PRAISE FOR

THE MAKING OF THE ATOMIC BOMB

"A great book. Mr. Rhodes has done a beautiful job, and I don't see how anyone can ever top it."

—LUIS W. ALVAREZ,
Nobel Laureate for Physics, 1968

". . . what I read already impressed me with the author's knowledge of much of the history of the science which led to the development of nuclear energy and nuclear bombs and of the personalities which contributed in the U.S. to the development of these. I was particularly impressed by his realization of the importance of Leo Szilard's contributions which are almost always underestimated but which he fully realizes and perhaps even overestimates. I hope the book will find a wide readership."

—EUGENE P. WIGNER,
Nobel Laureate for Physics, 1963

"I found *The Making of the Atomic Bomb* well written, interesting and one of the best in the great family of books on the subject. It is fascinating as a novel, and I have learned from it many things I did not know. Mr. Rhodes has done his homework conscientiously and intelligently."

—EMILIO SEGRÈ,
Nobel Laureate for Physics, 1959

"Mr. Rhodes gives careful attention to the role which chemists played in developing the bomb. *The Making of the Atomic Bomb* strikes me as the most complete account of the Manhattan Project to date.

—GLENN T. SEABORG,
Nobel Laureate for Chemistry, 1951

"*The Making of the Atomic Bomb* is an epic worthy of Milton. Nowhere else have I seen the whole story put down with such elegance and gusto and in such revealing detail and simple language which carries the reader through wonderful and profound scientific discoveries and their application.

The great figures of the age, scientific, military, and political, come to life when confronted with the fateful and awesome decisions which faced them in this agonizing century. This great book dealing with the most profound problems of the 20th century can help us to apprehend the opportunities and pitfalls that face the world in the 21st."

—I. I. RABI,
Nobel Laureate for Physics, 1944

BOOKS BY
RICHARD RHODES

NONFICTION

The Making of the Atomic Bomb

Looking for America

The Inland Ground

FICTION

Sons of Earth

The Last Safari

Holy Secrets

The Ungodly

Richard Rhodes

THE MAKING OF THE ATOMIC BOMB

A TOUCHSTONE BOOK
Published by Simon & Schuster
New York London Toronto Sydney Tokyo Singapore

TOUCHSTONE
Rockefeller Center
1230 Avenue of the Americas
New York, NY 10020

First Touchstone Edition 1988

Manufactured in the United States of America

10 9 8 7 6 5 4 3 2

Library of Congress Cataloging-in-Publication Data
Rhodes, Richard.
The making of the atomic bomb
Bibliography: p.
Includes index.
1. Atomic bomb—History. I. Title
QC773.R46 1986 623.4'5119'09 86-15445
ISBN 0-671-44133-7
ISBN 0-684-81378-5 (Pbk)

The author is grateful for permission to reprint excerpts from:

Reminiscences of Los Alamos, 1943–1945 by Lawrence Badash, et al., copyright © 1980 by D. Reidel Publishing Company, Dordrecht, Holland.

Energy and Conflict by Stanley A. Blumberg and Gwinn Owens, copyright © 1976. Published by G. P. Putnam's Sons and reprinted by permission of Ann Elmo Agency.

Rutherford by A. S. Eve, copyright 1939. Reprinted by permission of Cambridge University Press.

Atoms in the Family by Laura Fermi, copyright 1954. Reprinted by permission of University of Chicago Press.

What Little I Remember by Otto Frisch, copyright © 1979. Reprinted by permission of Cambridge University Press.

Now It Can Be Told by Leslie R. Groves, copyright © 1962 by Leslie R. Groves. Reprinted by permission of Harold Ober Associates, Inc.

Hiroshima Diary by Michihiko Hachiya, translated and edited by Warner Wells, M.D., copyright 1955. Reprinted by permission of University of North Carolina Press.

The Uranium People by Leona Marshall Libby, copyright © 1979. Reprinted by permission of Charles Scribner's Sons.

Death in Life by Robert Jay Lifton, copyright © 1982 by Robert Jay Lifton. Reprinted by permission of Basic Books, Inc. Publishers.

Children of the Atomic Bomb by Arata Osada, copyright © 1967. Midwest Publishers.

Niels Bohr by Stefan Rozental, copyright © 1967. Reprinted by permission of North-Holland Physics Publishing, Amsterdam.

Enrico Fermi, Physicist by Emilio Segrè, copyright © 1970. Reprinted by permission of University of Chicago Press.

Robert Oppenheimer: Letters and Recollections by Alice Kimball Smith and Charles Weiner, copyright © 1980 by Alice Kimball Smith and Charles Weiner. Reprinted by permission of Harvard University Press; also reprinted by permission of Spencer R. Weart at the American Institute of Physics and for quotes from the Bridgeman Papers, Harvard University Archives.

Adventures of a Mathematician by Stanislaw Ulam, copyright © 1977 by S. M. Ulam. Reprinted by permission of Charles Scribner's Sons.

Leo Szilard: His Version of the Facts by Spencer R. Weart and Gertrude Weiss Szilard, copyright © 1978. Reprinted by permission of the MIT Press.

All in Our Time by Jane Wilson, copyright © 1975 by the Educational Foundation for Nuclear Science, Chicago, Ill., 60637. Reprinted by permission of the Bulletin of the Atomic Scientists, a magazine of science and world affairs.

In memory
John Cushman
1926–1984

The author acknowledges with gratitude the support of the Ford Foundation and the Alfred P. Sloan Foundation in the research and writing of this book.

Taken as a story of human achievement, and human blindness, the discoveries in the sciences are among the great epics.

Robert Oppenheimer

In an enterprise such as the building of the atomic bomb the difference between ideas, hopes, suggestions and theoretical calculations, and solid numbers based on measurement, is paramount. All the committees, the politicking and the plans would have come to naught if a few unpredictable nuclear cross sections had been different from what they are by a factor of two.

Emilio Segrè

Contents

PART ONE

PROFOUND AND NECESSARY TRUTH

It is a profound and necessary truth that the deep things in science are not found because they are useful; they are found because it was possible to find them.

Robert Oppenheimer

It is still an unending source of surprise for me to see how a few scribbles on a blackboard or on a sheet of paper could change the course of human affairs.

Stanislaw Ulam

1
Moonshine

In London, where Southampton Row passes Russell Square, across from the British Museum in Bloomsbury, Leo Szilard waited irritably one gray Depression morning for the stoplight to change. A trace of rain had fallen during the night; Tuesday, September 12, 1933, dawned cool, humid and dull. Drizzling rain would begin again in early afternoon. When Szilard told the story later he never mentioned his destination that morning. He may have had none; he often walked to think. In any case another destination intervened. The stoplight changed to green. Szilard stepped off the curb. As he crossed the street time cracked open before him and he saw a way to the future, death into the world and all our woe, the shape of things to come.

Leo Szilard, the Hungarian theoretical physicist, born of Jewish heritage in Budapest on February 11, 1898, was thirty-five years old in 1933. At five feet, six inches he was not tall even for the day. Nor was he yet the "short fat man," round-faced and potbellied, "his eyes shining with intelligence and wit" and "as generous with his ideas as a Maori chief with his wives," that the French biologist Jacques Monod met in a later year. Midway between trim youth and portly middle age, Szilard had thick, curly, dark hair and an animated face with full lips, flat cheekbones and dark brown eyes. In photographs he still chose to look soulful. He had reason.

His deepest ambition, more profound even than his commitment to science, was somehow to save the world.

The Shape of Things to Come was H. G. Wells' new novel, just published, reviewed with avuncular warmth in *The Times* on September 1. "Mr. Wells' newest 'dream of the future' is its own brilliant justification," *The Times* praised, obscurely. The visionary English novelist was one among Szilard's network of influential acquaintances, a network he assembled by plating his articulate intelligence with the purest brass.

In 1928, in Berlin, where he was a *Privatdozent* at the University of Berlin and a confidant and partner in practical invention of Albert Einstein, Szilard had read Wells' tract *The Open Conspiracy.* The Open Conspiracy was to be a public collusion of science-minded industrialists and financiers to establish a world republic. Thus to save the world. Szilard appropriated Wells' term and used it off and on for the rest of his life. More to the point, he traveled to London in 1929 to meet Wells and bid for the Central European rights to his books. Given Szilard's ambition he would certainly have discussed much more than publishing rights. But the meeting prompted no immediate further connection. He had not yet encountered the most appealing orphan among Wells' Dickensian crowd of tales.

Szilard's past prepared him for his revelation on Southampton Row. He was the son of a civil engineer. His mother was loving and he was well provided for. "I knew languages because we had governesses at home, first in order to learn German and second in order to learn French." He was "sort of a mascot" to classmates at his *Gymnasium,* the University of Budapest's famous Minta. "When I was young," he told an audience once, "I had two great interests in life; one was physics and the other politics." He remembers informing his awed classmates, at the beginning of the Great War, when he was sixteen, how the fortunes of nations should go, based on his precocious weighing of the belligerents' relative political strength:

> I said to them at the time that I did of course not know who would win the war, but I did know how the war ought to end. It ought to end by the defeat of the central powers, that is the Austro-Hungarian monarchy and Germany, and also end by the defeat of Russia. I said I couldn't quite see how this could happen, since they were fighting on opposite sides, but I said that this was really what ought to happen. In retrospect I find it difficult to understand how at the age of sixteen and without any direct knowledge of countries other than Hungary, I was able to make this statement.

He seems to have assembled his essential identity by sixteen. He believed his clarity of judgment peaked then, never to increase further; it "perhaps even declined."

His sixteenth year was the first year of a war that would shatter the political and legal agreements of an age. That coincidence—or catalyst—by itself could turn a young man messianic. To the end of his life he made dull men uncomfortable and vain men mad.

He graduated from the Minta in 1916, taking the Eötvös Prize, the Hungarian national prize in mathematics, and considered his further education. He was interested in physics but "there was no career in physics in Hungary." If he studied physics he could become at best a high school teacher. He thought of studying chemistry, which might be useful later when he picked up physics, but that wasn't likely either to be a living. He settled on electrical engineering. Economic justifications may not tell all. A friend of his studying in Berlin noticed as late as 1922 that Szilard, despite his Eötvös Prize, "felt that his skill in mathematical operations could not compete with that of his colleagues." On the other hand, he was not alone among Hungarians of future prominence in physics in avoiding the backwater science taught in Hungarian universities at the time.

He began engineering studies in Budapest at the King Joseph Institute of Technology, then was drafted into the Austro-Hungarian Army. Because he had a *Gymnasium* education he was sent directly to officers' school to train for the cavalry. A leave of absence almost certainly saved his life. He asked for leave ostensibly to give his parents moral support while his brother had a serious operation. In fact, he was ill. He thought he had pneumonia. He wanted to be treated in Budapest, near his parents, rather than in a frontier Army hospital. He waited standing at attention for his commanding officer to appear to hear his request while his fever burned at 102 degrees. The captain was reluctant; Szilard characteristically insisted on his leave and got it, found friends to support him to the train, arrived in Vienna with a lower temperature but a bad cough and reached Budapest and a decent hospital. His illness was diagnosed as Spanish influenza, one of the first cases on the Austro-Hungarian side. The war was winding down. Using "family connections" he arranged some weeks later to be mustered out. "Not long afterward, I heard that my own regiment," sent to the front, "had been under severe attack and that all of my comrades had disappeared."

In the summer of 1919, when Lenin's Hungarian protégé Bela Kun and his Communist and Social Democratic followers established a short-lived Soviet republic in Hungary in the disordered aftermath of Austro-Hungarian defeat, Szilard decided it was time to study abroad. He was twenty-one years old. Just as he arranged for a passport, at the beginning of August, the Kun regime collapsed; he managed another passport from the right-wing regime of Admiral Nicholas Horthy that succeeded it and left Hungary around Christmastime.

Still reluctantly committed to engineering, Szilard enrolled in the Technische Hochschule, the technology institute, in Berlin. But what had seemed necessary in Hungary seemed merely practical in Germany. The physics faculty of the University of Berlin included Nobel laureates Albert Einstein, Max Planck and Max von Laue, theoreticians of the first rank. Fritz Haber, whose method for fixing nitrogen from the air to make nitrates for gunpowder saved Germany from early defeat in the Great War, was only one among many chemists and physicists of distinction at the several government- and industry-sponsored Kaiser Wilhelm Institutes in the elegant Berlin suburb of Dahlem. The difference in scientific opportunity between Budapest and Berlin left Szilard physically unable to listen to engineering lectures. "In the end, as always, the subconscious proved stronger than the conscious and made it impossible for me to make any progress in my studies of engineering. Finally the ego gave in, and I left the Technische Hochschule to complete my studies at the University, some time around the middle of '21."

Physics students at that time wandered Europe in search of exceptional masters much as their forebears in scholarship and craft had done since medieval days. Universities in Germany were institutions of the state; a professor was a salaried civil servant who also collected fees directly from his students for the courses he chose to give (a *Privatdozent,* by contrast, was a visiting scholar with teaching privileges who received no salary but might collect fees). If someone whose specialty you wished to learn taught at Munich, you went to Munich; if at Göttingen, you went to Göttingen. Science grew out of the craft tradition in any case; in the first third of the twentieth century it retained—and to some extent still retains—an informal system of mastery and apprenticeship over which was laid the more recent system of the European graduate school. This informal collegiality partly explains the feeling among scientists of Szilard's generation of membership in an exclusive group, almost a guild, of international scope and values.

Szilard's good friend and fellow Hungarian, the theoretical physicist Eugene Wigner, who was studying chemical engineering at the Technische Hochschule at the time of Szilard's conversion, watched him take the University of Berlin by storm. "As soon as it became clear to Szilard that physics was his real interest, he introduced himself, with characteristic directness, to Albert Einstein." Einstein was a man who lived apart—preferring originality to repetition, he taught few courses—but Wigner remembers that Szilard convinced him to give them a seminar on statistical mechanics. Max Planck was a gaunt, bald elder statesman whose study of radiation emitted by a uniformly heated surface (such as the interior of a kiln) had led him to discover a universal constant of nature. He followed the canny tradition among leading scientists of accepting only the most promising students for tutelage; Szilard won his attention. Max von Laue, the handsome director of the university's Institute for Theoretical Physics, who founded the science of X-ray crystallography and created a popular sensation by thus making the atomic lattices of crystals visible for the first time, accepted Szilard into his brilliant course in relativity theory and eventually sponsored his Ph.D. dissertation.

The postwar German infection of despair, cynicism and rage at defeat ran a course close to febrile hallucination in Berlin. The university, centrally located between Dorotheenstrasse and Unter den Linden due east of the Brandenburg Gate, was well positioned to observe the bizarre effects. Szilard missed the November 1918 revolution that began among mutinous sailors at Kiel, quickly spread to Berlin and led to the retreat of the Kaiser to Holland, to armistice and eventually to the founding, after bloody riots, of the insecure Weimar Republic. By the time he arrived in Berlin at the end of 1919 more than eight months of martial law had been lifted, leaving a city at first starving and bleak but soon restored to intoxicating life.

"There was snow on the ground," an Englishman recalls of his first look at postwar Berlin in the middle of the night, "and the blend of snow, neon and huge hulking buildings was unearthly. You felt you had arrived somewhere totally strange." To a German involved in the Berlin theater of the 1920s "the air was always bright, as if it were peppered, like New York late in autumn: you needed little sleep and never seemed tired. Nowhere else did you fail in such good form, nowhere else could you be knocked on the chin time and again without being counted out." The German aristocracy retreated from view, and intellectuals, film stars and journalists took its place; the major annual social event in the city where an imperial palace stood empty was the Press Ball, sponsored by the Berlin Press Club, which drew as many as six thousand guests.

Ludwig Mies van der Rohe designed his first glass-walled skyscraper in postwar Berlin. Yehudi Menuhin made his precocious debut, with Einstein in the audience to applaud him. George Grosz sorted among his years of savage observation on Berlin's wide boulevards and published *Ecce Homo.* Vladimir Nabokov was there, observing "an elderly, rosy-faced beggar woman with legs cut off at the pelvis . . . set down like a bust at the foot of a wall and . . . selling paradoxical shoelaces." Fyodor Vinberg, one of the Czar's departed officers, was there, publishing a shoddy newspaper, promoting *The Protocols of the Elders of Zion,* which he had personally introduced into Germany from Russia—a new German edition of that pseudo-Machiavellian, patently fraudulent fantasy of world conquest sold more than 100,000 copies—and openly advocating the violent destruction of the Jews. Hitler was not there until the end, because he was barred from northern Germany after his release from prison in 1924, but he sent rumpelstiltskin Joseph Goebbels to stand in for him; Goebbels learned to break heads and spin propaganda in an open, lusty, jazz-drunk city he slandered in his diary as "a dark and mysterious enigma."

In the summer of 1922 the rate of exchange in Germany sank to 400 marks to the dollar. It fell to 7,000 to the dollar at the beginning of January 1923, the truly terrible year. One hundred sixty thousand in July. One million in August. And 4.2 *trillion* marks to the dollar on November 23, 1923, when adjustment finally began. Banks advertised for bookkeepers good with zeros and paid out cash withdrawals by weight. Antique stores filled to the ceiling with the pawned treasures of the bankrupt middle class. A theater seat sold for an egg. Only those with hard currency—mostly foreigners—thrived at a time when it was possible to cross Germany by first-class railroad carriage for pennies, but they also earned the enmity of starving Germans. "No, one did not feel guilty," the visiting Englishman crows, "one felt it was perfectly normal, a gift from the gods."

The German physicist Walter Elsasser, who later emigrated to the United States, worked in Berlin in 1923 during an interlude in his student years; his father had agreed to pay his personal expenses. He was no foreigner, but with foreign help he was able to live like one:

> In order to make me independent of [inflation], my father had appealed to his friend, Kaufmann, the banker from Basle, who had established for me an account in American dollars at a large bank. . . . Once a week I took half a day off to go downtown by subway and withdrew my allowance in marks; and it was more each time, of course. Returning to my rented room, I at once bought enough food staples to last the week, for within three days, all the prices would have risen appreciably, by fifteen percent, say, so that my allowance would have run short and would not have permitted such pleasures as an excursion to Potsdam or to the lake country on Sundays. . . . I was too young, much too callous, and too inexperienced to understand what this galloping inflation must have meant—actual starvation and misery—to people who had to live on pensions or other fixed incomes, or even to wage earners, especially those with children, whose pay lagged behind the rate of inflation.

So must Szilard have lived, except that no one recalls ever seeing him cook for himself; he preferred the offerings of delicatessens and cafés. He would have understood what inflation meant and some of the reasons for its extremity. But though Szilard was preternaturally observant—"During a long life among scientists," writes Wigner, "I have met no one with more imagination and originality, with more independence of thought and opinion"—his recollections and his papers preserve almost nothing of these Berlin days. Germany's premier city at the height of its postwar social, political and intellectual upheaval earns exactly one sentence from Szilard: "Berlin at that time lived in the heyday of physics." That was how much physics, giving extraordinary birth during the 1920s to its modern synthesis, meant to him.

Four years of study usually preceded a German student's thesis work. Then, with a professor's approval, the student solved a problem of his own conception or one his professor supplied. "In order to be acceptable," says Szilard, it "had to be a piece of really original work." If the thesis found favor, the student took an oral examination one afternoon and if he passed he was duly awarded a doctorate.

Szilard had already given a year of his life to the Army and two years to engineering. He wasted no time advancing through physics. In the summer of 1921 he went to Max von Laue and asked for a thesis topic. Von Laue apparently decided to challenge Szilard—the challenge may have been friendly or it may have been an attempt to put him in his place—and gave him an obscure problem in relativity theory. "I couldn't make any headway with it. As a matter of fact, I was not even convinced that this was a problem that could be solved." Szilard worked on it for six months, until the Christmas season, "and I thought Christmastime is not a time to work, it is a time to loaf, so I thought I would just think whatever comes to my mind."

What he thought, in three weeks, was how to solve a baffling inconsistency in thermodynamics, the branch of physics that concerns relationships between heat and other forms of energy. There are two thermodynamic theories, both highly successful at predicting heat phenomena. One, the phenomenological, is more abstract and generalized (and therefore more useful); the other, the statistical, is based on an atomic model and corresponds more closely to physical reality. In particular, the statistical theory depicts thermal equilibrium as a state of random motion of atoms. Einstein, for example, had demonstrated in important papers in 1905 that Brownian motion—the continuous, random motion of particles such as pollen suspended in a liquid—was such a state. But the more useful phenomenological theory treated thermal equilibrium as if it were static, a state of no change. That was the inconsistency.

Szilard went for long walks—Berlin would have been cold and gray, the grayness sometimes relieved by days of brilliant sunshine—"and I saw something in the middle of the walk; when I came home I wrote it down; next morning I woke up with a new idea and I went for another walk; this crystallized in my mind and in the evening I wrote it down." It was, he thought, the most creative period of his life. "Within three weeks I had produced a manuscript of something which was really quite original. But I didn't dare to take it to von Laue, because it was not what he had asked me to do."

He took it instead to Einstein after a seminar, buttonholed him and said he would like to tell him about something he had been doing.

"Well, what have you been doing?" Szilard remembers Einstein saying.

Szilard reported his "quite original" idea.

"That's impossible," Einstein said. "This is something that cannot be done."

"Well, yes, but I did it."

"How did you do it?"

Szilard began explaining. "Five or ten minutes" later, he says, Einstein understood. After only a year of university physics, Szilard had worked out a rigorous mathematical proof that the random motion of thermal equilibrium could be fitted within the framework of the phenomenological theory in its original, classical form, without reference to a limiting atomic model—"and [Einstein] liked this very much."

Thus emboldened, Szilard took his paper—its title would be "On the extension of phenomenological thermodynamics to fluctuation phenomena"—to von Laue, who received it quizzically and took it home. "And next morning, early in the morning, the telephone rang. It was von Laue. He said, 'Your manuscript has been accepted as your thesis for the Ph.D. degree.' "

Six months later Szilard wrote another paper in thermodynamics, "On the decrease of entropy in a thermodynamic system by the intervention of intelligent beings," that eventually would be recognized as one of the important foundation documents of modern information theory. By then he had his advanced degree; he was Dr. Leo Szilard now. He experimented with X-ray effects in crystals, von Laue's field, at the Kaiser Wilhelm Institute for Chemistry in Dahlem until 1925; that year the University of Berlin accepted his entropy paper as his *Habilitationsschrift,* his inaugural dissertation, and he was thereupon appointed a *Privatdozent,* a position he held until he left for England in 1933.

One of Szilard's sidelines, then and later, was invention. Between 1924 and 1934 he applied to the German patent office individually or jointly with his partner Albert Einstein for twenty-nine patents. Most of the joint applications dealt with home refrigeration. "A sad newspaper story. . .

caught the attention of Einstein and Szilard one morning," writes one of Szilard's later American protégés: "It was reported in a Berlin newspaper that an entire family, including a number of young children, had been found asphyxiated in their apartment as a result of their inhalation of the noxious fumes of the [chemical] that was used as the refrigerant in their primitive refrigerator and that had escaped in the night through a leaky pump valve." Whereupon the two physicists devised a method of pumping metallicized refrigerant by electromagnetism, a method that required no moving parts (and therefore no valve seals that might leak) except the refrigerant itself. A.E.G., the German General Electric, signed Szilard on as a paid consultant and actually built one of the Einstein-Szilard refrigerators, but the magnetic pump was so noisy compared to even the noisy conventional compressors of the day that it never left the engineering lab.

Another, oddly similar invention, also patented, might have won Szilard world acclaim if he had taken it beyond the patent stage. Independently of the American experimental physicist Ernest O. Lawrence and at least three months earlier, Szilard worked out the basic principle and general design of what came to be called, as Lawrence's invention, the cyclotron, a device for accelerating nuclear particles in a circular magnetic field, a sort of nuclear pump. Szilard applied for a patent on his device on January 5, 1929; Lawrence first thought of the cyclotron on about April 1, 1929, producing a small working model a year later—for which he won the 1939 Nobel Prize in Physics.

Szilard's originality stopped at no waterline. Somewhere along the way from sixteen-year-old prophet of the fate of nations to thirty-one-year-old open conspirer negotiating publishing rights with H. G. Wells, he conceived an Open Conspiracy of his own. He dated his social invention from "the mid-twenties in Germany." If so, then he went to see Wells in 1929 as much from enthusiasm for the Englishman's perspicacity as for his vision. C. P. Snow, the British physicist and novelist, writes of Leo Szilard that he "had a temperament uncommon anywhere, maybe a little less uncommon among major scientists. He had a powerful ego and invulnerable egocentricity: but he projected the force of that personality outward, with beneficent intention toward his fellow creatures. In that sense, he had a family resemblance to Einstein on a reduced scale." Beneficent intention in this instance is a document proposing a new organization: *Der Bund*—the order, the confederacy, or, more simply, the band.

The Bund, Szilard writes, would be "a closely knit group of people whose inner bond is pervaded by a religious and scientific spirit":

> If we possessed a magical spell with which to recognize the "best" individuals of the rising generation at an early age . . . then we would be able to train

> them to independent thinking, and through education in close association we could create a spiritual leadership class with inner cohesion which would renew itself on its own.

Members of this class would not be awarded wealth or personal glory. To the contrary, they would be required to take on exceptional responsibilities, "burdens" that might "demonstrate their devotion." It seemed to Szilard that such a group stood a good chance of influencing public affairs even if it had no formal structure or constitutional position. But there was also the possibility that it might "take over a more direct influence on public affairs as part of the political system, next to government and parliament, or in the place of government and parliament."

"The Order," Szilard wrote at a different time, "was not supposed to be something like a political party . . . but rather it was supposed to represent the state." He saw representative democracy working itself out somehow within the cells of thirty to forty people that would form the mature political structure of the Bund. "Because of the method of selection [and education] . . . there would be a good chance that decisions at the top level would be reached by fair majorities."

Szilard pursued one version or another of his Bund throughout his life. It appears as late as 1961, by then suitably disguised, in his popular story "The Voice of the Dolphins": a tankful of dolphins at a "Vienna Institute" begin to impart their compelling wisdom to the world through their keepers and interpreters, who are U.S. and Russian scientists; the narrator slyly implies that the keepers may be the real source of wisdom, exploiting mankind's fascination with superhuman saviors to save it.

A wild burst of optimism—or opportunism—energized Szilard in 1930 to organize a group of acquaintances, most of them young physicists, to begin the work of banding together. He was convinced in the mid-1920s that "the parliamentary form of democracy would not have a very long life in Germany" but he "thought that it might survive one or two generations." Within five years he understood otherwise. "I reached the conclusion something would go wrong in Germany . . . in 1930." Hjalmar Schacht, the president of the German Reichsbank, meeting in Paris that year with a committee of economists called to decide how much Germany could pay in war reparations, announced that Germany could pay none at all unless its former colonies, stripped from it after the war, were returned. "This was such a striking statement to make that it caught my attention, and I concluded that if Hjalmar Schacht believed he could get away with this, things must be rather bad. I was so impressed by this that I wrote a letter to my bank and transferred every single penny I had out of Germany into Switzerland."

A far more organized Bund was advancing to power in Germany with another and more primitive program to save the world. That program, set out arrogantly in an autobiographical book—*Mein Kampf*—would achieve a lengthy and bloody trial. Yet Szilard in the years ahead would lead a drive to assemble a Bund of sorts; submerged from view, working to more urgent and more immediate ends than utopia, that "closely knit group of people" would finally influence world events more enormously even than Nazism.

Sometime during the 1920s, a new field of research caught Szilard's attention: nuclear physics, the study of the nucleus of the atom, where most of its mass—and therefore its energy—is concentrated. He was familiar with the long record of outstanding work in the general field of radioactivity of the German chemist Otto Hahn and the Austrian physicist Lise Meitner, who made a productive team at the Kaiser Wilhelm Institute for Chemistry. No doubt he was also alert as always to the peculiar tension in the air that signaled the possibility of new developments.

The nuclei of some light atoms could be shattered by bombarding them with atomic particles; that much the great British experimental physicist Ernest Rutherford had already demonstrated. Rutherford used one nucleus to bombard another, but since both nuclei were strongly positively charged, the bombarded nucleus repelled most attacks. Physicists were therefore looking for ways to accelerate particles to greater velocities, to force them past the nucleus' electrical barrier. Szilard's design of a cyclotron-like particle accelerator that could serve such a purpose indicates that he was thinking about nuclear physics as early as 1928.

Until 1932 he did no more than think. He had other work and nuclear physics was not yet sufficiently interesting to him. It became compelling in 1932. A discovery in physics opened the field to new possibilities while discoveries Szilard made in literature and utopianism opened his mind to new approaches to world salvation.

On February 27, 1932, in a letter to the British journal *Nature,* physicist James Chadwick of the Cavendish Laboratory at Cambridge University, Ernest Rutherford's laboratory, announced the possible existence of a neutron. (He confirmed the neutron's existence in a longer paper in the *Proceedings of the Royal Society* four months later, but Szilard would no more have doubted it at the time of Chadwick's first cautious announcement than did Chadwick himself; like many scientific discoveries, it was obvious once it was demonstrated, and Szilard could repeat the demonstration in Berlin if he chose.) The neutron, a particle with nearly the same mass as the positively charged proton that until 1932 was the sole certain component of the atomic nucleus, had no electric charge, which meant it could pass through the surrounding electrical barrier and enter into the nucleus. The neutron would open the atomic nucleus to examination. It might even be a way to force the nucleus to give up some of its enormous energy.

Just then, in 1932, Szilard found or took up for the first time that appealing orphan among H. G. Wells' books that he had failed to discover before: *The World Set Free.* Despite its title, it was not a tract like *The Open Conspiracy.* It was a prophetic novel, published in 1914, before the beginning of the Great War. Thirty years later Szilard could still summarize *The World Set Free* in accurate detail. Wells describes, he says:

> . . . the liberation of atomic energy on a large scale for industrial purposes, the development of atomic bombs, and a world war which was apparently fought by an alliance of England, France, and perhaps including America, against Germany and Austria, the powers located in the central part of Europe. He places this war in the year 1956, and in this war the major cities of the world are all destroyed by atomic bombs.

More personal discoveries emerged from Wells' visionary novel—ideas that anticipated or echoed Szilard's utopian plans, responses that may have guided him in the years ahead. Wells writes that his scientist hero, for example, was "oppressed, he was indeed scared, by his sense of the immense consequences of his discovery. He had a vague idea that night that he ought not to publish his results, that they were premature, that some secret association of wise men should take care of his work and hand it on from generation to generation until the world was riper for its practical application."

Yet *The World Set Free* influenced Szilard less than its subject matter might suggest. "This book made a very great impression on me, but I didn't regard it as anything but fiction. It didn't start me thinking of whether or not such things could in fact happen. I had not been working in nuclear physics up to that time."

By his own account, a different and quieter dialogue changed the direction of Szilard's work. The friend who had introduced him to H. G. Wells returned in 1932 to the Continent:

> I met him again in Berlin and there ensued a memorable conversation. Otto Mandl said that now he really thought he knew what it would take to save mankind from a series of ever-recurring wars that could destroy it. He said that Man has a heroic streak in himself. Man is not satisfied with a happy idyllic life: he has the need to fight and to encounter danger. And he concluded that what mankind must do to save itself is to launch an enterprise aimed at leaving the earth. On this task he thought the energies of mankind

> could be concentrated and the need for heroism could be satisfied. I remember very well my own reaction. I told him that this was somewhat new to me, and that I really didn't know whether I would agree with him. The only thing I could say was this: that if I came to the conclusion that this was what mankind needed, if I wanted to contribute something to save mankind, then I would probably go into nuclear physics, because only through the liberation of atomic energy could we obtain the means which would enable man not only to leave the earth but to leave the solar system.

Such must have been Szilard's conclusion; that year he moved to the Harnack House of the Kaiser Wilhelm Institutes—a residence for visiting scientists sponsored by German industry, a faculty club of sorts—and approached Lise Meitner about the possibility of doing experimental work with her in nuclear physics. Thus to save mankind.

He always lived out of suitcases, in rented rooms. At the Harnack House he kept the keys to his two suitcases at hand and the suitcases packed. "All I had to do was turn the key and leave when things got too bad." Things got bad enough to delay a decision about working with Meitner. An older Hungarian friend, Szilard remembers—Michael Polanyi, a chemist at the Kaiser Wilhelm Institutes with a family to consider—viewed the German political scene optimistically, like many others in Germany at the time. "They all thought that civilized Germans would not stand for anything really rough happening." Szilard held no such sanguine view, noting that the Germans themselves were paralyzed with cynicism, one of the uglier effects on morals of losing a major war.

Adolf Hitler was appointed Chancellor of Germany on January 30, 1933. On the night of February 27 a Nazi gang directed by the head of the Berlin SA, Hitler's private army, set fire to the imposing chambers of the Reichstag. The building was totally destroyed. Hitler blamed the arson on the Communists and bullied a stunned Reichstag into awarding him emergency powers. Szilard found Polanyi still unconvinced after the fire. "He looked at me and said, 'Do you really mean to say that you think that [Minister] of the Interior [Hermann Göring] had anything to do with this?' and I said, 'Yes, this is precisely what I mean.' He just looked at me with incredulous eyes." In late March, Jewish judges and lawyers in Prussia and Bavaria were dismissed from practice. On the weekend of April 1, Julius Streicher directed a national boycott of Jewish businesses and Jews were beaten in the streets. "I took a train from Berlin to Vienna on a certain date, close to the first of April, 1933," Szilard writes. "The train was empty. The same train the next day was overcrowded, was stopped at the frontier, the people had to get out, and everybody was interrogated by the Nazis.

This just goes to show that if you want to succeed in this world you don't have to be much cleverer than other people, you just have to be one day earlier."

The Law for the Restoration of the Career Civil Service was promulgated throughout Germany on April 7 and thousands of Jewish scholars and scientists lost their positions in German universities. From England, where he landed in early May, Szilard went furiously to work to help them emigrate and to find jobs for them in England, the United States, Palestine, India, China and points between. If he couldn't yet save all the world, he could at least save some part of it.

He came up for air in September. By then he was living at the Imperial Hotel in Russell Square, having transferred £1,595 from Zurich to his bank in London. More than half the money, £854, he held in trust for his brother Béla; the rest would see him through the year. Szilard's funds came from his patent licenses, refrigeration consulting and *Privatdozent* fees. He was busy finding jobs for others and couldn't be bothered to seek one himself. He had few expenses in any case; a week's lodging and three meals a day at a good London hotel cost about £5.5; he was a bachelor most of his life and his needs were simple.

"I was no longer thinking about this conversation [with Otto Mandl about space travel], or about H. G. Wells' book either, until I found myself in London about the time of the British Association [meeting]." Szilard's syntax slips here: the crucial word is *until.* He had been too distracted by events and by rescue work to think creatively about nuclear physics. He had even been considering going into biology, a radical change of field but one that a number of able physicists have managed, in prewar days and since. Such a change is highly significant psychologically and Szilard was to make it in 1946. But in September 1933, a meeting of the British Association for the Advancement of Science, an annual assembly, intervened.

If on Friday, September 1, lounging in the lobby of the Imperial Hotel, Szilard read *The Times'* review of *The Shape of Things to Come,* then he noticed the anonymous critic's opinion that Wells had "attempted something of the sort on earlier occasions—that rather haphazard work, 'The World Set Free,' comes particularly to mind—but never with anything like the same continuous abundance and solidity of detail, or indeed, the power to persuade as to the terrifying probability of some of the more immediate and disastrous developments." And may have thought again of the atomic bombs of Wells' earlier work, of Wells' Open Conspiracy and his own, of Nazi Germany and its able physicists, of ruined cities and general war.

Without question Szilard read *The Times* of September 12, with its provocative sequence of headlines:

THE BRITISH ASSOCIATION

BREAKING DOWN
THE ATOM

TRANSFORMATION OF
ELEMENTS

Ernest Rutherford, *The Times* reported, had recited a history of "the discoveries of the last quarter of a century in atomic transmutation," including:

THE NEUTRON
NOVEL TRANSFORMATIONS

All of which made Szilard restive. The leading scientists in Great Britain were meeting and he wasn't there. He was safe, he had money in the bank, but he was only another anonymous Jewish refugee down and out in London, lingering over morning coffee in a hotel lobby, unemployed and unknown.

Then, midway along the second column of *The Times'* summary of Rutherford's speech, he found:

> HOPE OF TRANSFORMING ANY ATOM
>
> What, Lord Rutherford asked in conclusion, were the prospects 20 or 30 years ahead?
>
> High voltages of the order of millions of volts would probably be unnecessary as a means of accelerating the bombarding particles. Transformations might be effected with 30,000 or 70,000 volts. . . . He believed that we should be able to transform all the elements ultimately.
>
> We might in these processes obtain very much more energy than the proton supplied, but on the average we could not expect to obtain energy in this way. It was a very poor and inefficient way of producing energy, and anyone who looked for a source of power in the transformation of the atoms was talking moonshine.

Did Szilard know what "moonshine" meant—"foolish or visionary talk"? Did he have to ask the doorman as he threw down the newspaper and stormed out into the street? "Lord Rutherford was reported to have said that whoever talks about the liberation of atomic energy on an industrial scale is talking moonshine. Pronouncements of experts to the effect that something cannot be done have always irritated me."

"This sort of set me pondering as I was walking in the streets of London, and I remember that I stopped for a red light at the intersection of Southampton Row. . . . I was pondering whether Lord Rutherford might not prove to be wrong."

"It occurred to me that neutrons, in contrast to alpha particles, do not ionize [i.e., interact electrically with] the substance through which they pass.

"Consequently, neutrons need not stop until they hit a nucleus with which they may react."

Szilard was not the first to realize that the neutron might slip past the positive electrical barrier of the nucleus; that realization had come to other physicists as well. But he was the first to imagine a mechanism whereby more energy might be released in the neutron's bombardment of the nucleus than the neutron itself supplied.

There was an analogous process in chemistry. Polanyi had studied it. A comparatively small number of active particles—oxygen atoms, for example—admitted into a chemically unstable system, worked like leaven to elicit a chemical reaction at temperatures much lower than the temperature that the reaction normally required. Chain reaction, the process was called. One center of chemical reaction produces thousands of product molecules. One center occasionally has an especially favorable encounter with a reactant and instead of forming only one new center, it forms two or more, each of which is capable in turn of propagating a reaction chain.

Chemical chain reactions are self-limiting. Were they not, they would run away in geometric progression: 1, 2, 4, 8, 16, 32, 64, 128, 256, 512, 1024, 2048, 4096, 8192, 16384, 32768, 65536, 131072, 262144, 524288, 1048576, 2097152, 4194304, 8388608, 16777216, 33554432, 67108868, 134217736 . . .

"As the light changed to green and I crossed the street," Szilard recalls, "it . . . suddenly occurred to me that if we could find an element which is split by neutrons and which would emit *two* neutrons when it absorbs *one* neutron, such an element, if assembled in sufficiently large mass, could sustain a nuclear chain reaction.

"I didn't see at the moment just how one would go about finding such an element, or what experiments would be needed, but the idea never left me. In certain circumstances it might be possible to set up a nuclear chain reaction, liberate energy on an industrial scale, and construct atomic bombs."

Leo Szilard stepped up onto the sidewalk. Behind him the light changed to red.

2
Atoms and Void

Atomic energy requires an atom. No such beast was born legitimately into physics until the beginning of the twentieth century. The atom as an idea—as an invisible layer of eternal, elemental substance below the world of appearances where things combine, teem, dissolve and rot—is ancient. Leucippus, a Greek philosopher of the fifth century B.C. whose name survives on the strength of an allusion in Aristotle, proposed the concept; Democritus, a wealthy Thracian of the same era and wider repute, developed it. " 'For by convention color exists,' " the Greek physician Galen quotes from one of Democritus' seventy-two lost books, " 'by convention bitter, by convention sweet, but in reality atoms and void.' " From the seventeenth century onward, physicists postulated atomic models of the world whenever developments in physical theory seemed to require them. But whether or not atoms really existed was a matter for continuing debate.

Gradually the debate shifted to the question of what kind of atom was necessary and possible. Isaac Newton imagined something like a miniature billiard ball to serve the purposes of his mechanical universe of masses in motion: "It seems probable to me," he wrote in 1704, "that God in the beginning formed matter in solid, massy, hard, impenetrable, movable particles, of such sizes and figures, and with such other properties, and in such proportion to space, as most conduced to the end to which he formed them." The Scottish physicist James Clerk Maxwell, who organized the founding of the Cavendish Laboratory, published a seminal *Treatise on Electricity and Magnetism* in 1873 that modified Newton's purely mechanical universe of particles colliding in a void by introducing into it the idea of an electromagnetic field. The field permeated the void; electric and magnetic energy propagated through it at the speed of light; light itself, Clerk Maxwell demonstrated, is a form of electromagnetic radiation. But despite his modifications, Clerk Maxwell was as devoted as Newton to a hard, mechanical atom:

> Though in the course of ages catastrophes have occurred and may yet occur in the heavens, though ancient systems may be dissolved and new systems evolved out of their ruins, the [atoms] out of which [the sun and the planets] are built—the foundation stones of the material universe—remain unbroken and unworn. They continue this day as they were created—perfect in number and measure and weight.

Max Planck thought otherwise. He doubted that atoms existed at all, as did many of his colleagues—the particulate theory of matter was an English invention more than a Continental, and its faintly Britannic odor made it repulsive to the xenophobic German nose—but if atoms did exist he was sure they could not be mechanical. "It is of paramount importance," he confessed in his *Scientific Autobiography,* "that the outside world is something independent from man, something absolute, and the quest for laws which apply to this absolute appeared to me as the most sublime scientific pursuit in life." Of all the laws of physics, Planck believed that the thermodynamic laws applied most basically to the independent "outside world" that his need for an absolute required. He saw early that purely mechanical atoms violated the second law of thermodynamics. His choice was clear.

The second law specifies that heat will not pass spontaneously from a colder to a hotter body without some change in the system. Or, as Planck himself generalized it in his Ph.D. dissertation at the University of Munich in 1879, that "the process of heat conduction cannot be completely reversed by any means." Besides forbidding the construction of perpetual-motion machines, the second law defines what Planck's predecessor Rudolf Clausius named *entropy:* because energy dissipates as heat whenever work is done—heat that cannot be collected back into useful, organized form—the universe must slowly run down to randomness. This vision of increasing disorder means that the universe is one-way and not reversible; the second law is the expression in physical form of what we call time. But the equations of mechanical physics—of what is now called classical physics—

theoretically allowed the universe to run equally well forward or backward. "Thus," an important German chemist complained, "in a purely mechanical world, the tree could become a shoot and a seed again, the butterfly turn back into a caterpillar, and the old man into a child. No explanation is given by the mechanistic doctrine for the fact that this does not happen. . . . The actual irreversibility of natural phenomena thus proves the existence of phenomena that cannot be described by mechanical equations; and with this the verdict on scientific materialism is settled." Planck, writing a few years earlier, was characteristically more succinct: "The consistent implementation of the second law . . . is incompatible with the assumption of finite atoms."

A major part of the problem was that atoms were not then directly accessible to experiment. They were a useful concept in chemistry, where they were invoked to explain why certain substances—elements—combine to make other substances but cannot themselves be chemically broken down. Atoms seemed to be the reason gases behaved as they did, expanding to fill whatever container they were let into and pushing equally on all the container's walls. They were invoked again to explain the surprising discovery that every element, heated in a laboratory flame or vaporized in an electric arc, colors the resulting light and that such light, spread out into its rainbow spectrum by a prism or a diffraction grating, invariably is divided into bands by characteristic bright lines. But as late as 1894, when Robert Cecil, the third Marquis of Salisbury, chancellor of Oxford and former Prime Minister of England, catalogued the unfinished business of science in his presidential address to the British Association, whether atoms were real or only convenient and what structure they hid were still undecided issues:

> What the atom of each element is, whether it is a movement, or a thing, or a vortex, or a point having inertia, whether there is any limit to its divisibility, and, if so, how that limit is imposed, whether the long list of elements is final, or whether any of them have any common origin, all these questions remain surrounded by a darkness as profound as ever.

Physics worked that way, sorting among alternatives: all science works that way. The chemist Michael Polanyi, Leo Szilard's friend, looked into the workings of science in his later years at the University of Manchester and at Oxford. He discovered a traditional organization far different from what most nonscientists suppose. A "republic of science," he called it, a community of independent men and women freely cooperating, "a highly simplified example of a free society." Not all philosophers of science, which is what Polanyi became, have agreed. Even Polanyi sometimes called science an "orthodoxy." But his republican model of science is powerful in the same way successful scientific models are powerful: it explains relationships that have not been clear.

Polanyi asked straightforward questions. How were scientists chosen? What oath of allegiance did they swear? Who guided their research—chose the problems to be studied, approved the experiments, judged the value of the results? In the last analysis, who decided what was scientifically "true"? Armed with these questions, Polanyi then stepped back and looked at science from outside.

Behind the great structure that in only three centuries had begun to reshape the entire human world lay a basic commitment to a naturalistic view of life. Other views of life dominated at other times and places—the magical, the mythological. Children learned the naturalistic outlook when they learned to speak, when they learned to read, when they went to school. "Millions are spent annually on the cultivation and dissemination of science by the public authorities," Polanyi wrote once when he felt impatient with those who refused to understand his point, "who will not give a penny for the advancement of astrology or sorcery. In other words, our civilization is deeply committed to certain beliefs about the nature of things; beliefs which are different, for example, from those to which the early Egyptian or the Aztec civilizations were committed."

Most young people learned no more than the orthodoxy of science. They acquired "the established doctrine, the dead letter." Some, at university, went on to study the beginnings of method. They practiced experimental proof in routine research. They discovered science's "uncertainties and its eternally provisional nature." That began to bring it to life.

Which was not yet to become a scientist. To become a scientist, Polanyi thought, required "a full initiation." Such an initiation came from "close personal association with the intimate views and practice of a distinguished master." The practice of science was not itself a science; it was an art, to be passed from master to apprentice as the art of painting is passed or as the skills and traditions of the law or of medicine are passed. You could not learn the law from books and classes alone. You could not learn medicine. No more could you learn science, because nothing in science ever quite fits; no experiment is ever final proof; everything is simplified and approximate.

The American theoretical physicist Richard Feynman once spoke about his science with similar candor to a lecture hall crowded with undergraduates at the California Institute of Technology. "What do we mean by 'understanding' something?" Feynman asked innocently. His amused sense of human limitation informs his answer:

> We can imagine that this complicated array of moving things which constitutes "the world" is something like a great chess game being played by the gods, and we are observers of the game. We do not know what the rules of the game are; all we are allowed to do is to *watch* the playing. Of course, if we watch long enough, we may eventually catch on to a few of the rules. *The rules of the game* are what we mean by *fundamental physics.* Even if we know every rule, however . . . what we really can explain in terms of those rules is very limited, because almost all situations are so enormously complicated that we cannot follow the plays of the game using the rules, much less tell what is going to happen next. We must, therefore, limit ourselves to the more basic question of the rules of the game. If we know the rules, we consider that we "understand" the world.

Learning the feel of proof; learning judgment; learning which hunches to play; learning which stunning calculations to rework, which experimental results *not* to trust: these skills admitted you to the spectators' benches at the chess game of the gods, and acquiring them required sitting first at the feet of a master.

Polanyi found one other necessary requirement for full initiation into science: belief. If science has become the orthodoxy of the West, individuals are nevertheless still free to take it or leave it, in whole or in part; believers in astrology, Marxism and virgin birth abound. But "no one can become a scientist unless he presumes that the scientific doctrine and method are fundamentally sound and that their ultimate premises can be unquestioningly accepted."

Becoming a scientist is necessarily an act of profound commitment to the scientific system and the scientific world view. "Any account of science which does not explicitly describe it as something we believe in is essentially incomplete and a false pretense. It amounts to a claim that science is essentially different from and superior to all human beliefs that are not scientific statements—and this is untrue." Belief is the oath of allegiance that scientists swear.

That was how scientists were chosen and admitted to the order. They constituted a republic of educated believers taught through a chain of masters and apprentices to judge carefully the slippery edges of their work.

Who then guided that work? The question was really two questions: who decided which problems to study, which experiments to perform? And who judged the value of the results?

Polanyi proposed an analogy. Imagine, he said, a group of workers faced with the problem of assembling a very large, very complex jigsaw puzzle. How could they organize themselves to do the job most efficiently?

Each worker could take some of the pieces from the pile and try to fit them together. That would be an efficient method if assembling a puzzle was like shelling peas. But it wasn't. The pieces weren't isolated. They fitted together into a whole. And the chance of any one worker's collection of pieces fitting together was small. Even if the group made enough copies of the pieces to give every worker the entire puzzle to attack, no one would accomplish as much alone as the group might if it could contrive a way to work together.

The best way to do the job, Polanyi argued, was to allow each worker to keep track of what every other worker was doing. "Let them work on putting the puzzle together in the sight of the others, so that every time a piece of it is fitted in by one [worker], all the others will immediately watch out for the next step that becomes possible in consequence." That way, even though each worker acts on his own initiative, he acts to further the entire group's achievement. The group works independently together; the puzzle is assembled in the most efficient way.

Polanyi thought science reached into the unknown along a series of what he called "growing points," each point the place where the most productive discoveries were being made. Alerted by their network of scientific publications and professional friendships—by the complete openness of their communication, an absolute and vital freedom of speech—scientists rushed to work at just those points where their particular talents would bring them the maximum emotional and intellectual return on their investment of effort and thought.

It was clear, then, who among scientists judged the value of scientific results: every member of the group, as in a Quaker meeting. "The authority of scientific opinion remains *essentially mutual;* it is established *between* scientists, not *above* them." There were leading scientists, scientists who worked with unusual fertility at the growing points of their fields; but science had no ultimate leaders. Consensus ruled.

Not that every scientist was competent to judge every contribution. The network solved that problem too. Suppose Scientist M announces a new result. He knows his highly specialized subject better than anyone in the world; who is competent to judge him? But next to Scientist M are Scientists L and N. Their subjects overlap M's, so they understand his work well enough to assess its quality and reliability and to understand where it fits into science. Next to L and N are other scientists, K and O and J and P, who know L and N well enough to decide whether to trust their judgment about M. On out to Scientists A and Z, whose subjects are almost completely removed from M's.

"This network is the seat of scientific opinion," Polanyi emphasized; "of an opinion which is not held by any single human brain, but which, split

into thousands of different fragments, is held by a multitude of individuals, each of whom endorses the other's opinion at second hand, by relying on the consensual chains which link him to all the others through a sequence of overlapping neighborhoods." Science, Polanyi was hinting, worked like a giant brain of individual intelligences linked together. That was the source of its cumulative and seemingly inexorable power. But the price of that power, as both Polanyi and Feynman are careful to emphasize, is voluntary limitation. Science succeeds in the difficult task of sustaining a political network among men and women of differing backgrounds and differing values, and in the even more difficult task of discovering the rules of the chess game of the gods, by severely limiting its range of competence. "Physics," as Eugene Wigner once reminded a group of his fellows, "does not even try to give us complete information about the events around us—it gives information about the *correlations* between those events."

Which still left the question of what standards scientists consulted when they passed judgment on the contributions of their peers. Good science, original work, always went beyond the body of received opinion, always represented a dissent from orthodoxy. How, then, could the orthodox fairly assess it?

Polanyi suspected that science's system of masters and apprentices protected it from rigidity. The apprentice learned high standards of judgment from his master. At the same time he learned to trust his *own* judgment: he learned the possibility and the necessity of dissent. Books and lectures might teach rules; masters taught controlled rebellion, if only by the example of their own original—and in that sense rebellious—work.

Apprentices learned three broad criteria of scientific judgment. The first criterion was plausibility. That would eliminate crackpots and frauds. It might also (and sometimes did) eliminate ideas so original that the orthodox could not recognize them, but to work at all, science had to take that risk. The second criterion was scientific value, a composite consisting of equal parts accuracy, importance to the entire system of whatever branch of science the idea belonged to, and intrinsic interest. The third criterion was originality. Patent examiners assess an invention for originality according to the degree of surprise the invention produces in specialists familiar with the art. Scientists judged new theories and new discoveries similarly. Plausibility and scientific value measured an idea's quality by the standards of orthodoxy; originality measured the quality of its dissent.

Polanyi's model of an open republic of science where each scientist judges the work of his peers against mutually agreed upon and mutually supported standards explains why the atom found such precarious lodging in nineteenth-century physics. It was plausible; it had considerable scientific value, especially in systematic importance; but no one had yet made any surprising discoveries about it. None, at least, sufficient to convince the network of only about one thousand men and women throughout the world in 1895 who called themselves physicists and the larger, associated network of chemists.

The atom's time was at hand. The great surprises in basic science in the nineteenth century came in chemistry. The great surprises in basic science in the first half of the twentieth century would come in physics.

In 1895, when young Ernest Rutherford roared up out of the Antipodes to study physics at the Cavendish with a view to making his name, the New Zealand he left behind was still a rough frontier. British nonconformist craftsmen and farmers and a few adventurous gentry had settled the rugged volcanic archipelago in the 1840s, pushing aside the Polynesian Maori who had found it first five centuries before; the Maori gave up serious resistance after decades of bloody skirmish only in 1871, the year Rutherford was born. He attended recently established schools, drove the cows home for milking, rode horseback into the bush to shoot wild pigeons from the berry-laden branches of virgin miro trees, helped at his father's flax mill at Brightwater where wild flax cut from aboriginal swamps was retted, scutched and hackled for linen thread and tow. He lost two younger brothers to drowning; the family searched the Pacific shore near the farm for months.

It was a hard and healthy childhood. Rutherford capped it by winning scholarships, first to modest Nelson College in nearby Nelson, South Island, then to the University of New Zealand, where he earned an M.A. with double firsts in mathematics and physical science at twenty-two. He was sturdy, enthusiastic and smart, qualities he would need to carry him from rural New Zealand to the leadership of British science. Another, more subtle quality, a braiding of country-boy acuity with a profound frontier innocence, was crucial to his unmatched lifetime record of physical discovery. As his protégé James Chadwick said, Rutherford's ultimate distinction was "his genius to be astonished." He preserved that quality against every assault of success and despite a well-hidden but sometimes sickening insecurity, the stiff scar of his colonial birth.

His genius found its first occasion at the University of New Zealand, where Rutherford in 1893 stayed on to earn a B.Sc. Heinrich Hertz's 1887 discovery of "electric waves"—radio, we call the phenomenon now—had impressed Rutherford wonderfully, as it did young people everywhere in the world. To study the waves he set up a Hertzian oscillator—electrically charged metal knobs spaced to make sparks jump between metal plates—in a dank basement cloakroom. He was looking for a problem for his first independent work of research.

He located it in a general agreement among scientists, pointedly including Hertz himself, that high-frequency alternating current, the sort of current a Hertzian oscillator produced when the spark radiation surged rapidly back and forth between the metal plates, would not magnetize iron. Rutherford suspected otherwise and ingeniously proved he was right. The work earned him an 1851 Exhibition scholarship to Cambridge. He was spading up potatoes in the family garden when the cable came. His mother called the news down the row; he laughed and jettisoned his spade, shouting triumph for son and mother both: "That's the last potato I'll dig!" (Thirty-six years later, when he was created Baron Rutherford of Nelson, he sent his mother a cable in her turn: "Now Lord Rutherford, more your honour than mine.")

"Magnetization of iron by high-frequency discharges" was skilled observation and brave dissent. With deeper originality, Rutherford noticed a subtle converse reaction while magnetizing iron needles with high-frequency current: needles already saturated with magnetism became partly *demagnetized* when a high-frequency current passed by. His genius to be astonished was at work. He quickly realized that he could use radio waves, picked up by a suitable antenna and fed into a coil of wire, to induce a high-frequency current into a packet of magnetized needles. Then the needles would be partly demagnetized and if he set a compass beside them it would swing to show the change.

By the time he arrived on borrowed funds at Cambridge in September 1895 to take up work at the Cavendish under its renowned director, J. J. Thomson, Rutherford had elaborated his observation into a device for detecting radio waves at a distance—in effect, the first crude radio receiver. Guglielmo Marconi was still laboring to perfect his version of a receiver at his father's estate in Italy; for a few months the young New Zealander held the world record in detecting radio transmissions at a distance.

Rutherford's experiments delighted the distinguished British scientists who learned of them from J. J. Thomson. They quickly adopted Rutherford, even seating him one evening at the Fellows' high table at King's in the place of honor next to the provost, which made him feel, he said, "like an ass in a lion's skin" and which shaded certain snobs on the Cavendish staff green with envy. Thomson generously arranged for a nervous but exultant Rutherford to read his third scientific paper, "A magnetic detector of electrical waves and some of its applications," at the June 18, 1896, meeting of the Royal Society of London, the foremost scientific organization in the world. Marconi only caught up with him in September.

Rutherford was poor. He was engaged to Mary Newton, the daughter of his University of New Zealand landlady, but the couple had postponed marriage until his fortunes improved. Working to improve them, he wrote his fiancée in the midst of his midwinter research: "The reason I am so keen on the subject [of radio detection] is because of its practical importance. . . . If my next week's experiments come out as well as I anticipate, I see a chance of making cash rapidly in the future."

There is mystery here, mystery that carries forward all the way to "moonshine." Rutherford was known in later years as a hard man with a research budget, unwilling to accept grants from industry or private donors, unwilling even to ask, convinced that string and sealing wax would carry the day. He was actively hostile to the commercialization of scientific research, telling his Russian protégé Peter Kapitza, for example, when Kapitza was offered consulting work in industry, "You cannot serve God and Mammon at the same time." The mystery bears on what C. P. Snow, who knew him, calls the "one curious exception" to Rutherford's "infallible" intuition, adding that "no scientist has made fewer mistakes." The exception was Rutherford's refusal to admit the possibility of usable energy from the atom, the very refusal that irritated Leo Szilard in 1933. "I believe that he was fearful that his beloved nuclear domain was about to be invaded by infidels who wished to blow it to pieces by exploiting it commercially," another protégé, Mark Oliphant, speculates. Yet Rutherford himself was eager to exploit radio commercially in January 1896. Whence the dramatic and lifelong change?

The record is ambiguous but suggestive. The English scientific tradition was historically genteel. It generally disdained research patents and any other legal and commercial restraints that threatened the open dissemination of scientific results. In practice that guard of scientific liberty could molder into clubbish distaste for "vulgar commercialism." Ernest Marsden, a Rutherford-trained physicist and an insightful biographer, heard that "in his early days at Cambridge there were some few who said that Rutherford was not a cultured man." One component of that canard may have been contempt for his eagerness to make a profit from radio.

It seems that J. J. Thomson intervened. A grand new work had abruptly offered itself. On November 8, 1895, one month after Rutherford arrived at Cambridge, the German physicist Wilhelm Röntgen discovered X rays radiating from the fluorescing glass wall of a cathode-ray tube. Röntgen reported his discovery in December and stunned the world. The strange radiation was a new growing point for science and Thomson began studying it almost immediately. At the same time he also continued his experiments with cathode rays, experiments that would culminate in 1897 in his identification of what he called the "negative corpuscle"—the electron, the first atomic particle to be identified. He must have needed help. He would also have understood the extraordinary opportunity for original research that radiation offered a young man of Rutherford's skill at experiment.

To settle the issue Thomson wrote the grand old man of British science, Lord Kelvin, then seventy-two, asking his opinion of the commercial possibilities of radio—"before tempting Rutherford to turn to the new subject," Marsden says. Kelvin after all, vulgar commercialism or not, had developed the transoceanic telegraph cable. "The reply of the great man was that [radio] might justify a captial expenditure of a £100,000 Company on its promotion, but no more."

By April 24 Rutherford has seen the light. He writes Mary Newton: "I hope to make both ends meet somehow, but I must expect to dub out my first year. . . . My scientific work at present is progressing slowly. I am working with the Professor this term on Röntgen Rays. I am a little full up of my old subject and am glad of a change. I expect it will be a good thing for me to work with the Professor for a time. I have done one research to show I can work by myself." The tone is chastened and not nearly convinced, as if a ghostly, parental J. J. Thomson were speaking through Rutherford to his fiancée. He has not yet appeared before the Royal Society, where he was hardly "a little full up" of his subject. But the turnabout is accomplished. Hereafter Rutherford's healthy ambition will go to scientific honors, not commercial success.

It seems probable that J. J. Thomson sat eager young Ernest Rutherford down in the darkly paneled rooms of the Gothic Revival Cavendish Laboratory that Clerk Maxwell had founded, at the university where Newton wrote his great *Principia,* and kindly told him he could not serve God and Mammon at the same time. It seems probable that the news that the distinguished director of the Cavendish had written the Olympian Lord Kelvin about the commercial ambitions of a brash New Zealander chagrined Rutherford to the bone and that he went away from the encounter feeling grotesquely like a parvenu. He would never make the same mistake again, even if it meant strapping his laboratories for funds, even if it meant driving away the best of his protégés, as eventually it did. Even if it meant that energy from his cherished atom could be nothing more than moonshine. But if Rutherford gave up commercial wealth for holy science, he won the atom in exchange. He found its constituent parts and named them. With string and sealing wax he made the atom real.

The sealing wax was blood red and it was the Bank of England's most visible contribution to science. British experimenters used Bank of England sealing wax to make glass tubes airtight. Rutherford's earliest work on the atom, like J. J. Thomson's work with cathode rays, grew out of nineteenth-century examination of the fascinating effects produced by evacuating the air from a glass tube that had metal plates sealed into its ends and then connecting the metal plates to a battery or an induction coil. Thus charged with electricity, the emptiness inside the sealed tube glowed. The glow emerged from the negative plate—the cathode—and disappeared into the positive plate—the anode. If you made the anode into a cylinder and sealed the cylinder into the middle of the tube you could project a beam of glow—of cathode rays—through the cylinder and on into the end of the tube opposite the cathode. If the beam was energetic enough to hit the glass it would make the glass fluoresce. The cathode-ray tube, suitably modified, its all-glass end flattened and covered with phosphors to increase the fluorescence, is the television tube of today.

In the spring of 1897 Thomson demonstrated that the beam of glowing matter in a cathode-ray tube was not made up of light waves, as (he wrote drily) "the almost unanimous opinion of German physicists" held. Rather, cathode rays were negatively charged particles boiling off the negative cathode and attracted to the positive anode. These particles could be deflected by an electric field and bent into curved paths by a magnetic field. They were much lighter than hydrogen atoms and were identical "whatever the gas through which the discharge passes" if gas was introduced into the tube. Since they were lighter than the lightest known kind of matter and identical regardless of the kind of matter they were born from, it followed that they must be some basic constituent *part* of matter, and if they were a part, then there must be a whole. The real, physical electron implied a real, physical atom: the particulate theory of matter was therefore justified for the first time convincingly by physical experiment. They sang J. J.'s success at the annual Cavendish dinner:

The corpuscle won the day
And in freedom went away
And became a cathode ray.

Armed with the electron, and knowing from other experiments that what was left when electrons were stripped away from an atom was a much more massive remainder that was positively charged, Thomson went on in the next decade to develop a model of the atom that came to be called the "plum pudding" model. The Thomson atom, "a number of negatively-electrified corpuscles enclosed in a sphere of uniform positive electrification" like raisins in a pudding, was a hybrid: particulate electrons and diffuse remainder. It served the useful purpose of demonstrating mathematically that electrons could be arranged in stable configurations within an atom and that the mathematically stable arrangements could account for the similarities and regularities among chemical elements that the periodic table of the elements displays. It was becoming clear that electrons were responsible for chemical affinities between elements, that chemistry was ultimately electrical.

Thomson just missed discovering X rays in 1894. He was not so unlucky in legend as the Oxford physicist Frederick Smith, who found that photographic plates kept near a cathode-ray tube were liable to be fogged and merely told his assistant to move them to another place. Thomson noticed that glass tubing held "at a distance of some feet from the discharge-tube" fluoresced just as the wall of the tube itself did when bombarded with cathode rays, but he was too intent on studying the rays themselves to pursue the cause. Röntgen isolated the effect by covering his cathode-ray tube with black paper. When a nearby screen of fluorescent material still glowed he realized that whatever was causing the screen to glow was passing through the paper and the intervening air. If he held his hand between the covered tube and the screen, his hand slightly reduced the glow on the screen but in dark shadow he could *see its bones.*

Röntgen's discovery intrigued other researchers besides J. J. Thomson and Ernest Rutherford. The Frenchman Henri Becquerel was a third-generation physicist who, like his father and grandfather before him, occupied the chair of physics at the Musée d'Histoire Naturelle in Paris; like them also he was an expert on phosphorescence and fluorescence—in his case, particularly of uranium. He heard a report of Röntgen's work at the weekly meeting of the Académie des Sciences on January 20, 1896. He learned that the X rays emerged from the fluorescing glass, which immediately suggested to him that he should test various fluorescing materials to see if they also emitted X rays. He worked for ten days without success, read an article on X rays on January 30 that encouraged him to keep working and decided to try a uranium salt, uranyl potassium sulfate.

His first experiment succeeded—he found that the uranium salt emitted radiation—but misled him. He had sealed a photographic plate in black paper, sprinkled a layer of the uranium salt onto the paper and "exposed the whole thing to the sun for several hours." When he developed the photographic plate "I saw the silhouette of the phosphorescent substance in black on the negative." He mistakenly thought sunlight activated the effect, much as cathode rays released Röntgen's X rays from the glass.

The story of Becquerel's subsequent serendipity is famous. When he tried to repeat his experiment on February 26 and again on February 27 Paris was gray. He put the covered photographic plate away in a dark drawer, uranium salt in place. On March 1 he decided to go ahead and develop the plate, "expecting to find the images very feeble. On the contrary, the silhouettes appeared with great intensity. I thought at once that the action might be able to go on in the dark." Energetic, penetrating radiation from inert matter unstimulated by rays or light: now Rutherford had his subject, as Marie and Pierre Curie, looking for the pure element that radiated, had their backbreaking work.

Between 1898, when Rutherford first turned his attention to the phenomenon Henri Becquerel found and which Marie Curie named *radioactivity,* and 1911, when he made the most important discovery of his life, the young New Zealand physicist systematically dissected the atom.

He studied the radiations emitted by uranium and thorium and named two of them: "There are present at least two distinct types of radiation—one that is very readily absorbed, which will be termed for convenience the α [alpha] radiation, and the other of a more penetrative character, which will be termed the β [beta] radiation." (A Frenchman, P. V. Villard, later discovered the third distinct type, a form of high-energy X rays that was named gamma radiation in keeping with Rutherford's scheme.) The work was done at the Cavendish, but by the time he published it, in 1899, when he was twenty-seven, Rutherford had moved to Montreal to become professor of physics at McGill University. A Canadian tobacco merchant had given money there to build a physics laboratory and to endow a number of professorships, including Rutherford's. "The McGill University has a good name," Rutherford wrote his mother. "£500 is not so bad [a salary] and as the physical laboratory is the best of its kind in the world, I cannot complain."

In 1900 Rutherford reported the discovery of a radioactive gas emanating from the radioactive element thorium. Marie and Pierre Curie soon discovered that radium (which they had purified from uranium ores in 1898) also gave off a radioactive gas. Rutherford needed a good chemist to help him establish whether the thorium "emanation" was thorium or something else; fortunately he was able to shanghai a young Oxford man at McGill, Frederick Soddy, of talent sufficient eventually to earn a Nobel Prize. "At the beginning of the winter [of 1900]," Soddy remembers, "Ernest Rutherford, the Junior Professor of Physics, called on me in the laboratory and told me about the discoveries he had made. He had just returned with his bride from New Zealand . . . but before leaving Canada for his trip he had discovered what he called the thorium emanation. . . . I was, of course, intensely interested and suggested that the chemical character of the [substance] ought to be examined."

The gas proved to have no chemical character whatsoever. That, says Soddy, "conveyed the tremendous and inevitable conclusion that the element thorium was slowly and spontaneously transmuting itself into [chemically inert] argon gas!" Soddy and Rutherford had observed the spontaneous disintegration of the radioactive elements, one of the major discoveries of twentieth-century physics. They set about tracing the way uranium, radium and thorium changed their elemental nature by radiating away part of their substance as alpha and beta particles. They discovered that each different radioactive product possessed a characteristic "half-life," the time required for its radiation to reduce to half its previously measured intensity. The half-life measured the transmutation of half the atoms in an element into atoms of another element or of a physically variant form of the same element—an "isotope," as Soddy later named it. Half-life became a way to detect the presence of amounts of transmuted substances—"decay products"—too small to detect chemically. The half-life of uranium proved to be 4.5 billion years, of radium 1,620 years, of one decay product of thorium 22 minutes, of another decay product of thorium 27 days. Some decay products appeared and transmuted themselves in minute fractions of a second—in the twinkle of an eye. It was work of immense importance to physics, opening up field after new field to excited view, and "for more than two years," as Soddy remembered afterward, "life, scientific life, became hectic to a degree rare in the lifetime of an individual, rare perhaps in the lifetime of an institution."

Along the way Rutherford explored the radiation emanating from the radioactive elements in the course of their transmutation. He demonstrated that beta radiation consisted of high-energy electrons "similar in all respects to cathode rays." He suspected, and later in England conclusively proved, that alpha particles were positively charged helium atoms ejected during radioactive decay. Helium is found captured in the crystalline spaces of uranium and thorium ores; now he knew why.

An important 1903 paper written with Soddy, "Radioactive change," offered the first informed calculations of the amount of energy released by radioactive decay:

> It may therefore be stated that the total energy of radiation during the disintegration of one gram of radium cannot be less than 10^8 [i.e., 100,000,000] gram-calories, and may be between 10^9 and 10^{10} gram-calories. . . . The union of hydrogen and oxygen liberates approximately 4×10^3 [i.e., 4,000] gram-calories per gram of water produced, and this reaction sets free more energy for a given weight than any other chemical change known. The energy of radioactive change must therefore be at least twenty-thousand times, and may be a million times, as great as the energy of any molecular change.

That was the formal scientific statement; informally Rutherford inclined to whimsical eschatology. A Cambridge associate writing an article on radioactivity that year, 1903, considered quoting Rutherford's "playful suggestion that, could a proper detonator be found, it was just conceivable that a wave of atomic disintegration might be started through matter, which would indeed make this old world vanish in smoke." Rutherford liked to quip that "some fool in a laboratory might blow up the universe unawares." If atomic energy would never be useful, it might still be dangerous.

Soddy, who returned to England that year, examined the theme more seriously. Lecturing on radium to the Corps of Royal Engineers in 1904, he speculated presciently on the uses to which atomic energy might be put:

> It is probable that all heavy matter possesses—latent and bound up with the structure of the atom—a similar quantity of energy to that possessed by radium. If it could be tapped and controlled what an agent it would be in shaping the world's destiny! The man who put his hand on the lever by which a parsimonious nature regulates so jealously the output of this store of energy would possess a weapon by which he could destroy the earth if he chose.

Soddy did not think the possibility likely: "The fact that we exist is a proof that [massive energetic release] did not occur; that it has not occurred is the best possible assurance that it never will. We may trust Nature to guard her secret."

H. G. Wells thought Nature less trustworthy when he read similar statements in Soddy's 1909 book *Interpretation of Radium.* "My idea is taken from Soddy," he wrote of *The World Set Free.* "One of the good old scientific romances," he called his novel; it was important enough to him that he interrupted a series of social novels to write it. Rutherford's and Soddy's discussions of radioactive change therefore inspired the science-fiction novel that eventually started Leo Szilard thinking about chain reactions and atomic bombs.

In the summer of 1903 the Rutherfords visited the Curies in Paris. Mme. Curie happened to be receiving her doctorate in science on the day of their arrival; mutual friends arranged a celebration. "After a very lively evening," Rutherford recalled, "we retired about 11 o'clock in the garden, where Professor Curie brought out a tube coated in part with zinc sulphide and containing a large quantity of radium in solution. The luminosity was brilliant in the darkness and it was a splendid finale to an unforgettable day." The zinc-sulfide coating fluoresced white, making the radium's ejection of energetic particles on its progess down the periodic table from uranium to lead visible in the darkness of the Paris evening. The light was bright enough to show Rutherford Pierre Curie's hands, "in a very inflamed and painful state due to exposure to radium rays." Hands swollen with radiation burns was another object lesson in what the energy of matter could do.

A twenty-six-year-old German chemist from Frankfurt, Otto Hahn, came to Montreal in 1905 to work with Rutherford. Hahn had already discovered a new "element," radiothorium, later understood to be one of thorium's twelve isotopes. He studied thorium radiation with Rutherford; together they determined that the alpha particles ejected from thorium had the same mass as the alpha particles ejected from radium and those from another radioactive element, actinium. The various particles were probably therefore identical—one conclusion along the way to Rutherford's proof in 1908 that the alpha particle was inevitably a charged helium atom. Hahn went back to Germany in 1906 to begin a distinguished career as a discoverer of isotopes and elements; Leo Szilard encountered him working with physicist Lise Meitner at the Kaiser Wilhelm Institute for Chemistry in the 1920s in Berlin.

Rutherford's research at McGill unraveling the complex transmutations of the radioactive elements earned him, in 1908, a Nobel Prize—not in physics but in chemistry. He had wanted that prize, writing his wife when she returned to New Zealand to visit her family in late 1904, "I may have a chance if I keep going," and again early in 1905, "They are all following on my trail, and if I am to have a chance for a Nobel Prize in the next few years I must keep my work moving." The award for chemistry rather than for physics at least amused him. "It remained to the end a good joke against him," says his son-in-law, "which he thoroughly appreciated, that he was thereby branded for all time as a chemist and no true physicist."

An eyewitness to the ceremonies said Rutherford looked ridiculously young—he was thirty-seven—and made the speech of the evening. He announced his recent confirmation, only briefly reported the month before, that the alpha particle was in fact helium. The confirming experiment was typically elegant. Rutherford had a glassblower make him a tube with extremely thin walls. He evacuated the tube and filled it with radon gas, a fertile source of alpha particles. The tube was gastight, but its thin walls allowed alpha particles to escape. Rutherford surrounded the radon tube with another glass tube, pumped out the air between the two tubes and sealed off the space. "After some days," he told his Stockholm audience triumphantly, "a bright spectrum of helium was observed in the outer vessel." Rutherford's experiments still stun with their simplicity. "In this Rutherford was an artist," says a former student. "All his experiments had style."

In the spring of 1907 Rutherford had left Montreal with his family—

by then including a six-year-old daughter, his only child—and moved back to England. He had accepted appointment as professor of physics at Manchester, in the city where John Dalton had first revived the atomic theory almost exactly a century earlier. Rutherford bought a house and went immediately to work. He inherited an experienced German physicist named Hans Geiger who had been his predecessor's assistant. Years later Geiger fondly recalled the Manchester days, Rutherford settled in among his gear:

> I see his quiet research room at the top of the physics building, under the roof, where his radium was kept and in which so much well-known work on the emanation was carried out. But I also see the gloomy cellar in which he had fitted up his delicate apparatus for the study of the alpha rays. Rutherford loved this room. One went down two steps and then heard from the darkness Rutherford's voice reminding one that a hot-pipe crossed the room at head-level, and to step over two water-pipes. Then finally, in the feeble light one saw the great man himself seated at his apparatus.

The Rutherford house was cheerier; another Manchester protégé liked to recall that "supper in the white-painted dining room on Saturdays and Sundays preceded pow-wows till all hours in the study on the first floor; tea on Sundays in the drawing room often followed a spin on the Cheshire roads in the motor." There was no liquor in the house because Mary Rutherford did not approve of drinking. Smoking she reluctantly allowed because her husband smoked heavily, pipe and cigarettes both.

Now in early middle age he was famously loud, a "tribal chief," as a student said, fond of banter and slang. He would march around the lab singing "Onward Christian Soldiers" off key. He took up room in the world now; you knew he was coming. He was ruddy-faced with twinkling blue eyes and he was beginning to develop a substantial belly. The diffidence was well hidden: his handshake was brief, limp and boneless; "he gave the impression," says another former student, "that he was shy of physical contact." He could still be mortified by condescension, blushing bright red and turning aside dumbstruck. With his students he was quieter, gentler, solid gold. "He was a man," pronounces one in high praise, "who never did dirty tricks."

Chaim Weizmann, the Russian-Jewish biochemist who was later elected the first president of Israel, was working at Manchester on fermentation products in those days. He and Rutherford became good friends. "Youthful, energetic, boisterous," Weizmann recalled, "he suggested anything but the scientist. He talked readily and vigorously on every subject under the sun, often without knowing anything about it. Going down to the refectory for lunch I would hear the loud, friendly voice rolling up the corridor." Rutherford had no political knowledge at all, Weizmann thought, but excused him on the grounds that his important scientific work took all his time. "He was a kindly person, but he did not suffer fools gladly."

In September 1907, his first term at Manchester, Rutherford made up a list of possible subjects for research. Number seven on the list was "Scattering of alpha rays." Working over the years to establish the alpha particle's identity, he had come to appreciate its great value as an atomic probe; because it was massive compared to the high-energy but nearly weightless beta electron, it interacted vigorously with matter. The measure of that interaction could reveal the atom's structure. "I was brought up to look at the atom as a nice hard fellow, red or grey in colour, according to taste," Rutherford told a dinner audience once. By 1907 it was clear to him that the atom was not a hard fellow at all but was substantially empty space. The German physicist Philipp Lenard had demonstrated as much in 1903 by bombarding elements with cathode rays. Lenard dramatized his findings with a vivid metaphor: the space occupied by a cubic meter of solid platinum, he said, was as empty as the space of stars beyond the earth.

But if there was empty space in atoms—void within void—there was something else as well. In 1906, at McGill, Rutherford had studied the magnetic deflection of alpha particles by projecting them through a narrow defining slit and passing the resulting thin beam through a magnetic field. At one point he covered half the defining slit with a sheet of mica only about three thousandths of a centimeter thick, thin enough to allow alpha particles to go through. He was recording the results of the experiment on photographic paper; he found that the edges of the part of the beam covered with the mica were blurred. The blurring meant that as the alpha particles passed through, the atoms of mica were deflecting—scattering—many of them from a straight line by as much as two degrees of angle. Since an intense magnetic field scattered the uncovered alpha particles only a little more, something unusual was happening. For a particle as comparatively massive as the alpha, moving at such high velocity, two degrees was an enormous deflection. Rutherford calculated that it would require an electrical field of about 100 million volts per centimeter of mica to scatter an alpha particle so far. "Such results bring out clearly," he wrote, "the fact that the atoms of matter must be the seat of very intense electrical forces." It was just this scattering that he marked down on his list to study.

To do so he needed not only to count but also to *see* individual alpha particles. At Manchester he accepted the challenge of perfecting the necessary instruments. He worked with Hans Geiger to develop an electrical device that clicked off the arrival of each individual alpha particle into a counting chamber. Geiger would later elaborate the invention into the familiar Geiger counter of modern radiation studies.

There was a way to make individual alpha particles visible using zinc sulfide, the compound that coated the tube of radium solution Pierre Curie had carried into the night garden in Paris in 1903. A small glass plate coated with zinc sulfide and bombarded with alpha particles briefly fluoresced at the point where each particle struck, a phenomenon known as "scintillation" from the Greek word for spark. Under a microscope the faint scintillations in the zinc sulfide could be individually distinguished and counted. The method was tedious in the extreme. It required sitting for at least thirty minutes in a dark room to adapt the eyes, then taking counting turns of only a minute at a time—the change signaled by a timer that rang a bell—because focusing the eyes consistently on a small, dim screen was impossible for much longer than that. Even through the microscope the scintillations hovered at the edge of visibility; a counter who expected an experiment to produce a certain number of scintillations sometimes unintentionally saw imaginary flashes. So the question was whether the count was generally accurate. Rutherford and Geiger compared the observation counts with matched counts by the electric method. When the observation method proved reliable they put the electric counter away. It could count, but it couldn't see, and Rutherford was interested first of all in locating an alpha particle's position in space.

Geiger went to work on alpha scattering, aided by Ernest Marsden, then an eighteen-year-old Manchester undergraduate. They observed alpha particles coming out of a firing tube and passing through foils of such metals as aluminum, silver, gold and platinum. The results were generally consistent with expectation: alpha particles might very well accumulate as much as two degrees of total deflection bouncing around among atoms of the plum-pudding sort. But the experiment was troubled with stray particles. Geiger and Marsden thought molecules in the walls of the firing tube might be scattering them. They tried eliminating the strays by narrowing and defining the end of the firing tube with a series of graduated metal washers. That proved no help.

Rutherford wandered into the room. The three men talked over the problem. Something about it alerted Rutherford's intuition for promising side effects. Almost as an afterthought he turned to Marsden and said, "See if you can get some effect of alpha particles directly reflected from a metal surface." Marsden knew that a negative result was expected—alpha particles shot *through* thin foils, they did not bounce *back* from them—but that missing a positive result would be an unforgivable sin. He took great care to prepare a strong alpha source. He aimed the pencil-narrow beam of alphas at a forty-five degree angle onto a sheet of gold foil. He positioned his scintillation screen on the same side of the foil, beside the alpha beam, so that a particle bouncing back would strike the screen and register as a scintillation. Between firing tube and screen he interposed a thick lead plate so that no direct alpha particles could interfere.

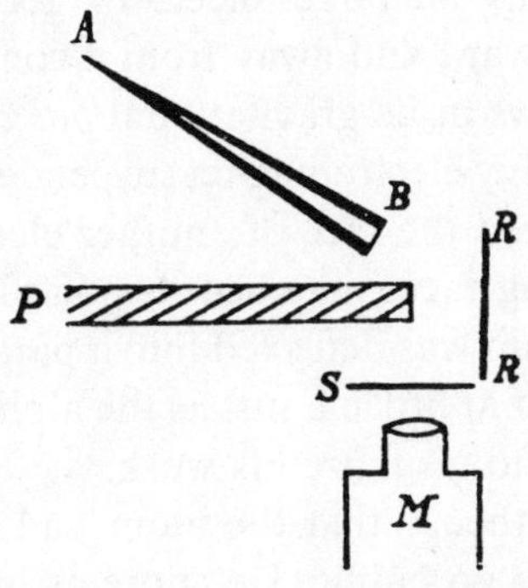

Arrangement of Ernest Marsden's experiment: A-B, alpha particle source. R-R, gold foil. P, lead plate. S, zinc sulfide scintillation screen. M, microscope.

Immediately, and to his surprise, he found what he was looking for. "I remember well reporting the result to Rutherford," he wrote, ". . .when I met him on the steps leading to his private room, and the joy with which I told him."

A few weeks later, at Rutherford's direction, Geiger and Marsden formulated the experiment for publication. "If the high velocity and mass of the α-particle be taken into account," they concluded, "it seems surprising that some of the α-particles, as the experiment shows, can be turned within a layer of 6×10^{-5} [i.e., .00006] cm. of gold through an angle of 90°, and even more. To produce a similar effect by magnetic field, the enormous field of 10^9 absolute units would be required." Rutherford in the meantime went off to ponder what the scattering meant.

He pondered, in the midst of other work, for more than a year. He had a first quick intuition of what the experiment portended and then lost it. Even after he announced his spectacular conclusion he was reluctant to promote it. One reason for his reluctance might be that the discovery contradicted the atomic models J. J. Thomson and Lord Kelvin had postulated earlier. There were physical objections to his interpretation of Marsden's discovery that would require working out as well.

Rutherford had been genuinely astonished by Marsden's results. "It was quite the most incredible event that has ever happened to me in my life," he said later. "It was almost as incredible as if you fired a 15-inch shell at a piece of tissue paper and it came back and hit you. On consideration I realised that this scattering backwards must be the result of a single collision, and when I made calculations I saw that it was impossible to get anything of that order of magnitude unless you took a system in which the greatest part of the mass of the atom was concentrated in a minute nucleus."

"Collision" is misleading. What Rutherford had visualized, making calculations and drawing diagrammatic atoms on large sheets of good paper, was exactly the sort of curving path toward and away from a compact, massive central body that a comet follows in its gravitational *pas de deux* with the sun. He had a model made, a heavy electromagnet suspended as a pendulum on thirty feet of wire that grazed the face of another electromagnet set on a table. With the two grazing faces matched in polarity and therefore repelling each other, the pendulum was deflected into a parabolic path according to its velocity and angle of approach, just as the alpha particles were deflected. He needed as always to visualize his work.

When further experiment confirmed his theory that the atom had a small, massive nucleus, he was finally ready to go public. He chose as his forum an old Manchester organization, the Manchester Literary and Philosophical Society—"largely the general public," says James Chadwick, who attended the historic occasion as a student on March 7, 1911, ". . . people interested in literary and philosophical ideas, largely business people."

The first item on the agenda was a Manchester fruit importer's report that he had found a rare snake in a consignment of Jamaica bananas. He exhibited the snake. Then it was Rutherford's turn. Only an abstract of the announcement survives, but Chadwick remembers how it felt to hear it: it was "a most shattering performance to us, young boys that we were. . . . We realized this was obviously the truth, this was it."

Rutherford had found the nucleus of his atom. He did not yet have an arrangement for its electrons. At the Manchester meeting he spoke of "a central electric charge concentrated at a point and surrounded by a uniform spherical distribution of opposite electricity equal in amount." That was sufficiently idealized for calculation, but it neglected the significant physical fact that the "opposite electricity" must be embodied in electrons. Somehow they would have to be arranged around the nucleus.

Another mystery. A Japanese theoretical physicist, Hantaro Nagaoka, had postulated in 1903 a "Saturnian" model of the atom with flat rings of electrons revolving like Saturn's rings around a "positively charged particle." Nagaoka adapted the mathematics for his model from James Clerk Maxwell's first triumphant paper, published in 1859, "On the stability of motion of Saturn's rings." All Rutherford's biographers agree that Rutherford was unaware of Nagaoka's paper until March 11, 1911—after the Manchester meeting—when he heard about it by postcard from a physicist friend: "Campbell tells me that Nagaoka once tried to deduce a big positive centre in his atom in order to account for optical effects." He thereupon looked up the paper in the *Philosophical Magazine* and added a discussion of it to the last page of the full-length paper, "The scattering of α and β particles by matter and the structure of the atom," that he sent to the same magazine in April. He described Nagaoka's atom in that paper as being "supposed to consist of a central attracting mass surrounded by rings of rotating electrons."

But it seems that Nagaoka had recently visited him, because the Japanese physicist wrote from Tokyo on February 22, 1911, thanking him "for the great kindness you showed me in Manchester."* Yet the two physicists seem not to have discussed atomic models, or Nagaoka would probably have continued the discussion in his letter and Rutherford, a totally honest man, would certainly have acknowledged it in his paper.

One reason Rutherford was unaware of Nagaoka's Saturnian model of the atom is that it had been criticized and abandoned soon after Nagaoka introduced it because it suffered from a severe defect, the same theoretical defect that marred the atom Rutherford was now proposing. The rings of Saturn are stable because the force operating between the particles of debris that make them up—gravity—is attractive. The force operating between the electrons of Nagaoka's Saturnian electron rings, however—negative electric charge—was repulsive. It followed mathematically that whenever two or more electrons equally spaced on an orbit rotated around the nucleus, they would drift into modes of oscillation—instabilities—that would quickly tear the atom apart.

What was true for Nagaoka's Saturnian atom was also true, theoretically, for the atom Rutherford had found by experiment. It the atom operated by the mechanical laws of classical physics, the Newtonian laws that govern relationships within planetary systems, then Rutherford's model should not work. But his was not a merely theoretical construct. It was the result of real physical experiment. And work it clearly did. It was as stable as the ages and it bounced back alpha particles like cannon shells.

Someone would have to resolve the contradiction between classical physics and Rutherford's experimentally tested atom. It would need to be someone with qualities different from Rutherford's: not an experimentalist but a theoretician, yet a theoretician rooted deeply in the real. He would need at least as much courage as Rutherford had and equal self-confidence. He would need to be willing to step through the mechanical looking glass

*Nagaoka indicates indirectly that the visit took place sometime prior to July 1910—after Marsden's 1909 discovery and before Rutherford's announcement to Geiger at Christmastime 1910 that he had worked out an explanation.

into a strange, nonmechanical world where what happened on the atomic scale could not be modeled with planets or pendulums.

As if he had been called to the cause, such a person abruptly appeared in Manchester. Writing to an American friend on March 18, 1912, Rutherford announced the arrival: "Bohr, a Dane, has pulled out of Cambridge and turned up here to get some experience in radioactive work." "Bohr" was Niels Henrick David Bohr, the Danish theoretical physicist. He was then twenty-seven years old.

3

Tvi

"There came into the room a slight-looking boy," Ernest Rutherford's McGill colleague and biographer A. S. Eve recalls of Manchester days, "whom Rutherford at once took into his study. Mrs. Rutherford explained to me that the visitor was a young Dane, and that her husband thought very highly indeed of his work. No wonder, it was Niels Bohr!" The memory is odd. Bohr was an exceptional athlete. The Danes cheered his university soccer exploits. He skied, bicycled and sailed; he chopped wood; he was unbeatable at Ping-Pong; he routinely took stairs two at a time. He was also physically imposing: tall for his generation, with "an enormous domed head," says C. P. Snow, a long, heavy jaw and big hands. He was thinner as a young man than later and his shock of unruly, combed-back hair might have seemed boyish to a man of Eve's age, twelve years older than Rutherford. But Niels Bohr was hardly "slight-looking."

Something other than Bohr's physical appearance triggered Eve's dissonant memory: probably his presence, which could be hesitant. He was "much more muscular and athletic than his cautious manner suggested," Snow confirms. "It didn't help that he spoke with a soft voice, not much above a whisper."All his life Bohr talked so quietly—and yet indefatigably—that people strained to hear him. Snow knew him as "a talker as hard to get to the point as Henry James in his later years," but his speech differed dramatically between public and private and between initial exploration of a subject and eventual mastery. Publicly, according to Oskar Klein, a student of Bohr's and then a colleague, "he took the greatest care to get the most accurately shaded formulation of the matter." Albert Einstein admired Bohr for "uttering his opinions like one perpetually groping and never like one who [believed himself to be] in the possession of definite truth." If Bohr groped through the exploratory phases of his deliberations, with mastery "his assurance grew and his speech became vigorous and full of vivid images," Lise Meitner's physicist nephew Otto Frisch noted. And privately, among close friends, says Klein, "he would express himself with drastic imagery and strong expressions of admiration as well as criticism."

Bohr's manner was as binary as his speech. Einstein first met Bohr in Berlin in the spring of 1920. "Not often in life," he wrote to Bohr afterward, "has a human being caused me such joy by his mere presence as you did," and he reported to their mutual friend Paul Ehrenfest, an Austrian physicist at Leiden, "I am as much in love with him as you are." Despite his enthusiasm Einstein did not fail to observe closely his new Danish friend; his verdict in Bohr's thirty-fifth year is similar to Eve's in his twenty-eighth: "He is like an extremely sensitive child who moves around the world in a sort of trance." At first meeting—until Bohr began to speak—the theoretician Abraham Pais thought the long, heavy face "gloomy" in the extreme and puzzled at that momentary impression when everyone knew "its intense animation and its warm and sunny smile."

Bohr's contributions to twentieth-century physics would rank second only to Einstein's. He would become a scientist-statesman of unmatched foresight. To a greater extent than is usually the case with scientists, his sense of personal identity—his hard-won selfhood and the emotional values he grounded there—was crucial to his work. For a time, when he was a young man, that identity was painfully divided.

Bohr's father, Christian Bohr, was professor of physiology at the University of Copenhagen. In Christian Bohr's case the Bohr jaw extended below a thick mustache and the face was rounded, the forehead not so high. He may have been athletic; he was certainly a sports enthusiast, who encouraged and helped finance the Akademisk Boldklub for which his sons would one day play champion soccer (Niels' younger brother Harald at the 1908 Olympics). He was progressive in politics; he worked for the emancipation of women; he was skeptical of religion but nominally conforming, a solid bourgeois intellectual.

Christian Bohr published his first scientific paper at twenty-two, took a medical degree and then a Ph.D. in physiology, studied under the distinguished physiologist Carl Ludwig at Leipzig. Respiration was his special subject and he brought to that research the practice, still novel in the early 1880s, of careful physical and chemical experiment. Outside the laboratory, a friend of his explains, he was a "keen worshipper" of Goethe; larger issues of philosophy intrigued him.

One of the great arguments of the day was vitalism versus mechanism, a disguised form of the old and continuing debate between those, including the religious, who believe that the world has purpose and those who believe it operates automatically and by chance or in recurring unprogressive cycles. The German chemist who scoffed in 1895 at the "purely mechanical world" of "scientific materialism" that would allow a butterfly to turn back into a caterpillar was disputing the same issue, an issue as old as Aristotle. In Christian Bohr's field of expertise it emerged in the question whether organisms and their subsystems—their eyes, their lungs—were assembled to preexisting purpose or according to the blind and unbreathing laws of chemistry and of evolution. The extreme proponent of the mechanistic position in biology then was a German named Ernst Heinrich Haeckel, who insisted that organic and inorganic matter were one and the same. Life arose by spontaneous generation, Haeckel argued; psychology was properly a branch of physiology; the soul was not immortal nor the will free. Despite his commitment to scientific experiment Christian Bohr chose to side against Haeckel, possibly because of his worship of Goethe. He had then the difficult work of reconciling his practice with his views.

Partly for that reason, partly to enjoy the company of friends, he began stopping at a café for discussions with the philosopher Harald Høffding after the regular Friday sessions of the Royal Danish Academy of Sciences and Letters, of which they were both members. The congenial physicist C. Christensen, who spent his childhood as a shepherd, soon added a third point of view. The men moved from café meetings to regular rotation among their homes. The philologist Vilhelm Thomsen joined them to make a formidable foursome: a physicist, a biologist, a philologist, a philosopher. Niels and Harald Bohr sat at their feet all through childhood.

As earnest of his commitment to female emancipation Christian Bohr taught review classes to prepare women for university study. One of his students was a Jewish banker's daughter, Ellen Adler. Her family was cultured, wealthy, prominent in Danish life; her father was elected at various times to both the lower and upper houses of the Folketing, the Danish parliament. Christian Bohr courted her; they were married in 1881. She had a "lovable personality" and great unselfishness, a friend of her sons would say. Apparently she submerged her Judaism after her marriage. Nor did she matriculate at the university as she must originally have planned.

Christian and Ellen Bohr began married life in the Adler family townhouse that faced, across a wide street of ancient cobbles, Christiansborg Palace, the seat of the Folketing. Niels Bohr was born in that favorable place on October 7, 1885, second child and first son. When his father accepted an appointment at the university in 1886 the Bohr family moved to a house beside the Surgical Academy, where the physiology laboratories were located. There Niels and his brother Harald, nineteen months younger, grew up.

As far back as Niels Bohr could remember, he liked to dream of great interrelationships. His father was fond of speaking in paradoxes; Niels may have discovered his dreaming in that paternal habit of mind. At the same time the boy was profoundly literal-minded, a trait often undervalued that became his anchoring virtue as a physicist. Walking with him when he was about three years old, his father began pointing out the balanced structure of a tree—the trunk, the limbs, the branches, the twigs—assembling the tree for his son from its parts. The literal child saw the wholeness of the organism and dissented: if it wasn't like that, he said, it wouldn't be a tree. Bohr told that story all his life, the last time only days before he died, seventy-eight years old, in 1962. "I was from first youth able to say something about philosophical questions," he summarized proudly then. And because of that ability, he said, "I was considered something of a different character."

Harald Bohr was bright, witty, exuberant and assumed at first to be the smarter of the two brothers. "At a very early stage, however," says Niels Bohr's later collaborator and biographer Stefan Rozental, "Christian Bohr took the opposite view; he realized Niels' great abilities and special gifts and the extent of his imagination." The father phrased his realization in what would have been a cruel comparison if the brothers had been less devoted. Niels, he pronounced, was "the special one in the family."

Assigned in the fifth grade to draw a house, Niels produced a remarkably mature drawing but counted the fence pickets first. He liked carpentry and metalworking; he was household handyman from an early age. "Even as a child [he] was considered the thinker of the family," says a younger colleague, "and his father listened closely to his views on fundamental problems." He almost certainly had trouble learning to write and always had trouble writing. His mother served loyally as his amanuensis: he dictated his schoolwork to her and she copied it down.

He and Harald bonded in childhood close as twins. "There runs like a leitmotif above all else," Rozental notices, "the inseparability that characterized the relationship between the two brothers." They spoke and thought *"à deux,"* recalls one of their friends. "In my whole youth," Bohr reminisced, "my brother played a very large part. . . . I had very much to do with my brother. He was in all respects more clever than I." Harald in his turn told whoever asked that he was merely an ordinary person and his brother pure gold, and seems to have meant it.

Speech is a clumsiness and writing an impoverishment. Not language but the surface of the body is the child's first map of the world, undifferentiated between subject and object, coextensive with the world it maps until awakening consciousness divides it off. Niels Bohr liked to show how a stick used as a probe—a blind man's cane, for example—became an extension of the arm. Feeling seemed to move to the end of the stick, he said. The observation was one he often repeated—it struck his physicist protégés as wondrous—like the story of the boy and the tree, because it was charged with emotional meaning for him.

He seems to have been a child of deep connection. That is a preverbal gift. His father, with his own Goethesque yearnings for purpose and wholeness—for natural unity, for the oceanic consolations of religion without the antique formalisms—especially sensed it. His overvalued expectation burdened the boy.

Religious conflict broke early. Niels "believed literally what he learnt from the lessons on religion at school," says Oskar Klein. "For a long time this made the sensitive boy unhappy on account of his parents' lack of faith." Bohr at twenty-seven, in a Christmastime letter to his fiancée from Cambridge, remembered the unhappiness as paternal betrayal: "I see a little boy in the snow-covered street on his way to church. It was the only day his father went to church. Why? So the little boy would not feel different from other little boys. He never said a word to the little boy about belief or doubt, and the little boy believed with all of his heart."

The difficulty with writing was a more ominous sign. The family patched the problem over by supplying him with his mother's services as a secretary. He did not compose mentally while alone and then call in his helper. He composed on the spot, laboriously. That was the whispering that reminded C. P. Snow of the later Henry James. As an adult Bohr drafted and redrafted even private letters. His reworking of scientific papers in draft and then repeatedly in proof became legendary. Once after continued appeals to Zurich for the incomparable critical aid of the Austrian theoretical physicist Wolfgang Pauli, who knew Bohr well, Pauli responded warily, "If the last proof is sent away, then I will come." Bohr collaborated first with his mother and with Harald, then with his wife, then with a lifelong series of younger physicists. They cherished the opportunity of working with Bohr, but the experience could be disturbing. He wanted not only their attention but also their intellectual and emotional commitment: he wanted to convince his collaborators that he was right. Until he succeeded he doubted his conclusions himself, or at least doubted the language of their formulation.

Behind the difficulty with writing lay another, more pervasive difficulty. It took the form of anxiety that without the extraordinary support of his mother and his brother would have been crippling. For a time, it was.

It may have emerged first as religious doubt, which appeared, according to Klein, when Niels was "a young man." Bohr doubted as he had believed, "with unusual resolution." By the time he matriculated at the University of Copenhagen in the autumn of 1903, when he was eighteen, the doubt had become pervasive, intoxicating him with terrifying infinities.

Bohr had a favorite novel. Its author, Poul Martin Møller, introduced *En Dansk Students Eventyr (The Adventures of a Danish Student)* as a reading before the University of Copenhagen student union in 1824. It was published posthumously. It was short, witty and deceptively lighthearted. In an important lecture in 1960, "The Unity of Human Knowledge," Bohr described Møller's book as "an unfinished novel still read with delight by the older as well as the younger generation in [Denmark]." It gives, he said, "a remarkably vivid and suggestive account of the interplay between the various aspects of our position [as human beings]." After the Great War the Danish government helped Bohr establish an institute in Copenhagen. The most promising young physicists in the world pilgrimaged to study there. "Every one of those who came into closer contact with Bohr at the Institute," writes his collaborator Léon Rosenfeld, "as soon as he showed himself sufficiently proficient in the Danish language, was acquainted with the little book: it was part of his initiation."

What magic was contained in the little book? It was the first Danish novel with a contemporary setting: student life, and especially the extended conversations of two student cousins, one a "licentiate"—a degree candidate—the other a "philistine." The philistine is a familiar type, says Bohr, "very soberly efficient in practical affairs"; the licentiate, more exotic, "is addicted to remote philosophical meditations detrimental to his social activities." Bohr quotes one of the licentiate's "philosophical meditations":

> [I start] to think about my own thoughts of the situation in which I find myself. I even think that I think of it, and divide myself into an infinite retrogressive sequence of "I's" who consider each other. I do not know at which "I" to stop as the actual, and in the moment I stop at one, there is indeed again an "I" which stops at it. I become confused and feel a dizziness as if I were looking down into a bottomless abyss.

"Bohr kept coming back to the different meanings of the word 'I,' " Robert Oppenheimer remembered, "the 'I' that acts, the 'I' that thinks, the 'I' that studies itself."

Other conditions that trouble the licentiate in Møller's novel might be taken from a clinical description of the conditions that troubled the young Niels Bohr. This disability, for example:

> Certainly I have seen thoughts put on paper before; but since I have come distinctly to perceive the contradiction implied in such an action, I feel completely incapable of forming a single written sentence. . . . I torture myself to solve the unaccountable puzzle, how one can think, talk, or write. You see, my friend, a movement presupposes a direction. The mind cannot proceed without moving along a certain line; but before following this line, it must already have thought it. Therefore one has already thought every thought before one thinks it. Thus every thought, which seems the work of a minute, presupposes an eternity. This could almost drive me to madness.

Or this complaint, on the fragmentation of the self and its multiplying duplicity, which Bohr in later years was wont to quote:

> Thus on many occasions man divides himself into two persons, one of whom tries to fool the other, while a third one, who is in fact the same as the other two, is filled with wonder at this confusion. In short, thinking becomes dramatic and quietly acts the most complicated plots with itself and for itself; and the spectator again and again becomes actor.

"Bohr would point to those scenes," Rosenfeld notes, "in which the licentiate describes how he loses the count of his many egos, or [discourses] on the impossibility of formulating a thought, and from these fanciful antinomies he would lead his interlocutor . . . to the heart of the problem of unambiguous communication of experience, whose earnestness he thus dramatically emphasized." Rosenfeld worshiped Bohr; he failed to see, or chose not to report, that for Bohr the struggles of the licentiate were more than "fanciful antinomies."

Ratiocination—that is the technical term for what the licentiate does, the term for what the young Bohr did as well—is a defense mechanism against anxiety. Thought spirals, panicky and compulsive. Doubt doubles and redoubles, paralyzing action, emptying out the world. The mechanism is infinitely regressive because once the victim knows the trick, he can doubt anything, even doubt itself. Philosophically the phenomenon could be interesting, but as a practical matter ratiocination is a way of stalling. If work is never finished, its quality cannot be judged. The trouble is that stalling postpones the confrontation and adds that guilt to the burden.

Anxiety increases; the mechanism accelerates its spiraling flights; the self feels as if it will fragment; the multiplying "I" dramatizes the feeling of impending breakup. At that point madness reveals its horrors; the image that recurred in Bohr's conversation and writing throughout his life was the licentiate's "bottomless abyss." We are "suspended in language," Bohr liked to say, evoking that abyss; and one of his favorite quotations was two lines from Schiller:

Nur die Fülle führt zur Klarheit,
Und im Abgrund wohnt die Wahrheit.

Only wholeness leads to clarity,
And truth lies in the abyss.

But it was not in Møller that Bohr found solid footing. He needed more than a novel, however apposite, for that. He needed what we all need for sanity: he needed love and work.

"I took a great interest in philosophy in the years after my [high school] examination," Bohr said in his last interview. "I came especially in close connection with Høffding." Harald Høffding was Bohr's father's old friend, the other charter member of the Friday-night discussion group. Bohr had known him from childhood. Born in 1843, he was twelve years older than Christian Bohr, a profound, sensitive and kindly man. He was a skillful interpreter of the work of Søren Kierkegaard and of William James and a respected philosopher in his own right: an anti-Hegelian, a pragmatist interested in questions of perceptive discontinuity. Bohr became a Høffding student. It seems certain he also turned personally to Høffding for help. He made a good choice. Høffding had struggled through a crisis of his own as a young man, a crisis that brought him, he wrote later, near "despair."

Høffding was twelve years old when Søren Kierkegaard died of a lung infection in chill November 1855, old enough to have heard of the near-riot at the grave a somber walk outside the city walls, old enough for the strange, awkward, fiercely eloquent poet of multiple pseudonyms to have been a living figure. With that familiarity as a point of origin Høffding later turned to Kierkegaard's writings for solace from despair. He found it especially in *Stages on Life's Way,* a black-humorous dramatization of a dialectic of spiritual stages, each independent, disconnected, bridgeable only by an irrational leap of faith. Høffding championed the prolific and difficult Dane in gratitude; his second major book, published in 1892, would help establish Kierkegaard as an important philosopher rather than merely a literary stylist given to outbursts of raving, as Danish critics had first chosen to regard him.

Kierkegaard had much to offer Bohr, especially as Høffding interpreted him. Kierkegaard examined the same states of mind as had Poul Martin Møller. Møller taught Kierkegaard moral philosophy at the university and seems to have been a guide. After Møller's death Kierkegaard dedicated *The Concept of Dread* to him and referred to him in a draft of the dedication as "my youth's enthusiasm, my beginning's *confidant,* mighty trumpet of my awakening, my departed friend." From Møller to Kierkegaard to Høffding to Bohr: the line of descent was direct.

Kierkegaard notoriously suffered from a proliferation of identities and doubts. The doubling of consciousness is a central theme in Kierkegaard's work, as it was in Møller's before him. It would even seem to be a hazard of long standing among the Danes. The Danish word for despair, *Fortvivlelse,* carries lodged at its heart the morpheme *tvi,* which means "two" and signifies the doubling of consciousness. *Tvivl* in Danish means "doubt"; *Tvivlesyg* means "skepticism"; *Tvetydighed,* "ambiguity." The self watching itself is indeed a commonplace of puritanism, closely akin to the Christian conscience.

But unlike Møller, who jollies the licentiate's *Tvivl* away, Kierkegaard struggled to find a track through the maze of mirrors. Høffding, in his *History of Modern Philosophy,* which Bohr would have read as an undergraduate, summarizes the track he understood Kierkegaard to have found: "His leading idea was that the different possible conceptions of life are so sharply opposed to one another that we must make a choice between them, hence his catchword *either-or;* moreover, it must be a choice which each particular person must make for himself, hence his second catchword, *the individual.*" And, following: "Only in the world of possibilities is there continuity; in the world of reality decision always comes through a breach of continuity." Continuity in the sense that it afflicted Bohr was the proliferating stream of doubts and "I's" that plagued him; a breach of that continuity—decisiveness, function—was the termination he hoped to find.

He turned first to mathematics. He learned in a university lecture about Riemannian geometry, a type of non-Euclidean geometry developed by the German mathematician Georg Riemann to represent the functions of complex variables. Riemann showed how such multivalued functions (a number, its square root, its logarithm and so on) could be represented and related on a stack of coincident geometric planes that came to be called Riemann surfaces. "At that time," Bohr said in his last interview, "I really thought to write something about philosophy, and that was about this analogy with multivalued functions. I felt that the various problems in psychology—which were called the big philosophical problems, of the free will and such things—that one could really reduce them when one considered how one really went about them, and that was done on the analogy to multivalued functions." By then he thought the problem might be one of language, of the ambiguity—the multiple values, as it were—between different meanings of the word "I." Separate each different meaning on a different plane and you could keep track of what you were talking about. The confusion of identities would resolve itself graphically before one's eyes.

The scheme was too schematic for Bohr. Mathematics was probably too much like ratiocination, leaving him isolated within his anxiety. He thought of writing a book about his mathematical analogies but leapt instead to work that was far more concrete. But notice that the mathematical analogy begins to embed the problem of doubt within the framework of language, identifying doubt as a specialized form of verbal ambiguity, and notice that it seeks to clarify ambiguities by isolating their several variant meanings on separate, disconnected planes.

The solid work Bohr took up, in February 1905, when he was nineteen years old, was a problem in experimental physics. Each year the Royal Danish Academy of Sciences and Letters announced problems for study against a two-year deadline, after which the academy awarded gold and silver medals for successful papers. In 1905 the physics problem was to determine the surface tension of a number of liquids by measuring the waves produced in those liquids when they were allowed to run out through a hole (the braided cascade of a garden hose demonstrates such waves). The method had been proposed by the British Nobelist John William Strutt, Lord Rayleigh, but no one had yet tried it out. Bohr and one other contestant accepted the challenge.

Bohr went to work in the physiology laboratory where he had watched and then assisted his father for years, learning the craft of experiment. To produce stable jets he decided to use drawn-out glass tubes. Because the method required large quantities of liquid he limited his experiment to water. The tubes had to be flattened on the sides to make an oval cross section; that gave the jet of water the shape it needed to evolve braidlike waves. All the work of heating, softening and drawing out the tubes Bohr did himself; he found it hypnotic. Rosenfeld says Bohr "took such delight in this operation that, completely forgetting its original purpose, he spent hours passing tube after tube through the flame."

Each separate experimental determination of the surface-tension value took hours. It had to be done at night, when the lab was unoccupied, because the jets were easily disturbed by vibration. Slow work, but Bohr also dawdled. The academy had allowed two years. Toward the end of that time Christian Bohr realized his son was procrastinating to the point where he might not finish his paper before the deadline. "The experiments had no end," Bohr told Rosenfeld some years later on a bicycle ride in the country; "I always noticed new details that I thought I had first to understand. At last my father sent me out here, away from the laboratory, and I had to write up the paper."

"Out here" was Naerumgaard, the Adler country estate north of Copenhagen. There, away from the temptations of the laboratory, Niels wrote and Harald transcribed an essay of 114 pages. Niels submitted it to the academy on the day of deadline, but even then it was incomplete; three days later he turned in an eleven-page addendum that had been accidentally left off.

The essay, Bohr's first scientific paper, determined the surface tension only of water but also uniquely extended Rayleigh's theory. It won a gold medal from the academy. It was an outstanding achievement for someone so young and it set Bohr's course for physics. Unlike mathematicized philosophy, physics was anchored solidly in the real world.

In 1909 the Royal Society of London accepted the surface-tension paper in modified form for its *Philosophical Transactions.* Bohr, who was still only a student working toward his master's degree when the essay appeared, had to explain to the secretary of the society, who had addressed him by his presumed academic title, that he was "not a professor."

Retreating to the country had helped him once. It might help again. Naerumgaard ceased to be available when the Adler family donated it for use as a school. When the time came to study for his master's degree examinations, between March and May 1909, Bohr traveled to Vissenbjerg, on the island of Funen, the next island west from Copenhagen's Zealand, to stay at the parsonage of the parents of Christian Bohr's laboratory assistant. Niels procrastinated on Funen by reading *Stages on Life's Way.* The day he finished it he enthusiastically mailed the book to Harald. "This is the only thing I have to send," he wrote his younger brother; "nevertheless, I don't think I could easily find anything better. . . . It is something of the finest I have ever read." At the end of June, back in Copenhagen, again on deadline day, Bohr turned in his master's thesis, copied out in his mother's hand.

Harald had sprinted ahead of him by then, having won his M.Sc. in April and gone off to the Georgia-Augusta University in Göttingen, Germany, the center of European mathematics, to study for his Ph.D. He received that degree in Göttingen in June 1910. Niels wrote his younger brother tongue-in-cheek that his "envy would soon be growing over the rooftops," but in fact he was happy with his progress on his own doctoral dissertation despite having spent "four months speculating about a silly question about some silly electrons and [succeeding] only in writing circa fourteen more or less divergent rough drafts." Christensen had posed Bohr a problem in the electron theory of metals for his master's thesis; the subject interested Bohr enough to continue pursuing it as his doctoral work.

He was specializing in theoretical studies now; to try to do experimental work too, he explained, was "unpractical."

He returned to the parsonage at Vissenbjerg in the autumn of 1910. His work slowed. He may have recalled the licentiate's dissertation problems, for he again turned to Kierkegaard. "He made a powerful impression on me when I wrote my dissertation in a parsonage in Funen, and I read his works night and day," Bohr told his friend and former student J. Rud Nielsen in 1933. "His honesty and his willingness to think the problems through to their very limit is what is great. And his language is wonderful, often sublime. There is of course much in Kierkegaard that I cannot accept. I ascribe that to the times in which he lived. But I admire his intensity and perseverance, his analysis to the utmost limit, and the fact that through these qualities he turned misfortune and suffering into something good."

He finished his Ph.D. thesis, "Studies in the electron theory of metals," by the end of January 1911. On February 3, suddenly, at fifty-six, his father died. He dedicated his thesis "in deepest gratitude to the memory of my father." He loved his father; if there had been a burden of expectation he was free of that burden now.

As was customary, he publicly defended his thesis in Copenhagen on May 13. "Dr. Bohr, a pale and modest young man," the Copenhagen newspaper *Dagbladet* reported under a crude drawing of the candidate standing in white tie and tails at a heavy lectern, "did not take much part in the proceedings, whose short duration is a record." The small hall was crowded to overflowing. Christiansen, one of the two examiners, said simply that hardly anyone in Denmark was well enough informed on the subject to judge the candidate's work.

Before he died Christian Bohr had helped arrange a fellowship from the Carlsberg Foundation for his son for study abroad. Niels spent the summer sailing and hiking with Margrethe Nørland, the sister of a friend, a beautiful young student whom he had met in 1910 and to whom, shortly before his departure, he became engaged. Then he went off in late September to Cambridge. He had arranged to study at the Cavendish under J. J. Thomson.

> 29 Sept. 1911
> Eltisley Avenue 10,
> Newnham, Cambridge
>
> Oh Harald!
>
> Things are going so well for me. I have just been talking to J. J. Thomson and have explained to him, as well as I could, my ideas about radiation, magnetism, etc. If you only knew what it meant to me to talk to such a man. He was extremely nice to me, and we talked

> about so much; and I do believe that he thought there was some sense in what I said. He is now going to read [my dissertation] and he invited me to have dinner with him Sunday at Trinity College; then he will talk with me about it. You can imagine that I am happy. . . . I now have my own little flat. It is at the edge of town and is very nice in all respects. I have two rooms and eat all alone in my own room. It is very nice here; now, as I am sitting and writing to you, it blazes and rumbles in my own little fireplace.

Niels Bohr was delighted with Cambridge. His father's Anglophilia had prepared him to like English settings; the university offered the tradition of Newton and Clerk Maxwell and the great Cavendish Laboratory with its awesome record of physical discovery. Bohr found that his schoolboy English needed work and set out reading *David Copperfield* with an authoritative new dictionary at hand, looking up every uncertain word. He discovered that the laboratory was crowded and undersupplied. On the other hand, it was amusing to have to go about in cap and gown (once he was admitted to Trinity as a research student) "under threat of high fines," to see the Trinity high table "where they eat so much and so first-rate that it is quite unbelievable and incomprehensible that they can stand it," to walk "for an hour before dinner across the most beautiful meadows along the river, with the hedges flecked with red berries and with isolated windblown willow trees—imagine all this under the most magnificent autumn sky with scurrying clouds and blustering wind." He joined a soccer club; called on physiologists who had been students of his father; attended physics lectures; worked on an experiment Thomson had assigned him; allowed the English ladies, "absolute geniuses at drawing you out," to do their duty by him at dinner parties.

But Thomson never got around to reading his dissertation. The first meeting had not, in fact, gone so well. The new student from Denmark had done more than explain his ideas; he had shown Thomson the errors he found in Thomson's electron-theory work. "I wonder," Bohr wrote Margrethe soon after, "what he will say to my disagreement with his ideas." And a little later: "I'm longing to hear what Thomson will say. He's a great man. I hope he will not get angry with my silly talk."

Thomson may or may not have been angry. He was not much interested in electrons anymore. He had turned his attention to positive rays—the experiment he assigned Bohr concerned such rays and Bohr found it distinctly unpromising—and in any case had very little patience with theoretical discussions. "It takes half a year to get to know an Englishman," Bohr said in his last interview. ". . . It was the custom in England that they would be polite and so on, but they wouldn't be interested to see anybody. . . . I went Sundays to the dinner in Trinity College. . . . I was sitting there, and nobody spoke to me ever in many Sundays. But then they understood that I was not more eager to speak to them than they were to speak to me. And then we were friends, you see, and then the whole thing was different." The insight is generalized; Thomson's indifference was perhaps its first specific instance.

Then Rutherford turned up at Cambridge.

He "came down from Manchester to speak at the annual Cavendish Dinner," says Bohr. "Although on this occasion I did not come into personal contact with [him], I received a deep impression of the charm and power of his personality by which he had been able to achieve almost the incredible wherever he worked. The dinner"—in December—"took place in a most humorous atmosphere and gave the opportunity for several of Rutherford's colleagues to recall some of the many anecdotes which already then were attached to his name." Rutherford spoke warmly of the recent work of the physicist C. T. R. Wilson, the inventor of the cloud chamber (which made the paths of charged particles visible as lines of water droplets hovering in supersaturated fog) and a friend from Cambridge student days. Wilson had "just then," says Bohr, photographed alpha particles in his cloud chamber scattering from interactions with nuclei, "the phenomenon which only a few months before had led [Rutherford] to his epoch-making discovery of the atomic nucleus."

Bohr had matters on his mind that he would soon relate to the problem of the nucleus and its theoretically unstable electrons, but it was Rutherford's enthusiastic informality that most impressed him at the annual dinner. Remembering this period of his life long afterward, he would single out for special praise among Rutherford's qualities "the patience to listen to every young man when he felt he had any idea, however modest, on his mind." In contrast, presumably, to J. J. Thomson, whatever Thomson's other virtues.

Soon after the dinner Bohr went up to Manchester to visit "one of my recently deceased father's colleagues who was also a close friend of Rutherford," whom Bohr wanted to meet. The close friend brought them together. Rutherford looked over the young Dane and liked what he saw despite his prejudice against theoreticians. Someone asked him later about the discrepancy. "Bohr's different," Rutherford roared, disguising affection with bluster. "He's a football player!" Bohr was different in another regard as well; he was easily the most talented of all Rutherford's many students—and Rutherford trained no fewer than eleven Nobel Prize winners during his life, an unsurpassed record.

Bohr held up his decision between Cambridge and Manchester until he could go over everything with Harald, who visited him in Cambridge in January 1912 for the purpose. Then Bohr eagerly wrote Rutherford for permission to study at Manchester, as they had discussed in December. Rutherford had advised him then not to give up on Cambridge too quickly—Manchester is always here, he told him, it won't run away—and so Bohr proposed to arrive for spring term, which began in late March. Rutherford gladly agreed. Bohr felt he was being wasted at Cambridge. He wanted substantial work.

His first six weeks in Manchester he spent following "an introductory course on the experimental methods of radioactive research," with Geiger and Marsden among the instructors. He continued pursuing his independent studies in electron theory. He began a lifelong friendship with a young Hungarian aristocrat, George de Hevesy, a radiochemist with a long, sensitive face dominated by a towering nose. De Hevesy's father was a court councillor, his mother a baroness; as a child he had hunted partridge in the private game park of the Austro-Hungarian emperor Franz Josef next to his grandfather's estate. Now he was working to meet a challenge Rutherford had thrown at him one day to separate radioactive decay products from their parent substances. Out of that work he developed over the next several decades the science of using radioactive tracers in medical and biological research, one more useful offspring of Rutherford's casual but fecund paternity.

Bohr learned about radiochemistry from de Hevesy. He began to see connections with his electron-theory work. His sudden burst of intuitions then was spectacular. He realized in the space of a few weeks that radioactive properties originated in the atomic nucleus but chemical properties depended primarily on the number and distribution of electrons. He realized—the idea was wild but happened to be true—that since the electrons determined the chemistry and the total positive charge of the nucleus determined the number of electrons, an element's position on the periodic table of the elements was exactly the nuclear charge (or "atomic number"): hydrogen first with a nuclear charge of 1, then helium with a nuclear charge of 2 and so on up to uranium at 92.

De Hevesy remarked to him that the number of known radioelements already far outnumbered the available spaces on the periodic table and Bohr made more intuitive connections. Soddy had pointed out that the radioelements were generally not new elements, only variant physical forms of the natural elements (he would soon give them their modern name, isotopes). Bohr realized that the radioelements must have the same atomic number as the natural elements with which they were chemically identical. That enabled him to rough out what came to be called the radioactive dis-

1 H 1.0080																	2 He 4.0026
3 Li 6.941	4 Be 9.0122											5 B 10.81	6 C 12.011	7 N 14.0067	8 O 15.9994	9 F 18.9984	10 Ne 20.179
11 Na 22.9898	12 Mg 24.305											13 Al 26.9815	14 Si 28.086	15 P 30.9738	16 S 32.06	17 Cl 35.453	18 Ar 39.948
19 K 39.102	20 Ca 40.08	21 Sc 44.956	22 Ti 47.90	23 V 50.941	24 Cr 51.996	25 Mn 54.9380	26 Fe 55.847	27 Co 58.9332	28 Ni 58.71	29 Cu 63.54	30 Zn 65.37	31 Ga 69.72	32 Ge 72.59	33 As 74.9216	34 Se 78.96	35 Br 79.909	36 Kr 83.80
37 Rb 85.467	38 Sr 87.62	39 Y 88.906	40 Zr 91.22	41 Nb 92.906	42 Mo 95.94	43 Tc (99)	44 Ru 101.07	45 Rh 102.906	46 Pd 106.4	47 Ag 107.870	48 Cd 112.40	49 In 114.82	50 Sn 118.69	51 Sb 121.75	52 Te 127.60	53 I 126.9045	54 Xe 131.30
55 Cs 132.906	56 Ba 137.34	57 La 138.906	72 Hf 178.49	73 Ta 180.948	74 W 183.85	75 Re 186.2	76 Os 190.2	77 Ir 192.2	78 Pt 195.09	79 Au 196.967	80 Hg 200.59	81 Tl 204.37	82 Pb 207.2	83 Bi 208.981	84 Po (210)	85 At (210)	86 Rn (222)
87 Fr (223)	88 Ra (226)	89 Ac (227)															

Lanthanide Series

58 Ce 140.12	59 Pr 140.908	60 Nd 144.24	61 Pm (147)	62 Sm 150.4	63 Eu 151.96	64 Gd 157.25	65 Tb 158.925	66 Dy 162.50	67 Ho 164.930	68 Er 167.26	69 Tm 168.934	70 Yb 173.04	71 Lu 174.97

Actinide Series

90 Th (232)	91 Pa (231)	92 U (238)	93 Np (237)	94 Pu (242)

Periodic table of the elements. The lanthanide series ("rare earths"), beginning with lanthanum (57), and the actinide series, which begins with actinium (89) and includes thorium (90) and uranium (92), are chemically similar. Other families of elements read vertically down the table—at the far right, for example, the noble gases: helium, neon, argon, krypton, xenon, radon.

placement law: that when an element transmutes itself through radioactive decay it shifts its position on the periodic table two places to the left if it emits an alpha particle (a helium nucleus, atomic number 2), one place to the right if it emits a beta ray (an energetic electron, which leaves behind in the nucleus an extra positive charge).

All these first rough insights would be the work of other men's years to anchor soundly in theory and experiment. Bohr ran them in to Rutherford. To his surprise, he found the discoverer of the nucleus cautious about his own discovery. "Rutherford . . . thought that the meagre evidence [so far obtained] about the nuclear atom was not certain enough to draw such consequences," Bohr recalled. "And I said to him that I was sure that it would be the final proof of his atom." If not convinced, Rutherford was at least impressed; when de Hevesy asked him a question about radiation one day Rutherford responded cheerfully, "Ask Bohr!"

Rutherford was well prepared for surprises, then, when Bohr came to see him again in mid-June. Bohr told Harald what he was on to in a letter on June 19, after the meeting:

> It could be that I've perhaps found out a little bit about the structure of atoms. You must not tell anyone anything about it, otherwise I certainly could not write you this soon. If I'm right, it would not be an indication of the nature of a possibility . . . but perhaps a little piece of reality. . . . You understand that I may yet be wrong, for it hasn't been worked out fully yet (but I don't think so); nor do I believe that Rutherford thinks it's completely wild; he is the right kind of man and would never say that he was convinced of something that was not entirely worked out. You can imagine how anxious I am to finish quickly.

Bohr had caught a first glimpse of how to stabilize the electrons that orbited with such theoretical instability around Rutherford's nucleus. Rutherford sent him off to his rooms to work it out. Time was running short; he planned to marry Margrethe Nørland in Copenhagen on August 1. He wrote Harald on July 17 that he was "getting along fairly well; I believe I have found out a few things; but it is certainly taking more time to work them out than I was foolish enough to believe at first. I hope to have a little paper ready to show to Rutherford before I leave, so I'm busy, so busy; but the unbelieveable heat here in Manchester doesn't exactly help my diligence. How I look forward to talking to you!" By the following Wednesday, July 22, he had seen Rutherford, won further encouragement, and was making plans to meet Harald on the way home.

Bohr married, a serene marriage with a strong, intelligent and beautiful woman that lasted a lifetime. He taught at the University of Copenhagen through the autumn term. The new model of the atom he was struggling to develop continued to tax him. On November 4 he wrote Rutherford that he expected "to be able to finish the paper in a few weeks." A few weeks passed; with nothing finished he arranged to be relieved of his university teaching and retreated to the country with Margrethe. The old system worked; he produced "a very long paper on all these things." Then an important new idea came to him and he broke up his original long paper and began rewriting it into three parts. "On the constitution of atoms and molecules," so proudly and bravely titled—Part I mailed to Rutherford on March 6, 1913, Parts II and III finished and published before the end of the year—would change the course of twentieth-century physics. Bohr won the 1922 Nobel Prize in Physics for the work.

As far back as Bohr's doctoral dissertation he had decided that some of the phenomena he was examining could not be explained by the mechanical laws of Newtonian physics. "One must assume that there are forces in nature of a kind completely different from the usual mechanical sort," he wrote then. He knew where to look for these different forces: he looked to the work of Max Planck and Albert Einstein.

Planck was the German theoretician whom Leo Szilard would meet at the University of Berlin in 1921; born in 1858, Planck had taught at Berlin since 1889. In 1900 he had proposed a revolutionary idea to explain a persistent problem in mechanical physics, the so-called ultraviolet catastrophe. According to classical theory there should be an infinite amount of light (energy, radiation) inside a heated cavity such as a kiln. That was because classical theory, with its continuity of process, predicted that the particles in the heated walls of the cavity which vibrated to produce the light would vibrate to an infinite range of frequencies.

Obviously such was not the case. But what kept the energy in the cavity from running off infinitely into the far ultraviolet? Planck began his effort to find out in 1897 and pursued it for three hard years. Success came with a last-minute insight announced at a meeting of the Berlin Physical Society on October 19, 1900. Friends checked Planck's new formula that very night against experimentally derived values. They reported its accuracy to him the next morning. "Later measurements, too," Planck wrote proudly in 1947, at the end of his long life, "confirmed my radiation formula again and again—the finer the methods of measurement used, the more accurate the formula was found to be."

Planck solved the radiation problem by proposing that the vibrating particles can only radiate at certain energies. The permitted energies would be determined by a new number—"a universal constant," he says, "which I called *h*. Since it had the dimension of action (energy × time), I gave it the name, *elementary quantum of action.*" (*Quantum* is the neuter form of the Latin word *quantus,* meaning "how great.") Only those limited and finite energies could appear which were whole-number multiples of $h\nu$: of the frequency ν times Planck's *h*. Planck calculated *h* to be a very small number, close to the modern value of 6.63×10^{-27} erg-seconds. Universal *h* soon acquired its modern name: *Planck's constant.*

Planck, a thoroughgoing conservative, had no taste for pursuing the radical consequences of his radiation formula. Someone else did: Albert Einstein. In a paper in 1905 that eventually won for him the Nobel Prize,

Einstein connected Planck's idea of limited, discontinuous energy levels to the problem of the photoelectric effect. Light shone on certain metals knocks electrons free; the effect is applied today in the solar panels that power spacecraft. But the energy of the electrons knocked free of the metal does not depend, as common sense would suggest, on the brightness of the light. It depends instead on the *color* of the light—on its frequency.

Einstein saw a quantum condition in this odd fact. He proposed the heretical possibility that light, which years of careful scientific experiment had demonstrated to travel in waves, actually traveled in small individual packets—particles—which he called "energy quanta." Such photons (as they are called today), he wrote, have a distinctive energy $h\nu$ and they transfer most of that energy to the electrons they strike on the surface of the metal. A brighter light thus releases more electrons but not more energetic electrons; the energy of the electrons released depends on $h\nu$ and so on the frequency of the light. Thus Einstein advanced Planck's quantum idea from the status of a convenient tool for calculation to that of a possible physical fact.

With these advances in understanding Bohr was able to confront the problem of the mechanical instability of Rutherford's model of the atom. In July, at the time of the "little paper ready to show to Rutherford," he already had his central idea. It was this: that since classical mechanics predicted that an atom like Rutherford's, with a small, massive central nucleus surrounded by orbiting electrons, would be unstable, while in fact atoms are among the most stable of systems, classical mechanics was inadequate to describe such systems and would have to give way to a quantum approach. Planck had introduced quantum principles to save the laws of thermodynamics; Einstein had extended the quantum idea to light; Bohr now proposed to lodge quantum principles within the atom itself.

Through the autumn and early winter, back in Denmark, Bohr pursued the consequences of his idea. The difficulty with Rutherford's atom was that nothing about its design justified its stability. If it happened to be an atom with several electrons, it would fly apart. Even if it were a hydrogen atom with only one (mechanically stable) electron, classical theory predicted that the electron would radiate light as it changed direction in its orbit around the nucleus and therefore, the system losing energy, would spiral into the nucleus and crash. The Rutherford atom, from the point of view of Newtonian mechanics—as a miniature solar system—ought to be impossibly large or impossibly small.

Bohr therefore proposed that there must be what he called "stationary states" in the atom: orbits the electrons could occupy without instability, without radiating light, without spiraling in and crashing. He worked the numbers of this model and found they agreed very well with all sorts of experimental values. Then at least he had a plausible model, one that explained in particular some of the phenomena of chemistry. But it was apparently arbitrary; it was not more obviously a real picture of the atom than other useful models such as J. J. Thomson's plum pudding.

Help came then from an unlikely quarter. A professor of mathematics at King's College, London, J. W. Nicholson, whom Bohr had met and thought a fool, published a series of papers proposing a quantized Saturnian model of the atom to explain the unusual spectrum of the corona of the sun. The papers were published in June in an astronomy journal; Bohr didn't see them until December. He was quickly able to identify the inadequacies of Nicholson's model, but not before he felt the challenge of other researchers breathing down his neck—and not without noticing Nicholson's excursion into the jungle of spectral lines.

Oriented toward chemistry, communicating back and forth with George de Hevesy, Bohr had not thought of looking at spectroscopy for evidence to support his model of the atom. "The spectra was a very difficult problem," he said in his last interview. ". . . One thought that this is marvelous, but it is not possible to make progress there. Just as if you have the wing of a butterfly, then certainly it is very regular with the colors and so on, but nobody thought that one could get the basis of biology from the coloring of the wing of a butterfly."

Taking Nicholson's hint, Bohr now turned to the wings of the spectral butterfly.

Spectroscopy was a well-developed field in 1912. The eighteenth-century Scottish physicist Thomas Melvill had first productively explored it. He mixed chemical salts with alcohol, lit the mixtures and studied the resulting light through a prism. Each different chemical produced characteristic patches of color. That suggested the possibility of using spectra for chemical analysis, to identify unknown substances. The prism spectroscope, invented in 1859, advanced the science. It used a narrow slit set in front of a prism to limit the patches of light to similarly narrow lines; these could be directed onto a ruled scale (and later onto strips of photographic film) to measure their spacing and calculate their wavelengths. Such characteristic patterns of lines came to be called line spectra. Every element had its own unique line spectrum. Helium was discovered in the chromosphere of the sun in 1868 as a series of unusual spectral lines twenty-three years before it was discovered mixed into uranium ore on earth. The line spectra had their uses.

But no one understood what produced the lines. At best, mathematicians and spectroscopists who liked to play with wavelength numbers were able to find beautiful harmonic regularities among sets of spectral lines. Johann Balmer, a nineteenth-century Swiss mathematical physicist, identified in 1885 one of the most basic harmonies, a formula for calculating the wavelengths of the spectral lines of hydrogen. These, collectively called the Balmer series, look like this:

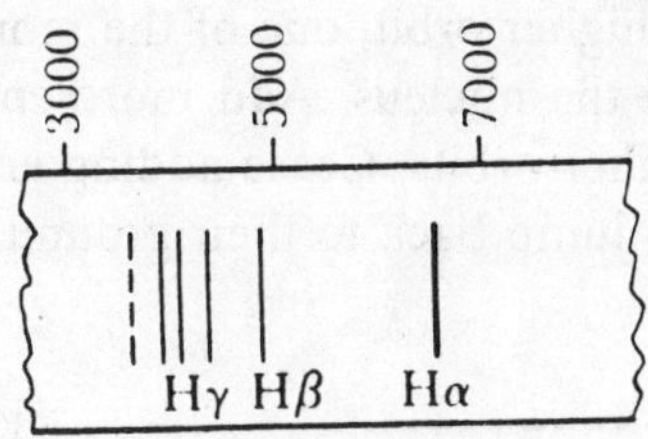

Balmer series

It is not necessary to understand mathematics to appreciate the simplicity of the formula Balmer derived that predicts a line's location on the spectral band to an accuracy of within one part in a thousand, a formula that has only one arbitrary number:

$$\lambda = 3645.6\left(\frac{n^2}{n^2 - 4}\right)$$

(the Greek letter λ, lambda, stands for the wavelength of the line; *n* takes the values 3, 4, 5 and so on for the various lines). Using his formula, Balmer was able to predict the wavelengths of lines to be expected for parts of the hydrogen spectrum not yet studied. They were found where he said they would be.

A Swedish spectroscopist, Johannes Rydberg, went Balmer one better and published in 1890 a general formula valid for a great many different line spectra. The Balmer formula then became a special case of the more general Rydberg equation, which was built around a number called the Rydberg constant. That number, subsequently derived by experiment and one of the most accurately known of all universal constants, takes the precise modern value of 109,677 cm^{-1}.

Bohr would have known these formulae and numbers from undergraduate physics, especially since Christensen was an admirer of Rydberg and had thoroughly studied his work. But spectroscopy was far from Bohr's field and he presumably had forgotten them. He sought out his old friend and classmate, Hans Hansen, a physicist and student of spectroscopy just returned from Göttingen. Hansen reviewed the regularity of line spectra with him. Bohr looked up the numbers. "As soon as I saw Balmer's formula," he said afterward, "the whole thing was immediately clear to me."

What was immediately clear was the relationship between his orbiting electrons and the lines of spectral light. Bohr proposed that an electron bound to a nucleus normally occupies a stable, basic orbit called a ground state. Add energy to the atom—heat it, for example—and the electron responds by jumping to a higher orbit, one of the more energetic stationary states farther away from the nucleus. Add more energy and the electron continues jumping to higher orbits. Cease adding energy—leave the atom alone—and the electrons jump back to their ground states, like this:

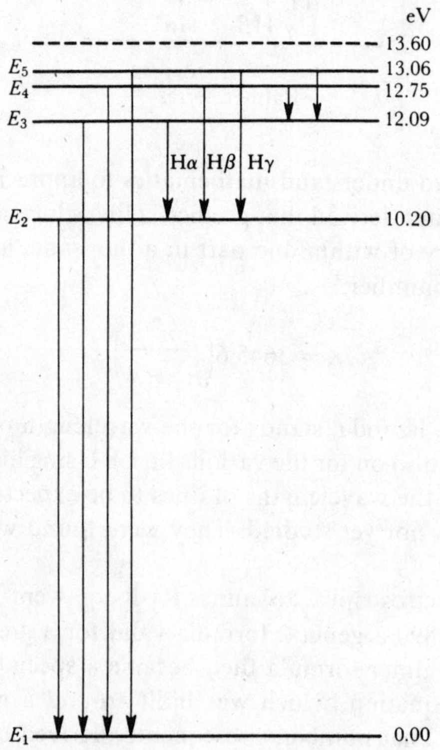

With each jump, each electron emits a photon of characteristic energy. The jumps, and so the photon energies, are limited by Planck's constant. Subtract the value of a lower-energy stationary state W_2 from the value of a higher energy stationary state W_1 and you get exactly the energy of the light as *hv*. So here was the physical mechanism of Planck's cavity radiation.

From this elegant simplification, $W_1 - W_2 = hv$, Bohr was able to derive the Balmer series. The lines of the Balmer series turn out to be exactly the energies of the photons that the hydrogen electron emits when it jumps down from orbit to orbit to its ground state.

Then, sensationally, with the simple formula

$$R = \frac{2\pi^2 me^4}{h^3}$$

(where *m* is the mass of the electron, *e* the electron charge and *h* Planck's constant—all fundamental numbers, not arbitrary numbers Bohr made up) Bohr produced Rydberg's constant, calculating it within 7 percent of its experimentally measured value! "There is nothing in the world which impresses a physicist more," an American physicist comments, "than a numerical agreement between experiment and theory, and I do not think that there can ever have been a numerical agreement more impressive than this one, as I can testify who remember its advent."

"On the constitution of atoms and molecules" was seminally important to physics. Besides proposing a useful model of the atom, it demonstrated that events that take place on the atomic scale are quantized: that just as matter exists as atoms and particles in a state of essential graininess, so also does process. Process is discontinuous and the "granule" of process—of electron motions within the atom, for example—is Planck's constant. The older mechanistic physics was therefore imprecise; though a good approximation that worked for large-scale events, it failed to account for atomic subtleties.

Bohr was happy to force this confrontation between the old physics and the new. He felt that it would be fruitful for physics. Because original work is inherently rebellious, his paper was not only an examination of the physical world but also a political document. It proposed, in a sense, to begin a reform movement in physics: to limit claims and clear up epistemological fallacies. Mechanistic physics had become authoritarian. It had outreached itself to claim universal application, to claim that the universe and everything in it is rigidly governed by mechanistic cause and effect. That was Haeckelism carried to a cold extreme. It stifled Niels Bohr as biological Haeckelism had stifled Christian Bohr and as a similar authoritarianism in philosophy and in bourgeois Christianity had stifled Søren Kierkegaard.

When Rutherford saw Bohr's Part I paper, for example, he immediately found a problem. "There appears to me one grave difficulty in your hypothesis," he wrote Bohr on March 20, "which I have no doubt you fully realise, namely, how does an electron decide what frequency it is going to vibrate at when it passes from one stationary state to the other? It seems to me that you would have to assume that the electron knows beforehand where it is going to stop." Einstein showed in 1917 that the physical answer to Rutherford's question is statistical—any frequency is possible, and the ones that turn up happen to have the best odds. But Bohr answered the question in a later lecture in more philosophical and even anthropomorphic terms: "Every change in the state of an atom should be regarded as an individual process, incapable of more detailed description, by which the atom goes over from one so-called stationary state to another. . . . We are here so far removed from a causal description that an atom in a stationary state may in general even be said to possess a free choice between various possible transitions." The "catchwords" here, as Harald Høffding might say, are *individual* and *free choice.* Bohr means the changes of state within individual atoms are not predictable; the catchwords color that physical limitation with personal emotion.

In fact the 1913 paper was deeply important emotionally to Bohr. It is a remarkable example of how science works and of the sense of personal authentication that scientific discovery can bestow. Bohr's emotional preoccupations sensitized him to see previously unperceived regularities in the natural world. The parallels between his early psychological concerns and his interpretation of atomic processes are uncanny, so much so that without the great predictive ability of the paper its assumptions would seem totally arbitrary.

Whether or not the will is free, for example, was a question that Bohr took seriously. To identify a kind of freedom of choice within the atom itself was a triumph for his carefully assembled structure of beliefs. The separate, distinct electron orbits that Bohr called stationary states recall Kierkegaard's stages. They also recall Bohr's attempt to redefine the problem of free will by invoking separate, distinct Riemann surfaces. And as Kierkegaard's stages are discontinuous, negotiable only by leaps of faith, so do Bohr's electrons leap discontinuously from orbit to orbit. Bohr insisted as one of the two "principal assumptions" of his paper that the electron's whereabouts between orbits cannot be calculated or even visualized. Before and after are completely discontinuous. In that sense, each stationary state of the electron is complete and unique, and in that wholeness is stability. By contrast, the continuous process predicted by classical mechanics, which Bohr apparently associated with the licentiate's endless ratiocination, tears the atom apart or spirals it into radiative collapse.

Bohr may have found his way through his youthful emotional crisis in part by calling up his childhood gift of literal-mindedness. He famously insisted on anchoring physics in fact and refused to carry argument beyond physical evidence. He was never a system-builder. "Bohr characteristically avoids such a word as 'principle,' " says Rosenfeld; "he prefers to speak of 'point of view' or, better still, 'argument,' i.e. line of reasoning; likewise, he rarely mentions the 'laws of nature,' but rather refers to 'regularities of the phenomena.' " Bohr was not displaying false humility with his choice of terms; he was reminding himself and his colleagues that physics is not a grand philosophical system of authoritarian command but simply a way, in his favorite phrase, of "asking questions of Nature." He apologized similarly for his tentative, rambling habit of speech: "I try not to speak more clearly than I think."

"He points out," Rosenfeld adds, "that the idealized concepts we use in science must ultimately derive from common experiences of daily life which cannot themselves be further analysed; therefore, whenever any two such idealizations turn out to be incompatible, this can only mean that some mutual limitation is imposed upon their validity." Bohr had found a solution to the spiraling flights of doubt by stepping out of what Kierkegaard called "the fairyland of the imagination" and back into the real world. In the real world material objects endure; their atoms cannot, then, ordinarily be unstable. In the real world cause and effect sometimes seem to limit our freedom, but at other times we know we choose. In the real world it is meaningless to doubt existence; the doubt itself demonstrates the existence of the doubter. Much of the difficulty was language, that slippery medium in which Bohr saw us inextricably suspended. "It is wrong," he told his colleagues repeatedly, "to think that the task of physics is to find out how nature *is*"—which is the territory classical physics had claimed for itself. "Physics concerns what we can *say* about nature."

Later Bohr would develop far more elaborately the idea of mutual limitations as a guide to greater understanding. It would supply a deep philosophical basis for his statecraft as well as for his physics. In 1913 he first demonstrated its resolving power. "It was clear," he remembered at the end of his life, "and that was *the* point about the Rutherford atom, that we had something from which we could not proceed at all in any other way than by radical change. And that was the reason then that [I] took it up so seriously."

4

The Long Grave Already Dug

Otto Hahn cherished the day the Kaiser came to visit. The official dedication of the first two Kaiser Wilhelm Institutes, one for chemistry, one for physical chemistry, on October 23, 1912—Bohr in Copenhagen was approaching his quantized atom—was a wet day in the suburb of Dahlem southwest of Berlin. The Kaiser, Wilhelm II, Victoria's eldest grandson, wore a raincloak to protect his uniform, the dark collar of his greatcoat turned out over the lighter shawl of the cloak. The officials who walked the requisite paces behind him, his scholarly friend Adolf von Harnack and the distinguished chemist Emil Fischer foremost among them, made do with dark coats and top hats; those farther back in the procession who carried umbrellas kept them furled. Schoolboys, caps in hand, lined the curbs of the shining street like soldiers on parade. They stood at childish attention, awe dazing their dreamy faces, as this corpulent middle-aged man with upturned dark mustaches who believed he ruled them by divine right passed in review. They were thirteen, perhaps fourteen years old. They would be soldiers soon enough.

Officials in the Ministry of Culture had encouraged His Imperial Majesty to support German science. He responded by donating land for a research center on what had been a royal farm. Industry and government then lavishly endowed a science foundation, the Kaiser Wilhelm

Society, to operate the proposed institutes, of which there would be seven by 1914.

The society began its official life early in 1911 with von Harnack, a theologian who was the son of a chemist, as its first president. The imperial architect, Ernst von Ihne, went briskly to work. The Kaiser came to Dahlem to dedicate the first two finished buildings, and the Institute for Chemistry especially must have pleased him. It was set back on a broad lawn at the corner of Thielallee and Faradayweg: three stories of cut stone filigreed with six-paned windows, a steep, gabled slate roof and at the roofline high above the entrance a classical pediment supported by four Doric columns. A wing angled off paralleling the cross street. Fitted between the main building and the wing like a hinge, a round tower rose up dramatically four stories high. Von Ihne had surmounted the tower with a dome. Apparently the dome was meant to flatter the Kaiser's taste. A sense of humor was not one of Wilhelm II's strong points and no doubt it did. The dome took the form of a giant *Pickelhaube,* the comic-opera spiked helmet that the Kaiser and his soldiers wore.

Leaving Ernest Rutherford in Montreal in 1906 Hahn had moved to Berlin to work with Emil Fischer at the university. Fischer was an organic chemist who knew little about radioactivity, but he understood that the field was opening to importance and that Hahn was a first-rate man. He made room for Hahn in a wood shop in the basement of his laboratories and arranged Hahn's appointment as a *Privatdozent,* which stirred less forward-looking chemists on the faculty to wonder aloud at the deplorable decline in standards. A chemist who claimed to identify new elements with a gold-foil electroscope must be at least an embarrassment, if not in fact a fraud.

Hahn found the university's physicists more congenial than its chemists and regularly attended the physics colloquia. At one colloquium at the beginning of the autumn term in 1907 he met an Austrian woman, Lise Meitner, who had just arrived from Vienna. Meitner was twenty-nine, one year older than Hahn. She had earned her Ph.D. at the University of Vienna and had already published two papers on alpha and beta radiation. Max Planck's lectures in theoretical physics had drawn her to Berlin for postgraduate study.

Hahn was a gymnast, a skier and a mountain climber, boyishly good-looking, fond of beer and cigars, with a Rhineland drawl and a warm, self-deprecating sense of humor. He admired attractive women, went out of his way to cultivate them and stayed friends with a number of them throughout his happily married life. Meitner was petite, dark and pretty, if also morbidly shy. Hahn befriended her. When she found she had free time she decided to experiment. She needed a collaborator. So did Hahn. A physicist and a radiochemist, they would make a productive team.

They required a laboratory. Fischer agreed that Meitner could share the wood shop on condition that she never show her face in the laboratory upstairs where the students, all male, worked. For two years she observed the condition strictly; then, with the liberalization of the university, Fischer relented, allowed women into his classes and Meitner above the basement. Vienna had been only a little more enlightened. Meitner's father, an attorney—the Meitners were assimilated Austrian Jews, baptized all around—had insisted that she acquire a teacher's diploma in French before beginning to study physics so that she would always be able to support herself. Only then could she prepare for university work. With the diploma out of the way Meitner crammed eight years of *Gymnasium* preparation into two. She was the second woman ever to earn a Ph.D at Vienna. Her father subsidized her research in Berlin until at least 1912, when Max Planck, by now a warm supporter, appointed her to an assistantship. "The German Madame Curie," Einstein would call her, characteristically lumping the Germanic peoples together and forgetting her Austrian birth.

"There was no question," says Hahn, "of any closer relationship between us outside the laboratory. Lise Meitner had had a strict, lady-like upbringing and was very reserved, even shy." They never ate lunch together, never went for a walk, met only in colloquia and in the wood shop. "And yet we were really close friends." She whistled Brahms and Schumann to him to pass the long hours taking timed readings of radioactivity to establish identifying half-lives, and when Rutherford came through Berlin in 1908 on his way back from the Nobel Prize ceremonies she selflessly accompanied Mary Rutherford shopping while the two men indulged themselves in long talks.

The close friends moved together to the new institute in 1912 and worked to prepare an exhibit for the Kaiser. In his first venture into radiochemistry, in London before he went to Montreal, Hahn had spied out what he took to be a new element, radiothorium, that was one hundred thousand times as radioactive as its modest namesake. At McGill he found a third substance intermediate between the other two; he named it "mesothorium" and it was later identified as an isotope of radium. Mesothorium compounds glow in the dark at a different level of faint illumination from radiothorium compounds. Hahn thought the difference might amuse his sovereign. On a velvet cushion in a little box he mounted an unshielded sample of mesothorium equivalent in radiation intensity to 300 milligrams of radium. He presented his potent offering to the Kaiser and asked him to compare it to "an emanating sample of radiothorium that produced in the dark very nice luminous moving shapes on [a] screen." No one warned His Majesty of the radiation hazard because no safety standards for radiation exposure had yet been set. "If I did the same thing today," Hahn said fifty years later, "I should find myself in prison."

The mesothorium caused no obvious harm. The Kaiser passed on to the second institute, half a block up Faradayweg northwest beyond the angled wing. Two senior chemists managed the Chemistry Institute where Hahn and Meitner worked, but the Institute for Physical Chemistry and Electrochemistry, to give it its full name, was established specifically for the man who was its first director, a difficult, inventive German-Jewish chemist from Breslau named Fritz Haber. It was a reward of sorts. A German industrial foundation paid for it and endowed it because in 1909 Haber had succeeded in developing a practical method of extracting nitrogen from the air to make ammonia. The ammonia would serve for artificial fertilizer, replacing Germany's and the world's principal natural source, sodium nitrate dug from the bone-dry northern desert of Chile, an expensive and insecure supply. More strategically, the Haber process would be invaluable in time of war to produce nitrates for explosives; Germany had no nitrates of its own.

Kaiser Wilhelm enlarged at the dedication on the dangers of firedamp, the explosive mixture of methane and other gases that accumulates in mines. He urged his chemists to find some early means of detection. That was a task, he said, "worthy of the sweat of noble brows." Haber, noble brow—he shaved his bullet head, wore round horn-rimmed glasses and a toothbrush mustache, dressed well, wined and dined in elegance but suffered bitter marital discord—set out to invent a firedamp whistle that would sound a different pitch when dangerous gases were present. With a fine modern laboratory uncontaminated by old radioactivity Hahn and Meitner went to work at radiochemistry and the new field of nuclear physics. The Kaiser returned from Dahlem to his palace in Berlin, happy to have lent his name to yet another organ of burgeoning German power.

In the summer of 1913 Niels Bohr sailed with his young wife to England. He followed the second and third parts of his epochal paper, which he had sent ahead by mail to Rutherford; he wanted to discuss them before releasing them for publication. In Manchester he met his friend George de Hevesy again and some of the other research men. One he met, probably for the first time, was Henry Gwyn Jeffreys Moseley, called Harry, an Eton boy and an Oxford man who had worked for Rutherford as a demonstrator, teaching undergraduates, since 1910. Harry Moseley at twenty-six was poised for great accomplishment. He needed only the catalyst of Bohr's visit to set him off.

Moseley was a loner, "so reserved," says A. S. Russell, "that I could neither like him nor not like him," but with the unfortunate habit of allowing no loose statement of fact to pass unchallenged. When he stopped work long enough to take tea at the laboratory he even managed to inhibit Ernest Rutherford. Rutherford's other "boys" called him "Papa." Moseley respected the boisterous laureate but certainly never honored him with any such intimacy; he rather thought Rutherford played the stage colonial.

Harry came from a distinguished line of scientists. His great-grandfather had operated a lunatic asylum with healing enthusiasm but without benefit of medical license, but his grandfather was chaplain and professor of natural philosophy and astronomy at King's College and his father had sailed as a biologist on the three-year voyage of H.M.S. *Challenger* that produced a fifty-volume pioneering study of the world ocean. Henry Moseley—Harry had his father's first name—won the friendly praise of Charles Darwin for his one-volume popular account, *Notes by a Naturalist on the 'Challenger';* Harry in his turn would work with Darwin's physicist grandson Charles G. Darwin at Manchester.

If he was reserved to the point of stuffiness he was also indefatigable at experiment. He would go all out for fifteen hours, well into the night, until he was exhausted, eat a spartan meal of cheese sometime before dawn, find a few hours for sleep and breakfast at noon on fruit salad. He was trim, carefully dressed and conservative, fond of his sisters and his widowed mother, to whom he regularly wrote chatty and warmly devoted letters. Hay fever threw off his final honors examinations at Oxford; he despised teaching the Manchester undergraduates—many were foreigners, "Hindoos, Burmese, Jap, Egyptian and other vile forms of Indian," and he recoiled from their "scented dirtiness." But finally, in the autumn of 1912, Harry found his great subject.

"Some Germans have recently got wonderful results by passing X rays through crystals and then photographing them," he wrote his mother on October 10. The Germans, at Munich, were directed by Max von Laue. Von Laue had found that the orderly, repetitive atomic structure of a crystal produces monochromatic interference patterns from X rays just as the mirroring, slightly separated inner and outer surfaces of a soap bubble produce interference patterns of color from white light. X-ray crystallography was the discovery that would win von Laue the Nobel Prize. Moseley and C. G. Darwin set out with a will to explore the new field. They acquired the necessary equipment and worked through the winter. By May 1913 they had advanced to using crystals as spectroscopes and were finishing up a first solid piece of work. X rays are energetic light of extremely short wavelength. The atomic lattices of crystals spread out their spectra much as a prism does visible light. "We find," Moseley wrote his mother on May 18, "that an X ray bulb with a platinum target gives out a sharp line spectrum of five wavelengths. . . . Tomorrow we search for the spectra of other elements. There is here a whole new branch of spectroscopy, which is sure to tell one much about the nature of the atom."

Then Bohr arrived and the question they discussed was Bohr's old insight that the order of the elements in the periodic table ought to follow the atomic number rather than, as chemists thought, the atomic weight. (The atomic number of uranium, for example, is 92; the atomic weight of the commonest isotope of uranium is 238; a rarer isotope of uranium has an atomic weight of 235 and the same atomic number.) Harry could look for regular shifts in the wavelengths of X-ray line spectra and prove Bohr's contention. Atomic number would make a place in the periodic table for all the variant physical forms that had been discovered and that would soon be named isotopes; atomic number, emphasizing the charge on the nucleus as the determiner of the number of electrons and hence of the chemistry, would strongly confirm Rutherford's nuclear model of the atom; the X-ray spectral lines would further document Bohr's quantized electron orbits. The work would be Moseley's alone; Darwin by then had withdrawn to pursue other interests.

Bohr and the patient Margrethe went on to Cambridge to vacation and polish Bohr's paper. Rutherford left near the end of July with Mary on an expedition to the idyllic mountains of the Tyrol. Moseley stayed in "unbearably hot and stuffy" Manchester, blowing glass. "Even now near midnight," he wrote his mother two days after Rutherford's departure, "I discard coat and waistcoat and work with windows and door open to try to get some air. I will come to you as soon as I can get my apparatus to work before ever I start measurements." On August 13 he was still at it. He wrote his married sister Margery to explain what he was after:

> I want in this way to find the wave-lengths of the X ray spectra of as many elements as possible, as I believe they will prove much more important and fundamental than the ordinary light spectra. The method of finding the wave-lengths is to reflect the X rays which come from a target of the element investigated [when such a target is bombarded with cathode rays]. . . . I have then merely to find at which angles the rays are reflected, and that gives the wave-lengths. I aim at an accuracy of at least one in a thousand.

The Bohrs returned to Copenhagen, the Rutherfords from the Tyrol, and now it was September and time for the annual meeting of the British

Association, this year in Birmingham. Bohr had not planned to attend, especially after lingering overlong in Cambridge, but Rutherford thought he should: his quantized atom with its stunning spectral predictions would be the talk of the conference. Bohr relented and rushed over. Birmingham's hotels were booked tight. He slept the first night on a billiard table. Then the resourceful de Hevesy found him a berth in a girls' college. "And that was very, very practical and wonderful," Bohr remembered afterward, adding quickly that "the girls were away."

Sir Oliver Lodge, president of the British Association, mentioned Bohr's work in his opening address. Rutherford touted it in meetings. James Jeans, the Cambridge mathematical physicist, allowed wittily that "the only justification at present put forward for these assumptions is the very weighty one of success." A Cavendish physicist, Francis W. Aston, announced that he had succeeded in separating two different weights of neon by tediously diffusing a large sample over and over again several thousand times through pipe clay—"a definite proof," de Hevesy noted, "that elements of different atomic weight can have the same chemical properties." Marie Curie came across from France, "shy," says A. S. Eve, "retiring, self-possessed and noble." She fended off the bulldog British press by praising Rutherford: "great developments," she predicted, were "likely to transpire" from his work. He was "the one man living who promises to confer some inestimable boon on mankind."

Harald Bohr reported to his brother that autumn that the younger men at Göttingen "do not dare to believe that [your paper] can be objectively right; they find the assumptions too 'bold' and 'fantastic.' " Against the continuing skepticism of many European physicists Bohr heard from de Hevesy that Einstein himself, encountered at a conference in Vienna, had been deeply impressed. De Hevesy passed along a similar tale to Rutherford:

> Speaking with Einstein on different topics we came to speak on Bohr's theory, he told me that he had once similar ideas but he did not dare to publish them. "Should Bohr's theory be right, it is of the greatest importance." When I told him about the [recent discovery of spectral lines where Bohr's theory had predicted they should appear] the big eyes of Einstein looked still bigger and he told me "Then it is one of the greatest discoveries."
>
> I felt very happy hearing Einstein saying so.

So did Bohr.

Moseley labored on. He had trouble at first making sharp photographs of his X-ray spectra, but once he got the hang of it the results were outstanding. The important spectral lines shifted with absolute regularity as he went up the periodic table, one step at a time. He devised a little staircase of strips of film by matching up the lines. He wrote to Bohr on November 16: "During the last fortnight or so I have been getting results which will interest you. . . . So far I have dealt with the K [spectral line] series from Calcium to Zinc. . . . The results are exceedingly simple and largely what you would expect. . . . K = N − 1, very exactly, N being the atomic number." He had calcium at 20, scandium at 21, titanium at 22, vanadium at 23, chromium at 24 and so on up to zinc at 30. He concludes that his results "lend great weight to the general principles which you use, and I am delighted that this is so, as your theory is having a splendid effect on Physics." Harry Moseley's crisp work gave experimental confirmation of the Bohr-Rutherford atom that was far more solidly acceptable than Marsden's and Geiger's alpha-scattering experiments. "Because you see," Bohr said in his last interview, "actually the Rutherford work was not taken seriously. We cannot understand today, but it was not taken seriously at all. . . . The great change came from Moseley."

Otto Hahn was called upon once more to demonstrate his radioactive preparations. In the early spring of 1914 the Bayer Dye Works at Leverkusen, near Cologne in the Rhineland, gave a reception to celebrate the opening of a large lecture hall. Germany's chemical industry led the world and Bayer was the largest chemical company in Germany, with more than ten thousand employees. It manufactured some two thousand different dyestuffs, large tonnages of inorganic chemicals, a range of pharmaceuticals. The firm's managing director, Carl Duisberg, a chemist who preferred industrial management along American lines, had invited the *Oberpräsident* of the Rhineland to attend the reception; he then invited Hahn to add a glow to the proceedings.

Hahn lectured to the dignitaries on radioactivity. Near the beginning of the lecture he wrote Duisberg's name on a sealed photographic plate with a small glass tube filled with strong mesothorium. Technicians developed the plate while he spoke; at the end Hahn projected the radiographic signature onto a screen to appreciative applause.

The high point of the celebration at the vast 900-acre chemical complex came in the evening. "In the evening there was a banquet," Hahn remembered with nostalgia; "everything was exquisite. On each of the little tables there was a beautiful orchid, brought from Holland by air." Orchids delivered by swift biplane might be adequate symbols of German prosperity and power in 1914, but the managing director wanted to demonstrate German technological superiority as well, and found exotic statement: "At many of the tables," says Hahn, evoking an unrecognizably futuristic past, "the wine was cooled by means of liquid air in thermos vessels."

When war broke out Niels and Harald Bohr were hiking in the Austrian Alps, covering as much as twenty-two miles a day. "It is impossible to describe how amazing and wonderful it is," Niels had written to Margrethe along the way, "when the fog on the mountains suddenly comes driving down from all the peaks, initially as quite small clouds, finally to fill the whole valley." The brothers had planned to return home August 6; the war suddenly came driving down like the mountain fog and they rushed across Germany before the frontiers closed. In October Bohr would sail with his wife from neutral Denmark to teach for two years at Manchester. Rutherford, his boys off to war work, needed help.

Harry Moseley was in Australia with his mother at the beginning of August, attending the 1914 British Association meeting, in his spare time searching out the duck-billed platypus and picturesque silver mines. The patriotism of the Australians, who immediately began mobilizing, triggered his own Etonian spirit of loyalty to King and country. He sailed for England as soon as he could book passage. By late October he had gingered up a reluctant recruiting officer to arrange his commission as a lieutenant in the Royal Engineers ahead of the waiting list.

Chaim Weizmann, the tall, sturdy, Russian-born Jewish biochemist who was Ernest Rutherford's good friend at Manchester, was a passionate Zionist at a time when many, including many influential British Jews, believed Zionism to be at least visionary and naïve if not wrongheaded, fanatic, even a menace. But if Weizmann was a Zionist he was also deeply admiring of British democracy, and one of his first acts after the beginning of the war was to cut himself off from the international Zionist organization because it proposed to remain neutral. Its European leaders hated Czarist Russia, England's ally, and so did Weizmann, but unlike them he did not believe that Germany in cultural and technological superiority would win the war. He believed that the Western democracies would emerge victorious and that Jewish destiny lay with them.

He, his wife and his young son had been en route to a holiday in Switzerland at the outbreak of the war. They worked their way back to Paris, where he visited the elderly Baron Edmond de Rothschild, financial mainstay of the pioneering Jewish agricultural settlements in Palestine. To Weizmann's astonishment Rothschild shared his optimism about the eventual outcome of the war and its possibilities for Jewry. Though Weizmann had no official position in the Zionist movement, Rothschild urged him to seek out and talk to British leaders.

That matched his own inclinations. His hope of British influence had deep roots. He was the third child among fifteen of a timber merchant who assembled rafts of logs and floated them down the Vistula to Danzig for milling and export. The Weizmanns lived in that impoverished western region of Russia cordoned off for the Jews known as the Pale of Settlement. When Chaim was only eleven he had written a letter that prefigured his work in the war. "The eleven-year-old boy," reports his biographer Isaiah Berlin, "says that the kings and nations of the world are plainly set upon the ruin of the Jewish nation; the Jews must not let themselves be destroyed; England alone may help them to return and rise again in their ancient land of Palestine."

Young Weizmann's conviction drove him inexorably west. At eighteen he floated on one of his father's rafts to West Prussia, worked his way to Berlin and studied at the Technische Hochschule. In 1899 he took his Ph.D. at the University of Fribourg in Switzerland, then sold a patent to Bayer that considerably improved his finances. He moved to England in 1904, a move he thought "a deliberate and desperate step. . . . I was in danger of degenerating into a *Luftmensch* [literally, an "air-man"], one of those well-meaning, undisciplined and frustrated 'eternal students.' " Chemical research would save him from that fate; he settled in Manchester under the sponsorship of William Henry Perkin, Jr., the head of the chemistry department there, whose father had established the British coal-tar dye industry by isolating aniline blue, the purple dye after which the Mauve Decade was named.

Returning to Manchester from France in late August 1914, Weizmann found a circular on his desk from the British War Office "inviting every scientist in possession of any discovery of military value to report it." He possessed such a discovery and forthwith offered it to the War Office "without remuneration." The War Office chose not to reply. Weizmann went on with his research. At the same time he began the approach to British leaders that he and Rothschild had discussed that would elaborate into some two thousand interviews before the end of the war.

Weizmann's discovery was a bacillus and a process. The bacillus was *Clostridium acetobutylicum Weizmann,* informally called B-Y ("bacillus-Weizmann"), an anerobic organism that decomposes starch. He was trying to develop a process for making synthetic rubber when he found it, on an ear of corn. He thought he could make synthetic rubber from isoamyl alcohol, which is a minor byproduct of alcoholic fermentation. He went looking for a bacillus—millions of species and subspecies live in the soil and on plants—that converted starch to isoamyl alcohol more efficiently than known strains. "In the course of this investigation I found a bacterium which produced considerable amounts of a liquid smelling very much like isoamyl alcohol. But when I distilled it, it turned out to be a mixture of acetone and butyl alcohol in very pure form. Professor Perkins advised me to pour the stuff down the sink, but I retorted that no pure chemical is useless or ought to be thrown away."

That creature of serendipity was B-Y. Mixed with a mash of cooked corn it fermented the mash into a solution of water and three solvents—one part ethyl alcohol, three parts acetone, six parts butyl alcohol (butanol). The three solvents could then be separated by straightforward distillation. Weizmann tried developing a process for making synthetic rubber from butanol and succeeded. In the meantime, in the years just prior to the beginning of the war, the price of natural rubber fell and the clamor for synthetic rubber stilled.

Pursuing his efforts toward a Jewish homeland, Weizmann acquired in Manchester a loyal and influential friend, C. P. Scott, the tall, elderly, liberal editor of the *Manchester Guardian.* Among his many connections, Scott was David Lloyd George's most intimate political adviser. Weizmann found himself having breakfast one Friday morning in January 1915 with the vigorous little Welshman who was then Chancellor of the Exchequer and who would become Prime Minister in the middle of the war. Lloyd George had been raised on the Bible. He respected the idea of a Jewish return to Palestine, especially when Weizmann eloquently compared rocky, mountainous, diminutive Palestine with rocky, mountainous, diminutive Wales. Besides Lloyd George, Weizmann was surprised to find interest in Zionism among such men as Arthur Balfour, the former Prime Minister who would serve as Foreign Secretary in Lloyd George's cabinet, and Jan Christiaan Smuts, the highly respected Boer who joined the British War Cabinet in 1917 after serving behind the scenes previously. "Really messianic times are upon us," Weizmann wrote his wife during this period of early hope.

Weizmann had cultured B-Y primarily for its butanol. He happened one day to tell the chief research chemist of the Scottish branch of the Nobel explosives company about his fermentation research. The man was impressed. "You know," he said to Weizmann, "you may have the key to a very important situation in your hands." A major industrial explosion prevented Nobel from developing the process, but the company let the British government know.

"So it came about," writes Weizmann, "that one day in March [1915], I returned from a visit to Paris to find waiting for me a summons to the British Admiralty." The Admiralty, of which Winston Churchill, at forty-one exactly Weizmann's age, was First Lord, faced a severe shortage of acetone. That acrid solvent was a crucial ingredient in the manufacture of cordite, a propellant used in heavy artillery, including naval guns, that takes its name from the cordlike form in which it is usually extruded. The explosive material that hurled the heavy shells of the British Navy's big guns from ship to ship and ship to shore across miles of intervening water was a mixture of 64 parts nitrocellulose and 30.2 parts nitroglycerin stabilized with 5 parts petroleum jelly and softened—gelatinized—with 0.8 percent acetone. Cordite could not be manufactured without acetone, and without cordite the guns would need to be extensively rebuilt to accommodate hotter propellants that would otherwise quickly erode their barrels.

Weizmann agreed to see what he could do. Shortly he was brought into the presence of the First Lord. As Weizmann remembered the experience of meeting the "brisk, fascinating, charming and energetic" Winston Churchill:

> Almost his first words were: "Well, Dr. Weizmann, we need thirty thousand tons of acetone. Can you make it?" I was so terrified by this lordly request that I almost turned tail. I answered: "So far I have succeeded in making a few hundred cubic centimeters of acetone at a time by the fermentation process. I do my work in a laboratory. I am not a technician, I am only a research chemist. But, if I were somehow able to produce a ton of acetone, I would be able to multiply that by any factor you chose." . . . I was given carte blanche by Mr. Churchill and the department, and I took upon myself a task which was to tax all my energies for the next two years.

That was part one of Weizmann's acetone experience. Part two came in early June. The British War Cabinet had been shuffled in May because of the enlarging disaster of the Dardanelles campaign at Gallipoli; Herbert Asquith, the Prime Minister, had required Churchill's resignation as First Lord of the Admiralty and replaced him with Arthur Balfour; Lloyd George had moved from Chancellor of the Exchequer to Minister of Munitions. Lloyd George thus immediately inherited the acetone problem in the wider context of Army as well as Navy needs. Scott of the *Manchester Guardian* alerted him to Weizmann's work and the two men met on June 7. Weizmann told him what he had told Churchill previously. Lloyd George was impressed and gave him larger carte blanche to scale up his fermentation process.

In six months of experiments at the Nicholson gin factory in Bow, Weizmann achieved half-ton scale. The process proved efficient. It fermented 37 tons of solvents—about 11 tons of acetone—from 100 tons of grain. Weizmann began training industrial chemists while the government took over six English, Scottish and Irish distilleries to accommodate them. A shortage of American corn—German submarines strangled British shipping in the First War as in the Second—threatened to shut down the operations. "Horse-chestnuts were plentiful," notes Lloyd George in his *War Memoirs,* "and a national collection of them was organised for the purpose of using their starch content as a substitute for maize." Eventually acetone production was shifted to Canada and the United States and back to corn.

"When our difficulties were solved through Dr. Weizmann's genius," continues Lloyd George, "I said to him: 'You have rendered great service to the State, and I should like to ask the Prime Minister to recommend you to His Majesty for some honour.' He said, 'There is nothing I want for myself.' 'But is there nothing we can do as a recognition of your valuable assistance to the country?' I asked. He replied: 'Yes, I would like you to do something for my people.' . . . That was the fount and origin of the famous declaration about the National Home for Jews in Palestine."

The "famous declaration" came to be called the Balfour Declaration, a commitment by the British government in the form of a letter from Arthur Balfour to Baron Edmond de Rothschild to "view with favour the establishment in Palestine of a national home for the Jewish people" and to "use their best endeavours to facilitate the achievement of this object." That document originated far more complexly than in simple payment for Weizmann's biochemical services. Other spokesmen and statesmen were at work as well and Weizmann's two thousand interviews need to be counted in. Smuts identified the relationship long after the war when he said that Weizmann's "outstanding war work as a scientist had made him known and famous in high Allied circles, and his voice carried so much the greater weight in pleading for the Jewish National Home."

But despite these necessary qualifications, Lloyd George's version of the story deserves better than the condescension historians usually accord it. A letter of one hundred eighteen words signed by the Foreign Secretary committing His Majesty's government to a Jewish homeland in Palestine at some indefinite future time, "it being clearly understood that nothing shall be done which may prejudice the civil and religious rights of existing non-Jewish communities in Palestine," can hardly be counted an unseemly reward for saving the guns of the British Army and Navy from premature senility. Chaim Weizmann's experience was an early and instructive example of the power of science in time of war. Government took note. So did science.

A heavy German artillery bombardment preceded the second battle of Ypres that began on April 22, 1915. Ypres was (or had been: it hardly existed anymore) a modest market town in southeastern Belgium about eight miles north of the French border and less than thirty miles inland from the

French port of Dunkirk. Around Ypres spread shell-cratered, soggy downland dominated by unpromising low hills—the highest of them, Hill 60 on the military maps, volcanically contested, only 180 feet elevation. A line of Allied and, parallel northeastward, of German trenches curved through the area, emplaced since the previous November.

Before then, the German attacking and the British defending, the two armies had run a race to the sea. The Germans had hoped to win the race to turn the flank of the Allies. Not yet fully mobilized for war, they even threw in Ersatz Corps of ill-trained high school and university students to bolster their numbers and took 135,000 casualties in what the German people came to call the *Kindermord,* the murder of the children. But at the price of 50,000 lives the British held the narrow flank. The war that was supposed to be surgically brief—a quick march through Belgium, France's capitulation, home by Christmas—turned to a stagnant war of opposing trenches, in the Ypres salient as everywhere along the battle line from the Channel to the Alps.

The April 22 bombardment, the beginning of a concerted German attempt at breakthrough, had driven the Canadians and French Africans holding the line at Ypres deep into their trenches. At sunset it lifted. German troops moved back from the front line along perpendicular communication trenches, leaving behind only newly trained *Pioniere*—combat engineers. A German rocket signal went up. The *Pioniere* set to work opening valves. A greenish-yellow cloud hissed from nozzles and drifted on the wind across no-man's-land. It blanketed the ground, flowed into craters, over the rotting bodies of the dead, through wide brambles of barbed wire, drifted then across the sandbagged Allied parapets and down the trench walls past the firesteps, filled the trenches, found dugouts and deep shelters: and men who breathed it screamed in pain and choked. It was chlorine gas, caustic and asphyxiating. It smelled as chlorine smells and burned as chlorine burns.

Masses of Africans and Canadians stumbled back in retreat. Other masses, surprised and utterly uncomprehending, staggered out of their trenches into no-man's-land. Men clawed at their throats, stuffed their mouths with shirttails or scarves, tore the dirt with their bare hands and buried their faces in the earth. They writhed in agony, ten thousand of them, serious casualties; and five thousand others died. Entire divisions abandoned the line.

Germany achieved perfect surprise. All the belligerents had agreed under the *Hague Declaration of 1899 Concerning Asphyxiating Gases* "to abstain from the use of projectiles the sole object of which is the diffusion of asphyxiating or deleterious gases." None seemed to think tear gas covered by this declaration, though tear gases are more toxic than chlorine in sufficient concentration. The French used tear gas in the form of rifle grenades as early as August 1914; the Germans used it in artillery shells fired against the Russians at Bolimow at the end of January 1915 and on the Western Front first against the British at Nieuport in March. But the chlorine attack at Ypres was the first major and deliberate poison-gas attack of the war.

As later with other weapons of unfamiliar effect, the chlorine terrorized and bewildered. Men threw down their rifles and decamped. Medical officers at aid stations were suddenly overwhelmed with casualties the cause of whose injuries was unknown. Chemists among the men who survived the attack recognized chlorine quickly enough, however, and knew how easy it was to neutralize; within a week the women of London had sewn 300,000 pads of muslin-wrapped cotton for soaking in hyposulfite—the first crude gas masks.

Even though the German High Command allowed the use of gas at Ypres, it apparently doubted its tactical value. It had massed no reserve troops behind the lines to follow up. Allied divisions quickly closed the gap. Nothing came of the attack except agony.

Otto Hahn, a lieutenant in the infantry reserve, helped install the gas cylinders, 5,730 of them containing 168 tons of chlorine, originally at a different place in the line. Shovel crews dug them into the forward walls of the trenches at firestep level and sandbagged them thickly to protect them from shellfire. To work them you connected lead pipe to their valves, ran the pipe over the parapet into no-man's-land, waited for a rocket to signal a start and opened the valves for a predetermined time. Chlorine boils at −28.5° F unpressurized; it boiled out eagerly when released. But the prevailing winds had been wrong where Hahn's team of *Pioniere* first installed the chlorine cylinders. By the time the High Command decided to remove them to Ypres along a four-mile front where the wind blew more favorably, Hahn had been sent off to investigate gas-attack conditions in the Champagne.

In January he was ordered to German-occupied Brussels to see Fritz Haber. Haber had just been promoted from reserve sergeant major to captain, an unprecedented leap in rank in the aristocratic Germany Army. He needed the rank, he told Hahn, to accomplish his new work. "Haber informed me that his job was to set up a special unit for gas-warfare." It seems that Hahn was shocked. Haber offered reasons. They were reasons that would be heard again in time of war:

> He explained to me that the Western fronts, which were all bogged down, could be got moving again only by means of new weapons. One of the weap-

> ons contemplated was poison gas. . . . When I objected that this was a mode of warfare violating the Hague Convention he said that the French had already started it—though not to much effect—by using rifle-ammunition filled with gas. Besides, it was a way of saving countless lives, if it meant that the war could be brought to an end sooner.

Hahn followed Haber to work on gas warfare. So did the physicist James Franck, head of the physics department at Haber's institute, who, like Haber and Hahn, would later win the Nobel Prize. So did a crowd of industrial chemists employed by I.G. Farben, a cartel of eight chemical companies assembled in wartime by the energetic Carl Duisberg of Bayer. The plant at Leverkusen with the new lecture hall turned up hundreds of known toxic substances, many of them dye precursors and intermediates, and sent them off to the Kaiser Wilhelm Institute for Physical Chemistry and Electrochemistry for study. Berlin acquired depots for gas storage and a school where Hahn instructed in gas defense.

He also directed gas attacks. In Galicia on the Eastern Front in mid-June 1915, "the wind was favourable and we discharged a very poisonous gas, a mixture of chlorine and phosgene, against the [Russian] enemy lines. . . . Not a single shot was fired. . . . The attack was a complete success."

Because of its massive chemical industry, which supplied the world before the war, Germany was far ahead of the Allies in the production of chemicals for gas warfare. Early in the war the British had even been reduced to buying German dyestuffs (not for gas, for dyeing) through neutral countries; when the Germans discovered the subterfuge they proposed, with what compounding of cynicism and labored Teutonic humor the record does not reveal, to trade dyestuffs for scarce rubber and cotton. But France and Britain went immediately to work. By the end of the war at least 200,000 tons of chemical warfare agents had been manufactured and used, half by Germany, half by the several Allies together.

Abrogating the Hague Convention opened an array of new ecological niches, so to speak, in weaponry. Types of gas and means of delivery then proceeded to diversify like Darwin's finches. Germany introduced phosgene next after chlorine, mixing it with chlorine for cloud-gas attacks like Hahn's because of its slow rate of evaporation. The French retaliated in early 1916 with phosgene artillery shells. Phosgene then became a staple of the war, dispensed from cylinders, artillery shells, trench mortars, canisters fired from mortarlike "projectors" and bombs. It smelled like new-mown hay but it was by far the most toxic gas used, ten times as toxic as chlorine, fatal in ten minutes at a concentration of half a milligram per liter of air. At higher concentrations one or two breaths killed in a matter of hours. Phosgene—carbonyl chloride—hydrolyzed to hydrochloric acid in contact with water; that was its action in the water-saturated air deep in the delicate bubbled tissue of the human lung. It caused more than 80 percent of the war's gas fatalities.

Chlorpicrin—the British called it vomiting gas, the Germans called it *Klop*—a vicious compound of picric acid and bleaching powder, came along next. German engineers used it against Russian soldiers in August 1916. Its special virtue was its chemical inertness. It did not react with the several neutralizing chemicals packed in gas-mask canisters; only the modest layer of activated charcoal in the canisters removed it from the air by adsorption. So a high concentration could saturate the charcoal and get through. It worked like tear gas but induced nausea, vomiting and diarrhea as well. Men raised their masks to vomit; if the *Klop* had been mixed with phosgene, as it frequently was, they might then be lethally exposed. Chlorpicrin's other advantage was that it was simple and cheap to make.

The most horrible gas of the war, the gas that started a previously complacent United States developing a chemical-warfare capacity of its own, was dichlorethyl sulfide, known for its horseradish- or mustard-like smell as mustard gas. The Germans first used it on the night of July 17, 1917, in an artillery bombardment against the British at Ypres. The attack came as a complete surprise and caused thousands of casualties. Defense in the form of effective masks and efficient gas discipline had caught up with offense by the summer of 1917; Germany introduced mustard gas to break the deadlock, just as it had introduced chlorine before. Shells marked with yellow crosses rained down on the men at Ypres. At first they experienced not much more than sneezing and many put away their masks. Then they began vomiting. Their skin reddened and began to blister. Their eyelids inflamed and swelled shut. They had to be led away blinded to aid stations, more than fourteen thousand of them over the next three weeks.

Though the gas smelled like mustard in dense concentrations, in low concentrations, still extremely toxic, it was hardly noticeable. It persisted for days and even weeks in the field. A gas mask alone was no longer sufficient protection. Mustard dissolved rubber and leather; it soaked through multiple layers of cloth. One man might bring enough back to a dugout on the sole of his boot to blind temporarily an entire nest of his mates. Its odor could also be disguised with other gases. The Germans sometimes chose to disguise mustard with xylyl bromide, a tear gas that smells like lilac, and so it came to pass in the wartime spring that men ran in terror from a breeze scented with blossoming lilac shrubs.

These are not nearly all the gases and poisons developed in the boisterous, vicious laboratory of the Great War. There were sneezing gases and arsenic powders and a dozen tear gases and every combination. The French loaded artillery shells with cyanide—to no point except hatred, as it turned out, because the resulting vapors were lighter than air and immediately lofted away. By 1918 a typical artillery barrage locomoting east or west over the front lines counted nearly as many gas shells as high-explosive. Germany, always logical at war to the point of inhumanity, blamed the French and courted a succession of increasingly desperate breakthroughs. The chemists, like bargain hunters, imagined they were spending a pittance of tens of thousands of lives to save a purseful more. Britain reacted with moral outrage but capitulated in the name of parity.

It was more than Fritz Haber's wife could bear. Clara Immerwahr had been Haber's childhood sweetheart. She was the first woman to win a doctorate in chemistry from the University of Breslau. After she married Haber and bore him a son, a neglected housewife with a child to raise, she withdrew progressively from science and into depression. Her husband's work with poison gas triggered even more desperate melancholy. "She began to regard poison gas not only as a perversion of science but also as a sign of barbarism," a Haber biographer explains. "It brought back the tortures men said they had forgotten long ago. It degraded and corrupted the discipline [i.e., chemistry] which had opened new vistas of life." She asked, argued, finally adamantly demanded that her husband abandon gas work. Haber told her what he had told Hahn, adding for good measure, patriot that he was, that a scientist belongs to the world in times of peace but to his country in times of war. Then he stormed out to supervise a gas attack on the Eastern Front. Dr. Clara Immerwahr Haber committed suicide the same night.

The Allied campaign at Gallipoli began on April 25, 1915. The rough, southward-descending Gallipoli Peninsula looked westward toward the Aegean; eastward, across the narrow strait known as the Dardanelles—to the ancients and to Lord Byron, the Hellespont—it faced Turkish Asia. Capture the peninsula; control the Dardanelles, then the Sea of Marmara above, then the narrow Bosporus Strait that divides Europe from Asia, then Constantinople, and you might control the Black Sea, into which the Danube drains—a vast flanking movement against the Central Powers. Such were the ambitions of the War Cabinet, chivvied by Winston Churchill, for the Dardanelles campaign. The Turks, whose land it was, backed by the Germans, opposed the operation with machine guns and howitzers.

One Australian, one New Zealand, one French colonial and two British divisions landed at Gallipoli to establish narrow beachheads. The water of one beachhead bay churned as white at first as a rapid, the Turks pouring down ten thousand rounds a minute from the steep cliffs above; then it bloomed thick and red with blood. Geography, error and six Turkish divisions under a skillful German commander forestalled any effective advance. By early May, when a British Gurkha and a French division arrived to replace the Allied depletions, both sides had chiseled trenches in the stony ground.

The standoff persisted into summer. Sir Ian Hamilton, the Allied commander, Corfu-born, literary, with a Boer-stiffened right arm and the best of intentions, appealed for reinforcements. The War Cabinet had reorganized itself and expelled Churchill; it assented with reluctance to Hamilton's appeal and shipped out five divisions more.

Harry Moseley shipped out among them. He was a signaling officer now, 38th Brigade, 13th Infantry Division, one of Lord Kitchener's New Army batches made up of dedicated but inexperienced civilian volunteers. At Gibraltar on June 20 he signaled his mother "Our destination no longer in doubt." At Alexandria on June 27 he made his will, leaving everything, which was £2,200, to the Royal Society strictly "to be applied to the furtherance of experimental research in Pathology Physics Physiology Chemistry or other branches of science but not in pure mathematics astronomy or any branch of science which aims merely at describing cataloguing or systematizing."

Alexandria was "full of heat flies native troops and Australians" and after a week they sailed on to Cape Helles on the southern extremity of the Gallipoli Peninsula, a relatively secure bay behind the trench lines. There they could ease into combat in the form of artillery shells lobbed over the Dardanelles to Europe, as it were, from Turkish batteries in Asia. If men were bathing in the bay a lookout on the heights blew a trumpet blast to announce a round coming in. Centipedes and sand, Harry dispensing chlorodyne to his men to cure them of the grim amebic dysentery everyone caught from the beaches, Harry in silk pajamas sharing out the glorious Tiptree blackberry jam his mother sent. "The one real interest in life is the flies," he wrote her. "No mosquitoes, but flies by day and flies by night, flies in the water, flies in the food."

Toward the end of July the divisions crossed to Lemnos to stage for the reinforcing invasion. That was to divide the peninsula, gain the heights and outflank the Turkish lines toward Helles. Hamilton secreted twenty thousand men by the dark of the moon into the crowded trenches at a beach called Anzac halfway up the peninsula and the Turks were none the wiser. The remainder, some seventeen thousand New Army men, came ashore on the night of August 6, 1915, at Sulva Bay north of Anzac, to very little opposition.

When the Turks learned of the invasion they moved new divisions down the peninsula by forced march. The objective of the 38th Brigade, what was left of it toward the end, after days and nights of continuous marching and fighting, was an 850-foot hill, Chanuk Bair, inland a mile and a half from Anzac. To the west of Chanuk Bair and lower down was another hill with a patch of cultivated ground: the Farm. Moseley's column, commanded by Brigadier A. H. Baldwin, struggling up an imprisoning defile a yard wide and six hundred feet deep, found its way blocked by a descending train of mules loaded with ammunition. That was scabby passage and the brigadier in a fury of frustration led off north toward the Farm "over ghastly country in the pitch dark," says the brigade machine gunner, the men "falling headlong down holes and climbing up steep and slippery inclines." But they reached the Farm.

Baldwin's force then held the far left flank of the line of five thousand British, Australians and New Zealanders precariously dug into the slopes below the heights of Chanuk Bair, which the Turks still commanded from trenches.

The Turkish reinforcements arrived at night and crowded into the Chanuk trenches, thirty thousand strong. They launched their assault at dawn on August 10 with the sun breaking blindingly at their backs. John Masefield, the British poet, was there and lived to report: "They came on in a monstrous mass, packed shoulder to shoulder, in some places eight deep, in others three or four deep." On the left flank "the Turks got fairly in among our men with a weight which bore all before it, and what followed was a long succession of British rallies to a tussle body to body, with knives and stones and teeth, a fight of wild beasts in the ruined cornfields of The Farm." Harry Moseley, in the front line, lost that fight.

When he heard of Moseley's death, the American physicist Robert A. Millikan wrote in public eulogy that his loss alone made the war "one of the most hideous and most irreparable crimes in history."

Six miles below Dover down the chalk southeastern coast of England the old resort and harbor town of Folkestone fills a small valley which opens steeply to the strait. Hills shelter the town to the north; the chalk cliff west sustains a broad municipal promenade of lawns and flower beds. The harbor, where Allied soldiers embarked in great numbers for France, offers the refuge of a deep-water pier a third of a mile long with berths for eight steamers. The town remembers William Harvey, the seventeenth-century physician who discovered the circulation of the blood, as its most distinguished native son.

At Folkestone on a sunny, warm Friday afternoon, May 25, 1917, housewives came out in crowds to shop for the Whitsun weekend. A few miles away at Shorncliffe camp, Canadian troops mustered on the parade ground. There was bustle and enthusiasm in town and camp alike. It was payday.

Without warning the shops and streets exploded. A line of waiting housewives crumpled outside a greengrocer's. A wine merchant returned to the front of his shop to find his only customer decapitated. Blast felled passersby in a narrow passage between two old buildings. Horses slumped dead between the shafts of carriages. Finely shattered glass suddenly iced a section of street, a conservatory shed its windows, a crater obliterated a tennis court. Fires bloomed from damaged stores.

Only after the first explosions did the people of Folkestone notice the sound of engines beating the air. They hardly understood what they heard. They screamed "Zepps! Zepps!" for until then Zeppelin dirigibles had been the only mechanism of air attack they knew. "I saw two aeroplanes," a clergyman remembered who ran outside amid the clamor, "not Zeppelins, emerging from the disc of the sun almost overhead. Then four more, or five, in a line and others, all light bright silver insects hovering against the blue of the sky. . . . There was about a score in all, and we were charmed with the beauty of the sight." Charmed because aircraft of any kind were new to the British sky and these were white and large. The results were less charming: 95 killed, 195 injured. The parade ground at Shorncliffe camp was damaged but no one was hurt.

Folkestone was the little Guernica of the Great War. German Gotha bombers—oversized biplanes—had attacked England for the first time, bringing with them the burgeoning concept of strategic bombing. The England Squadron had been headed for London but had met a solid wall of clouds inland from Gravesend. Twenty-one aircraft turned south then and searched for alternative targets. Folkestone and its nearby army camp answered the need.

A Zeppelin bombed Antwerp early in the war as the Germans pushed through Belgium. Churchill sent Navy fighters to bomb Zeppelin hangars at Düsseldorf. Gothas bombed Salonika and a British squadron bombed the fortress town of Maidos in the Dardanelles during the campaign for Gallipoli. But the Gothas that attacked Folkestone in 1917 began the first effective and sustained campaign of strategic civilian bombardment. It fitted Prussian military strategist Karl von Clausewitz's doctrine of total war in much the same way that submarine attack did, carrying fear and horror directly to the enemy to weaken his will to resist. "You must not suppose that we set out to kill women and children," a captured Zeppelin commander told the British authorities, another rationalization that would echo. "We have higher military aims. You would not find one officer in the German Army or Navy who would go to war to kill women and children. Such things happen accidentally in war."

At first the Kaiser, thinking of royal relatives and historic buildings, kept London off the bombing list. His naval staff pressed him to relent, which he did by stages, first allowing the docks to be bombed from naval airships, then reluctantly enlarging permission westward across the city. But the hydrogen-filled airships of Count Ferdinand von Zeppelin were vulnerable to incendiary bullets; when British pilots learned to fire them the stage was set for the bombers.

They came on in irregular numbers, dependent in those later years of the war not only on the vagaries of weather but also on the vagaries, enforced by the British blockade, of substandard engine parts and inferior fuel. A squadron flew against London by daylight on June 13, nineteen days after Folkestone, dropped almost 10,000 pounds of bombs and caused the most numerous civilian bombing casualties of the war, 432 injured and 162 killed, including sixteen horribly mangled children in the basement of a nursery school. London was nearly defenseless and at first the military saw no reason to change that naked condition; the War Minister, the Earl of Derby, told the House of Lords that the bombing was without military significance because not a single soldier had been killed.

So the Gothas continued their attacks. They crossed the Channel from bases in Belgium three times in July, twice in August, and averaged two raids a month through the autumn and winter and spring for a total of twenty-seven in all, first by day and then increasingly, as the British improved their home defenses, by night. They dropped almost a quarter of a million pounds of bombs, killing 835 people, injuring 1,972 more.

Lloyd George, by then Prime Minister, appealed to the brilliant, reliable Smuts to develop an air program, including a system of home defense. Early-warning mechanisms were devised: oversized binaural gramophone horns connected by stethoscope to keen blind listeners; sound-focusing cavities carved into sea cliffs that could pick up the *wong-wong* of Gotha engines twenty miles out to sea. Barrage balloons raised aprons of steel cable that girdled London's airspace; enormous white arrows mounted on the ground on pivots guided the radioless defenders in their Sopwith Camels and Pups toward the invading German bombers. The completed defense system around London was primitive but effective and it needed only technological improvement to ready it for the next war.

At the same time the Germans explored strategic offense. They extended the range of their Gothas with extra fuel tanks. When daylight bombing became too risky they learned to fly and bomb at night, navigating by the stars. They produced a behemoth new four-engine bomber, the Giant, a biplane with a wingspan of 138 feet, unmatched until the advent of the American B-29 Superfortress more than two decades later. Its effective range approached 300 miles. A Giant dropped the largest bomb of the war on London on February 16, 1918, a 2,000-pounder that was thirteen feet long; it exploded on the grounds of the Royal Hospital in Chelsea. As they came to understand strategic bombing, the Germans turned from high explosives to incendiaries, reasoning presciently that fires might cause more damage by spreading and coalescing than any amount of explosives alone. By 1918 they had developed a ten-pound incendiary bomb of almost pure magnesium, the Elektron, that burned at between 2000° and 3000° and that water could not dowse. Only hope of a negotiated peace restrained Germany from attempting major incendiary raids on London in the final months of the war.

The Germans bombed to establish "a basis for peace" by destroying "the morale of the English people" and paralyzing their "will to fight." They succeeded in making the British mad enough to think strategic bombing through. "The day may not be far off," Smuts wrote in his report to Lloyd George, "when aerial operations with their devastation of enemy lands and destruction of industrial and populous centres on a vast scale may become the principal operations of the war, to which the older forms of military and naval operations may become secondary and subordinate."

The United States Army was slow to respond to gas warfare because it assumed that masks would adequately protect U.S. troops. The civilian Department of the Interior, which had experience dealing with poison gases in mines, therefore took the lead in chemical warfare studies. The Army quickly changed its mind when the Germans introduced mustard gas in July 1917. Research contracts for poison-gas development went out to Cornell, Johns Hopkins, Harvard, MIT, Princeton, Yale and other universities. With what a British observer could now call "the great importance attached in America to this branch of warfare," Army Ordnance began construction in November 1917 of a vast war-gas arsenal at Edgewood, Maryland, on waste and marshy land.

The plant, which cost $35.5 million—a complex of 15 miles of roads, 36 miles of railroad track, waterworks and power plants and 550 buildings for the manufacture of chlorine, phosgene, chlorpicrin, sulfur chloride and mustard gas—was completed in less than a year. Ten thousand military and civilian workers staffed it. By the end of the war it was capable of filling 1.1 million 75-mm gas shells a month as well as several million other sizes and types of shells, grenades, mortar bombs and projector drums.

"Had the war lasted longer," the British observer notes, "there can be no doubt that this centre of production would have represented one of the most important contributions by America to the world war."

Gas in any case was far less efficient at maiming and killing men than were artillery and machine-gun fire. Of a total of some 21 million battle casualties gas caused perhaps 5 percent, about 1 million. It killed at least 30,000 men, but at least 9 million died overall. Gas may have evoked special horror because it was unfamiliar and chemical rather than familiar and mechanical in its effects.

The machine gun forced the opposing armies into trenches; artillery carried the violence over the parapets once they were there. So the general staffs learned to calculate that they would lose 500,000 men in a six-month offensive or 300,000 men in six months of "ordinary" trench warfare. The British alone fired off more than 170 million artillery rounds, more than 5 million tons, in the course of the war. The shells, if they were not loaded with shrapnel in the first place, were designed to fragment when they exploded on impact; they produced by far the most horrible mutilations and dismemberings of the war, faces torn away, genitals torn away, a flying debris of arms and legs and heads, human flesh so pulped into the earth that the filling of sandbags with that earth was a repulsive punishment. Men cried out against the monstrousness on all sides.

The machine gun was less mutilating but far more efficient, the basic slaughtering tool of the war. "Concentrated essence of infantry," a military theorist daintily labeled it. Against the criminally stubborn conviction of the professional officer corps that courage, *élan* and naked steel must carry the day the machine gun was the ultimate argument. "I go forward," a British soldier writes of his experience in an attacking line of troops, ". . . up and down across ground like a huge ruined honeycomb, and my wave melts away, and the second wave comes up, and also melts away, and then the third wave merges into the ruins of the first and second, and after a while the fourth blunders into the remnants of the others." He was describing the Battle of the Somme, on July 1, 1916, when at least 21,000 men died in the first hour, possibly in the first few minutes, and 60,000 the first day.

Americans invented the machine gun: Hiram Stevens Maxim, a Yankee from Maine; Colonel Isaac Lewis, a West Pointer, director of the U.S. Army coast artillery school; William J. Browning, a gunmaker and businessman; and their predecessor Richard Jordan Gatling, who correctly located the machine gun among automated systems. "It bears the same relation to other firearms," Gatling noted, "that McCormack's Reaper does to the sickle, or the sewing machine to the common needle." The military historian John Keegan writes:

> For the most important thing about a machine-gun is that it is a *machine,* and one of quite an advanced type, similar in some respects to a high-precision lathe, in others to an automatic press. Like a lathe, it requires to be set up, so that it will operate within desired and predetermined limits; this was done on the Maxim gun . . . by adjusting the angle of the barrel relative to its fixed firing platform, and tightening or loosening its traversing screw. Then, like an automatic press, it would, when actuated by a simple trigger, begin and continue to perform its functions with a minimum of human attention, supplying its own power and only requiring a steady supply of raw material and a little routine maintenance to operate efficiently throughout a working shift.

The machine gun mechanized war. Artillery and gas mechanized war. They were the hardware of the war, the tools. But they were only proximately the mechanism of the slaughter. The ultimate mechanism was a method of organization—anachronistically speaking, a software package. "The basic lever," the writer Gil Elliot comments, "was the *conscription law,* which made vast numbers of men available for military service. The civil machinery which ensured the carrying out of this law, and the *military organization* which turned numbers of men into battalions and divisions, were each founded on a bureaucracy. The *production* of resources, in particular guns and ammunition, was a matter for civil organization. The *movement* of men and resources to the front, and the trench system of defence, were military concerns." Each interlocking system was logical in itself and each system could be rationalized by those who worked it and moved through it. Thus, Elliot demonstrates, "It is reasonable to obey the law, it is good to organize well, it is ingenious to devise guns of high technical capacity, it is sensible to shelter human beings against massive firepower by putting them in protective trenches."

What was the purpose of this complex organization? Officially it was supposed to save civilization, protect the rights of small democracies, demonstrate the superiority of Teutonic culture, beat the dirty Hun, beat the arrogant British, what have you. But the men caught in the middle came to glimpse a darker truth. "The War had become undisguisedly mechanical and inhuman," Siegfried Sassoon allows a fictional infantry officer to see. "What in earlier days had been drafts of volunteers were now droves of victims." Men on every front independently discovered their victimization. Awareness intensified as the war dragged on. In Russia it exploded in revolution. In Germany it motivated desertions and surrenders. Among the

French it led to mutinies in the front lines. Among the British it fostered malingering.

Whatever its ostensible purpose, the end result of the complex organization that was the efficient software of the Great War was the manufacture of corpses. This essentially industrial operation was fantasized by the generals as a "strategy of attrition." The British tried to kill Germans, the Germans tried to kill British and French and so on, a "strategy" so familiar by now that it almost sounds normal. It was not normal in Europe before 1914 and no one in authority expected it to evolve, despite the pioneering lessons of the American Civil War. Once the trenches were in place, the long grave already dug (John Masefield's bitterly ironic phrase), then the war stalemated and death-making overwhelmed any rational response. "The war machine," concludes Elliot, "rooted in law, organization, production, movement, science, technical ingenuity, with its product of six thousand deaths a day over a period of 1,500 days, was the permanent and realistic factor, impervious to fantasy, only slightly altered by human variation."

No human institution, Elliot stresses, was sufficiently strong to resist the death machine. A new mechanism, the tank, ended the stalemate. An old mechanism, the blockade, choked off the German supply of food and matériel. The increasing rebelliousness of the foot soldiers threatened the security of the bureaucrats. Or the death machine worked too well, as against France, and began to run out of raw material. The Yanks came over with their sleeves rolled up, an untrenched continent behind them where the trees were not hung with entrails. The war putrified to a close.

But the death machine had only sampled a vast new source of raw material: the civilians behind the lines. It had not yet evolved equipment efficient to process them, only big guns and clumsy biplane bombers. It had not yet evolved the necessary rationale that old people and women and children are combatants equally with armed and uniformed young men. That is why, despite its sickening squalor and brutality, the Great War looks so innocent to modern eyes.

5

Men from Mars

The first subway on the European continent was dug not in Paris or Berlin but in Budapest. Two miles long, completed in 1896, it connected the thriving Hungarian capital with its northwestern suburbs. During the same year the rebuilding of the grand palace of Franz Josef I, in one of his Dual-Monarchial manifestations King of Hungary, enlarged that structure to 860 rooms. Across the wide Danube rose a grandiose parliament, its dimensions measured in acres, six stories of Victorian mansard-roofed masonry bristling with Neo-Gothic pinnacles set around an elongated Renaissance dome braced by flying buttresses. The palace in hilly, quiet Buda confronted the parliament eastward in flat, bustling Pest. "Horse-drawn droshkies," Hungarian physicist Theodor von Kármán remembers of that time, carried "silk-gowned women and their Hussar counts in red uniforms and furred hats through the ancient war-scarred hills of Buda." But "such sights hid deeper social currents," von Kármán adds.

From the hills of Buda you could look far beyond Pest onto the great Hungarian plain, the Carpathian Basin enclosed 250 miles to the east by the bow of the Carpathian Mountains that the Magyars had crossed to found Hungary a thousand years before. Pest expanded within rings of boulevards on the Viennese model, its offices busy with banking, brokering, lucrative trade in grain, fruit, wine, beef, leather, timber and industrial production only lately established in a country where more than 96 percent of the population had lived in settlements of fewer than 20,000 persons as recently as fifty years before. Budapest, combining Buda, Óbuda and Pest, had grown faster than any other city on the Continent in those fifty years, rising from seventeenth to eighth in rank—almost a million souls. Now coffeehouses, "the fountain of illicit trading, adultery, puns, gossip and poetry," a Hungarian journalist thought, "the meeting places for the intellectuals and those opposed to oppression," enlivened the boulevards; parks and squares sponsored a cavalry of equestrian bronzes; and peasants visiting for the first time the Queen City of the Danube gawked suspiciously at blocks of mansions as fine as any in Europe.

Economic take-off, the late introduction of a nation rich in agricultural resources to the organizing mechanisms of capitalism and industrialization, was responsible for Hungary's boom. The operators of those mechanisms, by virtue of their superior ambition and energy but also by default, were Jews, who represented about 5 percent of the Hungarian population in 1910. The stubbornly rural and militaristic Magyar nobility had managed to keep 33 percent of the Hungarian people illiterate as late as 1918 and wanted nothing of vulgar commerce except its fruits. As a result, by 1904 Jewish families owned 37.5 percent of Hungary's arable land; by 1910, although Jews comprised only 0.1 percent of agricultural laborers and 7.3 percent of industrial workers, they counted 50.6 percent of Hungary's lawyers, 53 percent of its commercial businessmen, 59.9 percent of its doctors and 80 percent of its financiers. The only other significant middle class in Hungary was a vast bureaucracy of impoverished Hungarian gentry that came to vie with the Jewish bourgeoisie for political power. Caught between predominantly Jewish socialists and radicals on one side and the entrenched bureaucracy on the other, both sides hostile, the Jewish commercial elite allied itself for survival with the old nobility and the monarchy; one measure of that conservative alliance was the dramatic increase in the early twentieth century of ennobled Jews.

George de Hevesy's prosperous maternal grandfather, S. V. Schossberger, became in 1863 the first unconverted Jew ennobled since the Middle Ages, and in 1895 de Hevesy's entire family was ennobled. Max Neumann, the banker father of the brilliant mathematician John von Neumann, was elevated in 1913. Von Kármán's father's case was exceptional. Mór Kármán, the founder of the celebrated Minta school, was an educator rather than a wealthy businessman. In the last decades of the nineteenth century he reorganized the haphazard Hungarian school system along German lines, to its great improvement—and not incidentally wrested control of education from the religious institutions that dominated it and passed that control to the state. That won him a position at court and the duty of planning the education of a young archduke, the Emperor's cousin. As a result, writes von Kármán:

> One day in August 1907, Franz Joseph called him to the Palace, and said he wished to reward him for his fine job. He offered to make my father an Excellency.
>
> My father bowed slightly and said: "Imperial Majesty, I am very flattered. But I would prefer something which I could hand down to my children."
>
> The Emperor nodded his agreement and ordained that my father be given a place in the hereditary nobility. To receive a predicate of nobility, my father had to be landed. Fortunately he owned a small vineyard near Budapest, so the Emperor bestowed upon him the predicate "von Szolloskislak" (small grape). I have shortened it to von, for even to me, a Hungarian, the full title is almost unpronounceable.

Jewish family ennoblements in the hundred years prior to 1900 totaled 126; in the short decade and a half between 1900 and the outbreak of the Great War the insecure conservative alliance bartered 220 more. Some thousands of men in these 346 families were ultimately involved. They were thus brought into political connection, their power of independent action siphoned away.

Out of the prospering but vulnerable Hungarian Jewish middle class came no fewer than seven of the twentieth century's most exceptional scientists: in order of birth, Theodor von Kármán, George de Hevesy, Michael Polanyi, Leo Szilard, Eugene Wigner, John von Neumann and Edward Teller. All seven left Hungary as young men; all seven proved unusually versatile as well as talented and made major contributions to science and technology; two among them, de Hevesy and Wigner, eventually won Nobel Prizes.

The mystery of such a concentration of ability from so remote and provincial a place has fascinated the community of science. Recalling that "galaxy of brilliant Hungarian expatriates," Otto Frisch remembers that his friend Fritz Houtermans, a theoretical physicist, proposed the popular theory that "these people were really visitors from Mars; for them, he said, it was difficult to speak without an accent that would give them away and therefore they chose to pretend to be Hungarians whose inability to speak any language without accent is well known; except Hungarian, and [these] brilliant men all lived elsewhere." That was amusing to colleagues and flattering to the Hungarians, who liked the patina of mystery that romanticized their pasts. The truth is harsher: the Hungarians came to live elsewhere because lack of scientific opportunity and increasing and finally violent anti-Semitism drove them away. They took the lessons they learned in Hungary with them into the world.

They all began with talent, variously displayed and remembered. Von Kármán at six stunned his parents' party guests by quickly multiplying six-figure numbers in his head. Von Neumann at six joked with his father in classical Greek and had a truly photographic memory: he could recite entire chapters of books he had read. Edward Teller, like Einstein before him, was exceptionally late in learning—or choosing—to talk. His grandfather warned his parents that he might be retarded, but when Teller finally spoke, at three, he spoke in complete sentences.

Von Neumann too wondered about the mystery of his and his compatriots' origins. His friend and biographer, the Polish mathematician Stanislaw Ulam, remembers their discussions of the primitive rural foothills on both sides of the Carpathians, encompassing parts of Hungary, Czechoslovakia and Poland, populated thickly with impoverished Orthodox villages. "Johnny used to say that all the famous Jewish scientists, artists and writers who emigrated from Hungary around the time of the first World War came, either directly or indirectly, from those little Carpathian communities, moving up to Budapest as their material conditions improved." Progress, to people of such successful transition, could be a metaphysical faith. "As a boy," writes Teller, "I enjoyed science fiction. I read Jules Verne. His words carried me into an exciting world. The possibilities of man's improvement seemed unlimited. The achievements of science were fantastic, and they were good."

Leo Szilard, long before he encountered the novels of H. G. Wells, found another visionary student of the human past and future to admire. Szilard thought in maturity that his "addiction to the truth" and his "predilection for 'Saving the World' " were traceable first of all to the stories his mother told him. But apart from those, he said, "the most serious influence on my life came from a book which I read when I was ten years old. It was a Hungarian classic, taught in the schools, *The Tragedy of Man.*"

A long dramatic poem in which Adam, Eve and Lucifer are central characters, *The Tragedy of Man* was written by an idealistic but disillusioned young Hungarian nobleman named Imre Madach in the years after the failed Hungarian Revolution of 1848. A modern critic calls the work "the most dangerously pessimistic poem of the 19th century." It runs Adam through history with Lucifer as his guide, rather as the spirits of Christmas lead Ebenezer Scrooge, enrolling Adam successively as such real historical personages as Pharaoh, Miltiades, the knight Tancred, Kepler. Its pessimism resides in its dramatic strategy. Lucifer demonstrates to Adam the pointlessness of man's faith in progress by staging not imaginary experiences, as in *Faust* or *Peer Gynt,* but real historical events. Pharaoh frees his slaves and they revile him for leaving them without a dominating god; Miltiades returns from Marathon and is attacked by a murderous crowd of citizens his enemies have bribed; Kepler sells horoscopes to bejewel his faithless wife. Adam sensibly concludes that man will never achieve his ultimate ideals but ought to struggle toward them anyway, a conclusion that Szilard continued to endorse as late as 1945. "In [Madach's] book," he said then, "the devil shows Adam the history of mankind, [ending] with the sun dying down. Only a few Eskimos are left and they worry chiefly because there are too many Eskimos and too few seals [the last scene before Adam returns to the beginning again]. The thought is that there remains a rather narrow margin of hope after you have made your prophecy and it is pessimistic."

Szilard's qualified faith in progress and his liberal political values ultimately set him apart from his Hungarian peers. He believed that group was shaped by the special environment of Budapest at the turn of the century, "a society where economic security was taken for granted," as a historian paraphrases him, and "a high value was placed on intellectual achievement." The Minta that Szilard and Teller later attended deeply gratified von Kármán when he went there in the peaceful 1890s. "My father [who founded the school]," he writes, "was a great believer in teaching everything—Latin, math, and history—by showing its connection with everyday living." To begin Latin the students wandered the city copying down inscriptions from statues and museums; to begin mathematics they looked up figures for Hungary's wheat production and made tables and drew graphs. "At no time did we memorize rules from a book. Instead we sought to develop them ourselves." What better basic training for a scientist?

Eugene Wigner, small and trim, whose father managed a tannery and who would become one of the leading theoretical physicists of the twentieth century, entered the Lutheran *Gimnásium* in 1913; John von Neumann followed the next year. "We had two years of physics courses, the last two years," Wigner remembers. "And it was very interesting. Our teachers were just enormously good, but the mathematics teacher was fantastic. He gave private classes to Johnny von Neumann. He gave him private classes because he realized that this would be a great mathematician."

Von Neumann found a friend in Wigner. They walked and talked mathematics. Wigner's mathematical talent was exceptional, but he felt less than first-rate beside the prodigious banker's son. Von Neumann's brilliance impressed colleagues throughout his life. Teller recalls a truncated syllogism someone proposed to the effect that (a) Johnny can prove anything and (b) anything Johnny proves is correct. At Princeton, where in 1933 von Neumann at twenty-nine became the youngest member of the newly established Institute for Advanced Study, the saying gained currency that the Hungarian mathematician was indeed a demigod but that he had made a thorough, detailed study of human beings and could imitate them perfectly. The story hints at a certain manipulative coldness behind the mask of bonhomie von Neumann learned to wear, and even Wigner thought his friendships lacked intimacy. To Wigner he was nevertheless the only authentic genius of the lot.

These earlier memories of *Gimnásium* days contrast sharply with the turmoil that Teller experienced. Part of the difference was personal. Teller was bored in first-year math at the Minta and quickly managed to insult his mathematics teacher, who was also the principal of the school, by improving on a proof. The principal took the classroom display unkindly. "So you are a genius, Teller? Well, I don't like geniuses." But whatever Teller's personal difficulties, he was also confronted directly, as a schoolboy of only eleven years, with revolution and counterrevolution, with riots and violent bloodletting, with personal fear. What had been usually only implicit for the Martians who preceded him was made explicit before his eyes. "I think this was the first time I was deeply impressed by my father," he told his biographers. "He said anti-Semitism was coming. To me, the idea of anti-Semitism was new, and the fact that my father was so serious about it impressed me."

Von Kármán studied mechanical engineering at the University of Budapest before moving on to Göttingen in 1906; de Hevesy tried Budapest in 1903 before going to the Technische Hochschule in Berlin in 1904 and on to work with Fritz Haber and then with Ernest Rutherford; Szilard had studied at the Technology Institute in Budapest and served in the Army before the post-Armistice turmoil made him decide to leave. In contrast, Wigner, von Neumann and particularly Teller experienced the breakdown of Hungarian society as adolescents—Teller at the impressionable beginning of puberty—and at first hand.

"The Revolution arrived as a hurricane," an eyewitness to the Hungarian Revolution of October 1918 recalls. "No one prepared it and no one arranged it; it broke out by its own irresistible momentum." But there were antecedents: a general strike of half a million workers in Budapest and other Hungarian industrial centers in January 1918; another general strike of similar magnitude in June. In the autumn of that year masses of soldiers, students and workers gathered in Budapest. This first brief revolution began with anti-military and nationalistic claims. By the time the Hungarian National Council had been formed under Count Mihály Károli ("We can't even manage a revolution without a count," they joked in Budapest), in late October, there was expectation of real democratic reform: the council issued a manifesto calling for Hungarian independence, an end to the war, freedom of the press, a secret ballot and even female suffrage.

The Austro-Hungarian Dual Monarchy collapsed in November. Austrian novelist Robert Musil explained that collapse as well as anybody in a dry epitaph: *Es ist passiert* ("It sort of happened"). Hungary won a new government on October 31 and ecstatic crowds filled the streets of Budapest waving chrysanthemums, which had become the symbol of the revolution, and cheering the truckloads of soldiers and workers that pushed through.

The victory was not easy after all. The revolution hardly extended beyond Budapest. The new government was unable to negotiate anything better than a national dismembering. The founding of the Republic of Hungary, proclaimed on November 16, 1918, was shadowed by another founding on November 20: of the Hungarian Communist Party, by soldiers returning from Russian camps where they had been radicalized as prisoners of war. On March 21, 1919, four months after it began, the Republic of Hungary bloodlessly metamorphosed into the Hungarian Soviet Republic, its head a former prisoner of war, disciple of Lenin, journalist, Jew born in the Carpathians of Transylvania: Béla Kun. Arthur Koestler, a boy of fourteen then in Budapest, heard for the first time "the rousing tunes of the *Marseillaise* and of the *Internationale* which, during the hundred days of the Commune, drowned the music-loving town on the Danube in a fiery, melodious flood."

It was a little more than a hundred days: 133. They were days of confusion, hope, fear, comic ineptitude and some violence. Toward the end of the war von Kármán had returned to Budapest from aeronautics work with the Austro-Hungarian Air Force, where he had participated in the development of an early prototype of the helicopter. De Hevesy had also returned. Von Kármán helped reorganize and modernize the university in the brief days of the Republic and even served as undersecretary for universities during the Kun regime. He remembered its naïveté more than its violence: "So far as I can recall, there was no terrorism in Budapest during the one hundred days of the Bolsheviks, although I did hear of some sadistic excesses." Lacking a qualified physicist, the university hired de Hevesy as a lecturer on experimental physics during the winter of 1918–19. Undersecretary von Kármán appointed him to a newly established professorship of physical chemistry in March, but de Hevesy found Commune working conditions unsatisfactory and went off in May to Denmark to visit Bohr. The two old friends agreed he would join Bohr's new institute in Copenhagen as soon as it was built.

Arthur Koestler remembers that food was scarce, especially if you tried to buy it with the regime's ration cards and nearly worthless paper money, but for some reason the same paper would purchase an abundance of Commune-sponsored vanilla ice cream, which his family therefore consumed for breakfast, lunch and dinner. He mentions this curiosity, he remarks, "because it was typical of the happy-go-lucky, dilettantish, and even surrealistic ways in which the Commune was run." It was, Koestler thought, "all rather endearing—at least when compared to the lunacy and savagery which was to descend upon Europe in years to come."

The Hungarian Soviet Republic affected von Neumann and Teller far more severely. They were not admirers like young Koestler nor yet members of the intellectual elite like de Hevesy and von Kármán. They were children of businessmen—Max Teller was a prosperous attorney. Max von Neumann took his family and fled to Vienna. "We left Hungary," his son testified many years later, "very soon after the Communists seized power. . . . We left essentially as soon as it was feasible, which was about 30 or 40 days later, and we returned about 2 months after the Communists had been put down." In Vienna the elder von Neumann joined the group of Hungarian financiers working with the conservative nobility to overthrow the Commune.

Lacking protective wealth, the Tellers stuck it out grimly in Budapest, living with their fears. They made forays into the country to barter with the peasants for food. Teller heard of corpses hung from lampposts, though as with von Kármán's "sadistic excesses" he witnessed none himself. Faced with an overcrowded city, the Commune had socialized all housing. The day came for the Koestlers as for the Tellers when soldiers charged with requisitioning bourgeois excesses of floor space and furniture knocked on their doors. The Koestlers, who occupied two threadbare rooms in a boarding house, were allowed to keep what they had, Arthur discovering in the meantime that working people were interesting and different. The Tellers acquired two soldiers who slept on couches in Max Teller's two office rooms, connected to the Teller apartment. The soldiers were courteous; they sometimes shared their food; they urinated on the rubber plant; but because they searched for hoarded money (which was safely stashed in the cover linings of Max Teller's law books) or simply because the Tellers felt generally insecure, their alien presence terrified.

Yet it was not finally Hungarian communism that frightened Edward Teller's parents most. The leaders of the Commune and many among its officials were Jewish—necessarily, since the only intelligentsia Hungary had evolved up to that time was Jewish. Max Teller warned his son that anti-Semitism was coming. Teller's mother expressed her fears more vividly. "I shiver at what my people are doing," she told her son's governess in the heyday of the Commune. "When this is over there will be a terrible revenge."

In the summer of 1919, as the Commune faltered, eleven-year-old Edward and his older sister Emmi were packed off to safety at their maternal grandparents' home in Rumania. They returned in the autumn; by then Admiral Nicholas Horthy had ridden into Budapest on a white horse behind a new national army to install a violent fascist regime, the first in Europe. The Red Terror had come and gone, resulting in some five hundred deaths by execution. The White Terror of the Horthy regime was of another order of magnitude: at least 5,000 deaths and many of those sadistic; secret torture chambers; a selective but unrelenting anti-Semitism that drove tens of thousands of Jews into exile. A contemporary observer, a socialist equally biased against either extreme, wrote that he had "no desire whatever to palliate the brutalities and atrocities of the proletarian dictatorship; its harshness is not to be denied, even if its terrorists operated more with insults and threats than with actual deeds. But the tremendous difference between the Red and the White Terror is beyond all question." A friend of the new regime, Max von Neumann brought his family home.

In 1920 the Horthy regime introduced a *numerus clausus* law restricting university admission which required "that the comparative numbers of the entrants correspond as nearly as possible to the relative population of the various races or nationalities." The law, which would limit Jewish admissions to 5 percent, a drastic reduction, was deliberately anti-Semitic. Though he was admitted to the University of Budapest and might have stayed, von Neumann chose instead to leave Hungary at seventeen, in 1921, for Berlin, where he came under the influence of Fritz Haber and studied first for a chemical engineering degree, awarded at the Technical Institute of Zürich in 1925. A year later he picked up a Ph.D. *summa cum laude* in mathematics at Budapest; in 1927 he became a *Privatdozent* at the University of Berlin; in 1929, at twenty-five, he was invited to lecture at Princeton. He was professor of mathematics at Princeton by 1931 and accepted lifetime appointment to the Institute for Advanced Study in 1933.

Von Neumann experienced no personal violence in Hungary, only upheaval and whatever anxiety his parents communicated. He nevertheless felt himself scarred. His discussion with Stanislaw Ulam went on more ominously from identifying Carpathian villages as the ultimate places of origin of Hungary's talented expatriates. "It will be left to historians of science," Ulam writes, "to discover and explain the conditions which catalyzed the emergence of so many brilliant individuals from that area. . . . Johnny used to say that it was a coincidence of some cultural factors which he could not make precise: an external pressure on the whole society of this part of Central Europe, a feeling of extreme insecurity in the individuals, and the necessity to produce the unusual or else face extinction."

Teller was too young to leave Hungary during the worst of the Horthy years. This was the adolescent period, as *Time* magazine paraphrased Teller later, when Max Teller "dinned into his son two grim lessons: 1) he would have to emigrate to some more favorable country when he grew up and 2) as a member of a disliked minority he would have to excel the average just to stay even." Teller added a lesson of his own. "I loved science," he told an interviewer once. "But also it offered a possibility for escaping this doomed society." Von Kármán embeds in his autobiography a similarly striking statement about the place of science in his emotional life. When the Hungarian Soviet Republic collapsed he retreated to the home of a wealthy friend, then found his way back to Germany. "I was glad to get out of Hungary," he writes of his state of mind then. "I felt I had had enough of politicians and government upheavals. . . . Suddenly I was enveloped in the feeling that only science is lasting."

That science can be a refuge from the world is a conviction common among men and women who turn to it. Abraham Pais remarks that Einstein "once commented that he had sold himself body and soul to science, being in flight from the 'I' and the 'we' to the 'it.' " But science as a means of escaping from the familiar world of birth and childhood and language when that world mounts an overwhelming threat—science as a way out, a portable culture, an international fellowship and the only abiding certitude—must become a more desperate and therefore a more total dependency. Chaim Weizmann gives some measure of that totality in the harsher world of the Russian Pale when he writes that "the acquisition of knowledge was not for us so much a normal process of education as the storing up of weapons in an arsenal by means of which we hoped later to be able to hold our own in a hostile world." He remembers painfully that "every division of one's life was a watershed."

Teller's experience in Hungary before he left it in 1926, at seventeen, for the Technical Institute at Karlsruhe was far less rigorous than Weizmann's in the Pale. But external circumstance is no sure measure of internal wounding, and there are not many horrors as efficient for the generation of deep anger and terrible lifelong insecurity as the inability of a father to protect his child.

"In the last few years," Niels Bohr wrote the German theoretical physicist Arnold Sommerfeld at Munich in April 1922, "I have often felt myself scientifically very lonesome, under the impression that my effort to develop the principles of the quantum theory systematically to the best of my ability has been received with very little understanding." Through the war years Bohr had struggled to follow, wherever it might lead, the "radical change" he had introduced into physics. It led to frustration. However stunning Bohr's prewar results had been, too many older European scientists still thought his inconsistent hypotheses *ad hoc* and the idea of a quantized atom repugnant. The war itself stalled advance.

Yet he persisted, groping his way forward in the darkness. "Only a rare and uncanny intuition," writes the Italian physicist Emilio Segrè, "saved Bohr from getting lost in the maze." He guided himself delicately by what he called the correspondence principle. As Robert Oppenheimer once explained it, "Bohr remembered that physics was physics and that Newton described a great part of it and Maxwell a great part of it." So Bohr assumed that his quantum rules must approximate, "in situations where the actions involved were large compared to the quantum, to the classical rules of Newton and of Maxwell." That correspondence between the reliable old and the unfamiliar new gave him an outer limit, a wall to feel his way along.

Bohr built his Institute for Theoretical Physics with support from the University of Copenhagen and from Danish private industry, occupying it on January 18, 1921, after more than a year of delay—he struggled with the architect's plans as painfully as he struggled with his scientific papers. The city of Copenhagen ceded land for the institute on the edge of the Faelledpark, broad with soccer fields, where a carnival annually marks the Danish celebration of Constitution Day. The building itself was modest gray stucco with a red tile roof, no larger than many private homes, with four floors inside that looked like only three outside because the lowest floor was built partly below grade and the top floor, which served the Bohrs at first as an apartment, extended into the space under the peaked roof (later, as Bohr's family increased to five sons, he built a house next door and the apartment served as living quarters for visiting students and colleagues). The institute included a lecture hall, a library, laboratories, offices and a popular Ping-Pong table where Bohr often played. "His reactions were very fast and accurate," says Otto Frisch, "and he had tremendous will power and stamina. In a way those qualities characterized his scientific work as well."

In 1922, the year his Nobel Prize made him a Danish national hero, Bohr accomplished a second great theoretical triumph: an explanation of the atomic structure that underlies the regularities of the periodic table of the elements. It linked chemistry irrevocably to physics and is now standard in every basic chemistry text. Around the nucleus, Bohr proposed, atoms are built up of successive orbital shells of electrons—imagine a set of nested spheres—each shell capable of accommodating up to a certain number of electrons and no more. Elements that are similar chemically are similar because they have identical numbers of electrons in their outermost shells, available there for chemical combination. Barium, for example, an alkaline earth, the fifty-sixth element in the periodic table, atomic weight 137.34, has electron shells filled successively by 2, 8, 18, 18, 8 and 2 electrons. Radium, another alkaline earth, the eighty-eighth element, atomic weight 226, has electron shells filled successively by 2, 8, 18, 32, 18, 8 and 2 electrons. Because the outer shell of each element has two valence electrons, barium and radium are chemically similar despite their considerable difference in atomic weight and number. "That [the] insecure and contradictory foundation [of Bohr's quantum hypotheses]," Einstein would say, "was sufficient to enable a man of Bohr's unique instinct and perceptiveness to discover the major laws of spectral lines and of the electron shells of the atom as well as their significance for chemistry appeared to me like a miracle. . . . This is the highest form of musicality in the sphere of thought."

Confirming the miracle, Bohr predicted in the autumn of 1922 that element 72 when discovered would not be a rare earth, as chemists expected and as elements 57 through 71 are, but would rather be a valence 4 metal like zirconium. George de Hevesy, now settled in at Bohr's institute, and a newly arrived young Dutchman, Dirk Coster, went to work using X-ray spectroscopy to look for the element in zircon-bearing minerals. They had not finished their checking when Bohr went off with Margrethe in early December to claim his Nobel Prize. They called him in Stockholm the night before his Nobel lecture, only just in time: they had definitely identified element 72 and it was chemically almost identical to zirconium. They named the new element hafnium after Hafnia, the old Roman name for Copenhagen. Bohr announced its discovery with pride at the conclusion of his lecture the next day.

Despite his success with it, quantum theory needed a more solid foundation than Bohr's intuition. Arnold Sommerfeld in Munich was an early contributor to that work; after the war the brightest young men, searching out the growing point of physics, signed on to help. Bohr remembered the period as "a unique cooperation of a whole generation of theoretical physicists from many countries," an "unforgettable experience." He was lonesome no more.

Sommerfeld brought with him to Göttingen in the early summer of 1922 his most promising student, a twenty-year-old Bavarian named Werner Heisenberg, to hear Bohr as visiting lecturer there. "I shall never forget the first lecture," Heisenberg wrote fifty years later, the memory still textured with fine detail. "The hall was filled to capacity. The great Danish physicist . . . stood on the platform, his head slightly inclined, and a friendly but somewhat embarrassed smile on his lips. Summer light flooded in through the wide-open windows. Bohr spoke fairly softly, with a slight Danish accent. . . . Each one of his carefully formulated sentences revealed a long chain of underlying thoughts, of philosophical reflections, hinted at but never fully expressed. I found this approach highly exciting."

Heisenberg nevertheless raised pointed objection to one of Bohr's statements. Bohr had learned to be alert for bright students who were not afraid to argue. "At the end of the discussion he came over to me and asked me to join him that afternoon on a walk over the Hain Mountain," Heisenberg remembers. "My real scientific career only began that afternoon." It is the memory of a conversion. Bohr proposed that Heisenberg find his way to Copenhagen eventually so that they could work together. "Suddenly, the future looked full of hope." At dinner the next evening Bohr was startled to be challenged by two young men in the uniforms of the Göttingen police. One of them clapped him on the shoulder: "You are arrested on the charge of kidnapping small children!" They were students, genial frauds. The small child they guarded was Heisenberg, boyish with freckles and a stiff brush of red hair.

Heisenberg was athletic, vigorous, eager—"radiant," a close friend says. "He looked even greener in those days than he really was, for, being a member of the Youth Movement . . . he often wore, even after reaching man's estate, an open shirt and walking shorts." In the Youth Movement young Germans on hiking tours built campfires, sang folk songs, talked of knighthood and the Holy Grail and of service to the Fatherland. Many were idealists, but authoritarianism and anti-Semitism already bloomed poisonously among them. When Heisenberg finally got to Copenhagen at Eastertime in 1924 Bohr took him off on a hike through north Zealand and asked him about it all. " 'But now and then our papers also tell us about more ominous, anti-Semitic, trends in Germany, obviously fostered by demagogues,' " Heisenberg remembers Bohr questioning. " 'Have you come across any of that yourself?' " That was the work of some of the old officers embittered by the war, Heisenberg said, "but we don't take these groups very seriously."

Now, as part of the "unique cooperation" Bohr would speak of, they went freshly to work on quantum theory. Heisenberg seems to have begun with a distaste for visualizing unmeasurable events. As an undergraduate, for example, he had been shocked to read in Plato's *Timaeus* that atoms had geometric forms: "It saddened me to find a philosopher of Plato's critical acumen succumbing to such fancies." The orbits of Bohr's electrons were similarly fanciful, Heisenberg thought, and Max Born and Wolfgang

Pauli, his colleagues at Göttingen, concurred. No one could see inside an atom. What was known and measurable was the light that came out of the atomic interior, the frequencies and amplitudes associated with spectral lines. Heisenberg decided to reject models entirely and look for regularities among the numbers alone.

He returned to Göttingen as a *Privatdozent* working under Born. Toward the end of May 1925 his hay fever flared; he asked Born for two weeks' leave of absence and made his way to Heligoland, a stormy sliver of island twenty-eight miles off the German coast in the North Sea, where very little pollen blew. He walked; he swam long distances in the cold sea; "a few days were enough to jettison all the mathematical ballast that invariably encumbers the beginning of such attempts, and to arrive at a simple formulation of my problem." A few days more and he glimpsed the system he needed. It required a strange algebra that he cobbled together as he went along where numbers multiplied in one direction often produced different products from the same numbers multiplied in the opposite direction. He worried that his system might violate the basic physical law of the conservation of energy and he worked until three o'clock in the morning checking his figures, nervously making mistakes. By then he saw that he had "mathematical consistency and coherence." And so often with deep physical discovery, the experience was elating but also psychologically disturbing:

> At first, I was deeply alarmed. I had the feeling that, through the surface of atomic phenomena, I was looking at a strangely beautiful interior, and felt almost giddy at the thought that I now had to probe this wealth of mathematical structures nature had so generously spread out before me. I was far too excited to sleep, and so, as a new day dawned, I made for the southern tip of the island, where I had been longing to climb a rock jutting out into the sea. I now did so without too much trouble, and waited for the sun to rise.

Back in Göttingen Max Born recognized Heisenberg's strange mathematics as matrix algebra, a mathematical system for representing and manipulating arrays of numbers on matrices—grids—that had been devised in the 1850s and that Born's teacher David Hilbert had extended in 1904. In three months of intensive work Born, Heisenberg and their colleague Pascual Jordan then developed what Heisenberg calls "a coherent mathematical framework, one that promised to embrace all the multifarious aspects of atomic physics." Quantum mechanics, the new system was called. It fit the experimental evidence to a high degree of accuracy. Pauli managed with heroic effort to apply it to the hydrogen atom and derive in a consistent way the same results—the Balmer formula, Rydberg's constant—that Bohr had derived from inconsistent assumptions in 1913. Bohr was delighted. At Copenhagen, at Göttingen, at Munich, at Cambridge, the work of development went on.

The bow of the Carpathians as they curve around northwestward begins to define the northern border of Czechoslovakia. Long before it can complete that service the bow bends down toward the Austrian Alps, but a border region of mountainous uplift, the Sudetes, continues across Czechoslovakia. Some sixty miles beyond Prague it turns southwest to form a low range between Czechoslovakia and Germany that is called, in German, the Erzgebirge: the Ore Mountains. The Erzgebirge began to be mined for iron in medieval days. In 1516 a rich silver lode was discovered in Joachimsthal (St. Joachim's dale), in the territory of the Count von Schlick, who immediately appropriated the mine. In 1519 coins were first struck from its silver at his command. *Joachimsthaler,* the name for the new coins, shortened to *thaler,* became "dollar" in English before 1600. Thereby the U.S. dollar descends from the silver of Joachimsthal.

The Joachimsthal mines, ancient and cavernous, shored with smoky timbers, offered up other unusual ores, including a black, pitchy, heavy, nodular mineral descriptively named pitchblende. A German apothecary and self-taught chemist, Martin Heinrich Klaproth, who became the first professor of chemistry at the University of Berlin when it opened its doors in 1810, succeeded in 1789 in extracting a grayish metallic material from a sample of Joachimsthal pitchblende. He sought an appropriate name. Eight years previously Sir William Herschel, the German-born English astronomer, had discovered a new planet and named it Uranus after the earliest supreme god of Greek mythology, son and husband of Gaea, father of Titans and Cyclopes, whose son Chronus with Gaea's help castrated him and from whose wounded blood, falling then on Earth, the three vengeful Furies sprang. To honor Herschel's discovery Klaproth named his new metal *uranium.* It was found to serve, in the form of sodium and ammonium diuranates, as an excellent coloring agent of ceramic glazes, giving a good yellow at 0.006 percent and with higher percentages successively orange, brown, green and black. Uranium mining for ceramics, once begun, continued modestly at Joachimsthal into the modern era. It was from Joachimsthal pitchblende residues that Marie and Pierre Curie laboriously separated the first samples of the new elements they named radium and polonium. The radioactivity of the Erzgebirge ores thus lent glamour to the region's several spas, including Carlsbad and Marienbad, which could now announce that their waters were not only naturally heated but dispersed tonic radioactivity as well.

In the summer of 1921 a wealthy seventeen-year-old American student, a recent graduate of the Ethical Culture School of New York, made his way to Joachimsthal on an amateur prospecting trip. Young Robert Oppenheimer had begun collecting minerals when his grandfather, who lived in Hanau, Germany, had given him a modest starter collection on a visit there when Robert was a small boy, before the Great War. He dated his interest in science from that time. "This was certainly at first a collector's interest," he told an interviewer late in life, "but it began to be also a bit of a scientist's interest, not in historical problems of how rocks and minerals came to be, but really a fascination with crystals, their structure, birefringence, what you saw in polarized light, and all the canonical business." The grandfather was "an unsuccessful businessman, born himself in a hovel, really, in an almost medieval German village, with a taste for scholarship." Oppenheimer's father had left Hanau for America at seventeen, in 1898, worked his way to ownership of a textile-importing company and prospered importing lining fabrics for men's suits at a time when ready-made suits were replacing hand tailoring in the United States. The Oppenheimers—Julius; his beautiful and delicate wife Ella, artistically trained, from Baltimore; Robert, born April 22, 1904; and Frank, Robert's sidekick brother, eight years younger—could afford to summer in Europe and frequently did so.

Julius and Ella Oppenheimer were people of dignity and some caution, nonpracticing Jews. They lived in a spacious apartment on Riverside Drive near 88th Street overlooking the Hudson River and kept a summer house at Bay Shore on Long Island. They dressed with tailored care, practiced cultivation, sheltered themselves and their children from real and imagined harm. Ella Oppenheimer's congenitally unformed right hand, hidden always in a prosthetic glove, was not discussed, not even by the boys out of earshot among their friends. She was loving but formal: in her presence only her husband presumed to raise his voice. Julius Oppenheimer, according to one of Robert's friends a great talker and social arguer, according to another was "desperately amiable, anxious to be agreeable," but also essentially kind. He belonged to Columbia University educator Felix Adler's Society for Ethical Culture, of which Robert's school was an extension, which declared that "man must assume responsibility for the direction of his life and destiny": man, as opposed to God. Robert Oppenheimer remembered himself as "an unctuous, repulsively good little boy." His childhood, he said, "did not prepare me for the fact that the world is full of cruel and bitter things. It gave me no normal, healthy way to be a bastard." He was a frail child, frequently ill. For that reason, or because she had lost a middle son shortly after birth, his mother did not encourage him to run in the streets. He stayed home, collected minerals and at ten years of age wrote poems but still played with blocks.

He was already working up to science. A professional microscope was a childhood toy. He did laboratory experiments in the third grade, began keeping scientific notebooks in the fourth, began studying physics in the fifth, though for many years chemistry would interest him more. The curator of crystals at the American Museum of Natural History took him as a pupil. He lectured to the surprised and then delighted members of the New York Mineralogical Club when he was twelve—from the quality of his correspondence the membership had assumed he was an adult.

When he was fourteen, to get him out of doors and perhaps to help him find friends, his parents sent him to camp. He walked the trails of Camp Koenig looking for rocks and discoursing with the only friend he found on George Eliot, emboldened by Eliot's conviction that cause and effect ruled human affairs. He was shy, awkward, unbearably precious and condescending and he did not fight back. He wrote his parents that he was glad to be at camp because he was learning the facts of life. The Oppenheimers came running. When the camp director cracked down on dirty jokes, the other boys, the ones who called Robert "Cutie," traced the censorship to him and hauled him off to the camp icehouse, stripped him bare, beat him up—"tortured him," his friend says—painted his genitals and buttocks green and locked him away naked for the night. Responsibly he held out to the end of camp but never went back. "Still a little boy," another childhood friend, a girl he liked more than she knew, remembers him at fifteen; ". . . very frail, very pink-cheeked, very shy, and very brilliant of course. Very quickly everybody admitted that he was different from all the others and very superior. As far as studies were concerned he was good in everything. . . . Aside from that he was physically—you can't say clumsy exactly—he was rather undeveloped, not in the way he behaved but the way he went about, the way he walked, the way he sat. There was something strangely childish about him."

He graduated as Ethical Culture's valedictorian in February 1921. In April he underwent surgery for appendicitis. Recovered from that, he traveled with his family to Europe and off on his side trip to Joachimsthal. Somewhere along the way he "came down with a heavy, almost fatal case of trench dysentery." He was supposed to enter Harvard in September, but "I was sick abed—in Europe, actually, at the time." Severe colitis following the bout of dysentery laid him low for months. He spent the winter in the family apartment in New York.

To round off Robert's convalescence and toughen him up, his father arranged for a favorite English teacher at Ethical Culture, a warm, supportive Harvard graduate named Herbert Smith, to take him out West for the summer. Robert was then eighteen, his face still boyish but steadied by arresting blue-gray eyes. He was six feet tall, on an extremely narrow frame; he never in his life weighed more than 125 pounds and at times of illness or stress could waste to 115. Smith guided his charge to a dude ranch, Los Piños, in the Sangre de Cristo Mountains northeast of Santa Fe, and Robert chowed down, chopped wood, learned to ride horses and live in rain and weather.

A highlight of the summer was a pack trip. It started in Frijoles, a village within sheer, pueblo-carved Cañon de los Frijoles across the Rio Grande from the Sangre de Cristos, and ascended the canyons and mesas of the Pajarito Plateau up to the Valle Grande of the vast Jemez Caldera above 10,000 feet. The Jemez Caldera is a bowl-shaped volcanic crater twelve miles across with a grassy basin inside 3,500 feet below the rim, the basin divided by mountainous extrusions of lava into several high valleys. It is a million years old and one of the largest calderas in the world, visible even from the moon. Northward four miles from the Cañon de los Frijoles a parallel canyon took its Spanish name from the cottonwoods that shaded its washes: Los Alamos. Young Robert Oppenheimer first approached it in the summer of 1922.

Like Eastern semi-invalids in frontier days, Oppenheimer's encounter with wilderness, freeing him from overcivilized restraints, was decisive, a healing of faith. From an ill and perhaps hypochondriac boy he weathered across a vigorous summer to a physically confident young man. He arrived at Harvard tanned and fit, his body at least in shape.

At Harvard he imagined himself a Goth coming into Rome. "He intellectually looted the place," a classmate says. He routinely took six courses for credit—the requirement was five—and audited four more. Nor were they easy courses. He was majoring in chemistry, but a typical year might include four semesters of chemistry, two of French literature, two of mathematics, one of philosophy and three of physics, these only the courses credited. He read on his own as well, studied languages, found occasional weekends for sailing the 27-foot sloop his father had given him or for all-night hikes with friends, wrote short stories and poetry when the spirit moved him but generally shied from extracurricular activities and groups. Nor did he date; he was still unformed enough to brave no more than worshiping older women from afar. He judged later that "although I liked to work, I spread myself very thin and got by with murder." The murder he got by with resulted in a transcript solid with A's sprinkled with B's; he graduated *summa cum laude* in three years.

There is something frantic in all this grinding, however disguised in traditional Harvard languor. Oppenheimer had not yet found himself—is that more difficult for Americans than for Europeans like Szilard or Teller, who seem all of a piece from their earliest days?—and would not manage to do so at Harvard. Harvard, he would say, was "the most exciting time I've ever had in my life. I really had a chance to learn. I loved it. I almost came alive." Behind the intellectual excitement there was pain.

He was always an intensely, even a cleverly, private man, but late in life he revealed himself to a group of sensitive friends, a revelation that certainly reaches back all the way to his undergraduate years. "Up to now," he told that group in 1963, "and even more in the days of my almost infinitely prolonged adolescence, I hardly took an action, hardly did anything or failed to do anything, whether it was a paper in physics, or a lecture, or how I read a book, how I talked to a friend, how I loved, that did not arouse in me a very great sense of revulsion and of wrong." His friends at Harvard saw little of this side—an American university is after all a safe-house—but he hinted of it in his letters to Herbert Smith:

> Generously, you ask what I do. Aside from the activities exposed in last week's disgusting note, I labor, and write innumerable theses, notes, poems, stories, and junk; I go to the math lib[rary] and read and to the Phil lib and divide my time between Meinherr [Bertrand] Russell and the contemplation of a most beautiful and lovely lady who is writing a thesis on Spinoza—charmingly ironic, at that, don't you think? I make stenches in three different labs, listen to Allard gossip about Racine, serve tea and talk learnedly to a few lost souls, go off for the weekend to distill the low grade energy into laughter and exhaustion, read Greek, commit faux pas, search my desk for letters, and wish I were dead. Voila.

Part of that exaggerated death wish is Oppenheimer making himself interesting to his counselor, but part of it is pure misery—considering its probable weight, rather splendidly and courageously worn.

Both of Oppenheimer's closest college friends, Francis Fergusson and Paul Horgan, agree that he was prone to baroque exaggeration, to making more of things than things could sustain on their own. Since that tendency would eventually ruin his life, it deserves to be examined. Oppenheimer was no longer a frightened boy, but he was still an insecure and uncertain young man. He sorted among information, knowledge, eras, systems, languages, arcane and apposite skills in the spirit of trying them on for size. Exaggeration made it clear that he knew you knew how awkwardly they fit (and self-destructively at the same time supplied the awkwardness). That was perhaps its social function. Deeper was worse. Deeper was self-loathing, "a very great sense of revulsion and of wrong." Nothing was yet his, nothing was original, and what he had appropriated through learning he thought stolen and himself a thief: a Goth looting Rome. He loved the loot but despised the looter. He was as clear as Harry Moseley was clear in his last will about the difference between collectors and creators. At the same time, intellectual controls were the only controls he seems to have found at that point in his life, and he could hardly abandon them.

He tried writing, poems and short stories. His college letters are those of a literary man more than of a scientist. He would keep his literary skills and they would serve him well, but he acquired them first of all for the access he thought they might open to self-knowledge. At the same time, he hoped writing would somehow humanize him. He read *The Waste Land,* newly published, identified with its *Weltschmerz* and began to seek the stern consolations of Hindu philosophy. He worked through the rigors of Bertrand Russell's and Alfred North Whitehead's three-volume *Principia Mathematica* with Whitehead himself, newly arrived—only one other student braved the seminar—and prided himself throughout his life on that achievement. Crucially, he began to find the physics that underlay the chemistry, as he had found crystals emerging in clarity from the historical complexity of rocks: "It came over me that what I liked in chemistry was very close to physics; it's obvious that if you were reading physical chemistry and you began to run into thermodynamical and statistical mechanical ideas you'd want to find out about them. . . . It's a very odd picture; I never had an elementary course in physics."

He worked in the laboratory of Percy Bridgman, many years later a Nobel laureate, "a man," says Oppenheimer, "to whom one wanted to be an apprentice." He learned much of physics, but haphazardly. He graduated a chemist and was foolhardy enough to imagine that Ernest Rutherford would welcome him at Cambridge, where the Manchester physicist had moved in 1919 to take over direction of the Cavendish from the aging J. J. Thomson. "But Rutherford wouldn't have me," Oppenheimer told a historian later. "He didn't think much of Bridgman and my credentials were peculiar and not impressive, and certainly not impressive to a man with Rutherford's common sense. . . . I don't even know why I left Harvard, but I somehow felt that [Cambridge] was more near the center." Nor would Bridgman's letter of recommendation, though well meant, have helped with Rutherford. Oppenheimer had a "perfectly prodigious power of assimilation," the Harvard physicist wrote, and "his problems have in many cases shown a high degree of originality in treatment and much mathematical power." But "his weakness is on the experimental side. His type of mind is analytical, rather than physical, and he is not at home in the manipulations of the laboratory." Bridgman said honestly that he thought

Oppenheimer "a bit of a gamble." On the other hand, "if he does make good at all, I believe that he will be a very unusual success." After another healing summer in New Mexico with Paul Horgan and old friends from the summer of 1921, Oppenheimer went off to Cambridge to attack the center where he could.

J. J. Thomson still worked at the Cavendish. He let Oppenheimer in. "I am having a pretty bad time," Oppenheimer wrote to Francis Fergusson at Oxford on November 1. "The lab work is a terrible bore, and I am so bad at it that it is impossible to feel that I am learning anything. . . . The lectures are vile." Yet he thought "the academic standard here would depeople Harvard overnight." He worked in one corner of a large basement room at the Cavendish (the Garage, it was called); Thomson worked in another. He labored painfully to make thin films of beryllium for an experiment he seems never to have finished—James Chadwick, who had moved down from Manchester and was now Rutherford's assistant director of research, later put them to use. "The business of the laboratory was really quite a sham," Oppenheimer recalled, "but it got me into the laboratory where I heard talk and found out a good deal of what people were interested in."

Postwar work on quantum theory was just then getting under way. It excited Oppenheimer enormously. He wanted to be a part of it. He was afraid he might be too late. All his learning had come easily before. At Cambridge he hit the wall.

It was as much an emotional wall as an intellectual, probably more. "The melancholy of the little boy who will not play because he has been snubbed," he described it three years later, after he broke through. The British gave him the same silent treatment they had given Niels Bohr, but he lacked Bohr's hard-earned self-confidence. Herbert Smith sensed the approaching disaster. "How is Robert doing?" he wrote Fergusson. "Is frigid England hellish socially and climatically, as you found it? Or does he enjoy its exoticism? I've a notion, by the way, that your ability to show him about should be exercised with great tact, rather than in royal profusion. Your [two] years' start and social adaptivity are likely to make him despair. And instead of flying at your throat . . . I'm afraid he'd merely cease to think his own life worth living." Oppenheimer wrote Smith in December that he had not been busy "making a career for myself. . . . Really I have been engaged in the far more difficult business of making myself for a career." It was worse than that. He was in fact, as he later said, "on the point of bumping myself off. This was chronic." He saw Fergusson at Christmastime in Paris and reported despair at his lab work and frustration with sexual ventures. Then, contradicting Smith's prediction, he flew at Fergusson's throat and tried to strangle him. Fergusson easily set him aside. Back at Cambridge Oppenheimer tried a letter of explanation. He wrote that he was sending Fergusson a "noisy" poem. "I have left out, and that is probably where the fun came in, just as I did in Paris, the awful fact of excellence; but as you know, it is that fact now, combined with my inability to solder two copper wires together, which is probably succeeding in getting me crazy."

The awful fact of excellence did not continue to elude him. As he approached a point of psychological crisis he also drove hard to extend himself, understanding deeply that his mind must pull him through. He was "doing a tremendous amount of work," a friend said, "thinking, reading, discussing things, but obviously with a sense of great inner anxiety and alarm." A crucial change that year was his first meeting with Bohr. "When Rutherford introduced me to Bohr he asked me what I was working on. I told him and he said, 'How is it going?' I said, 'I'm in difficulties.' He said, 'Are the difficulties mathematical or physical?' I said, 'I don't know.' He said, 'That's bad.' " But something about Bohr—his avuncular warmth at least, what C. P. Snow calls his simple and genuine kindness, his uninsipid "sweetness"—helped release Oppenheimer to commitment: "At that point I forgot about beryllium and films and decided to try to learn the trade of being a theoretical physicist."

Whether the decision precipitated the crisis or began to relieve it is not clear from the record. Oppenheimer visited a Cambridge psychiatrist. Someone wrote his parents about his problems and they hurried over as they had hurried to Camp Koenig years before. They pushed their son to see a new psychiatrist. He found one in London on Harley Street. After a few sessions the man diagnosed dementia praecox, the older term for what is now called schizophrenia, a condition characterized by early adult onset, faulty thought processes, bizarre actions, a tendency to live in an inner world, incapacity to maintain normal interpersonal relationships and an extremely poor prognosis. Given the vagueness of the symptomatology and Oppenheimer's intellectual dazzle and profound distress, the psychiatrist's mistake is easy enough to understand. Fergusson met Oppenheimer in Harley Street one day and asked him how it had gone. "He said . . . that the guy was too stupid to follow him and that he knew more about his troubles than the [doctor] did, which was probably true."

Resolution began before the consultations on Harley Street, in the spring, on a ten-day visit to Corsica with two American friends. What happened to bring Oppenheimer through is a mystery, but a mystery important enough to him that he deliberately emphasized it—tantalizingly and incompletely—to one of the more sensitive of his profilers, Nuel Pharr Davis. Corsica, Oppenheimer wrote his brother Frank soon after his visit, was "a great place, with every virtue from wine to glaciers, and from langouste to brigantines." To Davis, late in life, he emphasized that although the United States Government had assembled hundreds of pages of information about him across the years, so that some people said his entire life was recorded there, the record in fact contained almost nothing of real importance. To prove his point, he said, he would mention Corsica. "The [Cambridge] psychiatrist was a prelude to what began for me in Corsica. You ask whether I will tell you the full story or whether you must dig it out. But it is known to few and they won't tell. You can't dig it out. What you need to know is that it was not a mere love affair, not a love affair at all, but love." It was, he said, "a great thing in my life, a great and lasting part of it."

Whether a love affair or love, Oppenheimer found his vocation in Cambridge that year: that was the certain healing. Science saved him from emotional disaster as science was saving Teller from social disaster. He moved to Göttingen, the old medieval town in Lower Saxony in central Germany with the university established by George II of England, in the autumn of 1926, late Weimar years. Max Born headed the university physics department, newly installed in institute buildings on Bunsenstrasse funded by the Rockefeller Foundation. Eugene Wigner traveled to Göttingen to work with Born, as had Werner Heisenberg and Wolfgang Pauli and, less happily, the Italian Enrico Fermi, all future Nobel laureates. James Franck, having moved over from Haber's institute at the KWI, a Nobelist as of 1925, supervised laboratory classes. The mathematicians Richard Courant, Herman Weyl and John von Neumann collaborated. Edward Teller would show up later on an assistantship.

The town was pleasant, for visiting Americans at least. They could drink *frisches Bier* at the fifteenth-century Schwartzen Bären, the Black Bears, and sit to crisp, delicate *wiener Schnitzel* at the Junkernschänke, the Junkers' Hall, under a steel engraving of former patron Otto von Bismarck. The Junkernschänke, four hundred years old, occupied three stories of stained glass and flowered half-timber at the corner of Barefoot and Jew streets, which makes it likely that Oppenheimer dined there: he would have appreciated the juxtaposition. When a student took his doctorate at Göttingen he was required by his classmates to kiss the Goose Girl, a pretty, lifesize bronze maiden within a bronze floral arbor that decorates the fountain on the square in front of the medieval town hall. To reach the lips of the *Gänseliesel* required wading or leaping the fountain pool, the real point of the exercise, a baptism into professional distinction Oppenheimer must have welcomed.

The townspeople still suffered from the disaster of the war and the inflation. Oppenheimer and other American students lodged at the walled mansion of a Göttingen physician who had lost everything and was forced to take in boarders. "Although this society [at the university] was extremely rich and warm and helpful to me," Oppenheimer says, "it was parked there in a very miserable German mood . . . bitter, sullen, and, I would say, discontent and angry and with all those ingredients which were later to produce a major disaster. And this I felt very much." At Göttingen he first measured the depth of German ruin. Teller generalized it later from his own experience of lost wars and their aftermaths: "Not only do wars create incredible suffering, but they engender deep hatreds that can last for generations."

Two of Oppenheimer's papers, "On the quantum theory of vibration-rotation bands" and "On the quantum theory of the problem of the two bodies," had already been accepted for publication in the *Proceedings of the Cambridge Philosophical Society* when he arrived at Göttingen, which helped to pave the way. As he came to his vocation the papers multiplied. His work was no longer apprenticeship but solid achievement. His special contribution, appropriate to the sweep of his mind, was to extend quantum theory beyond its narrow initial ground. His dissertation, "On the quantum theory of continuous spectra," was published in German in the prestigious *Zeitschrift für Physik.* Born marked it "with distinction"—high praise indeed. Oppenheimer and Born jointly worked out the quantum theory of molecules, an important and enduring contribution. Counting the dissertation, Oppenheimer published sixteen papers between 1926 and 1929. They established for him an international reputation as a theoretical physicist.

He came home a far more confident man. Harvard offered him a job; so did the young, vigorous California Institute of Technology at Pasadena. The University of California at Berkeley especially interested him because it was, as he said later, "a desert," meaning it taught no theoretical physics yet at all. He decided to take Berkeley and Caltech both, arranging to lecture on the Bay Area campus in the autumn and winter and shift to Pasadena in the spring. But first he went back to Europe on a National Research Council fellowship to tighten up his mathematics with Paul Ehrenfest at Leiden and then with Pauli, now at Zurich, a mind more analytical and critical even than Oppenheimer's, a taste in physics more refined. After Ehrenfest Oppenheimer had wanted to work in Copenhagen with Bohr. Ehrenfest thought not: Bohr's "largeness and vagueness," in Oppenheimer's words, were not the proper astringent. "I did see a copy of the letter [Ehrenfest] wrote Pauli. It was clear that he was sending me there to be fixed up."

Before he left the United States for Leiden Oppenheimer visited the Sangre de Cristos with Frank. The two brothers found a cabin and a piece of land they liked—"house and six acres and stream," in Robert's terse description—up high on a mountain meadow. The house was rough-hewn timber chinked with caulk; it lacked even a privy. While Robert was in Europe his father arranged a long-term lease and set aside three hundred dollars for what Oppenheimer calls "restoration." A summer in the mountains was restoration for the celebrated young theoretician as well.

At the end of that summer of 1927 the Fascist government of Benito Mussolini convened an International Physical Congress at Como on the southwestern end of fjord-like Lake Como in the lake district of northern Italy. The congress commemorated the centennial of the death in 1827 of Alessandro Volta, the Como-born Italian physicist who invented the electric battery and after whom the standard unit of electrical potential, the volt, is named. Everyone went to Como except Einstein, who refused to lend his prestige to Fascism. Everyone went because quantum theory was beleaguered and Niels Bohr was scheduled to speak in its defense.

At issue was an old problem that had emerged in a new and more challenging form. Einstein's 1905 work on the photoelectric effect had demonstrated that light sometimes behaves as if it consists not of waves but of particles. Turning the tables, early in 1926 an articulate, cultured Viennese theoretical physicist named Erwin Schrödinger published a wave theory of matter demonstrating that matter at the atomic level behaves as if it consists of waves. Schrödinger's theory was elegant, accessible and completely consistent. Its equations produced the quantized energy levels of the Bohr atom, but as harmonics of vibrating matter "waves" rather than as jumping electrons. Schrödinger soon thereafter proved that his "wave mechanics" was mathematically equivalent to quantum mechanics. "In other words," says Heisenberg, ". . . the two were but different mathematical formulations of the same structure." That pleased the quantum mechanicists because it strengthened their case and because Schrödinger's more straightforward mathematics simplified calculation.

But Schrödinger, whose sympathies lay with the older classical physics, made more far-reaching claims for his wave mechanics. In effect, he claimed that it represented the reality of the interior of the atom, that not particles but standing matter waves resided there, that the atom was thereby recovered for the classical physics of continuous process and absolute determinism. In Bohr's atom electrons navigated stationary states in quantum jumps that resulted in the emission of photons of light. Schrödinger offered, instead, multiple waves of matter that produced light by the process known as constructive interference, the waves adding their peaks of amplitude together. "This hypothesis," says Heisenberg dryly, "seemed to be too good to be true." For one thing, Planck's quantized radiation formula of 1900, by now exhaustively proven experimentally, opposed it. But many traditional physicists, who had never liked quantum theory, greeted Schrödinger's work, in Heisenberg's words, "with a sense of liberation." Late in the summer, hoping to talk over the problem, Heisenberg turned up at a seminar in Munich where Schrödinger was speaking. He raised his objections. "Wilhelm Wien, [a Nobel laureate] who held the chair of experimental physics at the University of Munich, answered rather sharply that one must really put an end to quantum jumps and the whole atomic mysticism, and the difficulties I had mentioned would certainly soon be solved by Schrödinger."

Bohr invited Schrödinger to Copenhagen. The debate began at the railroad station and continued morning and night, says Heisenberg:

> For though Bohr was an unusually considerate and obliging person, he was able in such a discussion, which concerned epistemological problems which he considered to be of vital importance, to insist fanatically and with almost terrifying relentlessness on complete clarity in all arguments. He would not give up, even after hours of struggling, [until] Schrödinger had admitted that [his] interpretation was insufficient, and could not even explain Planck's law. Every attempt from Schrödinger's side to get round this bitter result was slowly refuted point by point in infinitely laborious discussions.

Schrödinger came down with a cold and took to his bed. Unfortunately he was staying at the Bohrs'. "While Mrs. Bohr nursed him and brought in tea and cake, Niels Bohr kept sitting on the edge of the bed talking at [him]: 'But you must surely admit that . . .' " Schrödinger approached desperation. "If one has to go on with these damned quantum jumps," he exploded, "then I'm sorry that I ever started to work on atomic theory." Bohr, always glad for conflicts that sharpened understanding, calmed his exhausted guest with praise: "But the rest of us are so grateful that you did, for you have thus brought atomic physics a decisive step forward." Schrödinger returned home discouraged but unconvinced.

Bohr and Heisenberg then went to work on the problem of reconciling the dualisms of atomic theory. Bohr hoped to formulate an approach that would allow matter and light to exist both as particle and as wave; Heisenberg argued consistently for abandoning models entirely and sticking to mathematics. In late February 1927, says Heisenberg, both of them "utterly exhausted and rather tense," Bohr went off to Norway to ski. The young Bavarian tried, using quantum-mechanical equations, to calculate something so seemingly simple as the trajectory of an electron in a cloud chamber and realized it was hopeless. Facing that corner, he turned around. "I began to wonder whether we might not have been asking the wrong sort of question all along."

Working late one evening in his room under the eaves of Bohr's institute Heisenberg remembered a paradox Einstein had thrown at him in a conversation about the value of theory in scientific work. "It is the theory which decides what we can observe," Einstein had said. The memory made Heisenberg restless; he went downstairs and let himself out—it was after midnight—and walked past the great beech trees behind the institute into the open soccer fields of the Faelledpark. It was early March and it would have been cold, but Heisenberg was a vigorous walker who did his best thinking outdoors. "On this walk under the stars, the obvious idea occurred to me that one should postulate that nature allowed only experimental situations to occur which could be described within the framework of the [mathematical] formalism of quantum mechanics." The bald statement sounds wondrously arbitrary; its test would be its consistent mathematical formulation and, ultimately, its predictive power for experiment. But it led Heisenberg immediately to a stunning conclusion: that on the extremely small scale of the atom, there must be inherent limits to how precisely events could be known. If you identified the position of a particle—by allowing it to impact on a zinc-sulfide screen, for example, as Rutherford did—you changed its velocity and so lost that information. If you measured its velocity—by scattering gamma rays from it, perhaps—your energetic gamma-ray photons battered it into a different path and you could not then locate precisely where it was. One measurement always made the other measurement uncertain.

Heisenberg climbed back to his room and began formulating his idea mathematically: the product of the uncertainties in the measured values of the position and momentum cannot be smaller than Planck's constant. So *h* appeared again at the heart of physics to define the basic, unresolvable granularity of the universe. What Heisenberg conceived that night came to be called the uncertainty principle, and it meant the end of strict determinism in physics: because if atomic events are inherently blurred, if it is impossible to assemble complete information about the location of individual particles in time and space, then predictions of their future behavior can only be statistical. The dream or bad joke of the Marquis de Laplace, the eighteenth-century French mathematician and astronomer, that if he knew at one moment the precise location in time and space of every particle in the universe he could predict the future forever, was thus answered late at night in a Copenhagen park: nature blurs that divine prerogative away.

Bohr ought to have liked Heisenberg's democratization of the atomic interior. Instead it bothered him: he had returned from his ski trip with a grander conception of his own, one that reached back for its force to his earliest understanding of doubleness and ambiguity, to Poul Martin Møller and Søren Kierkegaard. He was particularly unhappy that his Bavarian protégé had not founded his uncertainty principle on the dualism between particles and waves. He trained on him the "terrifying relentlessness" he had previously directed at Schrödinger. Oskar Klein, Bohr's amanuensis of the period, fortunately mediated. But Heisenberg was only twenty-six, however brilliant. He gave ground. The uncertainty principle, he agreed, was just a special case of the more general conception Bohr had devised. With that concession Bohr allowed the paper Heisenberg had written to go to the printer. And set to work composing his Como address.

At Como in pleasant September Bohr began with a polite reference to Volta, "the great genius whom we are here assembled to commemorate," then plunged in. He proposed to try to develop "a certain general point of view" which might help "to harmonize the apparently conflicting views taken by different scientists." The problem, Bohr said, was that quantum conditions ruled on the atomic scale but our instruments for measuring those conditions—our senses, ultimately—worked in classical ways. That inadequacy imposed necessary limitations on what we could know. An experiment that demonstrates that light travels in photons is valid within the limits of its terms. An experiment that demonstrates that light travels in waves is equally valid within its limits. The same is true of particles and waves of matter. The reason both could be accepted as valid is that "particles" and "waves" are words, are abstractions. What we know is not particles and waves but the equipment of our experiments and how that equipment changes in experimental use. The equipment is large, the interiors of atoms small, and between the two must be interposed a necessary and limiting translation.

The solution, Bohr went on, is to accept the different and mutually exclusive results as *equally valid* and stand them side by side to build up a composite picture of the atomic domain. *Nur die Fülle führt zur Klarheit:* only wholeness leads to clarity. Bohr was never interested in an arrogant reductionism. He called instead—the word appears repeatedly in his Como lecture—for "renunciation," renunciation of the godlike determinism of classical physics where the intimate scale of the atomic interior was concerned. The name he chose for this "general point of view" was *complementarity,* a word that derives from the Latin *complementum,* "that which fills up or completes." Light as particle and light as wave, matter as particle and matter as wave, were mutually exclusive abstractions that complemented each other. They could not be merged or resolved; they had to stand side by side in their seeming paradox and contradiction; but accepting that uncomfortably non-Aristotelian condition meant physics could know more than it otherwise knew. And furthermore, as Heisenberg's recently published uncertainty principle demonstrated within its limited context, the universe appeared to be arranged that way as far down as human senses would ever be able to see.

Emilio Segrè, who heard Bohr lecture at Como in 1927 as a young engineering student, explains complementarity simply and clearly in a history of modern physics he wrote in retirement: "Two magnitudes are complementary when the measurement of one of them prevents the accurate simultaneous measurement of the other. Similarly, two concepts are complementary when one imposes limitations on the other."

Carefully Bohr then examined the conflicts of classical and quantum physics one at a time and showed how complementarity clarified them. In conclusion he briefly pointed to complementarity's connection to philosophy. The situation in physics, he said, "bears a deep-going analogy to the general difficulty in the formation of human ideas, inherent in the distinction between subject and object." That reached back all the way to the licentiate's dilemma in *Adventures of a Danish Student,* and resolved it: the I who thinks and the I who acts are different, mutually exclusive, but complementary abstractions of the self.

In the years to come Bohr would extend the compass of his "certain general point of view" far into the world. It would serve him as a guide not only in questions of physics but in the largest questions of statesmanship as well. But it never commanded the central place in physics he hoped it would. At Como a substantial minority of the older physicists were predictably unpersuaded. Nor was Einstein converted when he heard. In 1926 he had written to Max Born concerning the statistical nature of quantum theory that "quantum mechanics demands serious attention. But an inner voice tells me that this is not the true Jacob. The theory accomplishes a lot, but it does not bring us closer to the secrets of the Old One. In any case, I am convinced that He does not play dice." Another physics conference, the annual Solvay Conference sponsored by a wealthy Belgian industrial chemist named Ernest Solvay, was held in Brussels a month after Como. Einstein attended, as did Bohr, Max Planck, Marie Curie, Hendrick Lorentz, Max Born, Paul Ehrenfest, Erwin Schrödinger, Wolfgang Pauli, Werner Heisenberg and a crowd of others. "We all stayed at the same hotel," Heisenberg remembers, "and the keenest arguments took place, not in the conference hall but during the hotel meals. Bohr and Einstein were in the thick of it all."

Einstein refused to accept the idea that determinism on the atomic level was forbidden, that the fine structure of the universe was unknowable, that statistics rule. " 'God does not throw dice' was a phrase we often heard from his lips in these discussions," writes Heisenberg. "And so he refused point-blank to accept the uncertainty principle, and tried to think up cases in which the principle would not hold." Einstein would produce a challenging thought experiment at breakfast, the debate would go on all day, "and, as a rule, by suppertime we would have reached a point where Niels Bohr could prove to Einstein that even his latest experiment failed to shake the uncertainty principle. Einstein would look a bit worried, but by next morning he was ready with a new imaginary experiment more complicated than the last." This went on for days, until Ehrenfest chided Einstein—they were the oldest of friends—that he was ashamed of him, that Einstein was arguing against quantum theory just as irrationally as his opponents had argued against relativity theory. Einstein remained adamant (he remained adamant to the end of his life where quantum theory was concerned).

Bohr, for his part, supple pragmatist and democrat that he was, never an absolutist, heard once too often about Einstein's personal insight into the gambling habits of the Deity. He scolded his distinguished colleague finally in Einstein's own terms. God does not throw dice? "Nor is it our business to prescribe to God how He should run the world."

6

Machines

After the war, under Ernest Rutherford's direction, the Cavendish thrived. Robert Oppenheimer suffered there largely because he was not an experimentalist; for experimental physicists, Cambridge was exactly the center that Oppenheimer had thought it to be. C. P. Snow trained there a little later, in the early 1930s, and in his first novel, *The Search,* published in 1934, celebrated the experience in the narrative of a fictional young scientist:

> I shall not easily forget those Wednesday meetings in the Cavendish. For me they were the essence of all the *personal* excitement in science; they were romantic, if you like, and not on the plane of the highest experience I was soon to know [of scientific discovery]; but week after week I went away through the raw nights, with east winds howling from the fens down the old streets, full of a glow that I had seen and heard and been close to the leaders of the greatest movement in the world.

More crowded than ever, the laboratory was showing signs of wear and tear. Mark Oliphant remembers standing in the hallway outside Rutherford's office for the first time and noticing "uncarpeted floor boards, dingy varnished pine doors and stained plastered walls, indifferently lit by a skylight with dirty glass." Oliphant also records Rutherford's appearance at that time, the late 1920s, when the Cavendish director was in his mid-fifties: "I was received genially by a large, rather florid man, with thinning fair hair and a large moustache, who reminded me forcibly of the keeper of the general store and post office in a little village in the hills behind Adelaide where I had spent part of my childhood. Rutherford made me feel welcome and at ease at once. He spluttered a little as he talked, from time to time holding a match to a pipe which produced smoke and ash like a volcano."

With simple experimental apparatus Rutherford continued to produce astonishing discoveries. The most important of them besides the discovery of the nucleus had come to fruition in 1919, shortly before he left Manchester for Cambridge—he sent off the paper in April. Afterward, at the Cavendish, he and James Chadwick followed through. The 1919 Manchester paper actually summarized a series of investigations Rutherford carried out in his rare moments of spare time during the four years of war, when he kept the Manchester lab going almost singlehandedly while doing research for the Admiralty on submarine detection. It appeared in four parts. The first three parts cleared the way for the fourth, "An anomalous effect in nitrogen," which was revolutionary.

Ernest Marsden, whose examination of alpha scattering had led Rutherford to discover the atomic nucleus, had found a similarly fruitful oddity in the course of routine experimental studies at Manchester in 1915. Marsden was using alpha particles—helium nuclei, atomic weight 4—emanating from a small glass tube of radon gas to bombard hydrogen atoms. He did that by fixing the radon tube inside a sealed brass box fitted at one end with a zinc-sulfide scintillation screen, evacuating the box of air and then filling it with hydrogen gas. The alpha particles emanating from the radon bounced off the hydrogen atoms (atomic weight approximately 1) like marbles, transferring energy to the H atoms and setting some of them in motion toward the scintillation screen; Marsden then measured their range by interposing pieces of absorbing metal foils behind the screen until the scintillations stopped. Predictably, the less massive H atoms recoiled farther as a result of their collisions with the heavier alpha particles than did the alphas—about four times as far, says Rutherford—just as smaller and larger marbles colliding in a marbles game do.

That was straightforward enough. But then Marsden noticed, Rutherford relates, while the box was evacuated, that the glass radon tube itself "gave rise to a number of scintillations like those from hydrogen." He tried a tube made of quartz, then a nickel disk coated with a radium compound, and found similarly bright, H-like scintillations. "Marsden concluded that there was strong evidence that hydrogen arose from the radioactive matter itself." This conjecture would have been stunning, if true—so far radioactive atoms had been found to eject only helium nuclei, beta electrons and gamma rays in the course of their decay—but it was not the only possible deduction. Nor was it one that Rutherford, who after all had discovered two of the three basic radiations and had never found hydrogen among them, was likely to accept out of hand. Marsden had returned to New Zealand in 1915 to teach; Rutherford pursued the strange anomaly. He had a good idea what he was after. "I occasionally find an odd half day to try a few of my own experiments," he wrote Bohr on December 9, 1917, "and have got I think results that will ultimately prove of great importance. I wish you were here to talk matters over with. I am detecting and counting the lighter atoms set in motion by [alpha] particles. . . . I am also trying to break up the atom by this method."

His equipment was similar to Marsden's, a small brass box fitted with stopcocks to admit and evacuate gases from its interior, with a scintillation screen mounted on one end. For an alpha source he used a beveled brass disk coated with a radium compound:

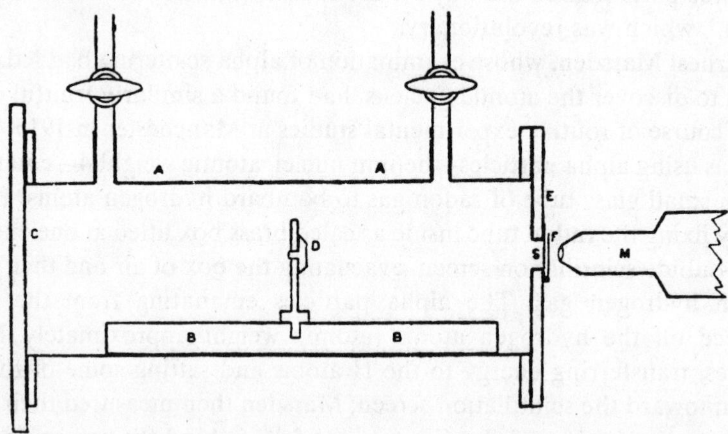

Arrangement of Ernest Rutherford's experiment: D, alpha source. S, zinc sulfide scintillation screen. M, microscope.

The likeliest explanation for Marsden's anomalous H atoms was contamination; hydrogen is light and chemically active and a minor component of the ubiquitous air. So Rutherford's problem was basically one of rigorous exclusion. He needed to narrow down the possible sources of hydrogen atoms in his box until he could conclusively prove their point of origin. He started by showing that they did not come from the radioactive materials alone. He established that they had the same mass and expected range as the H atoms that recoiled from alpha bombardment of hydrogen gas in Marsden's experiment. He admitted dry oxygen into the evacuated brass box, then carbon dioxide, and found in both cases that the H atoms coming off the radioactive source were slowed down by colliding with the atoms of those gases—fewer scintillations showed up on the screen.

Then he tried dry air. The result surprised him. Instead of decreasing the number of scintillations, as oxygen and carbon dioxide had done, dry air increased them—*doubled* them in fact.

These newfound scintillations "appeared to the eye to be about equal in brightness to H scintillations," Rutherford notes cautiously near the beginning of the revolutionary Part IV of his paper. He went after them. If they were H atoms, they still might be contaminants. He eliminated that possibility first. He showed that they could not be due merely to the hydrogen in water vapor (H_2O): drying the air even more thoroughly made little difference in their number. Dust might harbor H atoms like dangerous germs: he filtered the air he let into the box through long plugs of absorbent cotton but found little change.

Since the increase in H atoms occurred in air but not in oxygen or carbon dioxide, Rutherford deduced then that it "must be due either to nitrogen or to one of the other gases present in atmospheric air." And since air is 78 percent nitrogen, that gas appeared to be the likeliest candidate. He tested it simply, by comparing scintillations from air to scintillations from pure nitrogen. The test confirmed his hunch: "With pure nitrogen, the number of long-range scintillations under similar conditions was greater than in air." Finally, Rutherford established that the H atoms came in fact from the nitrogen and not from the radioactive source alone. And then he made his stunning announcement, couching it as always in the measured understatement of British science: "From the results so far obtained it is difficult to avoid the conclusion that the long-range atoms arising from collision of [alpha] particles with nitrogen are not nitrogen atoms but probably atoms of hydrogen. . . . If this be the case, we must conclude that the nitrogen atom is disintegrated." Newspapers soon published the discovery in plainer words: Sir Ernest Rutherford, headlines blared in 1919, had *split the atom.*

It was less a split than a transmutation, the first artificial transmutation ever achieved. When an alpha particle, atomic weight 4, collided with a nitrogen atom, atomic weight 14, knocking out a hydrogen nucleus (which Rutherford would shortly propose calling a *proton*), the net result was a new atom of oxygen in the form of the oxygen isotope O17: 4 plus 14 minus 1. There would hardly be enough O17 to breathe; only about one alpha particle in 300,000 crashed through the electrical barrier around the nitrogen nucleus to do its alchemical work.

But the discovery offered a new way to study the nucleus. Physicists had been confined so far to bouncing radiation off its exterior or measuring the radiation that naturally came out of the nucleus during radioactive decay. Now they had a technique for probing its insides as well. Rutherford and Chadwick soon went after other light atoms to see if they also could be disintegrated, and as it turned out, many of them—boron, fluorine, sodium, aluminum, phosphorus—could. But farther along the periodic table a barricade loomed. The naturally radioactive sources Rutherford used emitted relatively slow-moving alpha particles that lacked the power to penetrate past the increasingly formidable electrical barriers of heavier nuclei. Chadwick and others at the Cavendish began to talk of finding ways to accelerate particles to higher velocities. Rutherford, who scorned complex equipment, resisted. Particle acceleration was in any case difficult to do. For a time the newborn science of nuclear physics stalled.

Besides Rutherford's crowd of "boys," several individual researchers worked at the Cavendish, legatees of J. J. Thomson. One who pursued a different but related interest was a slim, handsome, athletic, wealthy experimentalist named Francis William Aston, the son of a Birmingham gunmaker's daughter and a Harborne metal merchant. As a child Aston made picric-acid bombs from soda-bottle cartridges and designed and launched huge tissue-paper fire balloons; as an adult, a lifelong bachelor, heir after 1908 to his father's wealth, he skied, built and raced motorcycles, played the cello and took elegant trips around the world, stopping off in Honolulu in 1909, at thirty-two, to learn surfing, which he thereafter declared to be the finest of all sports. Aston was one of Rutherford's regular Sunday partners at golf on the Gogs in Cambridge. It was he who had announced, at the 1913 meeting of the British Association, the separation of neon into two isotopes by laborious diffusion through pipe clay.

Aston trained originally as a chemist; Röntgen's discovery of X rays turned him to physics. J. J. Thomson brought him into the Cavendish in 1910, and it was because Thomson seemed to have separated neon into two components inside a positive-ray discharge tube that Aston took up the laborious work of attempting to confirm the difference by gaseous diffusion. Thomson found that he could separate beams of different kinds of atoms by subjecting his discharge tube to parallel magnetic and electrostatic fields. The beams he produced inside his tubes were not cathode rays; he was working now with "rays" repelled from the opposite plate, the positively charged anode. Such rays were streams of atomic nuclei stripped of their electrons: ionized. They could be generated from gas introduced into the tube. Or solid materials could be coated onto the anode plate itself, in which case ionized atoms of the material would boil off when the tube was evacuated and the anode was charged.

Mixed nuclei projected in a radiant beam through a magnetic field would bend into separated component beams according to their velocity, which gave a measure of their mass. An electrostatic field bent the component beams differently depending on their electrical charge, which gave a measure of their atomic number. "In this way," writes George de Hevesy, "a great variety of different atoms and atomic groupings were proved to be present in the discharge tube."

Aston thought hard about J. J.'s discharge tube while he worked during the war at the Royal Aircraft Establishment at Farnborough, southwest of London, developing tougher dopes and fabrics for aircraft coverings. He wanted to prove unequivocally that neon was isotopic—J. J. was still unconvinced—and saw the possibility of sorting the isotopes of other elements as well. He thought the positive-ray tube was the answer, but though it was good for general surveying, it was hopelessly imprecise.

By the time Aston returned to Cambridge in 1918 he had worked the problem out theoretically; he then began building the precision instrument he had envisioned. It charged a gas or a coating until the material ionized into its component electrons and nuclei and projected the nuclei through two slits that produced a knife-edge beam like the slit-narrowed beam of light in a spectrograph. It then subjected the beam to a strong electrostatic field; that sorted the different nuclei into separated beams. The separated beams proceeded onward through a magnetic field; that further sorted nuclei according to their mass, producing separated beams of isotopes. Finally the sorted beams struck the plateholder of a camera and marked their precise locations on a calibrated strip of film. How much the magnetic field bent the separated beams—where they blackened the strip of film—determined the mass of their component nuclei to a high degree of accuracy.

Aston called his invention a mass-spectrograph because it sorted elements and isotopes of elements by mass much as an optical spectrograph sorts light by its frequency. The mass-spectrograph was immediately and sensationally a success. "In letters to me in January and February, 1920," says Bohr, "Rutherford expressed his joy in Aston's work," which "gave such a convincing confirmation of Rutherford's atomic model." Of 281 naturally occurring isotopes, over the next two decades Aston identified 212. He discovered that the weights of the atoms of all the elements he measured, with the notable exception of hydrogen, were very nearly whole numbers, which was a powerful argument in favor of the theory that the elements were assembled in nature simply from protons and electrons—from hydrogen atoms, that is. Natural elements had not weighed up in whole numbers for the chemists because they were often mixtures of isotopes of different whole-number weights. Aston proved, for example, as he noted in a later lecture, "that neon consisted, beyond doubt, of isotopes 20 and 22, and that its atomic weight 20.2 was the result of these being present in the ratio of about 9 to 1." That satisfied even J. J. Thomson.

But why was hydrogen an exception? If the elements were built up from hydrogen atoms, why did the hydrogen atom itself, the elemental building block, weigh 1.008 alone? Why did it then shrink to 4 when it was packed in quartet as helium? Why not 4.032? And why was helium not exactly 4 but 4.002, or oxygen not exactly 16 but 15.994? What was the meaning of these extremely small, and varying, differences from whole numbers?

Atoms do not fall apart, Aston reasoned. Something very powerful holds them together. That glue is now called binding energy. To acquire it, hydrogen atoms packed together in a nucleus sacrifice some of their mass. This mass defect is what Aston found when he compared the hydrogen atom to the atoms of other elements following his whole-number rule. In addition, he said, nuclei may be more or less loosely packed. The density of their packing requires more or less binding energy, and that in turn requires more or less mass: hence the small variations. The difference between the measured mass and the whole number he expressed as a fraction, the packing fraction: roughly, the divergence of an element from its whole number divided by its whole number. "High packing fractions," Aston proposed, "indicate looseness of packing, and therefore low stability: low packing fractions the reverse." He plotted the packing fractions on a graph and demonstrated that the elements in the broad middle of the periodic table—nickel, iron, tin, for example—had the lowest packing fractions and were therefore the most stable, while elements at the extremes of the periodic table—hydrogen at the light end, for example, uranium at the heavy—had high packing fractions and were therefore the most unstable. Locked within all the elements, he said, but most unstably so in the case of those with high packing fractions, was mass converted to energy. Comparing helium to hydrogen, nearly 1 percent of the hydrogen mass was missing (4 divided by 4.032 = .992 = 99.2%). "If we were able to transmute [hydrogen] into [helium] nearly 1 percent of the mass would be annihilated. On the relativity equivalence of mass and energy now experimentally proved [Aston refers here to Einstein's famous equation $E = mc^2$], the quantity of energy liberated would be prodigious. Thus to change the hydrogen in a glass of water into helium would release enough energy to drive the 'Queen Mary' across the Atlantic and back at full speed."

Aston goes on in this lecture, delivered in 1936, to speculate about the social consequences of that energy release. Armed with the necessary knowledge, he says, "the nuclear chemists, I am convinced, will be able to synthesise elements just as ordinary chemists synthesise compounds, and it may be taken as certain that in some reactions sub-atomic energy will be liberated." And, continuing:

> There are those about us who say that such research should be stopped by law, alleging that man's destructive powers are already large enough. So, no doubt, the more elderly and ape-like of our prehistoric ancestors objected to the innovation of cooked food and pointed out the grave dangers attending the use of the newly discovered agency, fire. Personally I think there is no doubt that sub-atomic energy is available all around us, and that one day man will release and control its almost infinite power. We cannot prevent him from doing so and can only hope that he will not use it exclusively in blowing up his next door neighbor.

The mass-spectrograph Francis Aston invented in 1919 could not release the binding energy of the atom. But with it he identified that binding energy and located the groups of elements which in their comparative instability might be most likely to release it if suitably addressed. He was awarded the Nobel Prize in Chemistry in 1922 for his work. After accepting the award alongside Niels Bohr—"Stockholm has been the city of our dreams ever since," his sister, who regularly traveled with him, reminisces—he returned to the Cavendish to build larger and more accurate mass-spectrographs, operating them habitually at night because he "particularly detested," his sister says, "various human noises," including even conversations muffled through the walls of his rooms. "He was very fond of animals, especially cats and kittens, and would go to any amount of trouble to make their acquaintance, but he didn't like dogs of the barking kind." Although Aston respected Ernest Rutherford enormously, the Cavendish director's great boom must ever have been a trial.

The United States led the way in particle acceleration. The American mechanical tradition that advanced the factory and diversified the armory now extended into the laboratory as well. A congressman in 1914 had questioned a witness at an appropriations hearing, "What is a physicist? I was asked on the floor of the House what in the name of common sense a physicist is, and I could not answer." But the war made evident what a physicist was, made evident the value of science to the development of technology, including especially military technology, and government support and the support of private foundations were immediately forthcoming. Twice as many Americans became physicists in the dozen years between 1920 and 1932 as had in the previous sixty. They were better trained than their older counterparts, at least fifty of them in Europe on National Research Council or International Education Board or the new Guggenheim fellowships. By 1932 the United States counted about 2,500 physicists, three times as many as in 1919. The *Physical Review,* the journal that has been to American physicists what the *Zeitschrift für Physik* is to German, was considered a backwater publication, if not a joke, in Europe before the

1920s. It thickened to more than twice its previous size in that decade, increased in 1929 to biweekly publication, and began to find readers in Cambridge, Copenhagen, Göttingen and Berlin eager to scan it the moment it arrived.

Psychometricians have closely questioned American scientists of this first modern generation, curious to know what kind of men they were—there were few women among them—and from what backgrounds they emerged. Small liberal arts colleges in the Middle West and on the Pacific coast, one study found, were most productive of scientists then (by contrast, New England in the same period excelled at the manufacture of lawyers). Half the experimental physicists studied and fully 84 percent of the theoreticians were the sons of professional men, typically engineers, physicians and teachers, although a minority of experimentalists were farmers' sons. None of the fathers of the sixty-four scientists, including twenty-two physicists, in the largest of these studies was an unskilled laborer, and few of the fathers of physicists were businessmen. The physicists were almost all either first-born sons or eldest sons. Theoretical physicists averaged the highest verbal IQ's among all scientists studied, clustering around 170, almost 20 percent higher than the experimentalists. Theoreticians also averaged the highest spatial IQ's, experimentalists ranking second.

The sixty-four-man study which included twenty-two physicists among its "most eminent scientists in the U.S." produced this composite portrait of the American scientist in his prime:

> He is likely to have been a sickly child or to have lost a parent at an early age. He has a very high I.Q. and in boyhood began to do a great deal of reading. He tended to feel lonely and "different" and to be shy and aloof from his classmates. He had only a moderate interest in girls and did not begin dating them until college. He married late . . . has two children and finds security in family life; his marriage is more stable than the average. Not until his junior or senior year in college did he decide on his vocation as a scientist. What decided him (almost invariably) was a college project in which he had occasion to do some independent research—to find out things for himself. Once he discovered the pleasures of this kind of work, he never turned back. He is completely satisfied with his chosen vocation. . . . He works hard and devotedly in his laboratory, often seven days a week. He says his work is his life, and he has few recreations. . . . The movies bore him. He avoids social affairs and political activity, and religion plays no part in his life or thinking. Better than any other interest or activity, scientific research seems to meet the inner need of his nature.

Clearly this is close to Robert Oppenheimer. The group studied, like the American physics community then, was predominantly Protestant in origin with a disproportionate minority of Jews and no Catholics.

A psychological examination of scientists at Berkeley, using Rorschach and Thematic Apperception Tests as well as interviews, included six physicists and twelve chemists in a total group of forty. It found that scientists think about problems in much the same way artists do. Scientists and artists proved less similar in personality than in cognition, but both groups were similarly different from businessmen. Dramatically and significantly, almost half the scientists in this study reported themselves to have been fatherless as children, "their fathers dying early, or working away from home, or remaining so aloof and nonsupportive that their sons scarcely knew them." Those scientists who grew up with living fathers described them as "rigid, stern, aloof, and emotionally reserved." (A group of artists previously studied was similarly fatherless; a group of businessmen was not.)

Often fatherless and "shy, lonely," writes the psychometrician Lewis M. Terman, "slow in social development, indifferent to close personal relationships, group activities or politics," these highly intelligent young men found their way into science through a more personal discovery than the regularly reported pleasure of independent research. Guiding that research was usually a fatherly science teacher. Of the qualities that distinguished this mentor in the minds of his students, not teaching ability but "masterfulness, warmth and professional dignity" ranked first. One study of two hundred of these mentors concludes: "It would appear that the success of such teachers rests mainly upon their capacity to assume a father role to their students." The fatherless young man finds a masterful surrogate father of warmth and dignity, identifies with him and proceeds to emulate him. In a later stage of this process the independent scientist works toward becoming a mentor of historic stature himself.

The man who would found big-machine physics in America arrived at Berkeley one year before Oppenheimer, in 1928. Ernest Orlando Lawrence was three years older than the young theoretician and in many ways his opposite, an extreme of the composite American type. Both he and Oppenheimer were tall and both had blue eyes and high expectations. But Ernest Lawrence was an experimentalist, from prairie, small-town South Dakota; of Norwegian stock, the son of a superintendent of schools and teachers' college president; domestically educated through the Ph.D. at the Universities of South Dakota, Minnesota and Chicago and at Yale; with "almost an aversion to mathematical thought" according to one of his protégés, the later Nobel laureate Luis W. Alvarez; a boyish extrovert whose strongest expletives were "Sugar!" and "Oh fudge!" who learned to stand at ease among the empire builders of patrician California's Bohemian Grove; a master salesman who paid his way through college peddling aluminum kitchenware farm to farm; with a gift for inventing ingenious machines. Lawrence arrived at Berkeley from Yale in a Reo Flying Cloud with his parents and his younger brother in tow and put up at the faculty club. Fired compulsively with ambition—for physics, for himself—he worked from early morning until late at night.

As far back as his first year of graduate school, 1922, Lawrence had begun to think about how to generate high energies. His flamboyant, fatherly mentor encouraged him. William Francis Gray Swann, an Englishman who had found his way to Minnesota via the Department of Terrestrial Magnetism of the District of Columbia's private Carnegie Institution, took Lawrence along with him first to Chicago and then to Yale as he moved up the academic ladder himself. After Lawrence earned his Ph.D. and a promising reputation Swann convinced Yale to jump him over the traditional four years of instructorship to a starting position as assistant professor of physics. Swann's leaving Yale in 1926 was one reason Lawrence had decided to move West, that and Berkeley's offer of an associate professorship, a good laboratory, as many graduate-student assistants as he could handle and $3,300 a year, an offer Yale chose not to match.

At Berkeley, Lawrence said later, "it seemed opportune to review my plans for research, to see whether I might not profitably go into nuclear research, for the pioneer work of Rutherford and his school had clearly indicated that the next great frontier for the experimental physicist was surely the atomic nucleus." But as Luis Alvarez explains, "the tedious nature of Rutherford's technique . . . repelled most prospective nuclear physicists. Simple calculations showed that one microampere of electrically accelerated light nuclei would be more valuable than the world's total supply of radium—if the nuclear particles had energies in the neighborhood of a million electron volts."

Alpha particles or, better, protons could be accelerated by generating them in a discharge tube and then repelling or attracting them electrically. But no one knew how to confine in one place for any useful length of time, without electrical breakdown from sparking or overheating, the million volts that seemed to be necessary to penetrate the electrical barrier of the heavier nuclei. The problem was essentially mechanical and experimental; not surprisingly, it attracted the young generation of American experimental physicists who had grown up in small towns and on farms experimenting with radio. By 1925 Lawrence's boyhood friend and Minnesota classmate Merle Tuve, another protégé of W. F. G. Swann now installed at the Carnegie Institution and working with three other physicists, had managed brief but impressive accelerations with a high-voltage transformer submerged in oil; others, including Robert J. Van de Graaff at MIT and Charles C. Lauritsen at Caltech, were also developing machines.

Lawrence pursued more promising studies but kept the high-energy problem in mind. The essential vision came to him in the spring of 1929, four months before Oppenheimer arrived. "In his early bachelor days at Berkeley," writes Alvarez, "Lawrence spent many of his evenings in the library, reading widely. . . . Although he passed his French and German requirements for the doctor's degree by the slimmest of margins, and consequently had almost no facility with either language, he faithfully leafed through the back issues of the foreign periodicals, night after night." Such was the extent of Lawrence's compulsion. It paid. He was skimming the German *Arkiv für Elektrotechnik,* an electrical-engineering journal physicists seldom read, and happened upon a report by a Norwegian engineer named Rolf Wideröe, *Über ein neues Prinzip zur Herstellung hoher Spannungen:* "On a new principle for the production of higher voltages." The title arrested him. He studied the accompanying photographs and diagrams. They explained enough to set Lawrence off and he did not bother to struggle through the text.

Wideröe, elaborating on a principle established by a Swedish physicist in 1924, had found an ingenious way to avoid the high-voltage problem. He mounted two metal cylinders in line, attached them to a voltage source and evacuated them of air. The voltage source supplied 25,000 volts of high-frequency alternating current, current that changed rapidly from positive to negative potential. That meant it could be used both to push and to pull positive ions. Charge the first cylinder negatively to 25,000 volts, inject positive ions into one end, and the ions would be accelerated to 25,000 volts as they left the first cylinder for the second. Alternate the charge then—make the first cylinder positive and the second cylinder negative—and the ions would be pushed and pulled to further acceleration. Add more cylinders, each one longer than the last to allow for the increasing speed of the ions, and theoretically you could accelerate them further still, until such a time as they scattered too far outward from the center and crashed into the cylinder walls. Wideröe's important innovation was the use of a relatively small voltage to produce increasing acceleration. "This new idea," says Lawrence, "immediately impressed me as the real answer which I had been looking for to the technical problem of accelerating positive ions, and without looking at the article further I then and there made estimates of the general features of a linear accelerator for protons in the energy range above one million [volts]."

Lawrence's calculations momentarily discouraged him. The accelerator tube would be "some meters in length," too long, he thought, for the laboratory. (Linear accelerators today range in length up to two miles.) "And accordingly, I asked myself the question, instead of using a large number of cylindrical electrodes in line, might it not be possible to use two electrodes over and over again by sending the positive ions back and forth through the electrodes by some sort of appropriate magnetic field arrangement." The arrangement he conceived was a spiral. "It struck him almost immediately," Alvarez later wrote, "that one might 'wind up' a linear accelerator into a spiral accelerator by putting it in a magnetic field," because the magnetic lines of force in such a field guide the ions. Given a well-timed push, they would swing around in a spiral, the spiral becoming larger as the particles accelerated and were thus harder to confine. Then, making a simple calculation for the magnetic-field effects, Lawrence uncovered an unsuspected advantage to a spiral accelerator: in a magnetic field slow particles complete their smaller circuits in exactly the same time faster particles complete their larger circuits, which meant they could all be accelerated together, efficiently, with each alternating push.

Exuberantly Lawrence ran off to tell the world. An astronomer who was still awake at the faculty club was drafted to check his mathematics. He shocked one of his graduate students the next day by bombarding him with the mathematics of spiral accelerations but mustering no interest whatever in his thesis experiment. "Oh, that," Lawrence told the questioning student. "Well, you know as much on that now as I do. Just go ahead on your own." A faculty wife crossing the campus the next evening heard a startling "I'm going to be famous!" as the young experimentalist burst past her on the walk.

Lawrence then traveled East to a meeting of the American Physical Society and discovered that not many of his colleagues agreed. To less inspired mechanicians the scattering problem looked insurmountable. Merle Tuve was skeptical. Jesse Beams, a Yale colleague and a close friend, thought it was a great idea if it worked. Despite Lawrence's reputation as a go-getter—perhaps because no one encouraged him, perhaps because the idea was solid and sure in his head but the machine on the laboratory bench might not be—he kept putting off building his spiral particle accelerator. He was not the first man of ambition to find himself stalling on the summit ridge of a famous future.

Oppenheimer arrived in a battered gray Chrysler in the late summer of 1929 from another holiday at the Sangre de Cristos ranch with Frank—the ranch was named Perro Caliente now, "hot dog," Oppenheimer's cheer when he had learned the property could be leased. He put up at the faculty club and the two opposite numbers, he and Lawrence, became close friends. Oppenheimer saw "unbelievable vitality and love of life" in Lawrence. "Work all day, run off for tennis, and work half the night. His interest was so primarily active [and] instrumental and mine just the opposite." They rode horses together, Lawrence in jodhpurs and using an

English saddle in the American West—to distance himself, Oppenheimer thought, from the farm. When Lawrence could get away they went off on long recreational drives in the Reo to Yosemite and Death Valley.

A distinguished experimentalist from the University of Hamburg, Otto Stern, a Breslau Ph.D., forty-one that year and on his way to a Nobel Prize (though Lawrence would beat him), gave Lawrence the necessary boost. Sometime after the Christmas holidays the two men dined out in San Francisco, a pleasant ferry ride across the unbridged bay. Lawrence rehearsed again his practiced story of particles spinning to boundless energies in a confining magnetic field, but instead of coughing politely and changing the subject, as so many other colleagues had done, Stern produced a Germanic duplicate of Lawrence's original enthusiasm and barked at him to leave the restaurant immediately and go to work. Lawrence waited in decency until morning, cornered one of his graduate students and committed him to the project as soon as he had finished studying for his Ph.D. exam.

The machine that resulted looked, in top and side views, like this:

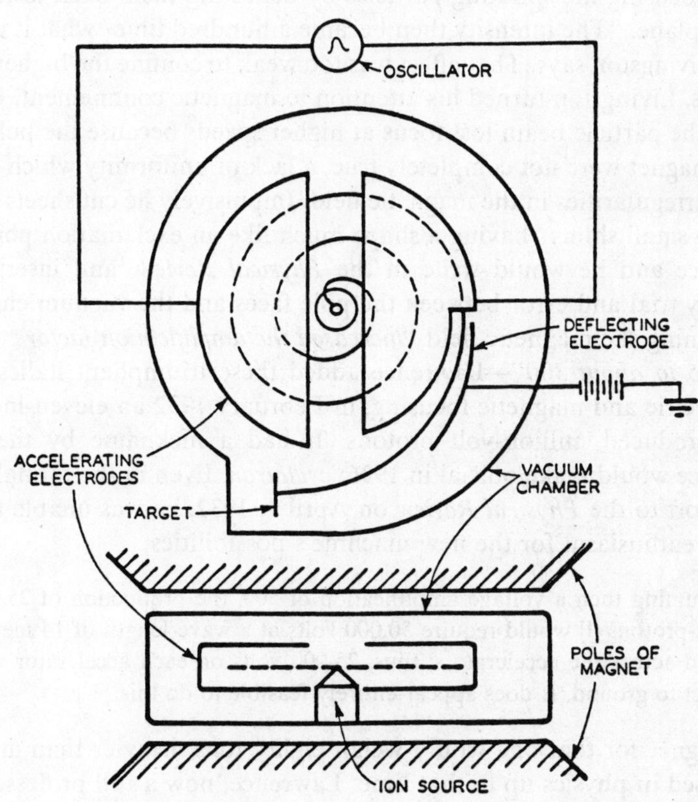

The two cylinders of the Wideröe accelerator have become two brass electrodes shaped like the cut halves of a cylindrical flask. These are contained completely within a vacuum tank and the vacuum tank is mounted between the round, flat poles of a large electromagnet.

In the space between the two electrodes (which came to be called *dees* because of their shape), at the center point, a hot filament and an outlet for hydrogen gas work together to produce protons which stream off into the magnetic field. The two dees, alternately charged, push and pull the protons as they come around. When they have been accelerated through about a hundred spirals the particles exit in a beam which can then be directed onto a target. With a 4.5-inch chamber and with less than 1,000 volts on the dees, on January 2, 1931, Lawrence and his student M. Stanley Livingston produced 80,000-volt protons.

The scattering problem solved itself at low accelerations when Livingston thought to remove the fine grid of wires installed in the gap between the dees that kept the accelerating electric field out of the drift space inside. The electric fields between the dee edges suddenly began functioning as lenses, focusing the spiraling particles by deflecting them back toward the middle plane. "The intensity then became a hundred times what it was before," Livingston says. That effect was too weak to confine the higher-speed particles. Livingston turned his attention to magnetic confinement. He suspected the particle beam lost focus at higher speeds because the pole faces of the magnet were not completely true, a lack of uniformity which in turn caused irregularities in the magnetic field. Impulsively he cut sheets of iron foil into small shims "having a shape much like an exclamation point," as Lawrence and he would write in the *Physical Review,* and inserted the shims by trial and error between the pole faces and the vacuum chamber. Thus tuning the magnetic field *"increased the amplification factor . . . from about 75 to about 300"*—Lawrence added these triumphant italics. With both electric and magnetic focusing, in February 1932 an eleven-inch machine produced million-volt protons. It had a nickname by then that Lawrence would make official in 1936: *cyclotron.* Even in the formal scientific report to the *Physical Review* on April 1, 1932, he was unable to contain his enthusiasm for the new machine's possibilities:

> Assuming then a voltage amplification of 500, the production of 25,000,000 volt-protons [!] would require 50,000 volts at a wave-length of 14 meters applied across the accelerators; thus, 25,000 volts on each accelerator with respect to ground. It does appear entirely feasible to do this.

The magnet for that one would weigh eighty tons, heavier than any machine used in physics up to that time. Lawrence, now a full professor, was already raising funds.

* * *

In his graduate-student days in Europe Robert Oppenheimer told a friend that he dreamed of founding a great school of theoretical physics in the United States—at Berkeley, as it happened, the second desert after New Mexico that he chose to colonize. Ernest Lawrence seems to have dreamed of founding a great laboratory. Both men coveted success and, each in his own way, the rewards of success, but they were differently driven.

Oppenheimer's youthful preciosity matured in Europe and the early Berkeley years into refinement that was usually admirable if still sometimes exquisite. Oppenheimer crafted that persona for himself at least in part from a distaste for vulgarity that probably originated in rebellion against his entrepreneurial father and that was not without elements of anti-Semitic self-hatred. Along the way he convinced himself that ambition and worldly success were vulgar, a conviction bolstered nicely by trust fund earnings to the extent of ten thousand dollars a year. Thereby he confounded his own strivings. The American experimental physicist I. I. Rabi would later question why "men of Oppenheimer's gifts do not discover everything worth discovering." His answer addresses one possible source of limitation:

> It seems to me that in some respects Oppenheimer was overeducated in those fields which lie outside the scientific tradition, such as his interest in religion, in the Hindu religion in particular, which resulted in a feeling for the mystery of the universe that surrounded him almost like a fog. He saw physics clearly, looking toward what had already been done, but at the border he tended to feel that there was much more of the mysterious and novel than there actually was. . . . Some may call it a lack of faith, but in my opinion it was more a turning away from the hard, crude methods of theoretical physics into a mystical realm of broad intuition.

But Oppenheimer's revulsion from what he considered vulgar, from just those "hard, crude methods" to which Rabi refers, must have been another and more directly punishing confusion. His elegant physics, so far as an outsider can tell—his scientific papers are nearly impenetrable to the nonmathematician and deliberately so—is a physics of bank shots. It works the sides and the corners and uses the full court but prefers not to drive relentlessly for the goal. Wolfgang Pauli and the hard, distant Cambridge theoretician Paul A. M. Dirac, Eugene Wigner's brother-in-law, both mathematicians of formidable originality, were his models. Oppenheimer first described the so-called tunnel effect whereby an uncertainly located particle sails through the electrical barrier around the nucleus on a light breeze of probability, existing—in particle terms—then ceasing to exist, then instantly existing again on the other side. But George Gamow, the antic

Russian, lecturing in Cambridge, devised the tunnel-effect equations that the experimenters used. Hans Bethe in the late 1930s first defined the mechanisms of carbon-cycle thermonuclear burning that fire the stars, work which won for him the Nobel Prize; Oppenheimer looked into the subtleties of the invisible cosmic margins, modeled the imploding collapse of dying suns and described theoretical stellar objects that would not be discovered for thirty and forty years—neutron stars, black holes—because the instruments required to detect them, radio telescopes and X-ray satellites, had not been invented yet. (Alvarez believes if Oppenheimer had lived long enough to see these developments he would have won a Nobel Prize for his work.) That was originality not so much ahead of its time as outside the frame.

Some of this psychological and creative convolution winds through a capsule essay on the virtues of discipline that Oppenheimer composed within a letter to his brother Frank in March 1932, when he was not quite twenty-eight years old. It is worth copying out at length; it hints of the long, self-punishing penance he expected to serve to cleanse any stain of crudity from his soul:

> You put a hard question on the virtue of discipline. What you say is true: I do value it—and I think that you do too—more than for its earthly fruit, proficiency. I think that one can give only a metaphysical ground for this evaluation; but the variety of metaphysics which gave an answer to your question has been very great, the metaphysics themselves very disparate: the bhagavad gita, Ecclesiastes, the Stoa, the beginning of the Laws, Hugo of St Victor, St Thomas, John of the Cross, Spinoza. This very great disparity suggests that the fact that discipline is good for the soul is more fundamental than any of the grounds given for its goodness. I believe that through discipline, though not through discipline alone, we can achieve serenity, and a certain small but precious measure of freedom from the accidents of incarnation, and charity, and that detachment which preserves the world which it renounces. I believe that through discipline we can learn to preserve what is essential to our happiness in more and more adverse circumstances, and to abandon with simplicity what would else have seemed to us indispensable; that we come a little to see the world without the gross distortion of personal desire, and in seeing it so, accept more easily our earthly privation and its earthly horror— But because I believe that the reward of discipline is greater than its immediate objective, I would not have you think that discipline without objective is possible: in its nature discipline involves the subjection of the soul to some perhaps minor end; and that end must be real, if the discipline is not to be factitious. Therefore I think that all things which evoke discipline: study, and our duties to men and to the commonwealth, war, and personal hardship, and even the need for subsistence, ought to be greeted by us with profound grati-

tude, for only through them can we attain to the least detachment; and only so can we know peace.

Lawrence, orders of magnitude less articulate than Oppenheimer, was also fiercely driven; the question is what drove him. A paragraph from a letter to his brother John, written at about the same time as Oppenheimer's essay, is revealing: "Interested to hear you have had a period of depression. I have them often—sometimes nothing seems to be OK—but I have gotten used to them now. I expect the blues and I endure them. Of course the best palliative is work, but sometimes it is hard to work under the circumstances." That work is only a "palliative," not a cure, hints at how blue the blues could be. Lawrence was a hidden sufferer, in some measure manic-depressive; he kept moving not to fall in.

To all these emotional troublings—Oppenheimer's and Lawrence's, as Bohr's and others' before and since—science offered an anchor: in discovery is the preservation of the world. The psychologist who studied scientists at Berkeley with Rorschach and TAT found that "uncommon sensitivity to experiences—usually sensory experiences" is the beginning of creative discovery in science. "Heightened sensitivity is accompanied in thinking by overalertness to relatively unimportant or tangential aspects of problems. It makes [scientists] look for and postulate significance in things which customarily would not be singled out. It encourages highly individualized and even autistic ways of thinking." Consider Rutherford playing his thoroughly unlikely hunch about alpha backscattering, Heisenberg remembering an obscure remark of Einstein's and concluding that nature only performed in consonance with his mathematics, Lawrence flipping compulsively through obscure foreign journals:

> Were this thinking not in the framework of scientific work, it would be considered paranoid. In scientific work, creative thinking demands seeing things not seen previously, or in ways not previously imagined; and this necessitates jumping off from "normal" positions, and taking risks by departing from reality. The difference between the thinking of the paranoid patient and the scientist comes from the latter's ability and willingness to test out his fantasies or grandiose conceptualizations through the systems of checks and balances science has established—and to give up those schemes that are shown not to be valid on the basis of these scientific checks. It is specifically because science provides such a framework of rules and regulations to control and set bounds to paranoid thinking that a scientist can feel comfortable about taking the paranoid leaps. Without this structuring, the threat of such unrealistic, illogical, and even bizarre thinking to overall thought and personality organization in general would be too great to permit the scientist the freedom of such fantasying.

At the leading edges of science, at the threshold of the truly new, the threat has often nearly overwhelmed. Thus Rutherford's shock at rebounding alpha particles, "quite the most incredible event that has ever happened to me in my life." Thus Heisenberg's "deep alarm" when he came upon his quantum mechanics, his hallucination of looking through "the surface of atomic phenomena" into "a strangely beautiful interior" that left him giddy. Thus also, in November 1915, Einstein's extreme reaction when he realized that the general theory of relativity he was painfully developing in the isolation of his study explained anomalies in the orbit of Mercury that had been a mystery to astronomers for more than fifty years. The theoretical physicist Abraham Pais, his biographer, concludes: "This discovery was, I believe, by far the strongest emotional experience in Einstein's scientific life, perhaps in all his life. Nature had spoken to him. He had to be right. 'For a few days, I was beside myself with joyous excitement.' Later, he told [a friend] that his discovery had given him palpitations of the heart. What he told [another friend] is even more profoundly significant: when he saw that his calculations agreed with the unexplained astronomical observations, he had the feeling that something actually snapped in him."

The compensation for such emotional risk can be enormous. For the scientist, at exactly the moment of discovery—that most unstable existential moment—the external world, nature itself, deeply confirms his innermost fantastic convictions. Anchored abruptly in the world, Leviathan gasping on his hook, he is saved from extreme mental disorder by the most profound affirmation of the real.

Bohr especially understood this mechanism and had the courage to turn it around and use it as an instrument of assay. Otto Frisch remembers a discussion someone attempted to deflect by telling Bohr it made him giddy, to which Bohr responded: "But if anybody says he can think about quantum problems without getting giddy, that only shows that he has not understood the first thing about them." Much later, Oppenheimer once told an audience, Bohr was listening to Pauli talking about a new theory on which he had recently been attacked. "And Bohr asked, at the end, 'Is this really crazy enough? The quantum mechanics was really crazy.' And Pauli said, 'I hope so, but maybe not quite.' " Bohr's understanding of how crazy discovery must be clarifies why Oppenheimer sometimes found himself unable to push alone into the raw original. To do so requires a sturdiness at the core of identity—even a brutality—that men as different as Niels Bohr and Ernest Lawrence had earned or been granted that he was unlucky enough to lack. It seems he was cut out for other work: for now, building that school of theoretical physics he had dreamed of.

* * *

On June 3, 1920, Ernest Rutherford delivered the Bakerian Lecture before the Royal Society of London. It was the second time he had been invited to fill the distinguished lectureship. He used the occasion to sum up present understanding of the "nuclear constitution" and to discuss his successful transmutation of the nitrogen atom reported the previous year, the usual backward glance of such formal public events. But unusually and presciently, he also chose to speculate about the possibility of a third major constituent of atoms besides electrons and protons. He spoke of "the possible existence of an atom of mass 1 which has zero nucleus charge." Such an atomic structure, he thought, seemed by no means impossible. It would not be a new elementary particle, he supposed, but a combination of existing particles, an electron and a proton intimately united, forming a single neutral particle.

"Such an atom," Rutherford went on with his usual perspicacity, "would have very novel properties. Its external [electrical] field would be practically zero, except very close to the nucleus, and in consequence it should be able to move freely through matter. Its presence would probably be difficult to detect by the spectroscope, and it may be impossible to contain it in a sealed vessel." Those might be its peculiarities. This would be its exceptional use: "On the other hand, it should enter readily the structure of atoms, and may either unite with the nucleus or be disintegrated by its intense field." A neutral particle, if such existed— a *neutron*—might be the most effective of all tools to probe the atomic nucleus.

Rutherford's assistant James Chadwick attended this lecture and found cause for disagreement. Chadwick was then twenty-nine years old. He had trained at Manchester and followed Rutherford down to Cambridge. He had accomplished much already—as a young man, two of his colleagues write, his output "was hardly inferior to that of Moseley"—but he had sat out the Great War in a German internment camp, to the detriment of his health and to his everlasting boredom, and he was eager to move the new work of nuclear physics along. A neutral particle would be a wonder, but Chadwick thought Rutherford had deduced it from flimsy evidence.

That winter he discovered his mistake. Rutherford invited him to participate in the work of extending the nitrogen transmutation results to heavier elements. Chadwick had improved scintillation counting by developing a microscope that gathered more light and by tightening up procedures. He also knew chemistry and might help eliminate hydrogen as a possible contaminant, a challenge to the nitrogen results that still bothered Rutherford. "But also, I think," said Chadwick many years later in a memorial lecture, "he wanted company to support the tedium of counting in the dark—and to lend an ear to his robust rendering of 'Onward, Christian Soldiers.' "

"Before the experiments," Chadwick once told an interviewer, "before we began to observe in these experiments, we had to accustom ourselves to the dark, to get our eyes adjusted, and we had a big box in the room in which we took refuge while Crowe, Rutherford's personal assistant and technician, prepared the apparatus. That is to say, he brought the radioactive source down from the radium room, put it in the apparatus, evacuated it, or filled it with whatever, put the various sources in and made the arrangements that we'd agreed upon. And we sat in this dark room, dark box, for perhaps half an hour or so, and naturally, talked." Among other things, they talked about Rutherford's Bakerian Lecture. "And it was then that I realized that these observations which I suspected were quite wrong, and which proved to be wrong later on, had nothing whatever to do with his suggestion of the neutron, not really. He just hung the suggestion on to it. Because it had been in his mind for some considerable time."

Most physicists had been content with the seemingly complete symmetry of two particles, the electron and the proton, one negative, one positive. Outside the atom—among the stripped, ionized matter beaming through a discharge tube, for example—two elementary atomic constituents might be enough. But Rutherford was concerned with how each element was assembled. "He had asked himself," Chadwick continues, "and kept on asking himself, how the atoms were built up, how on earth were you going to get—the general idea being at that time that protons and electrons were the constituents of an atomic nucleus . . . how on earth were you going to build up a big nucleus with a large positive charge? And the answer was a neutral particle."

From the lightest elements in the periodic table beyond hydrogen to the heaviest, atomic number—the nucleus' electrical charge and a count of its protons—differed from atomic weight. Helium's atomic number was 2 but its atomic weight was 4; nitrogen's atomic number was 7 but its atomic weight was 14; and the disparity increased farther along: silver, 47 but 107; barium, 56 but 137; radium, 88 but 226; uranium, 92 but 235 or 238. Theory at the time proposed that the difference was made up by additional protons in the nucleus closely associated with nuclear electrons that neutralized them. But the nucleus had a definite maximum size, well established by experiment, and as elements increased in atomic number and atomic weight there appeared to be less and less room in their nuclei for all the extra electrons. The problem worsened with the development in the 1920s of quantum theory, which made it clear that confining particles as light as electrons so closely would require enormous energies, energies that ought to show up when the nucleus was disturbed but never did. The only evidence for the presence of electrons in the nucleus was its occasional ejection of beta particles, energetic electrons. That was something to go on, but given the other difficulties with packing electrons into the nucleus it was not enough.

"And so," Chadwick concludes, "it was these conversations that convinced me that the neutron must exist. The only question was how the devil could one get evidence for it. . . . It was shortly after that I began to make experiments on the side when I could. [The Cavendish] was very busy, and left me little time, and occasionally Rutherford's interest would revive, but only occasionally." Chadwick would search for the neutron with Rutherford's blessing, but the frustrating work of experiment was usually his alone.

His temperament matched the challenge of discovering a particle that might leave little trace of itself in its passage through matter; he was a shy, quiet, conscientious, reliable man, something of a neutron himself. Rutherford even felt it necessary to scold him for giving the boys at the Cavendish too much attention, though Chadwick took their care and nurturing to be his primary responsibility. "It was Chadwick," remembers Mark Oliphant, "who saw that research students got the equipment they needed, within the very limited resources of the stores and funds at his disposal." If he seemed "dour and unsmiling" at first, with time "the kindly, helpful and generous person beneath became apparent." He tended, says Otto Frisch, "to conceal his kindness behind a gruff façade."

The façade was protective. James Chadwick was tall, wiry, dark, with a high forehead, thin lips and a raven's-beak nose. "He had," say his joint biographers, colleagues both, "a deep voice and a dry sense of humour with a characteristic chuckle." He was born in the village of Bollington, south of Manchester in Cheshire, in 1891. When he was still a small boy his father left their country home to start a laundry in Manchester; Chadwick's grandmother seems to have raised him. He sat for two scholarships to the University of Manchester at sixteen, an early age even in the English educational system, won them both, kept one and went off to the university.

He meant to read mathematics. The entrance interviews were held publicly in a large, crowded hall. Chadwick got into the wrong line. He had already begun to answer the lecturer's questions when he realized he was being questioned for a physics course. Since he was too timid to explain, he decided that the physics lecturer impressed him and he would read for physics. The first year he was sorry, his biographers report: "the physics classes were large and noisy." The second year he heard Rutherford lecture on his early New Zealand experiments and was converted. In his third year Rutherford gave him a research project. His timidity again confounded him, this time almost fatally for his career: he discovered a snag in the procedure Rutherford had recommended to him but could not bring himself to point it out. Rutherford thought he missed it. Man and boy found their way past that misunderstanding and Chadwick graduated from Manchester in 1911 with first-class honors.

He stayed on for his master's degree, working with A. S. Russell and following the research in those productive years of Geiger, Marsden, de Hevesy, Moseley, Darwin and Bohr. In 1913, taking his M.Sc., he won an important research scholarship that required him to change laboratories to broaden his training. By then Geiger had returned to Berlin; Chadwick followed. Which was a pleasure while it lasted—Geiger made a point of introducing Chadwick around, so that he became acquainted with Einstein, Hahn and Meitner, among others in Berlin—but the war intervened.

A reserve officer, Geiger was called up early. He fortified Chadwick with a personal check for two hundred marks before he left. Some of the young Englishman's German friends advised him to leave the country quickly, but others convinced him to wait to avoid the danger of encountering troop trains along the way. On August 2 Chadwick tried to buy a ticket home by way of Holland at the Cook's Tours office in Berlin. Cook's suggested going through Switzerland instead. That struck Chadwick's friends as risky. He again accepted their advice and settled in to wait.

Then it was too late. He was arrested along with a German friend for allegedly making subversive remarks—merely speaking English would have done the job in those first weeks of hysterical nationalism—and languished in a Berlin jail for ten days before Geiger's laboratory arranged his release. Once out he returned to the laboratory until chaos retreated behind order again and the Kaiser's government found time to direct that all Englishmen in Germany be interned for the duration of the war.

The place of internment was a race track at Ruhleben—the name means "quiet life"—near Spandau. Chadwick shared with five other men a box stall designed for two horses and must have thought of Gulliver. In the winter he had to stamp his feet till late morning before they thawed. He and other interns formed a scientific society and even managed to conduct experiments. Chadwick's cold, hungry, quiet life at Ruhleben continued for four interminable years. This was the time, he said later, making the best of it, when he really began to grow up. He returned to Manchester after the Armistice with his digestion ruined and £11 in his pocket. He was at least alive, unlike poor Harry Moseley. Rutherford took him in.

Some of the experiments Chadwick conducted at the Cavendish in the 1920s to look for the neutron, he says, "were so desperate, so far-fetched as to belong to the days of alchemy." He and Rutherford both thought of the neutron, as Rutherford had imagined it in his Bakerian Lecture, as a close union of proton and electron. They therefore conjured up various ways to torture hydrogen—blasting it with electrical discharges, searching out the effects on it of passing cosmic rays—in the hope that the H atom that had been stable since the early days of the universe would somehow agree to collapse into neutrality at their hands.

The neutral particle resisted their blandishments and the nucleus resisted attack. The laboratory, Chadwick remembers, "passed through a relatively quiet spell. Much interesting and important work was done, but it was work of consolidation rather than of discovery; in spite of many attempts the paths to new fields could not be found." It began to seem, he adds, that "the problem of the new structure of the nucleus might indeed have to be left to the next generation, as Rutherford had once said and as many physicists continued to believe." Rutherford "was a little disappointed, because it was so very difficult to find out anything really important." Quantum theory bloomed while nuclear studies stalled. Rutherford had felt optimistic enough in 1923 to shout at the annual meeting of the British Association, "We are living in the heroic age of physics!" By 1927, in a paper on atomic structure, he was a little less confident. "We are not yet able to do more than guess at the structure even of the lighter and presumably least complex atoms," he writes. He proposed a structure nonetheless, with electrons in the nucleus orbiting around nuclear protons, an atom within an atom.

They had other work. In hindsight, it was necessary preparation. The scintillation method of detecting radiation had reached its limit of effectiveness: it was unreliable if the counting rate was greater than 150 per minute or less than about 3 per minute, and both ranges now came into view in nuclear studies. A disagreement between the Cavendish and the Vienna Radium Institute convinced even Rutherford of the necessity of change. Vienna had reproduced the Cavendish's light-element disintegration experiments and published completely different results. Worse, the Vienna physicists attributed the discrepancy to inferior Cavendish equipment. Chadwick laboriously reran the experiments with a specially made microscope with zinc sulfide coated directly onto the lens of the microscope's objective, which greatly brightened the field. The results confirmed the Cavendish's earlier count. Chadwick then went to Vienna. "He found," write his biographers, "that the scintillation counting was done by three young women—it was thought that not only did women have better eyes than men but they were less likely to be distracted by thinking while counting!" Chadwick observed the young women at work and realized that because they understood what was expected of the experiments they produced the expected results, unconsciously counting nonexistent scintillations. To test the technicians he gave them, without explanation, an unfamiliar experiment; this time their counts matched his own. Vienna apologized.

Hans Geiger, among others, turned back to the electrical counter he had devised with Rutherford in 1908 and improved it. The result, the Geiger counter, was essentially an electrically charged wire strung inside a gas-filled tube with a thinly covered window that allowed charged particles to enter. Once inside the tube the charged particles ionized gas atoms; the electrons thus stripped from the gas atoms were drawn to the positively charged wire; that changed the current level in the wire; the change, in the form of an electrical pulse, could then be run through an amplifier and converted to a sound—typically a click—or shown as a jump in the sweep of a light beam on the television-like screen of an oscilloscope. The electrical counter could operate continuously and could count above and below the limits possible to fallible physicists peering at scintillation screens. But the early counters had a significant disadvantage: they were highly sensitive to gamma radiation, much more so than zinc sulfide, and the radium compounds the Cavendish used as alpha sources gave off plentiful gamma rays. Polonium, the radioactive element that Marie Curie had discovered in 1898 and named after her native Poland, could be an excellent alternative. It was a good alpha source and with a gamma-ray background 100,000 times less intense than radium it was much less likely to overload an electrical counter. Unfortunately, polonium was difficult to acquire. A ton of uranium ore contained only about 0.1 gram, too little for commercial separation. It was available practically only as a byproduct of the radioactive decay of radium, and radium too was scarce.

There was time in those years to recover from the bleakness of the war and get on with living. In 1925 Chadwick married Aileen Stewart-Brown, daughter of a family long established in business in Liverpool. He had been living at Gonville and Caius College; now he made plans for permanent residence. A year later, in the midst of house-building, when Rutherford asked him and another Cavendish man to take on part of the work of revising Rutherford's old textbook on radioactivity, he fitted in the duty at night, working bundled in an overcoat at a writing table moved close to the fireplace of a drafty temporary rental. When the fire burned low he even pulled on gloves.

At the end of the decade the Rutherfords suffered a personal tragedy. Their daughter Eileen, twenty-nine years old and the mother of three children—she was married to a theoretician, R. H. Fowler, who kept up that end of physics at the Cavendish—gave birth to a fourth; one week later, on

December 23, she was felled by a lethal blood clot. "The loss of his only child," writes A. S. Eve, "whom he loved and admired, aged Rutherford for a time; he looked older and stooped more. He continued his life and work with a manful purpose, and one of the delights of his life was his group of four grandchildren. His face always lit up when he spoke of them."

Rutherford was elevated to baron in the New Year's Honours List of 1931, the year he would turn sixty. A kiwi crested his armorial bearings; they were supported on the dexter side by a figure representing Hermes Trismegistus, the Egyptian god of wisdom who was supposed to have written alchemical books, and on the sinister side by a Maori holding a club; and the two crossed curves that quartered his escutcheon traced the matched growth and decay of activity that gives each radioactive element and isotope its characteristic half-life.

Around 1928 a German physicist, Walther Bothe, "a real physicist's physicist" to Emilio Segrè, and Bothe's student Herbert Becker began studying the gamma radiation excited by alpha bombardment of light elements. They surveyed the light elements from lithium to oxygen as well as magnesium, aluminum and silver. Since they were concentrating on gamma radiation excited from a target they wanted a minimum gamma background and used a polonium radiation source. "I don't know how [Bothe] got his sources," Chadwick puzzles, "but he did." Lise Meitner had generously sent polonium to Chadwick from the Kaiser Wilhelm Institutes, but it was too little to allow Chadwick to do the work Bothe was doing.

The Germans found gamma excitation with boron, magnesium and aluminum, as they had more or less expected, because alpha particles disintegrate those elements, but they also and unexpectedly found it with lithium and beryllium, which alphas in this reaction did not disintegrate. "Indeed," writes Norman Feather, one of Chadwick's colleagues at the Cavendish, "with beryllium, the intensity of the . . . radiation was nearly ten times as great as with any other element investigated." That was strange enough; equally strange was the oddity that beryllium emitted this intense radiation under alpha bombardment without emitting protons. Bothe and Becker reported their results briefly in August 1930, then more fully in December. The radiation they had excited from beryllium had more energy than the bombarding alpha particles. The principle of the conservation of energy required a source for the excess; they proposed that it came from nuclear disintegration despite the absence of protons.

Chadwick set one of his research students, an Australian named H. C. Webster, to work studying these unusual results. A French team began the same study a little later with better resources: Irène Curie, Mme. Curie's somber and talented daughter, then thirty-three, and her husband Frédéric Joliot, two years younger, a handsome, outgoing man trained originally as an engineer whose charm reminded Segrè of the French singer Maurice Chevalier.

Marie Curie's Radium Institute at the east end of the Rue Pierre Curie in the Latin Quarter, built just before the war with funds from the French government and the Pasteur Foundation, had the advantage in any studies that required polonium. Radon gas decays over time to three only mildly radioactive isotopes: lead 210, bismuth 210 and polonium 210, which thus become available for chemical separation. Medical doctors throughout the world then used radon sealed into glass ampules—"seeds"—for cancer treatment. When the radon decayed, which it did in a matter of days, the seeds no longer served. Many physicians sent them on to Paris as a tribute to the woman who discovered radium. They accumulated to the world's largest source of polonium.

The Joliot-Curies had worked independently for the two years since their marriage in 1927; in 1929 they decided to work in collaboration. They first developed new chemical techniques for separating polonium, and by 1931 had purified a volume of the element almost ten times more intense than any other existing source. With their powerful new source they turned their attention to the mystery of beryllium.

Chadwick's student H. C. Webster had progressed in the meantime, by the late spring of 1931, beyond recapitulation to discovery: he found, says Chadwick, "that the radiation from beryllium which was emitted in the same direction as the . . . alpha-particles was more penetrating than the radiation emitted in a backward direction." Gamma radiation, an energetic form of light, should be emitted equally in every direction from a point source such as a nucleus, just as visible light radiates equally from a lightbulb filament. A particle, on the other hand, would usually be bumped forward by an incoming alpha. "And that, of course," Chadwick adds, "was a point which excited me very much indeed, because I thought, 'Here's the neutron.' "

With twin daughters now, Chadwick had become a family man of regular habits. Among the most sacred of these was his annual June family vacation. The possibility of finding his long-sought neutron was not sufficient cause to change his plans. It might have been, but he thought he needed a cloud chamber for the next step in the search, and the one immediately available to him at the Cavendish was not in working order. He found a cloud chamber in other hands; its owner agreed to help Webster use it when he had finished using it himself. Still assuming that the neutron

1

1. English novelist H. G. Wells. His 1914 novel, *The World Set Free,* predicted atomic bombs, atomic war and world government.

2

2. As a young man, Hungarian physicist Leo Szilard dreamed of saving the world. "If we could find an element which is split by neutrons . . . "

3. Pierre and Marie Curie in their Paris laboratory, c. 1900. The elements they first isolated from pitchblende residues, polonium and radium, radiated far more energy than any chemical process could account for.

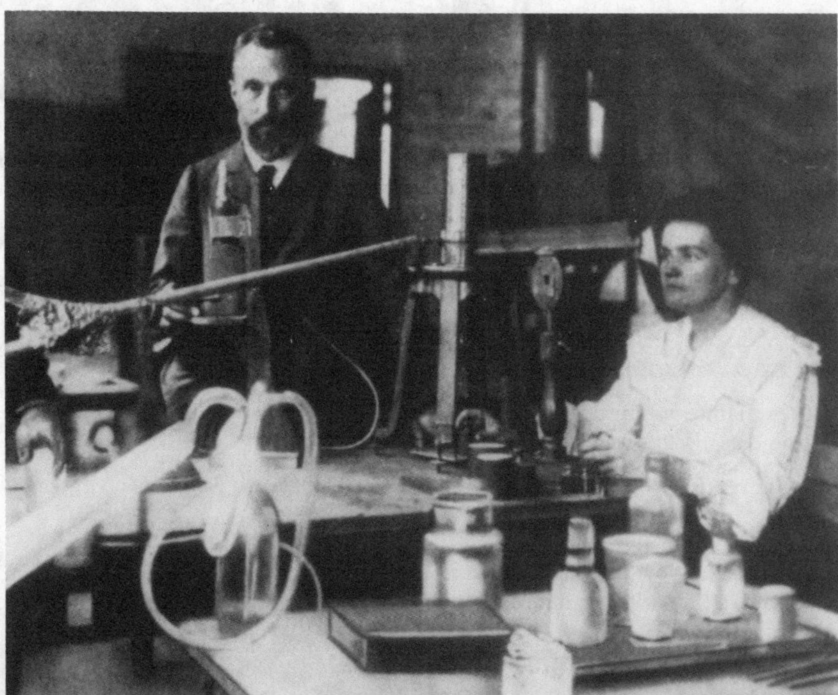

3

4

4. New Zealander Ernest Rutherford discovered the atomic nucleus. James Jeans called him "the Newton of atomic physics." C. 1902.

5

5. The Cavendish Laboratory in Cambridge, England, the world center of early-20th-century experimental physics.

6

6. Otto Hahn and Lise Meitner, chemist and physicist, made a productive team in Berlin.

7

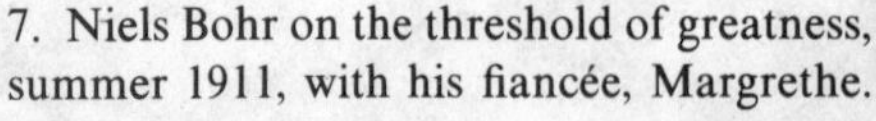

7. Niels Bohr on the threshold of greatness, summer 1911, with his fiancée, Margrethe.

8

8. October 1912: The Kaiser led the way to dedicate the new institute built on farmland he donated in the Berlin suburb of Dahlem.

9. The Kaiser Wilhelm Institute for Chemistry, another measure of burgeoning German power.

9

10

10. Chemist Fritz Haber (*left*) and theoretician Albert Einstein, c. 1914. Haber guided German development of poison gases in the Great War; Einstein spoke out for pacifism and pursued the general theory of relativity. He had already formulated the fateful mass-energy equivalence, $E = mc^2$.

12

11

11. Cambridge physicist Harry Moseley, killed at Gallipoli, 1915. A eulogist said his death alone made the war a "hideous" and "irreparable" crime.

12. American soldiers preparing for gas drill, c. 1917. "It was a way of saving countless lives," Otto Hahn remembers Fritz Haber arguing of poison gas, ". . . if it meant that the war could be brought to an end sooner."

13

13. Niels Bohr's new Institute for Theoretical Physics in Copenhagen, completed in 1921. The best young physicists in the world pilgrimaged here to work and to learn.

14

14. Niels Bohr in the 1920s.

15

15. At Como, Italy, in 1927, Enrico Fermi, Werner Heisenberg and Wolfgang Pauli (*l. to r.*) heard Bohr define complementarity.

16

17

16. Fermi and his group in Rome prepared through the early 1930s for major work and found it bombarding the elements with neutrons to induce artificial radioactivities previously unknown. Uranium was a complex puzzle. *L. to r.*, Emilio Segrè, Enrico Persico and Enrico Fermi at Ostia, 1927. 17. The Physics Institute on the Via Panisperna.

18. Cambridge physicist Francis Aston's mass-spectrograph sorted out isotopes by mass. Their whole-number weights led to an understanding of binding energy, the glue that holds atoms together. "Personally I think there is no doubt that sub-atomic energy is available all around us," Aston lectured, "and that one day man will release and control its almost infinite power."

18

19

19. The first anti-Jewish law Adolf Hitler promulgated, in April 1933, stripped "non-Aryan" academics of their posts. More than 100 physicists fled Germany.

20

20. With Europe in turmoil, Bohr's annual Copenhagen conferences became job forums. In the front row (*l. to r.*): Oskar Klein, Bohr, Heisenberg, Pauli, George Gamow, Lev Landau, Hendrik Kramers.

22

22. Identifying the third basic constituent of matter fell to Rutherford protégé James Chadwick. The discovery of the neutron in 1932 opened the atomic nucleus to detailed examination. Chadwick's colleagues hailed him as "the personification of the ideal experimentalist."

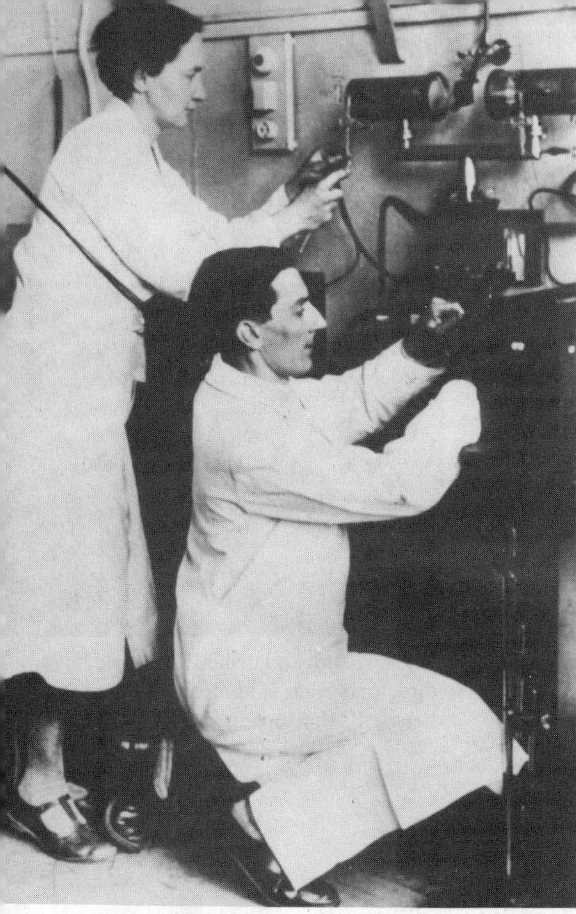

21

21. Frédéric and Irène Joliot-Curie at the Radium Institute in Paris discovered artificial radioactivity but missed the neutron. C. 1935.

23. At Berkeley in the 1930s theoretician Robert Oppenheimer (*left*) and experimentalist Ernest O. Lawrence built a great American school of physics.

23

24

24. Lawrence's Nobel Prize–winning cyclotron battered secrets from the nucleus and proved a potent source of neutrons. Here Lawrence examines the vacuum chamber of the 37-inch machine, completed in 1937.

25

25. Two distinguished Cavendish directors: J. J. Thomson (*left*) and Ernest Rutherford in the 1930s.

26

27

26. Mathematician John von Neumann departed Europe early for a lifetime appointment at the Institute for Advanced Study. 27. Leo Szilard, photographed by Gertrud Weiss at Oxford in 1936. The chain-reaction patent was already a British military secret.

28

29

28. After England, the physicists who escaped Nazi Germany emigrated in increasing numbers to the United States. Future Nobel laureate Hans Bethe won appointment at Cornell. 29. His Stuttgart professor's daughter Rose Ewald followed in 1936. "Rose was then twenty, and I fell in love with her."

30

30. The war against the Jews spread to Italy and threatened Laura Fermi. The 1938 Nobel Prize offered the couple escape with financial security; with their children Giulio and Nella they went on from Stockholm to New York. "We have founded the American branch of the Fermi family," Fermi mocked.

32

31

32. Otto Frisch, c. 1938. With Meitner, his aunt, he prized out the revolutionary meaning of the Hahn-Strassmann uranium discovery.

33

31. Lise Meitner at 59 in 1937. At Christmastime 1938 in Stockholm she heard from Otto Hahn of his stunning discovery with Fritz Strassmann that slow neutrons bombarding uranium made barium—the first evidence that the uranium atom split.

33. Otto Hahn at sixty in 1939. His "barium fantasy" would change the world.

34. One of Hahn's radiochemistry worktables at the Kaiser Wilhelm Institute for Chemistry.

34

35

36

35. The medieval fortress at Kungälv, Sweden, that looked down upon Frisch and Meitner as they worked.

36. Herbert Anderson at Columbia first demonstrated nuclear fission in the United States in January 1939.

37

37. At Munich in September 1938, British Prime Minister Neville Chamberlain agreed to Nazi demands to partition Czechoslovakia. "Peace with honour," he told the London crowds. "Complete surrender," Winston Churchill charged.

38

38. The APO target room at the Carnegie Institution's Department of Terrestrial Magnetism, Washington, D.C., after the demonstration of fission there on the night of January 28, 1939. *L. to r.*, Robert Meyer, Merle Tuve, Fermi, Richard Roberts, Léon Rosenfeld, Erik Bohr, Niels Bohr, Gregory Breit, John Fleming.

39

39. Albert Einstein's 1939 letter to President Franklin Roosevelt reporting the possibility of German atomic bomb research led FDR to appoint a Uranium Committee headed by ineffectual Bureau of Standards director Lyman J. Briggs (*left*).

was an electron-proton doublet with enough residual electrical charge to ionize a gas at least weakly, Chadwick wanted Webster to aim the beryllium radiation into the cloud chamber and see if he could photograph its ionizing tracks. He left his student to the work and went off on holiday.

"Of course," Chadwick said in retrospect of the neutron he was hunting for, "they should not have seen anything" in the cloud chamber, nor did they. "They wrote and told me what had happened, that they hadn't found anything, which disappointed me very much." When Webster moved on to the University of Bristol, Chadwick decided to take over the beryllium research himself.

First he had to shift his laboratory to a different part of the Cavendish building, and that delayed him; then he had to prepare a strong polonium source. In the matter of polonium he was lucky. Norman Feather had spent the 1929–30 academic year in Baltimore, in the physics department at Johns Hopkins, and there befriended an English physician who was in charge of the radium supply at Baltimore's Kelly Hospital. The physician had stored away several hundred used radon seeds; "together," Feather remembers, "they contained almost as much polonium as was available to Curie and Joliot in Paris." The hospital donated them to the Cavendish and Feather brought them home. Chadwick accomplished the dangerous chemical separation that autumn.

Irène Joliot-Curie reported her first results to the French Academy of Sciences on December 28, 1931. The beryllium radiation, she found, was even more penetrating than Bothe and Becker had reported. She standardized her measurements and put the energy of the radiation at three times the energy of the bombarding alpha particle.

The Joliot-Curies decided next to see if the beryllium radiation would knock protons out of matter as alpha particles did. "They fitted their ionization chamber with a thin window," explains Feather, "and placed various materials close to the window in the path of the radiation. They found nothing, except with materials such as paraffin wax and cellophane which already contained hydrogen in chemical combination. When thin layers of these substances were close to the window, the current in the ionization chamber was greater than usual. By a series of experimental tests, both simple and elegant, they produced convincing evidence that this excess ionization was due to protons ejected from the hydrogenous material." The Joliot-Curies understood then that what they were seeing were elastic collisions—like the collisions of billiard balls or marbles—between the beryllium radiation and the nuclei of H atoms.

But they were still committed to their previous conviction that the penetrating radiation from beryllium was gamma radiation. They had not thought about the possibility of a neutral particle. They had not read Rutherford's Bakerian Lecture because such lectures were invariably, in their experience, only recapitulations of previously reported work. Rutherford and Chadwick alone had thought seriously about the neutron.

On January 18, 1932, the Joliot-Curies reported to the Academy of Sciences their discovery that paraffin wax emitted high-velocity protons when bombarded by beryllium radiation. But that was not the title and the argument of the paper they wrote. They titled their paper "The emission of protons of high velocity from hydrogenous materials irradiated with very penetrating gamma rays." Which was as unlikely as if a marble should deflect a wrecking ball. Gamma rays could deflect electrons, a phenomenon known as the Compton effect after its discoverer, the American experimental physicist Arthur Holly Compton, but a proton is 1,836 times heavier than an electron and not easily moved.

At the Cavendish in early February Chadwick found the *Comptes Rendus*, the French physics journal, in his morning mail, discovered the Joliot-Curie paper and read it with widening eyes:

> Not many minutes afterward Feather came to my room to tell me about this report, as astonished as I was. A little later that morning I told Rutherford. It was a custom of long standing that I should visit him about 11 a.m. to tell him any news of interest and to discuss the work in progress in the laboratory. As I told him about the Curie-Joliot observation and their views on it, I saw his growing amazement; and finally he burst out "I don't believe it." Such an impatient remark was utterly out of character, and in all my long association with him I recall no similar occasion. I mention it to emphasize the electrifying effect of the Curie-Joliot report. Of course, Rutherford agreed that one must believe the observations; the explanation was quite another matter.

No further duty interposed itself between Chadwick and his destiny. He went fervently to work, starting on February 7, 1932, a Sunday: "It so happened that I was just ready to begin experiment [when he read of the Joliot-Curie discovery]. . . . I started with an open mind, though naturally my thoughts were on the neutron. I was reasonably sure that the Curie-Joliot observations could not be ascribed to a kind of Compton effect, for I had looked for this more than once. I was convinced that there was something quite new as well as strange."

His simple apparatus consisted of a radiation source and an ionization chamber, the chamber connected to a vacuum-tube amplifier and thence to an oscilloscope. The radiation source, an evacuated metal tube strapped to a rough-sawn block of pine, contained a one-centimeter silver disk coated with polonium mounted close behind a two-centimeter disk of pure beryllium, a silver-gray metal that is three times as light as aluminum. Alpha particles from the polonium striking beryllium nuclei knocked out the penetrating beryllium radiation, which, Chadwick found immediately, would pass essentially unimpeded through as much as two centimeters of lead.

The half-inch opening into the small ionization chamber that faced this radiation source was covered with aluminum foil. Within the shallow chamber, in an atmosphere of air at normal pressure, a small charged plate collected electrons ionized by incoming radiation and moved their pulses along to the amplifier and oscilloscope. "For the purpose at hand," explains Norman Feather, "such an arrangement was ideal. If the amplifier were carefully designed, it was possible to ensure that the magnitude of the oscillograph deflection was directly proportional to the amount of ionization produced in the chamber. . . . The energy of the recoil atom producing the ionization could thus be calculated directly from the size of the deflection on the oscillograph record."

Chadwick mounted a sheet of paraffin two millimeters thick in front of the aluminum-foil window into the ionization chamber; immediately, he wrote in his final report on the experiment, "the number of deflections recorded by the oscillograph increased markedly." That showed that particles ejected from the paraffin were entering the chamber. Then he began interposing sheets of aluminum foil between the wax and the chamber window until no more kicks appeared on the oscilloscope; by scaling the absorptions of aluminum compared to air he calculated the range of the particles as just over 40 centimeters in air; that range meant "it was obvious that the particles were protons."

Thus repeating the Joliot-Curie work prepared the way. Now Chadwick broke new ground. He removed the paraffin sheet. He wanted to study what happens to other elements bombarded directly by the beryllium radiation. Elements in the form of solids he mounted in front of the chamber window: "In this way lithium, beryllium, boron, carbon and nitrogen, as paracyanogen, were tested." Elements in the form of gases he simply pumped into the chamber to replace the ambient air: "Hydrogen, helium, nitrogen, oxygen, and argon were examined in this way." In every case the kicks increased on the oscilloscope; the powerful beryllium radiation knocked protons out of all the elements Chadwick tested. It knocked about the same number out of each element. And, most important for his conclusion, the energies of the recoiling protons were significantly greater than they could possibly be if the beryllium radiation consisted of gamma rays. "In general," Chadwick wrote, "the experimental results show that if the recoil atoms are to be explained by collision with a [gamma-ray photon], we must assume a larger and larger energy for the [photon] as the mass of the struck atom increases." Then, quietly, in what in fact is a devastating criticism of the Joliot-Curie thesis, invoking the basic physical rule that no more energy or momentum can come out of an event than went into it: "It is evident that we must either relinquish the application of the conservation of energy and momentum in these collisions or adopt another hypothesis about the nature of the radiation." When they read that sentence the Joliot-Curies were deeply and properly chagrined.

The hypothesis Chadwick proposed adopting should come as no surprise: "If we suppose that the radiation is not a [gamma] radiation, but consists of particles of mass very nearly equal to that of the proton, all the difficulties connected with the collisions disappear, both with regard to their frequency and to the energy transfer to different masses. In order to explain the great penetrating power of the radiation we must further assume that the particle has no net charge. . . . We may suppose it [to be] the 'neutron' discussed by Rutherford in his Bakerian Lecture of 1920."

Chadwick then worked the numbers to show that his hypothesis was the correct one to explain the facts.

"It was a strenuous time," he said afterward. From beginning to end the work took ten days and he kept up his Cavendish responsibilities besides. He averaged perhaps three hours of sleep a night, labored over the weekend of February 13–14 as well, finished probably on the seventeenth, a Wednesday, the day he sent off a first brief report to *Nature* to establish priority of discovery. He titled that report, published as a letter to the editor, "Possible existence of a neutron." "But there was no doubt whatever in my mind or I should not have written the letter."

"To [Chadwick's] great credit," writes Segrè in tribute, "when the neutron was not present [in earlier experiments] he did not detect it, and when it ultimately was there he perceived it immediately, clearly and convincingly. These are the marks of a great experimental physicist."

A young Russian, Peter Kapitza, had come up to Cambridge in 1921 to work at the Cavendish. He was solid, dedicated, charming and technically inventive and he soon made himself the apple of Rutherford's eye, the only one among all the boys, even including Chadwick, who could convince the frugal director to allow large sums of money to be spent for apparatus. In 1936 Rutherford would attack Chadwick angrily for encouraging the construction of a cyclotron at the Cavendish; but already in 1932 Kapitza had a separate laboratory in an elegant new brick building in the Cavendish courtyard for his expensive experiments with powerful magnetic fields. As Kapitza had settled in at Cambridge he had noticed what he considered to be an excessive and unproductive deference of British physics students to their seniors. He therefore founded a club, the Kapitza Club, devoted to open and unhierarchical discussion. Membership was limited and coveted. Members met in college rooms and Kapitza frequently opened discussions with deliberate howlers so that even the youngest would speak up to correct him, loosening the grip of tradition on their necks.

That Wednesday Kapitza wined and dined the exhausted Chadwick into what Mark Oliphant calls "a very mellow mood," then brought him along to a Kapitza Club meeting. "The intense excitement of all in the Cavendish, including Rutherford," Oliphant remembers, "was already remarkable, for we had heard rumors of Chadwick's results." Oliphant says Chadwick spoke lucidly and with conviction, not failing to mention the contributions of Bothe, Becker, Webster and the Joliot-Curies, "a lesson to us all." C. P. Snow, who was also present, remembers the performance as "one of the shortest accounts ever made about a major discovery." When tall and birdlike Chadwick finished speaking he looked over the assembly and announced abruptly, "Now I want to be chloroformed and put to bed for a fortnight."

He deserved his rest. He had discovered a new elementary particle, the third basic constituent of matter. It was this neutral mass that compounded the weight of the elements without adding electrical charge. Two protons and 2 neutrons made a helium nucleus; 7 protons and 7 neutrons a nitrogen; 47 protons and 60 neutrons a silver; 56 protons and 81 neutrons a barium; 92 protons and 146 (or 143) neutrons a uranium.

And because the neutron was as massive as a proton but carried no electrical charge, it was hardly affected by the shell of electrons around a nucleus; nor did the electrical barrier of the nucleus itself block its way. It would therefore serve as a new nuclear probe of surpassing power of penetration. "A beam of thermal neutrons," writes the American theoretical physicist Philip Morrison, "moving at about the speed of sound, which corresponds to a kinetic energy of only about a fortieth of an electron volt, produces nuclear reactions in many materials much more easily than a beam of protons of millions of volts energy, traveling thousands of times faster." Ernest Lawrence's cyclotron, spiraling protons to million-volt energies for the first time the same month that Chadwick made his fateful discovery, fortunately proved to be adaptable to the production of neutrons. More than any other development, Chadwick's neutron made practical the detailed examination of the nucleus. Hans Bethe once remarked that he considered everything before 1932 "the prehistory of nuclear physics, and from 1932 on the history of nuclear physics." The difference, he said, was the discovery of the neutron.

Word of the discovery reached Copenhagen in the midst of preparations for an amateur theatrical, a parody of Goethe's *Faust*, to celebrate the tenth anniversary of the opening of Bohr's Institute for Theoretical Physics. The postdoctoral dramatists gave the new particle the last word. They had cast Wolfgang Pauli, a corpulent man with a smooth, round face and protuberant, heavy-lidded eyes who resembled the actor Peter Lorre, as Mephistopheles, Bohr as The Lord. Eclectically they cast Chadwick in absentia as Wagner and an anonymous illustrator drew him into the script, "*the personification of the ideal experimentalist*" according to the stage directions, balancing a vastly magnified neutron on his finger:

In Copenhagen, as before in Cambridge, Chadwick reports his discovery briefly and succinctly:

The Neutron *has come to be.*
Loaded with Mass is he.
Of Charge, forever free.
Pauli, do you agree?

Pauli steps forward to dispense his Mephistophelean blessing:

That which experiment has found—
Though theory has no part in—
Is always reckoned more than sound
To put your mind and heart in. . . .

And a chorus of clowning, friendly physicists, Bohr's brilliant young crew, dances out to sing a finale and bring the curtain down:

Now a reality,
Once but a vision.
What classicality,
Grace and precision!
Hailed with cordiality,
Honored in song,
Eternal Neutrality
Pulls us along!

It was the last peaceful time many of them would know for years to come.

7
Exodus

"Antisemitism is strong here and political réaction is violent," Albert Einstein wrote Paul Ehrenfest from Berlin in December 1919. The letter coincides with Einstein's discovery by the popular press, the beginning of his years of international celebrity. "A new figure in world history," the *Berliner Illustrirte Zeitung* described him under a cover photograph on December 14, ". . .whose investigations signify a complete revision of our concepts of nature, and are on a par with the insights of a Copernicus, a Kepler, a Newton." Immediately the anti-Semites and fascists set to work on him.

Einstein was already, at forty-three, respected in the first rank of theoretical physicists. He had been nominated for the Nobel Prize in all but two years since 1910, the secondings increasing in number after 1917; Max Planck, who was not given to exaggeration, wrote the Nobel Committee in 1919 that Einstein "made the first step beyond Newton." The award might have come sooner than in 1922 (belatedly for 1921: the 1922 prize was Bohr's) had relativity been less paradoxical a revelation.

Physically Einstein was not yet the amused, grandfatherly notable of his later American years. His mustache was still dark and his thick black hair had only begun to gray. C. P. Snow would observe "a massive body, very heavily muscled." The Swabian-born physicist's friends thought his loud laugh boyish; his enemies thought it rude. "A powerful sensuality,"

Snow suspected, suspecting also that Einstein took his sensuality to be "one of the chains of personality that ought to be slipped off." Nor had he yet learned, in the psychoanalyst Erik Erikson's words, "to look into cameras as if he were meeting the eyes of the future beholders of his image." In the past year Einstein had endured a stomach ulcer, jaundice and a painful divorce; he had lost and partly regained fifty-six pounds; his mother was dying of cancer: fatigue stained his expressive face. Leopold Infeld, a young Polish physicist who knocked at his door in postwar Berlin seeking a letter of recommendation, found him "dressed in a morning coat and striped trousers with one important button missing." Infeld knew Einstein's face from magazines and newsreels. "But no picture could reproduce the shining glow of his eyes." They were large and dark brown, and the diffident young visitor was one of many—Leo Szilard was another—who found comfort in those cold days in their honest warmth.

The immediate occasion for world notice was an eclipse of the sun. Einstein had presented a paper to the Prussian Academy of Sciences in Berlin on November 25, 1915, "The field equations of gravitation," in which, he reported happily, "finally the general theory of relativity is closed as a logical structure." The paper stands as his first finished statement of the general theory. It was susceptible of proof. It explained mysterious anomalies in the orbit of Mercury—that confirmed prediction was the one which left Einstein feeling something had snapped in him. The general theory also predicted that starlight would be deflected, when it passed a massive body like the sun, through an angle equal to twice the value Newtonian theory predicts. The Great War delayed measurement of the Einstein value. A total eclipse of the sun (which would block the sun's glare and make the stars beyond it visible) due on May 29, 1919, offered the first postwar occasion. The British, not the Germans, followed through. Cambridge astronomer Arthur Stanley Eddington led an expedition to Principe Island, off the West African coast; the Greenwich Observatory sent another expedition to Sobral, inland from the coast of northern Brazil. A joint meeting of the Royal Society and the Royal Astronomical Society at Burlington House in London on November 6, under a portrait of Newton, confirmed the stunning results: the Einstein value, not the Newton value, held good. "One of the greatest achievements in the history of human thought," J. J. Thomson told the assembled worthies. "It is not the discovery of an outlying island but of a whole continent of new scientific ideas."

That was news. *The Times* headlined it REVOLUTION IN SCIENCE and the word spread. From that day forward Einstein was a marked man.

It rankled German chauvinists, including rightist students and some physicists, that the eyes of the world should turn to a Jew who had declared himself a pacifist during the bloodiest of nationalistic wars and who spoke out for internationalism now. When Einstein prepared to offer a series of popular lectures in the University of Berlin's largest hall—everyone was lecturing on relativity that winter—students complained of the expense for coal and electricity. The student body president challenged Einstein to hire his own hall. He ignored the insult and spoke in the university hall as scheduled, but at least one of his lectures, in February, was disrupted.

He was challenged more seriously the following August by an organization assembled under obscure leadership and extravagant but clandestine financing that called itself the Committee of German Scientists for the Preservation of Pure Scholarship. The 1905 Nobel laureate Philipp Lenard, seeing relativity hailed and Einstein come to fame, retreated into a vindictive anti-Semitism and lent his respectability to the Committee, which attacked relativity theory as a Jewish corruption and Einstein as a tasteless self-promoter. The organization held a well-attended public meeting in Berlin's Philharmonic Hall on August 20. Einstein went to listen—one speaker, as Leopold Infeld recalled, "said that uproar about the theory of relativity was hostile to the German spirit"—and stayed to scorn the crackpot talk with laughter and satiric applause.

The criticism nevertheless stung. Einstein mistakenly thought the majority of his German colleagues subscribed to it. Rashly he struck off an uncharacteristically defensive statement. It appeared in the *Berliner Tageblatt* three days after the Philharmonic Hall meeting. "My Answer to the Antirelativity Theory Company Ltd." shocked his friends, but it presciently identified the deeper issues of the Committee attack. "I have good reason to believe that motives other than a desire to search for truth are at the bottom of their enterprise," Einstein wrote. And parenthetically, leaving his implications unstated in elision: "(Were I a German national, with or without swastika, instead of a Jew of liberal, international disposition, then . . .)." A month later his sense of humor had returned; he asked Max Born not to be too hard on him: "Everyone has to sacrifice at the altar of stupidity from time to time . . . and this I have done with my article." But before then he had seriously considered leaving Germany.

It would not be the first time. Einstein had renounced German citizenship and departed the country once before, at the extraordinary age of sixteen. That earlier rejection, which he reversed two decades later, prepared him for the final one, after the Weimar interlude, when Adolf Hitler came to power.

Germany had been united in empire for only eight years when Einstein was born in Ulm on March 14, 1879. He grew up in Munich. He was slow to speak, but he was not, as legend has it, slow in his studies; he consistently earned the highest or next-highest marks in mathematics and Latin in school and *Gymnasium*. At four or five the "miracle" of a compass his father showed him excited him so much, he remembered, that he "trembled and grew cold." It seemed to him then that "there had to be something behind objects that lay deeply hidden." He would look for the something which objects hid, though his particular genius was to discover that there was nothing behind them to hide; that objects, as matter and as energy, were all; that even space and time were not the invisible matrices of the material world but its attributes. "If you will not take the answer too seriously," he told a clamorous crowd of reporters in New York in 1921 who asked him for a short explanation of relativity, "and consider it only as a kind of joke, then I can explain it as follows. It was formerly believed that if all material things disappeared out of the universe, time and space would be left. According to the relativity theory, however, time and space disappear together with the things."

The quiet child became a rebellious adolescent. He was working his own way through Kant and Darwin and mathematics while the *Gymnasium* pounded him with rote. He veered off into religion—Judaism—and came back bitterly disillusioned: "Through the reading of popular scientific books I soon reached the conviction that much of the stories in the Bible could not be true. . . . The consequence was a positively fanatic free-thinking coupled with the impression that youth is intentionally being deceived by the state through lies; it was a crushing impression. Suspicion against every kind of authority grew out of this experience, a sceptical attitude towards the convictions which were alive in any specific social environment."

His father stumbled in business, not for the first time. The family moved across the Alps to Milan to start again, but Albert stayed behind in a boardinghouse to complete his *Gymnasium* work. He was probably expelled from the *Gymnasium* before he could quit. He acquired a doctor's certificate claiming nervous disorders. It was not only the autocracy of his German school that he despised. "Politically," he wrote later, "I hated Germany from my youth." He had thought of renouncing his citizenship while his family was still in Munich, as a rebellious adolescent of fifteen. That began a long family debate. He won it after he moved from Milan to Zurich to try again to finish his schooling; his father wrote the German authorities on his behalf. Einstein renounced his German citizenship officially on January 28, 1896. The Swiss took him aboard in 1901. He liked their doughty democracy and was prepared to serve in their militia but was found medically unfit (because of flat feet and varicose veins); but one reason he quit Germany was to avoid the duty of Prussian conscription, *Kadavergehorsamkeit*, the obedience of the corpse.

The boy and the young man rebelled to protect the child within—the "victorious child," Erik Erikson has it in Einstein's case, the child with its uninhibited creativity preserved into adulthood. Einstein grazes the point in a letter to James Franck:

> I sometimes ask myself how it came about that I was the one to develop the theory of relativity. The reason, I think, is that a normal adult never stops to think about problems of space and time. These are things which he has thought of as a child. But my intellectual development was retarded, as a result of which I began to wonder about space and time only when I had already grown up.

"Relativity" was a misnomer. Einstein worked his way to a new physics by demanding consistency and greater objectivity of the old. If the speed of light is a constant, then something else must serve as the elastic between two systems at motion in relation to one another—even if that something else is time. If a body gives off an amount E of energy its mass minutely diminishes. But if energy has mass, then mass must have energy: the two must be equivalent: $E = mc^2$, $E/c^2 = m$. (I.e., an amount of energy E in joules is equal to an amount of mass m in kilograms multiplied by the square of the speed of light, an enormous number, 3×10^8 meters per second times 3×10^8 m/s $= 9 \times 10^{16}$ or 90,000,000,000,000,000 joules per kilogram. Dividing E by c^2 demonstrates how large an amount of energy is contained within even a small mass.)

Einstein came to that beautiful, harrowing equivalency in 1907, in a long paper published in the *Jahrbuch der Radioaktivität und Elektronik*. "It is possible," he wrote there, "that radioactive processes may become known in which a considerably larger percentage of the mass of the initial atom is converted into radiations of various kinds than is the case for radium." Like Soddy and Rutherford earlier in England, he saw the lesson of radium that there was vast energy stored in matter, though he was not at all sure that it could be released, even experimentally. "The line of thought is amusing and fascinating," he confided to a friend at the time, "but I wonder if the dear Lord laughs about it and has led me around by the nose." He had his Ph.D. then from the University of Zurich and Max Planck had begun to correspond with him, but he had not yet left the patent office where he worked as a technical expert from 1902 to 1909, the years of his first great burst of papers including those on Brownian motion, the photoelectric effect and special relativity.

He habilitated as a *Privatdozent* at the University of Bern in 1908 but held on to the patent-office job for another year for security. Finally in October 1909, *after* receiving his first honorary doctorate, he moved up to associate professor at the University of Zurich. A full professorship enticed him to isolated Prague—he was married now, with a wife and two sons to support—but happily the Polytechnic in Zurich drew him back a year later with a matching offer. The academic hesitations measure how radically new was his work. It was 1913 before Max Planck, Fritz Haber and a muster of German notables, recognizing the waste, offered him a triple appointment in Berlin: a research position under the aegis of the Prussian Academy of Sciences, a research professorship at the university and the directorship of the planned Kaiser Wilhelm Institute for Physics. After the Germans left, Einstein quipped to his assistant, Otto Stern, that they were "like men looking for a rare postage stamp."

He arrived in Berlin in April 1914. In the war years, separated from his first wife and living alone, he completed the general theory. To Max Born that "great work of art" was "the greatest feat of human thinking about nature, the most amazing combination of philosophical penetration, physical intuition, and mathematical skill" even though "its connections with experience were slender." Einstein's crowning achievement ameliorated for him the universal madness of the war:

> I begin to feel comfortable amid the present insane tumult, in conscious detachment from all things which preoccupy the crazy community. Why should one not be able to live contentedly as a member of the service personnel in the lunatic asylum? After all, one respects the lunatics as the people for whom the building in which one lives exists. Up to a point, you can make your own choice of institution—though the distinction between them is smaller than you think in your younger years.

Einstein raised funds for the Zionist cause of a Hebrew university in Palestine on a first trip to the United States, with Chaim Weizmann, in April and May 1921. He had seen the crowds of Eastern Jews stumbling into Berlin in the wake of war and revolution, watched the German incitement against them and decided to take their part. His guide to Zionist thinking was the eloquent spokesman and organizer Kurt Blumenfeld, who also served in that capacity to the young Hannah Arendt. It was Blumenfeld who convinced him to accompany Weizmann to America—his relations with the forceful, singleminded Weizmann, Einstein told Abraham Pais once, "were, as Freud would say, ambivalent." He lectured on relativity at Columbia, the City College of New York and Princeton, met Fiorello La Guardia and President Warren G. Harding, conceived "a new theory of eternity" sitting through formal speeches at the annual dinner of the National Academy of Sciences and spoke to crowds of enthusiastic American Jews.

Back home he wrote that he "first discovered the Jewish people" in America. "I have seen any number of Jews, but the Jewish people I have never met either in Berlin or elsewhere in Germany. This Jewish people which I found in America came from Russia, Poland, and Eastern Europe generally. These men and women still retain a healthy national feeling; it has not yet been destroyed by the process of atomization and dispersion." The statement implicitly criticizes the Jews of Germany, whose "undignified assimilationist cravings and strivings," Einstein wrote elsewhere, had "always . . . annoyed" him. Blumenfeld propounded a radical, post-assimilatory Zionism and had taught him well. A decade later Hannah Arendt would write that "in a society on the whole hostile to Jews . . . it is possible to assimilate only by assimilating to anti-Semitism also." Einstein specialized in driving assumptions to their logical conclusions: clearly he had arrived at a similar understanding of the "Jewish question."

He was now not only the most famous scientist in the world but also a known spokesman for Jewish causes. In Berlin on June 24, 1922, right-wing extremists gunned down Walther Rathenau, the Weimar Republic's first Foreign Minister, a physical chemist and industrialist friend of Einstein and a highly visible Jew. It appeared that Einstein might be next. "I am supposed to belong to that group of persons whom the people are planning to assassinate," he wrote Max Planck. "I have been informed independently by serious persons that it would be dangerous for me in the near future to stay in Berlin or, for that matter, to appear anywhere in public in Germany." He lived privately until October, then left with his second wife, Elsa, on a long trip to the Far East and Japan, receiving notice of his Nobel Prize en route. He spent twelve days in Palestine on the way back and stopped over in Spain. By the time he returned to Berlin, German preoccupation with politics had temporarily retreated behind preoccupation with the Dadaistic mark, then soaring toward 54,000 to the dollar. Einstein went on with his work, including the Einstein-Szilard refrigerator pump and his first efforts toward a unified field theory, but began frequently to travel abroad.

The anti-Semitism Einstein found strong in Berlin in December 1919 was rampant in Munich. Pale, thin, thirty-year-old Adolf Hitler sat down that month at the single battered table in the cramped office of the German Workers Party, formerly a taproom, to draft his party's platform. A grotesque wood carving served as inspiration. It would follow its master into history; a touring Australian academic encountered it again in 1936:

> I was being shown round a famous collection of [Nazi] Party relics in Munich. The curator was a mild old man, a student of the old German academic class.

After showing me everything, he led, almost with bated breath, to his *pièce de résistance.* He produced a small sculptured wooden gibbet from which was suspended a brutally realistic figure of a dangling Jew. This piece of humourless sadism, he said, decorated the table at which Hitler founded the Party, seventeen years ago.

His pale blue eyes shining, Hitler read out the twenty-five points of his party's program the following February in the Festsaal of Munich's Hofbräuhaus before nearly two thousand people, the largest crowd the little German Workers Party had yet attracted. "These points of ours," he had shouted in triumph the day he finished drafting them, "are going to rival Luther's placard on the doors of Wittenberg!" All or part of six of them applied specifically to Jews: that Jews were not countrymen "of German blood" and therefore could not be citizens; that only citizens could hold public office or publish German-language newspapers; that no more non-Germans might immigrate into the country and that all non-Germans admitted since the beginning of the Great War should be expelled. The twenty-five points were never officially declared the program of the Nationalsozialistiche Deutsche Arbeiterpartei, the Nazi Party, which the German Workers Party evolved to, but their power was felt nevertheless.

The Beer Hall Putsch on November 8, 1923, delivered Hitler to a comfortable, sunlit cell in Landsberg prison, where he dictated his personal and political testament to his bashful acolyte Rudolf Hess. *Mein Kampf* has much to say about the Jews. Across the nearly seven hundred pages of its two volumes it refers to Jewry more frequently than to any other subject except Marxism—and Hitler considered Marxism a Jewish invention and a Jewish "weapon."

Jews, the future Chancellor of Germany declares in *Mein Kampf*, are "no lovers of water." He "often grew sick to my stomach from [their] smell." Their dress is "unclean," their appearance "generally unheroic." "A foreign people," they have "definite racial characteristics"; they are "inferior being[s]," "vampires" with "poison fangs," "yellow fist[s]" and "repulsive traits." "The personification of the devil as the symbol of all evil assumes the living shape of the Jew."

The attributes of the Jew are legion, Hitler goes on. The Jew is "a garbage separator, splashing his filth in the face of humanity." Or he is a "scribbler . . . who poison[s] men's souls like germ-carriers of the worst sort." Or "the cold-hearted, shameless, and calculating director of this revolting vice traffic in the scum of the big city." "Was there any form of filth or profligacy," Hitler asks rhetorically, ". . . without at least one Jew involved in it? If you cut even cautiously into such an abscess, you found, like a maggot in a rotting body, often dazzled by the sudden light—a kike!"

The Jew is "no German." Jews are a "race of dialectical liars"; a "people which lives only for this earth"; "the great masters of the lie"; "traitors, profiteers, usurers, and swindlers"; a "world hydra"; "a horde of rats." "Alone in this world they would stifle in filth and offal."

"Without any true culture," the Jew is "a *parasite* in the body of other peoples," "a sponger who like a noxious bacillus keeps spreading as soon as a favorable medium invites him." "He lacks idealism in any form." He is an "eternal blood-sucker" of "diabolical purposes," "restrained by no moral scruples," who "poisons the blood of others, but preserves his own." He "systematically ruins women and girls": "With satanic joy on his face, the black-haired Jewish youth lurks in wait for the unsuspecting girl whom he defiles with his blood, thus stealing her from her people." He is "master over bastards and bastards alone" and "it was and is Jews who bring the Negroes into the Rhineland, always with the same secret thought and clear aim of ruining the hated white race by the necessarily resulting bastardization." Syphilis is a "Jewish disease," a "Jewification of our spiritual life and mammonization of our mating instinct [that] will sooner or later destroy our entire offspring." The Jew "makes a mockery of natural feelings, overthrows all concepts of beauty and sublimity, of the noble and the good, and instead drags men down into the sphere of his own base nature." "An apparition in a black caftan and black hair locks," responsible for "spiritual pestilence worse than the Black Death of olden times," the Jew is a "coward," a "plunderer," a "menace," a "foreign element," a "viper," a "tyrant," a "ferment of decomposition."

The sun shines in the wide windows of Hitler's cell at Landsberg. Boyish in lederhosen, he remembers that he was blinded by mustard gas below Ypres. He wrote a poem during the war, a poem out of a dream, before he took shrapnel in the thigh on the Somme, before Ypres:

I often go on bitter nights
To Wotan's oak in the quiet glade
With dark powers to weave a union—
The runic letters the moon makes with its magic spell
And all who are full of impudence during the day
Are made small by the magic formula!

. . . . .

Hitler's testament is almost finished. He dictates, his blanched face tumefying:

> If at the beginning of the War and during the War twelve or fifteen thousand of these Hebrew corrupters of the people had been held under poison gas, as happened to hundreds of thousands of our very best German workers in the field, the sacrifice of millions at the front would not have been in vain.

* * *

The dispersion of the Jewish people from Palestine—the Diaspora—began in the sixth century B.C. when Babylon conquered the southern Palestinian kingdom of Judah, destroyed Solomon's temple and carried a large body of Jews into captivity. By the beginning of the Christian era, under Roman hegemony, Jews had established communities in Egypt, in Greece, around the Mediterranean and on the shores of the Black Sea and there were Jewish slaves with the Roman legions on the Rhine. Conditions worsened again for the Jews when the Empire was Christianized in the fourth century A.D. with the conversion of the Emperor Constantine; Christianity and Judaism competed, in a Darwinian sense, for the same Holy Land and the same holy books. Under systematic persecution only a small remnant of the Jewish people remained in Judea. The fantasy of Jews as a brotherhood of evil was invented during this era when Christianity fought its missionary way to dominance.

In the disorder of the Dark Ages the Jews lost even their vestigial Roman citizenship. Those who sought protection won it from rulers like Charlemagne's son Louis the Pious who knew their worth as merchants and craftsmen, but the price of protection was that they became the ruler's property. Their rights were thus no longer inherent but chartered. Against that threatening insecurity Jews could count their gain of judicial autonomy: within their communities they were allowed to administer their own laws. In parts of Spain they had the power even of life and death.

The medieval Church, challenged by the spread of learning and the militancy of Islam to shore up its defenses against heresy, exercised its increasing power over the Jews balefully. The Lateran Councils of 1179 and 1215 made the baleful conflict visible by denying Jews authority over Christians, denying them Christian servants, relegating moneylending to Jews by forbidding it to Christians, forbidding Christians lodging in Jewish quarters and thus officially sanctioning the establishment of ghettos and, most onerously, requiring every Jew to wear a distinguishing badge—frequently, on local authority, the yellow Magen David that the Nazis later restored. Every Jew who ventured from the ghetto distinctively marked was a painted bird, exposed to attack.

The fantasy of Jews as a brotherhood of evil swelled in medieval times to a full-blown demonology. The Jewish Messiah became the Antichrist. The Jews became sorcerers of Satan who poisoned wells, tortured the consecrated Host and murdered Christian children to collect their blood for diabolic rites. When the Black Death struck in the fourteenth century, a supposedly demonic people who poisoned wells were obvious suspects: they needed only to have infiltrated some more vicious poison into the water supply. A quarter of Europe died of plague, and in that time of horror tens of thousands of Jews were burned, drowned, hanged or buried alive in retaliation. Massacre became endemic; 350 Jewish communities were decimated in German lands alone.

The English were the first to expel the Jews entirely. The Jews of England belonged to the Crown, which had systematically extracted their wealth through a special Exchequer to the Jews. By 1290 it had bled them dry. Edward I thereupon confiscated what little they had left and threw them out. They crossed to France, but expulsion from that country followed in 1392; from Spain, at the demand of the Inquisition, in 1492; from Portugal in 1497. Since Germany was a region of multiple sovereignties, German Jews could not be generally expelled. They had been fleeing eastward from bitter German persecution in any case since the twelfth century.

The Jews expelled from Western Europe fled to Poland, a large and thinly populated kingdom where elected monarchs welcomed them with generous charters. The medieval German of these emigrant Ashkenazim evolved to Yiddish; they founded villages and towns; they dispersed up and down the long eastern Polish frontier and lived in relative peace for two hundred years.

Twenty-five thousand at the end of the fifteenth century had increased at least tenfold by the middle of the seventeenth. Then, in violent wars with Russia and Sweden, Poland began to break up. Cossacks and their peasant allies murdered great numbers of Jews and sacked hundreds of their communities. The Ukraine was split in two; Poland lost the northern half to Russia. War and disorder continued into the eighteenth century with Prussia, Austria and Turkey variously joining battle. When Russia invaded Poland in 1768, Prussia proposed a three-way partition with Austria to forestall a complete takeover. That led to Poland's partial dismemberment in 1772. In 1795, after another Russian invasion, the country was completely partitioned and ceased to exist. (Much truncated, it was revived by the Congress of Vienna in 1814 as Congress Poland, joined to Russia by the linkage of Polish kingship for the Czar.) Its Jewish population had increased by then to more than one million souls. Prussia acquired about 150,000 but promptly expelled them eastward. Austria acquired about 250,000. Russia, which soon controlled more than three-fourths of what had been the Polish commonwealth, then also controlled the fates of most of the Eastern Jews. But while Poland had welcomed them, Russia despised them. Its economy was too primitive to need their commercial skills and it abhorred their religion. To Catherine the Great her one million new subjects were first and foremost "the enemies of Christ."

The enemies of Christ became Russia's "Jewish problem." In Russia's benighted intolerance it framed only two solutions: assimilation (by conversion to Christianity) or expulsion. For the interim it practiced quarantine. A decree of 1791 limited Jewish residence to the formerly Polish territories and the unpopulated steppes above the Black Sea, a region that extended north across 286,000 square miles of central Europe to the Baltic: the Pale of Settlement ("pale" in its old sense of "enclosed by a boundary"). The Ashkenazim numbered one-ninth of the Pale's total population, and might have prospered there, but they were burdened with further restrictions. They were heavily taxed, they could not live in the villages as they had done for generations, they could not keep the village inns or sell liquor to the peasants. Their traditional local governments, the *kehillot*, were stripped of legal authority but required to collect Jewish taxes. More horribly, under Nicholas I after 1825 the *kehillot* were charged to conscript twelve-year-old Jewish children for a lifetime of forced service in the Russian Army—six years of brutal "education" followed by twenty-five years in the ranks—a fate that befell between 40,000 and 50,000 Jewish sons before the requirement was relaxed in 1856. The memory of that cruelty would endure: Edward Teller's grandmother responded to his childhood misbehavior, he reminisced once with a friend, by warning him to be a good boy or the Russians would get him.

While Eastern Jews toiled to survive in Mother Russia, emancipation was proceeding in the West. Small Jewish communities had reestablished themselves, made up partly of nominal converts to Christianity who had escaped Spain and Portugal for Holland and England and America, partly of Eastern returnees. The Austrian emperor Joseph II issued an Edict of Tolerance in 1782.

The edicts of emperors were less important to the political future of the Jewish people than the temper of the Enlightenment with its religious skepticism and its faith in the self-evident rights of man. The time had come in the evolution of European forms of government when no single group or class any longer had the power to dominate all others as the nobility had previously done. The nation-state evolved in part to remove this impasse by investing power in the state itself. Such a mechanism made no distinction between Jew and Christian. American Jews thus became American citizens automatically with the Revolution and the Bill of Rights.

The French, remembering ghettos and expulsions, found the emancipation of the Jews of France more difficult. "The Jews should be denied everything as a nation," the Count of Clermont-Tonnerre argued in the French National Assembly, "but granted everything as individuals. . . . It is intolerable that [they] should become a separate political formation or class in the country. Every one of them must individually become a citizen." When a Jewish community contracted its loyalty to a monarch in exchange for his protection it only did what other medieval classes and orders had done. But the nation-state was secular and it considered the autonomous Jewish theocracies lodged within its borders in secular terms. In secular terms a separate political body, theocratic or not, to which citizens gave their first loyalty was potentially a rival and inherently subversive. Much monstrosity would devolve from that reification. In the meantime Liberty, Equality and Fraternity prevailed and the Jews of France became *citoyens* on a September Tuesday in 1791.

Emancipations as they progressed within less revolutionary states included Holland-Belgium, 1795; Sweden, 1848; Denmark and Greece, 1849; England by a gradual unmuddling completely in 1866; Austria, 1867; Spain by the withdrawal of its 1492 order of expulsion in 1868; the new German Empire, 1871. Though they were influential out of all proportion to their numbers, the emancipated Jews of Western Europe, many of whom moved directly to assimilate, were only a minute fraction of the Diaspora. The preponderance of the Jewish people, increased by 1850 to 2.5 million, by 1900 to 5 million, struggled in increasing misery in the Pale.

At his coronation in 1856, amid remissions and amnesties, Czar Alexander II abolished the special conscription of Jewish children. Other alleviations followed, all designed to encourage Jewish assimilation. "Useful" Jews—wealthy merchants, university graduates, craftsmen and medical assistants—were allowed residence in the interior of Russia, beyond the Pale. The universities were restored to autonomy and Jews allowed to attend. Within the Pale Jews received limited civil rights and became eligible for local councils. But the Czar who freed 30 million peasants from serfdom was dismayed to discover that reform after so many centuries of repression might lead not to expressions of gratitude but to revolutionary agitation and revolt, as it did in Congress Poland in 1863, and the liberalization of Russian life stalled.

Revolutionaries—a splinter group that called itself "The People's Will"—murdered Alexander on March 13, 1881, by lobbing a hail of small bombs into his open carriage in broad daylight on a main street of St. Petersburg as he drove home from reviewing the Imperial Guards. One member of The People's Will, not a bomber, was Jewish; that was pretext enough, in the confused aftermath of regicide, to blame the assassination on the Jews. A wave of pogroms—the curious Russian word refers to a violent riot by one group against another—began that continued until 1884. "Jewish disorders," the dogmatic new Czar, Alexander III, called these murderous raids of drunken mobs on Jewish quarters everywhere in the Pale. They erupted with the active participation or tacit consent of the authorities. More than two hundred Jewish communities were attacked. The first wave of pogroms—there would be more in later decades—left

20,000 Jews homeless and 100,000 ruined. Women were raped, families murdered. The government blamed the violence on anarchists and moved to expel even the "useful" Jews back into the ghettos of the Pale.

With the pogroms came the 1882 May Laws, revising or repealing previous reforms and imposing catastrophic new restrictions. Between 1881 and 1900 more than 1 million Jews emigrated from Russia and central Europe to the United States and another 1.5 million between 1900 and 1920. A much smaller number of emigrants, like Chaim Weizmann, chose Western Europe and England. Most found less opportunity there than their American counterparts and more virulent anti-Semitism.

One of the important sources of German anti-Semitism in the years after the Great War was the strange forgery known as *The Protocols of the Elders of Zion*. Adolf Hitler took the *Protocols* as a text, to the extent that National Socialism had a text, for world domination. "I have read *The Protocols of the Elders of Zion*," Hitler told one of his loyalists; "it simply appalled me. The stealthiness of the enemy, and his ubiquity! I saw at once that we must copy it—in our own way, of course." Heinrich Himmler confirmed that connection: "We owe the art of government to the Jews." To the *Protocols*, he meant, which "the Führer learned by heart."

The *Protocols* were Russian work. They link the Jewish experience in Russia with the Jewish experience in Germany, where so few Jews actually lived—only about 500,000 in 1933, less than 1 percent of the German population. If Russia's hostility to the Jews was rooted in part in religious conflict, German anti-Semitism, by contrast, needed a secular myth. A half-educated apostate autodidact like Hitler especially needed some structure on which to hang his anti-Semitic pathology. German anti-Semitism had plentiful German antecedents—Richard Wagner's foamings were high on Hitler's list—but the *Protocols* happened to arrive at the right time and place to earn a prominent position well forward. In the 1920s and 1930s millions of copies of various translations and editions were sold throughout the world.

The book is cast in the form of lectures and begins in midsentence, its scene unset, as if torn from the evil hands of its perpetrators. To supply the missing background, editors usually bound in explanatory material. A popular preliminary was a chapter from the novel *Biarritz*, the work of a minor German postal official, entitled "In the Jewish Cemetery in Prague." Editors offered this lurid fiction, like the fiction of the *Protocols* themselves, as fact. The historian Norman Cohn summarizes its setting:

> At eleven o'clock the gates of the cemetery creak softly and the rustling of long coats is heard, as they touch against the stones and shrubbery. A vague white figure passes like a shadow through the cemetery until it reaches a cer-

> tain tombstone; here it kneels down, touches the tombstone three times with its forehead and whispers a prayer. Another figure approaches; it is that of an old man, bent and limping; he coughs and sighs as he moves. The figure takes its place next to its predecessor and it too kneels down and whispers a prayer. . . . Thirteen times this procedure is repeated. When the thirteenth and last figure has taken its place a clock strikes midnight. From the grave there comes a sharp, metallic sound. A blue flame appears and lights up the thirteen kneeling figures. A hollow voice [the thirteenth figure] says, "I greet you, heads of the twelve tribes of Israel." It is the Devil speaking; and the figures dutifully reply, "We greet you, son of the accursed."

The *Protocols* follow. They are twenty-four in all—some eighty pages in book form. "What I am about to set forth, then," explains the speaker at the beginning of the first Protocol, "is our system from the two points of view, that of ourselves and that of the *goyim*." Much about the system set forth is incoherent, but the *Protocols* elaborate three main themes: a bitter attack on liberalism, the political methods of the Jewish world conspiracy and an outline of the world government the Elders expect soon to install.

The attack on liberalism would be comical if the *Protocols* had not found such vicious use. Liberalism "produced Constitutional States . . . and a constitution, as you well know, is nothing else but a school of discords, misunderstandings, quarrels, disagreements, fruitless party agitations, party whims. . . .We replaced the ruler by a caricature of a government—by a president, taken from the mob, from the midst of our puppet creatures, our slaves." A touching loyalty to the Russian *ancien régime* surfaces from time to time and must have given European readers pause:

> The principal guarantee of stability of rule is to confirm the aureole of power, and this aureole is attained only by such a majestic inflexibility of might as shall carry on its face the emblems of inviolability from mystical causes—from the choice of God. Such was, until recent times, the Russian autocracy, the one and only serious foe we had in the world, without counting the Papacy.

In brief, the Elders have stage-managed the invention and dissemination of modern ideas—of the modern world. Everything more recent than the Russian imperial system of czar, landed nobility and serfs is part and parcel of their diabolical work. Which helps explain how so obscure a study as physics came in Germany in the 1920s to be counted part of the Jewish conspiracy.

The Elders work to establish a world autocracy ruled by a leader who is a "patriarchial paternal" guardian. Liberalism will be rooted out, the masses led away from politics, censorship strict, freedom of the press abolished. A third of the population will be recruited for amateur spying ("It will then be no disgrace to be a spy and informer, but a merit") and a vast secret police will keep order. All these were Nazi strategies, and certainly Hitler's debt to the *Protocols* is evident in *Mein Kampf* and explicitly acknowledged.

Russia's contribution to German anti-Semitism was plagiarized from a work of political satire, *Dialogues from Hell Between Montesquieu and Machiavelli*, written by a French lawyer, Maurice Joly, and first published in Brussels in 1864. Montesquieu speaks for liberalism, Machiavelli for despotism. The concoction of the *Protocols* was probably the work of the head of the czarist secret police outside Russia, a Paris-based agent named Pyotr Ivanovich Rachkovsky. Borrowing and paraphrasing Machiavelli's speeches without even bothering to change their order and attributing them to a secret Jewish council, Rachkovsky was attempting to discredit Russian liberalism by showing it to be a Jewish plot. A St. Petersburg newspaper serialized the earliest version of the *Protocols* in 1903. It was one of three books belonging to the Czarina Alexandra Feodorovna—the other two were the Bible and *War and Peace*—found among her possessions at Ekaterinburg after the murder of the imperial family by Communist revolutionaries on July 17, 1918.

That coincidence returned the *Protocols* west. Fyodor Vinberg, who arranged the German translation and publication of the *Protocols* in Berlin in 1920, was a colonel in the Imperial Guard. The Czarina had been an honorary colonel of his regiment and he had worshiped her. He escaped to Germany at the end of the Great War convinced that her murderers had been Jews. Thereafter revenge on the Jews was the central fixation of his life. He was a friend to Hitler's advisers, particularly the Nazi Party "philosopher," Russian-born Alfred Rosenberg, who published a study of the *Protocols* in 1923.

The fiction of a Jewish world conspiracy had practical value for the Nazi Party. As it had done for earlier anti-Semitic parties, writes Hannah Arendt, who was on the scene as a student in Berlin in the 1920s, it "gave them the advantage of a domestic program, and conditions were such that one had to enter the arena of social struggle in order to win political power. They could pretend to fight the Jews exactly as the workers were fighting the bourgeoisie. Their advantage was that by attacking the Jews, who were believed to be the secret power behind governments, they could openly attack the state itself."

The fiction also served for propaganda, to reassure the German people: if the Jews could dominate the world, then so could the Aryans. Arendt continues: "Thus the Protocols presented world conquest as a practical possibility, implied that the whole affair was only a question of inspired or shrewd know-how, and that nobody stood in the way of a German victory over the entire world but a patently small people, the Jews, who ruled it without possessing instruments of violence—an easy opponent, therefore, once their secret was discovered and their method emulated on a larger scale."

But the scurrilities of *Mein Kampf*, which on the evidence of their incoherence are not calculated manipulations but violent emotional outbursts, demonstrate that Hitler pathologically feared and hated the Jews. In black megalomania he masked an intelligent, industrious and much-persecuted people with the distorted features of his own terror. And that would make all the difference.

A German journalist had the temerity in 1931 to ask Adolf Hitler where he would find the brains to run the country if he took it over. Hitler snapped that *he* would be the brains but went on contemptuously to enlist the help of the German class that still resisted voting the Nazis into power:

> Do you think perhaps that, in the event of a successful revolution along the lines of my party, we would not inherit the brains in droves? Do you believe that the German middle class, this flower of the intelligentsia, would refuse to serve us and place their minds at our disposal? The German middle class would take its stand on the famed ground of the accomplished fact; we will do what we like with the middle class.

But what about the Jews, the journalist persisted—those talented people, war heroes among them, Einstein among them? "Everything they have created has been stolen from us," Hitler charged. "Everything that they know will be used against us. They should just go and foment their unrest among other peoples. We do not need them."

At noon on January 30, 1933, Adolf Hitler, forty-three years old, gleefully accepted appointment as Chancellor of Germany. With the Reichstag fire and the subsequent suspension of constitutional liberties, with the Enabling Act of March 23 by which the Reichstag voluntarily gave over its powers to the Hitler cabinet, the Nazis began to consolidate their control. They moved immediately to legalize anti-Semitism and abolish the civil rights of German Jews. Meeting at his country retreat in Berchtesgaden with Joseph Goebbels, now his propaganda minister, Hitler decided on a boycott of Jewish businesses as an opening sally. The national boycott began on Saturday, April 1. Already during the previous week Jewish judges and lawyers had been dismissed from practice in Prussia and Bavaria. Now newspapers conveniently published business addresses and teams of Nazi storm troopers stationed themselves at storefronts to direct the mobs. Jews caught in the streets were beaten while the police looked on. The boycott was a nationwide German pogrom and it lasted through a violent weekend.

A month earlier, the evening after the Reichstag fire, Wolfgang Pauli had dropped in on a Göttingen group that included Edward Teller. The group had discussed Germany's political situation and Pauli had declared emphatically that the idea of a German dictatorship was *Quatsch*, Pauli's favorite dismissal: rubbish, mush, nonsense. "I have seen dictatorship in Russia," he told them. "In Germany it just couldn't happen." In Hamburg Otto Frisch had mustered similar optimism, as indeed had many Germans. "I didn't take Hitler at all seriously at first," Frisch told an interviewer later. "I had the feeling, 'Well, chancellors come and chancellors go, and he will be no worse than the rest of them.' Then things began to change." The Third Reich promulgated its first anti-Jewish ordinance on April 7. The Law for the Restoration of the Professional Civil Service, the harbinger of some four hundred anti-Semitic laws and decrees the Nazis would issue, changed Teller's life, Pauli's, Frisch's, the lives of their colleagues decisively, forever. It announced bluntly that "civil servants of non-Aryan descent must retire." A decree defining "non-Aryan" followed on April 11: anyone "descended from non-Aryan, especially Jewish, parents or grandparents." Universities were state institutions. Members of their faculties were therefore civil servants. The new law abruptly stripped a quarter of the physicists of Germany, including eleven who had earned or would earn Nobel Prizes, of their positions and their livelihood. It immediately affected some 1,600 scholars in all. Nor were academics dismissed by the Reich likely to find other work. To survive they would have to emigrate.

Some had already left, among them Einstein and the older Hungarians. Einstein read the signs correctly because he was Einstein and because he had borne the brunt of the attack since immediately after the war; the Hungarians had become connoisseurs by now of advancing fascism.

Theodor von Kármán departed first, from Aachen. He had pioneered aeronautical physics; the California Institute of Technology, then vigorously assembling its future reputation, wanted to include that specialty in its curriculum. Aviation philanthropist Daniel Guggenheim was prevailed upon to contribute. The Guggenheim Aeronautical Laboratory, with a ten-foot wind tunnel, began operation under von Kármán's direction in 1930.

Caltech also courted Einstein. So did Oxford and Columbia, but he was attracted to the cosmological work of the dean of Caltech graduate studies, a Massachusetts-born physicist of Quaker background named Richard Chace Tolman. Ongoing observations at Mount Wilson Observatory, above Pasadena, might confirm the last of the three original predictions of the general theory of relativity, the gravitational red-shifting of the light of high-density stars. Tolman sent a delegation to Berlin; Einstein agreed to visit Pasadena in 1931 as a research associate.

He did, twice, returning to Berlin between, dining in Southern California with Charlie Chaplin, viewing a rough cut of Sergei Eisenstein's death-obsessed film *Que Viva Mexico!* with its sponsor Upton Sinclair. As his second visit approached, in December, Einstein was ready to reassess his future: "I decided today," he wrote in his diary, "that I shall essentially give up my Berlin position and shall be a bird of passage for the rest of my life."

The bird of passage was not to nest in Pasadena. Abraham Flexner, the American educator, sought out Einstein at Caltech. Flexner was in the process of founding a new institution, not yet located or named, chartered in 1930 with a $5 million endowment. The two men strolled for most of an hour up and down the halls of the club where Einstein was staying. They met again at Oxford in May and once more at the Einsteins' summer house at Caputh, outside Berlin, in June. "We sat then on the veranda and talked until evening," Flexner recalled, "when Einstein invited me to stay to supper. After supper we talked until almost eleven. By that time it was perfectly clear that Einstein and his wife were prepared to come to America." They walked together to the bus stop. "*Ich bin Feuer und Flamme dafür*," Einstein told his guest as he put him on the bus: "I am fire and flame for it." The Institute for Advanced Study would be established in Princeton, New Jersey. Einstein was its first great acquisition. He had suggested a salary of $3,000 a year. His wife and Flexner negotiated a more respectable $15,000. It was what Caltech had been prepared to pay. But at Caltech, as in Zurich before, Einstein would have been expected to teach. At the Institute for Advanced Study his only responsibility was thought.

The Einsteins left Caputh in December 1932, scheduled to divide the coming year between Princeton and Berlin. Einstein knew better. "Turn around," he told his wife as they stepped off the porch of their house. "You will never see it again." She thought his pessimism foolish.

In mid-March the Nazi SA searched the empty house for hidden weapons. By then Einstein had spoken out publicly against Hitler and was returning to Europe to prepare to move. He settled temporarily at a resort town on the Belgian coast, Le Coq sur Mer, with his wife, his two stepdaughters, his secretary, his assistant and two Belgian guards: assassination threatened again. In Berlin his son-in-law arranged to have his furniture packed. The French obligingly transported his personal papers to Paris by diplomatic pouch. At the end of March 1933 the most original physicist of the twentieth century once again renounced his German citizenship.

Princeton University acquired John von Neumann and Eugene Wigner in 1930, in Wigner's puckish recollection, as a package deal. The university sought advice on improving its science from Paul Ehrenfest, who "recommended to them not to invite a single person but at least two . . . who already knew each other, who wouldn't feel suddenly put on an island where they have no intimate contact with anybody. Johnny's name was of course well known by that time the world over, so they decided to invite Johnny von Neumann. They looked: who wrote articles with John von Neumann? They found: Mr. Wigner. So they sent a telegram to me also." In fact, Wigner had already earned a high reputation in a recondite area of physics known as group theory, about which he published a book in 1931. He accepted the invitation to Princeton to look it over and perhaps to look America over as well. "There was no question in the mind of any person that the days of foreigners [in Germany], particularly with Jewish ancestry, were numbered. . . . It was so obvious that you didn't have to be perceptive. . . . It was like, 'Well, it will be colder in December.' Yes, it will be. We know it well."

Leo Szilard in Berlin debated his future in a musing letter to Eugene Wigner written on October 8, 1932. He was apparently still trying to organize his Bund: the knowledge had got into his blood that he had work to accomplish at the moment more noble than science, he wrote—bad luck, it couldn't be distilled out again. He understood he wasn't allowed to complain if such work commanded no office space in the world. He was considering a professorship in experimental physics in India since it would be essentially only a teaching post and he could therefore turn his creative energies elsewhere. Only the gods knew what might be available in Europe or on the American coast between Washington and Boston, places he might prefer, so he perforce might go to India. In any case, until he found a position he would at least be free to do science without feeling guilty.

Szilard promised to write Wigner again when he had an "actual program." He did not yet know that his actual program would be organizing the desperate rescue. He parked his bags at the Harnack House in Dahlem and sat down with Lise Meitner to talk about doing nuclear physics at the Kaiser Wilhelm Institute. She had Hahn, and Hahn was superb, but he was a chemist. She could use a jack-of-all-trades like Szilard. But the collaboration was not to be. Events moved too quickly. Szilard took his train from Berlin, the train that proved him, if not more clever than most people, at least a day earlier. That was "close to the first of April, 1933."

If Pauli, safe behind the lines in Zurich, had misread events before, he was clear enough once the new law was announced. Walter Elsasser, among the first to leave, chose neutral Switzerland, entrained for Zurich and homed on the physics building at the Polytechnic. "On entering the main door of this building one faces a broad and straight staircase leading directly to the second floor. Before I could take my first step on it, there appeared at the top of the stairs the moon-face of Wolfgang Pauli, who shouted down: 'Elsasser,' he said, 'you are the first to come up these stairs; I can see how in the months to come there will be many, many more to climb up here.' " The idea of a German dictatorship was no longer *Quatsch.*

Longstanding anti-Semitic discrimination in academic appointments weighted the civil service law dismissals in favor of the natural sciences, fields of study that had evolved more recently than the older disciplines of the liberal arts, that German scholarship had looked down upon as "materialistic" and that had therefore proved less impenetrable to Jews. Medicine incurred 423 dismissals, physics 106, mathematics 60—in the physical and biological sciences other than medicine, an immediate total of 406 scientists. The University of Berlin and the University of Frankfurt each lost a third of its faculty.

The promising young theoretical physicist Hans Bethe, then at Tübingen, first heard of his dismissal from one of his students, who wrote him to say he read of it in the papers and wondered what he should do. Bethe thought the question impertinent—it was he who had been dismissed, not the student—and asked for a copy of the news story. Hans Geiger was professor of experimental physics at Tübingen at the time, having moved there from Berlin. When Bethe joined the faculty as a theoretician in November 1932, "Geiger explained his experiments to me, and in other ways made a lot of me, so all seemed to be well on the personal level." Sensibly, then, Bethe wrote the vacationing Geiger for advice. "He wrote back a completely cold letter saying that with the changed situation it would be necessary to dispense with my further services—period. There was no kind word, no regret—nothing." A few days later the official notice arrived.

Bethe at twenty-seven was sturdy, indefatigable, a skier and mountain climber, exceptionally self-confident in physics if still socially diffident. His eyes were blue, his features Germanic; his thick, dark-brown hair, cut short, stood up on his head like a brush. His custom of plowing through difficulties eventually won Bethe comparison with a battleship, except that this particularly equable vessel usually boomed with laughter. He had already published important work.

Born in Strasbourg on July 2, 1906, Bethe moved during childhood to Kiel and then to Frankfurt as his father, a university physiologist, achieved increasing academic success. He did not think of himself as a Jew: "I was not Jewish. My mother was Jewish, and until Hitler came that made no difference whatever." His father's background was Protestant and Prussian; his mother was the daughter of a Strasbourg professor of medicine. He counted two Jewish grandparents, more than enough to trigger the Tübingen dismissal.

Bethe began university studies at Frankfurt in 1924. Two years later, recognizing his gift for theoretical work, his adviser sent him to Arnold Sommerfeld in Munich. Sommerfeld had trained nearly a third of the full professors of theoretical physics in the German-speaking world; his protégés included Max von Laue, Wolfgang Pauli and Werner Heisenberg. The American chemist Linus Pauling came to work with Sommerfeld while Bethe was there, as did the German Rudolf Peierls and Americans Edward U. Condon and I. I. Rabi. Edward Teller arrived from Karlsruhe in 1928, but before the relationship between the two young men could develop into friendship Teller was incapacitated in a streetcar accident, his right foot severed just above the ankle. By the time the amputation healed, Sommerfeld had gone off on a sixtieth-birthday trip around the world, leaving Bethe, who had just passed his doctoral examinations, to look for a job on his own; missing Sommerfeld, Teller chose to move on to Leipzig to study with Heisenberg. Bethe went to the Cavendish on a Rockefeller Fellowship, then to Rome, before accepting appointment at Tübingen.

Since Geiger refused to help challenge his Tübingen dismissal, Bethe appealed to Munich. "Sommerfeld immediately replied, 'You are most welcome here. I will have your fellowship again for you. Just come back.' " After a time in Munich Bethe was invited to Manchester, then to Copenhagen to work with Bohr. In the summer of 1934 Cornell University offered him an assistant professorship. One of his former students, now on the Ithaca physics faculty, had recommended him for the post. He accepted and shipped for America, arriving in early February 1935.

Teller took his Ph.D. under Heisenberg at Leipzig in 1930, stayed on there for another year as a research associate, then shifted to Göttingen to work in its Institute for Physical Chemistry. "His early papers," Eugene Wigner writes, "were entirely in the spirit of the times: the expanding world of the applications of quantum mechanics." Teller probed the more developed part of physics—chemical and molecular physics—with vigorous originality, producing some thirty papers between 1930 and 1936, most of them written with collaborators because he was sloppy at calculation and impatient with the detailed effort of following through.

"It was a foregone conclusion that I had to leave," Teller remembers. "After all, not only was I a Jew, I was not even a German citizen. I wanted to be a scientist. The possibility to remain a scientist in Germany and to have any chance of continuing to work had vanished with the coming of Hitler. I had to leave, as many others did, as soon as I could." The director of his institute, Arnold Eucken, "an old German nationalist," confirmed Teller's conclusion as they left on the same southbound train for spring vacation in March 1933. "I really want you here," Teller remembers Eucken equivocating, "but with this new situation, there is no point in your staying. I would like to help you, but you have no future in Germany." The problem then was where to go. Back in Göttingen after a tense confrontation with his parents in Budapest—they wanted him to stay in Hungary—Teller sat down to apply for a Rockefeller Fellowship to Copenhagen to work with Bohr.

In Hamburg Otto Frisch decided he would have to take Hitler seriously after all. Frisch, a personable young experimentalist with a gift for ingenious invention, worked for Otto Stern, the tubby Galician who apprenticed under Einstein and who had barked at Ernest Lawrence four years previously to get busy on his notion of a cyclotron. Stern was "quite shocked," Frisch writes, "to find that I was of Jewish origin, just as was he himself and another two of his four collaborators. He would have to leave and the three of us as well," although "the University of Hamburg—with the traditions of a Free Hansa city—was very reluctant to put the racial laws into effect, and I wasn't sacked until several months after the other universities had toed the line."

Before the Nazis promulgated the civil service law Frisch had applied for, and won, a Rockefeller Fellowship to work with Enrico Fermi in Rome. The program was designed to free promising young scientists from their immediate duties for a year of research abroad, after which they were expected to return to duty again. At a time of crisis the foundation unfortunately chose to enforce its rules narrowly. Frisch was soon "very disappointed and at first rather disgusted when [the foundation] told me that, the situation having changed because of the Hitler laws, they had to withdraw [their] offer of a grant because I no longer had a job to come back to."

In the meantime Bohr turned up in Hamburg. He was traveling throughout Germany to determine who needed help. "To me it was a great experience," Frisch writes, "to be suddenly confronted with Niels Bohr—an almost legendary name for me—and to see him smile at me like a kindly father; he took me by my waistcoat button and said: 'I hope you will come and work with us sometime; we like people who can carry out "thought experiments"!' " (Frisch had recently verified the prediction of quantum theory that an atom recoils when it emits a photon, a movement previously considered too slight to measure.) "That night I wrote home to my mother . . . and told her not to worry: the Good Lord himself had taken me by my waistcoat button and smiled at me. That was exactly how I felt."

Stern, secure personally in independent wealth and international reputation, set out to find places for his people. "Stern said he would go traveling," continues Frisch, "and see if he could sell his Jewish collaborators—I mean find places for them. And he said he would try to sell me to Madame Curie. So I said, 'Well, do what you can. I'll be very grateful for anything you can do. Just sell me to whoever wants to have me.' And when he came back [from visiting laboratories abroad] he said that Madame Curie had not bought me, but Blackett had." Patrick Maynard Stuart Blackett, London-born, tall, a Navy man, with a lean, vigorous face, was one of Rutherford's protégés and a future laureate. He had just departed the Cavendish for a workingmen's college in London, Birkbeck, after a furious argument over the extent of the Cavendish teaching load. "If physics laboratories have to be run dictatorially," Blackett had sworn, emerging white-faced from Rutherford's office, "I would rather be my own dictator." Birkbeck was a night school; experimenters could work at peace all day, except when Blackett's automatic cloud chamber, triggered by a passing cosmic ray, went off like a cannon in their midst. It was temporary duty. Frisch took it. When the appointment ran out the following year he crossed the North Sea to Copenhagen to work with the Good Lord.

He had the comfort of knowing that for the immediate future his aunt was safe. Lise Meitner was forbidden as of the following September to lecture at the University of Berlin, but because her citizenship was Austrian rather than German she was allowed to continue her work at the KWI. She had a subterfuge to confess, however. When Hahn, who had been lecturing on radiochemistry that spring at Cornell, returned hurriedly to salvage what he could from the wreckage of the Institutes' staff, Meitner sought him out. Her nephew explains:

> Lise Meitner had always kept quiet about her Jewish connection. She had never felt that she was in any way related to Jewish tradition. Although she was, racially speaking, a complete Jew, she had been baptized in her infancy and had never considered herself as anything but a Protestant who happened to have Jewish ancestors. And when all this [anti-Semitic] trouble began she felt, perhaps partly to let sleeping dogs lie and partly not to embarrass her friends, that she would keep quiet about it. It was rather an embarrassment when Hitler forced it all out into the open, so to say, and she had to go and tell Hahn, "You know, I am really Jewish and I am apt to be an embarrassment to you."

At Göttingen the Nobel laureate James Franck, a physical chemist, had a talk with Niels Bohr. Though Franck was Jewish, he was exempt from the civil service law because he had fought at the front in the Great War. He was no less outraged. The problem was deciding what to do. He listened to many people, but he told a friend long afterward that it was Bohr who persuaded him: Bohr insisted that individuals really were responsible for the political actions of their societies. Franck was director of Göttingen's Second Physical Institute. He resigned in protest on April 17 and made sure the newspapers knew.

Max Born shared Franck's convictions and admired his courage but disliked public confrontation. Placed on indefinite "leave of absence" as of April 25, but hearing from the university curator that arrangements might eventually be made to reinstate him, Born responded brusquely that he wanted no special treatment. "We decided to leave Germany at once," he writes. The Borns had already rented an apartment in an Alpine valley town for the summer; they slipped the possession date forward and went early. "Thus we left for the South Tyrol at the beginning of May." He passed the news to Einstein via Leiden. "Ehrenfest sent me your letter," Einstein responded on May 30 from Oxford, which was courting him. "I am glad that you have resigned your positions (you and Franck). Thank God there is no risk involved for either of you. But my heart aches at the thought of the young ones."

The young ones—the scientists and scholars just beginning to establish themselves, as yet unpublished, without international reputation—needed more than informal arrangements. They needed organized support.

Leo Szilard's early train delivered him to Vienna, where he put up at the Regina Hotel. The news of the Law for the Restoration of the Professional Civil Service reached him there, probably in the lobby, and he read the first list of dismissals. That outrage sent him into the street to walk. He encountered an old friend from Berlin, Jacob Marshack, an econometrician. Szilard insisted they had to do something to help. Together they went to see Gottfried Kuhnwald—"the old, hunchbacked Jewish adviser of the Christian Social party," a Szilard admirer explains. "Kuhnwald was a mysterious and shrewd man, very Austrian, with sideburns like Franz Josef. He agreed at once that there would be a great expulsion. He said that when it happened, the French would pray for the victims, the British would organize their rescue, and the Americans would pay for it."

Kuhnwald sent the conspirators to a German economist then visiting Vienna. He advised them in turn that Sir William Beveridge, the director of the London School of Economics, was also visiting Vienna at that time, working on the history of prices, and was registered at the Regina. Szilard bearded the Englishman in his room and found he had not yet thought further than the modest charity of appointing one dismissed economist to the school. That response was at least three orders of magnitude too timid for Szilard's taste and he prepared to assault Sir William with the truth.

Kuhnwald, Beveridge and Szilard met for tea and Szilard read out the list of academic dismissals. Beveridge then agreed, Szilard's admirer writes, "that as soon as he got back to England and got through the most important things on his agenda, he would try to form a committee to find places for the academic victims of Nazism; and he suggested that Szilard should come to London and occasionally prod him. If he prodded him long enough and frequently enough, he would probably be able to do something."

The busy economist required very little prodding. Szilard followed him to London and on a weekend at Cambridge in May Beveridge convinced Ernest Rutherford to head an Academic Assistance Council. The council announced itself on May 22, proposing "to provide a clearing house and centre of information" and to "seek to raise a fund." Among the distinguished academics who signed the announcement besides Beveridge and Rutherford were J. S. Haldane, Gilbert Murray, A. E. Housman, J. J. Thomson, G. M. Trevelyan and John Maynard Keynes.

At about the same time a similar response was building in the United States. John Dewey helped assemble a Faculty Fellowship Fund at Columbia University. There were other immediate private initiatives such as the hiring of Hans Bethe at Cornell. The major U.S. effort, the Emergency Committee in Aid of Displaced German Scholars, was organized under the auspices of the Institute for International Education.

Szilard beat the bushes that summer. He did not feel he could properly represent the Academic Assistance Council (though he ran its office for the month of August as an upaid volunteer), so he traveled and worked to coordinate existing groups and start new ones. A "long and satisfactory interview" early in May with Chaim Weizmann elicited support from English Jewry. Einstein had thought of creating a "university for exiles"; Szilard, working through Léon Rosenfeld, convinced him to devote his prestige to the common effort instead. In Switzerland he nudged the International Students' Service and the Intellectual Cooperation Section of the League of Nations; in Holland he nudged a nervous and disorganized Ehrenfest, who had a small fund available to support visiting theoretical physicists. The university rectors in Belgium were "sympathetic," Szilard reported back to Beveridge, but "war reminiscences make it difficult to establish in Belgium any organization for the helping of German scientists."

The Bohrs coordinated their own exhausting efforts with Szilard's. Bohr convened his usual summer conference in Copenhagen, but this time, writes Otto Frisch, "he proposed to use [it] as a sort of labour exchange." Frisch found it "a confusing affair, with so many people and so little time to sort them out."

It was Bohr with whom Edward Teller had hoped to work when he applied in Göttingen for a Rockefeller Fellowship. The foundation denied him an award on the same grounds it had removed Otto Frisch's: because he had no place of employment to return to. James Franck and Max Born interceded on Teller's behalf with the English, and shortly there arrived not one but two offers of temporary appointments. Teller accepted an assistantship in physics at University College, London. From there, at the beginning of 1934, with the Rockefeller to secure him, he shifted to Copenhagen.

Szilard had help from an American, a Columbia University man, a physicist named Benjamin Liebowitz who had invented a new kind of shirt collar and established himself in the business of shirt manufacturing. At forty-two, Liebowitz was seven years older than Szilard. The two men had met when Szilard had visited the United States briefly in early 1932 and had renewed their acquaintance afterward in Berlin. Like Szilard, Liebowitz had taken up unpaid relief work. The two threw in together, the New Yorker supplying Szilard with a useful American connection. Liebowitz characterized the German situation vividly in a letter back to New York in early May:

> It is impossible to describe the utter despair of all classes of Jews in Germany. The thoroughness with which they are being hounded out and stopped short in their careers is appalling. Unless help comes from the outside, there is no outlook for thousands, perhaps hundreds of thousands, except starvation or [suicide]. It is a gigantic "cold pogrom" and it is not only against Jews; Communists of course are included, but are not singled out racially; Social Democrats and Liberals generally are now or are coming under the ban, especially if they protest in the least against the Nazi movement. . . .
>
> Dr. Leo Szilard . . . proved to be the best prognosticator—he was able to foresee events better than anybody else I know. Weeks before the storm broke he began to formulate plans to provide some means of helping the scientists and scholars of Germany.

Szilard was becoming nervous about his own lack of anchorage. He had not, he wrote another friend in August, "dismissed the idea of going to India, neither has this idea grown stronger." He was not opposed to America, but he would very much prefer to live in England. Although he was "rather tired," he felt "very happy in England." His happiness darkened to gloom as soon as he looked ahead: "It is quite probable that Germany will rearm and I do not believe that this will be stopped by intervention of other powers within the next years. Therefore it is likely to have in a few years two heavily armed antagonistic groups in Europe, and the consequence will be that we shall get war automatically, probably against the wish of either of the parties."

That prepared him for that cool, humid, dull day in September when he would step off the Southampton Row curb and begin to shape the things to come.

Einstein crossed the Channel to England for the last time on September 9 and came under the flamboyant protection of a Naval Air Service commander, barrister and M.P. named Oliver Stillingfleet Locker-Lampson, who had the peculiar distinction of having been invited, while serving under the Grand Duke Nicholas of Russia, to murder Rasputin, an invitation which uncharacteristic discretion led him to decline. Locker-Lampson sent the distinguished physicist off the next morning to a vacation house isolated on moorlands on the east coast of England. Einstein had left Belgium at his wife's insistence: she feared for his life. While she organized their emigration he settled in at Roughton Heath, walking the moors "talking to the goats," he said. There he learned of the suicide of Paul Ehrenfest, one of his oldest and closest friends, on September 25; Ehrenfest had tried to kill his youngest son and blinded him and then killed himself.

The largest public event of the rescue was a mass meeting in Royal Albert Hall, the great circular auditorium in London below Kensington Gardens. Einstein was the featured speaker and therefore all the hall's ten thousand seats were filled and the aisles crowded. Ernest Rutherford came down from Cambridge to chair the event. Afterward Einstein packed his bags and left for America, joining his wife on the *Westernland* when it stopped at Southampton on its way from Antwerp to New York, on October 7.

The mass meeting had been meant to raise money. It raised very little. Cambridge physicist P. B. Moon remembers Rutherford's frustration:

> He did a very great deal for the refugees from Hitler's Germany, finding places for some of them in his laboratory and scraping together what money he could to keep them and their families going until they could find established posts. He told me that one of them had come to him and said he had discovered something or other. "I stopped him short and said 'plenty of people know that,' but you know, Moon, these chaps are living on the smell of an oil rag. They've *got* to push themselves forward."

With the possible exception of French prayer, in fact, Gottfried Kuhnwald's shrewd prediction held true for the first two years of the rescue effort: the British alone nearly equaled the rest of the world in temporary appointments, and American contributions, largely from foundations like the Rockefeller, matched the rest dollar for dollar. Then, as the Depression began to ease and the English academic system pinched, emigration increased to the United States. Under official Emergency Committee auspices thirty scientists and scholars arrived in 1933, thirty-two in 1934, only fifteen in 1935; but forty-three came in 1938, ninety-seven in 1939, fifty-nine in 1940, fifty in 1941. Nor were many of these physicists: with their international network of friendships and acquaintances the physicists were better able than most to provide for each other. About one hundred refugee physicists emigrated to the United States between 1933 and 1941.

Princeton, Einstein reported to his friend Elizabeth, the Queen of Belgium, "is a wonderful little spot, a quaint and ceremonious village of puny demigods on stilts. Yet, by ignoring certain social conventions, I have been able to create for myself an atmosphere conducive to study and free from distraction." Wigner noticed that von Neumann "fell in love with America on the first day. He thought: these are sane people who don't talk in these traditional terms which are meaningless. To a certain extent the materialism of the United States, which was greater than that of Europe, appealed to him." When Stanislaw Ulam arrived in Princeton in 1935 he found von Neumann comfortably ensconced in a "large and impressive house. A black servant let me in." The von Neumanns gave two or three parties a week. "These were not completely carefree," Ulam notes; "the shadow of coming world events pervaded the social atmosphere." Ulam's own enthusiasm for America, formulated a few years later when he was a Junior Fellow at Harvard, was tempered with a criticism of the extreme weather: "I used to tell my friends that the United States was like the little child in a fairy tale, at whose birth all the good fairies came bearing gifts, and only one failed to come. It was the one bringing the climate."

Leopold Infeld, riding the train through New Jersey from New York to Princeton, "was astonished at so many wooden houses; in Europe they are looked down upon as cheap substitutes which do not, like brick, resist the attack of passing time." Inevitably on that passage he noticed "old junked cars, piles of scrap iron." At Princeton the campus was deserted. He found a hotel and asked where all the students had gone. Perhaps to see Notre Dame, the clerk said. "Was I crazy?" Infeld asked himself. "Notre Dame is in Paris. Here is Princeton with empty streets. What does it all mean?" He soon found out. "Suddenly the whole atmosphere changed. It happened in a discontinuous way, in a split second. Cars began to run, crowds of people streamed through the streets, noisy students shouted and sang." Infeld arrived on a Saturday; in those days Princeton played Notre Dame at football.

His first night in the New World, Hans Bethe walked all over New York.

A chemist, Kurt Mendelssohn, vividly recalled the morning after his escape: "When I woke up the sun was shining in my face. I had slept deeply, soundly and long—for the first time in many weeks. [The previous night] I had arrived in London and gone to bed without fear that at 3 a.m. a car with a couple of S.A. men would draw up and take me away."

Before it is science and career, before it is livelihood, before even it is family or love, freedom is sound sleep and safety to notice the play of morning sun.

8

Stirring and Digging

The seventh Solvay Conference, held in Brussels in late October 1933, was George Gamow's ticket of escape from a Soviet Union rapidly becoming inhospitable to theoretical physicists who persisted in modern views. The previous summer the tall, blond, powerfully built Odessan and his wife Rho, also a physicist, had tried to escape by paddling a faltboat—a collapsible rubber kayak—170 miles south from the Crimea to Turkey across the Black Sea without benefit of a weather report. They took a pocket compass, carefully hoarded hard-boiled eggs, cooking chocolate, two bottles of brandy and a bag of fresh strawberries, set out in the morning ostensibly on a recreational excursion and paddled hard all day and into the night. The only document they carried was Gamow's Danish motorcycle-driver's license, souvenir of the 1930 winter he spent in Copenhagen after working with Rutherford at the Cavendish. Gamow planned to show the Turks the document, announce himself in Danish to be a Dane, head for the nearest Danish consulate and put himself long-distance in Bohr's capable hands. But the Black Sea is named for its storms. The wind thwarted the Gamows' escape, drenching them in heavy seas, exhausting them through a long, cold night and finally blowing them back to shore.

Back in Leningrad the following year Gamow received notice from his government that he was officially delegated to the Solvay Conference. "I could not believe my eyes," he writes in his autobiography. It was an easy way out of the country—except that Rho had not been included. Gamow determined to acquire a second passport or defiantly stay home. Through the Bolshevik economist Nikolai Bukharin, whom he knew, he arranged an interview with Party Chairman Vyacheslav Molotov at the Kremlin. Molotov wondered that the theoretician could not live for two weeks without his wife. Gamow feigned camaraderie:

> "You see," I said, "to make my request persuasive I should tell you that my wife, being a physicist, acts as my scientific secretary, taking care of papers, notes, and so on. So I cannot attend a large congress like that without her help. But this is not true. The point is that she has never been abroad, and after Brussels I want to take her to Paris to see the Louvre, the *Folies Bergère*, and so forth, and to do some shopping."

That Molotov understood. "I don't think this will be difficult to arrange," he told Gamow.

When the time arrived to collect the passports Gamow found that Molotov had changed his mind, preferring not to set an awkward precedent. Gamow stubbornly refused to cooperate. The passport office called him three times to pick up his passport and three times he insisted he would wait until there were two. The fourth time "the voice on the telephone informed me that both passports were ready. And indeed they were!" (After the conference the young defectors sailed to America. Gamow taught at the University of Michigan's summer school in pleasant Ann Arbor and from there moved to accept a professorship at George Washington University in Washington, D.C.)

The Solvay Conference, devoted for the first time to nuclear physics, drew men and women from the highest ranks of two generations: Marie Curie, Rutherford, Bohr, Lise Meitner among the older physicists; Heisenberg, Pauli, Enrico Fermi, Chadwick (eight men in all from Cambridge and no one from devastated Göttingen), Gamow, Irène and Frédéric Joliot-Curie, Patrick Blackett, Rudolf Peierls among the younger. Ernest Lawrence, his cyclotron humming, was the token American that year.

They debated the structure of the proton. Other topics they discussed may have seemed more far-reaching at the time. None would prove to be. On August 2, 1932, working with a carefully prepared cloud chamber, an American experimentalist at Caltech named Carl Anderson had discovered a new particle in a shower of cosmic rays. The particle was an electron with a positive instead of a negative charge, a "positron," the first indication that the universe consists not only of matter but of antimatter as well. (Its discovery earned Anderson the 1936 Nobel Prize.) Physicists everywhere immediately looked through their files of cloud-chamber photographs and identified positron tracks they had misidentified before (the Joliot-Curies, who had missed the neutron, saw that they had also missed the positron). The new particle raised the possibility that the positively charged proton might in fact be compound, might be not a unitary particle but a neutron in association with a positron. (It was not; there proved not to be room in the nucleus for electrons positive or negative.)

After they had identified the positrons they had missed before, the Joliot-Curies had started up their cloud chamber again and looked for the new particle in other experimental arrangements. They found that if they bombarded medium-weight elements with alpha particles from polonium, the targets ejected protons. Then they noticed that lighter elements, including in particular aluminum and boron, sometimes ejected a neutron and then a positron instead of a proton. That seemed evidence for a compound proton. They presented their evidence with enthusiasm as a report to the Solvay Conference.

Lise Meitner attacked the Joliot-Curies' report. She had performed similar experiments at the KWI and she was highly respected for the cautious precision of her work. In her experiments, she emphasized, she had been "unable to uncover a *single* neutron." Sentiment favored Meitner. "In the end, the great majority of the physicists present did not believe in the accuracy of our experiments," Joliot says. "After the meeting we were feeling rather depressed." Fortunately the theoreticians intervened. "But at that moment Professor Niels Bohr took us aside . . . and told us he thought our results were very important. A little later Pauli gave us similar encouragement." The Joliot-Curies returned to Paris determined to settle the issue once and for all.

Husband and wife were then thirty-three and thirty-six years old, with a small daughter at home. They sailed and swam together in summer, skied together in winter, worked together efficiently in the laboratory in the Latin Quarter on the Rue Pierre Curie. Irène had succeeded her mother as director of the Radium Institute in 1932: the long-widowed pioneer was mortally ill with leukemia induced by too many years of exposure to radiation.

It seemed likely that the appearance of neutrons and positrons rather than protons might depend on the energy of the alpha particles attacking the target. The Joliot-Curies could test that possibility by moving their polonium source away from the target, slowing the alphas by forcing them to batter their way through longer ranges of air. Joliot went to work. Without question he was seeing neutrons. When he shifted the polonium away from the aluminum-foil target "the emission of neutrons [ceased] altogether when a minimum velocity [was] reached." But something else happened then to surprise him. After neutron emission ceased, positron emission continued—not stopping abruptly but decreasing "only over a period of time, like the radiation . . . from a naturally radioactive element." What was going on? Joliot had been observing the particles with a cloud chamber, catching their ionizing tracks in its supersaturated fog. Now he switched to a Geiger counter and called in Irène. As he explained to a colleague the next day: "I irradiate this target with alpha rays from my source; you can hear the Geiger counter crackling. I remove the source: the crackling ought to stop, but in fact it continues." The strange activity declined to half its initial intensity in about three minutes. They would hardly yet have dared to think of that period as a half-life. It might merely mark the erratic performance of the Geiger counter.

A young German physicist who specialized in Geiger counters, Wolfgang Gentner, was working at the institute that year. Joliot asked him to check the lab instruments. The couple went off to a social evening they could find no excuse to avoid. "The following morning," writes the colleague to whom Joliot spoke that day, "the Joliots found on their desk a little hand-written note from Gentner, telling them that the Geiger counters were in perfect working order."

They were nearly certain then that they had discovered how to make matter radioactive by artificial means.

They calculated the probable reaction. An aluminum nucleus of 13 protons and 14 neutrons, capturing an alpha particle of 2 protons and 2 neutrons and immediately re-emitting 1 neutron must be converting itself into an unstable isotope of phosphorus with 15 protons and 15 neutrons (13 + 2 protons = 15; 14 + 2 - 1 neutrons = 15). The phosphorus then probably decayed to silicon (14 protons, 16 neutrons). The 3-minute period was the half-life of that decay.

They could not chemically trace the infinitesimal accumulation of silicon. Joliot explained why in 1935, when he and his wife accepted the Nobel Prize in Chemistry for their discovery: "The yield of these transmutations is very small, and the weights of elements formed . . . are less than 10^{-15} [grams], representing at most a few million atoms"—too few to find by chemical reaction alone. But they could trace the radioactivity of the phosphorus with a Geiger counter. If it did indeed signal the artificial transmutation of some of the aluminum to phosphorus, they should be able to separate the two different elements chemically. The radioactivity would go with the new phosphorus and leave the untransmuted aluminum behind. But they needed a definitive separation that could be carried out within three minutes, before the faint induced radioactivity faded below their Geiger counter's threshold.

The request perplexed a chemist in a nearby laboratory—"never having envisaged chemistry from that point of view," says Joliot—but he contrived the necessary procedure. The Joliot-Curies irradiated a piece of aluminum foil, dropped it into a container of hydrochloric acid and covered the container. The acid dissolved the foil, producing, by reaction, gaseous hydrogen, which should carry the phosphorus with it out of solution. They drew off the gas into an inverted test tube. The dissolved aluminum fell silent then but the gas made the Geiger counter chatter: whatever was radioactive had been carried along. A different chemical test proved that the radioactive substance was phosphorus. Joliot bounded like a boy.

The discovery might serve as an offering to Irène's ailing mother, who had prepared the daughter and sponsored the son-in-law:

> Marie Curie saw our research work and I will never forget the expression of intense joy which came over her when Irène and I showed her the first artificially radioactive element in a little glass tube. I can still see her taking in her fingers (which were already burnt with radium) this little tube containing the radioactive compound—as yet one in which the activity was very weak. To verify what we had told her she held it near a Geiger-Müller counter and she could hear the rate meter giving off a great many "clicks." This was doubtless the last great satisfaction of her life.

The Joliot-Curies reported their work—"one of the most important discoveries of the century," Emilio Segrè says in his history of modern physics—in the *Comptes Rendus* on January 15, 1934, and in a letter to *Nature* dated four days later. "These experiments give the first chemical proof of artificial transmutation," they concluded proudly. Rutherford wrote them within a fortnight: "I congratulate you both on a fine piece of work which I am sure will ultimately prove of much importance." He had tried a number of such experiments himself, he said, "but without any success"—high praise from the master of experiment.

They had demonstrated that it was possible not only to chip pieces off the nucleus, as Rutherford had done, but also to force it artificially to release some of its energy in radioactive decay. Joliot foresaw the potential consequences of that attack in his half of the joint Nobel Prize address. Given the progress of science, he said, "we are entitled to think that scientists, building up or shattering elements at will, will be able to bring about transmutations of an explosive type. . . . If such transmutations do succeed in spreading in matter, the enormous liberation of useful energy can be imagined." But he saw the possibility of cataclysm "if the contagion spreads to all the elements of our planet":

> Astronomers sometimes observe that a star of medium magnitude increases suddenly in size; a star invisible to the naked eye may become very brilliant and visible without any telescope—the appearance of a Nova. This sudden flaring up of the star is perhaps due to transmutations of an explosive character like those which our wandering imagination is perceiving now—a process that the investigators will no doubt attempt to realize while taking, we hope, the necessary precautions.

Leo Szilard received no invitation to the Solvay Conference. By October 1933 he had not accomplished any nuclear physics of note except within the well-equipped laboratory of his brain. In August he had written a friend that he was "spending much money at present for travelling about and earn of course nothing and cannot possibly go on with this for very long." The idea of a nuclear chain reaction "became a sort of obsession" with him. When he heard of the Joliot-Curies' discovery, in January, his obsession bloomed: "I suddenly saw that the tools were on hand to explore the possibility of such a chain reaction."

He moved to a less expensive hotel, the Strand Palace, near Trafalgar Square, and settled in to think. He had "a little money saved up" after all, "enough perhaps to live for a year in the style in which I was accustomed to live, and therefore I was in no particular hurry to look for a job"—the excitement of new ideas thus relieving his August urgency. The bath was down the hall. "I remember that I went into my bath . . . around nine o'clock in the morning. There is no place as good to think as a bathtub. I would just soak there and think, and around twelve o'clock the maid would knock and say, 'Are you all right, sir?' Then I usually got out and made a few notes, dictated a few memoranda."

One of the "memoranda" took the form of a patent application, filed March 12, 1934, relating to atomic energy. It was the first of several, that year and the next, all finally merged into one complete specification, "Improvements in or Relating to the Transmutation of Chemical Elements." (The same day Szilard applied for a patent, never issued, proposing the storage of books on microfilm.) Szilard had already realized—in September, in the context of inducing a chain reaction—that neutrons would be more efficient than alpha particles at bombarding nuclei. He applied that insight now to propose an alternative method for creating artificial radioactivity:

> In accordance with the present invention radio-active bodies are generated by bombarding suitable elements with neutrons. . . . Such uncharged nuclei penetrate even substances containing the heavier elements without ionization losses and cause the formation of radio-active substances.

That was a first step. It was also a cheeky piece of bravado. Szilard had only theoretical grounds for believing that neutrons might induce radioactivity artificially. He had not done the necessary experiments. Only the Joliot-Curies had carried out such experiments so far, and they used alpha particles. Szilard was pursuing more than artificial radioactivity. He was pursuing chain reactions, power generation, atomic bombs. He had not yet found patentable form for these excursions. He wondered which element or elements might emit two or more neutrons for each neutron captured. He decided at some point, he said later, "that the reasonable thing to do would be to investigate systematically all the elements. There were ninety-two of them. But of course this is a rather boring task, so I thought that I would get some money, have some apparatus built, and then hire somebody who would just sit down and go through one element after the other."

The task would hardly be boring. The truth is that Szilard lacked the resources for such work—access to a laboratory, a dedicated crew, sufficient financial support. "None of the physicists had any enthusiasm for this idea of a chain reaction," he would remember. Rutherford threw him out. Blackett told him, "Look, you will have no luck with such fantastic ideas in England. Yes, perhaps in Russia. If a Russian physicist went to the government and [said], 'We must make a chain reaction,' they would give him all the money and facilities which he would need. But you won't get it in England." Soaking in his bath against the London chill, Szilard turned back to mapping the future. The opportunity to explore the elements systematically for surprises by bombarding them with neutrons passed him by.

It fell instead to Enrico Fermi and his team of young colleagues in Rome. Fermi was prepared. He had all on hand that Szilard did not. He saw as soon as Szilard that the neutron would serve better than the alpha particle for nuclear bombardment. The point was not obvious. One used alphas to generate neutrons (as the Joliot-Curies had done along the way to chasing down their positrons). Since not all the alphas found targets, the neutral particles were correspondingly that much more scarce. As Otto Frisch would write: "I remember that my reaction and probably that of many others was that Fermi's was really a silly experiment because neutrons were much fewer than alpha particles. What that simple argument overlooked of course was that they are very much more effective."

Fermi was prepared because he had been organizing his laboratory for a major expedition into nuclear physics for more than four years. If Italy had been one of the hot centers of physical research he might have been too preoccupied to plan ahead so carefully. But Italian physics was a ruin as sere as Pompeii when he came to it. He had no choice but to push aside the debris and start fresh.

Both Fermi's biographers—his wife Laura and his protégé and fellow Nobel laureate Emilio Segrè—assign the beginning of his commitment to physics to the period of psychological trauma following the death of his older brother Giulio when Fermi was fourteen years old, in the winter of 1915. Only a year apart in age, the two boys had been inseparable; Giulio's death during minor surgery for a throat abscess left Enrico suddenly bereft.

That same winter young Enrico browsed on market day among the stalls of Rome's Campo dei Fiori, where a statue commemorates the philosopher Giordano Bruno, Copernicus' defender, who was burned at the stake there in 1600 by the Inquisition. Fermi found two used volumes in Latin, *Elementorum physicae mathematicae*, the work of a Jesuit physicist, published in 1840. The desolate boy used his allowance to buy the physics textbooks and carried them home. They excited him enough that he read them straight through. When he was finished he told his older sister Maria he had not even noticed they were written in Latin. "Fermi must have studied the treatise very thoroughly," Segrè would decide, looking through the old volumes many years later, "because it contains marginal notes, corrections of errors, and several scraps of paper with notes in Fermi's handwriting."

From that point forward Fermi's development as a physicist proceeded, with a single significant exception, rapidly and smoothly. A friend of his father, an engineer named Adolfo Amidei, guided his adolescent mathematical and physical studies, lending him texts in algebra, trigonometry, analytical geometry, calculus and theoretical mechanics between 1914 and 1917. When Enrico graduated from the *liceo* early, skipping his third year, Amidei asked him if he preferred mathematics or physics as a career and made a point of writing down, with emphasis, the young man's exact reply: "I studied mathematics with passion because I considered it necessary for the study of physics, *to which I want to dedicate myself exclusively*. . . . I've read all the best-known books of physics."

Amidei then advised Fermi to enroll not at the University of Rome but at the University of Pisa, because he could compete in Pisa to be admitted as a fellow to an affiliated Scuola Normale Superiore of international reputation that would pay his room and board. Among other reasons for the advice, Amidei told Segrè, he wanted to remove Fermi from his family home, where "a very depressing atmosphere prevailed . . . after Giulio's death."

When the Scuola Normale examiner saw Fermi's competition essay on the assigned theme "Characteristics of sound" he was stunned. It set forth, reports Segrè, "the partial differential equation of a vibrating rod, which Fermi solved by Fourier analysis, finding the eigenvalues and the eigenfrequencies . . . which would have been creditable for a doctoral examination." Calling in the seventeen-year-old *liceo* graduate, the examiner told him he was extraordinary and predicted he would become an important scientist. By 1920 Fermi could write a friend that he had reached the point of teaching his Pisa teachers: "In the physics department I am slowly becoming the most influential authority. In fact, one of these days I shall hold (in the presence of several magnates) a lecture on quantum theory, of which I'm always a great propagandist." He worked out his first theory of permanent value to physics while he was still a student in Pisa, a predictive deduction in general relativity.

The exception to his rapid progress came in the winter of 1923, when Fermi won a postdoctoral fellowship to travel to Göttingen to study under Max Born. Wolfgang Pauli was there then, and Werner Heisenberg and the brilliant young theoretician Pascual Jordan, but somehow Fermi's exceptional ability went unnoticed and he found himself ignored. Since he was, in Segrè's phrase, "shy, proud, and accustomed to solitude," he may have brought the ostracism on himself. Or the Germans may have been prejudiced against him by Italy's poor reputation in physics. Or, more dynamically, Fermi's visceral aversion to philosophy may have left him tongue-tied: he "could not penetrate Heisenberg's early papers on quantum mechanics, not because of any mathematical difficulties, but because the physical concepts were alien to him and seemed somewhat nebulous" and he wrote papers in Göttingen "he could just as well have written in Rome." Segrè has concluded that "Fermi remembered Göttingen as a sort of failure. He was there for a few months. He sat aside at his table and did his work. He didn't profit. They didn't recognize him." The following year Paul Ehrenfest sent along praise through the intermediary of a former student who looked up Fermi in Rome. A three-month fellowship then took the young Italian to Leiden for the traditional Ehrenfest tightening. After that Fermi could be sure of his worth.

He was always averse to philosophical physics; a rigorous simplicity, an insistence on concreteness, became the hallmark of his style. Segrè thought him inclined "toward concrete questions verifiable by direct experiment." Wigner noticed that Fermi "disliked complicated theories and avoided them as much as possible." Bethe remarked Fermi's "enlightening simplicity." Less generously, the sharp-tongued Pauli called him a "quantum engineer"; Victor Weisskopf, though an admirer, saw some truth in Pauli's canard, a difference in style from more philosophical originals like Bohr. "Not a philosopher," Robert Oppenheimer once sketched him.

"Passion for clarity. He was simply unable to let things be foggy. Since they always are, this kept him pretty active." An American physicist who worked with the middle-aged Fermi thought him "cold and clear. . . . Maybe a little ruthless in the way he would go directly to the facts in deciding any question, tending to disdain or ignore the vague laws of human nature."

Fermi's passion for clarity was also a passion to quantify. He seems to have attempted to quantify everything within reach, as if he was only comfortable when phenomena and relationships could be classified or numbered. "Fermi's thumb was his always ready yardstick," Laura Fermi writes. "By placing it near his left eye and closing his right, he would measure the distance of a range of mountains, the height of a tree, even the speed at which a bird was flying." His love of classification "was inborn," Laura Fermi concludes, "and I have heard him 'arrange people' according to their height, looks, wealth, or even sex appeal."

Fermi was born in Rome on September 29, 1901, into a family that had successfully made the transition during the nineteenth century from peasant agriculture in the Po Valley to career civil service with the Italian national railroad. His father was a *capo divisione* in the railroad's administration, a civil rank that corresponded to the military rank of brigadier general. In accord with a common Italian practice of the day, the infant Enrico was sent to live in the country with a wet nurse. So was his brother Giulio, but because Enrico's health was delicate he did not return to his mother and father until he was two and a half years old. Confronted then with a roomful of strangers purporting to be his family, and "perhaps," writes Laura Fermi, "missing the rough effusiveness of his nurse," he began to cry:

> His mother talked to him in a firm voice and asked him to stop at once; in this home naughty boys were not tolerated. Immediately the child complied, dried his tears, and fussed no longer. Then, as in later childhood, he assumed the attitude that there is no point in fighting authority. If *they* wanted him to behave that way, all right, he would; it was easier to go along with *them* than against.

In 1926, when he was twenty-five years old, Fermi was chosen under the Italian system of *concorsos*, national competitions, to become professor of theoretical physics at the University of Rome. An influential patron had seen to the creation of the new post, a Sicilian named Orso Mario Corbino, a short, dark, volatile man, forty-six when Fermi sought him out in 1921, the director of the university physics institute, an exceptional physicist and a Senator of the Kingdom. Since the old guard of Italian physicists resented Fermi's rapid promotion, he especially welcomed the protection of Corbino's patronage. Corbino found support for his efforts to improve Italian physics from the Fascist government of the bulletheaded former journalist Benito Mussolini, although the senator was not himself a party member.

In the later 1920s Corbino and his young professor agreed that the time was ripe for the small group they were assembling in Rome to colonize new territory on the frontier of physics. They chose as their territory the atomic nucleus, then finding description in quantum mechanics but not yet experimentally disassembled. Fermi's tall, erudite Pisa classmate Franco Rasetti signed on as Corbino's first assistant early in 1927. Rasetti and Fermi together recruited Segrè, who had been studying engineering, by taking him along to the Como conference and explaining the achievements of the assembled luminaries to him—by then, Segrè saw, Pauli and Heisenberg recognized Fermi's talents and included him among their friends. The son of the prosperous owner of a paper mill, Segrè contributed elegance to the group as well as brains.

Corbino added Edoardo Amaldi, the son of a mathematics professor at the University of Padua, by frankly raiding the engineering school. The group quickly nicknamed Fermi "the Pope" for his quantum infallibility; Corbino, like Rutherford at the Cavendish, called them all his "boys." Rasetti departed to Caltech, Segrè to Amsterdam, for seasoning. Fermi sent them out again in the early 1930s, after the decision to go into nuclear physics: Segrè to work with Otto Stern in Hamburg, Amaldi to Leipzig to the laboratory of the physical chemist Peter Debye, Rasetti to Lise Meitner at the KWI. By 1933, with a departmental budget above $2,000 a year, ten times the budget of most Italian physics departments, with a well-made cloud chamber and a nearby radium source and KWI training in the vagaries of Geiger counters, the group was ready to begin.

In the meantime, two months after the Solvay Conference, Fermi completed the major theoretical work of his life, a fundamental paper on beta decay. Beta decay, the creation and expulsion by the nucleus of high-energy electrons in the course of radioactive change, had needed a detailed, quantitative theory, and Fermi supplied it entire. He introduced a new type of force, the "weak interaction," completing the four basic forces known in nature: gravity and electromagnetism, which operate at long range, and the strong force and Fermi's weak force, which operate within nuclear dimensions. He introduced a new fundamental constant, now called the Fermi constant, determining it from existing experimental data. "A fantastic paper," Victor Weisskopf later praised it, ". . . a monument to Fermi's intuition." In London the editor of *Nature* rejected it on the grounds that it was too remote from physical reality, which Fermi found irritating but amusing; he published it instead in the little-known weekly journal of the Italian Research Council, *Ricerca Scientifica*, where Amaldi's wife Ginestra worked, and later in the *Zeitschrift für Physik*. With only minor adjustments Fermi's theory of beta decay continues to be definitive.

The *Comptes Rendus* reporting the Joliot-Curies' discovery of artificial radioactivity reached Rome shortly after Fermi returned from skiing in the Alps, in January 1934. "We had not yet found any [nuclear physics] problems to work on," Amaldi reminisces. ". . . Then came out the paper of Joliot, and Fermi immediately started to look for the radioactivity." Like Szilard, Fermi saw the advantages of using neutrons. I. I. Rabi catalogues those advantages in a lecture:

> Since the neutron carries no charge, there is no strong electrical repulsion to prevent its entry into nuclei. In fact, the forces of attraction which hold nuclei together may pull the neutron into the nucleus. When a neutron enters a nucleus, the effects are about as catastrophic as if the moon struck the earth. The nucleus is violently shaken up by the blow, especially if the collision results in the capture of the neutron. A large increase in energy occurs and must be dissipated, and this may happen in a variety of ways, all of them interesting.

When Fermi began his neutron-bombardment experiments he was thirty-three years old, short, muscular, dark, with thick black hair, a narrow nose and surprising gray-blue eyes. His voice was deep and he grinned easily. Marriage to the petitely beautiful Laura Capon, the daughter of a Jewish officer in the Italian Navy, had encouraged him in methodical habits: he worked for several hours privately at home, arrived at the physics institute at nine, worked until twelve-thirty, lunched at home, returned to the institute at four and continued work until eight in the evening, returning home then for dinner. With marriage he had also gained weight.

He and his team of young colleagues occupied the south section of the second floor of the institute, sharing the space with Corbino and with the chief physicist of Rome's Sanità Pubblica—its health department—a generous soul named G. C. Trabacchi who lent Corbino's boys some of the instruments and supplies they needed for their experiments (in return they cherished him, nicknaming him "Divine Providence"). Antonino Lo Sordo, a frustrated old-guard physicist, fended off the encroaching horde from an office at the north end of the floor. Corbino and his family lived above, the residence overlooking a private garden in back with a goldfish pond at its focus. The first floor served students; the basement held electrical generators and a lead-lined safe for the Sanità's gram of radium, worth 670,000 lire—about $34,000—in that year of its most historic use. Glass pipes passed through a wall of the special safe to carry radon, formed in the decay of radium, to a compact extraction plant, a modest refinery of glass-pipe towers that purified and dried the radioactive gas. The residential upper story of the institute, contracted above the longer lower floors to make room at one end for the dome of a small rotunda, was roofed with tile. "The location of the building in a small park on a hill near the central part of Rome was convenient and beautiful at the same time," Segrè recalls. "The garden, landscaped with palm trees and bamboo thickets, with its prevailing silence (except at dusk, when gatherings of sparrows populated the greenery), made the institute a most peaceful and attractive center of study." A gravel path that shone white in the golden Roman sun led down to the Via Panisperna.

As usual, Fermi hewed the neutron experiments by hand. In February and early March he personally assembled crude Geiger counters from aluminum cylinders acquired by cutting the bottoms off tubes of medicinal tablets. Wired, filled with gas, their ends sealed and leads attached, the counters were slightly smaller than rolls of breath mints and a hundred times less efficient than modern commercial units, but with Fermi to operate them they served. While he built Geiger counters he asked Rasetti to prepare a neutron source in the form of polonium evaporated onto beryllium. Since polonium emits relatively low-energy alpha particles, the resulting source emitted relatively few neutrons per second, and Fermi and Rasetti irradiated several samples without success.

At that point Rasetti, showing a surprising lack of eagerness for historic experiment, went off to Morocco for Easter vacation. Fermi cast about for some way to acquire a stronger neutron source. The rationale for using polonium in the first place, in Paris and Cambridge and Berlin as well as in Rome, had been that a stronger alpha emitter like radon also emitted strong beta and gamma radiation, which disturbed the instruments and interfered with measurements. Fermi realized suddenly that since he was trying to observe a *delayed* effect, he was measuring only *after* he removed the neutron source in any case—and therefore any beta and gamma radiation wouldn't matter and he could use radon. Trabacchi had the radon to spare and willingly dispensed it; with a half-life of only 3.82 days it was perishable in any case and his glowing gram of radium continually exhaled a fresh draft.

To the basement of the physics institute on the Via Panisperna, in his gray lab coat, in mid-March, Fermi thus carried a snippet of glass tubing no larger than the first joint of his little finger. It was flame-sealed at one end and partly filled with powdered beryllium. He set the sealed end of this capsule into a container of liquid air. The radon, directed from the outlet of the extraction plant into the capsule, condensed on its walls in the −200°C cold. Fermi then had to attempt quickly to heat and draw closed the other end of the capsule, without cracking the glass, before the radon evaporated and escaped. When he succeeded, he finished preparing the neutron source by dropping it into a two-foot length of glass tubing of larger diameter and sealing it into the far end so that it could be handled at a distance safe from dangerous exposure to its gamma rays. For all the tedious preparation its useful life was brief.

In the beginning Fermi worked alone. He intended eventually to irradiate most of the elements in the periodic table and he started methodically with the lightest. His source, he calculated, supplied him with more than 100,000 neutrons per second. "Small cylindrical containers filled with the substances tested," he would explain in his first report, "were subjected to the action of the radiation from this source during intervals of time varying from several minutes to several hours." Fermi first irradiated water—testing hydrogen and oxygen at the same time—then lithium, beryllium, boron and carbon without inducing them to radioactivity. Laura Fermi says he wavered then, discouraged by the lack of results, but Fermi seldom talked shop at home and doubt seems unlikely: he knew from the Joliot-Curie work that aluminum, a little farther along, reacted with alphas, and neutrons should prove even more effective.

In any case he succeeded on his next attempt, with fluorine: "Calcium fluoride, irradiated for a few minutes and rapidly brought into the vicinity of the counter, causes in the first few moments an increase of pulses; the effect decreases rapidly, reaching the half-value in about 10 seconds."

Soon he found a radioactivity in aluminum with a half-life of twelve minutes, different from the Joliot-Curies' discovery. Putting aluminum first to link his work with theirs, he reported his findings in a letter to the *Ricerca Scientifica* on March 25, 1934.

A Roman numeral *I* distinguishes that first report on "Radioactivity induced by neutron bombardment." The search was on. To move it along Fermi recruited Amaldi and Segrè and cabled Rasetti in Morocco to rush home. Segrè writes:

> We organized our activities this way: Fermi would do a good part of the experiments and calculations. Amaldi would take care of what we would now call the electronics, and I would secure the substances to be irradiated, the sources, etc. Now, of course, this division of labor was by no means rigid, and we all participated in all phases of the work, but we had a certain division of responsibility along these lines, and we proceeded at great speed. We needed all the help we could get, and we even enlisted the help of a younger brother of one of the students (probably 12 years old), persuading him that it was most interesting and important that he should prepare some neat paper cylinders in which we could irradiate our stuff.

The next letter that went to the *Ricerca Scientifica* (and in summary form to *Nature*) reported artificially induced radioactivity in iron, silicon, phosphorus, chlorine, vanadium, copper, arsenic, silver, tellurium, iodine, chromium, barium, sodium, magnesium, titanium, zinc, selenium, antimony, bromine and lanthanum. By then they had established a routine: they irradiated substances at one end of the second floor and tested them under the Geiger counters at the other end, down a long hall. That shielded the counters from stray radiation from the neutron source. But it also meant, whenever the half-life of an induced radioactivity was short, that someone had to run down the hall. "Amaldi and Fermi prided themselves on being the fastest runners," Laura Fermi notes, "and theirs was the task of speeding short-lived substances from one end of the corridor to the other. They always raced, and Enrico claims that he could run faster than Edoardo. But he is not a good loser." A dignified Spaniard showed up one day to confer with "His Excellency Fermi." Rome's young professor of theoretical physics, a dirty lab coat flying out behind him, nearly knocked the visitor down.

They came, finally, to uranium. They had roughly classified the effects they were seeing. Light elements generally transmuted to lighter elements by ejecting either a proton or an alpha particle. But the electrical barrier around the nucleus works against exits as well as entrances, and that barrier increases in strength with increasing atomic number. So heavy elements got heavier, not lighter: they captured the bombarding neutron, threw off its binding energy by emitting gamma radiation, and thus, with the addition of the neutron's mass, but with no added or subtracted charge, became a heavier isotope of themselves. Which then decayed by the delayed emission of a negative beta ray to an element with one more unit of atomic number. Uranium did the same; after a delay it emitted a beta electron. That should mean, Fermi realized, that bombarding uranium with neutrons was producing first a heavier isotope, uranium 239, and then a new, man-made transuranic element, atomic number 93, something never seen on earth before.

It was necessary to purify their uranium sample (uranium nitrate in solution, a light yellow liquid) of the obscuring beta activity its natural decay products gave off. (Uranium decays naturally through a series of fourteen complex steps down the periodic table to thorium, protactinium, radium, radon, polonium and bismuth to lead.) Trabacchi in his generosity had by then even lent the group a young chemist, Oscar D'Agostino, fresh from training in radiochemistry on the Rue Pierre Curie; D'Agostino accomplished the laborious purification in early May. They were using stronger sources now, up to 800 millicuries of radon, about a million neutrons per second. Irradiating the uranium nitrate gave "a very intense effect with several periods [of half-lives]: one period of about 1 minute, another of 13 minutes besides longer periods not yet exactly determined"—thus their May 10 report.

These several induced radioactivities were all beta emitters. They made whatever atom was emitting them heavier by one atomic number. It seemed to follow, then, that they were transmutations up the periodic table into the uncharted new region of man-made elements. To confirm that stunning possibility Fermi needed to demonstrate with chemical separations that the neutron bombardment was not unaccountably creating elements *lighter* than uranium. The one-minute half-life was too short to work with, so he concentrated on the thirteen-minute substance. D'Agostino diluted the irradiated uranium nitrate with 50 percent nitric acid, dissolved into the acid a small amount of manganese salt and set the solution to boil. By adding sodium chlorate to the boiling solution he precipitated crystals of manganese dioxide. When he filtered the crystals from the solution the radioactivity went with the manganese, much as the radioactivity the Joliot-Curies had induced in aluminum went off with the hydrogen gas. If the radioactivity could be precipitated out of the uranium solution along with a manganese carrier, then it must not be uranium anymore.

By adding other carriers and precipitating other compounds D'Agostino proved that the thirteen-minute substance was neither protactinium (91), thorium (90), actinium (89), radium (88), bismuth (83) nor lead (82). Its behavior excluded elements 87 (then known as ekacesium), and 86 (radon). Element 85 was unknown. Perhaps because the half-lives were different, Fermi made no attempt to check polonium (84). But he felt he had been sufficiently thorough. "This negative evidence about the identity of the 13 min-activity from a large number of heavy elements," he reported cautiously in *Nature* in June, "suggests the possibility that the atomic number of the element may be greater than 92."

Corbino injudiciously announced "a new element" at the annual convocation, the King of Italy in attendance, that closed the academic year, which set the press baying and gave Fermi a few sleepless nights. Having so splendidly accomplished Szilard's "rather boring task," the weary physicist was happy to depart after that with his wife and their small daughter Nella for a summer lecture tour sponsored by the Italian government through Argentina, Uruguay and Brazil.

Leo Szilard had emerged from his bath that spring of 1934 to pursue his favorite causes, not yet joined, of releasing the energy of the nucleus and of saving the world. In a late-April memorandum condemning the recent Japanese occupation of Manchuria he seemed to look ahead to a far future: "The discoveries of scientists," he wrote, "have given weapons to mankind which may destroy our present civilization if we do not succeed in avoiding further wars." He probably meant military aircraft; the horrors of strategic bombing and even its potential for deterrence through a balance of terror were much bruited at mid-decade. But almost certainly he was also thinking of atomic bombs.

Several weeks earlier, looking for a patron, he had sent Sir Hugo Hirst, the founder of the British General Electric Company, a copy of the first chapter of *The World Set Free*. "Of course," he wrote Sir Hugo with a touch of bitterness, still brooding on Rutherford's prediction, "all this is moonshine, but I have reason to believe that in so far as the industrial applications of the present discoveries in physics are concerned, the forecast of the writers may prove to be more accurate than the forecast of the scientists. The physicists have conclusive arguments as to why we cannot create at present new sources of energy for industrial purposes; I am not so sure whether they do not miss the point."

That Szilard saw beyond "energy for industrial purposes" to the possibility of weapons of war is evident in his next patent amendments, dated June 28 and July 4, 1934. Previously he had described "the transmutation of chemical elements"; now he added "the liberation of nuclear energy for power production and other purposes through nuclear transmutation." He proposed for the first time "a chain reaction in which particles which carry no positive charge and the mass of which is approximately equal to the proton mass or a multiple thereof [i.e., neutrons] form the links of the chain." He described the essential features of what came to be known as a "critical mass"—the volume of a chain-reacting substance necessary to make the chain reaction self-sustaining. He saw that the critical mass could be reduced by surrounding a sphere of chain-reacting substance with "some cheap heavy material, for instance lead," that would reflect neutrons back into the sphere, the basic concept for what came to be known (by analogy with the mud tamped into drill holes to confine conventional explosives) as "tamper." And he understood what would happen if he assembled a critical mass, spelling out the results simply on the fourth page of his application:

> If the thickness is larger than the critical value . . . I can produce an explosion.

As if to mark in some distant inhuman ledger the end of one age and the beginning of another, Marie Sklodowska Curie, born in Warsaw, Poland, on November 7, 1867, died that day of Szilard's filing, July 4, 1934, in

Savoy. Einstein's was the best eulogy: "Marie Curie is," he said, "of all celebrated beings, the only one whom fame has not corrupted."

There is nothing in the documentary record to indicate that Szilard was yet thinking of uranium. His June amendment describes a possible chain reaction using light, silvery beryllium, element number 4 on the periodic table.

To study that metal Szilard needed access to a laboratory and a source of radiation. The beryllium nucleus was so lightly bound he suspected he could knock neutrons out of it not only with alpha particles or neutrons but even with gamma rays or high-energy X rays. Radium emitted gamma rays and radium was available conveniently at the nearest large hospital. So Szilard, an unusually practical visionary, dropped in to see the director of the physics department at the medical college of St. Bartholomew's Hospital. Couldn't he use St. Bart's radium, "which was not much in use in summertime," for experiments? Something of value to medicine might emerge. The director thought he could if he teamed up with someone on the staff. "There was a very nice young Englishman, Mr. [T. A.] Chalmers, who was game, and so we teamed up and for the next two months we did experiments."

Their first experiment demonstrated a brilliantly simple method for separating isotopes of iodine by bombarding an iodine compound with neutrons. They then used this Szilard-Chalmers effect (as it came to be called), which was extremely sensitive, as a tool for measuring the production of neutrons in their second experiment: knocking neutrons out of beryllium using the gamma radiation from radium. "These experiments," Szilard reminisces wryly, "established me as a nuclear physicist, not in the eyes of Cambridge, but in the eyes of Oxford. [Szilard had in fact applied to Rutherford that spring to work at the Cavendish and Rutherford had turned him down.] I had never done work in nuclear physics before, but Oxford considered me an expert. . . . Cambridge . . . would never had made that mistake. For them I was just an upstart who might make all sorts of observations, but these observations could not be regarded as discoveries until they had been repeated at Cambridge and confirmed."

If Szilard's summer work helped establish his Oxford reputation, it was also a personal disappointment: beryllium proved an unsatisfactory candidate for chain reaction. The problem, not settled until 1935, lay with the established mass of helium. The one stable isotope of beryllium consists of two helium nuclei lightly bound by a neutron. Its apparently high mass, which was calculated from Francis Aston's measurements of the mass of helium, seemed to indicate that it should be unstable. But the mass spectrograph was a skittish instrument even in the hands of its inventor, and as

Bethe, Rutherford and others were about to demonstrate, Aston's measurements were inaccurate: he had set the mass of helium too high. One casualty of that error was beryllium's candidacy for chain reaction, for nuclear power and atomic bombs.

Emilio Segrè and Edoardo Amaldi pilgrimaged to Cambridge early in July, short on English but carrying with them a comprehensive report on the Rome neutron-bombardment investigations. They met Chadwick, Kapitza and the other regulars at the Cavendish; observed the retired J. J. Thomson making his rounds; noted Aston, says Amaldi innocently, "going on improving the accuracy of his measurements of atomic masses"; and had a memorable meeting with Rutherford, whose "strong personality dominated the whole laboratory."

The two young physicists had come to compare experiments with two of Rutherford's boys. An unanswered question hung over the neutron work, a question that called existing nuclear theory into doubt. The *Nature* paper they brought with them discussed the difficulty frankly. It concerned what is called "radiative capture," the typical reaction of the heavy elements to neutron bombardment: a nucleus captures a neutron, emits a photon of gamma radiation to stabilize itself energetically and thus becomes an isotope one mass unit heavier.

Theory at the time treated the nucleus as if it was one large particle. As such, it had a definite diameter, which was modest enough that a speeding neutron could go in one side and exit out the other in about 10^{-21} seconds, a billion times less than a trillionth of a second. Any capture process would have to work within that brief interval of time. Otherwise the neutron would be gone. Capturing a neutron means stopping it within a nucleus. To do that the nucleus has to absorb the neutron's energy of motion. The nucleus in turn has to get rid of the excess energy. Which it does: by emitting a gamma photon.

But the gamma-emission times Fermi's group had measured were different from what theory said they ought to be. The nuclei the Rome group had studied took at least 10^{-16} seconds to get around to gamma emission—one hundred thousand times too long. And that was unaccountable.

Definite proof of radiative capture would sharpen the challenge to theory. That required proving beyond doubt, by experiment, that a heavier isotope really forms when a heavy nucleus captures a neutron. The Cavendish team Segrè and Amaldi came to visit in the summer of 1934 accomplished the first part of the proof, using sodium, while the Italians were on hand. They then returned to Rome and enlisted D'Agostino's help to perform the confirming chemistry. In the heat of Roman August they looked for additional clear-cut examples and won a double prize: "We also found a second case of 'proven' radiative capture," Amaldi writes, "which was based on the discovery of a new radioisotope of [aluminum] with a lifetime of almost 3 minutes."

Fermi planned to stop off in London for an international physics conference on his way home from South America. His young colleagues sent him word of their aluminum discovery. He reported to the conference on the neutron work. (Szilard also attended, happy to hear praise for his summer experiments and well launched toward a paying fellowship at Oxford.) Fermi said his group had studied sixty elements so far and had induced radioactivity in forty of them. Discussing the radiative-capture problem he cited the Cavendish results "and those of Amaldi and Segrè on aluminium," which were both, he said, "to be considered particularly important." Segrè describes the tempestuous aftermath:

> Shortly afterwards I caught a cold and could not go to the laboratory for several days. Amaldi tried to repeat our experiments and found a different [half-life] for irradiated aluminum which showed that our so-called (n,γ) reaction [i.e., neutron in, gamma photon out] did not occur. This was hurriedly relayed to Fermi who resented having communicated a result which now looked to be in error. He strongly criticized us and did not conceal his displeasure. The whole business was becoming very troublesome because we could not find any fault with the various experiments which gave inconsistent results.

The chastened junior members had their work cut out for them. A new recruit joined them, a tall, broad-shouldered, handsome tennis champion from Pisa named Bruno Pontecorvo, as they set about polishing their first rough work. Neutron bombardment activated some elements more intensely than others. They had previously categorized that activation only generally as strong, medium or weak. Now they proposed to establish a quantitative scale of activibility. They needed some standard intensity against which to measure the intensity of other activations. They chose the convenient 2.3-minute half-life period that neutron bombardment induced in silver.

Amaldi and Pontecorvo got the assignment. They immediately found, to their surprise, that their silver cylinders activated differently in different parts of the laboratory. "In particular," writes Amaldi, "there were certain wooden tables near a spectroscope in a dark room which had miraculous properties, since silver irradiated on those tables gained much more activity than when it was irradiated on a marble table in the same room."

That was a mystery worth exploring. On October 18 they started a systematic investigation, a series of measurements made inside and outside a lead housing. By October 22 they were prepared to measure what might happen when only a lead wedge separated the neutron source from its target. But the experimenters had to give student examinations that morning and Fermi decided to go ahead on his own. He described the historic moment late in life to a colleague curious about the process of discovery in physics:

> I will tell you how I came to make the discovery which I suppose is the most important one I have made. We were working very hard on the neutron-induced radioactivity and the results we were obtaining made no sense. One day, as I came into the laboratory, it occurred to me that I should examine the effect of placing a piece of lead before the incident neutrons. Instead of my usual custom, I took great pains to have the piece of lead precisely machined. I was clearly dissatisfied with something: I tried every excuse to postpone putting the piece of lead in its place. When finally, with some reluctance, I was going to put it in its place, I said to myself: "No, I do not want this piece of lead here; what I want is a piece of paraffin." It was just like that with no advance warning, no conscious prior reasoning. I immediately took some odd piece of paraffin and placed it where the piece of lead was to have been.

The extraordinary result of substituting paraffin wax for a heavy element like lead was a dramatic increase in the intensity of the activation. "About noon," Segrè remembers, "everybody was summoned to watch the miraculous effects of the filtration by paraffin. At first I thought a counter had gone wrong, because such strong activities had not appeared before, but it was immediately demonstrated that the strong activation resulted from the filtering by the paraffin of the radiation that produced the radioactivity." Laura Fermi says "the halls of the physics building resounded with loud exclamations: 'Fantastic! Incredible! Black magic!' "

Not even his most important discovery kept Fermi from going home for lunch. He was alone; his wife and daughter would not return from a visit to the country until the following morning. He pondered in solitude and may have considered the difference between wood and marble tables as well as between paraffin and lead. When he returned in midafternoon he proposed an answer: the neutrons were colliding with the hydrogen nuclei in the paraffin and the wood. That slowed them down. Everyone had assumed that faster neutrons were better for nuclear bombardment because faster protons and alpha particles always had been better. But the analogy ignored the neutron's distinctive neutrality. A charged particle needed energy to push through the nucleus' electrical barrier. A neutron did not. Slowing down a neutron gave it more time in the vicinity of the nucleus, and that gave it more time to be captured.

The simple way to test Fermi's theory was to try some other material besides paraffin that contained hydrogen (other light nuclei would also work to slow neutrons down, but hydrogen would work best: its nuclei are protons, about the same size and mass as neutrons, and they therefore bounce hardest and soak up the most energy per collision). Down to the first floor and out the back door they marched with their silver cylinder and their neutron source extended in its long glass tube, to the pond in Corbino's garden where Rasetti had experimented with raising salamanders, where they had all caught the fad one summer of sailing candle-powered toy boats, where the dark, curving leaves and leathery gray drupes of an almond tree shaded the lively goldfish.

The hydrogen in water (and in goldfish) worked as well as paraffin. Back in the lab they quickly tested whatever they could lay hands on to irradiate: silicon, zinc, phosphorus, which did not seem to be affected by the slow neutrons; copper, iodine, aluminum, which did. They tried radon without beryllium to make sure the paraffin was affecting neutrons and not gamma rays. They replaced the paraffin with an oxygen compound and found much less increase in induced radioactivity.

They went home to dinner but met afterward at Amaldi's, whose wife had a typewriter, to prepare a first report. "Fermi dictated while I wrote," Segrè remembers. "He stood by me; Rasetti, Amaldi, and Pontecorvo paced the room excitedly, all making comments at the same time." Laura Fermi recreates the scene: "They shouted their suggestions so loudly, they argued so heatedly about what to say and how to say it, they paced the floor in such audible agitation, they left the Amaldis' house in such a state, that the Amaldis' maid timidly inquired whether the guests had all been drunk."

Ginestra Amaldi delivered the typed paper, "Influence of hydrogenous substances on the radioactivity produced by neutrons—I," to the director of the *Ricerca Scientifica* the next morning. Tucked away in its historic paragraphs was a quiet justification for the confusion over aluminum: "The case of aluminum is noteworthy. In water it acquires an activity showing a period slightly shorter than 3 minutes. . . . This activity under normal conditions is so weak that it almost disappears compared to other activities generated in the same element."

Amaldi and Segrè had not been wrong about aluminum. They had simply irradiated different samples of the element on different tables. The hydrogen in the wooden table had slowed down some of the neutrons and enhanced the almost-three-minute activity. As Hans Bethe once noted wittily, the efficiency of slow neutrons "might never have been discovered if Italy were not rich in marble. . . . A marble table gave different results from a wooden table. If it had been done [in America], it all would have been done on a wooden table and people would never have found out."

The discovery of slow-neutron radioactivity meant that Fermi's group had to work its way through the elements again looking for different and enhanced half-lives—which is to say, different isotopes and decay products.

While that work proceeded a paper appeared in the *Physical Review* criticizing the group's earlier study of uranium. The paper's primary author was Aristide von Grosse, who had been one of Otto Hahn's assistants at the KWI and who had purified the first substantial sample of protactinium, the element Hahn and Meitner had discovered in 1917. Von Grosse argued that when Fermi irradiated uranium he had created protactinium, atomic number 91, not a new transuranic element. The Rome group took the paper as a challenge to further experiment. At the same time Hahn and Meitner decided proprietarily to repeat Fermi's previous uranium work. "It was a logical decision," Hahn explains in his scientific autobiography; "having been the discoverers of protactinium, we knew its chemical characteristics." The increasing number of different half-lives that investigators in Berlin and Paris found when they irradiated uranium were puzzling; Hahn correctly felt that he was better qualified than anyone else in the world to accomplish the subtle radiochemistry necessary to sort everything out.

In January and February 1935, in the midst of other projects, Amaldi set to work looking for alpha-emitting reactions in uranium in addition to the beta reactions the group had originally found. If uranium emitted alpha particles when it captured neutrons it would be transmuting down the periodic table rather than up, which might indeed produce protactinium along the way. Amaldi chose to use an ionization chamber connected to a linear amplifier to capture and measure the radiation. "I began to irradiate some foil[s] of uranium," he writes, ". . . and put them immediately after irradiation in front of the thin-window ionization chamber." Nothing happened. Conceivably the half-lives were too brief for the run down the hall from the irradiation area to the ionization chamber. Amaldi decided to try irradiating his samples directly in front of the chamber. That required screening out unwanted radiation. The gamma rays from his neutron source, which would have disturbed the ionization chamber, he blocked by setting a piece of lead between the source and the chamber: the desirable neutrons would find the lead no obstacle.

He also wanted to filter out uranium's natural alpha background. To do that he took advantage of the basic law of radioactivity that shorter half-lives mean more energetic radiation. The half-life of natural uranium is about 4.5 billion years; its alphas are proportionately mild, mild enough to be blocked by a layer of aluminum foil. On the other hand, if there really were half-lives in his experiment so short that he had to irradiate directly in front of the ionization chamber to catch them, their alphas should be energetic enough to breeze easily through the aluminum and the chamber window and enter the chamber for counting. So Amaldi wrapped his uranium samples with aluminum foil. It did not occur to him that his shielding might also screen out other reaction products. In 1935, alpha, beta and gamma radiation were the only reaction products anyone knew. "The experiments," Amaldi concludes, "gave negative results." He found no artificially induced alphas from uranium.

The Italians thought it even more probable then that by irradiating uranium they were creating new, man-made elements. Hahn and Meitner reported they thought so too. Fermi's group rounded up its work in the *Proceedings of the Royal Society* in a paper Rutherford approvingly passed along to that journal on February 15:

> Through these experiments our hypothesis that the 13-minute and 100-minute induced activities of uranium are due to transuranic elements seems to receive further support. The simplest interpretation consistent with the known facts is to assume that the 15-second, 13-minute and 100-minute activities are chain products [i.e., one decays into the next], probably with atomic number 92, 93 and 94 respectively and atomic weight 239.

But the truth was, uranium was a confusion, and no one yet knew.

What else besides beryllium? Leo Szilard asked himself in London. Beryllium looked suspicious. What other elements might chain-react? He answered with an amended patent specification on April 9, 1935: "Other examples for elements from which neutrons can liberate multiple neutrons are uranium and bromine." He was guessing, and without research funds he saw no way to experiment. The physicists he talked to remained profoundly skeptical of his ideas. "So I thought, there is after all something called 'chain reaction' in chemistry. It doesn't resemble a nuclear chain reaction, but still it's a chain reaction. So I thought I would talk to a chemist." The chemist he thought he would talk to was someone even more skillful than Leo Szilard at raising funds: Chaim Weizmann, who now lived and worked in London. Weizmann received Szilard and "understood what I told him." He asked Szilard how much money he needed. Szilard said £2,000—about $10,000. Though he was certainly hard-pressed for funding himself, Weizmann said he would see what he could do. Szilard recalls:

> I didn't hear from him for several weeks, but then I ran into Michael Polanyi, who by that time had arrived in Manchester and was head of the chemistry

> department there. Polanyi told me that Weizmann had come to talk to him about my ideas for the possibility of a chain reaction, and he wanted Polanyi's advice on whether he should get me this money. Polanyi thought that this experiment should be done.

A decade passed before Szilard and Weizmann met again, a gulf of history. Weizmann had not neglected Szilard's request, he explained then in apology in late 1945; he had only not succeeded in raising the funds.

Since the beginning of his rescue work in England Szilard had been in occasional contact with the physicist Frederick Alexander Lindemann, who was professor of experimental philosophy at Oxford and director of the Clarendon Laboratory there. It was Lindemann, wealthy and well-connected, who was arranging a fellowship for Szilard, part of his continuing campaign to arm the decrepit Oxford science laboratory against its splendid Cambridge rival. Lindemann had made effective use in that campaign of the Nazi expulsion of the Jewish academics but had given as good as he got: immediately upon hearing of the civil service law he had gone to Imperial Chemical Industries and convinced its directors to establish a grant program, arguing that such an investment would be not charity but money well spent. ICI had already begun paying out its first grant on May 1, 1933, while Beveridge and Szilard were still laying plans. It was an ICI grant that Szilard missed winning the following August, perhaps because he had not yet accomplished his summer of impressive experiment at St. Bart's, but Lindemann was paying attention now.

The tall, handsome Englishman, forty-nine years old in 1935, had been born in Germany, at Baden-Baden, because his mother chose not to allow advanced pregnancy to interfere with a visit to that fashionable spa. To provide their son with an outstanding education his English parents had sent him to the *Gymnasium* in Darmstadt. As a student before the Great War at the Darmstadt Technische Hochschule, where he was a protégé of the physical chemist Walther Nernst (the 1920 chemistry Nobelist), he had enjoyed such exceptional family connections that he found himself at times playing tennis with the Kaiser or the Czar. Inevitably the war made suspect such golden afternoons. Lindemann was chagrined and angered in 1915 to find that the British Army, noting his German birth certificate and German-sounding name, was unwilling to extend him a commission.

The Army's decision injured him deeply and may have changed his life. He had served as a co-secretary to the 1911 Solvay Conference, standing up proudly with Nernst, Rutherford, Planck, Einstein, Mme. Curie, but even before that youthful apotheosis Nernst had predicted difficulty: "If your father were not such a rich man," the blunt German had said, "you would become a great physicist." When the Army questioned Lindemann's patriotism, writes a refugee colleague, "he became withdrawn to avoid exposing himself to slights and insults. Secretiveness about his personal life developed into a mania and he discouraged personal approaches by a stand-offishness which was easily mistaken for arrogance." Lindemann retreated from original work and became a talented administrator, "the Prof," an "unbending Victorian gentleman," always impeccable in bowler hat, summer gray suit, winter dark suit, rolled-up umbrella and long, dark coat. If he could not win a uniform he would adopt one of his own.

He worked for his country during the war at the Royal Aircraft Factory at Farnborough, designing what are now called avionics and doing aeronautical research. Tailspins were recognized maneuvers in air fighting by 1916, a good way to shake off an attacker. Lindemann was the first to study them scientifically. To do so he took flying lessons—only changing from civilian clothes to flying clothes on the runway beside the plane—then coolly flew spin after spin, memorizing his instrument readings as he plummeted and writing them down after he had recovered level flight.

After the war Lindemann accepted appointment to an Oxford still donnishly disdainful of science. He escaped from that further condescension, says his colleague, into "gracious living," enjoying weekends with the nobility that were seldom vouchsafed to less well-born Oxford dons. By then a Rolls-Royce was part of his regalia. In June 1921, on a weekend at the country estate of the Duke and Duchess of Westminster, Lindemann met Winston Churchill, twelve years his senior. "The two men, so different in background and character, took to each other immediately and their acquaintance soon turned into a close friendship." Churchill recalled that he "saw a great deal of Frederick Lindemann" during the 1930s. "Lindemann was already an old friend of mine. . . . We came much closer from 1932 onwards, and he frequently motored over from Oxford to stay with me at Chartwell. Here we had many talks into the small hours of the morning about the dangers which seemed to be gathering upon us. Lindemann . . . became my chief adviser on the scientific aspects of modern war."

To this illustrious personage, a vegetarian who daily consumed copious quantities of olive oil and Port Salut, Szilard turned in the early summer of 1935 to discuss "the question whether or not the liberation of nuclear energy . . . can be achieved in the immediate future." If "double neutrons" could be produced, Szilard wrote Lindemann on June 3, "then it is certainly less bold to expect this achievement in the immediate future than to believe the opposite." That meant trouble, Szilard thought, if Germany achieved a chain reaction first, and he argued for "an attempt, whatever small chance of success it may have . . . to control this development as long as possible." Secrecy was the way to achieve such control: first, by winning agreement from the scientists involved to restrict publication, and second, by taking out patents.

Michael Polanyi had cautioned Szilard late in 1934 that "there is an opposition to you on account of taking patents." The British scientific tradition that opposed patents assumed that those who filed them did so for mercenary purposes; Szilard explained his patents to Lindemann to clear his name:

> Early in March last year it seemed advisable to envisage the possibility that . . . the release of large amounts of energy . . . might be imminent. Realising to what extent this hinges on the "double neutron," I have applied for a patent along these lines. . . . Obviously it would be misplaced to consider patents in this field private property and pursue them with a view to commercial exploitation for private purposes. When the time is ripe some suitable body will have to be created to ensure their proper use.

For the time being, Szilard proposed to work at Oxford on finding his "double neutrons," possibly raising £1,000 on the side "from private persons" so that he could hire a helper or two. To bait Lindemann's Clarendon ambitions, he argued in conclusion that "this type of work could greatly accelerate the building up of nuclear physics at Oxford." As indeed, had it gone forward, it might have done.

When he learned, possibly from Lindemann, that he could keep his patents secret only by assigning them to some appropriate agency of the British government, Szilard offered them first to the War Office. Director of Artillery J. Coombes turned them down on October 8, noting that "there appears to be no reason to keep the specification secret so far as the War Department is concerned." If Lindemann heard of the rejection he must have remembered his own rejection by the Army in 1915. The following February 1936, he intervened on Szilard's behalf with the Admiralty, Churchill's old bailiwick, writing the head of the Department of Scientific Research and Development cannily:

> I daresay you remember my ringing you up about a man working here who had a patent which he thought ought to be kept secret. I enclose a letter from him on the subject as you suggested. I am naturally somewhat less optimistic about the prospects than the inventor, but he is a very good physicist and even if the chances were a hundred to one against it seems to me it might be worth keeping the thing secret as it is not going to cost the Government anything.

The patent, Szilard explained in the letter Lindemann enclosed, "contains information which could be used in the construction of explosive bodies . . . very many thousand times more powerful than ordinary bombs." He was concerned about "the disasters which could be caused by their use on the part of certain Powers which might attack this country." Wisely and withal inexpensively the Admiralty accepted the patent into its safekeeping.

Eight months in Copenhagen had suited Edward Teller. He met George Gamow on the Odessan's last visit there, after the Solvay Conference of the previous autumn; the two of them roared across Denmark and back during Easter vacation on Gamow's motorcycle, working over a problem in quantum mechanics. The Rockefeller Foundation did not approve of marriage during a fellowship period, but James Franck had interceded on his behalf and Teller had married his childhood sweetheart, Mici Harkanyi, in Budapest on February 26. He had also written an important paper. He returned to London with Mici in the summer of 1934 with his reputation enhanced and again took up his lectureship at University College. Assuming they would settle in England, the Tellers signed a nine-year lease just before Christmas on a pleasant three-room flat.

Two offers arrived in January, one of which changed Teller's mind. The first was from Princeton: a lectureship. The second was from Gamow: a full professorship at George Washington University. GWU wanted to strengthen its physics department; Gamow wanted company and liked Teller's verve.

Teller was twenty-six years old and a newlywed. He was less than sure about living in the United States, but a full professorship was not something he could sensibly refuse. His wife found someone to sublet the flat. The U.S. State Department refused nonquota immigration visas because Teller had only taught for one year—the Copenhagen time counted merely as a fellowship—and was required to have taught for two. He had not, however, tried for visas on the Hungarian immigration quota because he assumed the quota was full. In fact there was room. The Tellers followed the Gamows across the Atlantic in August 1935.

Niels Bohr celebrated his fiftieth birthday on October 7. "Bohr in those days seemed at the height of his powers, bodily and mentally," Otto Frisch observes. "When he thundered up the steep staircase [of the institute], two steps at a time, there were few of us younger ones that could keep pace with him. The peace of the library was often broken by a brisk game of ping-pong, and I don't remember ever beating Bohr at that game." To honor Denmark's leading physicist, George de Hevesy organized a fund-raising campaign; the Danish people contributed 100,000 kroner to buy Bohr 0.6

gram of radium for his birthday. De Hevesy divided the radium, in liquid solution, into six equal parts, mixed each with beryllium powder and allowed them to dry, making six potent neutron sources. He had them mounted on the ends of long rods and stored them in a dry well in the basement of the institute that had been dug originally to supply vibration-free housing for a spectrograph.

The institute's annual Christmas party continued to be held in the well room, Stefan Rozental recalls: "The lid of the well served as a table, a Christmas tree stood in the middle, and all the personnel were gathered, from the chief down to the youngest apprentice in the workshop, and served a modest meal of sausages and beer. During the party Niels Bohr used to make a speech in which he gave a sort of survey of the past year." Safely below the sausages, stuck in a gallon flask of carbon disulphide, the neutron sources silently transmuted sulfur to radioactive phosphorus for de Hevesy's biological radioisotope studies.

Bohr had won national distinction for his work and the enduring gratitude of refugees for his aid; he had also faced personal pain. In 1932 the Danish Academy offered him lifetime free occupancy of the Danish House of Honor, a palatial estate in Pompeiian style built originally for the founder of Carlsberg Breweries and subsequently reserved for Denmark's most distinguished citizen (Knud Rasmussen, the polar explorer, was its previous occupant). By then the institute buildings included a modest director's house, but the Bohrs shared it with five handsome sons. They moved to the mansion beside the brewery, the best address in Denmark after the King's.

Two years later an accident took the Bohrs' eldest son, Christian, nineteen years old. Father, son and two friends were sailing on the Öresund, the sea passage between Denmark and Sweden, when a squall blew up. Christian "was drowned by falling over[board] in a very rough sea from a sloop," Robert Oppenheimer reports, "and Bohr circled as long as there was light, looking for him." But the Öresund is cold. For a time Bohr retreated into grief. Exhausting as it was, the refugee turmoil helped him.

Everyone at the institute followed Fermi's neutron work with fascination. Frisch, the only physicist on hand who knew Italian, was drafted to translate the successive papers aloud as soon as each issue of the *Ricerca Scientifica* arrived. The Copenhagen group was puzzled that slow neutrons affected some elements more intensely than others; on the one-particle model of the nucleus even a slow neutron should almost always shoot completely through a nucleus without capture.

From Cornell Hans Bethe published a paper calculating the slim odds of neutron capture. They conflicted squarely with observation. Frisch remembers the colloquium in Copenhagen in 1935 when someone reported on Bethe's paper:

> On that occasion Bohr kept interrupting, and I was beginning to wonder, with some irritation, why he didn't let the speaker finish. Then, in the middle of a sentence, Bohr suddenly stopped and sat down, his face completely dead. We looked at him for several seconds, getting anxious. Had he been taken unwell? But then he suddenly got up and said with an apologetic smile, "Now I understand it."

What Bohr understood about the nucleus he embodied in a landmark lecture to the Danish Academy on January 27, 1936, subsequently published in *Nature*. "Neutron capture and nuclear constitution" exploited the phenomenon of neutron capture to propose a new model of the nucleus; once again, as he had with Rutherford's planetary model of the atom, Bohr stood on the solid ground of experiment to argue for radical theoretical change.

He visualized a nucleus made up of neutrons and protons closely packed together—a model now familiar—rather than a single particle. (Nuclear particles collectively are known as nucleons.) A neutron entering such a crowded nucleus would not pass through; it would collide with the nearest nucleons, surrender its kinetic energy (as a cue ball does at break in billiards) and be captured by the strong force that holds the nucleus together. The energy added by the neutron would agitate the nearby nucleons; they would collide in turn with other nucleons beyond; the net effect would be a more generally agitated, "hotter" nucleus but one where no single component could quickly acquire enough energy to push through the electrical barrier and escape. If the nucleus then radiated its excess energy by ejecting a gamma photon, "cooling off," none of its nucleons could accrue enough energy to escape. The result, already confirmed by Fermi's experiments, would be the creation of a heavier isotope of the original element being bombarded.

More violent assaults on the nucleus, Bohr thought, would still disperse their energies throughout the compound nucleus created by their capture. Subsequent reconcentration of the energy might allow the nucleus to eject several charged or uncharged particles. Bohr did not think his compound model of the nucleus boded well for harnessing nuclear energy:

> For still more violent impacts, with particles of energies of about a thousand million volts, we must even be prepared for the collision to lead to an explosion of the whole nucleus. Not only are such energies, of course, at present far beyond the reach of experiments, but it does not need to be stressed that such effects would scarcely bring us any nearer to the solution of the much discussed problem of releasing the nuclear energy for practical purposes. Indeed, the more our knowledge of nuclear reactions advances the remoter this goal seems to become.

Thus by the mid-1930s the three most original living physicists had each spoken to the question of harnessing nuclear energy. Rutherford had dismissed it as moonshine; Einstein had compared it to shooting in the dark at scarce birds; Bohr thought it remote in direct proportion to understanding. If they seem less perceptive in their skepticism than Szilard, they also had a better grasp of the odds. The essential future is always unforeseen. They were experienced enough not to long for it.

In his lecture Bohr preferred to state only general principles, but to trace "the consequences of the general argument here developed" he had a specific mathematical model in mind. He published a discussion of that model the following year, in 1937. It reached all the way back to his doctoral dissertation on the surface tension of fluids to demonstrate the usefulness of treating the atomic nucleus as if it were a liquid drop.*

The tendency of molecules to stick together gives liquids a "skin" of surface tension. A falling raindrop thus rounds itself into a small perfect sphere. But any force acting on a liquid drop deforms it (think of the wobbles of a water-filled balloon thrown into the air and caught). Surface tension and deforming forces work against each other in complex ways; the molecules of the liquid bump and collide; the drop wobbles and distorts. Eventually the added energy dissipates as heat, and the drop steadies again.

The nucleus, Bohr proposed, was similar. The force that stuck the nucleons together was the nuclear strong force. Counteracting that strong force was the common electrical repulsion of the positively charged nuclear protons. The delicate balance between the two fundamental forces made the nucleus liquidlike. Energy added from the outside by particle bombardment deformed it; it wobbled like a liquid drop, oscillating complexly just as the braided streams of water Bohr had studied for his dissertation had oscillated. Which meant he could use Rayleigh's classical formulae for the surface tension of liquids to understand the complex nuclear energy levels and exchanges that Fermi's work had revealed. "This 1937 paper had to close with many issues not cleared up," writes the American theoretical physicist John Archibald Wheeler, who helped Bohr clear up more of them later. The liquid-drop model proved useful, however, and Frisch in Copenhagen and Meitner in Berlin, among others, took it to heart.

One fine October Thursday in 1937 Ernest Rutherford, a vigorous sixty-six, went out into the garden of his house on the green Cambridge Backs to

*George Gamow had proposed such a model in Copenhagen in 1928. Bohr credited it to Gamow at the October 1933 Solvay conference, as did Heisenberg. Bohr and his student Fritz Kalkar subsequently developed the model and physicists customarily attribute it to him.

trim a tree. He took a bad fall. He was "seedy" later in the day, Mary Rutherford said—nausea and indigestion—and she arranged for a masseur. Rutherford vomited that night. In the morning he called his family doctor. He suffered from a slight umbilical hernia, which he confined with a truss; his doctor found a possible strangulation, consulted with a specialist and directed the Rutherfords to the Evelyn Nursing Home for emergency surgery. Rutherford told his wife along the way that his business and financial affairs were all in order. She said his illness wasn't serious and asked him not to worry.

Surgery that evening confirmed a partial strangulation, released the imprisoned portion of the small intestine and restored its circulation. Saturday Rutherford seemed to be recovering but he began vomiting again on Sunday and there were signs of infection, deadly in those days before antibiotics. Monday he was worse; his doctors consulted the surgeon, a Melbourne man, who advised against a second operation given the patient's age and symptoms. Rutherford was made comfortable with intravenous saline, six pints by Tuesday, and a stomach tube. Tuesday morning, October 19, he was slightly improved, but though his wife judged him "a wonderful patient [who] bears his discomforts splendidly" and believed she discovered "just a thread of hope," he began that afternoon to weaken. A bequest he decided late in the day suggests he found gratitude in those last hours reviewing his life. "I want to leave a hundred pounds to Nelson College," he told Mary Rutherford. "You can see to it." And again loudly a little later: "Remember, a hundred to Nelson College." He died that evening. "Heart and circulation failed" because of massive infection, his doctor wrote, "and the end came peacefully."

An international gathering of physicists in Bologna that week celebrated the 200th anniversary of the birth of Luigi Galvani; Cambridge cabled the news of Rutherford's death on the morning of October 20. Bohr was on hand and accepted the grim duty of announcement. "When the meeting scheduled for that morning assembled," writes Mark Oliphant, "Bohr went to the front, and with faltering voice and tears in his eyes informed the gathering of what had happened." They were shocked at the abruptness of the loss. Bohr had visited Rutherford at Cambridge a few weeks earlier; the Cavendish men had seen their leader in fine fettle only days ago.

Bohr "spoke from the heart," says Oliphant, recalling "the debt which science owed so great a man whom he was privileged to call both his master and his friend." For Oliphant it was "one of the most moving experiences of my life." Remembering Rutherford in a letter to Oppenheimer on December 20 Bohr balanced loss with hope, complementarily: "Life is poorer without him; but still every thought about him will be a lasting encouragement." And in 1958, in a memorial lecture, Bohr said simply that "to me he had almost been as a second father."

The sub-dean of Westminster immediately approved interment of Rutherford's ashes in the nave of Westminster Abbey, just west of Newton's tomb and in line with Kelvin's. Eulogizing Rutherford at a conference in Calcutta the following January, James Jeans identified his place in the history of science:

> Voltaire said once that Newton was more fortunate than any other scientist could ever be, since it could fall to only one man to discover the laws which governed the universe. Had he lived in a later age, he might have said something similar of Rutherford and the realm of the infinitely small; for Rutherford was the Newton of atomic physics.

Ernest Rutherford unknowingly wrote his own more characteristic epitaph in a letter to A. S. Eve from his country cottage on the first day of that last October. He reported of his garden what he had also done for physics, vigorous and generous work: "I have made a still further clearance of the blackberry patch and the view is now quite attractive."

In September 1934, in the wake of Fermi's June *Nature* article "Possible production of elements of atomic number higher than 92," a curious paper appeared in a publication seldom read by physicists, the *Zeitschrift für Angewandte Chemie*—the *Journal of Applied Chemistry*. Its author was a respected German chemist, Ida Noddack, co-discoverer with her husband (in 1925) of the hard, platinum-white metallic element rhenium, atomic number 75. The paper was titled simply "On element 93" and it severely criticized Fermi's work. His "method of proof" was "not valid," Noddack wrote bluntly. He had demonstrated that "his new beta emitter" was not protactinium and then distinguished it from several other elements descending down the periodic table to lead, but it was "not clear why he chose to stop at lead." The old view that the radioactive elements form a continuous series beginning at uranium and ending at lead, wrote Noddack, was exactly what the Joliot-Curies' discovery of artificial radioactivity had disproved. "Fermi therefore ought to have compared his new radioelement with all known elements."

The fact was, Noddack went on, any number of elements could be precipitated out of uranium nitrate with manganese. Instead of assuming the production of a new transuranic element, "one could assume equally well that when neutrons are used to produce nuclear disintegrations, some distinctly new nuclear reactions take place which have not been observed previously." In the past, elements have transmuted only into their near neighbors. But "when heavy nuclei are bombarded by neutrons, it is conceivable that the nucleus breaks up into several large fragments, which would of course be isotopes of known elements but would not be neighbors." They would be, rather, much lighter elements farther down the periodic table than lead.

Segrè remembers reading the Noddack paper. He knows, because he asked them, that Hahn in Berlin and Joliot in Paris read it. It made very little sense to anyone. "I think whatever chemists read it," Frisch reminisces, "probably thought that this was quite pointless, carping criticism, and the physicists possibly even more so if they read it, because they would say, 'What's the use of criticizing unless you give some reason why that criticism would be valid?' Nobody had ever found a nuclear disintegration creating far-removed elements." Which was a point Noddack had carefully addressed, but was clearly one reason for the paper's neglect. The summary report for *Nature* on artificial radioactivity that Amaldi and Segrè had delivered to Rutherford in midsummer 1934 makes the assumption explicit: "It is reasonable to assume that the atomic number of the active element should be close to the atomic number . . . of the bombarded element."

But Fermi seldom left anything to assumption, however reasonable. He would certainly not have left to assumption this issue, about which he was already acutely sensitive because of Corbino's ill-timed speech (Noddack rubbed salt into that wound by referring to "the reports found in the newspapers"). He sat down and performed the necessary calculations. He later told at least Teller, Segrè and his American protégé Leona Woods that he had done so. Teller is quite sure he knows what those calculations were:

> Fermi refused to believe [Noddack]. . . . He knew how to calculate whether or not uranium could break in two. . . . He performed the calculation Mrs. Noddack suggested, and found that the probability was extraordinarily low. He concluded that Mrs. Noddack's suggestion could not possibly be correct. So he forgot about it. His theory was right . . . but . . . it was based on the . . . wrong experimental information.

Here Teller indicts Aston's measurement of the mass of helium (the same that had misled Szilard to beryllium), which "introduced a systematic error into calculating the mass and energy of nuclei."

Segrè finds Teller's version of the story possible but not persuasive. The helium mass number problem would not necessarily have ruled out breaking up the uranium nucleus. "You know, occasionally Fermi would tell you things, then you asked him, 'But really, how? Show me.' And then he would say, 'Oh, well, I know this on *c.i.f.*' He spoke Italian. '*C.i.f.*' meant '*con intuito formidable*,' 'with formidable intuition.' So how he did it, I don't know. On the other hand, Fermi made a lot of calculations which he kept to himself."

Leona Woods' version sheds light on Teller's:

> Why was Dr. Noddack's suggestion ignored? The reason is that she was ahead of her time. Bohr's liquid-drop model of the nucleus had not yet been formulated, and so there was at hand no accepted way to calculate whether breaking up into several large fragments was energetically allowed.

If Noddack's physics was *avant garde*, her chemistry was sound. By 1938 her article was gathering dust on back shelves, but Bohr had promulgated the liquid-drop model of the nucleus and the confused chemistry of uranium increasingly preoccupied Lise Meitner and Otto Hahn.

9

An Extensive Burst

"I believe all young people think about how they would like their lives to develop," Lise Meitner wrote in old age, looking back; "when I did so I always arrived at the conclusion that life need not be easy provided only that it was not empty. And this wish I have been granted." Sixty years old in 1938, the Austrian physicist had earned wide respect by hard and careful work. When Wolfgang Pauli had wished to propose an elusive, almost massless neutral particle to explain the energy that seemed to disappear in beta decay—it came to be called the neutrino—he had made his proposal in a letter to Lise Meitner and Hans Geiger. James Chadwick was "quite convinced that she would have discovered the neutron if it had been firmly in her mind, if she had had the advantage of, say, living in the Cavendish for years, as I had done." "Slight in figure and shy by nature," as her nephew Otto Frisch describes her, she was nevertheless formidable.

During the Great War she had volunteered as an X-ray technician with the Austrian Army; "there," says Frisch, "she had to cope with streams of injured Polish soldiers, not understanding their language, and with her medical bosses who interfered with her work, not understanding X-rays." She arranged her leaves from duty to coincide with Otto Hahn's and hurried to the Kaiser Wilhelm Institute for Chemistry in Dahlem to work with him; that was when they identified the element next down from uranium that they named protactinium. After the war she did physics separately until 1934, when, challenged by Fermi's work, she "persuaded Otto Hahn to renew our direct collaboration" to explore the consequences of bombarding uranium with neutrons. Meitner headed the physics department at the institute then, of which Hahn had become the director. She had attained by middle age, Hahn remarks fondly, "not only the dignity of a German professor, but also one of his proverbial attributes, absentmindedness." At a scientific gathering "a male colleague greeted her by saying, 'We met on an earlier occasion.' Not remembering that earlier occasion, she replied in all seriousness, 'You probably mistake me for Professor Hahn.' " Hahn supposed she was thinking of the many papers they had published together.

If she hid her shyness behind formidable reserve, among friends, Frisch says, "she could be lively and cheerful, and an excellent storyteller." Her nephew thought her "totally lacking in vanity." She wore her thick dark hair, now graying, pulled back and coiled in a bun and her youthful beauty had muted to bright but darkly circled eyes, a thin mouth, a prominent nose. She ate lightly but drank quantities of strong coffee. Music moved her; she followed it as other people follow trends and fashions in art (a family cultivation—her sister, Frisch's mother, was a concert pianist). She made a duet at the piano on visits with her musical nephew, "though hardly anybody else knew that she could play." She lived in an apartment at the KWI and when there was time she took long walks, ten miles or more a day: "It keeps me young and alert." Her most holy commitment, Frisch thought, "the vision she never lost" that filled her life, was "of physics as a battle for final truth."

The truth she battled for through the later 1930s was hidden somewhere in the complexities of uranium. She and Hahn, and beginning in 1935 a young German chemist named Fritz Strassmann, worked to sort out all the substances into which the heaviest of natural elements transmuted under neutron bombardment. By early 1938 they had identified no fewer than ten different half-life activities, many more than Fermi had demonstrated in his first pioneering survey. They assumed the substances must be either isotopes of uranium or transuranics. "For Hahn," says Frisch, "it was like the old days when new elements fell like apples when you shook the tree; [but] Lise Meitner found [the energetic reactions necessary to produce such new elements] unexpected and increasingly hard to explain."

Meanwhile Irène Curie had begun looking into uranium with a visiting Yugoslav, Pavel Savitch. They described a 3.5-hour activity the Germans had not reported and suggested it might be thorium, element 90, with which Curie had years of experience. If true, the Curie-Savitch suggestion would mean that a slow neutron somehow acquired the energy to knock an energetic alpha particle out of the uranium nucleus. The KWI trio scoffed, looked for the 3.5-hour activity, failed to find it and wrote the Radium Institute suggesting a public retraction. The French team identified the activity again and discovered they could separate it from their uranium by carrier chemistry using lanthanum (element 57, a rare earth). They proposed therefore that it must be either actinium, element 89, chemically similar to lanthanum but even harder than thorium to explain, or else a new and mysterious element.

Either way, their findings called the KWI work into doubt. Hahn met Joliot in May at a chemistry congress in Rome and told the Frenchman cordially but frankly that he was skeptical of Curie's discovery and intended to repeat her experiment and expose her error. By then, as Joliot undoubtedly knew, his wife had already raised the stakes, had tried to separate the "actinium" from its lanthanum carrier and had found it would not separate. No one imagined the substance could actually be lanthanum: how could a slow neutron transmute uranium into a much lighter rare earth thirty-four places down the periodic table? "It seems," Curie and Savitch reported that May in the *Comptes Rendus*, "that this substance cannot be anything except a transuranic element, possessing very different properties from those of other known transuranics, a hypothesis which raises great difficulties for its interpretation."

In the course of this exotic debate Meitner's status changed. Adolf Hitler bullied the young chancellor of Austria to a meeting at the German dictator's Berchtesgaden retreat in Bavaria in mid-February. "Who knows," Hitler threatened him, "perhaps I shall be suddenly overnight in Vienna: like a spring storm." On March 14 he was, triumphantly parading; the day before, with the raw new German Wehrmacht occupying its capital, Austria had proclaimed itself a province of the Third Reich and its most notorious native son had wept for joy. The *Anschluss*—the annexation—made Meitner a German citizen to whom all the ugly anti-Semitic laws applied that the Nazi state had been accumulating since 1933. "The years of the Hitler regime . . . were naturally very depressing," she wrote near the end of her life. "But work was a good friend, and I have often thought and said how wonderful it is that by work one may be granted a long respite of forgetfulness from oppressive political conditions." After the spring storm of the *Anschluss* her grant was abruptly withdrawn.

Max von Laue sought her out then. He had heard that Heinrich Himmler, head of the Nazi SS and chief of German police, had issued an order forbidding the emigration of any more academics. Meitner feared she might be expelled from the KWI and left unemployed and exposed. She made contact with Dutch colleagues including Dirk Coster, the physicist who had worked in Copenhagen with George de Hevesy in 1922 to discover hafnium. The Dutchmen persuaded their government to admit Meitner to Holland without a visa on a passport that was nothing more now than a sad souvenir.

Coster traveled to Berlin on Friday, July 16, arriving in the evening, and went straight to Dahlem to the KWI. The editor of *Naturwissenschaften*, Paul Rosbaud, an old friend, showed up as well, and together with Hahn the men spent the night helping Meitner pack. "I gave her a beautiful diamond ring," Hahn remembers, "that I had inherited from my mother and which I had never worn myself but always treasured; I wanted her to be provided for in an emergency."

Meitner left with Coster by train on Saturday morning. Nine years later she remembered the grim passage as if she had traveled alone:

> I took a train for Holland on the pretext that I wanted to spend a week's vacation. At the Dutch border, I got the scare of my life when a Nazi military patrol of five men going through the coaches picked up my Austrian passport, which had expired long ago. I got so frightened, my heart almost stopped beating. I knew that the Nazis had just declared open season on Jews, that the hunt was on. For ten minutes I sat there and waited, ten minutes that seemed like so many hours. Then one of the Nazi officials returned and handed me back the passport without a word. Two minutes later I descended on Dutch territory, where I was met by some of my Holland colleagues.

She was safe then. She moved on to Copenhagen for the emotional renewal of rest at the Carlsberg House of Honor with the Bohrs. Bohr had found a place for her in Sweden at the Physical Institute of the Academy of Sciences on the outskirts of Stockholm, a thriving laboratory directed by Karl Manne Georg Siegbahn, the 1924 Physics Nobel laureate for work in X-ray spectroscopy. The Nobel Foundation provided a grant. She traveled to that far northern exile, to a country where she had neither the language nor many friends, as if to prison.

Leo Szilard was looking for a patron. Frederick Lindemann had arranged an ICI fellowship for him at Oxford beginning in 1935, and for a while Szilard worked there, but the possibility of war in Europe made him restless. From Oxford in late March 1936 he had written Gertrud Weiss in Vienna that she should consider emigrating to America; he appears to have applied his reasoning to his own case as well. Szilard had met Weiss in his Berlin years and subsequently advised and quietly courted her. Now she had graduated from medical school. At his invitation she came to Oxford to see him. They walked in the country; she photographed him standing at roadside before a weathered log barrier, rounding at thirty-eight but not yet rotund, with a budding young tree filigreed behind him. "He told me he would be surprised if one could work in Vienna in two years. He said Hitler would be there. And he was"—the *Anschluss*—"almost to the day."

Szilard had written in his letter that England was "a *very* likeable country, but it would certainly be a lot smarter if you went to America. . . . In America you would be a free human being and very soon would not even be a 'stranger.' " (Weiss went, and stayed to become a distinguished expert in public health and, late in their wandering years, Szilard's wife.) During the same period Szilard wrote Michael Polanyi he would "stay in England until one year before the war, at which time I would shift my residence to New York City." The letter provoked comment, Szilard enjoyed recalling; it was "very funny, because how can anyone say what he will do one year *before* the war?" As it turned out, his prognostication was off by only four months: he arrived in the United States on January 2, 1938.

Before then Szilard had located a possible patron there, a Jewish financier of Virginia background named Lewis Lichtenstein Strauss, his first and middle names honoring his East Prussian maternal grandfather, his last name softened in Southern fashion to *straws.* Forty-two years old in 1938, Lewis Strauss was a full partner at the New York investment-banking house of Kuhn, Loeb, a self-made millionaire, an adaptable, clever but thin-skinned and pompous man.

Strauss had dreamed as a boy of becoming a physicist. The recession of 1913–14 had staggered his family's Richmond business—wholesale shoes—and his father had called on him at seventeen to drum a four-state territory. He did well; by 1917 he had saved twenty thousand dollars and was once again preparing to pursue a physics career. This time the Great War intervened. A childhood accident had left Strauss with marginal vision in one eye. His mother doted on him. She allowed his younger brother to volunteer for military service but looked for some less dangerous contribution for her favorite son. It turned up when Woodrow Wilson appointed the celebrated mining engineer and Belgian relief administrator Herbert Hoover as Food Administrator to manage U.S. supplies during the war. The wealthy Hoover was serving in Washington without pay and assembling a prosperous, unpaid young staff, Rhodes scholars preferred. Rosa Lichtenstein Strauss sent her boy.

He was twenty-one, knew how to ingratiate himself, knew also how to work. Improbable as it appears against a field of Rhodes scholars, within a month Hoover appointed the high-school-graduate wholesale shoe drummer as his private secretary. After the Armistice young Strauss shifted with Hoover to Paris, hastily picked up French at tutoring sessions over lunch and helped organize the allocation of 27 million tons of food and supplies to twenty-three countries. On the side he assisted the Jewish Joint Distribution Committee in its work of relieving the suffering of the hundreds of thousands of Jewish refugees streaming from Eastern Europe in the wake of war.

Strauss believed God had planned his life, which contributed greatly to his self-confidence. God let him take up employment when he was twenty-three, in 1919, at Kuhn, Loeb, a distinguished house with a number of major railroads among its clients. Four years later he married Alice Hanauer, daughter of one of the partners. His salary and participation reached $75,000 a year in 1926; the following year it escalated to $120,000. In 1929 he became a partner himself and settled into prosperous gentility.

The 1930s brought him pain and grief. After resisting Chaim Weizmann's attempts to convert him to Zionism at a Jewish conference in London in 1933—"My boy, you are difficult," Weizmann told him; "we will have to grind you down"—he returned to the United States to discover his mother terminally ill with cancer. She died early in 1935; the disease took his father as well in the hot summer of 1937. Strauss looked for a suitable memorial. "I became aware," he reports in his memoirs, "of the inadequate supply of radium for the treatment of cancer in American hospitals." He established the Lewis and Rosa Strauss Memorial Fund and turned up a young refugee physicist from Berlin, Arno Brasch. Brasch had designed a capacitor-driven discharge tube for producing bursts of high-energy X rays, a "surge generator." When Leo Szilard was working at St. Bart's with Chalmers in the summer of 1934 he had arranged for Brasch and his colleagues in Berlin to break up beryllium with hard X rays; the experiment had been a success and Brasch and four other contributors had signed the report to *Nature* along with Chalmers and Szilard. If X rays could break up beryllium they might at least induce radioactivity in other elements. "An isotope of cobalt thus produced," writes Strauss, "would be radioactive and would emit gamma rays similar to the radiation produced by radium. . . . Radioactive cobalt could be made . . . at a cost of a few dollars a gram. Radium was then priced at about fifty thousand dollars a gram. . . . I foresaw the possibility of producing the isotope in quantity and of giving it to hospitals as a memorial to my parents."

Enter Leo Szilard, still in England:

August 30, 1937

Dear Mr. Strauss:

I understand that you are interested in the development of a surge generator with the view of using it for producing artificially radioactive elements. . . .

> At present . . . I am not in the position of [offering manufacturing rights under this patent]. It is possible, however, that at a later date . . . I shall obtain full liberty of action concerning this patent. If this happens I shall let you have a non-exclusive license, royalty free, *but* limited to the production of radioactive elements by means of high voltage generated by a surge generator.
>
> Yours very truly,
> Leo Szilard

Brasch and Szilard owned the patent in question jointly. Szilard's letter offers to give his interest away free of charge nonexclusively to Strauss, a politic salutation to a rich man. But not even Leo Szilard could live on air, and as Strauss makes clear in his memoirs, the two young physicists eventually "asked me to finance them in the construction of a 'surge generator.' " On the other hand, Szilard as usual seems to have sought no personal financial gain from the project beyond, perhaps, basic support. In the time he could spare from observing the developing disaster in Europe he was apparently trying to promote the building of equipment with which he might explore further the possibility of a chain reaction.

He crossed the Atlantic in late September to reconnoiter. A friend remembers discussing the feasibility of an atomic bomb with Szilard during this period. "In the same conversation he spoke of his ideas for preserving peaches in tins in such a way that they would retain the texture and taste of the fresh fruit." When the surge-generator negotiations bogged down in debates among the lawyers, the resourceful Szilard distracted Strauss with the idea of using radiation to preserve and protect the natural products of farm and field. The tobacco worm might be exterminated, for example. But would irradiation harm the tobacco? Among Szilard's surviving papers is lodged a fading letter from Dr. M. Lenz of the Montefiore Hospital for Chronic Diseases that reports the decisive experiment:

> On April 14, 1938, at 2:30 p.m., your six cigars were irradiated with 100 kv., a filter focus distance of 20 cm. with ten minutes in front and ten minutes over the back of each cigar. This gave them 1000 r. in front and 1500 r. in back of each cigar.
>
> I hope that your friend finds the taste unchanged.

Szilard also bought pork from a meat market on Amsterdam Avenue, saving the receipt, and arranged its irradiation to see if X rays might kill the parasitic worm of trichinosis. He even dispatched his brother Béla to Chicago to discuss the matter with Swift & Company, which reported it had in fact made similar experiments of its own.

The surge-generator project developed through the year, incidentally giving Strauss the opportunity to meet Ernest Lawrence, who dropped by to pitch the new sixty-inch cyclotron he was building—the pole pieces were sixty inches across, but the magnet would weigh nearly two hundred tons. Lawrence and his brother John, a physician, had arrested their mother's cancer with accelerator radiation and intended to use the big cyclotron to further that research. Strauss remained loyal to the surge generator.

Segrè encountered Strauss's Hungarian wizard in New York that summer. The elegant Italian was professor of physics at Palermo by then, married to a German woman who had fled Breslau to escape the Nazis, with a young son:

> I left Palermo with a return ticket, and I arrived in New York. I met Szilard. "Oh, what are you doing here?" He was a good friend of mine. I knew him quite well. "What are you doing here? What's going on?"
>
> I said, "I'm going to Berkeley to look at the short-lived isotopes of element 43," which was my plan. "I'll work there the summer, and then I'll go back to Palermo."
>
> He said, "You are not going back to Palermo. By this fall, God knows what will happen! You can't go back."
>
> I said, "Well, I have a return ticket. Let's hope for the best."
>
> But I had gotten a passport for my wife and my son before leaving, because I smelled that the situation was dangerous. So I took the train in New York, Grand Central, and I bought the newspaper in Chicago. I still remember it. I will remember it as long as I live. I opened the newspaper, and I found out that Mussolini had started the antisemitic campaign and had fired everybody. So there I was. So I had the ticket and went to Berkeley. I started to work on my short-lived isotopes of technetium, but at the same time I tried to get some job. Then I got my wife here.

The pall of racism had dropped over Italy.

The physicists at the institute on Via Panisperna had been alert to the darkening Italian prospect since at least the mid-1930s. Segrè remembers asking Fermi in the spring of 1935 why the group's mood seemed less happy. Fermi suggested he look for an answer on the big table in the center of the institute reading room. Segrè did and found a world atlas there. He picked it up; it fell open automatically to a map of Ethiopia, which Italy in a show of Fascist bravado was about to invade. By the time the invasion began all but Amaldi were examining their options.

Fermi went off to the University of Michigan's summer school in Ann Arbor, renewing an affiliation he had begun with Laura in the summer of

1930. He liked America. "He was attracted," Segrè notes with an ear for Fermi's priorities, "by the well-equipped laboratories, the eagerness he sensed in the new generation of American physicists, and the cordial reception he enjoyed in academic circles. Mechanical proficiency and practical gadgets in America counterbalanced to an extent the beauty of Italy. American political life and political ideals were immeasurably superior to fascism." Fermi swam in Michigan's cool lakes and learned to enjoy American cooking. But the pressure of events in Italy was not yet sufficiently extreme, and Laura, Roman to her fine bones, was more than reluctant to leave the city of plane trees and classical ruins where she was born. Nor was anti-Semitism yet an issue in Italy—Mussolini had even declared he did not propose to make it one.

There was less to hold the other men. Rasetti summered at Columbia University that year, 1935, and decided to stay on. Segrè had shifted to Palermo but began looking toward Berkeley. Pontecorvo moved to Paris. D'Agostino went to work for the Italian National Research Council. Amaldi and Fermi pushed on alone, Amaldi remembers, Fermi even jettisoning his daily routine for the distraction of experiment:

> We worked with incredible stubbornness. We would begin at eight in the morning and take measurements [they were examining the unaccountably differing absorption of neutrons by different elements], almost without a break, until six or seven in the evening, and often later. The measurements . . . were repeated every three or four minutes, according to need, and for hours and hours for as many successive days as were necessary to reach a conclusion on a particular point. Having solved one problem, we immediately attacked another. . . . "Physics as soma" was our description of the work we performed while the general situation in Italy grew more and more bleak, first as a result of the Ethiopian campaign and then as Italy took part in the Spanish Civil War.

Fermi taught a summer course in thermodynamics at Columbia University in 1936 as the civil war began in Spain that would last three years, claim a million lives and set Mussolini decisively at Hitler's side. The following January Corbino died unexpectedly of pneumonia at sixty-one and the hostile occupant at the north end of the institute's second floor, Antonino Lo Sordo, a good Fascist, was appointed to succeed him. "That was a sign that Fermi's fortunes were declining in Italy," Segrè notes. "America," he concludes of those depressing years, "looked like the land of the future, separated by an ocean from the misfortunes, follies, and crimes of Europe."

If the *Anschluss* was a test of Hitler's strength, it was also a test of

Mussolini's willingness to acquiesce to complicity in crime. He had posed as Austria's protector; on the night of the March 1938 invasion Hitler waited near hysteria at the Chancellery in Berlin for a response from Rome to a letter he had sent justifying his action. The call came at 10:25 P.M. and the Führer snatched up the phone. "I have just come back from the Palazzo Venezia," his representative reported. "The Duce accepted the whole thing in a very friendly manner. He sends you his regards. . . . Mussolini said that Austria would be immaterial to him." Hitler replied: "Then please tell Mussolini I will never forget him for this! Never, never, never, no matter what happens! . . . As soon as the Austrian affair has been settled I shall be ready to go with him through thick and thin—through anything!" The Führer visited Rome in triumph in May, parading into districts the Duce had ordered hastily face-lifted to conceal their decay. Fermi's circle repeated the verse passed around the city by word of mouth by which an indignant Roman poet greeted the Nazi dictator:

Rome of travertine splendor
Patched with cardboard and plaster
Welcomes the little housepainter
As her next lord and master.

Italy would only be saved, Fermi told Segrè bitterly, if Mussolini went crazy and crawled on all fours.

The summer of 1938, July 14, brought the anti-Semitic *Manifesto della Razza* of which Segrè read in the Chicago newspaper on his way from New York to Berkeley. Italians are Aryans, the manifesto claimed. But "Jews do not belong to the Italian race." In Germany the vicious distinction had been commonplace; in Italy it was shocking. Italian Jews, only one in a thousand, were largely assimilated. The Fermis' two children—Giulio, a son, had been born in 1936—might be exempted since they were Catholic, born of a nominally Catholic father. But Laura was a Jew. She was spending the summer with the children in the Dolomites, the South Tyrol district named for the magnesian limestone that rings broad basin meadows with the flat, sharp formations Italians call "shovels." Enrico came up preoccupied in August to the meadow of San Martino di Castrozza to break the news. When Mussolini pushed through the first anti-Semitic laws early in September the Fermis decided to emigrate as soon as they could arrange their affairs. Fermi wrote four American universities and to avoid suspicion mailed each letter from a different Tyrolese town. Five schools shot back invitations. In confidence he accepted a professorship at Columbia and went off to Copenhagen to Bohr's annual gathering of the brethren.

The previous month the International Congress of Anthropological and Ethnological Sciences had invited Bohr to address it at a special session in Helsingør, Shakespeare's Elsinore, on the coast of Zealand north of Copenhagen. In the Renaissance castle there Denmark's most prominent citizen used the occasion to challenge Nazi racism publicly before the world. It was a brave statement by a brave man. Bohr understood that the major Western democracies were not likely to rally to the defense of his small, unprotected nation when Hitler eventually turned to look its way. George Placzek, a Bohemian theoretician working in Copenhagen whose tongue was almost as sharp as Pauli's, had already encapsulated that cruel truth. "Why should Hitler occupy Denmark?" Placzek quipped to Frisch one day. "He can just telephone, can't he?"

Against the brutal romanticism of German Blood and Earth, Bohr set the subtle corrective of complementarity. He spoke of "the dangers, well known to humanists, of judging from our own standpoint cultures developed within other societies." Complementarity, he proposed, offered a way to cope with the confusion. Subject and object interact to obscure each other in cultural comparisons as in physics and psychology; "we may truly say that different human cultures are complementary to each other. Indeed, each such culture represents a harmonious balance of traditional conventions by means of which latent possibilities of human life can unfold themselves in a way which reveals to us new aspects of its unlimited richness and variety."

The German delegates walked out. Bohr went on to say that the common aim of all science was "the gradual removal of prejudices," a complementary restorative to the usual pious characterization of science as a quest for incontrovertible truth. To a greater extent than any other scientist of the twentieth century Bohr perceived the institution of science to which he dedicated his life to be a profoundly political force in the world. The purpose of science, he believed, was to set men free. Totalitarianism, in Hannah Arendt's powerful image, drove toward "destroying all space between men and pressing men against each other." It was entirely in character that Bohr, at a time of increasing danger, publicly opposed that drive with the individualistic and enriching discretions of complementarity.

It was also entirely in character, when Fermi came to Copenhagen, that Bohr should lead him aside, take hold of his waistcoat button and whisper the message that his name had been mentioned for the Nobel Prize, a secret traditionally never foretold. Did Fermi wish his name withdrawn temporarily, given the political situation in Italy and the monetary restrictions, or would he like the selection process to go forward? Which was the same as telling Fermi he could have the Prize that year, 1938, if he wanted it and was welcome to use it to escape a homeland that threatened now despite the distinction he brought it to tear his wife from citizenship.

Leo Szilard's Cambridge collaborator Maurice Goldhaber emigrated to the United States in the late summer of 1938 and took up residence as an assistant professor of physics at the University of Illinois. Szilard appeared at Goldhaber's new apartment in Champaign in September to finish work they had begun together in England and stayed to follow the Munich crisis, for which purpose his host went out and bought a radio. Szilard understood, as Winston Churchill also understood and told his consituents at the end of August, that "the whole state of Europe and of the world is moving steadily towards a climax which cannot long be delayed." Before deciding between residency in England or the United States, Szilard said later, "I just thought I would wait and see."

The Sudetes, the border region of mountainous uplift that continues across Czechoslovakia from the Carpathians to the Erzgebirge, sustained at that time a German-speaking urban and industrialized population of some 2.3 million, about one-third of the population of western Czechoslovakia, formerly Bohemia. Nazi agitation began early in the Sudetenland; by 1935 a surrogate Nazi organization had become the largest political party in the Czechoslovakian republic. Hitler wanted Czechoslovakia next after Austria to facilitate his dream of German expansion, *Lebensraum*, and to deny airfields and support to the Soviet Union in the war he was well along in planning. The Sudetenland was his key. Czechoslovakia had built fortifications against German invasion across the Sudetes; after 1933 it imposed restrictions on the Sudeten Germans in an effort to protect that flank from subversion. Hitler opened his Czechoslovakian campaign even before the *Anschluss*, asserting the Reich's duty to protect the Sudeten Germans. Through the summer of 1938 German pressure on Czechoslovakia increased while the Western democracies maneuvered to avoid confrontation.

By the time Szilard began listening to Maurice Goldhaber's new radio the Czech government had established full martial law in the Sudetenland but also offered autonomy to the region in excess of what the Sudeten German Party had demanded. These developments prompted the British Prime Minister, Neville Chamberlain, to propose a meeting with Hitler. Hitler was delighted. He invited the Prime Minister to Berchtesgaden. The last outcome he wanted was a Czechoslovakian settlement. He signaled the Sudeten Nazis to increase their demands. Chamberlain heard the extremist proclamation on the radio on September 16 as he rode out by train from Munich: a call for immediate annexation to the German Reich. Back in

London on September 17 he recommended the annexation. Hitler, he said, "was in a fighting mood."

"The British and French cabinets at this time," writes Churchill, "presented a front of two overripe melons crushed together; whereas what was needed was a gleam of steel. On one thing they were all agreed: there should be no consultation with the Czechs. These should be confronted with the decision of their guardians. The Babes in the Wood had no worse treatment." The two governments, citing "conditions essential to security," decided that Czechoslovakia should cede to Germany all areas of the country where the population was more than 50 percent German. France had treaty obligations to Czechoslovakia but chose not to honor them. Facing such isolation, the small republic capitulated on September 21.

The Anglo-French proposals invoked self-determination for the German-speaking areas they defined. Hitler had agreed to such self-determination when he saw Chamberlain on September 16. Now the Prime Minister met with the Chancellor again, this time at Bad Godesberg on the Rhine outside Bonn, near Remagen. Hitler escalated his demands. "He told me," Chamberlain reported immediately afterward to the House of Commons, "that he never for one moment supposed that I should be able to come back and say that the principle [of self-determination] was accepted." Hitler wanted Czech acquiescence without self-determination by September 28 or he would invade. Chamberlain did not believe, however, he informed the Commons, that Hitler was deliberately deceiving him. The Nazi leader also told the Prime Minister "that this was the last of his territorial ambitions in Europe and that he had no wish to include in the Reich people of other races than Germans."

The Czechs mobilized a million and a half men. The French partly mobilized their army. The British fleet went active. At the same time a secret struggle may have been taking place between Hitler and the German general staff, which resisted any further plunge toward war. The result should have been stalemate, but Chamberlain moved again to concession. "Appeasement" was at that time a popular and not a pejorative word.

"How horrible, fantastic, incredible it is," the Prime Minister admonished the British people by radio on September 27, the night before Hitler's deadline, "that we should be digging trenches and trying on gas-masks here because of a quarrel in a faraway country between people of whom we know nothing!" He volunteered "to pay even a third visit to Germany." He was, he said, "a man of peace to the depths of my soul." He made the offer of a visit to Hitler at the same time directly by letter, and the Führer took him up on it the following afternoon. Chamberlain, French Premier Edouard Daladier, Mussolini and Hitler met at Munich on the evening of

September 29. By 2 A.M. the following morning the four leaders had agreed to Czech evacuation of the Sudetenland without self-determination within ten days beginning October 1. At Chamberlain's suggestion he and Hitler then met privately and agreed further to "regard the Agreement signed last night . . . as symbolic of the desire of our two peoples never to go to war with one another again." Before he left Munich, closeted with Mussolini, the Führer discussed Italian participation in the eventual invasion of the British Isles.

Chamberlain flew home. He read the joint declaration to the crowd gathered at the airport in welcome. Back in London he waved the declaration from an upper window of the Prime Minister's residence. "This is the second time there has come back from Germany to Downing Street peace with honour," he told the multitude below. "I believe it is peace in our time."

A group of refugee scientists was gathered outside the Clarendon Laboratory at Oxford the next morning discussing the Munich agreement when Frederick Lindemann drove up. Churchill had described the Czechoslovakian partition as amounting to "the complete surrender of the Western Democracies to the Nazi threat of force." Lindemann, Churchill's intimate adviser, was equally disgusted. One of the refugees asked him if he thought Chamberlain had something up his sleeve. "No," the Prof snapped, "something down his pants."

A cable came along to Lindemann then:

> HAVE ON ACCOUNT OF INTERNATIONAL SITUATION WITH GREAT REGRET POSTPONED MY SAILING FOR AN INDEFINITE PERIOD STOP WOULD BE VERY GRATEFUL IF YOU COULD CONSIDER ABSENCE AS LEAVE WITHOUT PAY STOP WRITING STOP PLEASE COMMUNICATE MY SINCERELY FELT GOOD WISHES TO ALL IN THESE DAYS OF GRAVE DECISIONS
>
> SZILARD

Szilard and Goldhaber found time during the crisis to write up a series of experiments with indium that they had started in England in 1937 and that Goldhaber and an Australian student, R. D. Hill, had completed before leaving for the United States. Szilard had thought indium might be a candidate for chain reaction but the results indicated that the radioactivity in indium of which Szilard had been suspicious was caused by a new type of reaction process, inelastic neutron scattering without neutron capture or loss. Szilard was discouraged. "As my knowledge of nuclear physics increased," he said later, "my faith in the possibility of a chain reaction gradually decreased." If other kinds of radiation also induced radioactivity in indium without producing neutrons, then he would have no more candidates for neutron multiplication and he would have to give up his belief in the process he still nicknamed "moonshine." That final experiment would be worked by friends at the University of Rochester in upstate New York, where he would travel in early December.

Otto Hahn opened the September 1938 issue of the *Comptes Rendus* to a shock. Part two of the Curie-Savitch study of the elusive 3.5-hour activity of uranium appeared there; amid much conjecture its most challenging conclusion was: "Taken altogether, the properties of $R_{3.5h}$ are those of lanthanum, from which it is not possible to separate it except by fractionation."*

Curie and Savitch believed that their $R_{3.5h}$ activity could be at least partly separated from lanthanum. It apparently did not occur to them that what was crystallizing out of solution might be another activity with a similar half-life, leaving a 3.5-hour lanthanum activity behind. They still could not believe—nor could anyone else—that uranium bombardment might produce an element thirty-five steps away down the periodic table. A Canadian radiochemist then visiting Dahlem records their German critic's response: "You can readily imagine Hahn's astonishment. . . . His reaction was that it just could not be, and that Curie and Savitch were very muddled up."

Despite his threat to Joliot in May, Hahn had not yet repeated the Curie-Savitch work. Now he passed the *Comptes Rendus* along to Fritz Strassmann. Strassmann studied the French paper and speculated that the muddle might have a physical cause—two similar radioactivities mixed together in the same solution. He told Hahn. Hahn laughed; the conclusion seemed improbable. On second thought, it was worth examining. As the Czechoslovakian crisis broke across Europe the two men bombarded uranium in peaceful Dahlem. They used a lanthanum carrier to precipitate rare-earth elements such as actinium (if any), a barium carrier to precipitate alkaline-earth elements such as radium (if any). (Carrier chemicals made it possible to separate from the parent solution the few thousand atoms of daughter substances produced by neutron bombardment. A chemically similar daughter substance, traceable by its unique half-life, would

*Fractionation—fractional crystallization—was a technique of chemical analysis pioneered by Marie Curie in the course of purifying polonium and radium. Most substances are more soluble at a high temperature than a low. Make a strong boiling solution of a substance—for rock candy, for example, sugar in water—cool the solution, and at some point the substance will emerge out of solution to form pure crystals. Fractional crystallization further involves separating out of the same solution several different, chemically similar substances by taking advantage of their tendency to crystallize at different temperatures according to differences in their atomic weights, lighter elements crystallizing first.

lodge in the spaces of the carrier's crystals as those regular solids formed from solution by chemical precipitation and would thus be carried away. Which carrier accomplished the carrying gave a clue to the part of the periodic table to which the unknown daughter substance belonged. Then it became a matter of further separating the daughter substance from the carrier by fractional crystallization, following it as before by tracing its characteristic radioactivity.)

After a hard week's work Hahn and Strassmann succeeded in identifying no fewer than sixteen different activities. Their barium separations gave them their most startling results: three previously unknown isotopes which they believed to be radium. They reported their findings in November in *Naturwissenschaften*. The creation of radium, element 88, from uranium, they pointed out, "must be due to the emission of two successive alpha particles."

If the physicists had found it hard to swallow that slow-neutron bombardment might produce thorium (90) or actinium (89), they found it even harder to swallow that it might produce radium. Lise Meitner wrote in warning from Stockholm suggesting pointedly that the two chemists check and recheck their results. Bohr invited Hahn to Copenhagen to lecture on the strange findings and tried to concoct a sufficiently crazy explanation:

> Bohr was skeptical and asked me if it was not highly improbable. . . . I had to reply that there was no other explanation, for our artificial radium could be separated only with weighable quantities of barium as carrier-substance. So apart from the radium only barium was present, and it was out of the question that it was anything but radium. Bohr suggested that these new radium isotopes of ours might perhaps in the end turn out to be strange transuranic elements.

Of the sixteen activities they had identified in neutron-bombarded uranium Hahn and Strassmann therefore now turned their full attention to the three controversial activities carried out of solution by barium.

Laura Fermi woke to the telephone early on the morning of November 10. A call would be placed from Stockholm, the operator advised her. Professor Fermi could expect it that evening at six.

Instantly awake to his wife's message, Fermi estimated the probability at 90 percent that the call would announce his Nobel Prize. As always he had planned conservatively, not counting on the award. The Fermis had prepared to leave for the United States from Italy shortly after the first of the year. Ostensibly Fermi was to lecture at Columbia for seven months and then return. For stays of longer than six months the United States required immigrant rather than tourist visas, and because Fermi was an academic he and his family could be granted such visas outside the Italian quota list. The ruse of a lecture series was devised to evade a drastic penalty: citizens leaving Italy permanently could take only the equivalent of fifty dollars with them out of the country. But the plan required circumspection. The Fermis could not sell their household goods or entirely empty their savings account without risking discovery. So the money from the Nobel Prize would be a godsend.

In the meantime they invested surreptitiously in what Fermi called "the refugee's trousseau." Laura's new coat was beaver and they distracted themselves on the day of the Stockholm call shopping for expensive watches. Diamonds, which had to be registered, they chose not to risk.

Near six o'clock the phone rang. It was Ginestra Amaldi wondering if they had heard. Everyone had gathered at the Amaldis to wait for the call, she reported. The Fermis turned on the six o'clock news. Laura long remembered the news:

> Hard, emphatic, pitiless, the commentator's voice read the second set of racial laws. The laws issued that day limited the activities and the civil status of the Jews. Their children were excluded from public schools. Jewish teachers were dismissed. Jewish lawyers, physicians, and other professionals could practice for Jewish clients only. Many Jewish firms were dissolved. "Aryan" servants were not allowed to work for Jews or to live in their homes. Jews were to be deprived of full citizenship rights, and their passports would be withdrawn.

The passports of Jews had already been marked. Fermi had contrived to keep his wife's passport clear.

They probably heard the news from Germany as well: of a vast pogrom the previous night—*Kristallnacht*, the night of glass. A seventeen-year-old Polish Jewish student had attempted to assassinate Ernst vom Rath, third secretary in the Germany Embassy in Paris, on November 7, in reprisal for Polish mistreatment of the student's parents. Vom Rath died on November 9 and the assassination served as an excuse for general anti-Semitic riot. Mobs torched synagogues, destroyed businesses and stores, dragged Jewish families from their homes and beat them in the streets. At least one hundred people died. A volume of plate glass was shattered that night across the Third Reich equal to half the annual production of its original Belgian sources. The SS arrested some thirty thousand Jewish men—"especially rich ones," its order had specified—and packed them into the concentration camps at Buchenwald, Dachau and Sachsenhausen, from which they could be ransomed only at the price of immediate pauperized emigration.

Fermi took the Stockholm call. The Nobel Prize, undivided, would be awarded for "your discovery of new radioactive substances belonging to the entire race of elements and for the discovery you made in the course of this work of the selective power of slow neutrons." In security the Fermis could leave the madness behind.

Lise Meitner had written Otto Hahn of her worries a few days before the Fermis arrived. "Most of the time I feel like a wind-up doll running on automatic," she told her old friend, "smiling along happily and empty of real life. From that you can judge for yourself how productive my efforts are at work. And still in the end I'm thankful for it because it forces me to keep my thoughts together, which isn't always easy." She was sorry Hahn's rheumatism had returned and was afraid he wasn't taking care of himself; she asked after Planck and von Laue by their private Hahn-Meitner nicknames, Max Sr. and Max Jr.; she greeted Hahn's wife, Edith, and wondered what Christmas plans he had for his son. His uranium work was "really very interesting." She hoped he would write again soon.

She was living in a small hotel room—there was hardly space to unpack—and having trouble sleeping. People told her she was too thin. Worse, conditions at the Physical Institute were not what she had expected them to be. A Swedish friend, Eva von Bahr-Bergius, a physicist she knew from Berlin who had been a lecturer at the University of Uppsala, had helped with arrangements and was gradually breaking the bad news. Manne Siegbahn had not wanted to take Meitner on. He had no money for her, he had complained; he could give her a place to work but no more. Von Bahr-Bergius had pursued the Nobel Foundation grant. But it provided nothing for equipment or assistance. Meitner blamed herself: "Of course it's my fault; I should have prepared much better and much earlier for my leaving, should at least have had drawings made of the most important apparatus [she would need]."

She was a strong woman, but she was miserable and alone. Hahn responded with sympathy. At midmonth she thanked him for that "dear letter," then changed moods and charged him with indifference: "Concerning myself I sometimes suspect you don't understand my way of thinking. . . . Right now I really don't know if anyone cares about my affairs at all or if they will ever be taken care of."

Hahn was pursuing Meitner's affairs as well as his own. With her moody letter at hand he stormed down to the revenue office, which was responsible for inventorying her furniture and other property before allowing its release, and laid on what he called "a little seizure of my 'ecstasy,' " after which "the matter went somewhat better." That news he wrote to Meitner on Monday evening, December 19, from the KWI. Only then did he report why he had not yet left the laboratory:

> As much as I can through all of this I am working, and Strassmann is working untiringly, on the uranium activities. . . . It's almost 11 at night; Strassmann will return at 11:30 so that I can see about going home. The fact is, there's something so strange about the "radium isotopes" that for the time being we are mentioning it only to you. The half-lives of the three isotopes are quite precisely determined; they can be separated from *all* elements except barium; all the processes are in tune. Just one is not—unless there are extremely unusual coincidences: the fractionation doesn't work. Our radium isotopes act like *barium*.

Hahn and Strassmann worked in three rooms on the ground floor of the Kaiser Wilhelm Institute for Chemistry, the building with the *Pickelhaube* dome: Hahn's large personal chemistry laboratory north off the main lobby, a measurement room across the hall at the near end of the wing that extended northwest along Faradayweg and an irradiation room at the far end of the wing. They separated the three functions of irradiation, measurement and chemistry to avoid contaminating one with radiation from another. All the rooms were fitted with worktables of unfinished raw pine roughed out by a careful carpenter who took the trouble to add a graceful taper to the legs. On the table in the irradiation room rested cylinders of beeswax-colored paraffin like angelfood cakes drilled for the neutron sources, which were gram-strength radium salts mixed with beryllium powder. Handmade Geiger counters, fixed in hinged, hollowed-out bricks of lead shielding on the table in the measurement room, connected through thin coiling wires back to breadboard amplifiers worked by silvered vacuum tubes like inverted bud vases. The amplifiers actuated gleaming brass clockwork counters with numbers showing black through angled miniature windows on their spines. Kraftboard-covered 90-volt Pertrix dry batteries that powered the system packed a shelf below the table. Hahn's laboratory table held the brackets, beakers, flasks, funnels and filters of radiochemistry. The two men moved in their work from room to room on a regular schedule determined by the duration of the half-lives they were studying. There would have been a pungency of nitrates in the air, mingled with the aroma of Hahn's inevitable cigar.

In his fifty-ninth year Hahn stooped slightly but looked younger than his age. His hairline had receded and his eyebrows had grown bushy; he had trimmed back to the edge of his upper lip the waxed Prussian mustache of his youth; his brown eyes still sparkled with warmth. By now he was unquestionably the ablest radiochemist in the world. He needed all his forty years' experience to decode uranium.

He and Strassmann had begun their renewed examination of the three "radium" isotopes early in December by attempting a purer separation from uranium. Strassmann suggested using barium chloride as a carrier rather than the customary barium sulfate because the chloride, Hahn explains, "forms beautiful little crystals" of exceptional purity. They wanted to be sure their separations would be free of contamination from other bombardment products with similar half-lives, the difficulty that had muddled Curie and Savitch. The procedure for the 86-minute activity they were studying, which they called "Ra-III," required them to irradiate about fifteen grams of purified uranium for twelve hours, wait several hours for their more intense 14-minute "Ra-II" to retreat from the foreground by decaying, then add barium chloride as a carrier and accomplish the separation. The Ra-III came out of the uranium solution with the barium, but it refused then to remain behind during fractionation when the barium crystallized away. Instead it crystallized with the barium.

"The attempts to separate our artificial 'radium isotopes' from barium in this way were unsuccessful," Hahn would explain in his Nobel Prize lecture; "no enrichment of the 'radium' was obtained. It was natural to ascribe this lack of success to the exceptionally low intensity of our preparations. It was always a question of merely a few thousands of atoms, which could only be detected as individual particles by the Geiger-Müller counter. Such a small number of atoms could be carried away by the great excess of inactive barium without any increase or decrease being perceptible." To check that possibility they retrieved from storage a known radium isotope they often worked with, the isotope they called "mesothorium." They diluted it to match the pale radioactivity of their few thousand atoms of Ra-III, then ran it through barium precipitation and fractionation. It separated away cleanly from the barium. Their technique was not at fault.

On Saturday, December 17, the day after Hahn stormed the revenue office on behalf of Meitner's furniture, he and Strassmann carried out a further heroic check. They mixed Ra-III with dilute mesothorium and precipitated and fractionated the two substances *together*. Then the chemical evidence was certain, whatever it might mean in physical terms: the mesothorium remained in solution when the barium carrier crystallized out but Ra-III went off with the barium, distributing itself uniformly and indivisibly throughout the small pure crystals. Hahn wrote an enthusiastic note in his pocket appointment book to mark the day: "Exciting fractionation of radium/barium/mesothorium."

It seemed their "radium" isotopes must be barium, element 56, slightly more than half as heavy as uranium and with just over half its charge. Hahn and Strassmann could hardly believe it. They conceived an even more convincing experiment. If their "radium" was really radium, then by beta decay it ought to transform itself one step up the periodic table to actinium (89). If, on the other hand, it was barium (56), then by beta decay it ought to transform itself one step up to lanthanum (57). And lanthanum could be separated from actinium by fractionation. They were carrying out this definitive project late Monday night, December 19, when Hahn sent Meitner the news.

"Perhaps you can suggest some fantastic explanation," he wrote. "We understand that it really *can't* break up into barium. . . . So try to think of some other possibility. Barium isotopes with much higher atomic weights than 137? If you can think of anything that might be publishable, then the three of us would be together in this work after all. We don't believe this is foolishness or that contaminations are playing tricks on us."

He closed by wishing his friend a "somewhat bearable" Christmas. Fritz Strassmann added "very warm greetings and best wishes." Hahn posted the letter to Stockholm late at night on his way home.

The two men took time from their readings to attend the annual KWI Christmas party the next day, though Hahn had little joy of it with Meitner gone. They continued the actinium-lanthanum experiment even as they worked up the radium-barium findings. After the party the institute would close for Christmas; they kept a typist busy until the end but were unable to finish their report. Hahn had called Paul Rosbaud at *Naturwissenschaften*, told him the news and asked him to make space in the next issue. Rosbaud was willing to pull a less urgent paper from the journal but cautioned that the manuscript must be delivered no later than Friday, December 23. Hahn arranged for a laboratory assistant to serve as typist on Thursday. In the meantime he and Strassmann would carry on alone.

Meitner received Hahn's Monday-night letter in Stockholm on Wednesday, December 21. It was startling; if the results held she saw it meant the uranium nucleus must fracture and she immediately wrote him back:

> Your radium results are very amazing. A process that works with slow neutrons and leads to barium! . . . To me for the time being the hypothesis of such an extensive burst seems very difficult to accept, but we have experienced so many surprises in nuclear physics that one cannot say without hesitation about anything: "It's impossible."

She was traveling on Friday to the village of Kungälv in the west of Sweden for a week's vacation, she told Hahn; "if you write me in the meantime please address your letter there." She sent him and his family "warmest greetings . . . and much love and the very best for the New Year."

That day Hahn and Strassmann had finished the actinium-lanthanum experiment—and confirmed lanthanum from barium decay. In the late evening, after they turned off their counters, Hahn wrote his exiled colleague again. The paper was not yet finished; a phrase from the letter would be reworked to more cautious language for the final draft: "Our radium proofs convince us that as chemists we must come to the conclusion that the three carefully-studied isotopes are not radium, but, from the standpoint of the chemist, barium."

Hahn had hoped Meitner might quickly find some physical explanation for his unprecedented chemistry. That would strengthen his conclusion and also put Meitner's name on the paper, the best possible Christmas gift. With the lanthanum confirmation at hand he could no longer delay. As it was he had withheld the news from physicists on his own staff and at the new physics institute nearby. Someone else—Curie and Savitch, for example—might very well have made the same discovery. And whatever the explanation, the discovery was clearly of major importance, a reaction unlike any other yet found. "We cannot hush up the results," Hahn wrote Meitner, "even though they may be absurd in physical terms. You can see that you will be performing a good deed if you find an alternative [explanation]. When we finish tomorrow or the day after I will send you a copy of the manuscript. . . . The whole thing is not very well suited for *Naturwissenschaften*. But they will publish it quickly."

Hahn mailed the letter to Stockholm. He did not yet know about Meitner's Kungälv vacation.

Leo Szilard's work at the University of Rochester confirmed that no neutrons came out when indium was irradiated. On December 21, as Hahn and Meitner exchanged their excited letters, Szilard advised the British Admiralty by letter:

> Further experiments . . . have definitely cleared up the anomalies which I have observed in 1936. . . . In view of this new work it does not now seem necessary to maintain [my] patent . . . nor would the waiving of the secrecy of this patent serve any useful purpose. I beg therefore to suggest that the patent be withdrawn altogether.

Szilard's faith in the possibility of a chain reaction, as he said later, had "just about reached the vanishing point."

Hahn and Strassmann had originally titled their paper "On the radium isotopes produced by the neutron bombardment of uranium and their behavior." With their new data they realized "radium" would no longer do. They considered changing "radium" to "barium" throughout the paper. But most of it had been written before the lanthanum experiment firmed their convictions. They would have had to rewrite from beginning to end, "especially," says Hahn in retrospect, "since in view of this result its major portion was not especially interesting any more." Christmas and the journal deadline were upon them and they had no time. They decided to jury-rig what was on hand. The results would be no less effective for being inelegant. They substituted the noncommittal phrase "alkaline-earth metals" for "radium isotopes" in the title—both barium and radium are alkaline-earth metals, as are beryllium, magnesium, calcium and strontium. They went through the draft putting equivocal quotation marks around their many references to radium and actinium. Then they attached seven cautious paragraphs at the end.

"Now we still have to discuss some newer experiments," this final section began, "which we publish rather hesitantly due to their peculiar results." They then summarized their series of experiments:

> We wanted to identify beyond any doubt the chemical properties of the parent members of the radioactive series which were separated with the barium and which have been designated as "radium isotopes." We have carried out fractional crystallizations and fractional precipitations, a method which is well-known for concentrating (or diluting) radium in barium salt solutions. . . .
>
> When we made appropriate tests with radioactive barium samples which were free of any later decay products, *the results were always negative. The activity was distributed evenly among all the barium fractions.* . . . We come to the conclusion that our "radium isotopes" have the properties of barium. As chemists we should actually state that the new products are not radium, but rather barium itself. Other elements besides radium or barium are out of the question.

They discussed actinium then, distinguished their work from that of Curie and Savitch and pointed out that all so-called transuranics would have to be reexamined. Not quite prepared to usurp the prerogative of the physicists, they closed on a tentative note:

> As chemists we really ought to revise the decay scheme given above and insert the symbols Ba, La, Ce [cerium], in place of Ra, Ac, Th [thorium]. However as "nuclear chemists," working very close to the field of physics, we cannot bring ourselves yet to take such a drastic step which goes against all previous laws of nuclear physics. There could perhaps be a series of unusual coincidences which has given us false indications.

Promising further experiments, they prepared to release their news to the world. Hahn mailed the paper and then felt the whole thing to be so improbable "that I wished I could get the document back out of the mail box"; or Paul Rosbaud came around to the KWI the same evening to pick it up. Both stories survive Hahn's later recollection. Since Rosbaud knew the paper's importance and dated its receipt December 22, 1938, he probably picked it up. But Hahn also visited the mailbox that night, to send a carbon copy of the seminal paper to Lise Meitner in Stockholm. His misgivings at publishing without her—or some dawning glimmer of the fateful consequences that might follow his discovery—may have accounted for his remembered apprehension.

The Swedish village of Kungälv—the name means King's River—is located some ten miles above the dominant western harbor city of Göteborg and six miles inland from the Kattegat coast. The river, now called North River, descends from Lake Vänern, the largest freshwater lake in Western Europe; at Kungälv it has cut a sheer granite southward-facing bluff, the precipice of Fontin, 335 feet high. The modern village is built along a single cobblestone lane on the narrow talus between the bluff and the river, its back to the wall.

As Norwegian Kongahalla the village was founded at a less constricted place downstream around A.D. 800. But an island hill rises from the river at Kungälv and is thus guarded by a natural moat, a defensive geography which the precipice of Fontin reinforces. In 1308, to mark the border there between Norway and Sweden, the Norwegians began to build on that island hill a monumental granite fortress, Bohus' Fäste (i.e., King Bohus' Fort), sod-ridged block walls mazing inward and upward to a cylindrical tower of thick stone with a conical roof that dominates the entire coastal valley. An accident of placement of three of the deep windows that penetrate the tower—two open above, one centered below—transforms it into a face staring with hollow eyes toward the Fontin bluff. To soften the grimness of that face the people of the valley named the tower *Fars Hatt*, Father's Hat, as if it evoked a workman in a cap. Through four hundred years of occupation Bohus' Fäste was besieged fourteen times while the settlements in the valley were put to the torch and the graveyard filled on the island below its hard walls.

The village was ordered moved upriver onto the island in 1612. The Danes ruled Norway from the fifteenth century to the early nineteenth century; they ceded the Kungälv region, Bohuslän, to Sweden by the Treaty of Roskilde in 1658. Fire in 1676 burned the island village and its burghers shifted for safety to the narrow shore. They laid out their lane and strip of houses extending west and east from a cobblestone marketplace where the talus widened to make room. Despite its fortress Kungälv is peaceful, especially in winter with the river frozen and a depth of clean snow on the ground. Its snug wooden houses, painted pastel, enclose rooms cozy with ships' chests and china cabinets and lace curtains, warmed by corner fireplaces faced with decorative tile, aromatic with coffee and baking. Eva von Bahr-Bergius and her husband Niklas built a house there in 1927, larger than most Kungälv houses but constructed in the same style. In 1938 Lise Meitner was alone in Stockholm. Otto Frisch was alone in Copenhagen, his mother, Meitner's sister, beyond reach in Vienna, his father incarcerated at Dachau, a victim of *Kristallnacht*. The Bergiuses therefore considerately invited aunt and nephew to Kungälv for Christmas dinner.

Meitner left Stockholm Friday morning, two days before Christmas. Frisch took the train ferry across from Denmark. His aunt arrived before him and registered at a quiet inn on Västra gatan, West Street, where they both would stay, a pale green building much like its modest neighbors but with a café on the ground floor. It faced a shadowed strip of garden north across the lane; above the stunted garden trees the dark bluff loomed. The other way, behind the inn, the flat, snow-covered flood plain of the river extended into open woods. The Bergiuses' house was a short walk eastward past the marketplace and the white church. Tired from travel, Frisch and Meitner met only briefly in the evening when Frisch came in.

In Copenhagen that winter he had been studying the magnetic behavior of neutrons. To further his work he needed a strong, uniform magnetic field, and on his way to Kungälv he had sketched out a large magnet he meant to design and build. He came downstairs on the morning before Christmas prepared to interest his aunt in his plans. She was already at breakfast and had no intention of discussing magnets: she had brought Hahn's December 19 letter downstairs with her and insisted Frisch read it. He did. "Barium," he told her, "I don't believe it. There's some mistake." He tried to change the subject to his magnet; she changed it back to barium. "Finally," says Meitner, ". . . we both became absorbed in my problem." They decided to go for a walk to see what they could puzzle out.

Frisch had brought cross-country skis and wanted to use them. He was concerned that his aunt would be unable to keep up. She could walk as fast as he could ski on level ground, she told him. She could and did. He fetched his skis and they went out, probably eastward to the Kungälv marketplace, which gave onto the flood plain of the river, then across the frozen river and into the open woods beyond.

"But it's impossible," Frisch remembers them saying in their collective effort to understand. "You couldn't chip a hundred particles off a nucleus in one blow. You couldn't even cut it across. If you tried to estimate the nuclear forces, all the bonds you'd have to cut all at once—it's fantastic. It's quite impossible that a nucleus could do that." Thirty years afterward Frisch summarized their thinking in more formal terms:

> But how could barium be formed from uranium? No larger fragments than protons or helium nuclei (alpha particles) had ever been chipped away from nuclei, and the thought that a large number of them should be chipped off at once could be dismissed; not enough energy was available to do that. Nor was it possible that the uranium nucleus could have been cleaved right across. Indeed a nucleus was not like a brittle solid that could be cleaved or broken; Bohr had stressed that a nucleus was much more like a liquid drop.

The liquid-drop model made a division of the nucleus seem possible. They sat down on a log. Meitner found a scrap of paper and a pencil in her purse. She drew circles. "Couldn't it be this sort of thing?"

Frisch: "Now, she always rather suffered from an inability to visualize things in three dimensions, whereas I had that ability quite well. I had, in fact, apparently come around to the same idea, and I drew a shape like a circle squashed in at two opposite points."

"Well, yes," Meitner said, "that is what I mean." She had meant to draw what Frisch had drawn, a liquid drop elongated like a dumbbell, but had drawn it end-on, indicating with a smaller dashed circle inside a larger solid circle the dumbbell's waist.

Frisch: "I remember that I immediately at that instant thought of the fact that electric charge diminishes surface tension." The liquid drop is held together by surface tension, the nucleus by the analogous strong force. But the electrical repulsion of the protons in the nucleus works against the strong force, and the heavier the element, the more intense the repulsion. Frisch continues:

> And so I promptly started to work out by how much the surface tension of a nucleus would be reduced. I don't know where we got all our numbers from, but I think I must have had a certain feeling for the binding energies and could make an estimate of the surface tension. Of course we knew the charge and the size reasonably well. And so, as an order of magnitude, the result was that at a charge [i.e., an atomic number] of approximately 100 the surface tension of the nucleus disappears; and therefore uranium at 92 must be pretty close to that instability.

They had discovered the reason no elements beyond uranium exist naturally in the world: the two forces working against each other in the nucleus eventually cancel each other out.

They pictured the uranium nucleus as a liquid drop gone wobbly with the looseness of its confinement and imagined it hit by even a barely energetic slow neutron. The neutron would add its energy to the whole. The nucleus would oscillate. In one of its many random modes of oscillation it might elongate. Since the strong force operates only over extremely short distances, the electric force repelling the two bulbs of an elongated drop would gain advantage. The two bulbs would push farther apart. A waist would form between them. The strong force would begin to regain the advantage within each of the two bulbs. It would work like surface tension to pull them into spheres. The electric repulsion would work at the same time to push the two separating spheres even farther apart.

Eventually the waist would give way. Two smaller nuclei would appear where one large nucleus had been before—barium and krypton, for example:

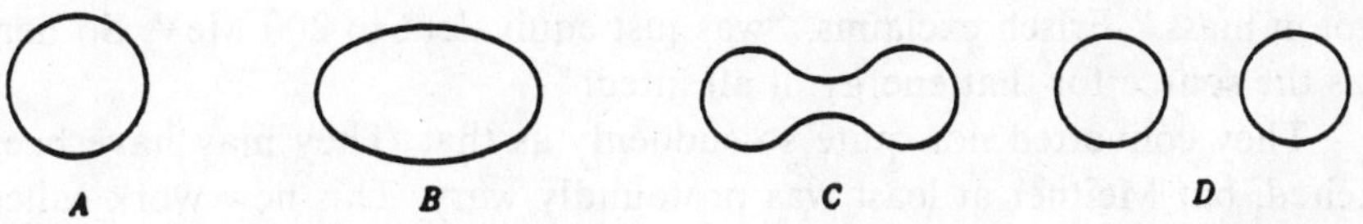

"Then," Frisch recalls, "Lise Meitner was saying that if you really do form two such fragments they would be pushed apart with great energy." They would be pushed apart by the mutual repulsion of their gathered protons at one-thirtieth the speed of light. Meitner or Frisch calculated that energy to be about 200 MeV: 200 million electron volts. An electron volt is the energy necessary to accelerate an electron through a potential difference of one volt. Two hundred million electron volts is not a large amount of energy, but it is an extremely large amount of energy from one atom. The most energetic chemical reactions release about 5 eV per atom. Ernest Lawrence was that year building a cyclotron with a nearly 200-ton magnet with which he hoped to accelerate particles by as much as 25 MeV. Frisch would calculate later that the energy from each bursting uranium nucleus would be sufficient to make a visible grain of sand visibly jump. In each mere gram of uranium there are about 2.5×10^{21} atoms, an absurdly large number, 25 followed by twenty zeros: 2,500,000,000,000,000,000,000.

They asked themselves what the source of all that energy could be. That was the crux of the problem and the reason no one had credited the possibility before. Neutron captures that had been observed before had involved much smaller energy releases.

When she was thirty-one, in 1909, Meitner had met Albert Einstein for the first time at a scientific conference in Salzburg. He "gave a lecture on the development of our views regarding the nature of radiation. At that time I certainly did not yet realize the full implications of his theory of relativity." She listened eagerly. In the course of the lecture Einstein used the theory of relativity to derive his equation $E = mc^2$, with which Meitner was then unfamiliar. Einstein showed thereby how to calculate the conversion of mass into energy. "These two facts," she reminisced in 1964, "were so overwhelmingly new and surprising that, to this day, I remember the lecture very well."

She remembered it in 1938, on the day before Christmas. She also "had the packing fractions in her head," says Frisch—she had memorized Francis Aston's numbers for the mass defects of nuclei. If the large uranium nucleus split into two smaller nuclei, the smaller nuclei would weigh less in total than their common parent. How much less? That was a calculation she could easily work: about one-fifth the mass of a proton less. Process one-fifth of the mass of a proton through $E = mc^2$. "One fifth of a proton mass," Frisch exclaims, "was just equivalent to 200 MeV. So here was the source for that energy; it all fitted!"

They converted not quite so suddenly as that. They may have been excited, but Meitner at least was profoundly wary. This new work called her previous four years' work with Hahn and Strassmann into doubt; if she was right about the one she was wrong about the other, just when she had escaped from Germany into the indifferent world of exile and needed most to confirm her reputation. "Lise Meitner sort of kept saying, 'We couldn't have seen it. This was so totally unexpected. Hahn is a good chemist and I trusted his chemistry to correspond to the elements he said they corresponded to. Who could have thought that it would be something so much lighter?' "

Christmas dinner at the Bergiuses' came and went. Frisch skied and Meitner walked. Nineteen thirty-eight was ticking to its end. With a week to pass in a small village they would certainly have visited the fortress and looked down from its ramparts onto the snow-covered valley, onto centuries of violent graves. Though they understood its energetics now, the discovery was still only physics to them; they did not yet imagine a chain reaction.

Hahn's letter of December 21, confirming lanthanum, was still not forwarded from Stockholm, nor was the carbon copy of the *Naturwissenschaften* paper. Hahn was eager to win Meitner's support and wrote Kungälv directly on the Wednesday after Christmas to woo her. Careful not to seem to usurp her place, he called the discovery his "barium fantasy" and questioned everything except the presence of barium and the absence of actinium, taking the humble chemist's part. "Naturally, I would be very interested to hear your frank opinion. Perhaps you could compute and publish something." He had continued to hold off telling other physicists, though he itched for physical confirmation of his chemistry. It was as though a maker of hand axes had discovered fire by striking flints while the sorcerers pondered how to harness lightning. He might hardly believe his luck and urgently seek their authentication even though he knew what burned his hand was real.

The letter reached Kungälv on Thursday; by return mail that day Meitner responded that the radium-barium finding was "very exciting. Otto R[obert] and I have already puzzled over it." But she let slip no answer to the puzzle and she asked about the lanthanum result.

Friday she sent Hahn a postcard: "Today the manuscript arrived." An important page was missing but it was all "very amazing." Nothing more; Hahn must have bitten his lip.

In Dahlem Rosbaud passed along the galley proofs. Hahn was more certain now of his findings. The manuscript had set the barium results "against all previous laws of nuclear physics." He moderated the phrase in proof to "against all previous experience."

But even with the carbon copy, the missing page and the December 21 letter finally at hand in Kungälv, Meitner hesitated to leap. On January 1, after conveying New Year's greetings to Hahn, she wrote: "We have read your work very thoroughly and consider it *perhaps* possible energetically after all that such a heavy nucleus bursts." She veered off to worry about their misbegotten transuranics, "not a good reference for my new start." Frisch added a New Year's wish of his own and a more genial reservation: "If your new findings are really true, it would certainly be of the greatest interest and I am very curious about further results."

Meitner returned to Stockholm later that day and Frisch to Copenhagen. He was "keen to submit our speculations—it wasn't really more at that time—to Bohr." The note of hesitancy in their letter to Hahn suggests they sought the authority of Bohr's blessing. Frisch saw him on January 3: "I had hardly begun to tell him, when he struck his forehead with his hand and exclaimed, 'Oh what idiots we have all been! Oh but this is wonderful! This is just as it must be!' " Their conversation lasted only a few minutes, Frisch wrote his aunt that day, "since Bohr immediately and in every respect was in agreement with us. . . . [He] still wants to consider this quantitatively this evening and to talk with me again about it tomorrow."

In Stockholm that day Meitner had received Hahn's revised proofs. Independently they quieted her doubt. She wrote Hahn emphatically: "I am fairly *certain* now that you really have a splitting towards barium and I consider it a wonderful result for which I congratulate you and Strassmann very warmly. . . . You now have a wide, beautiful field of work ahead of you. And believe me, even though I stand here very empty-handed at the moment, I am still happy about the marvelousness of these findings."

Now those findings needed interpretation. Aunt and nephew outlined a theoretical paper by long-distance telephone. Frisch drafted it Friday, January 6, and that evening took the trolley to the House of Honor to discuss it with Bohr, who was leaving for the United States the next morning for a term of work at the Institute for Advanced Study. There was time the next morning to type only part of the draft; Frisch delivered two pages to Bohr at the train station from which he and his nineteen-year-old son Erik were departing for Göteborg harbor. On the assumption that Frisch would immediately send the paper along to *Nature* Bohr promised not to mention it to their American colleagues until he heard from Frisch that it had been received and was in press. Among the notes he brought to that final discussion Frisch mentioned an experiment to confirm by physical means the Dahlem chemistry.

Hahn's and Strassmann's article had been published in Berlin on January 6. When it arrived in Copenhagen the next day Frisch thought to go over the whole business with George Placzek. Placzek was characteristically skeptical and characteristically witty about it. Uranium already suffered from alpha decay, Frisch remembers him scoffing; to think that it could be made to burst as well "was like dissecting a man killed by a falling brick and finding that he would have died of cancer." Placzek suggested that Frisch use a cloud chamber to look for energetic fragments that would prove the nucleus had split. The institute's radium-based neutron sources would fog a cloud-chamber photograph with gamma radiation, Frisch realized. But a simple ionization chamber would do. "One would expect fast-moving nuclei, of atomic number about 40-50 and atomic weight 100-150, and up to 100 MeV energy to emerge from a layer of uranium bombarded with neutrons," he explained his experiment in a subsequent report. "In spite of their high energy, these nuclei should have a range, in air, of a few millimetres only, on account of their high effective charge . . . which implies very dense ionization." In the course of their short passage his highly charged nuclear fragments would strip about 3 million electrons from the nuclei of air gases. They should be easy to find.

His chamber consisted of "two metal plates separated by a glass ring about 1 cm. high." The charged plates, which would collect the air ions, connected to a simple amplifier, which connected to an oscilloscope. To the bottom plate he attached a piece of uranium-coated foil. He set up the experiment in the basement of the institute and retrieved three of the neutron sources from the covered well. He placed the sources close to the foil and looked for the expected nuclei to emerge. Since they were highly energetic and strongly ionizing they would create quick, sharp, vertical pulses of the sweeping green beam of the oscilloscope.

Frisch started measurements on the afternoon of Friday, January 13, and "pulses at about the predicted amplitude and frequency (one or two per minute) were seen within a few hours." He ran checks with either the neutron sources or the uranium lining removed. He wrapped the sources with paraffin to slow the neutrons and "enhanced the effect by a factor of two." He continued measurements "until six in the morning to verify that the apparatus was working consistently." As had Werner Heisenberg before him, he lived upstairs at the institute; exhausted, he climbed the stairs to bed. He remembers thinking that 13 had proved once again to be his lucky number.

Even luckier than that: "At seven in the morning I was knocked out of bed by the postman who brought a telegram to say that my father had been released from concentration camp." His parents would move to Stockholm and share an apartment with his aunt, whose possessions, thanks to Hahn, were eventually shipped.

In "a state of slight confusion" Frisch spent the next day repeating the experiment for anyone who cared to see. One who came down in the morning to the basement laboratory was a black-haired, blue-eyed American biologist of Irish heritage named William A. Arnold who was studying on a Rockefeller Fellowship with George de Hevesy. Arnold was thirty-four, Frisch's age, on leave from the Hopkins Marine Station at Pacific Grove, California. He had made his way to Europe from San Francisco the previous September by freighter with his wife and young daughter. He could have gone to Berkeley to pick up radioisotope technique, but would have missed living in Copenhagen, learning from de Hevesy—would have missed contributing a coinage to the gamble that is history. Frisch showed the American the experiment and pointed out the pulses on the oscilloscope. "From the size of the spikes," Arnold recalls, "it was clear that they must represent 100-200 MeV, very much larger than the spikes from [uranium's natural background of] alpha particles."

> Later that day Frisch looked me up and said, "You work in a microbiology lab. What do you call the process in which one bacterium divides into two?" And I answered, "binary fission." He wanted to know if you could call it "fission" alone, and I said you could.

Frisch the sketch artist, good at visualizing as his aunt was not, had metamorphosed his liquid drop into a dividing living cell. Thereby the name for a multiplication of life became the name for a violent process of destruction. "I wrote home to my mother," says Frisch, "that I felt like someone who has caught an elephant by the tail."

Aunt and nephew conferred by telephone further over the weekend to prepare not one but two papers for *Nature*: a joint explanation of the reaction and Frisch's report of the confirming evidence of his experiment. Both reports—"Disintegration of uranium by neutrons: a new type of nuclear reaction" and "Physical evidence for the division of heavy nuclei under neutron bombardment"—used the new term "fission." Frisch finished the two papers on Monday evening, January 16, and posted them airmail to London the next morning. Since he and Bohr had already discussed the theoretical paper and since the experiment only confirmed the Hahn-Strassmann discovery, he did not hurry to let Bohr know.

Bohr sailed on the Swedish-American liner *Drottningholm* with his son Erik and the Belgian theoretician Léon Rosenfeld. "As we were boarding the ship," Rosenfeld recalls, "Bohr told me he had just been handed a note by Frisch, containing his and Lise Meitner's conclusions; we should 'try to understand it.' " That meant a working voyage; a blackboard was duly installed in Bohr's stateroom. The North Atlantic was stormy in that season; it made him "rather miserable, all the time on the verge of seasickness" but hardly stopped the work. The first question he wanted to answer was why, if the nucleus oscillated more or less randomly when it was bombarded, it seemed to prefer splitting into two parts rather than some other number. He was satisfied when he saw that the heaviest nuclei, because of their instability, require no more energy to split than they do to emit a single particle. It was a question of probabilities and two fragments were greatly more probable than a crowd.

The Fermis had arrived in New York on January 2, Laura feeling distinctly alien, Enrico announcing with his usual mock solemnity, "We have founded the American branch of the Fermi family." They put up temporarily at the King's Crown Hotel, opposite Columbia University, where Szilard was also living. George Pegram, the tall, soft-spoken Virginian who was chairman of the physics department and dean of graduate studies at Columbia, had met the Fermis as they debarked the *Franconia*; now in turn they waited at dockside to meet Bohr. The American theoretician John Archibald Wheeler, then twenty-nine years old, who had worked with Bohr in Copenhagen in the mid-1930s and would be working with him again at Princeton, joined them on the crowded West 57th Street pier. He had taught his regular Monday morning class, then caught a midday train.

As the *Drottningholm* berthed, at 1 P.M. on January 16, Laura Fermi saw Bohr on an upper deck leaning on the railing searching the crowd. She thought him worn when they met: "During the short time that had elapsed since our visit to his home, Professor Bohr seemed to have aged. For the last few months he had been extremely preoccupied about the political situation in Europe, and his worries showed on him. He stooped like a man carrying a heavy burden. His gaze, troubled and insecure, shifted from the one to the other of us, but stopped on none." No doubt Bohr was worried about Europe. He had also been seasick.

He had business in New York; he and Erik went off with the Fermis. Wheeler took Léon Rosenfeld along to Princeton. Keeping his promise to Frisch, Bohr had not mentioned the Hahn-Strassmann discovery and the Frisch-Meitner interpretation to either Fermi or Wheeler, but he had neglected to tell Rosenfeld of his pledge. Rosenfeld thought Frisch and Meitner had already sent off the paper that would give their work of interpretation priority. He passed on to Wheeler what Bohr had passed on to him. "In those days," Wheeler remembers, "I was in charge of the Monday evening journal club"—a weekly gathering of Princeton physicists to discuss the latest studies they found in physics journals, a way of keeping up. "It was the custom to get three things reported then, and here was something hot, as I had learned from Rosenfeld on the train." America first heard the news of the splitting of uranium—the term "fission" had not yet crossed the Atlantic—at the Princeton physics department journal club on the chill Monday evening of January 16, 1939. "The effect of my talk on the American physicists," says Rosenfeld ruefully, "was more spectacular than the fission phenomenon itself. They rushed about spreading the news in all directions."

Bohr arrived in Princeton the next day to take up residence and Rosenfeld casually mentioned the journal club talk. "I was immediately frightened," Bohr wrote his wife that night, "as I had promised Frisch I would wait until Hahn's note appeared and his own was sent off." It was more than a point of honor, though that would have been sufficient in itself to trigger the Bohr conscience. It was also that Meitner and Frisch were refugees who could use so spectacular a coup to establish themselves securely in exile. Bohr had at hand the work he and Rosenfeld had accomplished aboard the *Drottningholm*; for the next three days he labored to convert it into a letter to *Nature* that would give credit pointedly at the outset to Meitner and Frisch. Three days to produce a seven-hundred-word paper was for Niels Bohr great haste.

"Can you guess where I found out about [Bohr's news]?" asks Eugene Wigner. "In . . . the [Princeton] infirmary. Because I contracted jaundice and was in the infirmary for six weeks." Wigner and Princeton had not immediately got along; in 1936 "they said I should look for another job."

Princeton then, he thought, was "an ivory tower; people did not have any normal thinking about the facts of life and so forth and they looked down upon me." He sought another job and found one at the University of Wisconsin at Madison. "From the second day on I felt at home there. Somebody suggested we go to the track and we ran around the track and we were friends. We talked not only about the most difficult problems but about the daily events. We got down to earth almost." He met a young American woman in Wisconsin; they were quickly married. She became ill:

> I tried to conceal it from her that she had cancer and that there was no hope for her surviving. She was in a hospital in Madison and then she went to see her parents and I went with her but I didn't want to stay with her parents, of course, because I was, after all, a stranger to her parents. I went for a little while away to Michigan, Ann Arbor, and then I came back and saw her in her bed at her parents'. And then she told me essentially that she knows that she is close to death. She said, "Should I tell you where our suitcases are?" So she knew when she talked to me. I tried to conceal it from her because I felt that it would be better if a reasonably young person does not realize that she is doomed. Of course, we are all doomed.

He returned to Princeton in 1938, the university by then having more sensibly assessed his worth (a sophisticated and highly respected theoretician, Wigner shared the Nobel Prize in Physics in 1963 for his work on the structure of the nucleus).

After Bohr's arrival Szilard traveled down from New York to visit his sick friend and won a long-overdue surprise:

> Wigner told me of Hahn's discovery. Hahn found that uranium breaks into two parts when it absorbs a neutron. . . . When I heard this I immediately saw that these fragments, being heavier than corresponds to their charge, must emit neutrons, and if enough neutrons are emitted . . . then it should be, of course, possible to sustain a chain reaction. All the things which H. G. Wells predicted appeared suddenly real to me.

At Wigner's bedside in the Princeton infirmary the two Hungarians debated what to do.

In the meantime Bohr had sent his letter for *Nature* to Frisch in Copenhagen, asking him to forward it on "if, as I hope, Hahn's article has already been published and your and your aunt's note has already been submitted." He asked for the "latest news" on that front and wondered "how the experiments are proceeding." In a postscript he added that he had just seen the Hahn-Strassmann paper in *Naturwissenschaften*.

Ideas infect like viruses. The point of origin of the fission infection was Dahlem. From there it spread to Stockholm, to Kungälv, to Copenhagen. It crossed the Atlantic with Bohr and Rosenfeld. I. I. Rabi and the young California-born theoretician Willis Eugene Lamb, Jr., two Columbia men working at Princeton that week, both heard the news, Lamb perhaps from Wheeler, Rabi from Bohr himself. They returned to New York—"probably Friday night," Lamb thinks. Rabi says he told Fermi. In 1954 Fermi credited Lamb: "I remember one afternoon Willis Lamb came back very excited and said that Bohr had leaked out great news." Lamb recalls "spreading it around" but does not recall specifically telling Fermi. Possibly both men talked to the Italian laureate within a space of hours; it was information he of all physicists would most need to hear, since the Nobel lecture he had delivered only a month earlier, not yet printed, was now partly obsolete and an embarrassment. (Fermi confined revision to a footnote: "The discovery by Hahn and Strassmann . . . makes it necessary to reexamine all the problems of the transuranic elements, as many of them might be found to be products of a splitting of uranium." The many other radioactivities he and his group identified and his slow-neutron discovery still secured his Nobel Prize.)

Szilard also hoped to talk to Fermi: "I thought that if neutrons are in fact emitted in fission, this fact should be kept secret from the Germans. So I was very eager to contact Joliot and to contact Fermi, the two men who were most likely to think of this possibility." He had borrowed Wigner's apartment and had not yet left Princeton. "I got up one morning and wanted to go out. It was raining cats and dogs. I said, 'My God, I am going to catch cold!' Because at that time, the first years I was in America, each time I got wet I invariably caught a bad cold." He had to go out anyway. "I got wet and came home with a high fever, so I was not able to contact Fermi."

Fever or not, by January 25—Wednesday—Szilard had returned to New York, had seen the Hahn-Strassmann paper and was writing Lewis Strauss, whose patronage might now be more important than ever:

> I feel I ought to let you know of a very sensational new development in nuclear physics. In a paper . . . Hahn reports that he finds when bombarding uranium with neutrons the uranium breaking up. . . . This is entirely unexpected and exciting news for the average physicist. The Department of Physics at Princeton, where I spent the last few days, was like a stirred-up ant heap.
>
> Apart from the purely scientific interest there may be another aspect of this discovery, which so far does not seem to have caught the attention of those to whom I spoke. First of all it is obvious that the energy released in this new reaction must be very much higher than all previously known cases. . . .

> This in itself might make it possible to produce power by means of nuclear energy, but I do not think that this possibility is very exciting, for . . . the cost of investment would probably be too high to make the process worthwhile. . . .
>
> I see . . . possibilities in another direction. These might lead to large-scale production of energy and radioactive elements, unfortunately also perhaps to atomic bombs. This new discovery revives all the hopes and fears in this respect which I had in 1934 and 1935, and which I have as good as abandoned in the course of the last two years. At present I am running a high temperature and am therefore confined to my four walls, but perhaps I can tell you more about these new developments some other time.

The same day Fermi stepped into the office of John R. Dunning, a Columbia experimentalist whose specialty was neutrons, to propose an experiment. Dunning, his graduate student Herbert Anderson and others at Columbia had built a small cyclotron in the basement of Pupin Hall, the modern thirteen-story physics tower that faces downtown Manhattan from behind the library on the upper campus. A cyclotron was a potent source of neutrons; the two men talked about using it to perform an experiment similar to Frisch's experiment of January 13-14, of which they were as yet unaware. They discussed arrangements over lunch at the Columbia faculty club and afterward back at Pupin.

While Fermi was away from his desk Bohr arrived to tell him what he already knew. Finding an empty office, Bohr took the elevator to the basement, to the cyclotron area, where he turned up Herbert Anderson:

> He came right over and grabbed me by the shoulder. Bohr doesn't lecture you, he whispers in your ear. "Young man," he said, "let me explain to you about something new and exciting in physics." Then he told me about the splitting of the uranium nucleus and how naturally this fits in with the idea of the liquid drop. I was quite enchanted. Here was the great man himself, impressive in his bulk, sharing his excitement with me as if it were of the utmost importance for me to know what he had to say.

Bohr was en route to a conference in Washington on theoretical physics that would begin the next afternoon; he left to catch his train without seeing Fermi. As soon as Bohr was gone Anderson hunted up the Italian, who had returned to his office by now. "Before I had a chance to say anything," Anderson remembers, "he smiled in a friendly fashion and said, 'I think I know what you want to tell me. Let *me* explain it to you. . . .' I have to say that Fermi's explanation was even more dramatic than Bohr's."

Fermi helped Anderson and Dunning begin organizing the experiment he had discussed with Dunning earlier in the day. Anderson happened not long before to have built an ionization chamber and linear amplifier. "All we had to do was prepare a layer of uranium on one electrode and insert it into the chamber. That same afternoon we set up everything at the cyclotron. But the cyclotron was not working very well that day. Then I remembered some radon and beryllium which had been used as a source of neutrons in earlier experiments. It was a lucky thought." It came too late in the day; Fermi was also attending the Washington conference and had to leave. Anderson and Dunning closed up shop.

The Washington Conferences on Theoretical Physics, of which the 1939 meeting would be the fifth, were a George Gamow invention. He had stipulated their creation as a condition of his employment at George Washington University in 1934. He took Bohr's annual gathering in Copenhagen for a model; since there was no comparable assembly in the United States at the time, the Washington Conferences met with immediate success. At the instigation of Merle Tuve, Ernest Lawrence's boyhood friend and the driving force at the Department of Terrestrial Magnetism of the Carnegie Institution of Washington, the Carnegie Institution co-sponsored the conferences with GWU, though expenses were modest, for travel only, no more in total than five or six hundred dollars a year. People attended because they were interested. Edward Teller recalls the meetings as "in general small and exciting, thoroughly absorbing, and also a little tiring. Somehow, most of the running of the conferences Gamow left to me." The two men simply chose a topic and made up a list of invitees. Graduate students crowded in to listen. This year's topic was low-temperature physics.

Bohr sought out Gamow as soon as he arrived in Washington that evening. Gamow in turn called Teller: "Bohr has just come in. He has gone crazy. He says a neutron can split uranium." Teller thought of Fermi's experiments in Rome and the mess of radioactivities they produced and "suddenly understood the obvious." In Washington Fermi learned to his further disappointment from Bohr that Frisch was supposed to have done an experiment similar to the one left unfinished at Columbia. "Fermi . . . had no idea before that Frisch had made the experiment," Bohr wrote Margrethe a few days later. "I had no right to prevent others from experimentation, but I emphasized that Frisch had also spoken of an experiment in his notes. I said that it was all my fault that they all heard about Frisch and Meitner's explanation, and I earnestly asked them to wait [to make a public announcement] until I received a copy of Frisch's note to *Nature,* which I hoped would be waiting for me at Princeton [i.e., after the conference]." Fermi, understandably, seems to have argued against further delay.

Herbert Anderson returned to the basement of Pupin Hall that evening. He retrieved his neutron source. He calculated how many alpha particles the uranium oxide coated on a metal plate inside his ionization chamber would eject spontaneously in its normal process of radioactive decay: three thousand per minute. He calculated the probability of ten of those alphas appearing simultaneously to produce a spurious high-energy kick of the scanning beam of his oscilloscope: "practically never," he concluded in his laboratory notebook.

He set the neutron source beside the ionization chamber a little after 9 P.M. and began observing the effect on the oscilloscope. "Most kicks are due to .4 cm range α part[icles] [of approximately] .65 M[e]V," he noted. Then he saw what he was looking for: "Now large kicks which occur infrequently about 1 every 2 minutes." He counted them against the clock. In 60 minutes he had counted 33 large kicks. He removed the neutron source. "In 20 min" without a neutron source, he wrote, "0 counts." It was the first intentional observation of fission west of Copenhagen.

Dunning showed up later that evening, Anderson remembers, and "was very excited by the result I'd gotten." Anderson thought Dunning would telegraph Fermi immediately, but he seems not to have done so. Frisch, as he told Bohr later, had cabled no news of his confirming Copenhagen experiment because it seemed to him "just additional evidence of a discovery already made" and "cabling to you would have appeared unmodest to me." Dunning, despite his excitement at seeing the new phenomenon for himself, may have felt the same way.

Bohr woke to his dilemma. The conference would begin at two. As recently as three days previously he had written Frisch again, chiding him for not sending a copy of his and Meitner's *Nature* note. But he was less concerned now with that delay than he was with protecting the priority of Frisch's experiment, if any. Reluctantly he acceded to public announcement, stressing, he wrote Frisch afterward, "that no public account . . . could legitimately appear without mentioning your and your aunt's original interpretation of the Hahn results."

Fifty-one participants sat for a photograph in the course of the Fifth Washington Conference, and even a partial list of their names confirms the event's prestige. Otto Stern attended; Fermi; Bohr; Harold Urey of Columbia, who won the 1934 Nobel Prize in Chemistry for isolating a heavy form of hydrogen, deuterium, that carried a neutron in its nucleus; Gregory Breit, a waspish but inspired theoretician; Rabi; George Uhlenbeck, then at Columbia, who had been Paul Ehrenfest's assistant; Gamow; Teller; Hans Bethe down from Cornell; Léon Rosenfeld; Merle Tuve. Conspicuously absent was the Western crowd, probably because the two sponsoring institutions chose not to budget such long-distance travel.

Gamow opened the meeting by introducing Bohr. His news galvanized the room. A young physicist watching from the back saw an immediate application. Richard B. Roberts, Princeton-trained, worked with Tuve at the Department of Terrestrial Magnetism, the experimental section of the Carnegie Institution, located in a parklike setting in the Chevy Chase area of the capital. Roberts—thin, vigorous, with a strong jaw and wavy dark hair—still remembered the occasion vividly in 1979 in a draft autobiography:

> The Theo. Phys. Conference for 1939 was on the topic of low temperatures and I was not eager to attend. However, I went down to sit in the back row of the meeting. . . . Bohr and Fermi arrived and Bohr proceeded to reveal his news concerning the Hahn and Strassmann experiments. . . . He also told of Meitner's interpretation that the uranium had split. As usual he mumbled and rambled so there was little in his talk beyond the bare facts. Fermi then took over and gave his usual elegant presentation including all the implications.

Roberts noted in a letter to his father the Monday after the conference ended that "Fermi also . . . described an obvious experiment to test the theory"—Frisch's experiment, Fermi's, Dunning's and Anderson's experiment. "The remarkable thing is that this reaction results in 200 million volts of energy liberated and brings back the possibility of atomic power."

Bohr was calling the fission fragments "splitters." For the time being everyone borrowed that comical usage. Lawrence R. Hafstad, a longtime associate of Tuve, was sitting beside Roberts. When Fermi finished, the two men looked at each other, got up, left the meeting and lit out for the DTM. If "splitters" issued forth from uranium they intended to be among the first to see them.

In New York that day Szilard dragged himself to the nearest Western Union office and cabled the British Admiralty:

> KINDLY DISREGARD MY RECENT LETTER STOP WRITING

The secret patent had revived.

Naturwissenschaften reached Paris about January 16. One of Frédéric Joliot's associates recalls that "in a rather moving meeting [Joliot] made a report on this result to Madame Joliot and myself after having locked himself in for a few days and not talked to anybody." The Joliot-Curies were once again appalled to find they had barely missed a major discovery. In the next few days Joliot independently deduced the large energy release and considered the possibility of a chain reaction, as Szilard had thought he might. He tried to track down the neutrons from fission first, found that approach difficult, then set up an experiment somewhat like Frisch's. He detected fission fragments on January 26.

* * *

The newest building on the DTM grounds was the Atomic Physics Observatory, the working contents of which had just been brought on line two weeks before: a new 5 MV pressure Van de Graaff generator that Tuve, Roberts and their colleagues had built for $51,000 to extend their studies in the structure of the nucleus. The Van de Graaff was named for the Alabama-born physicist who invented it, but Tuve was the first—in 1932—to put it to practical use in experiment. It was essentially a monumental static-electricity generator, an insulated motor-driven pulley belt that picked up ions from discharge needles in its metal base, carried them up through an insulated support cylinder into a smooth metal storage sphere and deposited them on the sphere. As ions accumulated the sphere's voltage increased. The voltage could then be discharged as a spark—Van de Graaffs discharging lightning-bolt sparks have been staples of mad-scientist movies—or drawn off to power an accelerator tube. The new machine was built inside a pear-shaped pressure tank, as large as the tank of a water tower, that helped reduce accidental sparking.

When Tuve had first proposed the Van de Graaff to the zoning board of the prosperous Chevy Chase neighborhood the board had turned him down. Smashing atoms smacked of industrial process and the neighborhood had its property values to consider. Tuve noted the popularity of the Naval Observatory, across Connecticut Avenue a few miles west, and rechristened his project the Atomic Physics Observatory, which it was. As the APO it won approval.

Roberts and Hafstad chose to work with the APO. They had intended to use the old 1 MV Van de Graaff in the building next door to make neutrons for their splitter experiment, but that machine's ion-source filament was burned out. Although the APO's vacuum accelerator tube leaked, finding the leak looked to be less tedious than replacing the filament. In fact it needed two days. Hafstad went off Friday night on a ski weekend and another young Tuve protégé, R. C. Meyer, took his place.

Roberts' laboratory notebook entries summarize Saturday's work:

Sat 4:30 PM
Set up ionization chamber to try to detect

$U_{92}^{238} + n \rightarrow U_{92}^{239} \rightarrow Ba_{56}^{?} + Kr_{36}^{?}$

Neutrons from Li + D [accelerated deuterium nuclei bombarding lithium]
. . .
With uranium lined I.C. observed
α's [approximately] *1–2 mm and occasional 35 mm kicks* (*Ba + Kr?*)

The APO's target room was a small circular basement accessible down a steel ladder, a chilly kiva that smelled pleasantly of oil. As soon as Roberts saw the "tremendous pulses corresponding to very large energy release" he and Meyer ran every test they could think of. "We promptly tried the effect of paraffin (for slow neutrons) and then cadmium to remove the slow neutrons. We also tried all the other heavy elements available [to determine if they would split] and saw the same [i.e., fission] with thorium." Having made that original discovery (Frisch had made it independently in Copenhagen before them) they stopped to eat. "I told Tuve after supper and he immediately called Bohr and Fermi and they came out Saturday night."

Not only Bohr and Fermi came, in heavy, dark, pin-striped three-piece suits, Fermi swarthy with a day's growth of beard, but also Tuve; Rosenfeld; Teller; Erik Bohr, handsome in a heavy overcoat over a decorative Danish sweater; Gregory Breit, owlish in spectacles; and John A. Fleming, the conservative director of the DTM, who had the presence of mind to bring along a photographer. All except Teller posed in the target room with Meyer and Roberts for a historic photograph. The box of the ionization chamber in the foreground is stacked with disks of paraffin; Bohr holds the stub of an after-dinner cigar; Fermi's grin reveals the gap between his front teeth left by a baby tooth he shed late; Roberts looks into the camera weary but satisfied. Fermi had been amazed at the ionization pulses on the oscilloscope and had insisted they check for equipment malfunctions: he had never seen such pulses in Rome (they were captured by the aluminum foil Amaldi had wrapped around his uranium to block its alpha background). Bohr was still fretting. "I had to stand and look at the first [*sic*] experiment," he wrote Margrethe, "without knowing certainly if Frisch had done the same experiment and sent a note to *Nature*." Back at Princeton on Sunday he learned from other family letters that Frisch had. "There followed," Roberts concludes, "several days of excitement, press releases and phone calls."

Science reporter Thomas Henry had attended the conference; his story appeared in the *Washington Evening Star* on Saturday afternoon. The Associated Press picked it up. Shortened, it earned a place on an inside page of the Sunday *New York Times*. Dunning may have seen it there; he finally wired Fermi news that morning of the Columbia experiment. As Herbert Anderson remembers it, "Fermi . . . rushed back to Columbia and straightaway called me into his office. My notebook lists the experiments he felt we should do right away. The date was January 29, 1939." They had already agreed, says Anderson, that "I would teach him Americana, and he would teach me physics." Both lessons began in earnest.

The *San Francisco Chronicle* picked up the wire-service story. Luis W. Alvarez, Ernest Lawrence's tall, ice-blond protégé, a future Nobelist whose father was a prominent Mayo Clinic physician, read it at Berkeley sitting in a barber chair in Stevens Union having his hair cut. "So [I told] the barber to stop cutting my hair and I got right out of that barber chair and ran as fast as I could to the Radiation Lab . . . where my student Phil Abelson . . . had been [trying to identify] what transuranium elements were produced when neutrons hit uranium; he was so close to discovering fission that it was almost pitiful." Abelson still remembers the painful moment: "About 9:30 a.m. I heard the sound of running footsteps outside, and immediately afterward Alvarez burst into the laboratory. . . . When [he] told me the news, I almost went numb as I realized that I had come close but had missed a great discovery. . . . For nearly 24 hours I remained numb, not functioning very well. The next morning I was back to normal with a plan to proceed." By the end of the day Abelson found iodine as a decay product of tellurium from uranium irradiation, another way the nucleus could split (i.e., tellurium 52 + zirconium 40 = U 92).

Alvarez wired Gamow for details, learned of the Frisch experiment, then tracked down Oppenheimer:

> I remember telling Robert Oppenheimer that we were going to look for [ionization pulses from fission] and he said, "That's impossible" and gave a lot of theoretical reasons why fission couldn't really happen. When I invited him over to look at the oscilloscope later, when we saw the big pulses, I would say that in less than fifteen minutes Robert had decided that this was indeed a real effect and . . . he had decided that some neutrons would probably boil off in the reaction, and that you could make bombs and generate power, all inside of a few minutes. . . . It was amazing to see how rapidly his mind worked, and he came to the right conclusions.

The following Saturday Oppenheimer discussed the discovery in a letter to a friend at Caltech, outlining all the experiments Alvarez and others had accomplished during the week and speculating on applications:

> The U business is unbelievable. We first saw it in the papers, wired for more dope, and have had a lot of reports since. . . In how many ways does the U come apart? At random, as one might guess, or only in certain ways? And most of all, are there many neutrons that come off during the splitting, or from the excited pieces? If there are, then a 10 cm cube of U deuteride (one would need the D [deuterium, heavy hydrogen] to slow them without capture) should be quite something. What do you think? It is I think exciting, not in the rare way of positrons and mesotrons, but in a good honest practical way.

The next day, in a letter to George Uhlenbeck at Columbia, "quite something" became "might very well blow itself to hell." One of Oppenheimer's students, the American theoretical physicist Philip Morrison, recalls that "when fission was discovered, within perhaps a week there was on the blackboard in Robert Oppenheimer's office a drawing—a very bad, an execrable drawing—of a bomb."

Enrico Fermi made similar estimates. George Uhlenbeck, who shared an office with him in Pupin Hall, was there one day to overhear him. Fermi was standing at his panoramic office window high in the physics tower looking down the gray winter length of Manhattan Island, its streets alive as always with vendors and taxis and crowds. He cupped his hands as if he were holding a ball. "A little bomb like that," he said simply, for once not lightly mocking, "and it would all disappear."

PART TWO

A PECULIAR SOVEREIGNTY

The Manhattan District bore no relation to the industrial or social life of our country; it was a separate state, with its own airplanes and its own factories and its thousands of secrets. It had a peculiar sovereignty, one that could bring about the end, peacefully or violently, of all other sovereignties.

Herbert S. Marks

We must be curious to learn how such a set of objects—hundreds of power plants, thousands of bombs, tens of thousands of people massed in national establishments—can be traced back to a few people sitting at laboratory benches discussing the peculiar behavior of one type of atom.

Spencer R. Weart

10

Neutrons

At the end of January 1939, still ill with a feverish cold that had laid him low for more than a week but determined to prevent information on the possibility of a chain reaction in uranium from reaching physicists in Nazi Germany, Leo Szilard raised himself from his bed in the King's Crown Hotel on West 116th Street in Manhattan and went out into the New York winter to take counsel of his friend Isador Isaac Rabi. Rabi, no taller than Szilard but always a trimmer and cooler man, who would be the 1944 Nobel laureate in physics, was born in Galicia in 1898 and emigrated to the United States with his family as a small child. Yiddish had been his first language; he grew up on New York's Lower East Side, where his father worked in a sweatshop making women's blouses until he accumulated enough savings to open a grocery store. Because his family was Orthodox and fundamentalist in its Judaism, Rabi had not known that the earth revolved around the sun until he read it in a library book. A frightening vision of the vast yellow face of the rising moon seen as a child down a New York street had begun his turn toward science, as had his childhood reading of the cosmological first verses of the Book of Genesis. He was a man of abrupt and honest bluntness who did not easily tolerate fools. One reason for his impatience was certainly that it guarded from harm his deeply emotional commitment to science: he thought physics "infinite," he told a biographer in late middle age, and he was disappointed that young physicists of that later day, intent on technique, seemed to miss what he had found, "the mystery of it: how very different it is from what you can see, and how profound nature is."

Szilard learned from Rabi that Enrico Fermi had discussed the possibility of a chain reaction in his public presentation at the Fifth Washington Conference on Theoretical Physics that had met the week before. Szilard adjourned to Fermi's office but did not find him there. He went back to Rabi and asked him to talk to Fermi "and say that these things ought to be kept secret." Rabi agreed and Szilard returned to his sickbed.

He was recovering; a day or two later he again sought Rabi out:

> I said to him: "Did you talk to Fermi?" Rabi said, "Yes, I did." I said, "What did Fermi say?" Rabi said, "Fermi said 'Nuts!' " So I said, "Why did he say 'Nuts!'?" and Rabi said, "Well, I don't know, but he is in and we can ask him." So we went over to Fermi's office, and Rabi said to Fermi, "Look, Fermi, I told you what Szilard thought and you said 'Nuts!' and Szilard wants to know why you said 'Nuts!' " So Fermi said, "Well . . . there is the remote possibility that neutrons may be emitted in the fission of uranium and then of course perhaps a chain reaction can be made." Rabi said, "What do you mean by 'remote possibility'?" and Fermi said, "Well, ten per cent." Rabi said, "Ten per cent is not a remote possibility if it means that we may die of it. If I have pneumonia and the doctor tells me that there is a remote possibility that I might die, and it's ten percent, I get excited about it."

But despite Fermi's facility with American slang and Rabi's with probabilities Fermi and Szilard were unable to agree. For the time being they left the discussion there.

Fermi was not misleading Szilard. It was easy to estimate the explosive force of a quantity of uranium, as Fermi would do standing at his office window overlooking Manhattan, if fission proceeded automatically from mere assembly of the material; even journalists had managed that simple calculation. But such obviously was not the case for uranium in its natural form, or the substance would long ago have ceased to exist on earth. However energetically interesting a reaction, fission by itself was merely a laboratory curiosity. Only if it released secondary neutrons, and those in sufficient quantity to initiate and sustain a chain reaction, would it serve for anything more. "Nothing known then," writes Herbert Anderson, Fermi's young partner in experiment, "guaranteed the emission of neutrons. Neutron emission had to be observed experimentally and measured quantitatively." No such work had yet been done. It was, in fact, the new work Fermi had proposed to Anderson immediately upon returning from Washington. Which meant to Fermi that talk of developing fission into a weapon of war was absurdly premature.

Many years later Szilard succinctly summed up the difference between his position and Fermi's. "From the very beginning the line was drawn," he said. ". . . Fermi thought that the conservative thing was to play down the possibility that [a chain reaction] may happen, and I thought the conservative thing was to assume that it would happen and take all the necessary precautions."

Once he was well again Szilard had catching up to do. He cabled Oxford to ship him the cylinder of beryllium he had left behind at the Clarendon when he came to the United States, preliminary to mounting a neutron-emission experiment of his own. At Lewis Strauss's request he spent a day with the financier discussing the possible consequences of fission, which included, Strauss notes wistfully in his memoirs, making "the performance of our surge generator in Pasadena insignificant. The device had just been completed." The surge generator in which he had invested some tens of thousands of dollars had been cut down to size. The Strausses were scheduled to leave that evening by overnight train for a Palm Beach vacation; Szilard rode along as far as Washington to continue the discussion. He was massaging his patron: he needed to rent radium to combine with his beryllium to make a neutron source and hoped Strauss might be persuaded to support the expense.

Arriving late at Union Station in Washington, Szilard called the Edward Tellers. They were still recovering from the work of hosting the Washington Conference. Mici Teller protested the surprise visit, her husband remembers: "No! We are both much too tired. He must go to a hotel." They met Szilard anyway, whereupon to Teller's surprise Mici invited their countryman to stay with them:

> We drove to our home, and I showed Szilard to his room. He felt the bed suspiciously, then turned to me suddenly and said: "Is there a hotel nearby?" There was, and he continued: "Good! I have just remembered sleeping in this bed before. It is much too hard."
>
> But before he left, he sat on the edge of the hard bed and talked excitedly: "You heard Bohr on fission?"
>
> "Yes," I replied.
>
> Szilard continued: "You know what that means!"

What it meant to Szilard, Teller remembers, was that "Hitler's success could depend on it."

The next day Szilard discussed his plan for voluntary secrecy with Teller, then entrained for Princeton to pursue the same subject with Eugene Wigner, who was still drydocked in the infirmary with jaundice. Szilard was thus present in Princeton when yet another momentous insight struck Niels Bohr.

Bohr and Léon Rosenfeld were staying at the Nassau Club, the Princeton faculty center. On Sunday, February 5, George Placzek joined them at breakfast in the club dining room. The Bohemian theoretician had arrived in Princeton from Copenhagen the night before, another refugee from Nazi persecution. Talk turned to fission. "It is a relief that we are now rid of those transuranians," Rosenfeld remembers Bohr saying, referring to the confusing radioactivities Hahn, Meitner and Strassmann had found in the late 1930s that Bohr assumed could now be attributed to existing lighter elements—barium, lanthanum and the many other fission products researchers were beginning to identify.

Placzek was skeptical. "The situation is more confused than ever," he told Bohr. He began then to specify the sources of confusion. He was directly challenging the relevance of Bohr's liquid-drop model of the nucleus. The Danish laureate paid attention.

Physicists use a convenient measurement they call a "cross section" to indicate the probability that a particular nuclear reaction will or will not happen. The theoretical physicist Rudolf Peierls once explained the measurement with this analogy:

> For example, if I throw a ball at a glass window one square foot in area, there may be one chance in ten that the window will break, and nine chances in ten that the ball will just bounce. In the physicists' language, this particular window, for a ball thrown in this particular way, has a "disintegration cross-section" of 1/10 square foot and an "elastic cross-section" of 9/10 square foot.

Cross sections can be measured for many different nuclear reactions, and they are expressed not in square feet but in minute fractions of square centimeters, customarily 10^{-24}, because the diminutive nucleus is the target window of Peierls' analogy. The cross section that concerned Placzek in his discussion with Bohr was the capture cross section: the probability that a nucleus will capture an approaching neutron. In terms of Peierls' analogy, the capture cross section measures the chance that the window might be open when the ball arrives and might therefore admit the ball into the living room.

Nuclei capture neutrons of certain energies more frequently than they capture neutrons of other energies. They are naturally tuned, so to speak, to certain specific energy levels—as if Peierls' window opened more easily to balls thrown at only certain speeds. This phenomenon is known as resonance. The confusion Placzek delighted in reporting concerned a resonance in the capture cross sections of uranium and thorium.

Placzek pointed out that uranium and thorium both exhibit a capture resonance for neutrons with medium-range energies of about 25 electron volts. That meant, first of all, that although fission was one behavior uranium could exhibit under neutron bombardment, capture and subsequent transmutation continued to be another. Bohr was not ever to be rid of those inconvenient "transuranians." Some of them were real.

If a neutron penetrated a uranium nucleus, for example, the result might be fission. But if the neutron happened to be traveling at the appropriate energy when it penetrated—somewhere around 25 eV—the nucleus would probably capture it without fissioning. Beta decay would follow, increasing the nuclear charge by one unit; the result should be a new, as-yet-unnamed transuranic element of atomic number 93. That was one of Placzek's points. It would prove in time to be crucial.

The other source of confusion was more straightforward. It was also more immediately relevant to the question of how to harness nuclear energy. It concerned differences between uranium and thorium.

Thorium, element 90, a soft, heavy, lustrous, silver-white metal, was first isolated by the celebrated Swedish chemist Jöns Jakob Berzelius in 1828. Berzelius named the new element after Thor, the Norse god of thunder. Its oxide found commercial use beginning in the late nineteenth century as the primary component of the fragile woven mantles of gas lanterns: heat incandesces it a brilliant white. Because it is mildly radioactive, and radioactivity was once considered tonic, thorium was also for some years incorporated into a popular German toothpaste, Doramad. Auer, the company that made German gas mantles, also made the toothpaste. Hahn, Meitner and Strassmann, the Joliot-Curies and others had regularly studied thorium alongside uranium. Its behavior was often similar. Otto Frisch had first demonstrated that it fissioned. He bombarded it next after uranium in the course of his January experiment in Copenhagen, the experiment he had discussed with Bohr after he returned from Kungälv and Bohr had worked so hard in the United States to protect.

Frisch was then also the first to notice that the fission characteristics of thorium differed from those of uranium. Thorium did not respond to the magic of paraffin; it was unaffected by slow neutrons. Richard B. Roberts and his colleagues at the Department of Terrestrial Magnetism of the Carnegie Institution of Washington had just independently confirmed and extended Frisch's findings. With their 5 million volt Van de Graaff they could generate neutrons of several different, known energies. Continuing their experiments after their Saturday-night show for the Washington Conference group, they had compared uranium and thorium fission responses at varying energies as Frisch with his single neutron source could not. They found to their surprise (Frisch's paper had not yet appeared in *Nature*) that while both uranium and thorium fissioned under bombardment by fast neutrons, only uranium fissioned under bombardment by slow neutrons. Some energy between 0.5 MeV and 2.5 MeV marked a lower threshold for fast-neutron fission for both elements. (Bohr and John Wheeler, beginning work at Princeton on fission theory, had estimated the threshold energy to be about 1 MeV.) The slow neutrons that also fissioned uranium were effective at far lower energies. "From these comparisons," the DTM group concluded in a February paper, "it appears that the uranium fissions are produced by different processes for fast and slow neutrons."

Why, Placzek now prodded Bohr, should both uranium and thorium have similar capture resonances and similar fast-neutron thresholds but different responses to slow neutrons? If the liquid-drop model had any validity at all, the difference made no sense.

Bohr abruptly saw why and was struck dumb. Not to lose what he had only barely grasped, oblivious to courtesy, he pushed back his chair and strode from the room and from the club. Rosenfeld hurried to follow. "Taking a hasty leave of Placzek, I joined Bohr, who was walking silently, lost in deep meditation, which I was careful not to disturb." The two men tramped speechless through the snow across the Princeton campus to Fine Hall, the Neo-Gothic brick building where the Institute for Advanced Study was then lodged. They went in to Bohr's office, borrowed from Albert Einstein. It was spacious, with leaded windows, a fireplace, a large blackboard, an Oriental rug to warm the floor. No peripatetic like Bohr, Einstein had judged it too large and moved into a small secretarial annex nearby.

"As soon as we entered the office," Rosenfeld remembers, "[Bohr] rushed to the blackboard, telling me: 'Now listen: I have it all.' And he started—again without uttering a word—drawing graphs on the blackboard."

The first graph Bohr drew looked like this:

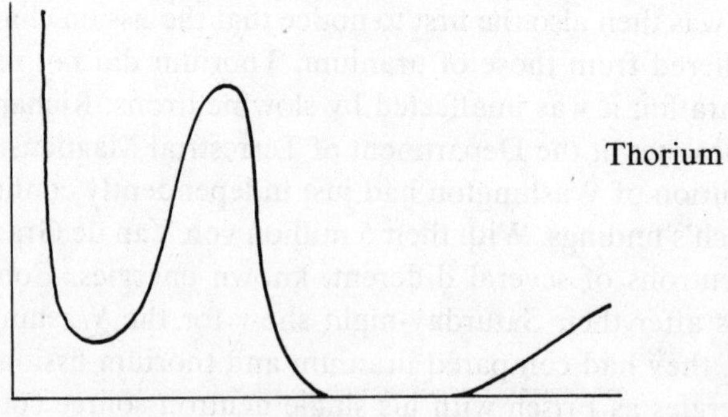

The horizontal axis plotted neutron energy left to right—low to high, slow to fast. The vertical axis charted cross sections—the probability of a particular nuclear reaction—and served a double purpose. The lazy S that filled most of the frame represented thorium's cross section for capture at different neutron energies, the steep central peak demonstrating the 25 eV resonance in the middle range. The tail that waved from the horizontal axis on the right side represented a different thorium cross section: its cross section for fission beginning at that high 1 MeV threshold. What Bohr had drawn was thus a visualization of thorium's changing response to bombardment by neutrons of increasing energy.

Bohr moved to the next section of blackboard and drew a second graph. He labeled it with the mass number of the isotope most plentiful in natural uranium. "He wrote the mass number 238 with very large figures," Rosenfeld says; "he broke several pieces of chalk in the process." Bohr's urgency marked the point of his insight. The second graph looked exactly like the first:

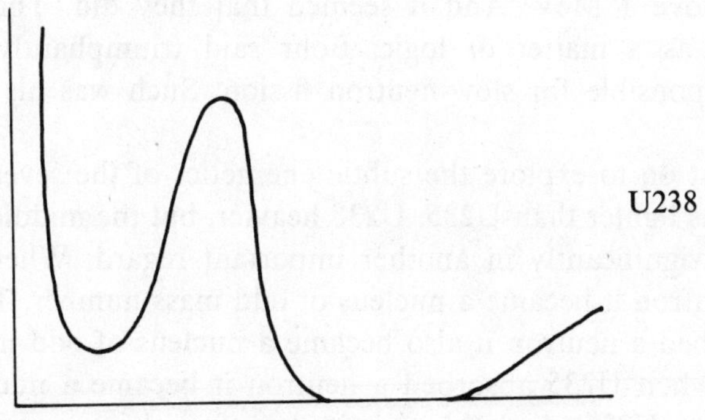

But a third graph was coming.

Francis Aston had found only U238 when he first passed uranium through his mass spectrograph at the Cavendish. In 1935, using a more powerful instrument, physicist Arthur Jeffrey Dempster of the University of Chicago detected a second, lighter isotope. "It was found," Dempster announced in a lecture, "that a few seconds' exposure was sufficient for the main component at 238 reported by Dr. Aston, but on long exposures a faint companion of mass number 235 was also present." Three years later a gifted Harvard postdoctoral fellow named Alfred Otto Carl Nier, the son of working-class German emigrants to Minnesota, measured the ratio of U235 to U238 in natural uranium as 1:139, which meant that U235 was present to the extent of about 0.7 percent. By contrast, thorium in its natural form is essentially all one isotope, Th232. And that natural difference in the composition of the two elements was the clue that set Bohr off. He drew his third graph. It depicted one cross section, not two:

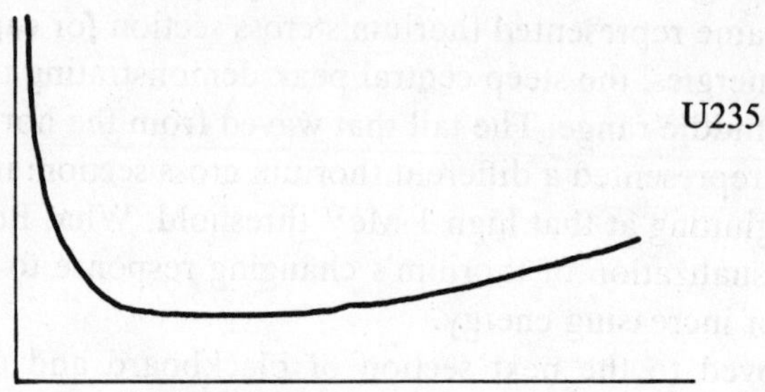

Having made a hard copy of his abrupt vision, Bohr was finally ready to explain himself.

Both thorium and U238 could be expected on theoretical grounds to behave similarly, he pointed out to Rosenfeld: to fission only with fast neutrons above 1 MeV. And it seemed that they did. That left U235. It followed as a matter of logic, Bohr said triumphantly, that U235 must be responsible for slow-neutron fission. Such was his essential insight.

He went on to explore the subtle energetics of the several reactions. Thorium was lighter than U235, U238 heavier, but the middle isotope differed more significantly in another important regard. When Th232 absorbed a neutron it became a nucleus of odd mass number, Th233. When U238 absorbed a neutron it also became a nucleus of odd mass number, U239. But when U235 absorbed a neutron it became a nucleus of even mass number, U236. And the vicissitudes of nuclear rearrangement are such, as Fermi would explain one day in a lecture, that "changing from an odd number of neutrons to an even number of neutrons released one or two MeV." Which meant that U235 had an inherent energetic advantage over its two competitors: it accrued energy toward fission simply by virtue of its change of mass; they did not.

Lise Meitner and Otto Frisch had realized in Kungälv that a certain amount of energy was necessary to agitate the nucleus to fission, but they had not considered in detail the energetics of that input. They were distracted by the enormous 200 MeV output. In fact, the uranium nucleus required an input of about 6 MeV to fission. That much energy was necessary to roil the nucleus to the point where it elongated and broke apart. The absorption of any neutron, regardless of its velocity, made available a binding energy of about 5.3 MeV. But that left U238 about 1 MeV short, which is why it needed fast neutrons of at least that threshold energy before it could fission.

U235 also earned 5.3 MeV when it absorbed a neutron. But it won Fermi's "one or two MeV" in addition simply by adjusting from an odd to an even mass. That put its total above 6 MeV. So *any neutron at all* would fission U235—slow, fast or in between. Which was what Bohr's third graph demonstrated: the probably continuous fission cross section of U235. From slow neutrons on the left only a fraction of an electron volt above zero energy, to fast neutrons on the right above 1 MeV that would also fission U238, any neutron an atom of U235 encountered would agitate it to fission. Natural uranium masked U235's continuous fissibility; the more abundant U238 captured most of the neutrons. Only by slowing the neutrons with paraffin below the U238 capture resonance at 25 eV had experimenters like Hahn, Strassmann and Frisch been able to coax the highly fissionable U235 out of hiding. In a burst of insight Bohr had answered Placzek's objections and replenished his liquid drop.

In January Bohr had produced a 700-word paper in three days to protect his European colleagues' priorities. Now, in his eagerness to spread the news of U235's special role in fission, he produced an 1,800-word paper in two days, mailing it to the *Physical Review* on February 7. "Resonance in uranium and thorium disintegrations and the phenomenon of nuclear fission" was nevertheless written with care, more care than it received in the reading. Everyone understood its basic hypothesis—that U235, not U238, is responsible for slow-neutron fission in uranium—though not everyone concurred without the confirmation of experiment. But probably because, as Fermi recalled, isotopes at that time "were considered almost magically inseparable," everyone overlooked its further implications. Szilard explained to Lewis Strauss that month that "slow neutrons seem to split a uranium isotope which is present in an abundance of about 1% in uranium." Richard Roberts at the DTM, in a 1940 draft report of considerable significance, asserted that "Bohr . . . ascribed the [slow] neutron reaction to U235 and the fast neutron reaction to U238." Roberts' misstatement was probably no more than a rough first approximation that he would have corrected in a polished report. Szilard's and Roberts' comments illustrate, however, that the slow-neutron fission of U235 preoccupied the physicists at first to the exclusion of a more ominous potentiality.

Bohr acknowledged it indirectly in his paper for the *Physical Review.* The slow-neutron fission of U235 occupied the foreground of his discussion because it explained the puzzling difference between uranium and thorium. But Bohr also considered U235's behavior under fast-neutron bombardment. "For fast neutrons," he wrote near the end of the paper, ". . .because of the scarcity of the isotope concerned, the fission yields will be much smaller than those obtained from neutron impacts on the abundant isotope." The statement implies but does not ask a pregnant question: what would the yields be for fast neutrons if U235 could be separated from U238?

The latest incarnation of Orso Corbino's garden fish pond in Rome was a tank of water three feet wide and three feet deep that Fermi and Anderson set up that winter in the basement of Pupin Hall. They planned to insert a radon-beryllium neutron source into the center of a five-inch spherical bulb and suspend the bulb in the middle of the tank. Neutrons from the beryllium would then diffuse through the surrounding water, which would slow them down. The neutrons would induce a characteristic 44-second half-life in strips of rhodium foil, Fermi's favorite neutron detector, set at various distances away from the bulb. Once he established a baseline of neutron activity using the Rn + Be source alone, Fermi intended to pack uranium oxide into the bulb around the source and make a second series of measurements. If more neutrons turned up in the water tank with uranium than without, he could deduce that uranium produced secondary neutrons when it fissioned and could roughly estimate their number. One neutron out for each neutron in was not enough to sustain a chain reaction, since inevitably some would be captured and others drift away: it needed something more than one secondary for each primary, preferably at least two.

Upstairs on the seventh floor Szilard discovered a different experiment in progress. Walter Zinn, a tall, blond Canadian postdoctoral research associate who taught at City College, was bombarding uranium with 2.5 MeV neutrons from a small accelerator. He had reasoned in terms of neutron energy rather than quantity; he was trying to demonstrate secondary neutron production by looking for neutrons faster than the 2.5 MeV's he supplied. So far he had managed only inconclusive results.

"Szilard watched my experiment with great interest," Zinn recalls, "and then suggested that perhaps it would be more successful if lower energy neutrons were available. I said, 'That's fine, but where do you get them?' Leo said, 'Just leave it to me, I'll get them.' "

Szilard meant to help Zinn, but he also coveted Zinn's ionization chamber. "All we needed to do," he said later, "was to get a gram of radium, get a block of beryllium, expose a piece of uranium to the neutrons which come from the beryllium, and then see by means of the ionization chamber which Zinn had built whether fast neutrons were emitted in the process. Such an experiment need not take more than an hour or two to perform, once the equipment has been built and if you have the neutron source. But of course we had no radium."

The problem was still money. The Radium Chemical Company of New York and Chicago, a subsidiary of the Union Minière du Haut-Katanga of Belgium, the dominant source of world radium supplies, was willing to rent a gram of radium for a minimum of three months for $125 a month. Szilard wrote Lewis Strauss at his Virginia farm on February 13 "to see whether you could sanction the expenditures" and presciently briefed the financier on the meaning of the latest developments. The letter's crucial paragraph addresses Bohr's new hypothesis that U235 is responsible for slow-neutron fission in natural uranium:

> If this isotope could be used for maintaining chain reactions, it would have to be separated from the bulk of uranium. This, no doubt, would be done if necessary, but it might take five to ten years before it can be done on a technical scale. Should small scale experiments show that the thorium and the bulk of uranium would not work, but the rare isotope of uranium would, we would have the task immediately to attack the question of concentrating the rare isotope of uranium.*

Strauss's surge-generator losses had inoculated him against further investment in the nuclear enterprise. He wanted to know, Szilard says, "just how sure I was that this would work." Since Szilard could offer no guarantees, Strauss offered no support. Szilard turned then to Benjamin Liebowitz. "He was not poor but he was not exactly wealthy. . . . I told him what this was all about, and he said, 'How much money do you need?' I said, 'Well, I'd like to borrow $2,000.' He took out his checkbook, he wrote out a check, I cashed the check, I rented the . . . radium, and in the meantime the beryllium block arrived from England."

The cylinder of beryllium, which Walter Zinn thought "a strange and unique object" and took for proof of Szilard's magic ways, arrived on February 18. The same day Szilard heard from Teller about significant work in

*The distinction between U235 and U238 had already fired a debate. "Fermi and a number of others," says John Dunning, "had considerable doubts about U-235 or even disagreed—they thought it was U-238 [that was responsible for slow-neutron fission]." The disagreement incensed Bohr, who told Léon Rosenfeld he was "outraged" that Fermi should question the logic of his argument that thorium and U238 stood on one side and U235 on the other. "It was both the strength and the weakness of Fermi," writes Rosenfeld, "to be so intent on following his own lines of thought that he was impervious to any outside influence. . . . He fancied there could be a different interpretation of the evidence discussed by Bohr, and that only experiment could decide." Dunning, on the other hand, "immediately accepted Bohr's argument." The important outcome was that Dunning began to think of isotope separation, while Fermi continued to pursue the possibility of a chain reaction in natural uranium. With unusual and uncharacteristically Fermian conservatism, so did Szilard.

Washington at the DTM. Richard Roberts and R. C. Meyer were preparing a letter to the *Physical Review* reporting the discovery of delayed neutrons from fission. These were not the instantaneous secondary neutrons the Columbia researchers were seeking, but they did confirm that the fission fragments had neutrons to spare and would give them up spontaneously.

The general excitement Teller found at the busy DTM laboratories impressed him more:

> As soon as I began taking interest in uranium, sharp discussion started on the practical significance. Tuve, Hafstad, and Roberts are entirely aware of what is involved. They also know of Fermi's experiments. Of course, I didn't say anything. The above-mentioned letter [to the *Physical Review*] cannot cause any harm. . . .
>
> I do not know their detailed plans, but I believe that urgent action [to maintain secrecy] is required. Very many people have discovered already what is involved. Those in Washington would like to persuade the Carnegie Institution that it should provide more money for U-research in view of the practical significance of the matter. . . . But right now this has no reality unless the [Carnegie] leadership becomes more interested than it has been so far. . . .
>
> I repeat that there is a chain-reaction mood in Washington. I only had to say "uranium" and then could listen for two hours to their thoughts.

The president of the Carnegie Institution was a New England Yankee, the grandson of two sea captains, an electrical engineer, inventor and former dean of the school of engineering at the Massachusetts Institute of Technology named Vannevar Bush. If Bush was initially less willing to invest in chain-reaction experiments than Teller would have liked him to be, he kept good company; neither Ernest Lawrence at Berkeley nor Otto Hahn in Dahlem nor Lise Meitner, visiting Copenhagen that February to work with Otto Frisch, chose to pursue moonshine. Only Columbia and Paris mounted early experiments, though the DTM would soon follow the Columbia lead.

Frédéric Joliot and two colleagues, a cultivated Austrian named Hans von Halban and a huge, keen Russian named Lew Kowarski, began an experiment similar to Fermi's the last week in February to identify secondary neutrons from fission. They also used a tank of water with a central neutron source but dissolved their uranium in the water rather than packing it around the source. More important to their priority of research, they had immediate access to the Radium Institute's ample radium supply.

Because Fermi's neutron source relied on radon rather than radium it induced an ambiguity into his experiment that Szilard caught and called to his attention: radon ejected much faster neutrons from beryllium than did radium; at least part of any increase in neutrons Fermi found in his tank might therefore result not from fission but from another, competing reaction in beryllium. Fermi thought the ambiguity trivial, but agreed, as Zinn had before, to repeat the experiment using a radium-beryllium source. Szilard generously offered his. But the radium to energize it was not yet in hand; Szilard was still negotiating its rental because his lack of official affiliation made the Radium Chemical Company nervous.

He got his radium, two grams sealed in a small brass capsule, early in March, after he arranged admission to the Columbia laboratories for three months as a guest researcher. He and Zinn immediately set up their experiment. They made an ingenious nest, like Chinese boxes, of its various components: a large cake of paraffin wax, the beryllium cylinder set at the bottom of a blind hole in the paraffin, the radium capsule fitted into the beryllium cylinder; resting on the beryllium, inside the paraffin, a box lined with neutron-absorbing cadmium filled with uranium oxide; pushed into that box, but shielded from the radium's gamma radiation by a lead plug, the ionization tube itself, which connected to an oscilloscope. With this arrangement, says Szilard, they could measure the flux of neutrons from the uranium with and without the cadmium shield:

> Everything was ready and all we had to do was to turn a switch, lean back, and watch the screen of a television tube. If flashes of light appeared on the screen, that would mean that neutrons were emitted in the fission process of uranium and this in turn would mean that the large-scale liberation of atomic energy was just around the corner. We turned the switch and saw the flashes. We watched them for a little while and then we switched everything off and went home.

They had made a rough estimate of neutron production: "We find the number of neutrons emitted per fission to be about two." With radium available merely by picking up the phone, the French team a week earlier had found "more than one neutron . . . produced for each neutron absorbed." Fermi and Anderson estimated "a yield of about two neutrons per each neutron captured." Szilard immediately alerted Wigner and Teller. Teller remembers the moment well:

> I was at my piano, attempting with the collaboration of a friend and his violin to make Mozart sound like Mozart, when the telephone rang. It was Szilard, calling from New York. He spoke to me in Hungarian, and he said only one thing: "I have found the neutrons."

Szilard also wired Lewis Strauss:

> PERFORMED TODAY PROPOSED EXPERIMENT WITH BERYLLIUM BLOCK WITH STRIKING RESULT. VERY LARGE NEUTRON EMISSION FOUND. ESTIMATE CHANCES FOR REACTION NOW ABOVE 50%.

Szilard had known what the neutrons would mean since the day he crossed the street in Bloomsbury: the shape of things to come. "That night," he recalled later, "there was very little doubt in my mind that the world was headed for grief."

Though he was still recovering from jaundice, Eugene Wigner responded vigorously to Szilard's disturbing news while a storm of betrayal broke over Central Europe. Hitler ordered the President and the Foreign Minister of Czechoslovakia to Berlin on March 14 and threatened to bomb Prague to rubble unless they surrendered their country. With the Nazi leader's encouragement the Slovaks formally seceded from the republic that day. Ruthenia, Czechoslovakia's narrow eastern extension along the Carpathians, also claimed independence as Carpatho-Ukraine, an exercise in grave-robbing abruptly terminated the following morning when the fascist Hungary of Admiral Horthy invaded the new nation with German endorsement. Hitler flew in triumph to Prague. On March 16 he decreed what was left of Czechoslovakia—Bohemia and Moravia—to be a German protectorate. The country that France and Great Britain had abandoned at Munich was partitioned without resistance.

Wigner caught the train to New York. On the morning of March 16 he met with Szilard, Fermi and George Pegram in Pegram's office. Since at least the end of January Szilard had been promoting a new version of his Bund—he called it the Association for Scientific Collaboration—to monitor research, collect and disburse funds and maintain secrecy, a civilian organization that might guide the development of atomic energy. He had discussed it with Lewis Strauss on the train to Washington, with Teller after the night of the hard bed, with Wigner in Princeton the weekend Bohr drew his graphs. As far as Wigner was concerned, the time for such amateurism was over. He "strongly appealed to us," says Szilard, "immediately to inform the United States government of these discoveries." It was "such a serious business that we could not assume responsibility for handling it."

At sixty-three George Braxton Pegram was a generation older than the two Hungarians and the Italian who debated in his office that morning. A South Carolinian who had earned his Ph.D. from Columbia in 1903 working with thorium, he had studied under Max Planck at the University of Berlin and corresponded with Ernest Rutherford when Rutherford was still progressing in fruitful exile at McGill. Pegram was tall and athletic, a champion at tennis well into his sixties, a canoeist when young who enjoyed paddling and sailing an eighteen-foot sponson around Manhattan Island. His interest in radioactivity may have been aroused by his father, a chemistry professor; "probably the most important problem before the physicist today," the senior Pegram told the North Carolina Academy of Sciences in 1911, "is that of making the enormous energy [within the atom] available for the world's work." The next year, as an associate professor of physics at Columbia, Pegram had written Albert Einstein encouraging him to come to New York to lecture on relativity theory. Pegram had brought Rabi and Fermi to Columbia, building the university's international reputation for nuclear research. He was gray now, with thinning hair, wire-rimmed glasses, protuberant ears, a strong, square, wide-chinned jaw. Radioactivity intrigued him still, but a university dean's well-worn conservatism counseled him to caution.

He knew someone in Washington, he told Wigner: Charles Edison, Undersecretary of the Navy. Wigner insisted Pegram immediately call the man. Pegram was willing to do so, but first the group should discuss logistics. Who would carry the news? Fermi was traveling to Washington that afternoon to lecture in the evening to a group of physicists; he could meet with the Navy the next day. His Nobel Prize should give him exceptional credibility. Pegram called Washington. Edison was unavailable; his office directed Pegram to Admiral Stanford C. Hooper, technical assistant to the Chief of Naval Operations. Hooper agreed to hear Fermi out. Pegram's call was the first direct contact between the physicists of nuclear fission and the United States government.

The next topic on the morning's agenda was secrecy. Fermi and Szilard had both written reports on their secondary-neutron experiments and were ready to send them to the *Physical Review.* With Pegram's concurrence they decided to go ahead and mail the reports to the *Review,* to establish priority, but to ask the editor to delay publishing them until the secrecy issue could be resolved. Both papers went off that day.

Pegram prepared a letter of introduction for Fermi to carry along to his appointment. It stated a hesitant case dense with hypotheticals:

> Experiments in the physics laboratory at Columbia University reveal that conditions may be found under which the chemical element uranium may be able to liberate its large excess of atomic energy, and that this might mean the possibility that uranium might be used as an explosive that would liberate a million times as much energy per pound as any known explosive. My own feeling is that the probabilities are against this, but my colleagues and I think that the bare possibility should not be disregarded.

Thus lightly armed, Fermi departed to engage the Navy.

The debate was hardly ended, nor Wigner's long day done. He returned to Princeton with Szilard in tow for an important meeting with Niels Bohr. It had been planned in advance; John Wheeler and Léon Rosenfeld would attend and Teller was making a special trip up from Washington. If Bohr could be convinced to swing his prestige behind secrecy, the campaign to isolate German nuclear physics research might work.

They met in the evening in Wigner's office. "Szilard outlined the Columbia data," Wheeler reports, "and the preliminary indications from it that at least two secondary neutrons emerge from each neutron-induced fission. Did this not mean that a nuclear explosive was certainly possible?" Not necessarily, Bohr countered. "We tried to convince him," Teller writes, "that we should go ahead with fission research but we should not publish the results. We should keep the results secret, lest the Nazis learn of them and produce nuclear explosions first. Bohr insisted that we would never succeed in producing nuclear energy and he also insisted that secrecy must never be introduced into physics."

Bohr's skepticism, says Wheeler, concerned "the enormous difficulty of separating the necessary quantities of U235." Fermi noted in a later lecture that "it was not very clear [in 1939] that the job of separating large amounts of uranium 235 was one that could be taken seriously." At the Princeton meeting, Teller remembers, Bohr insisted that "it can never be done unless you turn the United States into one huge factory."

More crucial for Bohr was the issue of secrecy. He had worked for decades to shape physics into an international community, a model within its limited franchise of what a peaceful, politically united world might be. Openness was its fragile, essential charter, an operational necessity, as freedom of speech is an operational necessity to a democracy. Complete openness enforced absolute honesty: the scientist reported *all* his results, favorable and unfavorable, where all could read them, making possible the ongoing correction of error. Secrecy would revoke that charter and subordinate science as a political system—Polanyi's "republic"—to the anarchic competition of the nation-states. No one was more anguished than Bohr by the menace of Nazi Germany; Laura Fermi remembers of this period, "two months after his landing in the United States," that "he spoke about the doom of Europe in increasingly apocalyptic terms, and his face was that of a man haunted by one idea." If U235 could be separated easily from U238, that misfortune might be cause for temporary compromise with principle in the interest of survival. Bohr thought the technology looked not even remotely accessible. The meeting dragged on inconclusively past midnight.

The next afternoon Fermi turned up at the Navy Department on

Constitution Avenue for his appointment with Admiral Hooper. He had probably planned a conservative presentation. The contempt of the desk officer who went in to announce him to the admiral encouraged that approach. "There's a wop outside," Fermi overheard the man say. So much for the authority of the Nobel Prize.

In what Lewis Strauss, by now a Navy volunteer, calls "a ramshackle old board room" Hooper assembled an audience of naval officers, officers from the Army's Bureau of Ordnance and two civilian scientists attached to the Naval Research Laboratory. One of the civilians, a bluff physicist named Ross Gunn, had watched Richard Roberts demonstrate fission in the target room of the 5 MV Van de Graaff at the DTM not long after Fermi passed through at the time of the Fifth Washington Conference. Gunn worked on submarine propulsion; he was eager to learn more about an energy source that burned no oxygen.

Fermi led his auditors through an hour of neutron physics. If the notes of one of the participants, a naval officer, are comprehensive, Fermi emphasized his water-tank measurements rather than Szilard's more direct ionization-chamber work. New experiments in preparation might confirm a chain reaction, Fermi explained. The problem then would be to assemble a sufficiently large mass of uranium to capture and use the secondary neutrons before they escaped through the surface of the material.

The officer taking notes interrupted. What might be the size of this mass? Would it fit into the breech of a gun?

Rather than look at physics down a gun barrel Fermi withdrew to the ultramundane. It might turn out to be the size of a small star, he said, smiling and knowing better.

Neutrons diffusing through a tank of water: it was all too vague. Except to alert Ross Gunn, the meeting came to nothing. "Enrico himself . . . doubted the relevance of his predictions," says Laura Fermi. The Navy reported itself interested in maintaining contact; representatives would undoubtedly visit the Columbia premises. Fermi smelled the condescension and cooled.

March 17 was a Friday; Szilard traveled down to Washington from Princeton with Teller; Fermi stayed the weekend. They got together, reports Szilard, "to discuss whether or not these things"—the *Physical Review* papers—"should be published. Both Teller and I thought that they should not. Fermi thought that they should. But after a long discussion, Fermi took the position that after all this was a democracy; if the majority was against publication, he would abide by the wish of the majority." Within a day or two the issue became moot. The group learned of the Joliot/von Halban/Kowarski paper, published in *Nature* on March 18. "From that moment on," Szilard notes, "Fermi was adamant that withholding publication made no sense."

The following month, on April 22, Joliot, von Halban and Kowarski published a second paper in *Nature* concerning secondary neutrons. This one, "Number of neutrons liberated in the nuclear fission of uranium," rang bells. Calculating on the basis of the experiment previously reported, the French team found 3.5 secondary neutrons per fission. "The interest of the phenomenon discussed here as a means of producing a chain of nuclear reactions," the three men wrote, "was already mentioned in our previous letter." Now they concluded that if a sufficient amount of uranium were immersed in a suitable moderator, "the fission chain will perpetuate itself and break up only after reaching the walls limiting the medium. Our experimental results show that this condition will most probably be satisfied." That is, uranium would most probably chain-react.

Joliot's was an authoritative voice. G. P. Thomson, J.J.'s son, who was professor of physics at Imperial College, London, heard it. "I began to consider carrying out certain experiments with uranium," he told a correspondent later. "What I had in mind was something rather more than a piece of pure research, for at the back of my thoughts there lay the possibility of a weapon." He applied forthwith to the British Air Ministry for a ton of uranium oxide, "ashamed of putting forward a proposal apparently so absurd."

More ominously, two initiatives originated simultaneously in Germany as a result of the French report. A physicist at Göttingen alerted the Reich Ministry of Education. That led to a secret conference in Berlin on April 29, which led in turn to a research program, a ban on uranium exports and provision for supplies of radium from the Czechoslovakian mines at Joachimsthal. (Otto Hahn was invited to the conference but arranged to be elsewhere.) The same week a young physicist working at Hamburg, Paul Harteck, wrote a letter jointly with his assistant to the German War Office:

> We take the liberty of calling to your attention the newest development in nuclear physics, which, in our opinion, will probably make it possible to produce an explosive many orders of magnitude more powerful than the conventional ones. . . . That country which first makes use of it has an unsurpassable advantage over the others.

The Harteck letter reached Kurt Diebner, a competent nuclear physicist stuck unhappily in the Wehrmacht's ordnance department studying high explosives. Diebner carried it to Hans Geiger. Geiger recommended pursuing the research. The War Office agreed.

A public debate in Washington on April 29 paralleled the secret conference in Berlin. The *New York Times* account accurately summarizes the divisions in the U.S. physics community at the time:

> Tempers and temperatures increased visibly today among members of the American Physical Society as they closed their Spring meeting with arguments over the probability of some scientist blowing up a sizable portion of the earth with a tiny bit of uranium, the element which produces radium.
>
> Dr. Niels Bohr of Copenhagen, a colleague of Dr. Albert Einstein at the Institute for Advanced Study, Princeton, N.J., declared that bombardment of a small amount of the pure Isotope U235 of uranium with slow neutron particles of atoms would start a "chain reaction" or atomic explosion sufficiently great to blow up a laboratory and the surrounding country for many miles.
>
> Many physicists declared, however, that it would be difficult, if not impossible, to separate Isotope 235 from the more abundant Isotope 238. The Isotope 235 is only 1 per cent of the uranium element.
>
> Dr. L. Onsager of Yale University described, however, a new apparatus in which, according to his calculations, the isotopes of elements can be separated in gaseous form in tubes which are cooled on one side and heated to high temperatures on the other.
>
> Other physicists argued that such a process would be almost prohibitively expensive and that the yield of Isotope 235 would be infinitesimally small. Nevertheless, they pointed out that, if Dr. Onsager's process of separation should work, the creation of a nuclear explosion which would wreck as large an area as New York City would be comparatively easy. A single neutron particle, striking the nucleus of a uranium atom, they declared, would be sufficient to set off a chain reaction of millions of other atoms.

The *Times* story assumes the truth of Bohr's argument in favor of U235, although even Bohr was apparently still emphasizing only a slow-neutron reaction. Fermi and others were not yet convinced of U235's role. The two uranium isotopes might not easily be separated in quantity, but it had occurred to John Dunning earlier in the month that they could be separated in microscopic amounts in Alfred Nier's mass spectrograph. Dunning had immediately written Nier a long, impassioned letter asking him, in effect, to resolve the dispute between Fermi and Bohr and push chain-reaction research dramatically forward. Nier, Dunning and Fermi all attended the American Physical Society meeting. In person Dunning urged Nier to try for a separation much as he had urged him in the key paragraph of his letter:

> There is one line of attack that deserves strong effort, and that is where we need your cooperation. . . . It is of the utmost importance to get some uranium

> isotopes separated in enough quantities for a real test. If you could separate effectively even tiny amounts of the two main isotopes [a third isotope, U234, is present in natural uranium to the trace extent of one part in 17,000], there is a good chance that [Eugene T.] Booth and I could demonstrate, by bombarding them with the cyclotron, which isotope is responsible. There is no other way to settle this business. If we could all cooperate and you aid us by separating some samples, then we could, by combining forces, settle the whole matter.

The important point for Dunning, the reason for his passion, was that if U235 was responsible for slow-neutron fission, then its fission cross section must be 139 times as large as the slow-neutron fission cross section of natural uranium, since it was present in the natural substance to the extent of only one part in 140. "By separating the 235 isotope," Herbert Anderson emphasizes in a memoir, "it would be much easier to obtain the chain reaction. More than this, with the separated isotope the prospect for a bomb with unprecedented explosive power would be very great."

Fermi urged Nier in similar terms; Nier recalls that he "went back and figured out how we might soup up our apparatus some in order to increase the output. . . . I did work on the problem, but at first it seemed like such a farfetched thing that I didn't work on it as hard as I might have. It was just one of a number of things I was trying to do."

Fermi in any case was more interested in pursuing a chain reaction in natural uranium than in attempting to separate isotopes. "He was not discouraged by the small cross-section for fission in the natural [element]," comments Anderson. " 'Stay with me,' he advised, 'we'll work with natural uranium. You'll see. We'll be the first to make the chain reaction.' I stuck with Fermi."

By mid-April Szilard had managed to borrow about five hundred pounds of black, grimy uranium oxide free of charge from the Eldorado Radium Corporation, an organization owned by the Russian-born Pregel brothers, Boris and Alexander. Boris had studied at the Radium Institute in Paris; Eldorado speculated in rare minerals and owned important uranium deposits at Great Bear Lake in the Northwest Territories of Canada.

Like Fermi's and Anderson's previous experiment, the new project involved measuring neutron production in a tank of liquid. For a more accurate reading the experimenters needed a longer exposure time than their customary rhodium foils activated to 44-second half-life would allow. They planned instead simply to fill the tank with a 10 percent solution of manganese, an ironlike metal that gives amethyst its purple color and that activates upon neutron bombardment to an isotope with a nearly 3-hour half-life. "The [radio]activity induced in manganese," they explained afterward in their report, "is proportional to the number of [slow] neutrons present." So the hydrogen in the water would serve to slow both the primary neutrons from the central neutron source and any secondary neutrons from fission, and the manganese in the water would serve to measure them—a nice economy of design.

Atoms on the surface of a mass of uranium are exposed to neutrons more efficiently than atoms deeper inside. Fermi and Szilard therefore decided not to bulk their five hundred pounds of uranium oxide into one large container but to distribute it throughout the tank by packing it into fifty-two cans as tall and narrow as lengths of pipe—two inches in diameter and two feet long.

Packing cans and mixing manganese solutions, which had to be changed and the manganese concentrated after each experimental run, was work. So was staying up half the night taking readings of manganese radioactivity. Fermi accepted the challenge with gusto. "He liked to work harder than anyone else," Anderson notes, "but everyone worked very hard." Except Szilard. "Szilard thought he ought to spend his time thinking." Fermi was insulted. "Szilard made a mortal sin," Segrè remembers, echoing Fermi. "He said, 'Oh, I don't want to work and dirty my hands like a painter's assistant.' " When Szilard announced that he had hired a stand-in, a young man whom Anderson remembers as "very competent," Fermi acceded to the arrangement without comment. But he never again pursued an experiment jointly with Szilard.

The arrangement as finally consummated looked like this:

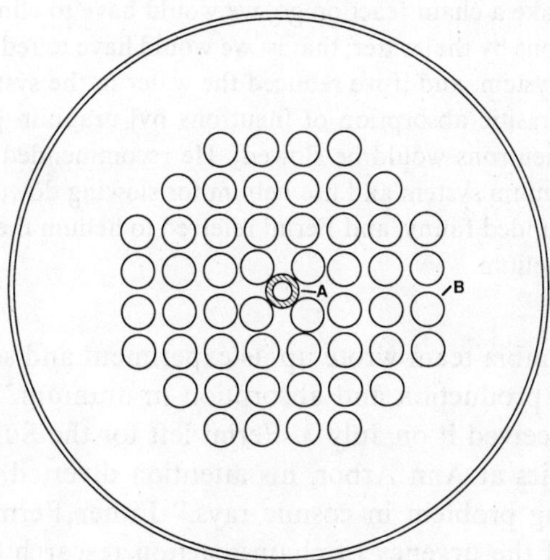

Szilard's Ra + Be source stands in the center of the tank, which holds 143 gallons of manganese solution; the fifty-two cans of UO_2 gather around.

It worked. The three physicists found neutron activity "about ten percent higher with uranium oxide than without it. This result shows that in our arrangement more neutrons are emitted by uranium than are absorbed by uranium." But the experiment raised puzzling questions. Resonance absorption, for example, was clearly a problem, capturing neutrons that might otherwise serve the chain reaction. The report estimates "an average emission [of secondary neutrons] of about 1.2 neutrons per thermal neutron" but notes that "this number should be increased, to perhaps 1.5," because some of the neutrons had obviously been captured without fissioning—demonstrating the big capture resonance around 25 eV that Bohr had attributed on his graphs to U238.

Another problem was the use of water as a moderator. As Fermi's team had discovered in Rome in 1934, hydrogen was more efficient than any other element at slowing down neutrons, and slow neutrons avoided the parasitic capture resonance of U238. But hydrogen itself also absorbed some slow neutrons, reducing further the number available for fission. And it was already clear that every possible secondary neutron would have to be husbanded carefully if a chain reaction was to be initiated in natural uranium. George Placzek came down from Cornell, where he had found a new home, for a visit, looked over the arrangement and insightfully foreclosed its future. As Szilard tells it:

> We were inclined to conclude that . . . the water-uranium system would sustain a chain reaction. . . . Placzek said that our conclusion was wrong because in order to make a chain reaction go, we would have to eliminate the absorption of [neutrons by the] water; that is, we would have to reduce the amount of water in the system, and if we reduced the water in the system, we would increase the parasitic absorption of [neutrons by] uranium [because with less water fewer neutrons would be slowed]. He recommended that we abandon the water-uranium system and use helium for slowing down the neutrons. To Fermi this sounded funny, and Fermi referred to helium thereafter invariably as Placzek's helium.

In June the Columbia team wrote up its experiment and sent the resulting paper, "Neutron production and absorption in uranium," to the *Physical Review,* which received it on July 3. Fermi left for the Summer School of Theoretical Physics at Ann Arbor, his attention diverted, says Anderson, "by an interesting problem in cosmic rays." Either Fermi did not share Szilard's sense of the urgency of chain-reaction research or he was withdrawing for a time from the Navy's indifference and Placzek's persuasive criticism of his uranium-water system; probably both. Anderson settled down to study resonance absorption in uranium, a project that would evolve into his doctoral dissertation.

Szilard remained in the steamy city: "I was left alone in New York. I still had no position at Columbia; my three months [of laboratory privileges] were up, but there were no experiments going on anyway and all I had to do was to think."

Szilard thought first about an alternative to water. The next common material up the periodic table that might work—that had a capture cross section considerably smaller than hydrogen's, that was cheap, that would be thermally and chemically stable—was carbon. The mineral form of carbon, chemically identical to diamond but the product of a different structure of crystallization, is graphite, a black, greasy, opaque, lustrous material that is the essential component of pencil lead. Although carbon slows neutrons much less rapidly than hydrogen, even that difference might be put to advantage by careful design.

Lewis Strauss was leaving for Europe the week of July 2. Hoping that the financier might coax support for uranium research from Belgium's Union Minière, Szilard sent Strauss a last-minute letter arguing that a chain reaction in uranium "is an immediate possibility" but chose not to mention his new uranium-graphite conception. Apparently he wanted to discuss it first with Fermi; the same day, July 3, he wrote the Italian laureate at length. "It seems to me now," he reported, "that there is a good chance that carbon might be an excellent element to use in place of hydrogen, and there is a strong temptation to gamble on this chance." He wanted to try "a large-scale experiment with a carbon-uranium-oxide mixture" as soon as they could acquire enough material. In the meantime he thought he would set up a small experiment to measure more accurately carbon's capture cross section, only the upper limit of which was then known. If carbon should prove unsuitable their "next best guess might be heavy water," rich in deuterium, though they would need "a few tons" of that scarce and expensive liquid. (Deuterium, H2, has a much smaller cross section for neutron capture than ordinary hydrogen.)

Across the one hundred sixty-third anniversary of the Declaration of Independence Szilard's ideas evolved rapidly. On July 5 he visited the National Carbon Company of New York to look into purchasing graphite blocks of high purity (because impurities such as boron with large capture cross sections would soak up too many neutrons). He wrote Fermi his finding the same day: "It seems that it will be possible to get sufficiently pure carbon at a reasonable price." He also mentioned arranging the uranium and carbon in layers.

Fermi sat down in Ann Arbor at the end of the week to respond to Szilard's first report. Independently he had arrived at a similar plan:

> Thank you for your letter. I was also considering the possibility of using carbon for slowing down the neutrons. . . . According to my estimates a possible recipe might be about 39,000 kg of carbon mixed with 600 kg of uranium. If it were really so the amounts of materials would certainly not be too large.
>
> Since however the amount of uranium that can be used, especially in a homogeneous mixture, is exceedingly small . . . *perhaps the use of thick layers of carbon separated by layers of uranium might allow use of a somewhat larger percentage of uranium.*

The idea of layering or in some other way separating the uranium from the graphite originated in calculations Fermi made in June for the manganese water-tank experiment. Fermi's calculations led both men to consider partitioning the oxide from the graphite in the new design they were independently evolving. Partitioning would give the fast secondary neutrons room to slow down, bouncing around in the moderator, before they encountered any U238 nuclei. Szilard's next letter, on July 8, mentions that "the carbon and the uranium oxide would not be mixed but built up in layers, or in any case used in some canned form." Both the July 5 and July 8 letters apparently crossed with Fermi's letter in the mail.

By the time he heard from Fermi, Szilard had seen still farther and realized that small spheres of uranium arranged within blocks of graphite would be "even more favorable from the point of view of a chain reaction than the system of plane uranium layers which was initially considered." The arrangement Szilard had in mind he called a "lattice." (A geodesic dome would represent such a lattice arrangement schematically if it were a complete sphere and if all its interior volume were filled like its surface with evenly spaced points.) His calculations indicated somewhat larger volumes of material than had Fermi's: "perhaps 50 tons of carbon and 5 tons of uranium." The entire experiment, he thought, would cost about $35,000.

If a chain reaction would work in graphite and uranium, Szilard assumed, then a bomb was probable. And if he had managed these conclusions, he further assumed, then so had his counterparts in Nazi Germany. He sought out Pegram in those early July days and tried to convince him of the urgent need for a large-scale experiment to settle the question. The dean resisted the assault: "He took the position that even though the matter appeared to be rather urgent, this being summer and Fermi being away there was really nothing that usefully could be done until the fall."

For several weeks Szilard had been trying on his own to raise funds from the U.S. military. In late May he had asked Wigner to contact the Army's Aberdeen Proving Ground, its weapons-development facility in Maryland. While he was thinking through the possibilities of a uranium-graphite system he had talked to Ross Gunn about Navy support. Now Fermi's letter of July 9 and a July 10 letter from Gunn arrived to discourage him. Fermi wrote of layering the carbon and uranium but calculated in terms of a homogeneous system—of graphite and uranium oxide crushed and mixed together. Szilard concluded he was being mocked: "I knew very well that Fermi . . . computed the homogeneous mixture only because it was the easiest to compute. This showed me that Fermi did not take this really seriously." Gunn in turn reported that "it seems almost impossible . . . to carry through any sort of an agreement [with the Navy] that would be really helpful to you. I regret this situation but see no escape."

Despite his Olympian ego not even Leo Szilard felt capable of saving the world entirely alone. He called on his Hungarian compatriots now for moral support. Edward Teller had moved to Manhattan for the summer to teach physics at Columbia; Eugene Wigner came up from Princeton to conspire with them. In later years Szilard would recount several different versions of how their conversation went, but a letter he wrote on August 15, 1939, offers reliable contemporary testimony: "Dr. Wigner is taking the stand that it is our duty to enlist the cooperation of the [Roosevelt] Administration. A few weeks ago he came to New York in order to discuss this point with Dr. Teller and me." Szilard had shown Wigner his uranium-graphite calculations. "He was impressed and he was concerned." Both Teller and Wigner, Szilard wrote in a background memorandum in 1941, "shared the opinion that no time must be lost in following up this line of development and in the discussion that followed, the opinion crystallized that an attempt ought to be made to enlist the support of the Government rather than that of private industry. Dr. Wigner, in particular, urged very strongly that the Government of the United States be advised."

But the discussion slipped away from that project into "worry about what would happen if the Germans got hold of large quantities of the uranium which the Belgians were mining in the Congo." Perhaps Szilard emphasized the futility of the government contacts that he and Fermi had already made. "So we began to think, through what channels could we approach the Belgian government and warn them against selling any uranium to Germany?"

It occurred to Szilard then that his old friend Albert Einstein knew the Queen of Belgium. Einstein had met Queen Elizabeth in 1929 on a trip to Antwerp to visit his uncle; thereafter the physicist and the sovereign maintained a regular correspondence, Einstein addressing her in plainspoken letters simply as "Queen."

The Hungarians were aware that Einstein was summering on Long Island. Szilard proposed visiting Einstein and asking him to alert Elizabeth of Belgium. Since Szilard owned no car and had never learned to drive he enlisted Wigner to deliver him. They called Einstein's office at the Institute for Advanced Study and learned he was staying at a summer house on Old Grove Pond on Nassau Point, the spit of land that divides Little from Great Peconic Bay on the northeastern arm of the island.

They called Einstein to arrange a day. At this time Szilard also furthered Wigner's proposal to contact the United States government by seeking advice from a knowledgeable emigré economist, Gustav Stolper, a Berliner resettled in New York who had once been a member of the Reichstag. Stolper offered to try to identify an influential messenger.

Wigner picked up Szilard on the morning of Sunday, July 16, and drove out Long Island to Peconic. They reached the area in early afternoon but had no luck soliciting directions to the house until Szilard thought to ask for it in Einstein's name. "We were on the point of giving up and going back to New York"—two world-class Hungarians lost among country lanes in summer heat—"when I saw a boy aged maybe seven or eight standing on the curb. I leaned out of the window and I said, 'Say, do you by any chance know where Professor Einstein lives?' The boy knew that and he offered to take us there."

C. P. Snow had visited Einstein at the same summer retreat two years before, also losing his way, and makes the scene familiar:

> He came into the sitting room a minute or two after we arrived. There was no furniture apart from some garden chairs and a small table. The window looked out on to the water, but the shutters were half closed to keep out the heat. The humidity was very high.
>
> At close quarters, Einstein's head was as I had imagined it: magnificent, with a humanizing touch of the comic. Great furrowed forehead; aureole of white hair; enormous bulging chocolate eyes. I can't guess what I should have expected from such a face if I hadn't known. A shrewd Swiss once said it had the brightness of a good artisan's countenance, that he looked like a reliable old-fashioned watchmaker in a small town who perhaps collected butterflies on a Sunday.
>
> What did surprise me was his physique. He had come in from sailing and was wearing nothing but a pair of shorts. It was a massive body, very heavily muscled: he was running to fat round the midriff and in the upper arms, rather like a footballer in middle-age, but he was still an unusually strong man. He was cordial, simple, utterly unshy. The large eyes looked at me, as though he was thinking: what had I come for, what did I want to talk about?

> . . . The hours went on. I have a hazy memory that several people drifted in and out of the room, but I do not remember who they were. Stifling heat. There appeared to be no set time for meals. He was already, I think, eating very little, but he was still smoking his pipe. Trays of open sandwiches—various kinds of wurst, cheese, cucumber—came in every now and then. It was all casual and Central European. We drank nothing but soda water.

Similarly settled, Szilard told Einstein about the Columbia secondary-neutron experiments and his calculations toward a chain reaction in uranium and graphite. Long afterward he would recall his surprise that Einstein had not yet heard of the possibility of a chain reaction. When he mentioned it Einstein interjected, *"Daran habe ich gar nicht gedacht!"*—"I never thought of that!" He was nevertheless, says Szilard, "very quick to see the implications and perfectly willing to do anything that needed to be done. He was willing to assume responsibility for sounding the alarm even though it was quite possible that the alarm might prove to be a false alarm. The one thing most scientists are really afraid of is to make fools of themselves. Einstein was free from such a fear and this above all is what made his position unique on this occasion."

Einstein hesitated to write Queen Elizabeth but was willing to contact an acquaintance who was a member of the Belgian cabinet. Wigner spoke up to insist again that the United States government should be alerted, pointing out, Szilard goes on, "that we should not approach a foreign government without giving the State Department an opportunity to object." Wigner suggested that they send the Belgian letter with a cover letter through State. All three men thought that made sense.

Einstein dictated a letter to the Belgian ambassador, a more formal contact appropriate to their State Department plan, and Wigner took it down in longhand in German. At the same time Szilard drafted a cover letter. Einstein's was the first of several such compositions—they served in succession as drafts—and the origin of most of the statements that ultimately found their way into the letter he actually sent.

Wigner carried the first Einstein draft back to Princeton, translated it into English and on Monday gave it to his secretary to type. When it was ready he mailed it to Szilard. Then he left Princeton to drive to California on vacation.

A message from Gustav Stolper awaited Szilard at the King's Crown. "He reported to me," Szilard wrote Einstein on July 19, "that he had discussed our problems with Dr. Alexander Sachs, a vice-president of the Lehman Corporation, biologist and national economist, and that Dr. Sachs wanted to talk to me about this matter." Eagerly Szilard arranged an appointment.

Alexander Sachs, born in Russia, was then forty-six years old. He had come to the United States when he was eleven, graduated from Columbia in biology at nineteen, worked as a clerk on Wall Street, returned to Columbia to study philosophy and then went on to Harvard with several prestigious fellowships to pursue philosophy, jurisprudence and sociology. He contributed economics text to Franklin Roosevelt's campaign speeches in 1932; beginning in 1933 he worked for three years for the National Recovery Administration, joining the Lehman Corporation in 1936. He had thick curls and a receding chin and looked and sounded like the comedian Ed Wynn. His associates at the NRA used to point him out to visiting firemen under that *nom de guerre* as ultimate proof, if the NRA itself was not sufficient, of Roosevelt's gift for radical innovation. Sachs communicated in dense, florid prose (he had been thinking that spring of writing a book entitled *The Inter-War Retreat from Reason as Exemplified in the Mis-history of the Recent Past and in the Contemporaneous Conduct of International Political and Economic Affairs by the United States and Great Britain*) but could coruscate in committee.

Sachs heard Szilard out. Then, as Szilard wrote Einstein, he "took the position, and completely convinced me, that these were matters which first and foremost concerned the White House and that the best thing to do, also from the practical point of view, was to inform Roosevelt. He said that if we gave him a statement he would make sure it reached Roosevelt in person." Among those who valued Sachs' opinions and called him from time to time for talks, it seems, was the President of the United States.

Szilard was stunned. The very boldness of the proposal won his heart after all the months when he had confronted caution and skepticism: "Although I have seen Dr. Sachs once," he told Einstein, "and really was not able to form any judgment about him, I nevertheless think that it could not do any harm to try this way and I also think that in this regard he is in a position to fulfill his promise."

Szilard met Sachs shortly after returning from Peconic—between Sunday and Wednesday. Unable at midweek to reach Wigner en route to California, he tracked down Teller, who thought Sachs' proposal preferable to the plan they had previously worked out. Drawing on the first Einstein draft, Szilard now prepared a draft letter to Roosevelt. He wrote it in German because Einstein's English was insecure, added a cover letter and mailed it to Long Island. "Perhaps you will be able to tell me over the telephone whether you would like to return the draft with your marginal comments by mail," he proposed in the cover letter, "or whether I should come out to discuss the whole thing once more with you." If he visited Peconic again, Szilard wrote, he would ask Teller to drive him, "not only because I believe his advice is valuable but also because I think you might enjoy getting to know him. He is particularly nice."

Einstein preferred to review a letter to the President in person. Teller therefore delivered Szilard to Peconic, probably on Sunday, July 30, in his sturdy 1935 Plymouth. "I entered history as Szilard's chauffeur," Teller aphorizes the experience. They found the Princeton laureate in old clothes and slippers. Elsa Einstein served tea. Szilard and Einstein composed a third text together, which Teller wrote down. "Yes, yes," Teller remembers Einstein commenting, "this would be the first time that man releases nuclear energy in a direct form rather than indirectly." Directly from fission, he meant, rather than indirectly from the sun, where a different nuclear reaction produces the copious radiation that reaches the earth as sunlight.

Einstein apparently questioned if Sachs was the best man to carry the news to Roosevelt. On August 2 Szilard wrote Einstein hoping "at long last" for a decision "upon whom we should try to get as middle man." He had seen Sachs in the interim; the economist, who certainly coveted the assignment of representing Albert Einstein to the President, had generously listed the financier Bernard Baruch or Karl T. Compton, the president of MIT, as possible alternates. On the other hand, he had strongly endorsed Charles Lindbergh, though he must have known that Roosevelt despised the famous aviator for his outspoken pro-German isolationism. Szilard wrote that he and Sachs had discussed "a somewhat longer and more extensive version" of the letter Einstein had written with Szilard at their second Peconic meeting; he now enclosed both the longer and shorter versions and asked Einstein to return his favorite along with a letter of introduction to Lindbergh.

Einstein opted for the longer version, which incorporated the shorter statement that had originated with him but carried additional paragraphs contributed by Szilard in consultation with Sachs. He signed both letters and returned them to Szilard in less than a week with a note hoping "that you will finally overcome your inner resistance; it's always questionable to try to do something too cleverly." That is, be bold and get moving. "We will try to follow your advice," Szilard rejoined on August 9, "and as far as possible overcome our inner resistances which, admittedly, exist. Incidentally, we are surely not trying to be too clever and will be quite satisfied if only we don't look too stupid."

Szilard transmitted the letter in its final form to Sachs on August 15 along with a memorandum of his own that elaborated on the letter's discussion of the possibilities and dangers of fission. He had not given up contacting Lindbergh—he drafted a letter to the aviator the following day—but he seems to have decided to try Sachs in the meantime, probably in the interest of moving the project on; he pointedly asked Sachs either to deliver the letter to Roosevelt or to return it.

One of the discussions Szilard had added to the longer draft that Einstein chose concerned who should serve as liaison between "the Administration and the group of physicists working on chain reactions in America." In his letter of transmittal to Sachs, Szilard now tacitly offered himself for that service. "If a man, having courage and imagination, could be found," he wrote, "and if such a man were put—in accordance with Dr. Einstein's suggestion—in the position to act with some measure of authority in this matter, this would certainly be an important step forward. In order that you may be able to see of what assistance such a man could be in our work, allow me please to give you a short account of the past history of the case." The short account that followed, an abbreviated and implicit *curriculum vitae,* essentially outlined Szilard's own role since Bohr's announcement of the discovery of fission seven crowded months earlier.

Szilard's offer was as innocent of American bureaucratic politics as it was bold. It was surely also the apotheosis of his drive to save the world. By this time the Hungarians at least believed they saw major humanitarian benefit inherent in what Eugene Wigner would describe in retrospect as "a horrible military weapon," explaining:

> Although none of us spoke much about it to the authorities [during this early period]—they considered us dreamers enough as it was—we did hope for another effect of the development of atomic weapons in addition to the warding off of imminent disaster. We realized that, should atomic weapons be developed, no two nations would be able to live in peace with each other unless their military forces were controlled by a common higher authority. We expected that these controls, if they were effective enough to abolish atomic warfare, would be effective enough to abolish also all other forms of war. This hope was almost as strong a spur to our endeavors as was our fear of becoming the victims of the enemy's atomic bombings.

From the horrible weapon which they were about to urge the United States to develop, Szilard, Teller and Wigner—"the Hungarian conspiracy," Merle Tuve was amused to call them—hoped for more than deterrence against German aggression. They also hoped for world government and world peace, conditions they imagined bombs made of uranium might enforce.

Alexander Sachs intended to read aloud to the President when he met with him. He believed busy people saw so much paper they tended to dismiss the printed word. "Our social system is such," he told a Senate committee in 1945, "that any public figure [is] punch-drunk with printer's ink. . . . This was a matter that the Commander in Chief and the head of the Nation must know. I could only do it if I could see him for a long stretch and read the material so it came in by way of the ear and not as a soft mascara on the eye." He needed a full hour of Franklin Delano Roosevelt's time.

History intervened to crowd the President's calendar. Having won the Rhineland, Austria and Czechoslovakia simply by taking them, having signed the Pact of Steel with Italy on May 22 and a ten-year treaty of non-aggression and neutrality with the USSR on August 23, Adolf Hitler ordered the invasion of Poland beginning at 4:45 A.M. on September 1, 1939, and precipitated the Second World War. The German invasion fielded fifty-six divisions against thirty Polish divisions strung thinly across the long Polish frontier; Hitler had ten times the aircraft, including plentiful squadrons of Stuka dive-bombers, and nine divisions of Panzer tanks against Polish horse cavalry armed with swords and spears. The assault was "a perfect specimen of the modern Blitzkrieg," writes Winston Churchill: "the close interaction on the battlefield of army and air force; the violent bombardment of all communications and of any town that seems an attractive target; the arming of an active Fifth Column; the free use of spies and parachutists; and above all, the irresistible forward thrusts of great masses of armour."

The mathematician Stanislaw Ulam had just returned from visiting Poland, bringing with him on a student visa his sixteen-year-old brother, Adam:

> Adam and I were staying in a hotel on Columbus Circle. It was a very hot, humid, New York night. I could not sleep very well. It must have been around one or two in the morning when the telephone rang. Dazed and perspiring, very uncomfortable, I picked up the receiver and the somber, throaty voice of my friend the topologist Witold Hurewicz began to recite the horrible tale of the start of war: "Warsaw has been bombed, the war has begun," he said. This is how I learned about the beginning of World War II. He kept describing what he had heard on the radio. I turned on my own. Adam was asleep; I did not wake him. There would be time to tell him the news in the morning. Our father and sister were in Poland, so were many other relatives. At that moment, I suddenly felt as if a curtain had fallen on my past life, cutting it off from my future. There has been a different color and meaning to everything ever since.

One of Roosevelt's first acts was to appeal to the belligerents to refrain from bombing civilian populations. Revulsion against the bombing of cities had grown in the United States since at least the Japanese bombing of

Shanghai in 1937. When Spanish Fascists bombed Barcelona in March 1938, Secretary of State Cordell Hull had condemned the atrocity publicly: "No theory of war can justify such conduct," he told reporters. ". . . I feel that I am speaking for the whole American people." In June the Senate passed a resolution condemning the "inhuman bombing of civilian populations." As war approached, revulsion began to give way to impulses of revenge; in the summer of 1939 Herbert Hoover could urge an international ban on the bombing of cities and still argue that "one of the impelling reasons for the unceasing building of bombing planes is to prepare reprisals." Bombing was bad because it was enemy bombing. *Scientific American* saw through to a darker truth: "Although . . . aerial bombing remains an unknown, indeterminate quantity, the world may be sure that the unwholesome atrocities which are happening today are but curtain raisers on insane dramas to come."

So although Roosevelt had asked Congress for increased funds for long-range bombers nine months before, in appealing to the belligerents on September 1, 1939, he could still articulate the moral indignation of millions of Americans:

> The ruthless bombing from the air of civilians in unfortified centers of population during the course of the hostilities which have raged in various quarters of the earth during the past few years, which has resulted in the maiming and in the death of thousands of defenseless men, women and children, has sickened the hearts of every civilized man and woman, and has profoundly shocked the conscience of humanity.
>
> If resort is had to this form of inhuman barbarism during the period of the tragic conflagration with which the world is now confronted, hundreds of thousands of innocent human beings who have no responsibility for, and who are not even remotely participating in, the hostilities which have now broken out, will lose their lives. I am therefore addressing this urgent appeal to every Government which may be engaged in hostilities publicly to affirm its determination that its armed forces shall in no event, and under no circumstances, undertake the bombardment from the air of civilian populations or of unfortified cities, upon the understanding that these same rules of warfare will be scrupulously observed by all of their opponents. I request an immediate reply.

Great Britain agreed to the President's terms the same day. Germany, busy bombing Warsaw, concurred on September 18.

The invasion of Poland brought Britain and France into the war on September 3. Abruptly Roosevelt's schedule filled to overflowing. In early September in particular he was working overtime with a reluctant Congress to revise the Neutrality Act to terms more favorable to Britain; Sachs was unable even to discuss arranging an interview until after the first week in September.

By September Kurt Diebner's new War Office department had consolidated German fission research under its authority. Diebner enlisted a young Leipzig theoretician named Erich Bagge and together the two physicists planned a secret conference to consider the feasibility of a weapons project. They had the authority to enlist the services of any German citizen they wished and they used it, sending out papers that left Hans Geiger, Walther Bothe, Otto Hahn and a number of other exceptional older men nervously uncertain if they were being invited to Berlin for consultation or ordered to active military service.

At the conference in Berlin on September 16 the physicists learned that German intelligence had discovered the beginnings of uranium research abroad—meaning, presumably, in the United States and Britain. They discussed the long, thorough theoretical paper by Niels Bohr and John Wheeler, "The mechanism of nuclear fission," that had been published in the September *Physical Review* and especially its conclusion, which Bohr and Wheeler had elaborated from Bohr's Sunday-morning graph work, that U235 was probably the isotope of uranium responsible for slow-neutron fission. Hahn like Bohr argued that isotope separation was difficult to the point of impossibility. Bagge proposed calling in Werner Heisenberg, his superior at Leipzig, to adjudicate.

Heisenberg therefore attended a second Berlin conference on September 26 and discussed two possible ways to harness the energy from fission: by slowing secondary neutrons with a moderator to make a "uranium burner" and by separating U235 to make an explosive. Paul Harteck, the Hamburg physicist who had written the War Office the previous April, traveled to the second conference armed with a paper he had just finished on the importance of layering uranium and moderator to avoid the U238 capture resonance—the same insight that had come independently to Fermi and Szilard in early July. Harteck's study, however, considered using heavy water as moderator, even though Harteck had worked with Rutherford at the Cavendish and knew from personal experience how expensive heavy-water production could be—water in which deuterium replaced hydrogen had to be tediously distilled from tons of ordinary H_2O.

Diebner and Bagge had outlined for the second conference a "Preparatory Working Plan for Initiating Experiments on the Exploitation of Nuclear Fission." Heisenberg would head up theoretical investigation. Bagge would measure deuterium's cross section for collision to establish how effectively heavy water might slow secondary neutrons. Harteck would look into isotope separation. Others would experiment to determine other significant nuclear constants. The War Office would take over the Kaiser Wilhelm Institute of Physics, finished in 1937 and beautifully equipped. Adequate funds would be forthcoming.

The German atomic bomb project was well begun.

It may have been no less complicated by humanitarian ambiguities than the project the Hungarians in the United States proposed. One young but highly respected German physicist involved in the work from near the beginning was Carl Friedrich von Weizsäcker, the son of the German Undersecretary of State. In a 1978 memoir von Weizsäcker remembers discussing the possibility of a bomb with Otto Hahn in the spring of 1939. Hahn opposed secrecy then partly on the grounds of scientific ethics but also partly because he "felt that if it were to be made, it would be worst for the entire world, even for Germany, if Hitler were to be the only one to have it." Like Szilard, Teller and Wigner, von Weizsäcker remembers realizing in discussions with a friend "that this discovery could not fail to radically change the political structure of the world":

> To a person finding himself at the beginning of an era, its simple fundamental structures may become visible like a distant landscape in the flash of a single stroke of lightning. But the path toward them in the dark is long and confusing. At that time [i.e., 1939] we were faced with a very simple logic. Wars waged with atom bombs as regularly recurring events, that is to say, nuclear wars as institutions, do not seem reconcilable with the survival of the participating nations. But the atom bomb exists. It exists in the minds of some men. According to the historically known logic of armaments and power systems, it will soon make its physical appearance. If that is so, then the participating nations and ultimately mankind itself can only survive if war as an institution is abolished.

Both sides might work from fear of the other. But some on both sides would be working also paradoxically believing they were preparing a new force that would ultimately bring peace to the world.

As September extended its violence Szilard grew impatient. He had heard nothing from Alexander Sachs. Pursuing Sachs' previous suggestions and his own leads, he arranged for Eugene Wigner to give him a letter of introduction to MIT president Karl T. Compton; recontacted a businessman of possible influence whom he had once interested in the Einstein-Szilard refrigerator pump; read a newspaper account of a Lindbergh speech and reported to Einstein that the aviator "is in fact not our man." Finally, the last week in September, he and Wigner visited Sachs and found to their dismay that the economist still held Einstein's letter. "He says he has spoken repeatedly with Roosevelt's secretary," Szilard reported to Einstein on October 3, "and has the impression that Roosevelt is so overburdened that it would be wiser to see him at a later date. He intends to go to Washington this week." The two Hungarians were ready to start over: "There is a distinct possibility that Sachs will be of no use to us. If this is the case, we must put the matter in someone else's hands. Wigner and I have decided to accord Sachs ten days' grace. Then I will write you again to let you know how matters stand."

But Alexander Sachs did indeed travel to Washington, not that week but the next, and on Wednesday, October 11, presented himself, probably in the late afternoon, at the White House. Roosevelt's aide, General Edwin M. Watson, "Pa" to Roosevelt and his intimates, sitting with his own executive secretary and military aide, reviewed Sachs' agenda. When he was convinced that the information was worth the President's time, Watson let Sachs into the Oval Office.

"Alex," Roosevelt hailed him, "what are you up to?"

Sachs liked to warm up the President with jokes. His sense of humor tended to learned parables. Now he told Roosevelt the story of the young American inventor who wrote a letter to Napoleon. The inventor proposed to build the emperor a fleet of ships that carried no sail but could attack England in any weather. He had it in his power to deliver Napoleon's armies to England in a few hours without fear of wind or storm, he wrote, and he was prepared to submit his plans. Napoleon scoffed: ships without sails? "Bah! Away with your visionists!"

The young inventor, Sachs concluded, was Robert Fulton. Roosevelt laughed easily; probably he laughed at that.

Sachs cautioned the President to listen carefully: what he had now to impart was at least the equivalent of the steamboat inventor's proposal to Napoleon. Not yet ready to listen, Roosevelt scribbled a message and summoned an aide. Shortly the aide returned with a treasure, a carefully wrapped bottle of Napoleon brandy that the Roosevelts had preserved in the family for years. The President poured two glasses, passed one to his visitor, toasted him and settled back.

Sachs had made a file for Roosevelt's reading of Einstein's letter and Szilard's memorandum. But neither document had suited his sense of how to present the information to a busy President. "I am an economist, not a scientist," he would tell friends, "but I had a prior relationship with the President, and Szilard and Einstein agreed I was the right person to make the relevant elaborate scientific material intelligible to Mr. Roosevelt. No scientist could sell it to him." Sachs had therefore prepared his own version of the fission story, a composite and paraphrase of the contents of the Einstein and Szilard presentations. Though he left those statements with Roosevelt, he read neither one of them aloud. He read not Einstein's subsequently famous letter but his own eight-hundred-word summation, the first authoritative report to a head of state of the possibility of using nuclear energy to make a weapon of war. It emphasized power production first, radioactive materials for medical use second and "bombs of hitherto unenvisaged potency and scope" third. It recommended making arrangements with Belgium for uranium supplies and expanding and accelerating experiment but imagined that American industry or private foundations would be willing to foot the bill. To that end it proposed that Roosevelt "designate an individual and a committee to serve as a liaison" between the scientists and the Administration.

Sachs had intentionally listed the peaceful potentials of fission first and second among its prospects. To emphasize the "ambivalence" of the discovery, he said later, the "two poles of good and evil" it embodied, he turned near the end of the discussion to Francis Aston's 1936 lecture, "Forty Years of Atomic Theory"—it had been published in 1938 as part of a collection, *Background to Modern Science,* which Sachs had brought along to the White House—where the English spectroscopist had ridiculed "the more elderly and apelike of our prehistoric ancestors" who "objected to the innovation of cooked food and pointed out the grave dangers attending the use of the newly discovered agency, fire." Sachs read the entire last paragraph of the lecture to Roosevelt, emphasizing the final sentences:

> Personally I think there is no doubt that sub-atomic energy is available all around us, and that one day man will release and control its almost infinite power. We cannot prevent him from doing so and can only hope that he will not use it exclusively in blowing up his next door neighbor.

"Alex," said Roosevelt, quickly understanding, "what you are after is to see that the Nazis don't blow us up."

"Precisely," Sachs said.

Roosevelt called in Watson. "This requires action," he told his aide.

Meeting afterward with Sachs, Watson went by the book. He proposed a committee consisting initially of the director of the Bureau of Standards, an Army representative and a Navy representative. The Bureau of Standards, established by Act of Congress in 1901, is the nation's physics laboratory, charged with applying science and technology in the national interest and for public benefit. Its director in 1939 was Dr. Lyman J. Briggs, a Johns Hopkins Ph.D. and a government scientist for forty-three years who had been nominated by Herbert Hoover and appointed by FDR. The military representatives were Lieutenant Colonel Keith F. Adamson and Commander Gilbert C. Hoover, both ordnance experts.

"Don't let Alex go without seeing me again," Roosevelt had directed Watson. Sachs met the same evening with Briggs, briefed him and proposed he and his committee of two get together with the physicists working on fission. Briggs agreed. Sachs saw the President again and declared himself satisfied. That was good enough for Roosevelt.

Briggs set a first meeting of the Advisory Committee on Uranium for October 21 in Washington, a Saturday. Sachs proposed to invite the emigrés; to counterbalance them Briggs invited Tuve, who found a schedule conflict and deputized Richard Roberts as his stand-in. Fermi, still nursing his Navy grievance, refused to attend but was willing to allow Teller to speak in his behalf. On the appointed day the Hungarian conspiracy breakfasted with Sachs at the Carleton Hotel, the out-of-towners having arrived the night before. From the hotel they proceeded to the Department of Commerce. The meeting then counted nine participants: Briggs, a Briggs assistant, Sachs, Szilard, Wigner, Teller, Roberts, Adamson for the Army and Hoover for the Navy.

Szilard began by emphasizing the possibility of a chain reaction in a uranium-graphite system. Whether such a system would work, he said, depended on the capture cross section of carbon and that was not yet sufficiently known. If the value was large, they would know that a large-scale experiment would fail. If the value was extremely small, a large-scale experiment would look highly promising. An intermediate value would necessitate a large-scale experiment to decide. He estimated the destructive potential of a uranium bomb to be as much as twenty thousand tons of high-explosive equivalent. Such a bomb, he had written in the memorandum Sachs carried to Roosevelt, would depend on fast neutrons and might be "too heavy to be transported by airplane," which meant he was still thinking of exploding natural uranium, not of separating U235.

Adamson, openly contemptuous, butted in. "In Aberdeen," Teller remembers him sneering, "we have a goat tethered to a stick with a ten-foot rope, and we have promised a big prize to anyone who can kill the goat with a death ray. Nobody has claimed the prize yet." As for twenty thousand tons of high explosive, the Army officer said, he'd been standing outside an ordance depot once when it blew up and it hadn't even knocked him down.

Restraining himself, Wigner spoke after Szilard, supporting his compatriot's argument.

Roberts raised serious objection. He was convinced that Szilard's optimism for a chain reaction was premature and his notion of a fast-neutron weapon made of natural uranium misguided. Roberts had co-authored a review of the subject just one month before. It agreed with Szilard that "there are not yet sufficient data to say definitely whether or not a uranium powerhouse is a possibility." But it also assessed—because the DTM had begun assessing—the question of the fast-neutron fission of natural uranium and found, because of resonance capture and extensive scattering of fast neutrons, that it was "very unlikely that the fast neutrons can produce a sufficient number of fissions to maintain a [chain] reaction."

The DTM physicist also pointed out that other lines of research might be more promising than a slow-neutron chain reaction in natural uranium. He meant isotope separation. At the University of Virginia Jesse Beams, formerly Ernest Lawrence's colleague at Yale, was applying to the task the high-speed centrifuges he was developing there. Roberts thought answers to these questions might require several years of work and that research should be left in the meantime to the universities.

Briggs spoke up to defend his committee. He argued vigorously that any assessment of the possibilities of fission at a time when Europe was at war had to include more than physics; it had to include the potential impact of the development on national defense.

Szilard was "astonished," as he told Pegram the next day, at Sachs' "active and enthusiastic" participation in the meeting. Sachs seconded Briggs and the Hungarians. "The issue was too important to wait," he recalled his argument, "and the important thing was to be helpful because if there was something to it there was danger of our being blown up. We had to take time by the forelock, and we had to be ahead."

Then it was Teller's turn. For himself, he announced in his deep, heavily accented voice, he strongly supported Szilard. But he had also been given the task of serving as messenger for Fermi and Tuve, who had discussed these issues in New York and had come to some agreement about them. "I said that this needed a little support. In particular we needed to acquire a good substance to slow down the neutrons, therefore we needed pure graphite, and this is expensive." Jesse Beams' centrifuge work also required support, Teller added.

"How much money do you need?" Commander Hoover wanted to know.

Szilard had not planned to ask for money. "The diversion of Government funds for such purposes as ours appears to be hardly possible," he explained to Pegram the next day, "and I have therefore myself avoided to make any such recommendation." But Teller answered Hoover promptly, probably speaking for Fermi: "For the first year of this research we need six thousand dollars, mostly in order to buy the graphite." ("My friends blamed me because the great enterprise of nuclear energy was to start with such a pittance," Teller reminisces; "they haven't forgiven me yet." Szilard, who would write Briggs on October 26 that the graphite alone for a large-scale experiment would cost at least $33,000, must have been appalled.)

Adamson had anticipated just such a raid on the public treasury. "At this point," says Szilard, "the representative of the Army started a rather longish tirade":

> He told us that it was naive to believe that we could make a significant contribution to defense by creating a new weapon. He said that if a new weapon is created, it usually takes two wars before one can know whether the weapon is any good or not. Then he explained rather laboriously that it is in the end not weapons which win the wars, but the morale of the troops. He went on in this vein for a long time until suddenly Wigner, the most polite of us, interrupted him. [Wigner] said in his high-pitched voice that it was very interesting for him to hear this. He always thought that weapons were very important and that this is what costs money, and this is why the Army needs such a large appropriation. But he was very interested to hear that he was wrong: it's not weapons but the morale which wins the wars. And if this is correct, perhaps one should take a second look at the budget of the Army, and maybe the budget could be cut.

"All right, all right," Adamson snapped, "you'll get your money."

The Uranium Committee produced a report for the President on November 1. It narrowly emphasized exploring a controlled chain reaction "as a continuous source of power in submarines." In addition, it noted, "If the reaction turns out to be explosive in character, it would provide a possible source of bombs with a destructiveness vastly greater than anything now known." The committee recommended "adequate support for a thorough investigation." Initially the government might undertake to supply four tons of pure graphite (this would allow Fermi and Szilard to measure the capture cross section of carbon) and, if justified later, fifty tons of uranium oxide.

Briggs heard from Pa Watson on November 17. The President had read the report, Watson wrote, and wanted to keep it on file. On file is where it remained, mute and inactive, well into 1940.

Even with Szilard and Fermi stalled, fission studies continued at many other American laboratories. Prodded by a late-October letter from Fermi, for example, Alfred Nier at the University of Minnesota finally began preparing to separate enough U235 from U238, using his mass spectroscope, to determine experimentally which isotope is responsible for slow-neutron fission. But to American physicists and administrators in and out of government a bomb of uranium seemed a remote possibility at best. However intense their sympathies, the war was still a European war.

11

Cross Sections

In the days before the war, Otto Frisch remembers, in Hamburg with Otto Stern, he used to run experiments by day and think intensely about physics well into the night. "I regularly came home," Frisch told an interviewer once, "had dinner at seven, had a quarter of an hour's nap after dinner, and then I sat down happily with a sheet of paper and a reading lamp and worked until about one o'clock at night—until I began to have hallucinations. . . . I began to see queer animals against the background of my room, and then I thought, 'Oh, well, better go to bed.' " The young Austrian's hypnagogic visions were "unpleasant feelings" but otherwise "it was an ideal life. I'd never had such a pleasant life, ever—this concentrated five hours work every night."

Through the spring of 1939, in contrast, after his early experiments with fission, Frisch found himself "in a state of complete doldrums. I had a feeling war was coming. What was the use of doing any research? I simply couldn't brace myself. I was in a pretty bad state, feeling, 'Nothing I start now is going to be any good.' " As his aunt, Lise Meitner, worried about her isolation in Stockholm, Frisch worried about his vulnerability in Copenhagen; when British colleagues visited he uncharacteristically campaigned among them:

> I first spoke to Blackett and then Oliphant when they passed through Copenhagen and said that I had a fear that Denmark would soon be overrun by

> Hitler, and if so, would there be a chance for me to go to England in time, because I'd rather work for England than do nothing or be compelled in some way or other to work for Hitler or be sent to a concentration camp.

Mark Oliphant directed the physics department at the University of Birmingham. Rather than initiate some complicated sponsorship he simply invited Frisch to visit him that summer to talk over the problem. "So I packed two small suitcases and traveled by ship and train, just like any tourist." The war overtook him safe in the English Midlands but with nothing more of his possessions on hand than the contents of his two small suitcases. His friends in Copenhagen had to store his belongings and arrange the repossession of the piano he was buying.

Oliphant found him work as an auxiliary lecturer. In that relative security he began to think about physics again. Fission still intrigued him. He lacked the neutron source he would need for direct attack. But he had followed Bohr's theoretical work: the distinction between the fissile characteristics of U235 and U238 in February; the major Bohr-Wheeler paper in September just as the German invasion of Poland brought war, "a great feeling of tense sobriety." He wondered if Bohr was right that U235 was the isotope responsible for slow-neutron fission. He conceived a way to find out: by preparing "a sample of uranium in which the proportions of the two isotopes were changed." That meant at least partly separating the isotopes, as Fermi and Dunning had encouraged Nier to do for the same reason. Frisch read up on methods. The simplest, he decided, was gaseous thermal diffusion, a technique developed by the German physical chemist Klaus Clusius. For equipment it required little more than a long tube standing on end with a heated rod inside running down its center. Fill the tube with some gaseous form of the material to be separated, cool the tube wall by flushing it with water, and "material enriched in the lighter isotope would accumulate near the top . . . while the heavier isotope would tend to go to the bottom."

Frisch set out to assemble his Clusius tube. Progress was slow. He planned to make the tube of glass, but the laboratory glassblower's first priority was Oliphant's secret war work, work about which Frisch, technically an enemy alien, was not supposed to know. Two physicists on Oliphant's staff, James Randall and H. A. H. Boot, were in fact developing the cavity magnetron, an electron tube capable of generating intense microwave radiation for ground and airborne radar—in C. P. Snow's assessment "the most valuable English scientific innovation in the Hitler war."

Meanwhile the British Chemical Society asked Frisch to write a review of advances in experimental nuclear physics for its annual report. "I managed to write that article in my bed-sitter where in daytime, with the gas fire going all day, the temperature rose to 42° Fahrenheit . . . while at night the water froze in the tumbler at my bedside." He wore his winter coat, set his typewriter on his lap and pulled his chair close to the fire. "The radiation from the gas fire stimulated the blood supply to my brain, and the article was completed on time."

Frisch's review article mentioned the possibility of a chain reaction only to discount it. He based that conclusion on Bohr's argument that the U238 in natural uranium would scatter fast neutrons, slowing them to capture-resonance energies; the few that escaped capture would not suffice, he thought, to initiate a slow-neutron chain reaction in the scarce U235. Slow neutrons in any case could never produce more than a modest explosion, Frisch pointed out; they took too long slowing down and finding a nucleus. As he explained later:

> That process would take times of the order of a sizeable part of a millisecond [i.e., a thousandth of a second], and for the whole chain reaction to develop would take several milliseconds; once the material got hot enough to vaporize, it would begin to expand and the reaction would be stopped before it got much further. So the thing might blow up like a pile of gunpowder, but no worse, and that wasn't worth the trouble.

Not long from Nazi Germany, Frisch found his argument against a violently explosive chain reaction reassuring. It was backed by the work of no less a theoretician than Niels Bohr. With satisfaction he published it.

It had seen the light of day before, most notably in an August 5, 1939, letter from Member of Parliament Winston Churchill to the British Secretary of State for Air. Concerned that Hitler might bluff Neville Chamberlain with threats of a new secret weapon, Churchill had collected a briefing from Frederick Lindemann and written to caution the cabinet not to fear "new explosives of devastating power" for at least "several years." The best authorities, the distinguished M.P. emphasized with a nod to Niels Bohr, held that "only a minor constituent of uranium is effective in these processes." That constituent would need to be laboriously extracted for any large-scale effects. "The chain process can take place only if the uranium is concentrated in a large mass," Churchill continued, slightly muddling the point. "As soon as the energy develops, it will explode with a mild detonation before any really violent effects can be produced. It might be as good as our present-day explosives, but it is unlikely to produce anything very much more dangerous." He concluded optimistically: "Dark hints will be dropped and terrifying whispers will be assiduously circulated, but it is to be hoped that nobody will be taken in by them."

40

40. The leaders of wartime American science, 1940. *L. to r,* Ernest Lawrence, Arthur Compton, Vannevar Bush, James Bryant Conant, Karl Compton, Alfred Loomis.

41. War came to Europe with the German invasion of Poland on September 1, 1939. Here Polish citizens in Warsaw study Nazi proclamations. Roosevelt appealed to the belligerents to refrain from bombing civilians.

41

42

42. Genia and Rudolf Peierls. While American efforts stalled, Peierls and Otto Frisch in England in 1940 worked out the essential theory of a fast-fission uranium bomb fueled with U235 and convinced his British colleagues that it was feasible.

43

43. Eugene T. Booth (*left*) and John Dunning (*right*) decided in 1940 to experiment with gaseous barrier diffusion to separate U235 from U238. The British took the same route.

44

44. Economist Alexander Sachs had carried the Einstein letter of warning to Roosevelt; he pushed the conservative Briggs committee without success for another year.

45

45. Nobel laureate theoretician Eugene P. Wigner, the third member of the "Hungarian conspiracy" with Szilard and Edward Teller. Szilard called him "the conscience of the project" from beginning to end.

46. Alfred O. C. Nier separated a sample of U235 with his mass-spectrograph; Columbia used it to confirm the rare isotope's responsibility for slow-neutron fission.

46

47

48

47. Australian Mark Oliphant visited the United States in 1941 and helped goad the American atomic-bomb program to commitment.

48. Glenn Seaborg, the codiscoverer of plutonium, with his bride-to-be, Helen Griggs, Los Angeles, 1942.

49

49. Strategic bombing soon bridged the barrier of the English Channel. Here: Coventry Cathedral, destroyed by German bombs.

50

50. The Japanese surprise attack on Pearl Harbor, December 7, 1941, finally precipitated the entry of the United States into the war against not only Japan but Germany and Italy as well. Immediately U.S. atomic bomb development accelerated.

51

51. Franklin Roosevelt saw the long-term potential and instinctively reserved nuclear-weapons policy to himself.

52

52. Louis B. Werner and Burris Cunningham in Chicago the day they isolated the first pure sample of plutonium, August 20, 1942.

53

53. Chicago Pile Number One, the first man-made nuclear reactor, under construction at the University of Chicago, November 1942. Lower layer holds uranium oxide pseudospheres, unfinished dead layer overlying. Note hammer in foreground for scale.

54

54. Oak Ridge Alpha I calutron racetrack for electromagnetic separation of U235. Silver-wound magnets protrude like ribs spaced by semicircular mass-spectrometer tanks. Spare tanks in left foreground.

55

55. K-25 gaseous-diffusion plant, Oak Ridge, Tennessee. Built to monumental scale, the structure is half a mile long with 42.6 acres under roof.

56

56. William S. "Deke" Parsons and Philip Abelson. Parsons directed ordnance development at Los Alamos; Abelson pioneered liquid thermal diffusion for uranium enrichment.

57

57. Abelson's liquid thermal diffusion rack. Steam circulated through an inner pipe, cooling water through an outer, causing U235 to diffuse inward and circulate upward. The resulting enriched material fed Ernest Lawrence's hungry calutrons.

58

59

58. U.S. plutonium-production complex on the Columbia River at Hanford, Washington. Twelve-hundred-ton graphite reactors drilled with 2,004 channels held uranium slugs; neutrons from fission transmuted 250 parts per million of U238 to plutonium. D pile in foreground between water tanks. 59. Pile face showing slug channels.

60

61

60. "Queen Mary" plutonium separation plant, Hanford. Dissolved irradiated slugs progressed by remote control through separation stages down the length of this 800-foot concrete building. 61. Interior showing processing cells.

62

63

62. The Norsk Hydro hydrogen electrolysis plant at Vemork, Norway, produced heavy water for German uranium research until disabled by Allied bombing. 63. The ferry *Hydro* on Lake Tinnsjö, Norway, sunk by commandos while carrying the last Norsk Hydro heavy water to Germany.

64

64. A secret laboratory was established in 1943 north of Santa Fe, New Mexico, on the forested Los Alamos mesa at 7,200 feet. Here scientists and engineers assembled to design and build the first atomic bombs. The Army Corps of Engineers constructed fourplex family apartments for housing.

65

65. Experiments at Los Alamos determined the critical masses of U235 and Pu239. Adding U235 cubes to a subcritical assembly within blocks of beryllium tamper measurably increased neutron flux. 66. The Los Alamos Tech Area.

66

67

67. The guillotine mechanism for studying supercritical assemblies (the Dragon experiment). 68. The first RaLa test. Note Army tanks for observers, lower left.

68

69

69. Niels Bohr learned of the U.S. program in 1943. The bomb, he foresaw, would end major war and challenge the nation-states to move toward an open world.

70

70. Polish mathematician Stanislaw Ulam calculated hydrodynamics at Los Alamos; in 1951 he conceived the essential breakthrough arrangement for a workable H-bomb.

71. Hungarian theorist Edward Teller (*left*) helped make the plutonium bomb work; Navy physicist Norris Bradbury directed its test assembly at Trinity. Teller guided H-bomb theoretical studies at Los Alamos.

71

72

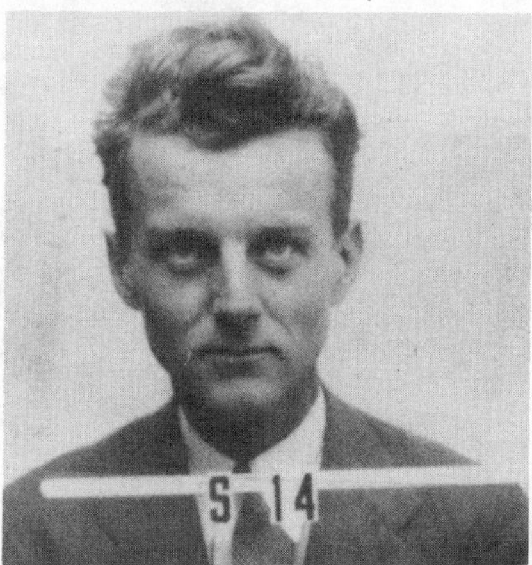

73

72. Seth Neddermeyer. His idea of using explosives to squeeze a nuclear core to criticality saved the plutonium bomb when impurities threatened its design. 73. Kitty Oppenheimer at Los Alamos with Peter.

74. The Los Alamos staff worked a six-day week; Sundays there was time for recreation. Shown here on a Sunday hike, *l. to r.*, standing, Emilio Segrè, Enrico Fermi, Hans Bethe, H. H. Staub, Victor Weisskopf; seated, Erika Staub, Elfriede Segrè.

74

75

75. The Normandy invasion in May 1944 led ultimately to Allied victory in Europe 12 months later. Supreme Commander Dwight D. Eisenhower visited the front lines.

76

76. Ferocious Japanese resistance claimed increasing U.S. casualties in the Pacific—30,000 of the 60,000 Americans committed on Iwo Jima, where 20,000 Japanese died.

77

77. At Los Alamos, Ukrainian chemist George Kistiakowsky (here riding Crisis) manufactured and tested the explosive lenses for the Fat Man bomb.

78

78. Early model Fat Man implosion bomb, upper segments removed to show interior. Overall diameter is about 5 feet.

79

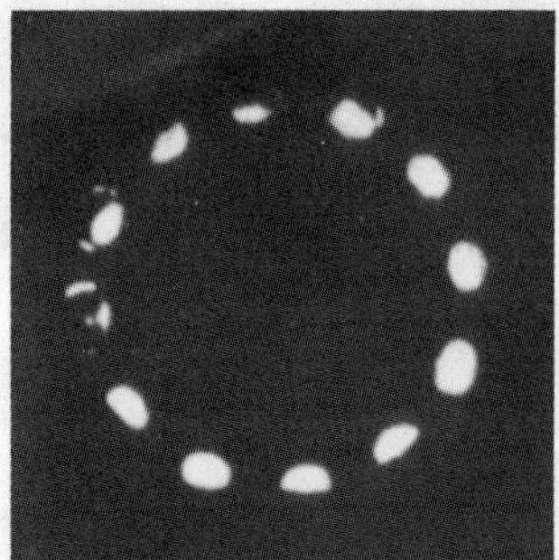

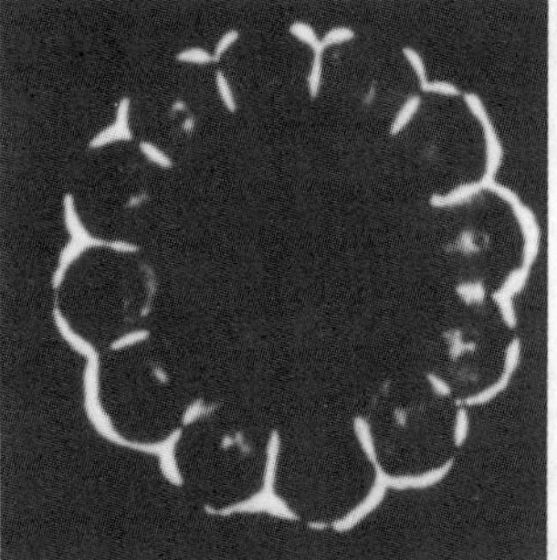

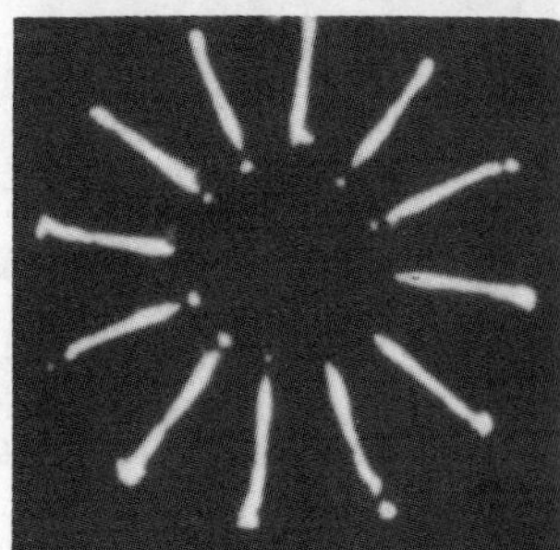

79. X-ray motion picture frames of implosion experiment. Note compression of core in final frames.

81

80. Shot tower at Trinity Site in the desert north of Alamogordo, N.M., where Los Alamos prepared in the spring of 1945 to test the plutonium bomb. 81. Base Camp.

82

82. After inserting the initiator into the core and mounting the assembly in a cylindrical plug of tamper, the crew delivered it to the tower for insertion into the bomb. 83. Firing and instrumentation bunkers.

84. Theoretician Philip Morrison (*left*), here with Ernest Lawrence, escorted the plutonium core to Trinity.

85

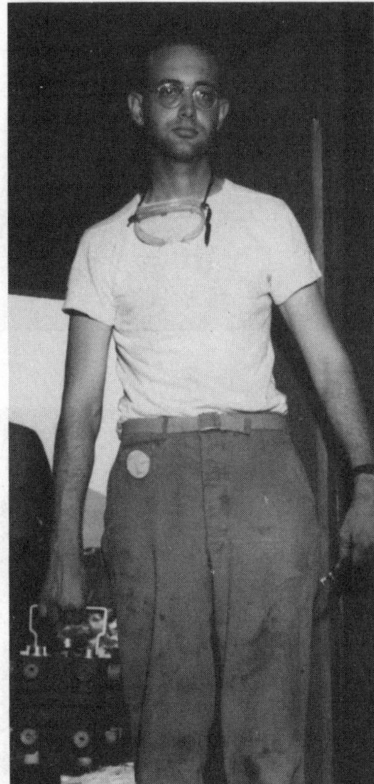

85. Sgt. Herbert Lehr delivered the core in its shock-mounted case to the McDonald Ranch assembly room at Trinity about 6 P.M., July 12, 1945. Assembly proceeded the following morning.

86

86. After inserting the initiator into the core and mounting the assembly in a cylindrical plug of tamper, the crew delivered it to the tower for insertion into the bomb. 87. The completely assembled Trinity bomb in its tower, with Norris Bradbury attending, July 15, 1945.

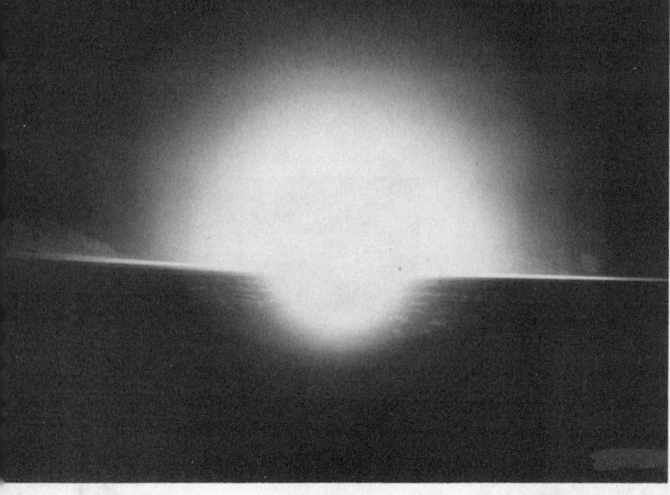

88–93. The first man-made nuclear explosion: Trinity, 0529:45 hours et seq., July 16, 1945. The sequence runs down this page and up the next. Note change of scale as the fireball expands. "This power of nature which we had first understood it to be," said I. I. Rabi, "—well, there it was."

88

89

90

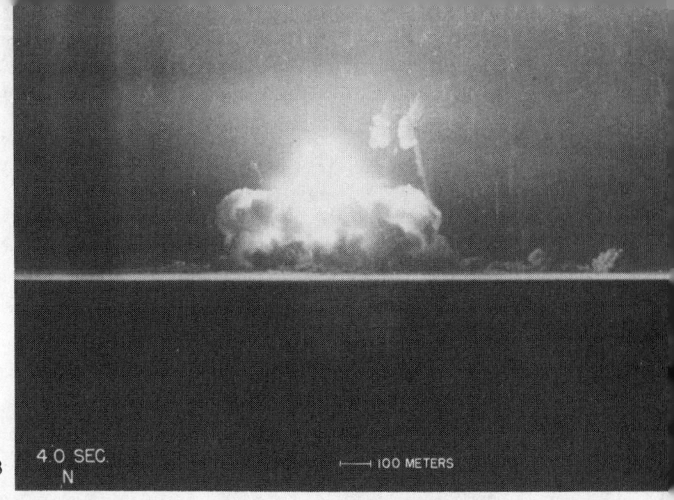

93

92

91

94

95

94. Twenty-four hours later Trinity, seen from the air, revealed a radioactive crater of green, glassy, fused desert sand. (Smaller crater to the south marks the 100-ton explosive test.)

95. Los Alamos director Robert Oppenheimer (*left*) subsequently visited the site with Manhattan Project commanding general Leslie R. Groves and found only the reinforcing rods of the tower footings left unvaporized.

96. In a final postwar celebration the British mission at Los Alamos pantomimed the war years. A stepladder stood in for the Trinity shot tower. Note Otto Frisch (third from left) in skirt playing housemaid.

96

Frisch found a friend that year in a fellow emigré at Birmingham, the theoretician Rudolf Peierls. A well-off Berliner, a slender man with a boyish face, a notable overbite and a mind of mathematical austerity, Peierls was born in 1907 and had arrived in England in 1933 on a Rockefeller Fellowship to Cambridge. With the Nazi purge of the German universities he chose to remain in England. He would be naturalized as a British citizen in February 1940, but until then he was technically an enemy alien. When Oliphant consulted with him from time to time on the mathematics of resonant cavities—important for microwave radar—both men were careful to pretend that the question was purely academic.

Peierls had already contributed significantly to the debate on the possible explosive effects of fission. The previous May one of Frédéric Joliot's associates in Paris, Francis Perrin, had published a first approximate formula for calculating the critical mass of uranium—the amount of uranium necessary to sustain a chain reaction. A lump smaller than a critical mass would be inert; a lump of critical size would explode spontaneously upon assembly.

The possibility of a critical mass is anchored in the fact that the surface area of a sphere increases more slowly with increasing radius than does the volume (as nearly r^2 to r^3). At some particular volume, depending on the density of the material and on its cross sections for scattering, capture and fission, more neutrons should find nuclei to fission than find surface to escape from; that volume is then the critical mass. Estimating the several cross sections of natural uranium, Francis Perrin put its critical mass at forty-four tons. A tamper around the uranium of iron or lead to bounce back neutrons might reduce the requirement, Perrin calculated, to only thirteen tons.

Peierls saw immediately that he could sharpen Perrin's formula. He did so in a theoretical paper he worked out in May and early June 1939 that the Cambridge Philosophical Society published in its *Proceedings* in October. Because a critical-mass formula based on slow-neutron fission would be mathematically complicated, requiring that the characteristics of the moderator be taken into account, Peierls proposed to consider "a simplified case": fission by unmoderated fast neutrons. Plugging in the fission cross section of natural uranium, which was essentially the fission cross section of U238, gave a critical mass, notes Peierls, "of the order of tons." As a weapon, an object of that size was too unwieldy to take seriously. "There was of course no chance of getting such a thing into any aeroplane, and the paper appeared to have no practical significance." Peierls was aware of the British and American concern for secrecy, but in this case he saw no reason not to publish.

The USSR opportunistically invaded Finland at the end of November. In the rest of Europe the strange standoff prevailed that isolationist Idaho senator William Borah would label the "phony war." The Peierlses moved to a larger house; early in the new year they generously invited Frisch to live with them. Genia Peierls, who was Russian, took the bachelor Austrian in hand. She "ran her house," writes Frisch, "with cheerful intelligence, a ringing Manchester voice and a Russian's sovereign disregard of the definite article. She taught me to shave every day and to dry dishes as fast as she washed them, a skill that has come in useful many times since." Life at the Peierlses was entertaining, but Frisch walked home through ominous blackouts so dark that he sometimes stumbled over roadside benches and could distinguish fellow pedestrians only by the glow of the luminous cards they had taken to wearing in their hatbands. Thus reminded of the continuing threat of German bombing, he found himself questioning his confident Chemical Society review: "Is that really true what I have written?"

Sometime in February 1940 he looked again. There had always been four possible mechanisms for an explosive chain reaction in uranium:

(1) slow-neutron fission of U238;
(2) fast-neutron fission of U238;
(3) slow-neutron fission of U235; and
(4) fast-neutron fission of U235.

Bohr's logical distinction between U238 and thorium on the one hand and U235 on the other ruled out (1): U238 was not fissioned by slow neutrons. (2) was inefficient because of scattering and the parasitic effects of the capture resonance of U238. (3) was possibly applicable to power production but too slow for a practical weapon. But what about (4)? Apparently no one in Britain, France or the United States had asked the question quite that way before.

If Frisch now glimpsed an opening into those depths he did so because he had looked carefully at isotope separation and had decided it could be accomplished even with so fugitive an isotope as U235. He was therefore prepared to consider the behavior of the pure substance unalloyed with U238, as Bohr, Fermi and even Szilard had not yet been. "I wondered—assuming that my Clusius separation tube worked well—if one could use a number of such tubes to produce enough uranium-235 to make a truly explosive chain reaction possible, not dependent on slow neutrons. How much of the isotope would be needed?"

He shared the problem with Peierls. Peierls had his critical-mass formula. In this case it required the cross section for fast-neutron fission of U235, a number no one knew because no one had yet separated a sufficient amount of the rare isotope to determine its cross section by experiment, the only way the number could be reliably known. Nevertheless, says Peierls, "we had read the paper of Bohr and Wheeler and had understood it, and it seemed to convince us that in those circumstances for neutrons in U235 the cross-section would be dominated by fission." Peierls could state simply what followed: "If a neutron hit the [U235] nucleus something was bound to happen."

What followed thus made the cross section intuitively obvious: it would be more or less the same as the familiar cross section that expressed the odds of hitting the uranium nucleus with a neutron at all—the geometric cross section, 10^{-23} square centimeters, an entire order of magnitude larger than the fission cross sections previously estimated for natural uranium that were small multiples of 10^{-24}.

"Just sort of playfully," Frisch writes, he plugged 10^{-23} cm^2 into Peierls' formula. "To my amazement" the answer "was very much smaller than I had expected; it was not a matter of tons, but something like a pound or two." A volume less than a golf ball for a substance so heavy as uranium.

But would that pound or two explode or fizzle? Peierls easily produced an estimate. The chain reaction would have to proceed faster than the vaporizing and swelling of the heating metal ball. Peierls calculated the time between neutron generations, between $1 \times 2 \times 4 \times 8 \times 16 \times 32 \times 64 \ldots$, to be about four millionths of a second, much faster than the several thousandths of a second Frisch had estimated for slow-neutron fission.

Then how destructive was the consequent explosion? Some eighty generations of neutrons—as many as could be expected to multiply before the swelling explosion separated the atoms of U235 enough to stop the chain reaction—still millionths of a second in total, gave temperatures as hot as the interior of the sun, pressures greater than the center of the earth where iron flows as a liquid. "I worked out the results of what such a nuclear explosion would be," says Peierls. "Both Frisch and I were staggered by them."

And finally, practically: could even a few pounds of U235 be separated from U238? Frisch writes:

> I had worked out the possible efficiency of my separation system with the help of Clusius's formula, and we came to the conclusion that with something like a hundred thousand similar separation tubes one might produce a pound of reasonably pure uranium-235 in a modest time, measured in weeks. At that point we stared at each other and realized that an atomic bomb might after all be possible.

"The cost of such a plant," Frisch adds for perspective, "would be insignificant compared with the cost of the war."

"Look, shouldn't somebody know about that?" Frisch then asked Peierls. They hastened their calculations to Mark Oliphant. "They convinced me," Oliphant testifies. He told them to write it all down.

They did, succinctly, in two parts, one part three typewritten pages, the other even briefer. Talking about it made them nervous, Peierls recalls (by then it was March and the exceptional cold had given way to warmer weather):

> I remember we were writing our memorandum . . . together in my room in the Physics Lab on the ground floor; it was a fine day and the window was open . . . and while we were discussing the wording a face suddenly appeared in the open window. And we were a little worried! It turned out that just underneath the window (which was facing south) people were growing some tomato plants, and somebody had been there bending down inspecting what these plants were doing.

The first of the two parts they titled "On the construction of a 'super-bomb'; based on a nuclear chain reaction in uranium." It was intended, they wrote, "to point out and discuss a possibility which seems to have been overlooked in . . . earlier discussions." They proceeded to cover the same ground they had previously covered together in private, noting that "the energy liberated by a 5 kg bomb would be equivalent to that of several thousand tons of dynamite." They described a simple mechanism for arming the weapon: making the uranium sphere in two parts "which are brought together first when the explosion is wanted. Once assembled, the bomb would explode within a second or less." Springs, they thought, might pull the two small hemispheres together. Assembly would have to be rapid or the chain reaction would begin prematurely, destroying the bomb but not much else. A byproduct of the explosion—about 20 percent of its energy, they thought—would be radiation, the equivalent of "a hundred tons of radium" that would be "fatal to living beings even a long time after the explosion." Effective protection from the weapon would be "hardly possible."

The second report, "Memorandum on the properties of a radioactive 'super-bomb,' " a less technical document, was apparently intended as an alternative presentation for nonscientists. This study explored beyond the technical questions of design and production to the strategic issues of possession and use; it managed at the same time both seemly innocence and extraordinary prescience:

> 1. As a weapon, the super-bomb would be practically irresistible. There is no material or structure that could be expected to resist the force of the explosion. . . .

2. Owing to the spreading of radioactive substances with the wind, the bomb could probably not be used without killing large numbers of civilians, and this may make it unsuitable as a weapon for use by this country. . . .

3. . . . It is quite conceivable that Germany is, in fact, developing this weapon. . . .

4. If one works on the assumption that Germany is, or will be, in the possession of this weapon, it must be realised that no shelters are available that would be effective and could be used on a large scale. The most effective reply would be a counter-threat with a similar weapon.

Thus in the first months of 1940 it was already clear to two intelligent observers that nuclear weapons would be weapons of mass destruction against which the only apparent defense would be the deterrent effect of mutual possession.

Frisch and Peierls finished their two reports and took them to Oliphant. He quizzed the men thoroughly, added a cover letter to their memoranda ("I have considered these suggestions in some detail and have had considerable discussion with the authors, with the result that I am convinced that the whole thing must be taken rather seriously, if only to make sure that the other side are not occupied in the production of such a bomb at the present time") and sent letter and documents off to Henry Thomas Tizard, an Oxford man, a chemist by training, the driving force behind British radar development, the civilian chairman of the Committee on the Scientific Survey of Air Defense—better known as the Tizard Committee—which was the most important British committee at the time concerned with the application of science to war.

"I have often been asked," Otto Frisch wrote many years afterward of the moment when he understood that a bomb might be possible after all, before he and Peierls carried the news to Mark Oliphant, "why I didn't abandon the project there and then, saying nothing to anybody. Why start on a project which, if it was successful, would end with the production of a weapon of unparalleled violence, a weapon of mass destruction such as the world had never seen? The answer was very simple. We were at war, and the idea was reasonably obvious; very probably some German scientists had had the same idea and were working on it."

Whatever scientists of one warring nation could conceive, the scientists of another warring nation might also conceive—and keep secret. That early in 1939 and early 1940, the nuclear arms race began. Responsible men who properly and understandably feared a dangerous enemy saw their own ideas reflected back to them malevolently distorted. Ideas that appeared defensive in friendly hands seen the other way around appeared aggressive. But they were the same ideas.

* * *

Werner Heisenberg sent his considered conclusions to the German War Office on December 6, 1939, while Fermi and Szilard waited for the $6,000 the Briggs Uranium Committee had allocated to them for graphite studies and Frisch prepared his pessimistic Chemical Society review. Heisenberg thought fission could lead to energy production even with ordinary uranium if a suitable moderator could be found. Water would not do, but "heavy water [or] very pure graphite would, on the other hand, suffice on present evidence." The surest method for building a reactor, Heisenberg wrote, "will be to enrich the uranium-235 isotope. The greater the degree of enrichment, the smaller the reactor can be made." Enrichment—increasing the proportion of U235 to U238—was also "the only method of producing explosives several orders of magnitude more powerful than the strongest explosives yet known." (The phrase indicates Heisenberg understood the possibility of fast-neutron fission even before Frisch and Peierls did.)

During the same period Paul Harteck in Hamburg was building a Clusius separation tube; in December he tested it by successfully separating isotopes of the heavy gas xenon. He traveled to Munich at Christmastime to discuss design improvements with Clusius, who was professor of physical chemistry at the university there. Auer, the thorium specialists, purveyors of gas mantles and radioactive toothpaste, delivered the first ton of pure uranium oxide processed from Joachimsthal ores to the War Office in January 1940. German uranium research was thriving.

Acquiring a suitable moderator looked more difficult. The German scientists favored heavy water, but Germany had no extraction plant of its own. Harteck calculated at the beginning of the year that a coal-fired installation would require 100,000 tons of coal for each ton of heavy water produced, an impossibility in wartime. The only source of heavy water in quantity in the world was an electrochemical plant built into a sheer 1,500-foot granite bluff beside a powerful waterfall at Vemork, near Rjukan, ninety miles west of Oslo in southern Norway. Norsk Hydro-Elektrisk Kvaelstofaktieselskab produced the rare liquid as a byproduct of hydrogen electrolysis for synthetic ammonia production.

I.G. Farben, the German chemical cartel assembled by Bayer's Carl Duisberg in the 1920s, owned stock in Norsk Hydro; learning of the War Office's need it approached the Norwegians with an offer to buy all the heavy water on hand, about fifty gallons worth some $120,000, and to order more at the rate of at least thirty gallons a month. Norsk Hydro was then producing less than three gallons a month, enough in the prewar years to glut the small physics-laboratory market. It wanted to know why Germany needed so vast a quantity. I.G. Farben chose not to say. In February the

Norwegian firm refused either to sell its existing stock or to increase production.

Heavy water also impressed the French team, a fact Joliot pased on to the French Minister of Armament, Raoul Dautry. When Dautry heard about the German bid for Norsk Hydro's supply he decided to win the water for France. A French bank, the Banque de Paris et des Pays Bas, controlled a majority interest in the Norwegian company and a former bank officer, Jacques Allier, was now a lieutenant in Dautry's ministry. Dautry briefed the balding, bespectacled Allier with Joliot on hand on February 20: the minister wanted the lieutenant to lead a team of French secret-service agents to Norway to acquire the heavy water.

Allier slipped into Oslo under an assumed name and met with the general manager of Norsk Hydro at the beginning of March. The French officer was prepared to pay up to 1.5 million kroner for the water and even to leave half for the Germans, but once the Norwegian heard what military purpose the substance might serve he volunteered his entire stock and refused payment. The water, divided among twenty-six cans, left Vemork by car soon afterward on a dark midnight. From Oslo Allier's team flew it to Edinburgh in two loads—German fighters forced down for inspection a decoy plane Allier had pretended to board at the time of the first loading—and then transported it by rail and Channel ferry to Paris, where Joliot prepared through the winter and spring of the phony war to use it in both homogeneous and heterogeneous uranium-oxide experiments.

Nuclear research in the Soviet Union during this period was limited to skillful laboratory work. Two associates of Soviet physicist Igor Kurchatov reported to the *Physical Review* in June 1940 that they had observed rare spontaneous fissioning in uranium. "The complete lack of any American response to the publication of the discovery," writes the American physicist Herbert F. York, "was one of the factors which convinced the Russians that there must be a big secret project under way in the United States." It was not yet big, but by then it had begun to be secret.

Japanese studies toward an atomic bomb began first within the military. The director of the Aviation Technology Research Institute of the Imperial Japanese Army, Takeo Yasuda, a lieutenant general and an alert electrical engineer, conscientiously followed the international scientific literature that related to his field; in the course of his reading in 1938 and 1939 he noticed and tracked the discovery of nuclear fission. In April 1940, foreseeing fission's possible consequences, he ordered an aide who was scientifically trained, Lieutenant Colonel Tatsusaburō Suzuki, to prepare a full report. Suzuki went to work with a will.

* * *

Niels Bohr had returned from Princeton to Copenhagen at the beginning of May 1939, preoccupied with the gathering European apocalypse. His friends had urged him to send for his family and remain in the United States. He had not been tempted. Refugees still escaping from Germany and now fleeing Central Europe as well needed him; his institute needed him; Denmark needed him. Hitler proposed on May 31 to compromise the neutrality of the Scandinavian countries with nonagression pacts. The pragmatic Danes alone accepted, fully aware the pact was worthless and even demeaning but unwilling to invite invasion for a paper victory. By autumn, when the John Wheelers offered to shelter one of Bohr's sons in Princeton for the duration of the conflict, Bohr reserved the offer against future need. "We are aware that a catastrophe might come any day," he wrote in the midst of Poland's agony.

Catastrophe for Denmark waited until April 1940 and came then with brutal efficiency. Bohr was lecturing in Norway. The British had announced their intention to mine Norwegian coastal waters against shipment of Norwegian iron ore to Nazi Germany. On the final evening of his lecture tour, April 8, Bohr dined with the King of Norway, Haakon VII, and found King and government lost in gloom at the prospect of a German attack. After dinner he boarded the night train for Copenhagen. A train ferry carried the cars across the Öresund at night to Helsingør while the passengers slept. Danish police pounding on compartment doors woke them to the news: the Germans had invaded not only Norway but Denmark as well. Two thousand German troops hidden in coal freighters moored near Langelinie, the Copenhagen pier of Hans Christian Andersen's Little Mermaid, had stormed ashore in the early morning, so unexpected a sight that night-shift workers bicycling home thought a motion picture was being filmed. A major German force had marched north through Schleswig-Holstein onto the Danish peninsula as well, crossing the border before dawn. German aircraft marked with black crosses dominated the air. German warships commanded the Kattegat and Skagerrak passages that open Denmark and southern Norway to the North Sea.

The Norwegians fought back, determined that their King, court and parliament must escape to exile. The Danes, in their flat country where Panzers might roll, did not. Rifle fire crackled in the streets of Copenhagen in the early morning, but King Christian X ordered an immediate ceasefire, which took effect at 6:25 A.M. By the time Bohr's train arrived in the capital city what Churchill would call "this ruthless *coup*" was complete, the streets littered with green surrender leaflets, the King preparing to receive the German chief of staff. Danish resistance would be dedicated and effective, but it would take less suicidal forms than open battle with the Wehrmacht.

The American Embassy quickly passed word that it could guarantee the Bohrs safe passage to the United States. Bohr again chose duty. His immediate concern was to burn the files of the refugee committee that had helped hundreds of emigrés to escape to exile. "It was characteristic of Niels Bohr," his collaborator, Stefan Rozental, writes, "that one of the first things he did was to contact the Chancellor of the University and other Danish authorities in order to protect those of the staff at the Institute whom the Germans might be expected to persecute." Those were Poles first of all, but Bohr also sought out government leaders to argue for concerted Danish resistance to any German attempt to install anti-Semitic laws in Denmark.

He even found time on the day of the occupation to worry about the large gold Nobel Prize medals that Max von Laue and James Franck had given him for safekeeping. Exporting gold from Germany was a serious criminal offense and their names were engraved on the medals. George de Hevesy devised an effective solution—literally: he dissolved the medals separately in acid. As solutions of black liquid in unmarked jars they sat out the war innocently on a laboratory shelf. Afterward the Nobel Foundation recast them and returned them to their owners.

Norsk Hydro was a prime German objective and there was heavy fighting around Rjukan, which held out until May 3, the last town in southern Norway to surrender. Then a management under duress reported to Paul Harteck that its heavy-water facility, the Vemork High Concentration Plant, could be expanded to increase production of the ideal neutron moderator to as much as 1.5 tons per year.

"What I should like," Henry Tizard wrote Mark Oliphant after he had studied the Frisch-Peierls memoranda, "would be to have quite a small committee to sit soon to advise what ought to be done, who should do it, and where it should be done, and I suggest that you, Thomson, and say Blackett, would form a sufficient nucleus for such a committee." Thomson was G. P. Thomson, J.J.'s son, the Imperial College physicist who had ordered up a ton of uranium oxide the previous year to study and felt ashamed at the absurdity. He had concluded after neutron-bombardment experiments that a chain reaction in natural uranium was unlikely and a war project therefore impractical. Tizard, who had been skeptical to begin with and had taken Thomson's conclusions as support for his skepticism, appointed Thomson chairman of the small committee; James Chadwick, now at Liverpool, his assistant P. B. Moon and Rutherford protégé John

Douglas Cockcroft were added to the list. Blackett was busy with other war work, although he would join the committee later. The group met informally for the first time on April 10 in the Royal Society's quarters at Burlington House.

It probably met as much to hear a visitor, the ubiquitous Jacques Allier of the Banque de Paris and the French Ministry of Armament, as to discuss the Frisch-Peierls work. Allier warned the British physicists about the German interest in heavy-water production and bid for collaboration on nuclear research between Britain and France. Only then, Thomson notes in the minutes he kept, did they consider "the possibility of separating isotopes . . . and it was agreed that the prospects were sufficiently good to justify small-scale experiments on uranium hexafluoride [a gaseous uranium compound]." They proposed rather ungenerously to remind Frisch to avoid "any possible leakage of news in view of the interest shown by the Germans." They were willing to inform him that his memorandum was being considered but not to supply details. (Peierls' name seems not yet to have made an impression on Thomson, and Tizard apparently retained the second Frisch-Peierls memorandum in his files.) "We entered the project with more scepticism than belief," the committee would report later, "though we felt it was a matter which had to be investigated." Thomson's minutes make that skepticism evident. Tizard for his part wrote Lindemann's brother Charles, a science adviser to the British Embassy in Paris, that he considered the French "unnecessarily excited" about the perils of German nuclear research. "I still . . . think that [the] probability of anything of real military significance is very low," he estimated in a note written the same week to the British War Cabinet staff.

It might have been as unpromising a start as the first meeting of the Briggs Uranium Committee had been, but the men on the Thomson committee were active, competent physicists, not military ordnance specialists, and whatever their initial skepticism they understood where the numbers Frisch and Peierls had used came from and what they might mean. At a second meeting on April 24 Thomson recorded laconically that "Dr. Frisch produced some notes to show that the uranium bomb was feasible." Many years later Oliphant recalled a more expansive response: "The Committee generally was electrified by the possibility." Chadwick's good opinion helped. He had just begun exploring fast-neutron fission himself with his new Liverpool cyclotron, the first in England, when he saw the Frisch-Peierls memorandum. At the April 24 meeting he awarded the emigrés' work chagrined confirmation: he "was embarrassed," says Oliphant, "confessing that he had reached similar conclusions, but did not feel justified in reporting them until more was known about the neutron cross sections from experiments. Peierls and Frisch had used calculated values. However, this confirmatory evidence led the Committee to pay great attention to the development of techniques for . . . separation."

Chadwick agreed to undertake the necessary studies. For several more weeks, until their protests through Oliphant registered with Thomson, Frisch and Peierls would be walled off from their own secrets. But work toward a bomb of chain-reacting uranium was now fairly begun, and this time it had found the right—fast—track.

Szilard chafed. The months after the first Uranium Committee meeting became "the most curious period of my life." No one called. "We heard nothing from Washington at all. . . . I had assumed that once we had demonstrated that in the fission of uranium neutrons are emitted, there would be no difficulty in getting people interested; but I was wrong." The Uranium Committee's November 1 report had in fact been languishing in Roosevelt's files; Watson finally decided on his own in early February 1940 to bring it up again. He asked Lyman Briggs if he had anything to add. Briggs reported the transfer, finally, of the $6,000 for Fermi's work on neutron absorption in graphite. That was "a crucial undertaking," Briggs said; he imagined it would determine "whether or not the undertaking has a practical application." He proposed to wait for results.

Something other than Briggs' penurious methodology triggered a new burst of activity from Szilard. He had spent the winter preparing a thorough theoretical study, "Divergent chain reactions in systems composed of uranium and carbon"—divergent in this case meaning chain reactions that continue to multiply once begun (the document's first footnote, numbered zero, cited "H. G. Wells, The World Set Free [1913]"). Early in the new year Joliot's group reported a uranium-water experiment that "seemed to come so close to being chain-reacting," says Szilard, "that if we improved the system somewhat by replacing water with graphite, in my opinion we should have gotten over the hump." He arranged lunch with Fermi to discuss the French paper. "I asked him, 'Did you read Joliot's paper?' He said he did. I asked him, 'What did you think of it?' and Fermi said, 'Not much.' " Szilard was furious. "At which point I saw no reason to continue the conversation and went home."

He traveled again to Princeton to see Einstein. They worked up another letter and sent it under Einstein's signature to Sachs. It emphasized the secret German uranium research at the Kaiser Wilhelm Institutes, about which they had learned from the physical chemist Peter Debye, the 1936 Nobel laureate in chemistry and director of the physics institute at Dahlem, who had been expelled recently to the United States, ostensibly on leave of absence, when he refused to give up Dutch citizenship and join the Nazi Reich. Sachs sent the Einstein letter on to Pa Watson for FDR. But Watson thought it sensible to check first with the Uranium Committee. Adamson responded, echoing Briggs: everything depended on the graphite measurements at Columbia. Watson proposed to wait for the official report. Sachs may have rebutted; Roosevelt wrote the gadfly economist on April 5 emphasizing that the Briggs committee was "the most practical method of continuing this research" but also calling for another committee meeting that Sachs might attend. Briggs dutifully scheduled it for Saturday afternoon, April 27.

In the meantime another development intervened. Alfred Nier at the University of Minnesota had gone to work, after Fermi wrote urging him again to do so, to prepare to separate measurable samples of U235 and U238. John Dunning sent him uranium hexafluoride, a highly corrosive compound that is a white solid at room temperature but volatilizes to a gas when heated to 140°F. "I worked with this for a couple of months in late 1939," Nier remembers. Unfortunately the gas was too volatile; it dispersed through Nier's three-foot glass spectrometer tube despite the best efforts of his vacuum pump to clear it and contaminated the collector plates:

> Finally I said, "This won't do." A new instrument was built in about 10 days in February, 1940. Our glass blower bent the horseshoe-shaped mass spectrometer tube for me; I made the metal parts myself. As a source of uranium, I used the less volatile uranium tetrachloride and tetrabromide left over from [his earlier] Harvard experiments. The first separation of U-235 and U-238 was actually accomplished on February 28 and 29, 1940. It was a leap year, and on Friday afternoon, February 29, I pasted the little samples [collected on nickel foil] on the margin of a handwritten letter and delivered them to the Minneapolis Post Office at about six o'clock. The letter was sent by airmail special delivery and arrived at Columbia University on Saturday. I was aroused early Sunday morning by a long-distance telephone call from John Dunning [who had worked through the night bombarding the samples with neutrons from the Columbia cyclotron]. The Columbia test of the samples clearly showed that U-235 was responsible for the slow neutron fission of uranium.

The demonstration vindicated Bohr's hypothesis, but it also led Briggs to even greater suspicion of the value of natural uranium; it was "very doubtful," he reported to Watson on April 9 "whether a chain reaction can be established without separating 235 from the rest of the uranium." Nier, Dunning and their collaborators Eugene T. Booth and Aristide von Grosse had written much the same thing in the *Physical Review* on March 15:

"These experiments emphasize the importance of uranium isotope separation on a larger scale for the investigation of chain reaction possibilities in uranium." But isotope separation was Dunning's approach to the problem in the first place and his enthusiasm as well; the slow-neutron finding hardly ruled out the Fermi-Szilard system. More misleading may have been the measurements Nier and the Columbia team published on April 15 using larger (but still microscopic) samples: "Furthermore, the number of fissions/microgram of U^{238} observed under these neutron intensity conditions, is sufficient to account for practically all the fast neutron fission observed in unseparated U." The statement was correct within the limits of measurement for such small samples, but its wording seems to deprecate U235 fast-neutron fission. In fact, Nier had not collected enough U235 to allow Columbia to measure that possibility. All anyone knew by then was that the U235 cross section for fast-neutron fission was less than the isotope's cross section for slow-neutron fission. But that cross section, as the first Nier/Columbia paper reported, was a whopping 400 to 500×10^{-24} cm^2.

Predictably, then, when the Uranium Committee met on April 27, with Sachs, Pegram, Fermi, Szilard and Wigner in attendance, it listened to the renewed debate, squared its shoulders at Sachs' exhortation to plunge ahead—and never wavered in its adamant conviction that a large-scale uranium-graphite experiment should await the outcome of Fermi's graphite measurements.

Now that the $6,000 had been paid, Columbia was able to buy the graphite Szilard had tracked down for Fermi's use. "Cartons of carefully-wrapped graphite bricks began to arrive at the Pupin Laboratory," Herbert Anderson remembers, four tons in all. "Fermi returned to the chain reaction problem with enthusiasm. This was the kind of physics he liked best. Together we stacked the graphite bricks in a neat pile. We cut narrow slots in some of the bricks for the rhodium foil detectors we wanted to insert, and soon we were ready to make measurements."

"So the physicists on the seventh floor of Pupin Laboratories started looking like coal miners," adds Fermi, "and the wives to whom these physicists came back tired at night were wondering what was happening."

The arrangement was designed to determine how far neutrons from a radon-beryllium source set in paraffin on the floor under the graphite column would diffuse up the column through the graphite after first slowing down in scattering collisions: the farther the neutrons traveled, the smaller was carbon's absorption cross section and therefore the better moderator it would be. The Pupin seventh floor became a racetrack like the second floor of the institute in Rome. Anderson describes the scene:

> A precise schedule was followed for each measurement. With the rhodium in place in the graphite, the source was inserted in its position inside the pile and removed after a one-minute exposure. To get the rhodium foil under the Geiger counter in the allotted 20 seconds [because its induced half-life is only 44 seconds] took coordination and some fast legwork. The division of labor was typical. I removed the source on signal; Fermi, stopwatch in hand, grabbed the rhodium and raced down the hall at top speed. He had just enough time to place the foil carefully into position, close the lead shield and, at the prescribed moment, start the count. Then with obvious satisfaction at seeing everything go right, he would watch the flashing lights on the scaler, tapping his fingers on the bench in time with the clicking of the register. Such a display of the phenomenon of radioactivity never failed to delight him.

The absorption cross section, as Fermi and Anderson subsequently calculated it, proved usefully small: 3×10^{-27} cm^2. And could be made smaller still, they thought, with purer graphite. The measurement strongly supported Fermi's and Szilard's plan to attempt to induce a slow-neutron chain reaction in natural uranium.

But while such a plan might demonstrate a potential future source of power, the American scientists and administrators who were advising Briggs could not yet identify any military use. In April the British Thomson committee asked A. V. Hill, a scientific adviser to the British Embassy in Washington, to find out what the Americans were doing about fission. According to the official history of the British atomic energy program, Hill talked to unidentified "scientists of the Carnegie Institution," whose opinions he reported pungently:

> It is not inconceivable that practical engineering applications and war use may emerge in the end. But I am assured by American colleagues that there is no sign of them at present and that it would be a sheer waste of time for people busy with urgent matters in England to turn to uranium as a war investigation. If anything likely to be of war value emerges they will certainly give us a hint of it in good time. A large number of American physicists are working on or interested in the subject; they have excellent facilities and equipment: they are extremely well disposed towards us: and they feel that it is much better that they should be pressing on with this than that our people should be wasting their time on what is scientifically very interesting, but for present practical needs probably a wild goose chase.

The opinion from the Carnegie may have been hardheaded, but it was based on more than prejudice. Roberts, Hafstad and fellow DTM physicist Norman P. Heydenburg had improved their measurements of cross sections for fast-neutron fission, scattering and capture in natural uranium.

Using their numbers, Edward Teller in one of the many calculations he made during this period arrived at a critical mass in excess of thirty tons, the same order of magnitude as Perrin and Peierls had calculated before him. With only slightly more pessimistic assumptions Roberts concluded that "the cross-section for capture [in natural uranium] is sufficiently large that it now seems impossible for a fast-neutron chain reaction to occur, even in an infinitely large block of pure uranium." By the spring of 1940 experiments at Columbia and the DTM had thus ruled out both slow- and significant fast-neutron fission in U238 and ruled in slow-neutron fission in U235. The asymmetry might have been a clue. No one picked it up.

Since at least the time of Einstein's first letter to FDR, Edward Teller had debated within himself the morality of weapons work. His life had twice been cruelly uprooted by totalitarianism. He understood Germany's frightening technological advantages at the outset of the war. "I came to the United States in 1935," he notes. ". . . The handwriting was on the wall. At that time, I believed that Hitler would conquer the world unless a miracle happened." But pure science still pacified him. "To deflect my attention from physics, my full-time job which I liked, to work on weapons, was not an easy matter. And for quite a time I did not make up my mind."

The accidental juxtaposition of two events led him to decision. "In the spring of 1940 it was announced that President Roosevelt would speak to a Pan American Scientific Congress in Washington, and as one of the professors of George Washington University I was invited. I did not intend to go." The other event of that crucial day, May 10, 1940, reversed his intention: the phony war abruptly ended. With seventy-seven divisions and 3,500 aircraft Germany without declaration or warning invaded Belgium, the Netherlands and Luxembourg to make way for the invasion of France. Teller thought Roosevelt might speak to that outrage. In his voluntary prewar isolation he had never bothered, Teller says, to visit the Capitol or listen to one of FDR's radio talks or otherwise involve himself in the political life of his adopted country, but he wanted now to see the President of the United States in person.

Alone among the scientists at the congress Teller knew about the Einstein letter. It was a direct link, he was an emotional man and the encounter with Roosevelt was eerily personal: "We had never met, but I had an irrational feeling he was talking to me." The President mentioned the German invasion, its challenge to "the continuance of the type of civilization" the people of the Americas valued, the distances of the modern world shortened by modern technology to timetables that removed the "mystic immunity" Americans once felt from European war. "Then he started to talk about the role of the scientist," Teller recalls, "who has been accused of inventing deadly weapons. He concluded: 'If the scientists in the free countries will not make weapons to defend the freedom of their countries, then freedom will be lost.' " Teller believed Roosevelt was not proposing what scientists *may* do "but something that was our duty and that we *must* do—to work out the military problems, because without the work of the scientists the war and the world would be lost."

Teller's memory of Roosevelt's speech differs from its text. The President said that most people abhor "conquest and war and bloodshed." He said that the search for truth was a great adventure but that "in other parts of the world, teachers and scholars are not permitted" that search—an observation of which Teller had personal knowledge. And then, cannily, Roosevelt offered absolution in advance for war work:

> You who are scientists may have been told that you are in part responsible for the debacle of today . . . but I assure you that it is not the scientists of the world who are responsible. . . . What has come about has been caused solely by those who would use, and are using, the progress that you have made along lines of peace in an entirely different cause.

"My mind was made up," Teller reports, "and it has not changed since."

Vannevar Bush made a similar choice that spring. The sharp-eyed Yankee engineer, who looked like a beardless Uncle Sam, had left his MIT vice presidency for the Carnegie Institution in the first place to position himself closer to the sources of government authority as war approached. Karl Compton had offered to move up to chairman of the MIT corporation and give him the presidency to keep him, but Bush had larger plans.

As a young man, with a doctorate in engineering behind him jointly from MIT and Harvard earned in one intense year, Bush in 1917 had gone patriotically to work for a research corporation developing a magnetic submarine detector. The device was effective, and one hundred sets got built; but because of bureaucratic confusion they were never put to use against German submarines. "That experience," Bush writes in a memoir, "forced into my mind pretty solidly the complete lack of proper liaison between the military and the civilian in the development of weapons in time of war, and what that lack meant."

In Washington after the invasion of Poland the Carnegie president gathered with a group of fellow science administrators—Frank Jewett, president of Bell Telephone Laboratories and the National Academy of Sciences; James Bryant Conant, the young president of Harvard, a distinguished chemist; Richard Tolman of Caltech, the theoretician who had wooed Einstein; Karl Compton—to worry about the approaching conflict:

> It was during the period of the "phony" war. We were agreed that the war was bound to break out into an intense struggle, that America was sure to get into it in one way or another sooner or later, that it would be a highly technical struggle, that we were by no means prepared in this regard, and finally and most importantly, that the military system as it existed . . . would never fully produce the new instrumentalities which we would certainly need.

They devised a national organization to do the job. Bush had learned his way around Washington and took the lead. The organization Bush wanted needed independent authority. He thought it should report directly to the President rather than through military channels and should have its own source of funds. He drafted a proposal. Then he arranged an introduction to Harry Hopkins.

A small-town Iowa boy, idealistic and energetic, Harry Lloyd Hopkins had fallen into New York social work after four years at Grinnell and won appointment at the beginning of the Depression administering emergency state relief. When the governor of New York was elected President, Hopkins moved with Roosevelt to Washington to help out with the New Deal. He ran the vast Works Progress Administration, then took over as Secretary of Commerce. His performance moved him closer and closer to the President, who picked up talent wherever he could find it; as war approached, Roosevelt invited Hopkins to dinner at the White House one evening and moved the man in for the duration as his closest adviser and aide. Hopkins was tall, a chain smoker and emaciated to the point of cachexia, his ghastly health the result of cancer surgery that took most of his stomach and left him unable to absorb much protein and therefore slowly starving to death. He kept an office in the White House basement but usually worked out of a cluttered bedroom suite—the Lincoln Bedroom—down the hall from FDR's.

When Bush met Hopkins, though the presidential aide was a liberal Democrat and the Carnegie president an admirer of Herbert Hoover and a self-styled Tory, "something meshed," writes Bush, "and we found we spoke the same language." Hopkins had a scheme for an Inventors Council. Bush countered with his more comprehensive National Defense Research Council. "Each of us was trying to sell something to the other." Bush won. Hopkins liked his plan.

In early June Bush made the rounds of Washington touching bases: the Army, the Navy, Congress, the National Academy of Sciences. On June

12 "Harry and I then went in to see the President. It was the first time I had met Franklin D. Roosevelt. . . . I had the plan for N.D.R.C. in four short paragraphs in the middle of a sheet of paper. The whole audience lasted less than ten minutes (Harry had no doubt been there before me). I came out with my 'OK-FDR' and all the wheels began to turn."

The National Defense Research Council immediately absorbed the Uranium Committee. That had been part of its purpose. Briggs was a cautious and frugal man, but his committee had also lacked the authority of a source of funds independent of the military. The white-haired director of the National Bureau of Standards would continue to be responsible for fission work. He would report now to James Bryant Conant, Harvard's wiry president, boyish in appearance but in practice cool and reserved, whom Bush had enlisted as soon as FDR authorized the new council.

The NDRC gave research in nuclear fission an articulate lobby within the executive branch. But though Bush and Conant felt challenged by German science—"the threat of a possible atomic bomb," writes Bush, "was in all our minds"—both men, concerned about scarce scientific resources, were initially more interested in proving the impossibility of such a weapon than in rushing to build one: the Germans could not do what could not be done. When Briggs wrapped up his pre-NDRC committee work in a report to Bush on July 1 he asked for $140,000, $40,000 of it for research on cross sections and other fundamental physical constants, $100,000 for the Fermi-Szilard large-scale uranium-graphite experiment (the military had decided to grant $100,000 on its own through the Naval Research Laboratory to isotope-separation studies). Bush allotted Briggs only the $40,000. Once again Fermi and Szilard were left to bide their time.

Winston Churchill had accepted George VI's invitation to form a government upon Neville Chamberlain's resignation the day Germany invaded the Lowlands; he shouldered the prime ministership calmly but felt the somber weight of office. C. P. Snow recalls a more paradoxical mood:

> I remember—I shall not forget it while I live—the beautiful, cloudless, desperate summer of 1940. . . . Oddly enough, most of us were very happy in those days. There was a kind of collective euphoria over the whole country. I don't know what we were thinking about. We were very busy. We had a purpose. We were living in constant excitement, usually, if we examined the true position, of an unpromising kind. In one's realistic moments, it was difficult to see what chance we had. But I doubt if most of us had many realistic moments, or thought much at all. We were all working like mad. We were sustained by a surge of national emotion, of which Churchill was both symbol and essence, evocator and voice.

Not only native-born Englishmen felt that surge. So did the emigré scientists whom Britain had sheltered. Franz Simon, an outstanding chemist whom Frederick Lindemann had extracted from Germany in 1933 for the Clarendon, wrote his old friend Max Born on the eve of the Battle of France that he longed to "use my whole force in the struggle for this country." Though he may not yet have realized it, Simon's opportunity had already arrived. Early in the year, when Frisch and Peierls were first beginning to discuss the ideas that would lead to their important memoranda, Peierls had consulted Simon about methods of isotope separation. Frisch had chosen to work with gaseous thermal diffusion—his Clusius tube—because it seemed to him the simplest method, but Simon had begun then to think about other systems. Half a dozen approaches had been tried in the past. You couldn't spit on the floor without separating isotopes, Simon joked; the problem was to collect them. He wanted to find a method adaptable to mass production, because with a 1:139 isotope ratio, uranium separation would have to proceed on a vast scale, as Frisch's calculation of 100,000 Clusius tubes demonstrated. Frisch dramatized the difficulty with a simile: "It was like getting a doctor who had after great labour made a minute quantity of a new drug and then saying to him: 'Now we want enough to pave the streets.' "

The surge of national emotion sustained Mark Oliphant as well, and in that mood he found even less patience than usual for obstructive rules. When P. B. Moon questioned the assumption that gaseous thermal diffusion was the method of choice for isotope separation, he won no encouragement from the Thomson committee, but back in Birmingham Oliphant simply told him to go ahead and talk it over with Peierls. "Within a week or two," writes Moon, "Peierls identified ordinary diffusion as a logically superior process and wrote directly to Thomson on the matter." Peierls proposed that the Thomson committee consult with Simon, the best man around. The committee hesitated, even though Simon was a naturalized citizen. Oliphant then authorized Peierls out of hand to visit Simon at Oxford.

Simon in the meantime had been working to convert a skeptical Lindemann. At Simon's suggestion Peierls had written to Lindemann on June 2. Together at Oxford later in June they approached Lindemann in person. "I do not know him sufficiently well to translate his grunts correctly," Peierls reported of the meeting. But he felt sure he had "convinced him that the whole thing ought to be taken seriously."

Like Peierls, Simon had settled on "ordinary" gaseous diffusion (as opposed to gaseous *thermal* diffusion) as the best method of isotope separation after winnowing through the alternatives. Gases diffuse through porous materials at rates that are determined by their molecular weight, lighter gases diffusing faster than heavier gases. Francis Aston had applied this principle in 1913 when he separated two isotopes of neon by diffusing a mixed sample several thousand times over and over through pipe clay—that is, unglazed bisque of the sort used to make clay pipes. Thick materials like pipe clay worked too slowly to be effective at factory scale; Simon sought a more efficient mechanism and concluded that a metal foil punctured with millions of microscopic holes would work faster. Divide a cylinder down its length with such a foil barrier, pump a gas of mixed isotopes into one side of the divided cylinder, and gas would diffuse through the barrier as it flowed from one end of the cylinder to the other. Compared to the gas left behind, the gas that diffused through the barrier would be selectively enriched in lighter isotopes. In the case of uranium hexafluoride the enrichment factor would be slight, 1.0043 under ideal conditions. But with enough repetitions of the process any degree of enrichment was possible, up to nearly 100 percent.

The immediate problem, Simon saw, was barrier material. The smaller the holes, the higher the pressures a separation system could sustain, and the higher the pressure, the smaller the equipment could be. Whatever the material, it would have to resist corrosion by uranium hexafluoride—"hex," they were beginning to call it, not necessarily in tribute to its evil contrarities—or the gas would clog its microscopic pores.

One morning that June, inspired, Simon took a hammer to a wire strainer he found in his kitchen. He carried the results to the Clarendon and called together two of his assistants—a Hungarian, Nicholas Kurti, and a big Rhodes scholar from Idaho, H. S. Arms. "Arms, Kurti," Simon announced, holding up the strainer, "I think we can now separate the isotopes." He had hammered the wires flat in demonstration, reducing the spaces between to pinholes.

"The first thing we used," Kurti recalls, "was 'Dutch cloth,' as I think it is called—a very fine copper gauze which has many hundreds of holes to the inch." The assistants hammered the holes even finer by hand. They tested the copper barrier not with hex but with a mixture of water vapor and carbon dioxide, "in other words something much like ordinary soda-water"—the first in an urgent series of experiments carried out through the summer and fall to study materials, pore size, pressures and other basic parameters preliminary to any equipment design.

In late June G. P. Thomson gave his committee a new name to disguise its activities: MAUD. The initials appear to form an acronym but do not. They arrived as a mysterious word in a cable from Lise Meitner to an English friend: MET NIELS AND MARGRETHE RECENTLY BOTH WELL BUT UNHAPPY ABOUT EVENTS PLEASE INFORM COCKCROFT AND MAUD RAY KENT.

Meitner's friend passed the message to Cockcroft, who decided, he wrote Chadwick, that MAUD RAY KENT was "an anagram for 'radium taken.' This agrees with other information that the Germans are getting hold of all the radium they can." Thomson borrowed the first word of Cockcroft's mysterious anagram for a suitably misleading name. The committee members did not learn until 1943 that Maud Ray was the governess who had taught Bohr's sons English; she lived in Kent.

The war crossed the Channel first in the air. As a result of the German bombing of Warsaw in the autumn of 1939, an act Germany represented as tactical because the Polish city was heavily fortified, the British Air Ministry had repudiated its pledge to refrain from strategic bombing. But neither belligerent was eager to exchange bombing raids, and although nightly blackouts added inconvenience and apprehension to the wartime burden of the people of both nations, the implicit truce held until mid-May 1940. Then within a week two events triggered British action. German raiders targeted for French airfields at Dijon lost their way and bombed the southern German city of Freiburg instead, killing fifty-seven people; the German Ministry of Propaganda brazenly denounced the bombing as British or French and threatened fivefold retaliation. Blacker and more violent non sequitur destroyed the city center of Rotterdam. Dutch forces were holding out stubbornly as late as May 14 in the northern section of that old Netherlands port. The German commanding general ordered a "short but devastating air raid" that he hoped might decide the battle. Negotiations with the Dutch advanced, the air raid was canceled, but the abort message arrived too late to stop half the hundred Heinkel 111's ordered into action from dropping 94 tons of bombs. The bombs started massive fires in stores of fats and margarine. The first official Dutch statement, issued from the embassy in Washington, placed casualties in the devastated city at 30,000, and the Western democracies responded with outrage. Actual deaths totaled about 1,000; some 78,000 people went homeless.

The British retaliated on May 15 by dispatching ninety-nine bombers to attack railway centers and supply depots in the Ruhr. Busy with the Battle of France, Hitler did not immediately strike back, but he issued a directive that prepared the way. He ordered the Luftwaffe "to undertake a full-scale offensive against the British homeland as soon as sufficient forces are available."

The initial German air attack, the Battle of Britain, began in mid-August: a month of ferocious daylight contests between the Luftwaffe and British Fighter Command for air supremacy in advance of Operation Sea Lion, Germany's planned cross-Channel invasion. It was not yet an attack on cities. British airfields and aircraft factories were primary targets. Hitler had reserved for himself the decision to bomb London, just as the Kaiser had done before him. Cities would soon go on the targeting list, however; the Luftwaffe was scheduled to raid Liverpool at night on August 28. Accident again intervened: German bombers aiming for oil storage tanks along the Thames overflew their targets on August 24 and bombed central London instead.

Churchill immediately retaliated, hurling four bombing raids in one week at Berlin. They accomplished little physical damage but incited Hitler to hysterical revenge:

> And if the British air force drops two or three or four thousand kilograms of bombs, then we will drop in a single night 150,000, 180,000, 230,000, 300,000, 400,000, a million kilograms. If they announce that they will attack our cities on a large scale, then we shall wipe their cities out!

The Luftwaffe was losing the Battle of Britain in any case, taking unacceptable losses—some 1,700 German aircraft compared to about 900 British. Night bombing would alleviate the losses, curtaining the bombers in dark asylum. But night bombing was notably less accurate than daylight bombing in those days before effective radar and required correspondingly larger targets. Cities and their civilian populations thus fell victim partly by default, because the technology necessary for more accurate targeting was not yet at hand. In any case terror was a weapon that Hitler especially prized, the destruction of what he called the enemy's "will-to-resist," and early in September he told his Sea Lion planners that "a systematic and long-drawn-out bombardment of London might produce an attitude in the enemy which will make Sea Lion unnecessary." He ordered the bombardment. Since it rained from the skies for months, it was hardly *Blitzkrieg,* lightning war, but the citizens exposed beneath it were not in the mood for fine distinctions, and they soon named it the Blitz.

Gresham's Law operated with air raid shelters as it operates with good and bad money: the basements of better department stores like Dickens and Jones, where clerks carried around refreshments—chocolates and ice cream—filled up first. Because the bombing followed regularly, night after night, Londoners had time to get used to it, but adjustment could go either way, the confident beginner slowly unraveling, the frightened beginner moving beyond fear.

More Londoners by far lived out the dangerous raids in their homes than in shelters: 27 percent fled to corrugated-iron Anderson shelters in back gardens, 9 percent to street shelters, only 4 percent into the Tube. By mid-November 13,700 tons of high explosives had fallen and 12,600 tons of incendiary canisters, an average of 201 tons per night; for the entire Blitz, September to May, the total tonnage reached 18,800—18.8 kilotons by modern measure, spread across nine months. London civilian deaths in 1940 and 1941 totaled 20,083, civilian deaths elsewhere in Britain 23,602, for a total death by Blitz in the second and third year of the war (about which the United States was still officially neutral) of 43,685. After that the bombing went the other way. Only twenty-seven Londoners lost their lives to bombs in 1942.

At Oxford in December 1940, Franz Simon, now officially working for the MAUD Committee, produced a report nearly as crucial to the future of uranium-bomb development as the original Frisch-Peierls memoranda had been. It was titled "Estimate of the size of an actual separation plant." Its aim, Simon wrote, was "to provide data for the size and costs of a plant which separates 1 kg per day of ^{235}U from the natural product." He estimated such a plant would cost about £5,000,000 and outlined its necessities in careful detail.

Simon had never trusted the mails. He trusted them even less at the height of the Blitz. He duplicated some forty copies of his report, accumulated enough rationed gasoline for a round trip and shortly before Christmas drove from Oxford into bomb-threatened London to deliver the fruit of half a year's hard work, his whole force in the struggle for his country, to G. P. Thomson.

The Germans may have been collecting radium, as Cockcroft thought MAUD RAY KENT signaled. They were certainly laying in industrial stocks of uranium. In June 1940, about the time Simon was hammering out his kitchen strainer, Auer ordered sixty tons of refined uranium oxide from the Union Minière in occupied Belgium. Paul Harteck in Hamburg tried that month to measure neutron multiplication in an ingenious arrangement of uranium oxide and dry ice—frozen carbon dioxide, a source of carbon free from any impurity other than oxygen—but was unable to convince Heisenberg to lend him enough uranium to guarantee unambiguous results. Heisenberg had larger plans. He had allied himself with von Weizsäcker at the KWI. In July they began designing a wooden laboratory building to be constructed on the grounds of the Kaiser Wilhelm Institute for Biology and Virus Research, next to the physics institute. To discourage the curious they named the building the Virus House. They intended to build a subcritical uranium burner there.

Germany had access to the world's only heavy-water factory and to thousands of tons of uranium ore in Belgium and the Belgian Congo. It had chemical plants second to none and competent physicists, chemists and engineers. It lacked only a cyclotron for measuring nuclear constants. The Fall of France—Paris was occupied June 14, an armistice signed June 22—filled that need. Kurt Diebner, the War Office's resident nuclear physics expert, rushed to Paris. Perrin, von Halban and Kowarski, he found, had escaped to England and taken Allier's twenty-six cans of heavy water with them, but Joliot had chosen to remain in France. (The French laureate would become president of the Directing Committee of the National Front, the largest Resistance organization of the war.)

German officers interrogated Joliot at length when he returned to his laboratory after the occupation began. Their interpreter, sent along from Heidelberg, turned out to be Wolfgang Gentner, the former Radium Institute student who had confirmed that Joliot's Geiger counter was working properly when Joliot discovered artificial radioactivity in 1933. Gentner arranged a secret meeting one evening at a student café and warned Joliot that the cyclotron he was building might be seized and shipped to Germany. Rather than allow that outrage Joliot negotiated a compromise: the cyclotron would stay but German physicists could use it for purely scientific experiments; Joliot would be allowed in turn to continue as laboratory director.

The Virus House was finished in October. Besides a laboratory the structure contained a special brick-lined pit, six feet deep, a variant of Fermi's water tank for neutron-multiplication studies. By December Heisenberg and von Weizsäcker had prepared the first of several such experiments. With water in the pit to serve as both reflector and radiation shield they lowered down a large aluminum canister packed with alternating layers of uranium oxide and paraffin. A radium-beryllium source in the center of the canister supplied neutrons, but the German physicists were able to measure no neutron multiplication at all. The experiment confirmed what Fermi and Szilard had already demonstrated: that ordinary hydrogen, whether in the form of water or paraffin, would not work with natural uranium to sustain a chain reaction.

That understanding left the German project with two possible moderator materials: graphite and heavy water. In January a misleading measurement reduced that number to one. At Heidelberg Walther Bothe, an exceptional experimentalist who would eventually share a Nobel Prize with Max Born, measured the absorption cross section of carbon using a 3.6-foot sphere of high-quality graphite submerged in a tank of water. He found a cross section of 6.4×10^{-27} cm^2, more than twice Fermi's value, and concluded that graphite, like ordinary water, would absorb too many neutrons to sustain a chain reaction in natural uranium. Von Halban and Kowarski, now at Cambridge and in contact with the MAUD Committee, similarly overestimated the carbon cross section—the graphite in both experiments was probably contaminated with neutron-absorbing impurities such as boron—but their work was eventually checked against Fermi's. Bothe could make no such check. The previous fall Szilard had assaulted Fermi with another secrecy appeal:

> When [Fermi] finished his [carbon absorption] measurement the question of secrecy again came up. I went to his office and said that now that we had this value perhaps the value ought not to be made public. And this time Fermi really lost his temper; he really thought this was absurd. There was nothing much more I could say, but next time when I dropped in his office he told me that Pegram had come to see him, and Pegram thought that this value should not be published. From that point the secrecy was on.

It was on just in time to prevent German researchers from pursuing a cheap, effective moderator. Bothe's measurement ended German experiments on graphite. Nothing in the record indicates the overestimate was deliberate, but it is worth noting that Walther Bothe, a protégé of Max Planck, had been hounded from the directorship of the physics institute of the University of Heidelberg in 1933 because he was anti-Nazi. "These galling fights so affected my health," he wrote later in a brief unpublished memoir, "that I had to spend a long period in a Badenweiler sanitorium." When Bothe was well again Planck appointed him to the Kaiser Wilhelm Society's Heidelberg physics institute, but "the Nazis continued to harass me, even to the accusation of scientific fraud."

At nearly the same time—early 1941—Harteck learned at Hamburg what Otto Frisch had recently learned at Liverpool. Frisch had moved to the industrial port city in the northwest of England to work with Chadwick and Chadwick's cyclotron. He built a Clusius tube there with a student assistant Chadwick assigned him—they moved in such energetic coordination through the laboratory that they won the nickname "Frisch and Chips"—and discovered, says Frisch, that "uranium hexafluoride is one of the gases for which the Clusius method does not work." The discovery set the British program back not at all, since Simon was already hard at work on gaseous barrier diffusion. But the German researchers had placed such faith in thermal diffusion that they had not bothered to develop alternatives. They quickly began doing so and identified several promising methods; oddly enough, barrier diffusion was not among them. Restudying the separation problem made it even clearer that U235 and U238 could only be separated by brute-force methods and at great expense.

When Harteck reported to the War Office in March 1941, following a conference with his colleagues, he stressed their consensus that isotope separation would be feasible "only for special applications in which cheapness is but a secondary consideration." Only for a bomb, he meant—so he told the historian David Irving after the war. The German physicists gave "special applications" second place on their list; they recommended urgent work first of all on the production of heavy water. Like Fermi and Szilard, they opted initially for a slow-neutron chain reaction in natural uranium. Make that reaction work and "special applications" might follow. Knowing no more than they knew, they hardly had a choice.

Lieutenant Colonel Suzuki reported back to Lieutenant General Yasuda in October 1940. He confined his report to a basic issue: the availability to Japan of uranium deposits. He looked beyond Japan to Korea and Burma and concluded that his country had access to sufficient uranium. A bomb was therefore possible.

Yasuda turned then to the director of Japan's Physical and Chemical Research Institute, who passed the problem on to his country's leading physicist, Yoshio Nishina. Nishina, born late in the Meiji era and fifty years old in 1940, known for theoretical work on the Compton Effect, had studied with Niels Bohr in Copenhagen, where he was remembered as a cosmopolitan and exceptional man. He had built a small cyclotron at his Tokyo laboratory, the Riken, and with help from an assistant who had trained at Berkeley was building in 1940 a 60-inch successor with a 250-ton magnet, the plans for which had been donated by Ernest Lawrence. More than one hundred young Japanese scientists, the cream of the crop, worked under Nishina at the Riken; to them he was *Oyabun,* "the old man," and he ran his laboratory Western-style with warmth and informality.

The Riken began measuring cross sections in December. In April 1941 the official order came through: the Imperial Army Air Force authorized research toward the development of an atomic bomb.

Leo Szilard was known by now throughout the American physics community as the leading apostle of secrecy in fission matters. To his mailbox, late in May 1940, came a puzzled note from a Princeton physicist, Louis A. Turner. Turner had written a Letter to the Editor of the *Physical Review,* a copy of which he enclosed. It was entitled "Atomic energy from U^{238}" and he wondered if it should be withheld from publication. "It seems as if it was wild enough speculation so that it could do no possible harm," Turner told Szilard, "but that is for someone else to say."

Turner had published a masterly twenty-nine-page review article on nuclear fission in the January *Reviews of Modern Physics* citing nearly one hundred papers that had appeared since Hahn and Strassmann reported their discovery twelve months earlier; the number of papers indicates the impact of the discovery on physics and the rush of physicists to explore it. Turner had also noted the recent Nier/Columbia report confirming the attribution of slow-neutron fission to U235. (He could hardly have missed it; the *New York Times* and other newspapers publicized the story widely. He wrote Szilard irritably or ingenuously that he found it "a little difficult to figure out the guiding principle [of keeping fission research secret] in view of the recent ample publicity given to the separation of isotopes.") His reading for the review article and the new Columbia measurements had stimulated him to further thought; the result was his *Physical Review* letter.

Since U235 is responsible for slow-neutron fission, the letter pointed out, and ordinary uranium contains only one part in 140 of that isotope, "it is natural to conclude that only 1/140 of any quantity of U can be considered as a possible source of atomic energy if slow neutrons are to be used." But the truth may be otherwise, Turner went on. The fission energy of most of the U238, if it could not be used directly, might yet find indirect release.

Turner was referring to the possibility that bombarding uranium with neutrons converted some of the uranium to transuranic elements, the transuranics that Bohr had hoped might have been banished by the discovery of fission. When an atom of U238 captured a neutron it became the isotope U239. That substance itself might fission, Turner suggested. But whether or not U239 did so, it was energetically unstable and would probably decay by beta emission to new elements heavier than uranium. And one or more of those new elements might be fissionable by slow neutrons—which would thereby indirectly put U238 to work.

The next element up the periodic table from uranium would be element 93. Turner selected as the likeliest candidate for fission not ${}_{93}X^{239}$, however, but the element next along, the element that 93 would probably decay to, ${}_{94}X^{239}$, which he called "eka-osmium."* And ${}_{94}\text{EkaOs}^{239}$, Turner proposed, changing from an odd to an even number of neutrons when it absorbed a neutron preparatory to fissioning (239 nucleons − 94 protons = 145 neutrons + 1 = 146) just as U235 changed to U236, ought to be even more fissionable than the lighter uranium isotope: "In ${}_{94}\text{EkaOs}^{240}$ . . . the excess energy would be even larger than in ${}_{92}U^{236}$ and a large cross section for fission would be expected."

* Although Bohr had speculated many years earlier that the transuranic elements, if any, would probably be chemically similar to uranium, researchers still commonly assumed that the transuranics would be chemically similar to the series of metals in the periodic table that begins with rhenium and osmium and includes platinum and gold. "Eka" is an old prefix meaning "beyond."

While Turner was thinking these theories through, two Berkeley men, Edwin M. McMillan and Philip M. Abelson, were moving independently toward demonstrating them. McMillan, a slim, freckled, California-born experimentalist, had been one of the men most responsible in the 1930s for improving Ernest Lawrence's cyclotrons to the point where they worked steadily and produced reliable results. Soon after the news of the discovery of fission reached Berkeley in late January 1939 he had devised an elegantly simple experiment to explore the phenomenon. "When a nucleus of uranium absorbs a neutron and fission takes place," McMillan told an audience later, "the two resulting fragments fly apart with great violence, sufficient to propel them through the air, or other matter, for some distance. This distance, called the 'range,' is a quantity of some interest, and I undertook to measure it." He did so first with thin sheets of aluminum foil "like the pages of a book" stacked on a layer of uranium oxide backed with filter paper. He bombarded the uranium with slow neutrons. Some of the fission fragments recoiled up into the stack of foils; each fragment embedded itself in a single sheet of foil at the end of its range, which depended on its mass; McMillan could then simply check successive sheets of foil in an ionization chamber, look for the characteristic half-lives of various fission products and read out the range (the uranium nucleus splits in many different ways, producing many different lighter-element nuclei).

But aluminum itself is activated by neutron bombardment, which made half-life measurements difficult. So McMillan replaced the foils with a stack of cigarette papers previously treated with acid to remove any trace of minerals that might develop radioactivity under bombardment. "Nothing very interesting about the fission fragments came out of this," he comments. The uranium coating on the filter paper under the stack of cigarette papers, on the other hand, "showed something very interesting." It showed two half-life activities different from those of the fission products that had recoiled away. And since whatever had remained in the uranium layer had not recoiled, the two different activities were probably not fission products. They were probably radioactivities induced in the uranium by captured neutrons. McMillan suspected that one of the two activities, the one with a half-life of 23 minutes, was one that Hahn, Meitner and Strassmann had identified in the 1930s as U239, "a uranium isotope produced by resonance neutron capture." The other activity left behind in the uranium layer had a longer half-life, about 2 days. In his report on his foil and cigarette-paper experiments McMillan chose not to speculate on what that second activity might be, but privately, he remembers, he thought "the two-day period could . . . be the product of the beta-decay of U-239, and therefore an isotope of [transuranic] element 93; in fact, this was the most reasonable explanation."

To check that explanation McMillan needed some hint of the substance's chemical identity. He expected that element 93 would behave chemically like the metal rhenium, element 75, next to osmium on the periodic table—would be "eka-rhenium" in the old terminology. He bombarded a larger uranium sample and enlisted the aid of Emilio Segrè, who was now working as a research associate at Berkeley. "Segrè was very familiar with the chemistry of [rhenium], since he and his co-workers [studying rhenium] had discovered [a similar element], now called technetium, in 1937." Segrè began a chemical analysis of the irradiated uranium; in the meantime McMillan sharpened his half-life measurement to 2.3 days. Segrè, says McMillan, "showed that the 2.3-day material had none of the properties of rhenium, and indeed acted like a rare earth instead." The rare earths, elements 57 (lanthanum) to 71 (lutetium), form a chemically closely related and odd series between barium and hafnium. Because of their middle-table atomic weights near barium, they often turn up as fission products. When Segrè found the 2.3-day activity acting not like rhenium, as expected, but like a rare earth, McMillan assumed that was what it was: "Since rare earths are prominent among the fission products, this discovery seemed at the time to end the story." Segrè even published a paper on his work titled "An unsuccessful search for transuranic elements."

McMillan might have left it there, but the fact that the 2.3-day substance did not recoil away from the uranium layer nagged at him. "As time went on and the fission process became better understood, I found it increasingly difficult to believe that one fission product should behave in a way so different from the rest, and early in 1940 I returned to the problem." The 60-inch cyclotron, with a massive rectangular-framed magnet spacious enough to shelter Lawrence's entire crew between its poles for a photograph—twenty-seven men, two rows seated on the lower jaw of the beast, Lawrence prominent at center, and a third row standing inside its maw—was up and running by then; McMillan used it to study the 2.3-day activity in more detail. He studied the activity chemically as well and managed the significant observation that it did not always fractionally crystallize out of solution as a rare earth would.

"By now it was the spring of 1940," McMillan continues, "and Dr. Philip Abelson came to Berkeley for a short vacation." Abelson was the young experimentalist for whose benefit Luis Alvarez had vacated his Berkeley barber chair half-shorn to pass along the news of the discovery of fission. He had finished his Berkeley Ph.D. and signed on with Merle Tuve at the DTM. Like McMillan, he had become suspicious of the conclusion that the 2.3-day activity was merely another rare-earth fission product. He found time in April 1940 to begin sorting out its chemistry—although he was a physicist by graduate training, he had earned his B.S. in chemistry at

Washington State. But he needed a bigger sample of bombarded uranium than he could produce with DTM equipment. "When he arrived for his vacation," says McMillan, "and our mutual interest became known to one another, we decided to work together." McMillan made up a new batch of irradiated uranium. Abelson pursued its chemistry.

"Within a day," Abelson recalls, "I established that the 2.3-day activity had chemical properties different from those of any known element.... [It] behaved much like uranium." Apparently the transuranics were not metals like rhenium and osmium but were part of a new series of rare-earth-like elements similar to uranium. For a rigorous proof that they had found a transuranic the two men isolated a pure uranium sample with strong 23-minute U239 activity and demonstrated with half-life measurements that the 2.3-day activity increased in intensity as the 23-minute activity declined. If the 2.3-day activity was different chemically from any other element and was created in the decay of U239, then it must be element 93. McMillan and Abelson wrote up their results. McMillan had already thought of a name for the new element—neptunium, for the next planet out beyond Uranus—but they chose not to offer the name in their report. They mailed the report, "Radioactive element 93," to the *Physical Review* on May 27, 1940, the same day Louis Turner sent Szilard his transuranic theories: anticipation and discovery can cut that close in science.

Presumably Szilard did not yet know of the Berkeley work (published June 15) when he answered Turner's letter on May 30, since he makes no mention of it, but he recognized the logic of Turner's argument, told him "it might eventually turn out to be a very important contribution"—and proposed he keep it secret. Szilard saw beyond what Turner had seen. He saw that a fissile element bred in uranium could be chemically separated away: that the relatively easy and relatively inexpensive process of chemical separation could replace the horrendously difficult and expensive process of physical separation of isotopes as a way to a bomb. But unstable element 93, neptunium, was not yet that fissile element and Szilard did not yet realize how small a quantity of pure fissile material was needed to make a critical mass. (Turner was first with his observation, but he was not alone. The idea occurred independently to von Weizsäcker one day in July, before the June *Physical Review* reached him in Germany with the McMillan-Abelson news, while he was riding the Berlin subway, though he assumed element 93 would do the job; he offered the idea to the War Office in a five-page report. A British team at the Cavendish worked it out and presented it to the MAUD Committee early in 1941. But the Germans thought only heavy water could make a uranium burner go in which the new elements might breed, and the British had become optimistic about isotope separation. Neither group therefore pursued the Turner approach.)

After Abelson returned to Washington, McMillan pressed on. Unstable neptunium decayed by beta emission with a 2.3-day half-life; he suspected it decayed to element 94. By analogy with uranium, which emits alpha particles naturally, element 94 should also be a natural alpha emitter. McMillan therefore looked for alphas with ranges different from the uranium alphas coming off his mixed uranium-neptunium samples. By autumn he had identified them. He tried some chemical separations, "finding that the alpha-activity did not belong to an isotope of protactinium, uranium or neptunium." He was that close.

But American science, spurred on by British appeals, was finally gearing up for war. Churchill had sent over Henry Tizard in the late summer of 1940 with a delegation of experts and a black-enameled metal steamer trunk, the original black box, full of military secrets. The prize specimen among them was the cavity magnetron developed in Mark Oliphant's laboratory at Birmingham. John Cockcroft, a future Nobel laureate with a vital mission, traveled along to explain the high-powered microwave generator. The Americans had never seen anything like it before. Cockcroft got together one weekend in October with Ernest Lawrence and multimillionaire physicist-financier Alfred Loomis, the last of the gentlemen scientists, at Loomis' private laboratory in the elegant suburban New York colony of Tuxedo Park. That meeting laid the groundwork for a major new NDRC laboratory at MIT. To keep its work secret it was named the Radiation Laboratory, as if serious scientists might actually be pursuing applications so dubious as those bruited by visionaries from nuclear physics. Loomis wanted Lawrence to direct the new laboratory. Lawrence preferred to stay at Berkeley laying plans and raising funds for a new 184-inch cyclotron but was willing to encourage his best people to move to Cambridge. He convinced McMillan: "I left Berkeley in November 1940 to take part in the development of radar for national defense." Lawrence's and McMillan's priorities are a measure of the priorities of American science in late 1940. Peacetime cyclotrons and radar for air defense came first before superbombs. With a different perspective on the matter, James Chadwick at Liverpool was so uncharacteristically incensed by the publication of the McMillan-Abelson paper reporting element 93 that he asked for, and got, an official protest through the British Embassy. An attaché was duly dispatched to Berkeley to scold Ernest Lawrence, the 1939 Nobel laureate in physics, for giving away secrets to the Germans in perilous times.

Laura and Enrico Fermi and their two children had moved from a Manhattan apartment in the summer of 1939 across the George Washington Bridge and beyond the Palisades to the pleasant suburb of Leonia, New

Jersey. Harold Urey, a short, intense, enthusiastic man, was a resident along with other Columbia families and had convinced the Fermis to buy a house there, praising Leonia's "excellent public schools," Laura writes, and extolling "the advantages of living in a middle-class town where one's children may have all that other children have." Among much good advice Urey cautioned the Italian couple to wage eternal war on crabgrass. Fermi was a product of Roman apartments; he quickly identified *Digitaria sanguinalis* neutrally as "an unlicensed annual" and chose to ignore it. Laura prepared to do battle but was unable to distinguish crabgrass from sod. Urey dropped by one day to give her counsel and identified the problem. "D'you know what's wrong with your lawn, Laura?" the chemistry laureate asked her compassionately. "It's *all* crab grass." Life was pleasant in Leonia; Fermi practiced fitting in. Segrè remembers that his friend "purposely studied contemporary Americana and read the comic strips. . . . Among adult immigrants, I have never seen a comparably earnest effort toward Americanization."

Segrè traveled to Indiana toward the end of 1940 to interview at Purdue, perfunctory interviewing because he meant to stay at Berkeley—"the machine was so good, I could do these things that nowhere else could I do." He continued eastward to visit the Fermis in Leonia. Independently of Turner, Segrè recalls, both he and Fermi had been thinking about element 94. On December 15, he writes, "we had a long walk along the Hudson, in freezing weather, during which we spoke of the possibility that the isotope of mass 239 of element 94 . . . might be a slow neutron fissioner. If this proved to be true, [it] could substitute for ^{235}U as a nuclear explosive. Furthermore, a nuclear reactor fueled with ordinary uranium would produce [the new element]. This gave an entirely new perspective on the making of nuclear explosives, eliminating the need to separate uranium isotopes, at that time a truly scary problem."

Lawrence happened to be visiting New York. "Fermi, Lawrence, Pegram and I met in Dean Pegram's office at Columbia University and developed plans for a cyclotron irradiation that could produce a sufficient amount of [element 94]." After Christmas Segrè returned to Berkeley.

A young chemist there, Glenn T. Seaborg, had already begun working toward identifying and isolating element 94. Born in Michigan of Swedish-American parents, Seaborg had grown up in Los Angeles and taken his Ph.D. at Berkeley in chemistry in 1937, when he was twenty-five. He was exceptionally tall, thin, guarded in the Swedish way but gifted and comfortable at work. The published record of Otto Hahn's 1933 Cornell lectures, *Applied Radiochemistry,* had been his guidebook in graduate school: radiochemistry was his passion. He had been practicing it at Berkeley in

January 1939 when the news of fission arrived; like Philip Abelson, he was excited by the discovery and chagrined to have missed it and had walked the streets for hours the night he heard.

As early as the end of August he had bombarded a sample of uranium to produce neptunium and had assigned one of his second-year graduate students, Arthur C. Wahl, to study its chemistry. His other collaborator in the search for 94 was Joseph W. Kennedy, like Seaborg a Berkeley chemistry instructor. By late November the group had progressed through four more bombardments, unraveling enough of neptunium's chemistry to devise techniques for isolating highly purified samples. Seaborg then wrote McMillan at MIT, a letter he summarizes in a careful history he wrote later that he cast as a contemporary diary: "I suggested that since he has now left Berkeley . . . and is therefore not in a position to continue this work [of studying neptunium and looking for element 94], that we would be very glad to carry on in his absence as his collaborators." McMillan acceded in mid-December; by the time Segrè returned to Berkeley Seaborg had separated out significant fractions of material from his bombarded samples, including uranium, fission products, purified neptunium and a rare-earth fraction that might contain 94.

Two searches were thus to proceed simultaneously. Seaborg's team would follow one especially intense alpha emitter it had identified in the hope of demonstrating that it was an isotope of 94, chemically different from all other known elements. At the same time, Segrè and Seaborg would produce neptunium 239 in quantity, look for its decay product (which ought to be 94^{239}) and attempt to measure that substance's fissibility.

Segrè and Seaborg bombarded ten grams of a solid uranium compound, uranyl nitrate hexahydrate (UNH), for six hours in the 60-inch cyclotron on January 9. They bombarded five more grams for an hour the next morning. By afternoon they knew from ionization-chamber measurements that they could make 94 by cyclotron bombardment; one kilogram of UNH, they calculated, suitably irradiated, should produce about 0.6 microgram (one millionth of a gram) from neptunium after allowing time for beta decay.

Seaborg's team identified an alpha-emitting daughter of Np238 on January 20. Definitive proof that it was 94 required chemical separation, and that delicate, tedious work proceeded during February. The crucial breakthrough came at the beginning of a week when everyone routinely labored past midnight to pursue the difficult fractionations to their end. On Sunday afternoon, February 23, Wahl discovered he could precipitate the alpha emitter from acid solution using thorium as a carrier. But he was not then able to separate the alpha emitter from the thorium. He talked to a

Berkeley chemistry professor who suggested using a more powerful oxidizing agent.

That evening Seaborg and Segrè began bombarding 1.2 kilograms of UNH in the 60-inch cyclotron to transmute some of its uranium into neptunium. They packed the UNH into glass tubes, set the tubes in holes drilled into a 10-inch block of paraffin and set the paraffin in a wooden box. Then they arranged the wooden box behind the beryllium target of the big cyclotron, which battered copious quantities of neutrons from the beryllium with powerful 16 MeV deuterons—favorite cyclotron projectiles, deuterium nuclei from heavy water. With the UNH in place in the cyclotron Seaborg climbed the stairs to the third floor of Gilman Hall where Wahl brewed fractionations under the roof in a cramped room relieved by a small balcony. Wahl tried the new oxidation chemistry that evening with Seaborg at his side. It worked; the thorium precipitated from solution and the alpha emitter stayed behind, enough of it to read out about 300 kicks per minute on the linear amplifier. That, writes Seaborg, was the "key step in its discovery," but they still needed a precipitate of the alpha emitter and they pushed on through the night. Seaborg remembers noticing the new day—lightning over San Francisco to the west across the Bay—when he stepped out onto the balcony to clear his lungs of fumes. Working again past midnight on Tuesday, Wahl filtered out a precipitate cleared of thorium. "With this final separation from Th," Seaborg records with emphasis, "it has been demonstrated that our alpha activity can be separated from all known elements and thus it is now clear that our *alpha activity is due to the new element with the atomic number 94.*"

The bombardment of Segrè's and Seaborg's kilogram sample, interrupted from time to time by other experiments that commanded the cyclotron, continued for a week. The UNH was rendered more intensely radioactive; the radioactivity would increase dangerously as they concentrated the Np239 they had made. They began working with goggles and lead shielding, dissolving the uranium first in two liters of ether and then proceeding through a series of laborious precipitations.

Their fifth and sixth reprecipitations they finished on Thursday, March 6. From 1.2 kilograms of UNH they had now separated less than a millionth of a gram of pure Np239 mixed with sufficient carrier to stain a miniature platinum dish that measured two-thirds of an inch across and half an inch deep. When they had dried this speck of matter God had not welcomed at the Creation they simply snipped off the sides of the platinum dish, covered the sample with a protective layer of Duco Cement, glued the dish to a piece of cardboard labeled Sample A and set it aside until it decayed completely to 94^{239}.

On Friday, March 28 (of the week when Field Marshal Erwin Rommel, commander of the Afrika Korps, opened a major offensive in North Africa; when the British meat ration was reduced to six ounces per person per week; when British torpedo bombers successfully attacked the Italian fleet as it returned from the Aegean, a performance that greatly interested the Japanese), Seaborg recorded:

> This morning Kennedy, Segrè and I made our first test for the fissionability of 94^{239} using Sample A. . . .
>
> Kennedy has constructed during the past few weeks a portable ionization chamber and linear amplifier suitable for detecting fission pulses. . . . Sample A (estimated to contain 0.25 micrograms of 94^{239}) was placed near the screened window of the ionization chamber embedded in paraffin near the beryllium target of the 37-inch cyclotron. The neutrons produced by the irradiation of the beryllium target with 8 MeV deuterons give a fission rate of 1 count per minute per microampere. When the ionization chamber is surrounded by a cadmium shield, the fission rate drops to essentially zero. . . .
>
> *This gives strong indications that 94^{239} undergoes fission with slow neutrons.*

Not until 1942 would they officially propose a name for the new element that fissioned like U235 but could be chemically separated from uranium. But Seaborg already knew what he would call it. Consistent with Martin Klaproth's inspiration in 1789 to link his discovery of a new element with the recent discovery of the planet Uranus and with McMillan's suggestion to extend the scheme to Neptune, Seaborg would name element 94 for Pluto, the ninth planet outward from the sun, discovered in 1930 and named for the Greek god of the underworld, a god of the earth's fertility but also the god of the dead: *plutonium.*

Frisch and Peierls had calculated a small U235 critical mass on the basis of sensible theory. Through the winter Merle Tuve's group at the DTM had continued to refine its cross-section measurements; in March Tuve was able to send to England a measured U235 fast-fission cross section that the British used to confirm a critical mass somewhat larger than the Frisch-Peierls estimate: about eighteen pounds untamped, nine or ten pounds surrounded by a suitably massive and reflective tamper. "This first test of theory," Peierls wrote triumphantly that month, "has given a completely positive answer and there is no doubt that the whole scheme is feasible (provided the technical problems of isotope separation are satisfactorily solved) and that the critical size for a U sphere is manageable."

Chadwick had also made further cross-section measurements. He was already a sober man; when he saw the new numbers a more intense sobriety seized him. He described the change in 1969 in an interview:

> I remember the spring of 1941 to this day. I realized then that a nuclear bomb was not only possible—it was inevitable. Sooner or later these ideas could not be peculiar to us. Everybody would think about them before long, and some country would put them into action. And I had nobody to talk to. You see, the chief people in the laboratory were Frisch and [Polish experimental physicist Joseph] Rotblat. However high my opinion of them was, they were not citizens of this country, and the others were quite young boys. And there was nobody to talk to about it. I had many sleepless nights. But I did realize how very very serious it could be. And I had then to start taking sleeping pills. It was the only remedy. I've never stopped since then. It's 28 years, and I don't think I've missed a single night in all those 28 years.

12

A Communication from Britain

James Bryant Conant traveled to London in the winter of 1941 to open a liaison office between the British government and the National Defense Research Council. Conant was the first American scientist of administrative rank to visit the beleaguered nation following the ad hoc exchanges of the Tizard Mission and he came to count the trip "the most extraordinary experience of my life." "I was hailed as a messenger of hope," he writes in his autobiography. "I saw a stouthearted population under bombardment. I saw an unflinching government with its back against the wall. Almost every hour I saw or heard something that made me proud to be a member of the human race."

The Harvard president, who would be forty-seven late in March, was welcomed not only because of his university affiliation or his distinction as a member of the NDRC. He had been an outspoken opponent of American isolationism during the long months of the phony war and was therefore welcomed especially as a sign—with only the Prime Minister dissenting. Churchill was less than delighted at the prospect of lunching with the president of Harvard. "What shall I talk to him about?" he was heard to ask. "He thought you would be an old man with a white beard, exuding learning and academic formality," Brendon Bracken, Churchill's aide, told Conant afterward. But braced by the American's belligerently pro-British views and put at ease by the tweed suit he chose to wear, the Prime Minister eventually warmed over lunch in the bomb-shelter basement at 10 Downing Street, proffering a Churchillian monologue during which he repeated one of his choicer recent coinages: "Give us the tools, and we will finish the job."

In 1920, at twenty-seven, when Conant was courting the woman he would marry—she was the only child of the Nobel laureate Harvard chemist T. W. Richards, a pioneer in measuring atomic weights—he had shared hopes for a grand future with her that coming from a less able man might have sounded absurd. "I said that I had three ambitions. The first was to become the leading organic chemist in the United States; after that I would like to be president of Harvard; and after that, a Cabinet member, perhaps Secretary of the Interior." Those may not seem conjoint ambitions, but Conant managed a version of each in turn. He was born of a Massachusetts family that had resided in the state since 1623. After Roxbury Latin and Harvard College he had taken a double Ph.D. under his future father-in-law in organic and physical chemistry. He emerged from the Great War with the rank of major for his work in poison-gas research at Edgewood. In his autobiography, written late in life, he justified his participation:

> I did not see in 1917, and do not see in 1968, why tearing a man's guts out by a high-explosive shell is to be preferred to maiming him by attacking his lungs or skin. All war is immoral. Logically, the 100 percent pacifist has the only impregnable position. Once that is abandoned, as it is when a nation becomes a belligerent, one can talk sensibly only in terms of the violation of agreements about the way war is conducted, or the consequences of a certain tactic or weapon.

Like Vannevar Bush, Conant was a patriot who believed in the application of advanced technology to war.

"Conant achieved an international reputation in both natural products chemistry and in physical-organic chemistry," writes the Ukrainian-born Harvard chemist George B. Kistiakowsky. Natural products include chlorophyll and hemoglobin and Conant contributed to the unraveling of both those vital molecules. His studies also helped generalize the concept of acids and bases, a concept now considered fundamental. If not the leading American organic chemist of his day, he ranked among the leaders. When Caltech tried to lure him away with a large research budget Harvard topped the offer and refused to let him go.

Number two on Conant's youthful list, the presidency of his alma mater, he won in 1933. He told the members of the Harvard Corporation who approached him that he didn't want the job, which was apparently a prerequisite, but would serve if elected. He was forty at the time of his election. He created the modern Harvard of eminent scholarship and publish-or-perish, up-or-out.

Conant's third ambition achieved approximate fulfillment after the war in high, though less than cabinet-rank, appointment; his long span of voluntary government service began with the NDRC.

In England in the late winter of 1941 he met with the leaders of the British government, had an audience with the King, picked up an honorary degree at Cambridge and walked the Backs afterward to see the crocuses in bloom, made room for the NDRC mission among hostile U.S. military and naval attachés, lunched with Churchill again. His mission in Britain was diplomatic rather than technical. He discussed gas warfare and explosives manufacture but was unable to share in the intense exchange of information on radar because he knew very little about electronics. But although he was familiar with the work on uranium and it fell within his official NDRC responsibilities, secrecy and his "strong belief in the 'need to know' principle" kept Conant from learning what the British had learned about the possibility of a bomb.

He met a "French scientist" at Oxford, probably Hans von Halban, who complained of inaction on uranium–heavy water research. "Since his complaints were clearly 'out of channels,' I quickly terminated the conversation and forgot the incident." That reaction was understandable: Conant could hardly know what security arrangements the British might have made with the Free French. But he also shied from Lindemann. They were lunching alone at a London club. "He introduced the subject of the study of the fission of uranium atoms. I reacted by repeating the doubts I had expressed and heard expressed at NDRC meetings." Lindemann brushed them aside and pounced:

> "You have left out of consideration," said [Lindemann], "the possibility of the construction of a bomb of enormous power." "How would that be possible?" I asked. "By first separating uranium 235," he said, "and then arranging for the two portions of the element to be brought together suddenly so that the resulting mass would spontaneously undergo a self-sustaining reaction."

Remarkably, the chairman of the chemistry and explosives division of the NDRC adds that, as late as March 1941, "this was the first I had heard about even the remote possibility of a bomb." Nor did he pursue the matter. "I assumed, quite correctly, that if and when Bush wished to be in touch with the atomic energy work in England, he would do so through channels involving Briggs." No wonder the Hungarian conspirers continued to tear their hair.

Then for the first time a ranking American physicist joined the debate whose voice could not be ignored. Even before Seaborg and Segrè confirmed the fissibility of plutonium, Ernest Lawrence had measured the prevailing American skepticism and conservatism against the increasing enthusiasm of his British friends and responded with characteristic fervor. Ralph H. Fowler, Ernest Rutherford's widower son-in-law, had visited Berkeley during the 1930s and attended picnics and weekend parties with the inventor of the cyclotron. Fowler was British scientific liaison officer in Washington now and from that close vantage he urged Lawrence to get involved. So did Mark Oliphant, whom Lawrence had met and liked on a visit to the Cavendish after the 1933 Solvay Conference.

Lawrence had encouraged the search for plutonium partly because he saw little hope for isotope separation by any of the methods so far discussed—by centrifuge, thermal diffusion or barrier diffusion. But around the beginning of the year he began thinking about separating isotopes electromagnetically, by the process that had already worked on a microscopic scale for Alfred Nier. It occurred to Lawrence that he could modify his superseded 37-inch cyclotron into a big mass spectrometer. The fact that Nier thought electromagnetic separation on an industrial scale impossible only spurred the Berkeley laureate on. Lawrence lived from machine to machine, as it were; conceiving a machine to do the job of liberating U235 from its confinement within U238 (while Fermi's uranium-graphite reactor manufactured Berkeley-born plutonium) gave him something solid to fight for, a tangible program to push.

It assembled itself by stages. He was not yet ready emotionally to set aside his peacetime plans. Warren Weaver, the director of the division of natural sciences at the Rockefeller Foundation, visited Berkeley in February to see how construction was progressing on the 4,900-ton, 184-inch cyclotron for which the foundation had awarded a $1,150,000 grant less than twelve months earlier. Lawrence took time to complain about the Uranium Committee's sloth—Weaver worked with another division of the NDRC—but then drove up behind the university to the cyclotron site on the hillside and first irritated and then enthralled the Rockefeller administrator with visions of a superior and much larger machine.

Lawrence rehearsed his complaint again in March when Conant, back from London, traveled out to deliver an address. "Light a fire under the Briggs committee," the energetic Californian badgered the president of Harvard. "What if German scientists succeed in making a nuclear bomb before we even investigate possibilities?" That prepared Lawrence for a full assault. He launched it on March 17 when he met with Karl Compton and Alfred Loomis at MIT.

Loomis had turned to physics after a lucrative career in the law and investment banking. Compton was a physicist of distinction who had taught for fifteen years at Princeton, where he took his Ph.D., before becoming president of MIT in 1930. Both men understood the politics of organizations. Yet they were sufficiently seized with Lawrence's fervor that Compton telephoned Vannevar Bush almost as soon as Lawrence left the room and dictated a follow-up letter the same day. Briggs was "by nature slow, conservative, methodical and accustomed to operate at peacetime government bureau tempo," Compton wrote, conveying Lawrence's blunt complaints, and had been "following a policy consistent with these qualities and still further inhibited by the requirement of secrecy." The British were ahead even though America had "the most in number and the best in quality of the nuclear physicists of the world." The Germans were "very active." Briggs had invited only a very few U.S. nuclear physicists into the work. There were other possibilities in fission research besides the pursuit of a slow-neutron chain reaction for power, possibilities "capable, if successful, of far more important military usage."

Though they felt free thus to lecture Bush, both Loomis and Compton stood in awe of Lawrence—Loomis had recently contributed $30,000 to a private fund simply to make it easier for Lawrence to travel around the country—and thought Bush could do no better than to turn him loose: "I hasten to say that the idea of Ernest himself taking an active part in any reorganization was in no sense suggested by him or even in his mind, but I do believe that it would be an ideal solution."

Bush's ego was commensurate with his responsibilities, as Loomis and Compton ought to have known. It might have been politic to welcome Lawrence's campaign, especially since Loomis was a first cousin and close friend of Henry L. Stimson, the respected and influential Secretary of War; but Bush decided instead to take it as a challenge to his authority, the first the physics community had mounted since he invented the NDRC, welcoming a fight he knew he could win. He met Lawrence in New York two days after the MIT meeting and let fly:

> I told him flatly that I was running the show, that we had established a procedure for handling it, that he could either conform to that as a member of the NDRC and put in his kicks through the internal mechanism, or he could be utterly on the outside and act as an individual in any way that he saw fit. He got into line and I arranged for him to have with Briggs a series of excellent conferences. However, I made it very clear to Lawrence that I proposed to

> make available to Briggs the best advice and consultation possible, but that in the last analysis I proposed to back up Briggs and his committee in their decision unless there was some decidedly strong case for entering into it personally. I think this matter was thoroughly straightened out, therefore, but it left its trail behind.

By threatening to push Ernest Lawrence out into the cold with the emigrés Bush managed temporarily to confine the uranium problem. Confinement lasted less than a month.

In 1940 Lawrence had recruited a Harvard experimentalist named Kenneth Bainbridge, by trade a nuclear physicist—Bainbridge built the Harvard cyclotron—to work on radar at MIT. When Conant went to London to open the new NDRC office there, Bainbridge and others had followed, to work with the British each in his own field of competence. But since Bainbridge knew nuclear physics as well as radar and had even looked into isotope separation, the British allowed him also to attend a full-dress meeting of the MAUD Committee. To Bainbridge's surprise, the committee had "a very good idea of the critical mass and [bomb] assembly [mechanism], and urged the exchange of personnel. . . . Their estimate was that a minimum of three years would be required to solve all the problems involved in producing an atomic weapon." Bainbridge immediately contacted Briggs and suggested he send someone over to represent the United States in uranium matters.

Beneath Bush's organizational bristle lay genuine perplexity. "I am no atomic scientist," he writes candidly; "most of this was over my head." As he saw the situation that April, "it would be possible to spend a very large amount of money indeed, and yet there is certainly no clear-cut path to defense results of great importance lying open before us at the present time." But he felt the increasing pressure—Lawrence's prodding, Bainbridge's confirmation of British progress—and reached out now for help.

"It was Bush's strategy," writes the American experimental physicist Arthur Compton, Karl's younger brother, "as co-ordinator of the nation's war research, to use the National Academy [of Sciences] as the court of final appeal for important scientific problems." On a Tuesday in mid-April, after meeting with Briggs, Bush wrote Frank B. Jewett, the senior Bell Telephone engineer who was president of the National Academy. Briggs had heard from Bainbridge and alerted Bush; Bush and Briggs, "disturbed," had conferred. "The British are apparently doing fully as much as we are, if not more, and yet it seems as though, if the problem were of really great importance, we ought to be carrying most of the burden in this country." Bush wanted "an energetic but dispassionate review of the entire situation by a highly competent group of physicists." The men chosen ought to have "sufficient knowledge to understand and sufficient detachment to cold bloodedly evaluate."

At a regular Washington meeting of the National Academy the following Friday Jewett, Bush and Briggs recruited their review group. They put Lawrence on the committee and the recently retired director of the research laboratory at General Electric, a physical chemist named William D. Coolidge. Then they sought out Arthur Compton, a Nobel laureate and professor of physics at the University of Chicago, and proposed he head the review. Compton humbly questioned his "fitness for the task" and jumped at the chance.

Arthur Holly Compton was the son of a Presbyterian minister and professor of philosophy at the College of Wooster in Wooster, Ohio. Compton's Mennonite mother was dedicated to missionary causes and had been the 1939 American Mother of the Year. He followed his older brother Karl into science and surpassed him in achievement but preserved the family piety as well. "Arthur Compton and God were daily companions," notes Leona Woods, Enrico Fermi's young protégé at the University of Chicago. She judged Compton nevertheless "a fine scientist and a fine man. . . . He was remarkably handsome all his life and athletically spare and strong." Fermi had concluded, writes Woods, that "tallness and handsomeness usually were inversely proportional to intelligence," but "he excepted Arthur Compton . . . whose intelligence he respected enormously."

Compton's physics was first-rate, as Fermi's respect implies. He graduated from the College of Wooster and took his Ph.D. at Princeton. In 1919, the first year of the program, he was appointed a National Research Council fellow and used the appointment to study under Rutherford at the Cavendish. The difficult work he began there—examining the scattering and absorption of gamma rays—led directly to the discovery of what came to be called the Compton effect, for which he won the Nobel Prize.

In 1920, Compton writes, he accepted a professorship at Washington University in St. Louis, "a small kind of place," to get out of the mainstream of physics so that he could concentrate on his scattering studies, which he was then extending from gamma rays to X rays. He scattered X rays with a graphite block and caught them and measured their wavelengths Moseley-style with a calcite-crystal X-ray spectrograph. He found that the X rays scattered by the graphite came out with wavelengths longer than their wavelengths going in: as if a shout bounced off a distant wall came back bizarrely deepened to a lower pitch. If X rays—light—were only a motion of waves, then their wavelengths would not have changed; Compton had in fact demonstrated in 1923 what Einstein had postulated in

1905 in his theory of the photoelectric effect: that light was wave but also simultaneously particle, photon. An X-ray photon had collided elastically with an electron, as billiard balls collide, had bounced off and thereby given up some of its energy. The calcite crystal revealed the energy loss as a longer wavelength of X-ray light. Arnold Sommerfeld hailed the Compton effect—elastic scattering of a photon by an electron—as "probably the most important discovery which could have been made in the current state of physics" because it proved that photons exist, which hardly anyone in 1923 yet believed, and demonstrated clearly the dual nature of light as both particle and wave.

The subtle experimenter lost his subtlety when he shifted from doing science to proselytizing for God. Rigor slipped to Chautauqua logic and he perpetrated such howlers as the notion that Heisenberg's uncertainty principle somehow extends beyond the dimensions of the atom into the human world and confirms free will. Bohr heard Compton's Free Will lecture when he visited the United States in the early 1930s and scoffed. "Bohr spoke highly of Compton as a physicist and a man," a friend of the Danish laureate remembers, "but he felt that Compton's philosopohy was too primitive: 'Compton would like to say that for God there is no uncertainty principle. That is nonsense. In physics we do not talk about God but about what we can know. If we are to speak of God we must do so in an entirely different manner.' "

In 1941 war work had already been kind to Arthur Compton's brother, moving Karl to national prominence within the science community and winning an important secret laboratory for MIT. Arthur wanted as much or more. There was the problem of pacifism, his mother's Mennonite creed and a course much discussed at that time in American vestries, a churchly counterpart to isolationism:

> In 1940, my forty-eighth year, I began to feel strongly my responsibility as a citizen for taking my proper part in the war that was then about to engulf my country, as it had already engulfed so much of the world. I talked, among others, with my minister in Chicago. He wondered why I was not supporting his appeal to the young people of our church to take a stand as pacifists. I replied in this manner: "As long as I am convinced, as I am, that there are values worth more to me than my own life, I cannot in sincerity argue that it is wrong to run the risk of death or to inflict death if necessary in the defense of those values."

Arthur Compton was ready, then, "a short time later," when Bush and the National Academy asked him to serve.

The review committee met immediately with some of Briggs' associates in Washington. A week later, May 5, 1941, it met again in Cambridge to hear from other Uranium Committee members and from Bainbridge. "There followed," writes Compton, "two weeks spent in discussing the military possibilities of uranium with others who were actively interested." Compton worked quickly to complete a seven-page report and delivered it to Jewett on May 17.

The report began with the statement that the committee was concerned with "the matter of possible military aspects of atomic fission" and listed three of those possibilities: "production of violently radioactive materials . . . carried by airplanes to be scattered as bombs over enemy territory," "a power source on submarines and other ships" and "violently explosive bombs." Radioactive dust would need a year's preparation after "the first successful production of a chain reaction," which meant "not earlier than 1943." A power source would need at least three years after a chain reaction. Bombs required concentrating U235 or possibly making plutonium in a chain reaction, so "atomic bombs can hardly be anticipated before 1945."

And that was that: no mention of fast-neutron fission, or critical mass, or bomb assembly mechanisms. The bulk of the report discussed "progress toward securing a chain reaction" and considered uranium-graphite, uranium-beryllium and uranium–heavy water systems. The committee proposed giving Fermi all the money he needed for his intermediate experiment and beyond. It also, more originally, discovered and emphasized the decisive long-range challenge of the new field:

> It would seem to us unlikely that the use of nuclear fission can become of military importance within less than two years. . . . If, however, the chain reaction can be produced and controlled, it may rapidly become a determining factor in warfare. Looking, therefore, to a struggle which may continue for a decade or more, it is important that we gain the lead in this development. That nation which first produces and controls the process will have an advantage which will grow as its applications multiply.

Bush was in the process of reorganizing government science when he received the NAS report. The NDRC, empowered equally with the military laboratories and the National Advisory Committee for Aeronautics, had served for research but lacked the authority to pursue engineering development. Bush proposed a new umbrella agency with wide authority over all government science in the service of war, the Office of Scientific Research and Development. Its director—Bush—would report personally to Roosevelt. Bush prepared to move up to the OSRD by calling in Conant to take over the NDRC. "And only after it was clear that I should shortly have a new position," writes Conant, "did Bush begin to take me into his confidence as he pondered on what to do with the Briggs Committee."Against the background of his British experience Conant told Bush his reaction to Compton's report was "almost completely negative."

Jewett had delivered the report to Bush with a cover letter calling it "authoritative and impressive," but privately he cautioned Bush that he had "a lurking fear" that the report "might be over-enthusiastic in parts and not so well balanced." Jewett also passed it to several senior colleagues for comment, including the 1923 Nobel laureate in physics, Robert A. Millikan of Caltech, and sent their comments along to Bush in early June. Bush responded with exasperation compounded with astonishing confusion about the developments in Britain:

> This uranium business is a headache! I have looked over Millikan's comments, and it is quite clear that he wrote them without realizing the present situation. The British have apparently definitely established the possibility of a chain reaction with 238 [*sic*], which entirely changes the complexion of the whole affair. Millikan bases his comments on the conviction that only 235 holds promise. This is natural, since he has not been brought in touch with recent developments which the British have told us about in great confidence.

He agreed that the work "ought to be handled in a somewhat more vigorous form," but he was still profoundly skeptical of its promise:

> Even if the physicists get all that they expect, I believe that there is a very long period of engineering work of the most difficult nature before anything practical can come out of the matter, unless there is an explosive involved, which I very much doubt.

The OSRD director was not yet convinced despite new word of plutonium's remarkable fissibility. Segrè and Seaborg had continued working through the spring of 1941 to determine the man-made element's various cross sections. On Sunday, May 18, having finally prepared a sample thin enough for accurate measurement, they calculated plutonium's cross section for slow-neutron fission at 1.7 times that of U235. When Lawrence heard the news on Monday, says Seaborg, he swung into action:

> We told Lawrence about our definitive demonstration yesterday of the slow neutron fissionability of 94^{239} and he was quite excited. He immediately phoned the University of Chicago to give the news to Arthur H. Compton. . . . Compton made an immediate attempt to phone (unsuccessfully) and then sent a telegram to Vannevar Bush. . . . In his telegram Compton indicated that the demonstration . . . greatly increases the importance of the fission problem since the available material [i.e., U238 transmuted to plutonium] is thus in-

creased by over 100 times. . . . He said that Alfred Loomis and Ernest Lawrence accordingly have requested him to urge anew the vital importance of pushing the [uranium-graphite] work at Columbia.

* * *

Whenever the U.S. program bogged down in bureaucratic doubt Hitler and his war machine rescued it. That summer's massive escalation, code-named Operation Barbarossa, was the opening of the Eastern Front at dawn on the morning of Sunday, June 22, a surge eastward with 164 divisions, including Finnish and Rumanian components, toward *Blitzkrieg* invasion of the USSR. The Führer's ambitious intention, declared with emphasis in a secret directive six months earlier, was "*to crush Soviet Russia in a quick campaign* even before the conclusion of the war against England." Hitler meant to push all the way to the Urals before winter and commandeer the Soviet Union's industrial and agricultural base; by July Panzers had crossed the Dnieper and were threatening Kiev.

The effect on Conant of his London experiences and the widening war was paradoxically to increase his skepticism of the program he had just accepted assignment to administer:

> What worried me about Compton's first report, I told Bush, was the assumption that achieving a chain reaction was so important that a large expenditure of both money and manpower was justified. To me, the defense of the free world was in such a dangerous state that only efforts which were likely to yield results within a matter of months or, at most, a year or two were worthy of serious consideration. In that summer of 1941, with recollections of what I had seen and heard in England fresh in my mind, I was impatient with the arguments of some of the physicists associated with the Uranium Committee whom I met from time to time. They talked in excited tones about the discovery of a new world in which power from a uranium reactor would revolutionize our industrialized society. These fancies left me cold. I suggested that until Nazi Germany was defeated all our energies should be concentrated on one immediate objective.

Having experienced the London Blitz, Conant had developed a siege mentality; Bush, as Conant points out, "was faced with a momentous decision as to priorities." Both men wanted a hard, practical assessment. They decided Compton's report needed an injection of common sense in the form of engineering expertise. Compton discreetly retired from the line; W. D. Coolidge, the General Electric scientist, temporarily took his place. Conant added an engineer from Bell Laboratories and another from Westinghouse and early in July the enlarged committee reviewed the first review.

Briggs was a convincing witness. By then he had received the April 9 minutes of a MAUD technical subcommittee meeting where Peierls reported that cross-section measurements confirmed the feasibility of a fast-neutron bomb. Briggs had also just learned from Lawrence that plutonium had a cross section for fast fission some ten times that of U238. Lawrence even submitted a separate report on element 94 that emphasized for the first time in U.S. official deliberations the importance of fast fission over slow. But Briggs was still preoccupied with a slow-neutron chain reaction for power production and so was the second NAS report. "In the summer of 1941," John Dunning's associate Eugene Booth remembers, "Briggs visited us in the basement of Pupin at Columbia to see our experiment for the separation of U235 by [gaseous] diffusion of uranium hexafluoride. He was interested, blessed us, but sent us no money."

The American program was in danger for its life that summer, Compton thought: "The government's responsible representatives were . . . very close to dropping fission studies from the war program." He believed the program was saved because of Lawrence's proposal to use plutonium to make a bomb. The fissibility of 94 may have convinced Compton. It was not decisive for the government's responsible representatives. They were hard men and needed hard facts. Those began to arrive. "More significant than the arguments of Compton and Lawrence," writes Conant, "was the news that a group of physicists in England had concluded that the construction of a bomb made out of uranium 235 was entirely feasible."

The British had been trying all winter and spring to pass the word. In July they tried again. G. P. Thomson had assembled a draft final report for the MAUD Committee to consider on June 23, the day after Barbarossa exploded across the Balkans and eastern Poland. Charles C. Lauritsen of Caltech, a respected senior physicist, was beginning work for the NDRC developing rockets and happened to be in London conferring with the British at the time of the MAUD draft. The committee invited him to attend its July 2 meeting at Burlington House. Lauritsen listened carefully, took notes and afterward talked individually with eight of the twenty-four physicists now attached to the work. When he returned to the United States the following week he immediately reported the MAUD findings to Bush. "In essence," says Conant, "he summarized the 'draft report.' " The physicists Lauritsen had interviewed had all pushed for a U.S.-built gaseous-diffusion plant.

The British government would not officially transmit the final MAUD Report to the United States government until early October, but the committee approved it on July 15 (and thereupon promptly disbanded) and by then Bush had been passed a copy of the Thomson draft, which embodied the essential findings. The MAUD Report differed from the two National

Academy studies as a blueprint differs from an architect's sketch. It announced at the outset:

> We have now reached the conclusion that it will be possible to make an effective uranium bomb which, containing some 25 lb of active material, would be equivalent as regards destructive effect to 1,800 tons of T.N.T. and would also release large quantities of radioactive substances. . . . A plant to produce 2¼ lb (1 kg) per day [of U235] (or 3 bombs per month) is estimated to cost approximately £5,000,000. . . . In spite of this very large expenditure we consider that the destructive effect, both material and moral, is so great that every effort should be made to produce bombs of this kind. . . . The material for the first bomb could be ready by the end of 1943. . . . Even if the war should end before the bombs are ready the effort would not be wasted, except in the unlikely event of complete disarmament, since no nation would care to risk being caught without a weapon of such destructive capabilities.

Of conclusions and recommendations the report offered, crisply, three:

> (i) The committee considers that the scheme for a uranium bomb is practicable and likely to lead to decisive results in the war.
>
> (ii) It recommends that this work continue on the highest priority and on the increasing scale necessary to obtain the weapon in the shortest possible time.
>
> (iii) That the present collaboration with America should be continued and extended especially in the region of experimental work.

"With the news from Great Britain unofficially in hand," Conant concludes in a secret history of the project he drafted in 1943, ". . . it became clear to the Director of OSRD and the Chairman of NDRC that a major push along the lines outlined was in order."

They still did not immediately organize that push. Nor was Conant, to his postwar recollection, yet convinced that a uranium bomb would work as described. British research and considered judgment had at least proposed a clear-cut program of *military* development. Bush took it to Vice President Henry Wallace, his White House sounding board, who was the only scientist in the cabinet, a plant geneticist who had developed several varieties of hybrid corn. "During July," writes Conant, "Bush had a discussion with Vice President Wallace about the question of spending a large amount of government money on the uranium program." After which Bush apparently decided to wait for official transmittal of the final MAUD Report.

"If each necessary step requires ten months of deliberation," Leo Szilard had complained to Alexander Sachs in 1940, "then obviously it will not be possible to carry out this development efficiently." The American program was moving faster now than that, but not by much.

While Lawrence and Compton championed plutonium that summer, a big, rawboned, war-battered Austrian hiding out within the German physics establishment tried to keep the fissile new element out of sight. He was an old friend of Otto Frisch:

> Fritz Houtermans and I had met in Berlin, but in London [before the war] I saw a lot more of that impressive eagle of a man, half Jewish as well as a Communist who had narrowly escaped the Gestapo. His father had been a Dutchman, but he was very proud of his mother's Jewish origin and liable to counter anti-semitic remarks by retorting "When your ancestors were still living in the trees mine were already forging cheques!" He was full of brilliant ideas.

Houtermans had taken a Ph.D. in experimental physics at Göttingen but was strong in theory. One of his brilliant ideas, developed in the late 1920s at the University of Berlin with a visiting British astronomer, Robert Atkinson, concerned the production of energy in stars. Atkinson was familiar with recent estimates by his older colleague Arthur Eddington that the sun and other stars burn at temperatures of 10 million and more degrees and have life spans of billions of years—a prodigious and unexplained expenditure of energy. On a walking tour near Göttingen in the summer of 1927 the two men had wondered if nuclear transformations of the sort Rutherford was producing at the Cavendish might account for the enduring stellar fires. They quickly worked out a basic theory, as Hans Bethe later described it, "that at the high temperatures in the interior of a star, the nuclei in the star could penetrate into other nuclei and cause nuclear reactions, releasing energy." The energy would be released when hot (and therefore fast-moving) hydrogen nuclei collided with enough force to overcome their respective electrical barriers and fused together, making helium nuclei and giving up binding energy in the process. With George Gamow, Houtermans and Atkinson later named these events *thermonuclear* reactions because they proceeded at such high temperatures.

In 1933 Houtermans emigrated to the Soviet Union, "but fell victim," writes Frisch, "to one of Stalin's purges and spent a couple of years in prison; his wife with two small children managed to escape and get to the U.S.A. When Hitler made his temporary pact with Stalin in 1939 it included an exchange of prisoners, and Houtermans was handed back to the Gestapo." Max von Laue, whom Frisch celebrates as "one of the few German scientists with the prestige and courage to stand up against the Nazis," managed to free Houtermans and arranged for him to work with a wealthy German inventor, Baron Manfred von Ardenne, who had studied physics and who maintained a private laboratory in Lichterfelde, outside Berlin. Von Ardenne was pursuing uranium research independently of Heisenberg and the War Office; to raise funds for the work he had approached the German Post Office, which commanded a large and largely unused budget for research. The Minister of Posts, imagining himself handing Hitler the decisive secret weapon of the war, had funded the building of a million-volt Van de Graaff and two cyclotrons, all under construction in 1941. Until they came on line Houtermans turned his attention to theory.

By August he had independently worked out all the basic ideas necessary to a bomb. He discussed them in a thirty-nine-page report, "On the question of unleashing chain reactions," that considered fast-neutron chain reactions, critical mass, U235, isotope separation and element 94. Houtermans emphasized making 94. "Every neutron which, instead of fissioning uranium-235, is captured by uranium-238," he wrote, "creates in this way a new nucleus, fissionable by thermal neutrons." He discussed his ideas privately with von Weizsäcker and Heisenberg, but he saw to it that the Post Office kept his report in its safe secure from War Office eyes. He had learned to cooperate for survival in the Soviet Union, where the NKVD—the KGB of its day—had knocked out all his teeth and kept him in solitary confinement for months. But in Germany as in the USSR he withheld as much information as he dared. His private endorsement of 94, to be transmuted by chain reaction from natural uranium, probably contributed to the neglect of isotope separation in Germany. After the summer of 1941 the German bomb program depended entirely on uranium and Vemork heavy water.

The British, at least, knew where they were going. Tizard was skeptical of the MAUD Report and doubted that a bomb could be produced before the end of the war. Lindemann—he was Lord Cherwell now, a baron, courtesy of his friend the P.M.—did not. Cherwell had followed the MAUD work carefully. He respected Thomson; Simon was an old friend; Peierls had read his grunts correctly after all. He trusted their judgment and set to work to reduce the lengthy report to a memorandum for Churchill. Churchill liked his documents held to half a page. So important was this one that Cherwell allowed it to run on for two and a half pages. He thought research should continue for six months and then face further review. He thought an isotope-separation plant should be erected not in the United States but in England—despite manpower shortages and the risk of German bombing—or "at worst" in Canada. In that conclusion he differed from the

MAUD Committee. "The reasons in favor [of an English location]," he wrote, "are the better chance of maintaining secrecy . . . but above all the fact that whoever possesses such a plant should be able to dictate terms to the rest of the world. However much I may trust my neighbor and depend on him, I am very much averse to putting myself completely at his mercy. I would, therefore, not press the Americans to undertake this work." His summation narrowed the odds but decisively raised the stakes:

> People who are working on these problems consider the odds are ten to one on success within two years. I would not bet more than two to one against or even money. But I am quite clear that we must go forward. It would be unforgivable if we let the Germans defeat us in war or reverse the verdict after they had been defeated.

Churchill received Cherwell's recommendation on August 27. Three days later he minuted his military advisers, alluding ironically to the effects of the Blitz: "Although personally I am quite content with the existing explosives, I feel we must not stand in the path of improvement, and I therefore think that action should be taken in the sense proposed by Lord Cherwell."

The British chiefs of staff concurred on September 3.

Mark Oliphant helped goad the American program over the top. "If Congress knew the true history of the atomic energy project," Leo Szilard said modestly after the war, "I have no doubt but that it would create a special medal to be given to meddling foreigners for distinguished services, and Dr. Oliphant would be the first to receive one." Conant in his 1943 secret history thought the "most important" reason the program changed direction in the autumn of 1941 was that "the all-out advocates of a head-on attack on the uranium problem had become more vocal and determined" and mentioned Oliphant's influence first of all.

Oliphant flew to the United States in late August—he considered the Pan-American Clipper through Lisbon too slow and usually traveled by unheated bomber—to work with his NDRC counterparts on radar. But he was also charged with inquiring why the United States was ignoring the MAUD Committee's findings. "The minutes and reports . . . had been sent to Lyman Briggs . . . and we were puzzled to receive virtually no comment. . . . I called on Briggs in Washington, only to find that this inarticulate and unimpressive man had put the reports in his safe and had not shown them to members of his Committee." Oliphant was "amazed and distressed."

He met then with the Uranium Committee. Samuel K. Allison was a new committee member, a talented experimentalist, a protégé of Arthur Compton at the University of Chicago. Oliphant "came to a meeting," Allison recalls, ". . . and said '*bomb*' in no uncertain terms. He told us we must concentrate every effort on the bomb and said we had no right to work on power plants or anything but the bomb. The bomb would cost twenty-five million dollars, he said, and Britain didn't have the money or the manpower, so it was up to us." Allison was surprised. Briggs had kept the committee in the dark. "I thought we were making a power source for submarines."

In desperation Oliphant reached out to the most effective champion he knew in the United States. He wired Ernest Lawrence: "I'll even fly from Washington to meet at a convenient time in Berkeley." At the beginning of September he did.

Lawrence drove Oliphant up the hill behind the Berkeley campus to the site of the 184-inch cyclotron where they could talk without being overheard. Oliphant rehearsed the MAUD Report, which Lawrence had not yet seen. Lawrence in turn proclaimed the possibility of electromagnetic separation of U235 in converted cyclotrons and the virtues of plutonium. "How much I still admire the way in which things are done in your laboratory," Oliphant would write him after their meeting. "I feel quite sure that in your hands the uranium question will receive proper and complete consideration." Back in his office Lawrence called Bush and Conant and arranged for Oliphant to see them. From Oliphant he collected a written summary of the secret British report.

In Washington Conant took Oliphant to dinner and listened with interest. Bush met him in New York and gave him a barely courteous twenty minutes. Neither administrator admitted to knowledge of the MAUD Report. "Gossip among nuclear physicists on forbidden subjects," Conant characterizes Oliphant's peregrinations in his secret history.

Oliphant also stopped by to talk to Fermi. He found the Italian laureate more cautious than ever, "non-committal about the fast-neutron bomb and not altogether happy about the Bohr-Wheeler theory of fission."

Before or after his meetings in Washington and New York Oliphant visited William D. Coolidge, the temporary chairman who produced the second NAS report, at General Electric in Schenectady. That visit at least stirred something like indignation. Coolidge immediately wrote Jewett of Oliphant's news, emphasizing for pure U235 "that the chain reaction in this case would take place thru the direct action of *fast* neutrons. . . . This information, so far as I know, was not available in this country until after the National Academy Committee had sent in its second report. I think that Oliphant's story should be given serious consideration." Information had indeed been available in the United States—at least the MAUD minutes, including Peierls' April 9 statement—but Briggs had locked it away for safekeeping. Oliphant returned to Birmingham wondering if he had made any impression at all.

Lawrence was already moving. He called Arthur Compton in Chicago after Oliphant left Berkeley. "Certain developments made him believe it would be possible to make an atomic bomb," Compton paraphrases the conversation. "Such a bomb, if developed in time, might determine the outcome of the war. The activity of the Germans in this field made it seem to him a matter of great urgency for us to press its development." It was no more than Szilard had argued two years earlier. Lawrence was scheduled to speak in Chicago on September 25. Conant would be in town to receive an honorary degree. Compton proposed to invite both men together to his home. Lawrence could then press the NDRC chairman directly.

Following his decision for political commitment at the Pan American Scientific Conference, Edward Teller had continued teaching at George Washington University but sought work in fission research. In March 1941, with Merle Tuve as one of their sponsors, the Tellers swore allegiance to the United States and became American citizens. Hans Bethe, who was teaching at Columbia for the spring term on temporary leave from Cornell, took the oath the same month. At the end of the term Bethe recommended that Columbia invite Teller to replace him. To work more closely with Fermi and Szilard—and to adjudicate their disputes, which he did with sensitivity—Teller accepted and moved to Manhattan, to an apartment on Morningside Drive.

In the midst of experiment Fermi found time to theorize. He and Teller had lunch at the University Club one pleasant day in September. Afterward, walking back to Pupin—"out of the blue," Teller says—Fermi wondered aloud if an atomic bomb might serve to heat a mass of deuterium sufficiently to begin thermonuclear fusion. Such a mechanism, a bomb fusing hydrogen to helium, should be three orders of magnitude as energetic as a fission bomb and far cheaper in terms of equivalent explosive force. For Fermi the idea was a throwaway. Teller found it a surpassing challenge and took it to heart.

Teller liked to break new ground. When he understood something theoretically he usually moved on without waiting for experimental confirmation. He understood the atomic bomb. He moved on to consider the possibility of a hydrogen bomb. He made extensive calculations. They were disappointing. "I decided that deuterium could not be ignited by atomic bombs," he recalls. "Next Sunday, we went on a walk. The Fermis and the Tellers. And I explained to Enrico why a hydrogen bomb could never be made. And he believed me." For a while, Teller even believed himself.

Enrico Fermi and Edward Teller were not, however, the first to conceive of using a nuclear chain reaction to initiate a thermonuclear reaction in hydrogen. That distinction apparently belongs to Japanese physicist Tokutaro Hagiwara of the faculty of science of the University of Kyoto. Hagiwara had followed world fission research and had conducted studies of his own. In May 1941 he lectured on "Super-explosive U235," reviewing existing knowledge. He was aware that an explosive chain reaction depended on U235 and understood the necessity of isotope separation: "Because of the potential application of this explosive chain reaction a practical method of achieving this must be found. Immediately, it is very important that a means of manufacturing U-235 on a large scale from natural uranium be found." He then discussed the linkage he saw between nuclear fission and thermonuclear fusion: "If by any chance U-235 could be manufactured in a large quantity and of proper concentration, U-235 has a great possibility of becoming useful as the initiating matter for a quantity of hydrogen. We have great expectations for this."

But before the Japanese or the Americans could build a hydrogen bomb they would have to build an atomic bomb. And in neither country was major support yet secure.

"It was a cool September evening," Arthur Compton remembers. "My wife greeted Conant and Lawrence as they came into our home and gave each of us a cup of coffee as we gathered around the fireplace. Then she busied herself upstairs so the three of us might talk freely."

Lawrence spoke with passion. He was "very vigorous in his expression of dissatisfaction with the U.S. program," writes Conant. "Dr. Oliphant had seen him during the summer and by recounting the British hopes had further fired Lawrence's zeal for more action in this whole field." Conant knew all about the British hopes, knew talk was cheap and chose to play the devil's advocate, easily gulling Compton, who thought his arguments turned the tide:

> Conant was reluctant. As a result of the reports so far received he had concluded that the time had come to drop the support of nuclear research as a subject for wartime study. . . . We could not afford to spend either our scientific or our industrial effort on an atomic program of highly questionable military value when every ounce of our strength was needed for the nation's defense.
>
> I rallied to Lawrence's support. . . .
>
> Conant began to be convinced.

"I could not resist the temptation," says the Harvard president, "to cut behind [Lawrence's] rhetoric by asking if he was prepared to shelve his own research programs." Compton cranks Conant's challenge to high melodrama:

> "If this task is as important as you men say," [Conant] remarked, "we must get going. I have argued with Vannevar Bush that the uranium project be put in wraps for the war period. Now you put before me plans for making a definite, highly effective weapon. If such a weapon is going to be made, we must do it first. We can't afford not to. But I'm here to tell you, nothing significant will happen on such a job as this unless we get into it with everything we've got."
>
> He turned to Lawrence. "Ernest, you say you are convinced of the importance of these fission bombs. Are you ready to devote the next several years of your life to getting them made?"
>
> . . . The question brought Lawrence up with a start. I can still recall the expression in his eyes as he sat there with his mouth half open. Here was a serious personal decision. . . . He hesitated only a moment: "If you tell me this is my job, I'll do it."

Back in Washington Conant briefed Bush on what he calls "the results of the involuntary conference in Chicago to which [I] had been exposed." The two administrators decided to order up a third National Academy report, enlarging Compton's committee this time to include W. K. Lewis, a chemical engineer with an outstanding reputation for estimating the potential success at industrial scale of laboratory processes, and Conant's Harvard colleague George B. Kistiakowsky, the resident NDRC explosives expert.

Tall, big-boned, boisterous, with a flat Slavic face and abiding self-confidence, Kistiakowsky had volunteered at eighteen for the White Russian Army and fought in the Russian Revolution. "I grew up in a family in which the question of civil rights, human freedom, was an important one," he told an interviewer late in life. "My father was a professor of sociology and wrote articles and books on the subject and got into trouble with the Czar's regime, very substantial trouble. Mother was also politically oriented. I think both of them went through a short period of being Marxists and then rejected it. That's why I really joined the anti-Bolshevik armies in '18. It was certainly not because I loved Czarism. Of course, I got completely disgusted with the White Army long before it was all over." Kistiakowsky escaped to Germany and took his doctorate at the University of Berlin in 1925. He might have stayed, but his professor advised him to look elsewhere. "He told me that if I wanted to go into an academic career I should emigrate; I would never get a job in Germany—'Here you will always be a Russian.' " Princeton accepted the Ukrainian chemist on a fellowship and soon hired him for its faculty. Then Harvard discovered and courted him. In 1930 he moved, becoming professor of chemistry in 1938.

Conant had been among those who lured Kistiakowsky from Princeton to Harvard. He valued highly his friend and fellow chemist's opinion. "When I retailed to him the idea that a bomb could be made by the rapid assembly of two masses of fissionable material, his first remark was that of a doubting Thomas. 'It would seem to be a difficult undertaking on a battlefield,' he remarked." But it was Kistiakowsky's judgment that finally convinced Conant, as British hopes and physicists' entreaties had not:

> A few weeks later when we met, his doubts were gone. "It can be made to work," he said. "I am one hundred percent sold."
>
> My doubts about Briggs' project evaporated as soon as I heard George Kistiakowsky's considered verdict. I had known George for many years. . . . I had asked him to be head of the NDRC division on explosives. . . . I had complete faith in his judgment. If he was sold on Arthur Compton's program, who was I to have reservations?

Oliphant convinced Lawrence, Lawrence convinced Compton, Kistiakowsky convinced Conant. Conant says Compton's and Lawrence's attitudes "counted heavily with Bush." But "more significant" was the MAUD Report, which G. P. Thomson, now British scientific liaison officer in Ottawa, officially transmitted to Conant on October 3. On October 9, without waiting for the third National Academy of Sciences review, Bush carried the report directly to the President.

Franklin Roosevelt, Henry Wallace and the director of the OSRD met that Thursday at the White House. In a memorandum Bush wrote to Conant the same day he makes it clear that the MAUD Report was the basis for the discussion: "I told the conference of the British conclusions." He told the President and the Vice President that the explosive core of an atomic bomb might weigh twenty-five pounds, that it might explode with a force equivalent to some eighteen hundred tons of TNT, that a vast industrial plant costing many times as much as a major oil refinery would be necessary to separate the U235, that the raw material might come from Canada and the Belgian Congo, that the British estimated the first bombs might be ready by the end of 1943. Bush tried to explain that an atomic bomb plant would produce no more than two or three bombs a month but doubted if the President took in that "relatively low yield." He emphasized that he was basing his statements "primarily on calculation with some laboratory investigation, but not on a proved case" and therefore could not guarantee success.

Bush was presenting, essentially, British calculations and British conclusions. Such a presentation made it appear that Britain was further advanced in the field than America. The discussion therefore shifted to the question of how the United States was attached or might attach itself to the British program. "I told of complete interchange with Britain on technical matters, and this was endorsed." Bush explained that the "technical people" in Britain had also formulated policy—had proposed that the government develop the atomic bomb as a weapon of war—and had passed their formulations along directly to the War Cabinet. In the United States, Bush said, an NDRC section and an advisory committee considered technical matters and only he and Conant considered policy.

Policy was the President's prerogative. As soon as Bush exposed it to view Roosevelt seized it. Bush took that decision to be the most important outcome of the meeting and put it emphatically first in his memorandum to Conant. Roosevelt wanted policy consideration restricted to a small group (it came to be called the Top Policy Group). He named its members: Vice President Wallace, Secretary of War Henry L. Stimson, Army Chief of Staff George C. Marshall, Bush and Conant. Every man owed his authority to the President. Roosevelt had instinctively reserved nuclear weapons policy to himself.

Thus at the outset of the U.S. atomic energy program scientists were summarily denied a voice in deciding the political and military uses of the weapons they were proposing to build. Bush accepted the usurpation happily. To him it was simply a matter of who would run the show. It left him on top and inside and he put it to use immediately to shoulder the physics community into line. Within hours, as he wrote Frank Jewett in November, he had "emphasized to Arthur Compton and his people the fact that they are asked to report upon the techniques, and that consideration of general policy has not been turned over to them as a subject."

Significantly, Bush associated the reservation of policy with relief from criticism: "Much of the difficulty in the past has been due to the fact that Ernest Lawrence in particular had strong ideas in regard to policy, and talked about them generally. . . . I cannot . . . bring him into the discussions, as I am not authorized by the President to do so." He applied just this test—silence on policy—to measure Lawrence's and Compton's loyalty: "I think [Lawrence] now understands this, and I am sure that Arthur Compton does, and I think our difficulties in this regard are over."

A scientist could choose to help or not to help build nuclear weapons. That was his only choice. The surrender of any further authority in the matter was the price of admission to what would grow to be a separate, secret state with separate sovereignty linked to the public state through the person and by the sole authority of the President.

Patriotism contributed to many decisions, but a deeper motive among the physicists, by the measure of their statements, was fear—fear of German triumph, fear of a thousand-year Reich made invulnerable with atomic bombs. And deeper even than fear was fatalism. The bomb was latent in nature as a genome is latent in flesh. Any nation might learn to command its expression. The race was therefore not merely against Germany. As Roosevelt apparently sensed, the race was against time.

There are indications in Bush's memorandum that Roosevelt was concerned less with a German challenge than with the long-term consequences of acquiring so decisive a new class of destructive instruments. "We discussed at some length after-war control," Bush wrote Conant, "together with sources of raw material" (sources of raw material were then believed to be few and far between; whoever commanded them might well, it seemed, monopolize the bomb). Roosevelt was thinking beyond developing bombs for the war that the United States had not yet entered. He was thinking about a military development that would change the political organization of the world.

Bush, who was a successful administrator partly because he knew the limits of his charter, then suggested that a "broader program"—industrial production—ought to be handled when the time came by some larger organization than the OSRD. Roosevelt agreed. Summarizing his assignment, Bush told the President he understood he was to expedite in every possible way the necessary research but was "not [to] proceed with any definite steps on this expanded plan until further instructions from him. . . . He indicated that this was correct." The money, the President told him, "would have to come from a special source available for such an unusual purpose and . . . he could arrange this."

The United States was not yet committed to building an atomic bomb. But it was committed to exploring thoroughly whether or not an atomic bomb could be built. One man, Franklin Roosevelt, decided that commitment—secretly, without consulting Congress or courts. It seemed to be a military decision and he was Commander in Chief.

Bush and Conant proceeded to order up from Arthur Compton a third NAS review. Compton asked Samuel Allison for the name of someone who could help him calculate the critical mass of U235. Allison had been corresponding with Enrico Fermi on the subject of carbon absorption cross sections and recommended him highly. Compton "called on Fermi in his office at Columbia University. Stepping to the blackboard he worked out for me, simply and directly, the equation from which could be calculated the critical size of a chain-reacting sphere. He had at his fingertips the most recent experimental values of the constants. He discussed for me the reliability of the data. . . . Even the most conservative estimate showed that the amount of fissionable metal needed to effect a nuclear explosion could hardly be greater than a hundred pounds."*

Compton moved on to Harold Urey's office to look into isotope separation. Urey was the recognized world leader in the field as a result of his Nobel Prize–winning work with hydrogen isotopes; he had directed isotope separation studies for the Uranium Committee and the Naval Research Laboratory since the beginning. He personally investigated chemical separation of U235 (which turned out to be impossible given the chemical compounds of the day) and separation by centrifuge. Estimating that a centrifuge plant that would produce one kilogram of U235 per day would require 40,000 to 50,000 yard-long centrifuges and would cost about $100 million, he had recently contracted with Westinghouse in the name of the Uranium Committtee for a prototype unit.

Urey was initially skeptical of gaseous barrier diffusion. He and John Dunning were not compatible, perhaps because they were both enthusiasts, and only when centrifuge development was well under way, in late 1940, did Urey turn his attention to the process that Dunning and Eugene Booth were working hard at their own expense to develop. They had chosen gaseous diffusion at dinner one evening in 1940 on their way home from a trip to Schenectady by systematically ruling out other methods as unsuitable for large-scale production, much as Peierls and Simon had done. They were interested in nuclear power, Booth remembers, not bomb-making. "Our reasons for pursuing the isotope separation path toward power production were simple and general. If a chain reaction became possible with normal uranium, a smaller and probably cheaper power plant could be made with enriched uranium."

Dunning and Urey produced a joint appraisal of the gaseous-diffusion process in November 1940. Dunning's barrier material at the time was fritted glass—partially fused and therefore porous silica, the material from which porcelain is made—which uranium hexafluoride was likely to cor-

*Compton's memory errs toward more optimism than Fermi's calculations warranted. After Compton's visit Gregory Breit, Briggs' theoretician on the Uranium Committee, asked Fermi to work his formulae on paper. Fermi was busy with his uranium-graphite experiment and produced, on October 6, a sketchy set of notes. He guessed at the cross sections and came up with 130,000 grams—287 pounds. "One cannot," he added, "in my opinion, exclude the possibility that [the critical mass] may be as low as 20,000 grams [44 pounds] or as high as one or more tons."

rode. They estimated that a gaseous-diffusion plant would involve some five thousand separate barrier tanks—"stages"—but made no attempt to determine cost and power requirements.

By the autumn of 1941, without official support, Dunning and Booth had nevertheless made significant progress. They had switched to brass barriers from which the zinc had been etched (brass is an alloy of copper and zinc; etching away the zinc made the material porous). In November, the month after Compton's visit, they would successfully enrich a measurable quantity of uranium with their equipment.

Compton traveled next to Princeton to see Eugene Wigner, who had been working closely with Fermi. Wigner clarified for Compton the difference between fast- and slow-neutron fission. He endorsed the uranium-graphite system Fermi was developing as a method for producing 94. "He urged me," writes Compton, "almost in tears, to help get the atomic program rolling. His lively fear that the Nazis would make the bomb first was the more impressive because from his life in Europe he knew them so well."

Back in Chicago Compton talked to Glenn Seaborg, who had come east from Berkeley at Compton's request. Seaborg was confident he could devise a large-scale, remote-controlled technology for separating 94 chemically from uranium.

Armed with this new round of information Compton called a meeting of his committee for October 21 in Schenectady. He prepared for the meeting by writing a draft report. A letter came from Lawrence saying he wanted to bring along Robert Oppenheimer: "I have a great deal of confidence in Oppie, and I'm anxious to have the benefit of his judgment in our deliberation." Conant had scolded Lawrence at Compton's fireside when he learned that Lawrence had asked Oppenheimer, still an outsider, for help with theory, but now Lawrence's request was granted.

A dispute between Lawrence and Oppenheimer about what Lawrence called the theoretician's "leftwandering activities" almost excluded him from the atomic bomb project. Oppenheimer, married now to the former Katherine Puening, known as Kitty, with a six-month-old son, had begun to wish for assignment. "Many of the men I had known went off to work on radar and other aspects of military research," he testified later. "I was not without envy of them." He learned the price of admission when he invited Lawrence to an organizational meeting at his elegant new home on Eagle Hill for a professional union, the American Association of Scientific Workers, of which Arthur Compton, among others, was a senior member. Lawrence wanted no part in any "causes and concerns," as he called political activities, and barred his staff as well: "I don't think it's a good idea," he told them. "I don't want you to join it. I know nothing wrong with it, but we're planning big things in connection with the war effort, and it wouldn't be right. I want no occasion for somebody in Washington to find fault with us." Oppenheimer was not so easily put off; he debated Lawrence's point, arguing that humanity was everyone's responsibility and that the more fortunate should help "underdogs." The Nazis came first, Lawrence countered. He told Oppenheimer about Conant's scolding. Oppenheimer reserved judgment. The October 21 meeting, where he could measure the scientific leaders of the uranium program against his own formidable gifts, changed his mind. "It was not until my first connection with the rudimentary atomic-energy enterprise," he testifies, "that I began to see any way in which I could be of direct use." When he saw his way to war work he quickly sacrificed his underdogs, writing Lawrence on November 12:

> I . . . assure you that there will be no further difficulties at any time with the A.A.S.W. . . . I doubt very much whether anyone will want to start at this time an organization which could in any way embarrass, divide or interfere with the work we have in hand. I have not yet spoken to everyone involved, but all those to whom I have spoken agree with us: so you can forget it.

Lawrence opened the Schenectady meeting by reading Oliphant's summary of the MAUD Report. Compton followed with a review based on his October travels. Oppenheimer weighed in during the discussion of U235's critical mass with an estimate of 100 kilograms, 220 pounds, close to Fermi's estimate of 130,000 grams. "Kistiakowsky," writes Compton, "explained the great economic advantage of being able to deliver a heavy blow with a bomb carried by a single plane."

But Compton was distressed to discover he could not move the engineers on the review committee—the practical souls Bush had insisted be added to bring the NAS reviews down to earth—to estimate either how much time it would take to build a bomb or how much the enterprise would cost:

> With one accord they refused. . . . There weren't enough data. The fact was that they had before them all the relevant information that existed, and some kind of answer was needed, however rough it might be, for otherwise our recommendation could not be acted upon. After some discussion, I suggested a total time of between three and five years, and a total cost . . . of some hundreds of millions of dollars. None of the committee members objected.

So the American numbers came out of a scientist's hat, as the British numbers had. Atomic energy was still too new for engineering.

If Compton was distressed by the refusal of commitment, Lawrence was appalled. Within twenty-four hours he mailed the committee chairman a bracing challenge edged with threat:

> In our meeting yesterday, there was a tendency to emphasize the uncertainties, and accordingly the possibility that uranium will not be a factor in the war. This to my mind, was very dangerous. . . .
>
> It will not be a calamity if, when we get the answers to the uranium problem, they turn out negative from the military point of view, but if the answers are fantastically positive and we fail to get them first, the results for our country may well be tragic disaster. I feel strongly, therefore, that anyone who hesitates on a vigorous, all-out effort on uranium assumes a grave responsibility.

But Compton had already been threatened by an expert, Vannevar Bush, and knew his duty well, though he did not yet know that Bush was already committed to expedition and expansion. He had difficulty estimating "the destructiveness of the bomb." The calculation "involved problems of gas pressure, specific heats at hitherto unknown temperatures, the transmission of radiations and particles through the material, and forces of inertia." He asked Gregory Breit for help. Breit was even more obsessed with secrecy than Briggs. "No help was forthcoming," says Compton, gritting his teeth. He turned then to Oppenheimer. "I had known 'Oppie' for some fourteen years and had found him most competent in seeing the essentials of an intricate problem and in interpreting what he saw. So I was glad to get a letter from him with helpful suggestions." Through the end of October Compton worked on.

At Leipzig in September Werner Heisenberg received the first forty gallons of heavy water from Norsk Hydro and immediately prepared another chain-reaction experiment like the unsuccessful effort at the Virus House in Dahlem the year before: a thirty-inch aluminum sphere filled with alternating layers of heavy water and uranium oxide, more than three hundred pounds of it, arranged around a central neutron source, the sphere itself then immersed in water in a laboratory tank. This time Heisenberg found some increase in neutrons, enough to extrapolate eventual success. The German laureate knew now from the work of von Weizsäcker and Houtermans that a sustained chain reaction in natural uranium would breed element 94. "It was from September 1941," he remarks in consequence, "that we saw an open road ahead of us, leading to the atomic bomb."

He decided to talk to Bohr. To what end he thought Bohr might help him he never unambiguously explained. His wife Elisabeth believes "he was lonely in Germany. Niels Bohr had become a father figure to him. . . . He thought that he could talk about anything with Bohr. . . . The advice of an older friend, more experienced in human and political affairs, had always been important to him." He "saw himself confronted with the spectre of the atomic bomb," Elisabeth Heisenberg explains, "and he wanted to signal to Bohr that Germany neither would nor could build a bomb. . . . Secretly he even hoped that his message could prevent the use of an atomic bomb on Germany one day. He was constantly tortured by this idea. . . . This vague hope was probably the strongest motivation for his trip."

Heisenberg and von Weizsäcker attended a scientific meeting in Copenhagen at the end of October, a meeting Bohr routinely boycotted as he boycotted all joint Danish and German activities, to emphasize his refusal to collaborate. He was willing to see Heisenberg, however, and received him, according to the German physicist's wife, "with great warmth and hospitality."

Heisenberg saved his crucial conversation for a long evening walk with Bohr through the brewery district around the Carlsberg House of Honor. "Being aware that Bohr was under the surveillance of the German political authorities," he recalled after the war, "and that his assertions about me would probably be reported to Germany, I tried to conduct this talk in such a way as to preclude putting my life into immediate danger." Heisenberg remembers asking Bohr if it was right for physicists to work on "the uranium problem" in wartime when there was a possibility that such work could lead to "grave consequences in the technique of war." Bohr, who had returned from the United States convinced that a bomb was practically impossible, "understood the meaning of the question immediately, as I realized from his slightly frightened reaction." Heisenberg apparently thought Bohr was privy to American secrets and was reacting guiltily to implicit exposure. But Bohr's next response suggests that he had been, rather, stunned at Heisenberg's revelation: he asked Heisenberg if a bomb really was possible. Heisenberg says he answered that a "terrific technical effort" would be necessary, which he hoped could not be realized in the present war. "Bohr was shocked by my reply, obviously assuming that I had intended to convey to him that Germany had made great progress in the direction of manufacturing atomic weapons. Although I tried subsequently to correct this false impression I probably did not succeed. . . . I was very unhappy about the result of this conversation."

Thus Heisenberg's version of the evening walk. Bohr's is less detailed. His son Aage, a Nobel laureate in his turn and his father's successor as director of the Copenhagen institute, summarizes it in a memoir:

> The impression that in Germany great military importance was given to [atomic energy research] was strengthened by the visit to Copenhagen in the autumn of 1941 of Werner Heisenberg and C. F. von Weizsäcker.... In a private conversation with my father Heisenberg brought up the question of the military applications of atomic energy. My father was very reticent and expressed his scepticism because of the great technical difficulties that had to be overcome, but he had the impression that Heisenberg thought that the new possibilities could decide the outcome of the war if the war dragged on.... [Heisenberg's] account [of the meeting] has no basis in actual events.

Robert Oppenheimer, who also had the story direct from Bohr, condenses the meeting to the comment: "Heisenberg and von Weizsäcker came over from Germany, and so did others. Bohr had the impression that they came less to tell what they knew than to see if Bohr knew anything that they did not; I believe that it was a standoff."

The two accounts are not incompatible, but both leave out a crucial fact: that Heisenberg passed to Bohr a drawing of the experimental heavy-water reactor he was working to build. If he did so clandestinely he certainly risked his life. If he did so cynically and with Nazi approval to misdirect Allied intelligence he was certainly no longer attached to Bohr as a father figure, as Elisabeth Heisenberg writes. Whatever his intent, it had the wrong effect on Bohr. Elisabeth Heisenberg thinks "Bohr essentially heard only one single sentence: The Germans knew that atomic bombs could be built. He was deeply shaken by this, and his consternation was so great that he lost track of all else." But Aage Bohr's and Oppenheimer's accounts imply a further response from Bohr: indignation, even incredulity, that Heisenberg would think Bohr might be willing in any way, for any reason, to cooperate with Nazi Germany. Heisenberg, in turn, was aghast that Bohr would fail to see and credit his reservations, would not understand, as his wife writes, that his "bond to his country and its people was not tantamount to a bond to the regime." To the contrary, she adds, "Bohr told Heisenberg that he understood completely that one had to use all of one's abilities and energies for one's country in time of war." Not surprisingly, since it implied Bohr thought the worst of him—that he was willing to work for the Nazis—"Heisenberg was deeply shocked by Bohr's reply."

The meeting, and especially the drawing Heisenberg passed, gave Bohr more to worry about, but he continued to doubt that any nation could afford sufficient industrial capacity, especially in wartime, to pursue isotope separation. He must have been pained at what he took to be the treachery of a brilliant and formerly devoted protégé. Heisenberg for his part found himself, says his wife, in "a state of confusion and despair." Even at risk he had not convinced Bohr of his sincerity nor in any way begun a dialogue to avert possible catastrophe. In the absence of such dialogue he had only managed potentially to alarm Germany's most powerful enemy further with news of progress in approaching the chain reaction. That news must necessarily accelerate Allied efforts to build a bomb. As Rudolf Peierls writes of this period in Heisenberg's life, "he had agreed to sup with the devil, and perhaps he found that there was not a long enough spoon."

Arthur Compton sent draft copies of the third National Academy of Sciences report to Vannevar Bush and Frank Jewett before the weekend of November 1. The new report was brief—six double-spaced typewritten pages (with forty-nine pages of technical appendices and figures)—and finally and emphatically to the point: "The special objective of the present report is to consider *the possibilities of an explosive fission reaction with U235.*" Progress toward separating uranium isotopes, Compton wrote, made renewed consideration urgent (a rationale somewhat less than candid: British progress had spurred the change).

This time the report knew what it was about: "*A fission bomb of superlative destructive power will result from bringing quickly together a sufficient mass of element U235.* This seems to be as sure as any untried prediction based upon theory and experiment can be." On the second page an estimate of critical mass elicited for the first time among the three NAS reports a mention of fast fission: "*The mass of U235 required to produce explosive fission under appropriate conditions can hardly be less than 2 kg nor greater than 100 kg.* These wide limits reflect chiefly the experimental uncertainty in the capture cross-section of U235 for fast neutrons."

The NAS estimate of destructiveness was low compared to the MAUD Report estimate, some 30 tons of TNT equivalent per kilogram of U235 (for 25 pounds, 300 tons compared to MAUD's 1,800 tons), but the American report attempted to compensate for its doubts about the efficacy of an intense energy release from a small amount of matter by emphasizing that the destructive effects on life of a bomb's radioactivity "may be as important as those of the explosion itself."

The centrifuge and gaseous diffusion programs were noted to be "approaching the stage of practical test." Fission bombs might be available "in significant quantity within three or four years." Like its predecessors the report stressed not the German challenge but the long-term prospect: "The possibility must be seriously considered that within a few years the use of bombs such as described here, or something similar using uranium fission, may determine military superiority. Adequate care for our national defense seems to demand urgent development of this program."

In detailed appendices Compton calculated the critical mass of a bomb heavily constrained in tamper at no more than 3.4 kilograms; Kistiakowsky debated whether a fission explosion would be as destructive in terms of energy produced as the explosion of an equivalently energetic mass of TNT and confirmed the feasibility of firing together two pieces of uranium at a speed of several thousand feet per second; and a senior physicist on Compton's committee reported favorably on the isotope-separation systems then under consideration and recommended "the principle of *parallel development,*" meaning pursuing them all at once, an expensive way to save time in case one or more failed.

Notably missing from the third report was any mention of the uranium-graphite work going on at Columbia or of plutonium. Compton remembers that a U235 bomb looked "more straightforward and more certain of accomplishment" than a plutonium bomb, but the omission also measures the extent to which Briggs' judgment of priorities, and Briggs himself, had been set aside. Bush writing Jewett before he met with Compton had already mentioned "leaving Briggs in charge of a section devoted as it is at the present time to physical measurements"—small potatoes indeed—and constituting "a new group under a full-time head to handle development." He was considering Ernest Lawrence but still thought Lawrence talked too much: "The matter . . . would have to be handled under the strictest sort of secrecy. This is the reason that I hesitate at the name of Ernest Lawrence."

If the third and last NAS report only rationalized a previous presidential decision, it at least served to check the British findings independently and to commit the American physics community to the cause. The United States had finally set its wheels to the bomb track. Its inertia was proportional to the juggernaut of its scientific, engineering and industrial might. Acceleration overcoming inertia, it now began to roll.

No document Franklin Delano Roosevelt signed authenticates the fateful decision to expedite research toward an atomic bomb that Vannevar Bush reported in his October 9 memorandum to James Bryant Conant: the archives divulge no smoking gun. The closest the records come to a piece of paper that changed the world is a banality. Bush personally delivered the third National Academy of Sciences report to the President on November 27, 1941. Roosevelt returned it to him two months later with a note on White House stationery written in black ink with a broad-nibbed pen, a note that would communicate only a commonplace of the housekeeping of state secrets except for the authority of its first vernacular expression and the initials it bears:

THE WHITE HOUSE
WASHINGTON

Jan 19 —

V.B.

OK — returned — I
think you had best keep
this in your own
safe

FDR

Text reads: "Jan 19— V.B. OK —returned— I think you had best keep this in your own safe FDR"

Still orphaned was plutonium, which Lawrence and Compton believed so promising. Compton found his chance to speak for it in early December when Bush and Conant called the members of the Uranium Committee to Washington to announce the reorganization of their work. Harold Urey would develop gaseous diffusion at Columbia, Bush and Conant had decided. Lawrence would pursue electromagnetic separation at Berkeley. A young chemical engineer, Eger V. Murphree, the director of research for Standard Oil of New Jersey, would supervise centrifuge development and look into broader questions of engineering. Compton in Chicago would be responsible for theoretical studies and the actual design of the bomb. "The meeting adjourned," writes Compton, "with the understanding that we would meet again in two weeks to compare progress and shape our plans more firmly."

Bush, Conant and Compton went to lunch at the Cosmos Club on Lafayette Square. There the Chicago physicist spoke up for plutonium. He argued that the advantage of chemical extraction rather than isotope separation made element 94 "a worthy competitor." Bush was wary. Conant pointed out that the new element's chemistry was still largely unknown. Compton recalls their exchange:

> "Seaborg tells me that within six months from the time [plutonium] is formed [by chain reaction] he can have it available for use in the bomb," was my comment.
>
> "Glenn Seaborg is a very competent young chemist, but he isn't that good," said Conant.

How good a chemist Glenn Seaborg might be remained to be seen. Compton, Conant remembers, went on to argue that "the construction of a self-sustaining chain reaction [in natural uranium—Fermi's and Szilard's project] would be a magnificent achievement" even if plutonium flunked as bomb material; "it would prove that the measurements and theoretical calculations were correct":

> I never knew whether it was this near-certainty of demonstrating a slow-neutron reaction which settled the matter in Van's mind, or whether he was impressed with Compton's faith in the production of a plutonium bomb, against my lack of faith as a chemist. At all events, within a matter of weeks he agreed to Arthur Compton's setting up at Chicago a highly secret project.

Bush had called the Washington meeting on a weekend to accommodate busy men. They had assembled on Saturday, December 6, 1941. Almost immediately they found themselves busier yet.

At 7 A.M. Hawaiian time on Sunday, December 7, 1941, near Kahuku Point at the northernmost reach of the island of Oahu, two U.S. Army privates in the process of shutting down the Opana mobile radar station, an aircraft reconnaissance unit which they had manned since 4 A.M., noticed an unusual disturbance on their oscilloscope screen. They checked and confirmed no malfunction and decided the large merged blur of light "must be a flight of some sort." Their plotting board indicated a bearing out of the northeast at a distance of 132 miles. More than fifty planes appeared to be involved. One of the men called the information center at Fort Shafter, at the other end of the island, where radar and visual reconnaissance reports were combined on a tabletop map. The lieutenant who took the phone heard the radar operator call the sightings "the biggest . . . he had ever seen." The operator did not, however, report his estimate of their number.

Both the Army and the Navy had been warned of imminent danger of

Japanese attack. The Japanese had convinced themselves that dominance over East Asia was vital to their survival. The American reaction to militant Japanese expansion into Manchuria and China—as many as 200,000 men, women and children were brutally slaughtered by the Japanese Army in Shanghai in 1937—had been to embargo war materials and freeze Japanese assets in the United States. Aviation fuel, steel and scrap iron went on the embargo list in September 1940 when the Japanese moved into French Indochina with the timid approval of Vichy France. After that the Japanese estimated they could survive no more than eighteen months without access to Asian oil and iron ore. For some time they had prepared for war while continuing to negotiate. Now negotiations had collapsed.

Lieutenant General Walter C. Short, commander of the Army's Hawaiian Department, received a coded message on November 27 signed in the name of the Chief of Staff—George Marshall—that read in part:

> Negotiations with Japan appear to be terminated to all practical purposes with only the barest possibility that the Japanese Government might come back and offer to continue. Japanese future action unpredictable but hostile action possible at any moment. If hostilities cannot, repeat cannot be avoided the United States desires that Japan commit the first overt act. . . . Measures should be carried out so as not, repeat not, to alarm civil population or disclose intent.

Short had at option three levels of alert, escalating from "a defense against sabotage, espionage and subversive activities without any threat from the outside" to full defense against "an all-out attack." He thought it obvious that the War Department message "was written basically for General MacArthur in the Philippines" and chose the limited sabotage defense, Alert No. 1.

Admiral Husband E. Kimmel, Commander in Chief of the U.S. Pacific Fleet, which was based at Pearl Harbor west of Honolulu on the southern coast of Oahu, received a similar but even more pointed message from the Navy Department a few hours later:

> This dispatch is to be considered a war warning. Negotiations with Japan looking toward stabilization of conditions in the Pacific have ceased and an aggressive move by Japan is expected within the next few days. The number and equipment of Japanese troops and the organization of naval task forces indicates an amphibious expedition against either the Philippines, Thai or Kra Peninsula or possibly Borneo. Execute an appropriate defensive deployment preparatory to carrying out the tasks assigned.

Kimmel noted the references to other theaters of potential conflict. When he and Short exchanged messages he noted the "more cautious phrasing" of the Army warning. "Appropriate defensive deployment" meant, he thought, full security measures for ships at sea. A surprise submarine attack seemed possible and he ordered the depth-bombing of any submarines discovered in the waters around Oahu.

The Army lieutenant who took the Opana radar call therefore had no expectation of danger. He looked for a routine explanation of the unusual report and found it. The Army paid radio station KGMB in Honolulu to play Hawaiian music throughout the night whenever it ferried aircraft to the Islands, giving its navigators a signal to seek. The lieutenant had heard such music on the radio that morning on his way to the information center. He decided that the radar must be picking up a flight of B-17's. The heading plotted at Opanu was the usual direction of approach from California. "Well, don't worry about it," the lieutenant told the radar men.

Pearl Harbor is a shallow, compound basin sheltered inland through a narrow outer channel from the sea. A bulge of land, Pearl City, and a mid-basin island, Ford Island, canalize the main anchorage of the harbor into a loop of narrow inlets. In 1941 drydocks, oil storage tanks and a submarine base occupied the harbor's irregular eastern shore. Seven battleships rode at anchor immediately southeast of Ford Island that Sunday morning: *Nevada* anchored alone; *Arizona* inboard of the repair ship *Vestal; Tennessee* inboard of *West Virginia; Maryland* inboard of *Oklahoma; California* alone. An eighth battleship, *Pennsylvania*, wedged naked in drydock nearby.

Lieutenant Commander Mitsuo Fuchida of the Japanese Imperial Navy, thirty-nine years old, who wore a red shirt to disguise from his men any blood he might shed and a white *hachimaki* tied around his flight helmet brushed with the calligraphic characters for "Certain Victory," called out *"Tora! Tora! Tora!"* at 0753 hours as his pilot banked around Barber's Point southwest of Pearl: *"Tiger!"* three times invoked to announce to the listening Japanese Navy that his first wave of 183 planes had achieved complete surprise. The 43 fighters, 49 high-level bombers, 51 dive-bombers and 40 torpedo planes he commanded had flown from six carriers holding station 200 miles to the north, carriers formidably escorted by battleships, heavy cruisers, destroyers and submarines that had left Hitokappu Bay on the northern Japanese island of Etorofu on November 25 and sailed blacked out in radio silence across the stormy but empty northern Pacific for almost two weeks to achieve this stunning rendezvous.

The torpedo bombers divided into groups of twos and threes and dived. The aircrews had prepared themselves to ram the battleships if necessary, but nothing restrained their attack. At 0758 the Ford Island command center radioed its frantic message to the world: AIR RAID PEARL HARBOR. THIS IS NOT DRILL. Admiral Kimmel saw the attack begin from a neighbor's lawn—"in utter disbelief and completely stunned," the neighbor remembers, "as white as the uniform he wore." Torpedoes struck a light cruiser and a target ship, a minelayer, another light cruiser, then the battleships: *Arizona* lifted out of the water; *West Virginia* washed by a huge waterspout; *Oklahoma* hit by three torpedoes one after another and immediately listing steeply to port; the bottom blown out of *Arizona;* three torpedoes into *California;* two more into *West Virginia*; a fourth into *Oklahoma* that bounced the big ship and rolled it over bottom up; *Arizona* taking a bomb that detonated its forward explosive stores, ripped the ship apart, killed at least a thousand men and blew high into the air a grisly rain of bodies, hands, legs and heads; a torpedo tearing out *Nevada*'s port bow. Thick black smoke rolled up to foul the blue Hawaiian morning and in the water, burning, screaming men attempted to swim through a dense scum of burning oil. Japanese fighters and bombers destroyed aircraft on the ground and strafed soldiers and marines pouring out of barracks at Hickam Field and Ewa Field and Wheeler. An hour later a second wave of 167 more attack aircraft deployed to further destruction. The two raids accounted for eight battleships, three light cruisers, three destroyers and four other ships sunk, capsized or damaged and 292 aircraft damaged or wrecked, including 117 bombers. And 2,403 Americans, military and civilian, killed, 1,178 wounded, in unprovoked assaults that lasted only minutes. The following afternoon, Franklin Roosevelt, addressing Congress in joint session, requested and won a declaration of war against not only Japan but Germany and Italy as well.

The man who conceived and planned the surprise attack on Pearl Harbor, Admiral Isoroku Yamamoto, Commander in Chief of the Japanese Combined Fleet, had few illusions about the ultimate success of a war against the United States. He had studied at Harvard and served as a naval attaché in Washington and knew America's strength. But if war had to come he meant "to give a fatal blow to the enemy fleet" when it was least expected, at the outset. By that act he hoped he could win his country six months to a year during which it might establish its Greater East Asia Co-Prosperity Sphere and dig in.

The torpedoes had been a challenge. Pearl Harbor was only forty feet deep. Torpedoes dropped from planes routinely sank seventy feet or more before bobbing up to attack depth. The Japanese had to reduce that plunge signficantly or bury their weapons in the Pearl mud.

They found in repeated experiments that they could sometimes manage a shallower drop by flying only forty feet above the water and holding down their air speed—the maneuver demanded skilled flying—but further improvement required torpedo redesign, largely by trial and error. As late as mid-October Fuchida's flyers were still managing no better than sixty-foot plunges, still far too deep.

A new stabilizer fin, originally designed for aerial stability, saved the mission. Tested during September, it consistently held the torpedo to less than forty feet and steadied it as well. But the pilots still needed aiming practice. Only thirty of the modified weapons could be promised by October 15, another fifty by the end of the month and the last hundred on November 30, after the task force was scheduled to sail.

The manufacturer did better. Realizing the weapons were vital to a secret program of unprecedented importance, manager Yukiro Fukuda bent company rules, drove his lathe and assembly crews overtime and delivered the last of the 180 specially modified torpedoes by November 17. Mitsubishi Munitions contributed decisively to the success of the first massive surprise blow of the Pacific War by the patriotic effort of its torpedo factory on Kyushu, the southernmost Japanese island, three miles up the Urakami River from the bay in the old port city of Nagasaki.

13 The New World

Enrico Fermi's team at Columbia University had been hard at work through 1941 while the government deliberated. Fermi, Leo Szilard, Herbert Anderson and the young physicists who had joined them may never have known how close they came to orphanhood. The isolation of plutonium at Berkeley added a potential military application to their reasons for pursuing a slow-neutron chain reaction in uranium and graphite, but given the necessary resources Fermi at least would certainly have pursued the chain reaction anyway as a physical experiment of fundamental and historic worth. He had missed discovering fission by the thickness of a sheet of aluminum foil; he would not willingly leave to someone else the demonstration of atomic energy's first sustained release. Thanks largely to Arthur Compton his work found continued support, which may help explain why he admired the pious Woosterite's intelligence so extravagantly.

Szilard had finally gone on the Columbia payroll on November 1, 1940, when the $40,000 National Defense Research Committee contract came through for physical-constant measurements. To help Fermi without the friction the two men generated when they worked side by side, Szilard undertook to apply his special talent for enlightened cajolery to the problem of procuring supplies of purified uranium and graphite. The record is thick with his correspondence with American graphite manufacturers dismayed to discover that what they thought were the purest of materials were in fact hopelessly contaminated, usually with traces of boron. The cross section for neutron absorption of that light, ubiquitous, silicon-like element, number 5 on the periodic table, was tremendous and poisonous. "Szilard at that time took extremely decisive and strong steps to try to organize the early phases of production of pure materials," says Fermi. ". . . He did a marvelous job which later on was taken over by a more powerful organization than was Szilard himself. Although to match Szilard it takes a few able-bodied customers."

In August and September the Columbia team prepared to assemble the largest uranium-graphite lattice yet devised. A slow-neutron chain reaction in natural uranium, like its fast-neutron counterpart U235, requires a critical mass: a volume of uranium and moderator sufficient to sustain neutron multiplication despite the inevitable loss of neutrons from its outer surface. No one yet knew the specifications of that critical volume, but it was obviously vast—on the order of some hundreds of tons. One way to create a self-sustaining chain reaction might be simply to continue stacking uranium and graphite together. But so crude an experiment, if it worked at all, would teach the experimenter very little about controlling the resulting reaction and might culminate in a disastrous and lethal runaway. Fermi proposed to approach the problem by the more circumspect route of a series of subcritical experiments designed to determine the necessary quantities and arrangements and to establish methods of control.

As always, he built directly on previous experience. He and Anderson had calculated the absorption cross section of carbon by measuring the diffusion of neutrons from a neutron source up a column of graphite. The new experiments would enlarge that column to take advantage of the increased stocks of graphite available and to make room for regularly spaced inclusions of uranium oxide: simplicity itself, but in physical form a thick, black, grimy, slippery mass of some thirty tons of extruded bars of graphite confining eight tons of oxide. Fermi named the structure a "pile." "Much of the standard nomenclature in nuclear science was developed at this time," Segrè writes. ". . . I thought for a while that this term was used to refer to a source of nuclear energy in analogy with Volta's use of the Italian term *pila* to denote his own great invention of a source of electrical energy [i.e., the Voltaic battery]. I was disillusioned by Fermi himself, who told me that he simply used the common English word *pile* as synonymous with *heap.*" The Italian laureate was continuing to master the plainsong of American speech.

The exponential pile Fermi proposed to build (so called because an exponent entered into the calculation of its relationship to a full-scale reactor) would be too big for any of the laboratories in Pupin. He sought larger quarters:

> We went to Dean Pegram, who was then the man who could carry out magic around the university, and we explained to him that we needed a big room. And when we say big we meant a really big room. Perhaps he made a crack about a church not being the most suited place for a physics laboratory . . . but I think a church would have been just precisely what we wanted. Well, he scouted around the campus and we went with him to dark corridors and under various heating pipes and so on to visit possible sites for this experiment and eventually a big room, not a church, but something that might have been compared in size with a church was discovered in Schermerhorn [Hall].

There, Fermi goes on, they began to build "this structure that at that time looked again in order of magnitude larger than anything that we had seen before. . . . It was a structure of graphite bricks and spread through these graphite bricks in some sort of pattern were big cans, cubic cans, containing uranium oxide." The cans, 8 by 8 by 8 inches, 288 of them in all, were made of tinned iron sheet; each could hold about 60 pounds of uranium oxide. Each cubic "cell" of the uranium-graphite lattice—a can and its surrounding graphite—was 16 inches on a side. Spheres of uranium in an arrangement of spherical cells would have been more efficient. In these beginning experiments, with materials of doubtful purity, Fermi was pursuing order-of-magnitude estimates, a first rough mapping of new territory. "This structure was chosen because of its constructional simplicity," the experimenters wrote afterward, "since it could be assembled without cutting our graphite bricks of 4″ by 4″ by 12″. Although we did not expect that the structure would approach too closely the optimum proportions, we thought it desirable to obtain some preliminary information as soon as possible." Promising results might also win further NDRC support.

"We were faced with a lot of hard and dirty work," Herbert Anderson recalls. "The black uranium oxide powder had to be . . . heated to drive off undesired moisture and then packed hot in the containers and soldered shut. To get the required density, the filling was done on a shaking table. Our little group, which by that time included Bernard Feld, George Weil, and Walter Zinn, looked at the heavy task before us with little enthusiasm. It would be exhausting work." Then Pegram to the rescue in Fermi's telling:

> We were reasonably strong, but I mean we were, after all, thinkers. So Dean Pegram again looked around and said that seems to be a job a little bit beyond your feeble strength, but there is a football squad at Columbia that contains a

dozen or so of very husky boys who take jobs by the hour just to carry them through college. Why don't you hire them?

And it was a marvelous idea; it was really a pleasure for once to direct the work of these husky boys, canning uranium—just shoving it in—handling packs of 50 or 100 pounds with the same ease as another person would have handled three or four pounds.

"Fermi tried to do his share of the work," Anderson adds; "he donned a lab coat and pitched in to do his stint with the football men, but it was clear that he was out of his class. The rest of us found a lot to keep us busy with measurements and calibrations that suddenly seemed to require exceptional care and precision."

For this first exponential experiment and the many similar experiments to come, Fermi defined a single fundamental magnitude for assessing the chain reaction, "the reproduction factor k." k was the average number of secondary neutrons produced by one original neutron in a lattice of infinite size—in other words, if the original neutron had all the room in the world in which to drift on its way to encountering a uranium nucleus. One neutron in the zero generation would produce k neutrons in the first generation, k^2 neutrons in the second generation, k^3 in the third generation and so on. If k was greater than 1.0, the series would diverge, the chain reaction would go, "in which case the production of neutrons is infinite." If k was less than 1.0, the series would eventually converge to zero: the chain reaction would die out. k would depend on the quantity and quality of materials used in the pile and the efficiency of their arrangement.

The cubical lattice that the Columbia football squad stacked in Schermerhorn Hall in September 1941 extrapolated to a disappointing first k of 0.87. "Now that is by 0.13 less than one," Fermi comments—13 percent less than the minimum necessary to make a chain reaction go—"and it was bad. However, at the moment we had a firm point to start from, and we had essentially to see whether we could squeeze the extra 0.13 or preferably a little bit more." The cans were made of iron, and iron absorbs neutrons. "So, out go the cans." Cubes of uranium were less efficient than spheres; next time the Columbia group would press the oxide into small rounded lumps. The materials were impure. "So, now, what do these impurities do?—clearly they can do only harm. Maybe they make harm to the tune of 13 percent." Szilard would continue his quest for materials of higher purity. "There was some considerable gain to be made . . . there."

"Well," concludes Fermi, "this brings us to Pearl Harbor."

Arthur Compton had less than two weeks to throw together a program between his discussion with Vannevar Bush and James Bryant Conant at the

Cosmos Club luncheon on December 6 and the first meeting on December 18 of the new leaders of what was now to be called the S-1 program. (S-1 for Section One of the Office of Scientific Research and Development: Conant would administer S-1, but the National Defense Research Committee was no longer directly involved; the bomb program had advanced from research into development.) On December 18, Conant notes in the secret history of the project he wrote in 1943, "the atmosphere was charged with excitement—the country had been at war nine days, an expansion of the S-1 program was now an accomplished matter. Enthusiasm and optimism reigned." Compton offered his program to Bush, Conant and Briggs the next day and followed up on December 20 with a memorandum. The projects that had come under his authority were scattered across the country at Columbia, Princeton, Chicago and Berkeley. For the time being he proposed leaving them there.

With the arrival of war, not to breathe a word of the mysteries they were exploring, the project leaders had adopted an informal code: plutonium was "copper," U235 "magnesium," uranium generically in the nonsensical British coinage "tube alloy." "On the basis of the present data," Compton wrote, optimism reigning, "it appears that explosive units of copper need be only half the size of those using magnesium, and that premature explosions can be ruled out." Because of the difficulty of engineering a remotely controlled chemical plant to extract plutonium, however, he thought that "the production of useful quantities of copper will take longer than the production of magnesium." For a timetable he offered:

> Knowledge of conditions for chain reaction by June 1, 1942.
> Production of chain reaction by October 1, 1942.
> Pilot plant for using reaction for copper production, October 1, 1943.
> Copper in usable quantities by December 31, 1944.

His schedule was designed to show that plutonium might be produced in time to influence the outcome of the war, the standard which Conant was insisting upon after Pearl Harbor even more vehemently than before. But the uranium-graphite work had not yet won even Compton's full confidence. If graphite proved impractical and "copper production" had to wait for heavy water (of which Harold Urey was urging the extraction at an existing plant in Canada), Compton's schedule would slip by "from 6 months to 18 months." And that might be too late to make a difference.

For the next six months, Compton estimated, the pile studies at Columbia, Princeton and Chicago would cost $590,000 for materials and $618,000 for salaries and support. "This figure seemed big to me," he remembers modestly, "accustomed as I was to work on research that needed not more than a few thousand dollars per year."

He had met with Pegram and Fermi to prepare this part of his proposal and concluded that when metallic uranium became available the project should be concentrated at Columbia. Over Christmas and through the first weeks of January it fell to Herbert Anderson, the native son, to find a building in the New York City area large enough to house a full-scale chain-reacting pile. Not to be outdone in the matter of informal codes, the Columbia team had named that culmination "the egg-boiling experiment." Anderson stumped the wintry boroughs and turned up seven likely locations for boiling uranium eggs. He proposed them to Szilard on January 21; they included a Polo Grounds structure, an aircraft hangar on Long Island that belonged to Curtiss-Wright and the hangar Goodyear used to house its blimps.

But as Compton reviewed the work of the several groups that had come under his authority, bringing their leaders together in Chicago three times during January, their disagreements and duplications made it obvious that all the developmental work on the chain reaction and on plutonium chemistry should be combined at one location. Pegram offered Columbia. They considered Princeton and Berkeley and industrial laboratories in Cleveland and Pittsburgh. Compton offered Chicago. No one wanted to move.

The third meeting of the new year, on Saturday, January 24, Compton conducted from his sickbed in one of the sparsely furnished spare bedrooms on the third floor of his large University Avenue house: he had the flu. Risking infection, Szilard attended, Ernest Lawrence, Luis Alvarez—Lawrence and Alvarez sitting together on the next bed—and several other men. "Each was arguing the merits of his own location," Compton writes, "and every case was good. I presented the case for Chicago." He had already won the support of his university's administration. "We will turn the university inside out if necessary to help win this war," its vice president had sworn. That was Compton's first argument: he knew the management and had its support. Second, more scientists were available to staff the operation in the Midwest than on the coasts, where faculties and graduate schools had been "completely drained" for other war work. Third, Chicago was conveniently and centrally located for travel to other sites.

Which convinced no one. Szilard had forty tons of graphite on hand at Columbia and a going concern. The arguments continued. Compton, who was notoriously indecisive, suffered their brunt as long as he could bear it. "Finally, wearied to the point of exhaustion but needing to make a firm decision, I told them that Chicago would be [the project's] location."

Lawrence scoffed. "You'll never get a chain reaction going here," he baited his fellow laureate. "The whole tempo of the University of Chicago is too slow."

"We'll have the chain reaction going here by the end of the year," Compton predicted.

"I'll bet you a thousand dollars you won't."

"I'll take you on that," Compton says he answered, "and these men here are the witnesses."

"I'll cut the stakes to a five-cent cigar," Lawrence hedged.

"Agreed," said Compton, who never smoked a cigar in his life.

After the crowd left, Compton shuffled wearily to his study and called Fermi. "He agreed at once to make the move to Chicago," Compton writes. Fermi may have agreed, but he found the decision burdensome. He was preparing further experiment. His group was exactly the right size. He owned a pleasant house in a pleasant suburb. He and Laura had buried a cache of Nobel Prize money in a lead pipe under the concrete floor of their basement coal bin against the possibility that as enemy aliens their assets would be frozen. Laura Fermi "had come to consider Leonia as our permanent home," she writes, "and loathed the idea of moving again." She says her husband "was unhappy to move. *They* (I did not know who they were) had decided to concentrate all *that* work (I did not know what it was) in Chicago and to enlarge it greatly, Enrico grumbled. It was the work he had started at Columbia with a small group of physicists. There is much to be said for a small group. It can work quite efficiently." But the country was at war. Fermi traveled back and forth by train until the end of April, then camped in Chicago. Laura dug up their buried treasure and followed at the end of June.

To Szilard, the day after the sickbed meeting—he had returned promptly to New York—Compton sent a respectful telegram: THANK YOU FOR COMING TO PRESENT ABLY COLUMBIA'S SITUATION. NOW WE NEED YOUR HELP IN ORGANIZING THE METALLURGICAL LABORATORY OF O.S.R.D. IN CHICAGO. CAN YOU ARRIVE HERE WEDNESDAY MORNING WITH FERMI AND WIGNER . . . TO DISCUSS DETAILS OF MOVING AND ORGANIZATION. Unlike the Radiation Laboratory at MIT, the new Metallurgical Laboratory hardly disguised its purpose in its name. Who would imagine its goal was the transmutation of the elements to make baseball-sized explosive spheres of unearthly metal?

Before Fermi and his team moved to Illinois they built one more exponential pile, this one loaded with cylindrical lumps of pressed uranium oxide three inches long and three inches in diameter that weighed four pounds each, some two thousand in all, set in blind holes drilled directly into graphite. A new recruit, a handsome, dark-haired young experimentalist named John Marshall, located a suitable press for the work in a junkyard in Jersey City and set it up on the seventh floor of Pupin; Walter Zinn designed stainless steel dies; the powdered oxide bound together under pressure as medicinal tablets pressed from powder—aspirin, for example—do.

Fermi was concerned to free the pile as completely as possible of moisture to reduce neutron absorption. He had canned the oxide before; now he decided to can the entire nine-foot graphite cube. "There are no ready-made cans of the needed size," Laura Fermi says dryly, "so Enrico ordered one." That, writes Albert Wattenberg, who joined the group in January, "required soldering together many strips of sheet metal. We were very fortunate in getting a sheet metal worker who made excellent solder joints. It was, however, quite a challenge to deal with him, since he could neither read nor speak English. We communicated with pictures, and somehow he did the job." Laura Fermi picks up the story: "To insure proper assembly, they marked each section with a little figure of a man: if the can were put together as it should be, all men would stand on their feet, otherwise on their heads." The Columbia men preheated the oxide lumps to 480°F before loading. They heated the contents of the room-sized can to the boiling point of water and pumped down a partial vacuum. Their heroic efforts reduced the pile's moisture to 0.03 percent. With the same relatively impure uranium and graphite they had used before but with these improved conditions and arrangements they measured *k* at the end of April at an encouraging 0.918.

In Chicago in the meantime Samuel Allison had built a smaller seven-foot exponential pile and measured *k* for his arrangement at 0.94. The University of Chicago had long ago sacrificed football to scholarship; Compton took over the warren of disused rooms under the west stands of Stagg Field, which was conveniently located immediately north of the main campus, and made space available there to Allison. Below solid masonry façades set with Gothic windows and crenellated towers the stands concealed ball courts as well as locker areas. The unheated room Allison had used for his experiment, sixty feet long, thirty feet wide, twenty-six feet high and sunk half below street level, was a doubles squash court.

December 6, 1941, the day of the bomb program expansion, marked another tidal event: Soviet forces under General Georgi Zhukov counterattacked across a two-hundred-mile front against the German Army congealed in snow and −35°F cold only thirty miles outside Moscow. "Like the supreme military genius who had trod this road a century before him," Churchill writes, evoking Napoleon Bonaparte, "Hitler now discovered what Russian winter meant." Zhukov's hundred divisions came as a bitter surprise—"well-fed, warmly clad and fresh Siberians," a German general describes them, "fully equipped for winter fighting" as the Wehrmacht troops were not—and armies that had advanced half a thousand miles to push within sight of the Kremlin stumbled back toward Germany nearly in rout. For the first time since Hitler began his conquests *Blitzkrieg* had failed. "The winter had fallen," Churchill writes. "The long war was certain." Hitler relieved his Army commander in chief of duty and appropriated that office to himself. By the end of March his casualties in the East, counting not the sick but only the wounded, numbered nearly 1.2 million men.

It was clear in Berlin that the German economy had reached the limits of its expansion. Tradeoffs must follow. The Minister of Munitions installed a rule similar to the rule upon which Conant was insisting in the United States, and the director of Reich military research promulgated it to the physicists studying uranium: "The work . . . is making demands which can be justified in the current recruiting and raw materials crisis only if there is a certainty of getting some benefit from it in the near future." After considering the question the War Office decided to reduce the priority of uranium research by assigning most of it to the Ministry of Education under Bernhard Rust, the scientifically illiterate SS *Obergruppenführer* and former provincial schoolteacher who had refused to sanction Lise Meitner's emigration following the *Anschluss.* The academic physicists were happy to be out from under the Army but chagrined to be consigned to a backwater ministry run by a party hack. Rust delegated authority to the Reich Research Council. That organization was part of the Reich Bureau of Standards. The KWI physicists considered its physics section head, Abraham Esau, incompetent. In effect, the German uranium program had slipped in status to the level of the old U.S. Uranium Committee and now had its Briggs.

The Research Council decided to appeal directly to the highest levels of the Reich for support. It organized an elaborate presentation and invited such dignitaries as Hermann Göring, Martin Bormann, Heinrich Himmler, Navy commander in chief Admiral Erich Raeder, Field Marshal Wilhelm Keitel and Albert Speer, Hitler's admired patrician architect who was Minister of Armaments and War Production. Heisenberg, Hahn, Bothe, Geiger, Clusius and Harteck were scheduled to speak at the February 26 meeting, Rust presiding, and an "Experimental Luncheon" would be served offering entrées prepared from frozen foods basted with synthetic shortening and bread made with soy flour.

Unfortunately for the council's ambitious plans, the secretary assigned to send out invitations enclosed the wrong lecture program. A secret scientific conference under the auspices of Army Ordnance had been scheduled at the Kaiser Wilhelm Society's Harnack House for the same day. Its program listed twenty-five highly technical scientific papers. That was the program the leaders of the Reich mistakenly received. Himmler regretted: he would be away from Berlin that day. Keitel was "too busy at the moment." Raeder would send a representative. None of the leaders chose to attend.

What Heisenberg had to say might have surprised them. He emphasized atomic energy for power but also discussed military uses. "Pure uranium-235 is thus seen to be an explosive of quite unimaginable force," he told his staff-level auditors. "The Americans seem to be pursuing this line of research with particular urgency." Inside a uranium reactor "a new element is created [i.e., plutonium] . . . which is in all probability as explosive as pure uranium-235, with the same colossal force." At the same time at Harnack House, where Leo Szilard once lodged, bags packed, Army Ordnance was learning that "it would suffice to bring together two lumps of this explosive, weighing a total of ten to a hundred kilograms, for it to detonate."

Basic knowledge of one direct route to an atomic bomb—via plutonium—was at hand. What was lacking was money and materials. The February 26 meeting won over at least the Minister of Education. "The first time large funds were available in Germany," Heisenberg recalled at the end of the war, "was in the spring of 1942, after that meeting with Rust, when we convinced him that we had absolutely definite proof that it could be done." Heisenberg's "large" is relative to the modest funds that had been available before, however. Not Bernhard Rust but Albert Speer needed to be convinced of the military promise of atomic energy to swell the scale of funding anywhere near the billions of reichsmarks that production of even ten kilograms of U235 or plutonium would require.

Speer did not recall the February 26 invitation after the war. Atomic energy first came to his attention, he writes in his memoirs, at one of his regular private luncheons with General Friedrich Fromm, the commander of the Home Army. "In the course of one of these meetings, at the end of April 1942, [Fromm] remarked that our only chance of winning the war lay in developing a weapon with totally new effects. He said he had contacts with a group of scientists who were on the track of a weapon which could annihilate whole cities. . . . Fromm proposed that we pay a joint visit to these men." Speer also heard that spring from the president of the Kaiser Wilhelm Society, who complained of lack of support for uranium research. "On May 6, 1942, I discussed this situation with Hitler and proposed that Göring be placed at the head of the Reich Research Council—thus emphasizing its importance."

That shift to the obese Reichsmarshal who commanded the Luftwaffe and whom Hitler had designated to be his successor carried only symbolic promotion. More crucial was a June 4 conference at Harnack House that Speer, Fromm, automobile and tank designer Ferdinand Porsche and other military and industrial leaders attended. In February Heisenberg had devoted most of his lecture to nuclear power. This time he emphasized military prospects. The secretary of the Kaiser Wilhelm Society was surprised: "The word 'bomb' which was used at this conference was news not only to me but for many others present, as I could see from their reaction." It was not news to Speer. When Heisenberg took questions from the floor, one of Speer's deputies asked how large a bomb capable of destroying a city would have to be. Heisenberg cupped his hands as Fermi had done sighting down Manhattan Island from Pupin Hall. "As large as a pineapple," he said.

After the briefings Speer questioned Heisenberg directly. How could nuclear physics be applied to the manufacture of atomic bombs? The German laureate seems to have shied from committing himself. "His answer was by no means encouraging," Speer remembers. "He declared, to be sure, that the scientific solution had already been found. . . . But the technical prerequisites for production would take years to develop, two years at the earliest, even provided that the program was given maximum support." They were crippled by an absence of cyclotrons, Heisenberg said. Speer offered to build cyclotrons "as large as or larger than those in the United States." Heisenberg demurred that German physicists lacked experience building large cyclotrons and would have to start small. Speer "urged the scientists to inform me of the measures, the sums of money and the materials they would need to further nuclear research." A few weeks later they did, but their requests looked picayune to a Reichsminister accustomed to dealing in billions of marks. They requested "an appropriation of several hundred thousand marks and some small amounts of steel, nickel, and other priority metals. . . . Rather put out by these modest requests in a matter of such crucial importance, I suggested that they take one or two million marks and correspondingly larger quantities of materials. But apparently more could not be utilized for the present, and in any case I had been given the impression that the atom bomb could no longer have any bearing on the course of the war."

Speer saw Hitler regularly and duly reported the findings of the June conferences:

> Hitler had sometimes spoken to me about the possibility of an atom bomb, but the idea quite obviously strained his intellectual capacity. He was also

> unable to grasp the revolutionary nature of nuclear physics. In the twenty-two hundred recorded points of my conferences with Hitler, nuclear fission comes up only once, and then is mentioned with extreme brevity. Hitler did sometimes comment on its prospects, but what I told him of my conferences with the physicists confirmed his view that there was not much profit in the matter. Actually, Professor Heisenberg had not given any final answer to my question whether a successful nuclear fission could be kept under control with absolute certainty or might continue as a chain reaction. Hitler was plainly not delighted with the possibility that the earth under his rule might be transformed into a glowing star. Occasionally, however, he joked that the scientists in their unworldly urge to lay bare all the secrets under heaven might some day set the globe on fire. But undoubtedly a good deal of time would pass before that came about, Hitler said; he would certainly not live to see it.

Following that, according to Speer, "on the suggestion of the nuclear physicists we scuttled the project to develop an atom bomb . . . after I had again queried them about deadlines and been told that we could not count on anything for three or four years." Work on what Speer calls "an energy-producing uranium motor for propelling machinery"—the heavy-water pile—would continue. "In the upshot," Heisenberg wrote in *Nature* in 1947, summarizing the war years, German physicists "were spared the decision as to whether or not they should aim at producing atomic bombs. The circumstances shaping policy in the critical year of 1942 guided their work automatically toward the problem of the utilization of nuclear energy in prime movers." But the Allies had not yet been informed.

"We may be engaged in a race toward realization," Vannevar Bush wrote Franklin Roosevelt on March 9, 1942; "but, if so, I have no indication of the status of the enemy program, and have taken no definite steps toward finding out." Why Bush was not more curious remains a mystery. Conant, Lawrence and Compton, not to mention the emigrés, fretted continually about the possibility of a German bomb. It was their primary reason for urging an American bomb. It was not Bush's or Roosevelt's—to them the bomb offered offensive advantage first of all—but the two leaders were alert to the German danger and surprisingly indifferent to assessing it.

The report that accompanied Bush's letter stated that five to ten pounds of "active material" would be "fairly certain" to explode with a force equivalent to 2,000 tons of TNT, up from 600 tons in the third National Academy of Sciences report of the previous November 6. It recommended building a centrifuge plant at a cost of $20 million that could produce enough U235 for one bomb a month and estimated that such a plant could be completed by December 1943. A gaseous diffusion plant, its cost unspecified, might deliver by the end of 1944. An electromagnetic separation plant—Ernest Lawrence's project—won the most attention in the report: it might "offer a short-cut," wrote Bush, and deliver "fully practicable quantities of material by the summer of 1943, with a time saving of perhaps six months or even more." In summary, "present opinion indicates that successful use is possible, and that this would be very important and might be determining in the war effort. It is also true that if the enemy arrived at results first it would be an exceedingly serious matter. The best estimate indicates completion in 1944, if every effort is made to expedite."

Roosevelt responded two days later: "I think the whole thing should be pushed not only in regard to development, but also with due regard to time. This is very much of the essence." Time, not money, was becoming the limiting factor in atomic bomb development.

A meeting on May 23 brought all the program leaders together with Conant to decide which of several methods of making a bomb should be moved on to the pilot-plant and industrial engineering stages. The centrifuge, gaseous barrier diffusion, electromagnetic and graphite or heavy-water plutonium-pile approaches all looked equally promising. Given wartime scarcities and budget priorities, which should be advanced? Conant used an arms-race argument to identify the point of decision:

> While all five methods now appear to be about equally promising, clearly the time of production of a dozen bombs by the five routes will certainly not be the same but might vary by six months or a year because of unforeseen delays. Therefore, if one discards one or two or three of the methods now, one may be betting on the slower horse unconsciously. To my mind the decision as to how "all out" the effort should be might well turn on the military appraisal of what would occur if either side had a dozen or two bombs before the other.

To that point Conant reviewed the evidence for a German bomb program, including new indications of espionage activity: information from the British that the Germans had a ton of heavy water; Peter Debye's report when he arrived in the United States eighteen months earlier that his colleagues at the KWI were hard at work; and "the recently intercepted instruction to their agents in this country [that] shows they are interested in what we are doing." Conant thought this last evidence the best. "If they are hard at work, they cannot be far behind since they started in 1939 with the same initial facts as the British and ourselves. There are still plenty of competent scientists left in Germany. They may be ahead of us by as much as a year, but hardly more."

If time, not money, was the crucial issue—in Conant's words, "if the possession of the new weapon in sufficient quantities would be a determining factor in the war"—then "three months' delay might be fatal." It followed that all five methods should be pushed at once, even though "to embark on this Napoleonic approach to the problem would require the commitment of perhaps $500,000,000 and quite a mess of machinery."

Glenn Seaborg arrived in Chicago aboard the streamliner City of San Francisco at 9:30 A.M. Sunday, April 19, 1942, his thirtieth birthday. As he left the station he noticed first that Chicago was cold compared to Berkeley—forty degrees that spring morning. Then headlines at a newsstand caught him up on the developing Pacific war: the Japanese reported American aircraft had bombed Tokyo and three other Honshu cities, a surprise attack that neither Southwest Pacific commander General Douglas MacArthur nor Washington acknowledged (it was Jimmy Doolittle's morale raid of sixteen B-25 bombers launched one-way across Japan to landing fields in China from the U.S. aircraft carrier *Hornet*). "This day . . . marks a transition point in my life," Seaborg writes in his carefully documented diary-style memoir, "for tomorrow I will take on the added responsibility of the 94 chemistry group at the Metallurgical Laboratory on the University of Chicago campus, the central component of the Metallurgical Project."

Transmuting U238 to plutonium in a chain-reacting pile was one thing, extracting the plutonium from the uranium quite another. The massive production piles that Compton's people were already beginning to plan would create the new element at a maximum concentration in the uranium of about 250 parts per million—a volume, uniformly dispersed through each two tons of mingled uranium and highly radioactive fission products, equal to the volume of one U.S. dime. Seaborg's work was somehow to pull that dime's worth out.

He had made a good beginning at Berkeley, exploring plutonium's unusual chemistry. Oxidizing agents are chemicals that strip electrons from the outer shells of atoms. Reducing agents conversely add electrons to the outer shells of atoms. Plutonium, it seemed, precipitated differently when it was treated with oxidizing agents than when it was treated with reducing agents. In a +4 oxidation state, the Berkeley team had found, the man-made element could be precipitated out of solution using a rare-earth compound such as lanthanum fluoride as a carrier. Oxidize the same plutonium to a +6 oxidation state and the precipitation no longer worked; the carrier crystallized but the plutonium remained behind in solution. That gave Seaborg a basic approach to extraction:

> We conceived the principle of the oxidation-reduction cycle. . . . This principle applied to any process involving the use of a substance which carried plutonium in one of its oxidation states but not in another. . . . For example, a carrier could be used to carry plutonium in one oxidation state and thus to separate it from uranium and the fission products. Then the carrier and the plutonium [now solid crystals] could be dissolved, the oxidation state of the plutonium changed, and the carrier reprecipitated, leaving the plutonium in solution. The oxidation state of the plutonium could again be changed and the cycles repeated. With this type of procedure, only a contaminating element having a chemistry nearly identical with the plutonium itself would fail to separate if a large number of oxidation-reduction cycles were employed.

A two-day chemistry conference began on Wednesday, April 23, with Eugene Wigner, Harold Urey, Princeton theoretician John A. Wheeler and a number of chemists already assigned to the Met Lab on hand. The scientists discussed seven possible ways to extract plutonium from irradiated uranium. They favored four that seemed particularly adaptable to remote control, not including precipitation. Seaborg, the new man, disagreed: "I, however, expressed confidence in the use of precipitation." They would nevertheless investigate all seven methods proposed. That would require the full-time work of forty men. One of Seaborg's jobs for months to come was recruiting. It worried him: "Sometimes I feel a little apprehensive about inviting . . . people to give up their secure university positions and come to work at the Met Lab. They must gamble on the future of their careers, and how long they will be diverted from them nobody knows." But if no one knew how long the work would last, most of them came to believe it transcendently important: "There is a statement of rather common currency around here and Berkeley that goes something like this: 'No matter what you do with the rest of your life, nothing will be as important to the future of the World as your work on this Project right now.' "

So far Seaborg had studied plutonium by following the characteristic radioactivity of minute amounts vastly diluted in carrier, the same tracer chemistry that Hahn, Fermi and the Joliot-Curies had used. Chemical reactions often proceed differently at different dilutions, however. To prove that an extraction process would work at industrial scale, Seaborg knew he would have to demonstrate it at industrial-scale concentrations. In peacetime he might have waited until a pile large enough to transmute at least gram quantities of plutonium was built and operating. That normal procedure was a luxury the bomb program could not afford.

Seaborg looked instead for a way to make more plutonium without a pile and a way to work with concentrated solutions of the little he might make. The resources of the OSRD came to his aid in the first instance, his own imagination and ingenuity in the second. He commandeered the 45-inch cyclotron at Washington University in St. Louis, where Compton had once hidden out, and arranged to have 300-pound batches of uranium nitrate hexahydrate bombarded heroically with neutrons for weeks and months at a time. So long and intense a bombardment would give him microgram quantities of plutonium—several hundred millionths of a gram, amounts hardly visible to the naked eye. He then somehow had to devise techniques for mixing, measuring and analyzing them.

Visiting New York earlier that month to deliver a lecture, Seaborg had sought out a quaint soul named Anton Alexander Benedetti-Pichler, a professor at Queens College in Flushing who had pioneered ultramicrochemistry, a technology for manipulating extremely small quantities of chemicals. Benedetti-Pichler had briefed Seaborg thoroughly and promised to send a list of essential equipment. Seaborg hired one of Benedetti-Pichler's former students and together the two men planned an ultramicrochemistry laboratory. "We looked for a good spot that would be vibration-free for the microbalances and settled on Room 405 (a former darkroom) in Jones Laboratory which has a concrete bench." The former darkroom, hardly six feet by nine, was scaled to the work.

Another specialist in ultramicrochemistry, Paul Kirk, taught at Berkeley. Seaborg hired a recent Ph.D. whom Kirk had trained, Burris Cunningham, and a graduate student, Louis B. Werner. "I always thought I was tall," the chemistry laureate comments, but Werner at six feet seven topped him by four inches, "a tight fit" in the small laboratory.

With the special tools of ultramicrochemistry the young chemists could work on undiluted quantities of chemicals as slight as tenths of a microgram (a dime weighs about 2.5 grams— 2,500,000 micrograms). They would manage their manipulations on the mechanical stage of a binocular stereoscopic microscope adjusted to 30-power magnification. Fine glass capillary straws substituted for test tubes and beakers; pipettes filled automatically by capillary attraction; small hypodermic syringes mounted on micromanipulators injected and removed reagents from centrifuge microcones; miniature centrifuges separated precipitated solids from liquids. The first balance the chemists used consisted of a single quartz fiber fixed at one end like a fishing pole stuck into a riverbank inside a glass housing that protected it from the least breath of air. To weigh their Lilliputian quantities of material they hung a weighing pan, made of a snippet of platinum foil that was itself almost too small to see, to the free end of the quartz fiber and measured how much the fiber bent, a deflection which was calibrated against standard weights. A more rugged balance developed at Berkeley had double pans suspended from opposite ends of a quartz-fiber beam strung with microscopic struts. "It was said," notes Seaborg, "that 'invisible material was being weighed with an invisible balance.' "

In addition to his new Met Lab responsibilities Seaborg still coordinated basic scientific studies of uranium and plutonium at Berkeley. At the beginning of June he traveled to California to meet with "the fellows on the third floor of Gilman Hall" and to marry Ernest Lawrence's secretary. On June 6, returning to Chicago through Los Angeles, where Seaborg's parents lived, bride and groom prepared for a quick Nevada wedding. They got off the train in Caliente, Nevada, stored their bags with the telegraph operator at the station and asked directions to the city hall. "But to our vexation we learned there is no city hall here and in order to get our marriage license we would have to go to the county seat, a town called Pioche, some 25 miles to the north." Providentially the deputy sheriff who served as Caliente's travel adviser and all-around troubleshooter turned out to be a June graduate of the Berkeley chemistry department. He arranged for the professor and his bride, Helen Griggs, to ride to Pioche in a mail truck. "Our witnesses were a janitor whom we recruited and [a] friendly clerk. We returned to Caliente on the mail truck's 4:30 run and checked into the local hotel here for our overnight stay."

Arriving in Chicago on June 9 Seaborg delivered his wife to the apartment he had rented before he left for California and proceeded immediately to his office. His mail informed him that Edward Teller was joining the Chicago project to work in the theoretical group under Eugene Wigner.

Two days later Robert Oppenheimer turned up in Chicago and dropped by to see Seaborg; they were old friends but "it was more than just a social call." Gregory Breit, the Wisconsin-based theoretician on the Uranium Committee who had been responsible for fast-neutron studies, had resigned from the bomb project in protest over what he felt were serious violations of security. "I do not believe that secrecy conditions are satisfactory in Dr. Compton's project," he had written Briggs on May 18. His litany of examples approached paranoia. "Within the Chicago project there are several individuals strongly opposed to secrecy. One of the men, for example, coaxed my secretary there to give him some official reports out of my safe while I was away on a trip. . . . The same individual talks quite freely within the group. . . . I have heard him advocate the principle that all parts of the work are so closely interrelated that it is desirable to discuss them as a whole." The dangerous individual Breit chose not to name was Enrico Fermi, pushing to make the chain reaction go. Compton had appointed Oppenheimer to replace Breit and Oppenheimer was visiting Seaborg for a briefing on the fast-neutron studies Seaborg was coordinating at

Berkeley. Studying fast-neutron reactions, Seaborg notes, was "a prerequisite to the design of an atomic bomb." Oppenheimer had found a place for himself on the ground floor.

The Washington University cyclotron crew moved the first 300 pounds of uranium nitrate hexahydrate into position around the machine's beryllium target on June 17. The UNH was scheduled for a month's bombardment, 50,000 microampere-hours. Though the chain reaction had not yet been proved and no one had yet seen plutonium, the various Met Lab councils of which Seaborg was a member had already begun debating the design and location of the big 250,000-kilowatt production piles that would create pounds of the strange metal if all went well. Fermi thought plutonium production needed an area a mile wide and two miles long for safety. Compton proposed building piles of increasing power to work up to full-scale production and was considering alternative sites in the Lake Michigan Dunes area and in the Tennessee Valley.

A question that would eventually encompass many other issues, some of them profound, was how to cool the big piles. Early in the organization of the Met Lab Compton had appointed an engineering council to consider such questions; besides an engineer and an industrial chemist the council included Samuel Allison, Fermi, Seaborg, Szilard and John A. Wheeler among its membership. By late June its discussions had progressed to the point of tentative commitment. Helium was one prospective coolant, to be circulated at high pressure inside a sealed steel shell; its zero cross section for neutron absorption was only one of its several advantages. Water was another coolant possibility, the heat-exchange medium most familiar to engineers but corrosive to uranium. An exotic third was bismuth, a metal with a low 520°F melting point that serves as a watchful solid in fuses and automatic fire alarms. Melted to a liquid it would transfer heat far more efficiently than helium or water. Szilard championed a liquid-bismuth cooling system in part because the metal could be circulated through the pile with a scaled-up version of the magnetic pump he and Albert Einstein had invented for refrigerators, a mechanism that had no moving parts to leak or fail.

The engineering council ruled out liquid cooling, Seaborg writes, "because of potential chemical action, danger of leaks and difficulty in transferring heat from oxide. . . . There was general agreement to use helium." Eugene Wigner had not been invited onto the council despite his interest in its problems and his thorough knowledge of chemical engineering. Wigner strongly favored water cooling, says Szilard, because "a water cooled system could be built in a much shorter time." Seaborg corroborates Wigner's continuing desperate concern about a German bomb:

> Compton repeated a conversation that ensued between him and Wigner on a possible schedule of the Germans. Like us, they have had three years since the discovery of fission to prepare a bomb. Assuming they know about [plutonium], they could run a heavy water pile for two months at 100,000 kw and produce six kilograms of it; thus it would be possible for them to have six bombs by the end of this year [1942]. On the other hand, we don't plan to have bombs in production until the first part of 1944.

Compton encouraged Wigner's group to design a water-cooled pile but ordered up detailed engineering studies only of a system using helium.

The basic issue behind the technical dispute was control, which Szilard at least understood they were systematically signing away to the U.S. government. A meeting on June 27 intensified the conflict. Bush's latest status report to Roosevelt on June 17 had proposed dividing the work of development and ultimate production between the OSRD and the U.S. Army Corps of Engineers, bringing in the Army to build and run the factories as Bush had planned to do all along. Roosevelt initialed Bush's cover letter "OK. FDR." and returned it immediately. The same day the Chief of Engineers ordered Colonel James C. Marshall of the Syracuse Engineer District, a 1918 West Point graduate with experience building air bases, to report to Washington for duty. Marshall selected the Boston construction engineering corporation of Stone & Webster as principal contractor for the bomb project. To report the reorganization Compton called the June 27 meeting of his group leaders and planning board. Allison, Fermi, Seaborg, Szilard, Teller, Wigner and Zinn attended, among others.

"Compton opened the meeting with a pep talk," Seaborg remembers, "asking us to go ahead with all vigor possible. He said our aim the past half-year has been to investigate the possibilities of producing an atomic bomb—now we have the responsibility to proceed from the military point of view on the assumption it can be done and we can assume we have a project for the entire duration of the war." Compton was stealthily working his way to the new arrangements. He emphasized the program's secrecy. "Only about six men in the U.S. Army are permitted to know what is going on," Seaborg paraphrases him; those privileged few included Secretary of War Henry L. Stimson—heady company for men who had only recently been graduate students or obscure academics—and "two construction experts," generals whom Compton then named. He described the responsibilities of the "construction experts" and finally broke the news: "It is hoped to have a contractor assume responsibility for the production plant." A contractor already had.

Compton's announcement had the effect he seems to have feared, Seaborg goes on: "A number of the people present expressed great concern about working for an industrial contractor because of their fear that this would not be a compatible environment in which to work." They would not have to work *for* such a contractor, though they would obviously have to work *with* one, but to make the reorganization palatable Compton hinted at worse that might be yet to come: "There was considerable talk about our being absorbed into the Army [i.e., commissioned as officers] and what the advantages and disadvantages might be. There were vigorous objections from most of the people present."

The problem would fester all summer and burst through again in the fall. Szilard would define it precisely in a memorandum: "Stated in abstract form, the trouble at Chicago arises out of the fact that the work is organized along somewhat authoritative [*sic*: authoritarian] rather than democratic lines." The visionary Hungarian physicist did not believe science could function by fiat. "In 1939," he had already written Vannevar Bush passionately in late May, before the cooling-system and contractor debates, "the Government of the United States was given a unique opportunity by Providence; this opportunity was lost. Nobody can tell now whether we shall be ready before German bombs wipe out American cities. Such scanty information as we have about work in Germany is not reassuring and all one can say with certainty is that we could move at least twice as fast if our difficulties were eliminated."

Three hundred pounds of irradiated UNH—yellowish crystals like rock salt—arrived from St. Louis by truck on July 27, a Monday:

> The UNH was surrounded by a layer of lead bricks. [Truman] Kohman and [Elwin H.] Covey were detailed to unload the shipment and carry it up to our lab on the fourth floor for extraction of the 94^{239}. The UNH crystals came packaged in small boxes of various sizes, made to fit into the various niches around the cyclotron target. Some of the boxes were made of masonite, but most of them were of quarter inch plywood. Unfortunately, some of the seams and edges had cracked open, allowing crystals of hot [i.e., radioactive] UNH to creep out. We could not get hold of any instrument to measure the radioactivity. I told Kohman and Covey their best protection would be to wear rubber gloves and a lab coat. . . . Although they struggled for half the day to get all the boxes and lead bricks upstairs into the storage area, I think they were conscientious and kept their radiation exposure to a minimum.

While Seaborg's high-spirited crew of young chemists began attempting to extract plutonium 239 from the bulky St. Louis UNH, wrestling with carboys of ether and heavy three-liter separatory funnels held at arm's length from behind lead shields, Cunningham and Werner in narrow Room 405 started toward isolating plutonium as a pure compound. They first measured out a 15-milliliter solution of UNH irradiated earlier that summer in the 60-inch Berkeley cyclotron. They assumed their solution then contained about one microgram of plutonium 239. (*Pu*239, that is: Seaborg had chosen the abbreviation Pu rather than Pl partly to avoid confusion with platinum, Pt, but also "facetiously," he says, "to create attention"—P.U. the old slang for putrid, something that raises a stink.) Working with their ultramicrochemical equipment—slow, tedious operations via micromanipulator gearing down large motions to microscopically small—on August 15, a Saturday, they mixed the rare earths cerium and lanthanum into their solution as carriers, partially evaporated it and precipitated the carriers and the Pu as fluorides. They dissolved the precipitated crystals in a few drops of sulfuric acid and evaporated the resulting solution to a volume of about one milliliter, a thousandth of a liter, some twenty drops. They checked the larger volume of solution left behind and found essentially no alpha activity, evidence that the alpha-active Pu had crystallized out with the rare earths. That was a day's work and they stored the precipitate solution carefully for Monday and went home.

On Monday, August 17, Cunningham and Werner began by oxidizing their small volume of precipitate to change the oxidation state of its Pu. They repeated the oxidation and reduction cycles on the solution several times. At the end of the day their quartz centrifuge microcone contained a minute drop of liquid that radiated some 57,000 alpha particles per minute. They set it in a steam bath to concentrate it.

On Tuesday the two men transferred the concentrated solution to a shallow platinum dish to prepare to concentrate it further. It began creeping over the sides. Rather than lose it they moved it quickly to the only larger dish at hand, which was contaminated with lanthanum. Their misjudgment of volume condemned them to another day of repurifying. Upstairs in the attic and on the roof Seaborg's bulk UNH crew stirred large-volume extractions of ether and water. It was hot and heavy work.

Room 405 had a purified concentrate again to process Wednesday morning. It was still contaminated with a potassium compound and with silver. Cunningham and Werner diluted it and precipitated out the silver as a chloride. They added five micrograms of lanthanum and precipitated out the Pu along with the lanthanum carrier. They dissolved the precipitate, oxidized it once more to change over the Pu and precipitated out the lanthanum. That left pure plutonium in solution, one more morning's work to bring down.

Of Thursday, August 20, 1942, Seaborg writes:

> Perhaps today was the most exciting and thrilling day I have experienced since coming to the Met Lab. Our microchemists isolated pure element 94 for

> the first time! This morning Cunningham and Werner set about fuming . . . yesterday's 94 solution containing about one microgram of 94^{239}, added hydrofluoric acid whereupon the reduced 94 precipitated as the fluoride . . . free of carrier material. . . .
>
> This precipitate of 94, which was viewed under the microscope and which was also visible to the naked eye, did not differ visibly from the rare-earth fluorides. . . .
>
> It is the first time that element 94 . . . has been beheld by the eye of man.

By afternoon "a holiday spirit prevailed in our group." After several hours' exposure to air "the precipitated [plutonium] had taken on a pinkish hue." Someone photographed Cunningham and Werner at their crowded bench in the narrow, tile-walled room—trim, strong-jawed young men looking weary. The crew upstairs that muscled carboys and lead bricks shuffled in like clumsy shepherds to peer through the microscope at the miracle of the tiny pinkish speck.

In the summer of 1942 Robert Oppenheimer gathered together at Berkeley a small group of theoretical physicists he was amused to call the "luminaries." Their job was to throw light on the actual design of an atomic bomb.

Hans Bethe, now thirty-six and a highly respected professor of physics at Cornell, had resisted joining the bomb project because he doubted the weapon's feasibility. "I considered . . . an atomic bomb so remote," Bethe told a biographer after the war, "that I completely refused to have anything to do with it. . . . Separating isotopes of such a heavy element [as uranium] was clearly a very difficult thing to do, and I thought we would never succeed in any practical way." But Bethe may well have headed the list of luminaries Oppenheimer wanted to attract. By 1942 the Cornell physicist had established himself as a theoretician of the first rank. His most outstanding contribution, for which he would receive the 1967 Nobel Prize in Physics, was to elucidate the production of energy in stars, identifying a cycle of thermonuclear reactions involving hydrogen, nitrogen and oxygen that is catalyzed by carbon and culminates in the creation of helium. Among other important work during the 1930s Bethe had been principal author of three lengthy review articles on nuclear physics, the first comprehensive survey of the field. Bound together, the three authoritative studies came to be called "Bethe's Bible."

He had wanted to help oppose Nazism. "After the fall of France," he says, "I was desperate to do something—to make some contribution to the war effort." First he developed a basic theory of armor penetration. On the recommendation of Theodor von Kármán, whom he consulted at Caltech, he and Edward Teller in 1940 extended and clarified shock-wave theory. In

1942 he joined the Radiation Laboratory at MIT to work on radar. That was where Oppenheimer found him.

Oppenheimer cleared his plan with Lee A. DuBridge, the director of the Rad Lab, then set a senior American theoretician, John H. Van Vleck, professor of physics at Harvard, to snare Bethe for the Berkeley summer study. "The essential point," he counseled Van Vleck, "is to enlist Bethe's interest, to impress on him the magnitude of the job we have to do . . . and to try to convince him, too, that our present plans . . . are the appropriate machinery." Oppenheimer felt the weight of the work. "Every time I think about our problem a new headache appears," he told the Harvard professor. "We shall certainly have our hands full." Van Vleck arranged to meet Bethe conspiratorially in Harvard Yard and succeeded in convincing him he was needed. The prearranged signal to Oppenheimer was a Western Union Kiddygram, an inexpensive standardized telegram with a message like "Brush your teeth."

Oppenheimer also invited Edward Teller. In 1939 Bethe had married Rose Ewald, the attractive and intelligent daughter of his Stuttgart physics professor Paul Ewald; Edward and Mici Teller, "our best friends in this country," had attended the New Rochelle wedding. Setting out for Berkeley in early July 1942, the Bethes stopped over in Chicago to pick up the Tellers. Teller showed Bethe Fermi's latest exponential pile. "He had a setup under one of the stands in Stagg Field," Bethe remembers—"in a squash court—with tremendous stacks of graphite." A chain reaction that made plutonium would bypass the problem of isotope separation. "I then," says Bethe, "became convinced that the atomic-bomb project was real, and that it would probably work."

The other luminaries enlisted for the summer study were Van Vleck, the Swiss-born Stanford theoretician Felix Bloch, Oppenheimer's former student and close collaborator Robert Serber, a young Indiana theoretician named Emil Konopinski and two postdoctoral assistants. Konopinski and Teller had arrived at the Met Lab at about the same time earlier that year. "We were newcomers in the bustling laboratory," Teller writes in a memoir, "and for a few days we were given no specific jobs." Teller proposed that he and Konopinski review his calculations that seemed to prove the impossibility of using an atomic bomb to ignite a thermonuclear reaction in deuterium:

> Konopinski agreed, and we tackled the job of writing a report to show, once and for all, that it could not be done. . . . But the more we worked on our report, the more obvious it became that the roadblocks which I had erected for Fermi's idea were not so high after all. We hurdled them one by one, and concluded that heavy hydrogen actually could be ignited by an atomic bomb

> to produce an explosion of tremendous magnitude. By the time we were on our way to California . . . we even thought we knew precisely how to do it.

That was not news Edward Teller was likely to hide under a bushel, whatever Oppenheimer's official agenda. Bethe was ushered into the glare as the streamliner clicked west: "We had a compartment on the train to California, so we could talk freely. . . . Teller told me that the fission bomb was all well and good and, essentially, was now a sure thing. In reality, the work had hardly begun. Teller likes to jump to conclusions. He said that what we really should think about was the possibility of igniting deuterium by a fission weapon—the hydrogen bomb."

At Berkeley the luminaries began meeting in Oppenheimer's office, "in the northwest corner of the fourth floor of old LeConte [Hall]," an older colleague remembers. "Like all those rooms, it had French doors opening out onto a balcony, to which there was easy access from the roof. Accordingly a very strong wire netting was fastened securely over his balcony." Only Oppenheimer had a key. "If a fire had ever started . . . in Oppenheimer's absence, it would have been tragic." But the fires that summer were still only theoretical.

The theoreticians let Teller's bomb distract them. It was new, important and spectacular and they were men with a compulsion to know. "The theory of the fission bomb was well taken care of by Serber and two of his young people," Bethe explains. They "seemed to have it well under control so we felt we didn't need to do much." The essentials of fast-neutron fission were firm—it needed experiment more than theory. The senior men turned their collective brilliance to fusion. They had not yet bothered to name generic bombs of uranium and plutonium. But from the pre-anthropic darkness where ideas abide in nonexistence until minds imagine them into the light, the new bomb emerged already chased with the technocratic euphemism of art deco slang: the Super, they named it.

Rose Bethe, who was then twenty-four, understood instantly. "My wife knew vaguely what we were talking about," says Bethe, "and on a walk in the mountains in Yosemite National Park she asked me to consider carefully whether I really wanted to continue to work on this. Finally, I decided to do it." The Super "was a terrible thing." But the fission bomb had to come first in any case and "the Germans were presumably doing it."

Teller had examined two thermonuclear reactions that fuse deuterium nuclei to heavier forms and simultaneously release binding energy. Both required that the deuterium nuclei be hot enough when they collided—energetic enough, violently enough in motion—to overcome the nuclear electrical barrier that usually repels them. The minimum necessary energy was thought at the time to come to about 35,000 electron volts, which corresponds to a temperature of about 400 million degrees. Given that temperature—and on earth only an atomic bomb might give it—both thermonuclear reactions should occur with equal probability. In the first, two deuterium nuclei collide and fuse to helium 3 with the ejection of a neutron and the release of 3.2 million electron volts of energy. In the second the same sort of collision produces tritium—hydrogen 3, an isotope of hydrogen with a nucleus of one proton and two neutrons that does not occur naturally on earth—with the ejection of a proton and the release of 4.0 MeV of energy.

The D + D reactions' release of 3.6 MeV was slightly less by mass than fission's net of 170 MeV. But fusion was essentially a thermal reaction, not inherently different in its kindling from an ordinary fire; it required no critical mass and was therefore potentially unlimited. Once ignited, its extent depended primarily on the volume of fuel—deuterium—its designers supplied. And deuterium, Harold Urey's discovery, the essential component of heavy water, was much easier and less expensive to separate from hydrogen than U235 was from U238 and much simpler to acquire than plutonium. Each kilogram of heavy hydrogen equaled about 85,000 tons TNT equivalent. Theoretically, 12 kilograms of liquid heavy hydrogen—26 pounds—ignited by one atomic bomb would explode with a force equivalent to 1 million tons of TNT. So far as Oppenheimer and his group knew at the beginning of the summer, an equivalent fission explosion would require some 500 atomic bombs.

That reckoning alone would have been enough to justify devoting the summer to imagining the Super a little way out of the darkness. Teller found something else as well, or thought he did, and with his usual pellmell facility he scattered it before them. There are many other thermonuclear reactions besides the D + D reactions. Bethe had examined a number of them methodically when looking for those that energized massive stars. Now Teller offered several which a fission bomb or a Super might inadvertently trigger. He proposed to the assembled luminaries the possibility that their bombs might ignite the earth's oceans or its atmosphere and burn up the world, the very result Hitler occasionally joked about with Albert Speer.

"I didn't believe it from the first minute," Bethe scoffs. "Oppie took it sufficiently seriously that he went to see Compton. I don't think I would have done it if I had been Oppie, but then Oppie was a more enthusiastic character than I was. I would have waited until we knew more." Oppenheimer had other urgent business with Compton in any case: the Super itself. Not to risk their loss, the bomb-project leaders were no longer allowed to fly. Oppenheimer tracked Compton by telephone at the beginning of a

July weekend to a country store in northern Michigan where he had stopped to pick up the keys to his lakeside summer cottage, got directions and caught the next train east. In the meantime Bethe applied himself to Teller's calculations.

The Cornell physicist's instant skepticism gives perspective to Compton's melodramatic recollection of his meeting with Oppenheimer:

> I'll never forget that morning. I drove Oppenheimer from the railroad station down to the beach looking out over the peaceful lake. There I listened to his story. . . .
>
> Was there really any chance that an atomic bomb would trigger the explosion of the nitrogen in the atmosphere or the hydrogen in the ocean? This would be the ultimate catastrophe. Better to accept the slavery of the Nazis than to run a chance of drawing the final curtain on mankind!
>
> We agreed there could be only one answer. Oppenheimer's team must go ahead with their calculations.

Bethe already had. "I very soon found some unjustified assumptions in Teller's calculations which made such a result extremely unlikely, to say the least. Teller was very soon persuaded by my arguments." The arguments—Bethe's and others'—against a runaway explosion appear most authoritatively in a technical history of the bomb design program prepared under Oppenheimer's supervision immediately after the war:

> It was assumed that only the most energetic of several possible [thermonuclear] reactions would occur, and that the reaction cross sections were at the maximum values theoretically possible. Calculation led to the result that no matter how high the temperature, energy loss would exceed energy production by a reasonable factor. At an assumed temperature of three million electron volts [compare the 35,000 eV known for D + D] the reaction failed to be self-propagating by a factor of 60. This temperature exceeded the calculated initial temperature of the deuterium reaction by a factor of 100, and that of the fission bomb by a larger factor. . . . The impossibility of igniting the atmosphere was thus assured by science and common sense.

Oppenheimer returned to that good news and they proceeded with the Super. Teller recaptures the mood: "My theories were strongly criticized by others in the group, but together with new difficulties, new solutions emerged. The discussions became fascinating and intense. Facts were questioned and the questions were answered by still more facts. . . . A spirit of spontaneity, adventure, and surprise prevailed during those weeks in Berkeley, and each member of the group helped move the discussion toward a positive conclusion."

There was serious trouble with Teller's D + D Super. The reactions would proceed too slowly to reach ignition before the fission trigger blew the assembly apart. Konopinski came to the rescue: "Konopinski suggested that, in addition to deuterium, we should investigate the reactions of the heaviest form of hydrogen, tritium." This, Teller explains, was at that time "only . . . a conversational guess." One tritium reaction of obvious interest was the fusion of a deuterium nucleus with a tritium nucleus, D + T, which results in the formation of a helium nucleus with the ejection of a neutron and the release of 17.6 MeV of energy. The D + T reaction kindled at a mere 5,000 eV, which corresponds to a temperature of 40 million degrees. But since tritium does not exist on earth it would have to be created. Neutrons bombarding an isotope of lithium, Li6, would transmute some of that light metal to tritium much as neutrons made plutonium from U235, but the only obvious source of such necessarily copious quantities of neutrons was Fermi's unproven pile. The luminaries did, however, consider the possibility of making tritium within the Super itself by packing the bomb with a dry form of lithium, lithium deuteride. But lithium in its natural form, like uranium in its natural form, contained too little of the desired isotope; to be effective, the Li6 would have to be separated. But lithium—element number 3 on the periodic table—would be much easier to separate than uranium . . . So the arguments progressed across the pleasant Berkeley summer. "We were forever inventing new tricks," Bethe says, "finding ways to calculate, and rejecting most of the tricks on the basis of the calculations. Now I could see at first-hand the tremendous intellectual power of Oppenheimer who was the unquestioned leader of our group. . . . The intellectual experience was unforgettable."

At the end of the summer, merging the Serber subgroup's work with their own, the luminaries concluded that the development of an atomic bomb would require a major scientific and technical effort. Glenn Seaborg heard Oppenheimer's deduction from that outcome at a meeting of the Met Lab technical council in Chicago on September 29. "Fast neutron work has no home," Seaborg paraphrases the Berkeley theoretician "[and] may need one." "Oppenheimer has plans in mind for fast neutron work," Compton told the council. Oppenheimer was scouting a site where the bomb might be designed and assembled. He thought such an operation might find a home in Cincinnati or with the plutonium production piles in Tennessee.

James Bryant Conant heard the results of the Berkeley summer study at a meeting of the S-1 Executive Committee in late August 1942 and jotted down a page of notes under the heading "Status of the Bomb." The fission bomb, he wrote, would explode according to the luminaries with "150 times energy of previous calculation" but, bad news, would require a critical mass "6 times the previous [estimated] size[:] 30 kg U235." Twelve kilograms of U235 were enough to explode, Conant noted, but inefficiently with "only 2% of energy." News of the Super then startled the NDRC chairman to a slip of the pencil:

> To denotate [*sic*: detonate] 5–10 kg of heavy hydrogen liquid would require 30 kg U235
>
> If you use 2 or 3 Tons of liquid deuterium and 30 kg U235 this would be equivalent 10^8 [i.e., 100,000,000] tons of TNT.
>
> Estimate devastation area of 1000 sq. km [or] 360 sq miles. Radioactivity lethal over same area for a few days.

Conant then drew a bold line with a steady hand and initialed the file note "JBC." As an afterthought or at a later time he added: "S-1 Executive Committee thinks the above probable. Heavy water is being pushed as hard as it can. [First] 100 kg of D will be available by fall of 1943 before 60 kg of U235 will be ready!"

A formal status report went off immediately from the Executive Committee to Bush. It predicted enough fissionable material for a test in eighteen months—by March 1944. It estimated that a 30-kilogram bomb of U235 "should have a destructive effect equivalent to the explosion of over 100,000 tons of TNT," much more than the mere 2,000 tons estimated earlier. And it dramatically announced the Super:

> If this [U235] unit is used to detonate a surrounding mass of 400 kg of liquid deuterium, the destructiveness should be equivalent to that of more than 10,-000,000 tons of TNT. This should devastate an area of more than 100 square miles.

The committee—Briggs, Compton, Lawrence, Urey, Eger Murphree and Conant—concluded by judging the bomb project important beyond all previous estimates: "We have become convinced that success in this program before the enemy can succeed is necessary for victory. We also believe that success of this program will win the war if it has not previously been terminated."

On August 29 Bush bumped the status report up to the Secretary of War, noting that "the physicists of the Executive Committee are unanimous in believing that this large added factor [i.e., the Super] can be obtained. . . . The ultimate potential possibilities are now considered to be very much greater than at the time of the [last] report."

The hydrogen bomb was thus under development in the United States onward from July 1942.

The problem that Leo Szilard would call "the trouble at Chicago"—the problem of authority and responsibility for pile-cooling design and much more—erupted in a brief rebellion at the Met Lab in September. Stone & Webster, the construction engineers the Army had hired, had spent the summer studying plutonium production. "Classical engineers," Leona Woods calls them, "who knew bridges and structures, canals, highways, and the like, but who had a very weak grasp or none at all of what was needed in the new nuclear industry." The firm sent one of its best engineers to brief Met Lab leaders on production plans. "The scientists sat deadly still with curled lips. The briefer was ignorant; he enraged and frightened everyone."

An exasperated Compton protégé, Volney Wilson, an idealistic young physicist responsible for pile instrumentation, called a confrontation meeting soon afterward on a hot autumn evening. (As a student Wilson had analyzed the motions of swimming fish and invented the competition swimming style known as the Dolphin; with it he had won in Olympics tryouts in 1938 but then suffered disqualification because the style was new and thus unauthorized, a purblindness on the part of the Olympics judges which may have conditioned Wilson's attitude toward authority.) In his memoirs Compton mixes up the autumn meeting with the similar disagreement in June; Woods, who worked for Wilson, remembers it better:

> We (some 60 or 70 scientists) assembled quietly in the commons room at Eckhart Hall, open windows bringing hot, humid air in with an infinitesimal breeze. No one spoke—it was a Quaker meeting. Finally Compton entered carrying a Bible. . . .
>
> Compton thought that the issue of Wilson's meeting was whether the plutonium production should be undertaken by large-scale industry or should be carried out by the scientists of the Metallurgical Project, keeping control in their hands. Instead, it seemed to me that the primary issue was to get rid of Stone & Webster.

Compton vouchsafed a parable. Without introduction he opened his Bible to Judges 7: 5–7 and read to Leo Szilard and Enrico Fermi, to Eugene Wigner, to John Wheeler and threescore serious scientists the story of how the Lord helped Gideon sort among His people to find a few good men to fight the Midianites when there were too many volunteers at hand to demonstrate clearly that the victory would be entirely the work of the Lord. "When Compton finished reading," Woods remembers, "he sat down." Not surprisingly, "there was more Quaker-meeting silence." Or astonishment. Then Volney Wilson stood to direct "well-considered fire and brimstone . . . at the incompetence of Stone & Webster." Many others in the group spoke as well, all opposing the Boston engineers. "After a while, silence fell and finally everyone got up and disbanded." Compton had reduced the discussion to a demand that the Met Lab capitulate to his authority. Fortunately the assembly of scientists ignored him. The Army would soon move the responsibility for plutonium production into more experienced hands than Stone & Webster's. When the change was proposed Compton eagerly endorsed it.

Szilard responded to the struggles at the Met Lab with anger that by now, after four years of frustration, had begun to harden into stoicism. Late in September he drafted a long memorandum to his colleagues that addressed specific Met Lab problems but also considered the deeper issue of the responsibility of scientists for their work. In draft and more moderately in finished form his examination by turns compliments and savages Compton's leadership: "In talking to Compton I frequently have the feeling that I am overplaying a delicate instrument." Beyond personality Szilard pointed to a destructive abdication by those whom Compton led: "I have often thought . . . that things would have been different if Compton's authority had actually originated with our group, rather than with the OSRD." He elaborates in the finished memorandum:

> The situation might be different if Compton considered himself as our representative in Washington and asked in our name for whatever was necessary to make our project successful. He could then refuse to make a decision on any of the issues which affect our work until he had an opportunity fully to discuss the matter with us.
>
> Viewed in this light, it ought to be clear to us that we, and we alone, are to be blamed for the frustration of our work.

An authoritarian organization had moved in—had been allowed to move in—to take over work that had been democratically begun. "There is a sprinkling of democratic spots here and there, but they do not form a coherent network which could be functional." Szilard was convinced that authoritarian organization was no way to do science. So were Wigner and the more detached Fermi. "If we brought the bomb to them all ready-made on a silver platter," Szilard remembers hearing Fermi say, "there would still be a fifty-fifty chance that they would mess it up." But beyond debating the virtues of contractors and cooling systems only Szilard continued to rebel:

> We may take the stand that the responsibility for the success of this work has been delegated by the President to Dr. Bush. It has been delegated by Dr.

> Bush to Dr. Conant. Dr. Conant delegates this responsibility (accompanied by only part of the necessary authority) to Compton. Compton delegates to each of us some particular task and we can lead a very pleasant life while we do our duty. We live in a pleasant part of a pleasant city, in the pleasant company of each other, and have in Dr. Compton the most pleasant "boss" we could wish to have. There is every reason why we should be happy and since there is a war on, we are even willing to work overtime.
>
> Alternatively, we may take the stand that those who have originated the work on this terrible weapon and those who have materially contributed to its development have, before God and the World, the duty to see to it that it should be ready to be used at the proper time and in the proper way.
>
> I believe that each of us has now to decide where he feels that his responsibility lies.

The Army had been involved in the bomb project since June, but the Corps of Engineers' Colonel Marshall had been unable to drive the project ahead of other national military priorities. Divided between the OSRD and the Army it began to look as if it might lose its way. Bush thought he saw a solution in an authoritative new Military Policy Committee that would retain the project under partly civilian control but delegate direction to a dynamic Army officer and back him up. "From my own point of view," he wrote at the end of August 1942, "faced as I am with the unanimous opinion of a group of men that I consider to be among the greatest scientists in the world, joined by highly competent engineers, I am prepared to recommend that nothing should stand in the way of putting this whole affair through to conclusion . . . even if it does cause moderate interference with other war efforts."

Bush had discussed his problems with the general in charge of the Army Services of Supply, Brehon Somervell. Independently Somervell worked out a solution of his own: assigning entire responsibility to the Corps of Engineers, which was under his command. The program would need a stronger leader. He had a man in mind. In mid-September he sought him out.

"On the day I learned that I was to direct the project which ultimately produced the atomic bomb," Albany-born Leslie Richard Groves wrote later, "I was probably the angriest officer in the United States Army." The West Point graduate, forty-six years old in 1942, goes on to explain why:

> It was on September 17, 1942, at 10:30 a.m., that I got the news. I had agreed, by noon that day, to telephone my acceptance of a proposed assignment to duty overseas. I was then a colonel in the Army Engineers, with most of the headaches of directing ten billion dollars' worth of military construction in

the country behind me—for good, I hoped. I wanted to get out of Washington, and quickly.

Brehon B. Somervell . . . my top superior, met me in a corridor of the new House of Representatives Office Building when I had finished testifying about a construction project before the Military Affairs Committee.

"About that duty overseas," General Somervell said, "you can tell them no."

"Why?" I inquired.

"The Secretary of War has selected you for a very important assignment."

"Where?"

"Washington."

"I don't want to stay in Washington."

"If you do the job right," General Somervell said carefully, "it will win the war."

Men like to recall, in later years, what they said at some important or possibly historic moment in their lives. . . . I remember only too well what I said to General Somervell that day.

I said, "Oh."

As deputy chief of construction for the entire U.S. Army, Groves knew enough about the bomb project to recognize its dubious claim to decisive effect and be thoroughly disappointed. He had just finished building the Pentagon, the most visible work of his career. He had seen the S-1 budget; it amounted in total to less than he had been spending in a week. He wanted assignment commanding troops. But he was career Army and understood he hardly had a choice. He crossed the Potomac to the Pentagon office of Somervell's chief of staff, Brigadier General Wilhelm D. Styer, for a briefing. Styer implied the job was well along and ought to be easy. The two officers worked up an order for Somervell to sign authorizing Groves "to take complete charge of the entire . . . project." Groves discovered he would be promoted to brigadier—for authority and in compensation—in a matter of days. He proposed to delay official appointment until the promotion came through. "I thought that there might be some problems in dealing with the many academic scientists involved in the project," he remembers of his initial innocence, "and I felt that my position would be stronger if they thought of me from the first as a general instead of as a promoted colonel." Styer agreed.

Groves was one inch short of six feet tall, jowly, with curly chestnut hair, blue eyes, a sparse mustache and sufficient girth to balloon over his webbing belt above and below its brass military buckle. Leona Woods thought he might weigh as much as 300 pounds; he was probably nearer

250 then, though he continued to expand. He had graduated from the University of Washington in 1914, studied engineering intensely for two years at MIT and gone on to West Point, where he graduated fourth in his class in 1918. Years at the Army Engineer School, the Command and General Staff College and the Army War College in the 1920s and 1930s completed his extensive education. He had seen duty in Hawaii, Europe and Central America. His father was a lawyer who left the law for the ministry and served in a country parish and an urban, working-class church before Grover Cleveland's Secretary of War convinced him to enlist as an Army chaplain on the Western frontier. "Entering West Point fulfilled my greatest ambition," Groves testifies. "I had been brought up in the Army, and in the main had lived on Army posts all my life. I was deeply impressed with the character and outstanding devotion to duty of the officers I knew." The dynamic engineer was married, with a thirteen-year-old daughter and a plebe son at West Point.

"A tremendous lone wolf," one of his subordinates describes Groves. Another, whose immediate superior Groves was about to become, distills their years together into grudgingly admiring vitriol. Lieutenant Colonel Kenneth D. Nichols—balding, bespectacled, thirty-four in 1942, West Point, Ph.D. in hydraulic engineering at Iowa State—remembers Groves as

> the biggest sonovabitch I've ever met in my life, but also one of the most capable individuals. He had an ego second to none, he had tireless energy—he was a big man, a heavy man but he never seemed to tire. He had absolute confidence in his decisions and he was absolutely ruthless in how he approached a problem to get it done. But that was the beauty of working for him—that you never had to worry about the decisions being made or what it meant. In fact I've often thought that if I were to have to do my part all over again, I would select Groves as boss. I hated his guts and so did everybody else but we had our form of understanding.

Nichols' previous boss, Colonel Marshall, had worked out of an office in Manhattan (where in August he had disguised the project to build an atomic bomb behind the name Manhattan Engineer District). But decisions of priority and supply were made in wartime in hurly-burly Washington offices, not in Manhattan, and to fight those battles the colonel had chosen the capable Nichols. Groves therefore sought out Nichols next after Styer. And found the project in even worse condition than he had feared: "I was not happy with the information I received; in fact, I was horrified."

He took Nichols with him to the Carnegie Institution on P Street to confront Vannevar Bush. Somervell had overlooked clearing Groves' appointment with Bush and the OSRD director was infuriated. He evaded

Groves' questions brusquely, which puzzled Groves. Controlling his anger until Groves and Nichols left, Bush then paid Styer a visit, which he describes in a contemporary memorandum:

> I told him (1) that I still felt, as I had told him and General Somervell previously, that the best move was to get the military commission first, and then the man to carry out their policies second; (2) that having seen General Groves briefly, I doubted whether he had sufficient tact for such a job.
>
> Styer disagreed on (1) and I simply said I wanted to be sure he understood my recommendation. On (2) he agreed the man is blunt, etc., but thought his other qualities would overbalance. . . . I fear we are in the soup.

Bush changed his mind within days. Groves immediately tackled his worst problems and solved them.

One of the first issues the heavyweight colonel had raised with Nichols was ore supply: was there sufficient uranium on hand? Nichols told him about a recent and fortuitous discovery: some 1,250 tons of extraordinarily rich pitchblende—it was 65 percent uranium oxide—that the Union Minière had shipped to the United States in 1940 from its Shinkolobwe mine in the Belgian Congo to remove it beyond German reach. Frédéric Joliot and Henry Tizard had independently warned the Belgians of the German danger in 1939. The ore was stored in the open in two thousand steel drums at Port Richmond on Staten Island. The Belgians had been trying for six months to alert the U.S. government to its presence. On Friday, September 18, Groves sent Nichols to New York to buy it.

On Saturday Groves drafted a letter in the name of Donald Nelson, the civilian head of the War Production Board, assigning a first-priority AAA rating to the Manhattan Engineer District. Groves personally carried the letter to Nelson. "His reaction was completely negative; however, he quickly reversed himself when I said that I would have to recommend to the President that the project should be abandoned because the War Production Board was unwilling to co-operate with his wishes." Groves was bluffing but it was not the bluster that swayed Nelson; he had probably heard by then from Bush and Henry Stimson. He signed the letter. "We had no major priority difficulties," notes Groves, "for nearly a year."

The same day Groves approved a directive that had been languishing on his predecessor's desk throughout the summer for the acquisition of 52,000 acres of land along the Clinch River in eastern Tennessee. Site X, the Met Lab called it. District Engineer Marshall had thought to wait to buy the land at least until the chain reaction was proved.

On September 23, the following Wednesday, Groves' promotion to brigadier came through. He hardly had time to pin on his stars before attending a command performance in the office of the Secretary of War called to assemble Bush's outmaneuvered Military Policy Committee with Stimson, Army Chief of Staff George Marshall, Bush, Conant, Somervell, Styer and an admiral on hand. Groves described how he intended to operate. Stimson proposed a nine-man committee to supervise. Groves held out for a more workable three and won his point. Discussion continued. Abruptly Groves asked to be excused: he needed to catch a train to Tennessee, he explained, to inspect Site X. The startled Secretary of War agreed and Leslie Richard Groves, the new broom that would sweep the Manhattan Engineer District clean, departed for Union Station. "You made me look like a million dollars," Somervell praised Groves when he got back to Washington. "I'd told them that if you were put in charge, things would really start moving." They did.

Enrico Fermi began planning a full-scale chain-reacting pile in May 1942 when one of the exponential piles his team built in the west stands of Stagg Field indicated its k at infinity would muster 0.995. The Met Lab was searching out higher-quality graphite and sponsoring production of pure uranium metal, denser than oxide; those and other improvements should push k above 1.0. "I remember I talked about the experiment on the Indiana dunes," Fermi told his wife after the war, "and it was the first time I saw the dunes. . . . I liked the dunes: it was a clear day, with no fog to dim colors. . . . We came out of the water, and we walked along the beach."

As they began preparations that summer Leona Woods remembers swimming "in frigid Lake Michigan every afternoon at five o'clock, off the huge breakwater rocks at the 55th street promontory"—she, Herbert Anderson, Fermi. She was still a graduate student, twenty-two and shy. "One evening, Enrico gave a party, inviting Edward and [Mici] Teller, Helen and Robert Mulliken (my research professor), and Herb Anderson, John Marshall, and me." They played Murder, the parlor game then in fashion. "The second the lights went out on this particular evening, I shrank into a corner and listened with astonishment to these brilliant, accomplished, famous sophisticated people shrieking and poking and kissing each other in the dark like little kids." All nice people are shy, Fermi consoled her when he knew her better; he had always been dominated by shyness. She records his sly self-mockery: "As he frequently said, he was amazed when he thought how modest he was."

Woods was finishing her thesis work during the summer but sometimes helped Anderson scour Chicago for lumber. CP-1—Chicago Pile Number One—Fermi planned to build in the form of a sphere, the most efficient shape to maximize k. Since the pile's layers of graphite bricks would enlarge concentrically up to its equator, they would need external support, and wood framing was light and easy to shape and assemble. "I was the buyer for a lot of lumber," Anderson says. "I remember the Sterling Lumber Company, how amazed they were by the orders I gave them, all with double X priority. But they delivered the lumber with no questions asked. There was almost no constraint on money and priority to get what we wanted."

Horseback riding one Saturday afternoon in the Cook County Forest Preserve twenty miles southwest of Chicago, Arthur and Betty Compton found an isolated, scenic site for the pile building, a terminal moraine forested with hawthorne and scrub oak known as the Argonne Forest. The Army's Nichols negotiated with the county to use the land; Stone & Webster began planning construction.

The Fermis rented a house from a businessman moving to Washington for war work; since they were enemy aliens and not allowed to own a shortwave radio the man had to have his big all-band Capehart temporarily disabled of its long-distance frequencies, though it continued to supply dance music to the party room on the third floor. Fermi was angry to find his mail being opened and complained indignantly until the practice was stopped (or managed more surreptitiously). The Comptons gave a series of parties to welcome newcomers to the Met Lab. "At each of these parties," Laura Fermi writes, "the English film *Next of Kin* was shown. It depicted in dark tones the consequences of negligence and carelessness. A briefcase laid down on the floor in a public place is stolen by a spy. English military plans become known to the enemy. Bombardments, destruction of civilian homes, and an unnecessarily high toll of lives on the fighting front are the result. . . . Willingly we accepted the hint and confined our social activities to the group of 'metallurgists.' " Compton, who describes himself as "one of those who must talk over important problems with his wife," arranged uniquely to have Betty Compton cleared. None of the other wives was supposed to know about her husband's work. Laura Fermi found out, like many others, only at the end of the war.

In mid-August Fermi's group could report a probable *k* for a graphite–uranium oxide pile of "close to 1.04." They were working on control-rod design and testing the vacuum properties of both metal sheet and balloon cloth. The cloth was Anderson's idea, a possible alternative to canning the pile to exclude neutron-absorbing air. It proved serviceable and Anderson followed up: "For the balloon cloth enclosure I went to the Goodyear Rubber Company in Akron, Ohio. The company had a good deal of experience in building blimps and rubber rafts but a square balloon 25′ on a side seemed a bit odd to them." They made it anyway, "with no questions asked." It should be good for a 1 percent improvement in *k*.

Between September 15 and November 15 Anderson, Walter Zinn and their crews also built sixteen successive exponential piles in the Stagg Field west stands to measure the purity of the various shipments of graphite, uranium oxide and metal they had begun to receive in quantity. Not all the uranium was acceptable. But Mallinckrodt Chemical Works in St. Louis, specialists at handling the ether necessary for oxide extraction, began producing highly purified brown oxide at the rate of thirty tons a month, and the National Carbon Company and a smaller supplier, by using purified petroleum coke for raw material and doubling furnace time, significantly improved graphite supplies (graphite is molded as coke, then baked in a high-temperature electric-arc oven for long hours until it crystallizes and its impurities vaporize away). By September regular deliveries began to arrive in covered trucks. Physicists doubled as laborers to unload the bricks and cans and pass them into the west stands for finishing.

Walter Zinn took charge of preparing the materials for the pile. The graphite came in from various manufacturers as rough 4¼ by 4¼-inch bars in 17- to 50-inch lengths. So that the bars would fit closely together they had to be smoothed and cut to standard 16½-inch lengths. About a fourth of them also had to be drilled for the lumps of uranium they would hold. A few required slots machined through to make channels for control rods. The uranium oxide needed to be compressed into what the physicists called "pseudospheres"—stubby cylinders with round-shouldered ends—for which purpose the press from the Jersey City junkyard had been shipped to Chicago the previous winter.

For crew Zinn had half a dozen young physicists, a thoroughly able carpenter and some thirty high school dropouts earning pocket money until their draft notices came through. They were Back of the Yards boys from the tough neighborhood beyond the Chicago stockyards and Zinn improved the fluency of his swearing keeping them in line.

Machining the graphite was like sharpening thousands of giant pencils. Zinn used power woodworking tools. A jointer first made two sides of each graphite brick perpendicular and smooth; a planer finished the other two surfaces; a swing saw cut the bricks to length. That processing produced 14 tons of bricks a day; each brick weighed 19 pounds.

To drill the blind, round-bottomed 3¼-inch holes for the uranium pseudospheres, two to a brick, Zinn adapted a heavy lathe. He mounted a 3¼-inch spade bit in the headstock of the lathe, where the material to be turned would normally be mounted, and forced the graphite up against the tool with the lathe carriage. Dull bits caused problems. Zinn tried tough carballoy bits first, but they were tedious to resharpen. He began making bits from old steel files, sharpening them by hand whenever they dulled. One sharpening was good for 60 holes, about an hour's work. Before they were through they would shape and finish 45,000 graphite bricks and drill 19,000 holes.

General Groves made his first appearance at the Met Lab on October 5 and delivered his first pronouncement. The technical council was debating cooling systems again. "The War Department considers the project important," Seaborg paraphrases Groves' formula, which they would all learn by heart. "There is no objection to a wrong decision with quick results. If there is a choice between two methods, one of which is good and the other looks promising, then build both." Get the cooling-system decision into Compton's hands by Saturday night, Groves demanded. It was Monday. They had been debating for months.

Groves moved on to Berkeley more impressed with their work than his Met Lab auditors realized. "I left Chicago feeling that the plutonium process seemed to offer us the greatest chances for success in producing bomb material," he recalls. "Every other process . . . depended upon the physical separation of materials having almost infinitesimal differences in their physical properties." Transmutation by chain reaction was entirely new, but the rest of the plutonium process, chemical separation, "while extremely difficult and completely unprecedented, did not seem to be impossible."

At the beginning of the month, to Compton's great relief, the brigadier had convinced E. I. du Pont de Nemours, the Delaware chemical and explosives manufacturers, to take over building and running the plutonium production piles under subcontract to Stone & Webster. He meant to involve the industrial chemists more extensively than that—meant for them to take over the plutonium project in its entirety. Du Pont resisted the increasing encroachment. "Its reasons were sound," writes Groves: "the evident physical operating hazards, the company's inexperience in the field of nuclear physics, the many doubts about the feasibility of the process, the paucity of proven theory, and the complete lack of essential technical design data." Du Pont also suspected, once it had sent an eight-man review team to Chicago at the beginning of November, that the plutonium project was the least promising of the several then under development and might even fail, tarnishing the company's reputation. Nor was it happy at the prospect of identifying itself with a secret weapon of mass destruction; it still remembered the general condemnation it had received for selling munitions to Britain and France before the United States entered the First World War. Groves told the Du Pont executive committee that the Germans were probably hard at work and the only defense against a Nazi atomic bomb would be an American bomb. And added what he took to be a clinching argument: "If we were successful in time, we would shorten the war and thus save tens of thousands of American casualties." The second week in November Du Pont admitted the possibility of regular production by 1945 and accepted the assignment (limiting itself to a profit of one dollar to avoid arms-merchant stigma), but made its skepticism and reluctance clear.

By then Stone & Webster's construction workers had gone on strike. The pile building scheduled for completion by October 20 would be indefinitely delayed. Fermi lived with the problem only long enough to recalculate the risks of pile control. In early November he cornered Compton in his office and proposed an alternative site: the doubles squash court where his team had built its series of exponential piles. A *k* greater than 1.0 presented an entirely different order of risk from a *k* of less than 1.0, however; Compton had, in Seaborg's words, a "dreadful decision" to make. "We did not see how a true nuclear explosion, such as that of an atomic bomb, could possibly occur," Compton writes with more calm than he probably felt at the time. "But the amount of potentially radioactive material present in the pile would be enormous and anything that would cause excessive ionizing radiation in such a location would be intolerable." He asked for Fermi's analysis of the probability of control.

No doubt Fermi discussed the various hand and automatic control rods he planned for the pile. But even slow-neutron fission generations had been calculated to multiply in thousandths of a second, which might flash the pile to dangerous levels of heat and radiation before any merely mechanical control system could move into position. The "most significant fact assuring us that the chain reaction could be controlled," says Compton, was one of the Richard Roberts team's earliest discoveries at the Carnegie Institution's Department of Terrestrial Magnetism following Bohr's announcement of the discovery of fission in 1939—in Compton's words, that "a certain small fraction of the neutrons associated with the fission process are not emitted at once but come off a few seconds after fission occurs." With a pile operating at *k* only marginally above 1.0, such delayed neutrons would slow the response sufficiently to allow time for adjustment.

For once Compton made a quick decision: with control seemingly assured, he allowed Fermi to build CP-1 in the west stands. He chose not to inform the president of the University of Chicago, Robert Maynard Hutchins, reasoning that he should not ask a lawyer to judge a matter of nuclear physics. "The only answer he could have given would have been—no. And this answer would have been wrong. So I assumed the responsibility myself." The word *meltdown* had not yet entered the reactor engineer's vocabulary—Fermi was only then inventing that specialty—but that is what

Compton was risking, a small Chernobyl in the midst of a crowded city. Except that Fermi, as he knew, was a formidably competent engineer.

In mid-November Fermi reorganized his team into two twelve-hour shifts, a day crew under Walter Zinn (who continued to supervise materials production as well), a night crew under Herbert Anderson. Construction began on Monday morning, November 16, 1942. From the balcony of the doubles squash court in the west stands of Stagg Field Fermi directed the hanging of the cubical dark-gray Goodyear balloon as his men hauled it into place with block and tackle. It dominated the room: bottom panel smoothed on the floor, top and three sides secured to the ceiling and the walls, the fourth side facing the balcony furled up out of the way like an awning. Someone drew a circle on the floor panel to locate the first layer of graphite and without ceremony the crew began positioning the dark, slippery bricks. The first layer was "dead" graphite that carried no load of uranium: solid crystalline carbon to diffuse and slow the neutrons that fission would generate. Up the pile as it stacked, the crews would alternate one layer of dead graphite with two layers of bricks each drilled and loaded with two five-pound uranium pseudospheres. That created a cubic cell of neutron-diffusing graphite around every lump of uranium.

To build the wooden framing, Herbert Anderson recalls, "Gus Knuth, the millwright, would be called in. We would show him . . . what we wanted, he would take a few measurements, and soon the timbers would be in place. There were no detailed plans or blueprints for the frame or the pile." Since they had batches of graphite, oxide and metal of varying purity, they improvised the placement of materials as they went along. Fermi, says Anderson, "spent a good deal of time calculating the most effective location for the various grades of [material] on hand."

They were soon averaging not quite two layers a shift, handing the bricks along from their delivery skids, sliding them to the workers on the pile, singing together to pass the time. The bricks in the dead graphite layers alternated direction, three running east and west and the next three north and south. That gave support to the oxide layers, which all ran together from front to back except at the outer edges, where dead graphite formed an outer shell. The physicist bricklayers had to be careful to line up the slots for the ten control-rod channels that passed at widely distributed points completely through the pile. "A simple design for a control rod was developed," says Anderson, "which could be made on the spot: cadmium sheet nailed to a flat wood strip. . . . The [thirteen-foot] strips had to be inserted and removed by hand. Except when the reactivity of the pile was being measured, they were kept inside the pile and locked using a simple hasp and padlock, the only keys to which were kept by Zinn and myself." Cadmium, which has a gargantuan absorption cross section for slow neutrons, held the pile quiescent.

As it grew they assembled wooden scaffolding to stand on and ran loads of bricks up to the working face on a portable materials elevator. Before the arrival of the elevator, during the period when they were building large exponential piles, they had simply leaned over from the precarious 2 by 12-inch scaffolding and reached the bricks up from the men on the floor below. Groves walked in on them one day and dressed them down for risking their necks. The elevator appeared unbidden soon after.

When they achieved the fifteenth layer Zinn and Anderson began measuring neutron intensity at the end of each shift at a fixed point near the center of the pile with the control rods removed. They used a boron trifluoride counter Leona Woods had devised that worked much like a Geiger counter, clicking off the neutron count. Standard indium foils bombarded to radioactivity by pile neutrons gave daily checks on the boron counter's calibration. Fermi had complained to Segrè in October that he was doing physics by telephone; now he moved a little closer to the work. "Each day we would report on the progress of the construction to Fermi," Anderson notes, "usually in his office in Eckhart Hall. Then we would present our sketch of the layers that we had assembled and reach some agreement on what would be added during the following shifts." Fermi took the raw boron-counter and indium measurements and calculated a countdown. As the pile approached its slow-neutron critical mass the neutrons generated within it by spontaneous fission multiplied through more and more generations before they were absorbed. At $k = 0.99$, for example, each neutron would multiply through an average one hundred generations before its chain of generations died out. Fermi divided the square of the radius of the pile by a measure of the intensity of radioactivity the pile induced in indium and got a number that would decrease to zero as the pile approached criticality. At layer 15 the countdown stood at 390; at layer 19 it dropped to 320. It was 270 at layer 25 and down to 149 at layer 36.

As winter locked down, the unheated west stands turned bitterly cold. Graphite dust blackened walls, floors, hallways, lab coats, faces, hands. A black haze dispersed light in the floodlit air. White teeth shone. Every surface was slippery, hands and feet routine casualties of dropped blocks. The men building the pile, lifting tons of materials every shift, stayed warm enough, but the unlucky security guards stationed at doors and entrances froze. Zinn scavenged rakish makeshift to thaw them out:

> We tried charcoal fires in empty oil drums—too much smoke. Then we secured a number of ornamental, imitation log, gas-fired fireplaces. These were

> hooked up to the gas mains, but they gobbled up the oxygen and replaced it with fumes which burned the eyes. . . . The University of Chicago came to the rescue. Years before, big league football had been banned from the campus; we found in an old locker a supply of raccoon fur coats. Thus, for a time we had the best dressed collegiate-style guards in the business.

Fermi had originally designed his first full-scale pile as a 76-layer sphere. Some 250 tons of better graphite from National Carbon now promised to reduce neutron absorption below previous estimates; more than 6 tons of high-purity uranium metal in the form of 2¼-inch cylinders began arriving from Iowa State College at Ames, where one of the Met Lab's chemistry group leaders, Frank Spedding, had converted a laboratory to backyard mass production. "Spedding's eggs," dropped in place of oxide pseudospheres into drilled graphite blocks that were then stacked in spherical configuration close to the center of the CP-1 lattice, significantly increased the value of *k*. Adjusting for the improvements, Fermi saw that they would not need to seal the Goodyear balloon and evacuate the air from the pile and could eliminate some 20 layers: his countdown should converge to zero, $k = 1.0$, between layers 56 and 57. Instead of a sphere the pile would take the form of a doorknob as big as a two-car garage, a flattened rotational ellipsoid 25 feet wide at the equator and 20 feet high from pole to pole:

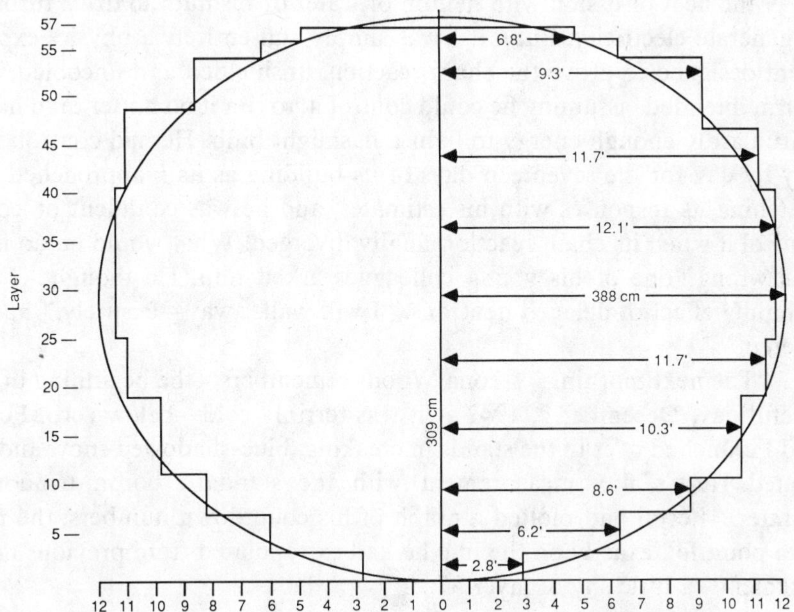

Anderson's crew assembled this final configuration on the night of December 1:

> That night the construction proceeded as usual, with all cadmium covered wood in place. When the 57th layer was completed, I called a halt to the work, in accordance with the agreement we had reached in the meeting with Fermi that afternoon. All the cadmium rods but one were removed and the neutron count taken following the standard procedure which had been followed on the previous days. It was clear from the count that once the only remaining cadmium rod was removed, the pile would go critical. I resisted great temptation to pull the final cadmium strip and be the first to make a pile chain react. However, Fermi had foreseen this temptation and extracted a promise from me to make the measurement, record the result, insert all cadmium rods, and lock them all in place.

Which Anderson dutifully did, and closed up the squash court and went home to bed.

The pile as it waited in the dark cold of Chicago winter to be released to the breeding of neutrons and plutonium contained 771,000 pounds of graphite, 80,590 pounds of uranium oxide and 12,400 pounds of uranium metal. It cost about $1 million to produce and build. Its only visible moving parts were its various control rods. If Fermi had planned it for power production he would have shielded it behind concrete or steel and pumped away the heat of fission with helium or water or bismuth to drive turbines to generate electricity. But CP-1 was simply and entirely a physics experiment designed to prove the chain reaction, unshielded and uncooled, and Fermi intended, assuming he could control it, to run it no hotter than half a watt, hardly enough energy to light a flashlight bulb. He had controlled it day by day for the seventeen days of its building as its *k* approached 1.0, matching its responses with his estimates, and he was confident he could control it when its chain reaction finally diverged. What would he do if he was wrong? one of his young colleagues asked him. He thought of the damping effect of delayed neutrons. "I will walk away—leisurely," he answered.

"The next morning," Leona Woods remembers—the beginning of the fateful day, December 2, 1942—"it was terribly cold—below zero. Fermi and I crunched over to the stands in creaking, blue-shadowed snow and repeated Herb's flux measurement with the standard boron trifluoride counter." Fermi had plotted a graph of his countdown numbers; the new data point fell exactly on the line he had extrapolated from previous measurements, a little shy of layer 57:

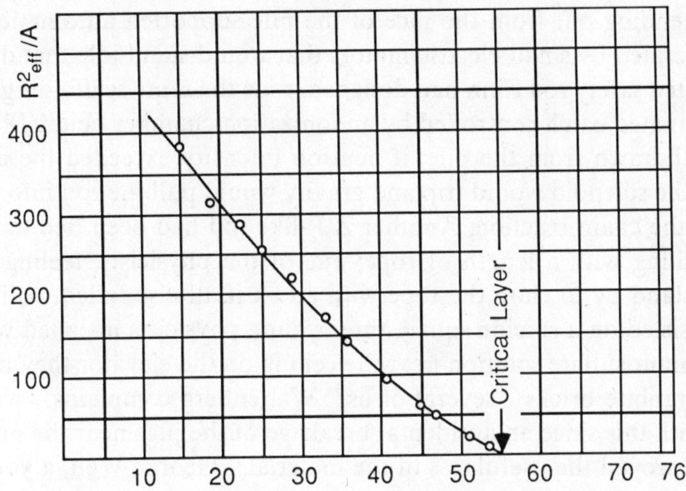

Fermi discussed a schedule for the day with Zinn and Volney Wilson, Woods continues; "then a sleepy Herb Anderson showed up. . . . Herb, Fermi and I went over to the apartment I shared with my sister (it was close to the stands) for something to eat. I made pancakes, mixing the batter so fast that there were bubbles of dry flour in it. When fried, these were somewhat crunchy between the teeth, and Herb thought I had put nuts in the batter."

Outside was raw wind. On the second day of gasoline rationing Chicagoans jammed streetcars and elevated trains, leaving almost half their usual traffic of automobiles at home. The State Department had announced that morning that two million Jews had perished in Europe and five million more were in danger. The Germans were preparing counterattack in North Africa; American marines and Japanese soldiers struggled in the hell of Guadalcanal.

> Back we mushed through the cold, creaking snow. . . . Fifty-seventh Street was strangely empty. Inside the hall of the west stands, it was as cold as outside. We put on the usual gray (now black with graphite) laboratory coats and entered the doubles squash court containing the looming pile enclosed in the dirty, grayish-black balloon cloth and then went up on the spectators' balcony. The balcony was originally meant for people to watch squash players, but now it was filled with control equipment and read-out circuits glowing and winking and radiating some gratefully received heat.

The instrumentation included redundant boron trifluoride counters for lower neutron intensities and ionization chambers for higher. A wooden pier extending out from the face of the pile supported automatic control rods operated by small electric motors that would stand idle that day. ZIP, a weighted safety rod Zinn had designed, rode the same scaffolding. A solenoid-actuated catch controlled by an ionization chamber held ZIP in position withdrawn from the pile; if neutron intensity exceeded the chamber setting the solenoid would trip and gravity would pull the rod into position to stop the chain reaction. Another ZIP-like rod had been tied to the balcony railing with a length of rope; one of the physicists, feeling foolish, would stand by to chop the rope with an ax if all else failed. Allison had even insisted on a suicide squad, three young physicists installed with jugs of cadmium-sulfate solution near the ceiling on the elevator they had used to lift graphite bricks; "several of us," Wattenberg complains, "were very upset with this since an accidental breakage of the jugs near the pile could have destroyed the usefulness of the material." George Weil, a young veteran of the Columbia days, took up position on the floor of the squash court to operate one of the cadmium control rods by hand at Fermi's order. Fermi had scalers that counted off boron trifluoride readings with loud clicks and a cylindrical pen recorder that performed a similar function silently, graphing pile intensities in ink on a roll of slowly rotating graph paper. For calculations he relied on his own trusted six-inch slide rule, the pocket calculator of its day.

Around midmorning Fermi began the crucial experiment. First he ordered all but the last cadmium rod removed and checked to see if the neutron intensity matched the measurement Anderson had made the night before. With that first comparison Volney Wilson's team working on the balcony took time to adjust its monitors. Fermi had calculated in advance the intensity he expected the pile to reach at each step of the way as George Weil withdrew the last thirteen-foot cadmium rod by measured increments.

When Wilson's team was ready, writes Wattenberg, "Fermi instructed Weil to move the cadmium rod to a position which was about half-way out. [The adjustment brought the pile to] well below critical condition. The intensity rose, the scalers increased their rates of clicking for a short while, and then the rate became steady, as it was supposed to." Fermi busied himself at his slide rule, calculating the rate of increase, and noted the numbers on the back. He called to Weil to move the rod out another six inches. "Again the neutron intensity increased and leveled off. The pile was still subcritical. Fermi had again been busy with his little slide rule and seemed very pleased with the results of his calculations. Every time the intensity leveled off, it was at the values he had anticipated for the position of the control rod."

The slow, careful checking continued through the morning. A crowd began to gather on the balcony. Szilard arrived, Wigner, Allison, Spedding whose metal eggs had flattened the pile. Twenty-five or thirty people accumulated on the balcony watching, most of them the young physicists who had done the work. No one photographed the scene but most of the spectators probably wore suits and ties in the genteel tradition of prewar physics and since it was cold in the squash court, near zero, they would have kept warm in coats and hats, scarves and gloves. The room was dingy with graphite dust. Fermi was calm. The pile rising before them, faced with raw 4 by 6-inch pine timbers up to its equator, domed bare graphite above, looked like an ominous black beehive in a bright box. Neutrons were its bees, dancing and hot.

Fermi called for another six-inch withdrawal. Weil reached up to comply. The neutron intensity leveled off at a rate outside the range of some of the instruments. Time passed, says Wattenberg, the watchers abiding in the cold, while Wilson's team again adjusted the electronics:

> After the instrumentation was reset, Fermi told Weil to remove the rod another six inches. The pile was still subcritical. The intensity was increasing slowly—when suddenly there was a very loud crash! The safety rod, ZIP, had been automatically released. Its relay had been activated by an ionization chamber because the intensity had exceeded the arbitrary level at which it had been set. It was 11:30 a.m., and Fermi said, "I'm hungry. Let's go to lunch." The other rods were put into the pile and locked.

At two in the afternoon they prepared to continue the experiment. Compton joined them. He brought along Crawford Greenewalt, the tall, handsome engineer who was the leader of the Du Pont contingent in Chicago. Forty-two people now occupied the squash court, most of them crowded onto the balcony.

Fermi ordered all but one of the cadmium rods again unlocked and removed. He asked Weil to set the last rod at one of the earlier morning settings and compared pile intensity to the earlier reading. When the measurements checked he directed Weil to remove the rod to the last setting before lunch, about seven feet out.

The closer *k* approached 1.0, the slower the rate of change of pile intensity. Fermi made another calculation. The pile was nearly critical. He asked that ZIP be slid in. That adjustment brought the neutron count down. "This time," he told Weil, "take the control rod out twelve inches." Weil withdrew the cadmium rod. Fermi nodded and ZIP was winched out as well. "This is going to do it," Fermi told Compton. The director of the plutonium project had found a place for himself at Fermi's side. "Now it will become self-sustaining. The trace [on the recorder] will climb and continue to climb; it will not level off."

Herbert Anderson was an eyewitness:

> At first you could hear the sound of the neutron counter, clickety-clack, clickety-clack. Then the clicks came more and more rapidly, and after a while they began to merge into a roar; the counter couldn't follow anymore. That was the moment to switch to the chart recorder. But when the switch was made, everyone watched in the sudden silence the mounting deflection of the recorder's pen. It was an awesome silence. Everyone realized the significance of that switch; we were in the high intensity regime and the counters were unable to cope with the situation anymore. Again and again, the scale of the recorder had to be changed to accommodate the neutron intensity which was increasing more and more rapidly. Suddenly Fermi raised his hand. "The pile has gone critical," he announced. No one present had any doubt about it.

Fermi allowed himself a grin. He would tell the technical council the next day that the pile achieved a *k* of 1.0006. Its neutron intensity was then doubling every two minutes. Left uncontrolled for an hour and a half, that rate of increase would have carried it to a million kilowatts. Long before so extreme a runaway it would have killed anyone left in the room and melted down.

"Then everyone began to wonder why he didn't shut the pile off," Anderson continues. "But Fermi was completely calm. He waited another minute, then another, and then when it seemed that the anxiety was too much to bear, he ordered 'ZIP in!' " It was 3:53 P.M. Fermi had run the pile for 4.5 minutes at one-half watt and brought to fruition all the years of discovery and experiment. Men had controlled the release of energy from the atomic nucleus.

The chain reaction was moonshine no more.

Eugene Wigner reports how they felt:

> Nothing very spectacular had happened. Nothing had moved and the pile itself had given no sound. Nevertheless, when the rods were pushed back in and the clicking died down, we suddenly experienced a let-down feeling, for all of us understood the language of the counter. Even though we had anticipated the success of the experiment, its accomplishment had a deep impact on us. For some time we had known that we were about to unlock a giant; still, we could not escape an eerie feeling when we knew we had actually done it. We felt as, I presume, everyone feels who has done something that he knows will have very far-reaching consequences which he cannot foresee.

Months earlier, realizing that the importation of Italian wine had been cut off by the war, Wigner had searched the liquor stores of Chicago for a celebratory *fiasca* of Chianti. He produced it now in a brown paper bag and

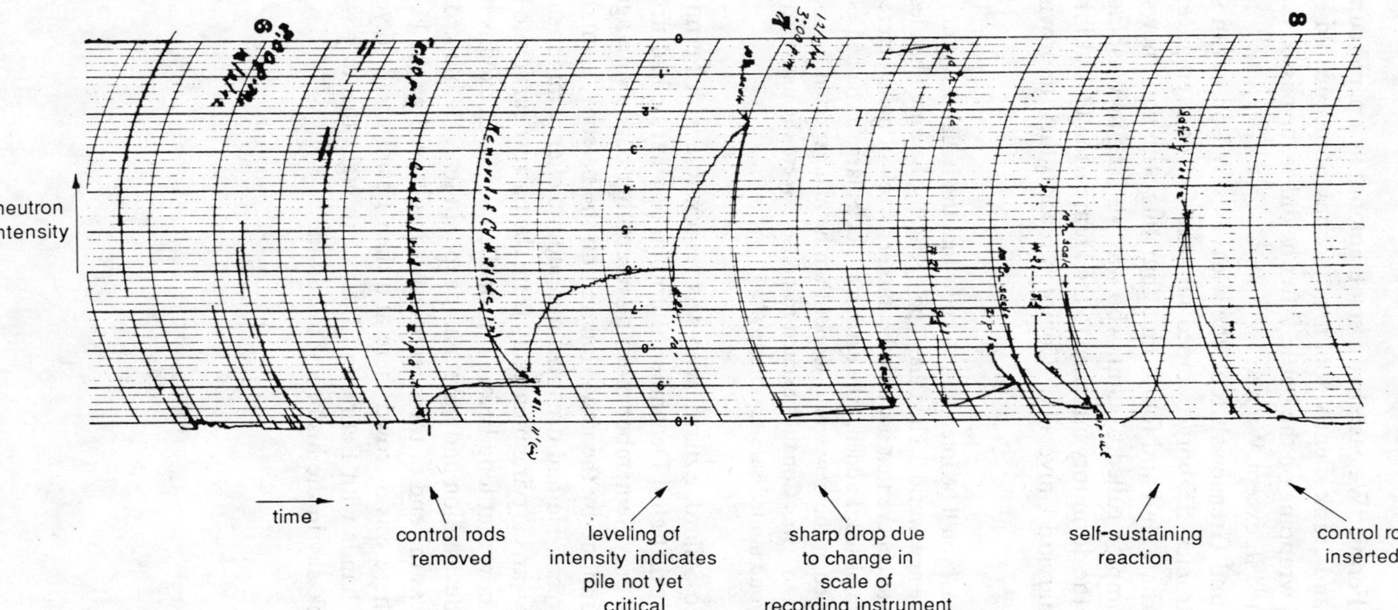

Neutron intensity in the pile as recorded by the chart recorder

presented it to Fermi. "We each had a small amount in a paper cup," Wattenberg says, "and drank silently, looking at Fermi. Someone told Fermi to sign the [straw] wrapping on the bottle. After he did so, he passed it around, and we all signed it, except Wigner."

Compton and Greenewalt took their leave as Wilson began shutting down the electronics. Seaborg bumped into the Du Pont engineer in the corridor of Eckhart Hall and found him "bursting with good news." Back in his office Compton called Conant, who was working in Washington "in my quarters in the dormitory attached to the Dumbarton Oaks Library and Collection of Harvard University." Compton records their improvised dialogue:

> "Jim," I said, "you'll be interested to know that the Italian navigator has just landed in the new world." Then, half apologetically, because I had led the S-1 Committee to believe that it would be another week or more before the pile could be completed, I added, "the earth was not as large as he had estimated, and he arrived at the new world sooner than he had expected."
>
> "Is that so," was Conant's excited response. "Were the natives friendly?"
>
> "Everyone landed safe and happy."

Except Leo Szilard. Szilard, who was responsible with Fermi for the accomplishment that chill December afternoon of what he had first imagined alone on a gray September morning in another country an age ago—the old world undone by the new—loitered on the balcony, a small round man in an overcoat. He had dreamed that atomic energy might substitute exploration for war, carrying men away from the narrow earth into the cosmos. He knew now that long before it propelled any such exodus it would increase war's devastation and mire man deeper in fear. He blinked behind his glasses. It was the end of the beginning. It might well be the beginning of the end. "There was a crowd there and then Fermi and I stayed there alone. I shook hands with Fermi and I said I thought this day would go down as a black day in the history of mankind."

14

Physics and Desert Country

Robert Oppenheimer was thirty-eight years old in 1942. He had done by then what Hans Bethe calls "massive scientific work." He was known and respected as a theoretician throughout the world of physics. Up to the time of the Berkeley summer study, however, few of his peers seem to have thought him capable of decisive leadership. Though he had matured deeply across the decade of the 1930s, his persistent mannerisms, especially his caustic tongue, may have screened his maturity from his colleagues' eyes. Yet the 1930s shaped Oppenheimer for the work that was now to challenge him.

His distinctive appearance sharpens the memory of an admiring new friend of that decade, a Berkeley professor and translator of French literature named Haakon Chevalier:

> [Oppenheimer] was tall, nervous and intent, and he moved with an odd gait, a kind of jog, with a great deal of swinging of his limbs, his head always a little to one side, one shoulder higher than the other. But it was the head that was the most striking: the halo of wispy black curly hair, the fine, sharp nose, and especially the eyes, surprisingly blue, having a strange depth and intensity, and yet expressive of a candor that was altogether disarming. He looked like a young Einstein, and at the same time like an overgrown choir boy.

Chevalier's portrait identifies Oppenheimer's youthfulness and sensitivity but misses the self-destructiveness: the chain-smoking, the persistent cough persistently ignored, the ravaged teeth, the usually empty stomach assaulted by highly praised martinis and highly spiced food. Oppenheimer's emaciation suggests he had an aversion to incorporating the world. His body embarrassed him and he seldom allowed himself to appear, as at the beach, undressed. At school he wore gray suits, blue shirts and well-polished black shoes. At home (a small spare apartment at first; later, after his marriage, the elegant house in the Berkeley hills he bought with a check the day he first toured it) he preferred jeans and blue chambray work shirts, the jeans hung on his narrow hips with a wide Western silver-buckled belt. It was not a common look in the 1930s—he had picked it up in New Mexico—and it was another detail that made him seem different.

Women thought him handsome and dashing. Before a party he might send gardenias not only to his own date but to his friends' dates as well. "He was great at a party," a female acquaintance of his later adulthood comments, "and women simply loved him." His unfailing attentiveness probably elicited that admiration: "He was always," writes Chevalier, "without seeming effort, aware of, and responsive to, everyone in the room, and was constantly anticipating unspoken wishes."

Men he could antagonize or amuse. Edward Teller first met Oppenheimer in 1937. The meeting, Teller says, was "painful but characteristic. On the evening I was to talk at a Berkeley colloquium, he took me out to a Mexican restaurant for dinner. I didn't have the practice in speaking that I've had since, and I was already a little nervous. The plates were so hot, and the spices were so hot—as you might suspect if you knew Oppenheimer—and his personality was so overpowering, that I lost my voice." Emilio Segrè notes that Oppenheimer "sometimes appeared amateurish and snobbish." Out of curiosity in 1940, while visiting Berkeley to deliver a lecture, Enrico Fermi attended a seminar one of Oppenheimer's protégés led in the master's style. "Emilio," Fermi joked afterward with Segrè, "I am getting rusty and old. I cannot follow the highbrow theory developed by Oppenheimer's pupils anymore. I went to their seminar and was depressed by my inability to understand them. Only the last sentence cheered me up; it was: 'and this is Fermi's theory of beta decay.' " Although Segrè found Oppenheimer "the fastest thinker I've ever met," with "an iron memory . . . brilliance and solid merits," he also saw "grave defects" including "occasional arrogance . . . [that] stung scientific colleagues where they were most sensitive." "Robert could make people feel they were fools," Bethe says simply. "He made me, but I didn't mind. Lawrence did. The two disagreed while they were both still at Berkeley. I think Robert would give Lawrence a feeling that he didn't know physics, and since that is what cyclotrons are for, Lawrence didn't like it." Oppenheimer recognized the habit without diagnosing it in a letter to his younger brother Frank: "But it is not easy—at least it is not easy for me—to be quite free of the desire to browbeat somebody or something." He called the behavior "beastliness." It did not win him friends.

Oppenheimer's mother died after a long battle with leukemia in late 1931; that was when he announced himself to Herbert Smith, his former Ethical Culture teacher, to be "the loneliest man in the world." His father died suddenly of a heart attack in 1937. The two deaths frame the beginning years of the unworldly physicist's discovery of the suffering in the world. Later he testified to the surprise of that discovery:

> My friends, both in Pasadena and in Berkeley, were mostly faculty people, scientists, classicists, and artists. I studied and read Sanskrit with Arthur Ryder. I read very widely, mostly classics, novels, plays, and poetry; and I read something of other parts of science. I was not interested in and did not read about economics or politics. I was almost wholly divorced from the contemporary scene in this country. I never read a newspaper or a current magazine like *Time* or *Harper's;* I had no radio, no telephone; I learned of the stock market crash in the fall of 1929 only long after the event; the first time I ever voted was in the Presidential election of 1936. To many of my friends, my indifference to contemporary affairs seemed bizarre, and they often chided me with being too much of a highbrow. I was interested in man and his experience; I was deeply interested in my science; but I had no understanding of the relations of man to his society. . . .
>
> Beginning in late 1936, my interests began to change.

Oppenheimer reports three reasons for the change. "I had had a continuing, smouldering fury about the treatment of the Jews in Germany," he mentions first. "I had relatives there, and was later to help in extricating them and bringing them to this country." They arrived only a few days after his father's death and he and Frank volunteered responsibility for them.

Second, says Oppenheimer, "I saw what the Depression was doing to my students." Philip Morrison, one of the wittiest of the young theoreticians, polio-crippled and poor, remembers in compensation the "very grave, very profound involvement in physics, the love of the whole thing, which we all had in those days." Oppenheimer could take his admiring students to dinner; he was unable to find them jobs. "And through them," he testifies, "I began to understand how deeply political and economic events could affect men's lives. I began to feel the need to participate more fully in the life of the community."

He had no framework yet. A woman would help him with that, her involvement the third reason he gives for his entry into the world: Jean Tatlock, the lithe, chiaroscuro daughter of an anti-Semite Berkeley medievalist. "In the autumn [of 1936], I began to court her, and we grew close to each other. We were at least twice close enough to marriage to think of ourselves as engaged." Tatlock was bright, passionate and compassionate, frequently depressed; their relationship was an ocean of storms. But so were Tatlock's other commitments. "She told me about her Communist Party memberships; they were on again, off again affairs, and never seemed to provide for her what she was seeking." The couple began to move together among what he calls "leftwing friends. . . . I liked the new sense of companionship, and at the time felt that I was coming to be part of the life of my time and country." He was taken with the causes of the Loyalists in the Spanish Civil War and the migrant workers in California, to both of which he contributed time and money. He read Engels and Feuerbach and all of Marx, finding their dialectics less rigorous than his taste: "I never accepted Communist dogma or theory; in fact, it never made sense to me."

He met his wife, Kitty, in the summer of 1939 in Pasadena. She was petite and dark, with a broad, high forehead, brown eyes, prominent cheekbones and a wide, expressive mouth. On the rebound she had married a young British physician, "Dr. [Stewart] Harrison, who was a friend and associate of the [Richard] Tolmans, [Charles C.] Lauritsens, and others of the California Institute of Technology faculty [Harrison was doing cancer research]. I learned of her earlier marriage to Joe Dallet, and of his death fighting in Spain. He had been a Communist Party official, and for a year or two during their brief marriage my wife was a Communist Party member. When I met her I found in her a deep loyalty to her former husband, a complete disengagement from any political activity, and a certain disappointment and contempt that the Communist Party was not in fact what she had once thought it was." The involvement was apparently immediate and intense.

Probably with his wife's encouragement, but certainly with his own growing good sense, Oppenheimer began to jettison political commitments that had come to seem parochial. "I went to a big Spanish relief party the night before Pearl Harbor," he testifies in example, "and the next day, as we heard the news of the outbreak of war, I decided that I had had about enough of the Spanish cause, and that there were other and more pressing crises in the world." He was willing similarly to abandon the American Association of Scientific Workers at Lawrence's insistence in order to help, as he supposed, to beat the Nazis to the atomic bomb.

By then, says Bethe, though Oppenheimer had been a poor teacher when he began, pitching quantum theory well above his students' untrained range, he had "created the greatest school of theoretical physics that the United States has ever known." Bethe's explanation for that evolution reveals the seedbed of Oppenheimer's later administrative leadership:

> Probably the most important ingredient he brought to his teaching was his exquisite taste. He always knew what were the important problems, as shown by his choice of subjects. He truly lived with those problems, struggling for a solution, and he communicated his concern to his group. . . . He was interested in everything, and in one afternoon [he and his students] might discuss quantum electrodynamics, cosmic rays, electron pair production and nuclear physics.

During the same period Oppenheimer's clumsiness with experiment evolved to appreciation and he consciously mastered experimental work—hands off. "He began to observe, not manipulate," a former student notes. "He learned to see the apparatus and to get a feeling of its experimental limitations. He grasped the underlying physics and had the best memory I know of. He could always see how far any particular experiment would go. When you couldn't carry it any farther, you could count on him to understand and to be thinking about the next thing you might want to try."

It remained for Oppenheimer to learn to control his "beastliness" and submerge his mannerisms. But he was always a quick study. Significantly, he was least convoluted, most direct, least mannered, most natural living simply at his unadorned ranch in the Pecos Valley high in the Sangre de Cristo Mountains of northern New Mexico.

Oppenheimer first met General Leslie R. Groves when Groves came to Berkeley from Chicago on his initial inspection tour early in October 1942. They attended a luncheon given by the president of the university; afterward they talked. Oppenheimer had already discussed the need for a fast-neutron laboratory at the Met Lab technical council meeting on September 29. He envisioned more responsibilities for that laboratory than basic fission studies, as he testified after the war:

> I became convinced, as did others, that a major change was called for in the work on the bomb itself. We needed a central laboratory devoted wholly to this purpose, where people could talk freely with each other, where theoretical ideas and experimental findings could affect each other, where the waste and

> frustration and error of the many compartmentalized experimental studies could be eliminated, where we could begin to come to grips with chemical, metallurgical, engineering, and ordnance problems that had so far received no consideration.

Memory compresses the laboratory's evolution here, however; Oppenheimer is not likely to have discussed eliminating Groves' cherished compartmentalization at their first meeting. To the contrary, he goes on to say, the two men first considered making the laboratory "a military establishment in which key personnel would be commissioned as officers," and he carried the idea far enough before he left Berkeley to visit a nearby military post to begin the process of commissioning.

Groves remembers that his "original impression gained from our first conversation in Berkeley" was that a central laboratory was a good idea; he felt strongly that "the work [of bomb design] should be started at once in order that one part of our operation, at any rate, could progress at what I hoped would be a comfortable pace." His immediate concern was leadership; he believed that the right man at the helm could sail even the most ungovernable boat. Ernest Lawrence would have been Groves' first choice, but the general doubted if anyone else could make electromagnetic isotope separation work. Compton had his hands full in Chicago. Harold Urey was a chemist. "Outside the project there may have been other suitable people, but they were all fully occupied on essential work, and none of those suggested appeared to be the equal of Oppenheimer." Groves had already sized up his man.

"It was not obvious that Oppenheimer would be [the new laboratory's] director," Bethe notes. "He had, after all, no experience in directing a large group of people. The laboratory would be devoted primarily to experiment and to engineering, and Oppenheimer was a theorist." Worse—in the eyes of the project leaders, Nobel laureates all—he had no Nobel Prize to distinguish him. There was also what Groves calls the "snag" of Oppenheimer's left-wing background, which "included much that was not to our liking by any means." Groves had not yet wrested control of Manhattan Project security from Army counterintelligence, and that organization adamantly refused to clear someone whose former fiancée, wife, brother and sister-in-law had all been members of the Communist Party once and perhaps, gone underground, still were.

The general wanted Oppenheimer anyway. "He's a genius," Groves told an interviewer off the record immediately after the war. "A real genius. While Lawrence is very bright he's not a genius, just a good hard worker. Why, Oppenheimer knows about everything. He can talk to you about anything you bring up. Well, not exactly. I guess there are a few things he doesn't know about. He doesn't know anything about sports."

Groves proposed Oppenheimer's name to the Military Policy Committee. It balked. "After much discussion I asked each member to give me the name of a man who would be a better choice. In a few weeks it became apparent that we were not going to find a better man; so Oppenheimer was asked to undertake the task." The physicist demurred later that he was chosen "by default. The truth is that the obvious people were already taken and that the Project had a bad name." Rabi would come to think that "it was a real stroke of genius on the part of General Groves, who was not generally considered to be a genius, to have appointed him," but at the time it seemed "a most improbable appointment. I was astonished." Groves on his way from Chicago to New York asked Oppenheimer on October 15, 1942, to ride on the train with him as far as Detroit to discuss the appointment. The two men met with Vannevar Bush in Washington on October 19. That long meeting was apparently decisive. Security questions would have to wait.

The next problem was where to locate the new laboratory. Already at his first meeting with Oppenheimer in Berkeley, Groves had stressed the need for isolation; however much or little the scientists who gathered at the new center would be allowed to talk to each other, the general intended to divide them away from the populace. "For this reason," Oppenheimer wrote his Illinois colleague John H. Manley in mid-October, "some rather far reaching geographical change in plans seems to be in the cards." (In the same letter Oppenheimer proposed "start[ing] now on a policy of absolutely unscrupulous recruiting of anyone we can lay hands on." He wanted the best he could get, and soon asked Groves for the likes of Bethe, Segrè, Serber and Teller.)

Site Y, as the hypothetical laboratory was initially called, needed good transportation, an adequate supply of water, a local labor force and a moderate climate for year-round construction and for experiment conducted outdoors. In his memoirs Groves lists safety as the primary reason he insisted on isolation— "so that nearby communities would not be adversely affected by any unforeseen results from our activities" —but the high steel fence topped with triple strands of barbed wire that eventually surrounded the laboratory was clearly not designed to confine explosions. Groves was in the midst of selecting sites for Manhattan Project production centers; the difference between his criteria for those locations and his criteria for Site Y was that at the bomb-design laboratory "we were faced with the necessity of importing a group of highly talented specialists, some of whom would be prima donnas, and of keeping them satisfied with their working and living conditions." If that in fact was Groves' intention, it was one of the few wartime goals he failed to achieve.

The general assigned the task of identifying a suitable location for the laboratory to Major John H. Dudley of the Manhattan Engineer District. Groves gave Dudley criteria more specific than satisfying prima donnas: room for 265 people, location at least two hundred miles from any international boundary but west of the Mississippi, some existing facilities, a natural bowl with the hills nearby that shaped the bowl so that fences might be strung on top and guarded. Traveling by air, rail, auto, jeep and horse through most of the American Southwest, Dudley found the perfect place: Oak City, Utah, "a delightful little oasis in south central Utah." But to claim it the Army would have had to evict several dozen families and remove a large area of farmland from production. Dudley thereupon recommended his second choice: Jemez Springs, New Mexico, a deep canyon about forty miles northwest of Santa Fe on the western slope of the Jemez Mountains— "a lovely spot," Oppenheimer thought in early November before he toured it, "and in every way satisfactory."

When the newly appointed director arrived on November 16 to inspect the Jemez Springs location with Dudley and Edwin McMillan, who was helping start the laboratory, he changed his mind. The canyon felt confining; Oppenheimer knew the region's grand scenic vistas and decided he wanted a laboratory with a view. McMillan also remembers expressing "considerable reservations about this site":

> We were arguing [with Dudley] when General Groves showed up. This had been planned. He would come in sometime in the afternoon and receive our report. As soon as Groves saw the site he didn't like it; he said, "This will never do." . . . At that point Oppenheimer spoke up and said "if you go on up the canyon you come out on top of the mesa and there's a boys' school there which might be a usable site."

Oppenheimer proposed the boys' school site, grouses Dudley, "as though it was a brand new idea." Dudley had already scouted the mesa twice, rejecting it because it failed to meet Groves' criteria. But a mesa is an inverted bowl, its perimeter similarly fencible. And the first requirement was to make the longhairs happy. "As I . . . knew the roads (or trails)," Dudley says sardonically, ". . . we drove directly there."

"The school was called Los Alamos," the daughter of its founder writes, "after the deep canyon which bordered the mesa to the south and which was groved with cottonwood trees along the sandy trickle of its stream." Ashley Pond, the founder, had been a sickly boarding-school boy sent West for his health, like Oppenheimer, who returned to New Mexico in later adulthood when his father died and left him with independent means. He opened the Los Alamos Ranch School on the 7,200-foot mesa in 1917. It was organized to invigorate pale scions, as Pond had been invigorated: boys slept on unheated porches of a chinked-log dormitory and wore shorts in winter snow; each was assigned a horse to ride and groom. It was, Emilio Segrè writes, "beautiful and savage country" : the dark Jemez Mountains to the west that formed the higher rim of the Jemez Caldera, the slumped cone of the old volcano of which Los Alamos was eroded tuffaceous spill; precipitously down from the mesa eastward the valley of the Rio Grande, "hot and barren" except for the green meander of the river, writes Laura Fermi, with "sand, cacti, a few piñon trees hardly rising above the ground, and space, immense, transparent, with no fog or moisture"; farther east the wall of the Rocky Mountains as that range extends south into New Mexico to form the Sangre de Cristo, reversing hue from green to red progressively at sunset. "I remember arriving [at Los Alamos]," McMillan continues of that first inspection, "and it was late in the afternoon. There was a slight snow falling. . . . It was cold and there were the boys and their masters out on the playing fields in shorts. I remarked that they really believed in hardening up the youth. As soon as Groves saw it, he said, in effect, 'This is the place.' "

"My two great loves are physics and desert country," Robert Oppenheimer had written a friend once; "it's a pity they can't be combined." Now they would be.

Leo Szilard, urban man, habitué of hotel lobbies, took a different view of the location when he heard about it. "Nobody could think straight in a place like that," he told his Met Lab colleagues. "Everybody who goes there will go crazy." The Corps of Engineers' appraisal prepared on November 21 describes a large forested site thirty-five miles by road northwest of Santa Fe with no gas or oil lines, one one-wire Forest Service telephone, average annual precipitation of 18.53 inches and an annual range of temperatures from −12° to 92°F. The land and improvements, including the boys' school with its sixty horses, two tractors, two trucks, fifty saddles, eight hundred cords of firewood, twenty-five tons of coal and sixteen hundred books, were worth $440,000. The school was willing to sell. The Manhattan Project acquired its scenic laboratory site.

Groves convinced the University of California to serve as contractor to operate the secret installation. Construction—of cheap, barracks-like buildings not intended to outlast the war, with coal-burning stoves and no sidewalks on which to escape the mire of spring and autumn mud—began almost immediately. "What we were trying to do," writes John Manley, the

University of Illinois physicist working with Oppenheimer then, "was build a new laboratory in the wilds of New Mexico with no initial equipment except the library of Horatio Alger books or whatever it was that those boys in the Ranch School read, and the pack equipment that they used going horseback riding, none of which helped us very much in getting neutron-producing accelerators." Robert R. Wilson, a young Berkeley Ph.D. teaching at Princeton, went up to Harvard for Oppenheimer and negotiated with Percy Bridgman for the Harvard cyclotron; Wisconsin would contribute two Van de Graaffs; from other laboratories, including Berkeley and the University of Illinois, Manley scavenged other gear. In the meantime Oppenheimer crisscrossed the country recruiting:

> The prospect of coming to Los Alamos aroused great misgivings. It was to be a military post; men were asked to sign up more or less for the duration; restrictions on travel and on the freedom of families to move about would be severe.... The notion of disappearing into the New Mexico desert for an indeterminate period and under quasi-military auspices disturbed a good many scientists, and the families of many more. But there was another side to it. Almost everyone realized that this was a great undertaking. Almost everyone knew that if it were completed successfully and rapidly enough, it might determine the outcome of the war. Almost everyone knew that it was an unparalleled opportunity to bring to bear the basic knowledge and art of science for the benefit of his country. Almost everyone knew that this job, if it were achieved, would be a part of history. This sense of excitement, of devotion and of patriotism in the end prevailed. Most of those with whom I talked came to Los Alamos.

One of the most tough-minded, I. I. Rabi, did not. His reasons are revealing. He continued developing radar at the Radiation Laboratory at MIT. "Oppenheimer wanted me to be the associate director," he told an interviewer many years later. "I thought it over and turned him down. I said, 'I'm very serious about this war. We could lose it with insufficient radar.' " The Columbia physicist thought radar more immediately important to the defense of his country than the distant prospect of an atomic bomb. Nor did he choose to work full time, he told Oppenheimer, to make "the culmination of three centuries of physics" a weapon of mass destruction. Oppenheimer responded that he would take "a different stand" if he thought the atomic bomb would serve as such a culmination. "To me it is primarily the development in time of war of a military weapon of some consequence." Either Oppenheimer had not yet thought his way through to a more millenarian view of the new weapon's implications or he chose to avoid discussing those implications with Rabi. He asked Rabi only to participate in an inaugural physics conference at Los Alamos in April 1943 and to help convince others, particularly Hans Bethe, to sign on. Eventually Rabi would come and go as a visiting consultant, one of the very few exceptions to Groves' compartmentalization and isolation rules.

Oppenheimer talked to the Bethes in Cambridge in snowy New England December; they questioned him at length about the life they would be asked to lead. Extracts from his letter of response sketch the invention of an instant community: "Laboratory . . . town . . . utilities, schools, hospitals . . . a sort of city manager . . . city engineer . . . teachers . . . M.P. camp . . . a laundry . . . two eating places . . . a recreation officer . . . libraries, pack trips, movies . . . bachelor apartments . . . a so-called Post Exchange . . . a vet . . . barbers and such like . . . a cantina where we can have beer and cokes and light lunches." The Bethes' best guarantee of satisfaction, Oppenheimer concluded, "is in the great effort and generosity that . . . Groves [has] brought to setting up this odd community and in [Groves'] evident desire to make a real success of it. In general [he is] not interested in saving money, but . . . in saving critical materials, in cutting down personnel, and in doing nothing which would attract Congressional attention to our hi-jinks." He chose not to mention the security arrangements, in the development of which he was participating: the perimeter fence, the pass controls, the virtual elimination of telephones ("Oppenheimer's idea was one telephone for himself," says Dudley, "one for the post commander, and any volume business would go out over a teletype."). By March Teller found Bethe taking "a very optimistic view, and there was no need whatever to persuade him to come."

Teller felt underemployed in Chicago and was eager to move to the new laboratory. John Manley asked him to write a prospectus to help with recruiting, which Teller sent to Oppenheimer in early January. During the Berkeley summer study the two men had begun what another participant judged a "mental love affair." Teller "liked and respected Oppie enormously. He kept wanting to talk about him with others who knew him, kept bringing up his name in conversation." Bethe noticed then and later that despite their many outward differences Teller and Oppenheimer were "fundamentally . . . very similar. Teller had an extremely quick understanding of things, so did Oppenheimer. . . . They were also somewhat alike in that their actual production, their scientific publications, did not measure up in any way to their capacity. I think Teller's mental capacity is very high, and so was Oppenheimer's but, on the other hand, their papers, while they included some very good ones, never reached really the top standards. Neither of them ever came up to the Nobel Prize level. I think you just cannot get to that level unless you are somewhat introverted." (Luis Alvarez, the 1968 physics Nobel laureate, disagrees, at least where Oppenheimer is concerned. He believes Oppenheimer would have won a Nobel Prize for his astrophysical work if he had lived long enough to see his predictions concerning exotic stellar objects—neutron stars, black holes—confirmed, as they have been, by discovery.) Both Oppenheimer and Teller wrote poetry; Oppenheimer pursued literature as Teller pursued music; and for a time in 1942 and 1943 the Hungarian apparently admired the older and socially more sophisticated New Yorker and hoped to count him for an ally.

As Oppenheimer traveled the country recruiting he discovered to his surprise that few of his colleagues were attracted to the notion of joining the Army. It fell to Rabi and his Rad Lab colleague Robert F. Bacher, during the weeks before Rabi decided to stay in Cambridge, to lead the revolt. The necessity of "scientific autonomy" was one crucial reason they cited for resisting militarization, Oppenheimer wrote Conant at the beginning of February 1943, and they insisted as a corollary that although "the execution of the security and secrecy measures should be in the hands of the military . . . the decision as to what measures should be applied must be in the hands of the Laboratory." On that point Oppenheimer concurred, "because I believe it is the only way to assure the cooperation and the unimpaired morale of the scientists." The stakes were higher than simply losing Rabi and Bacher, Oppenheimer told Conant: "I believe that the solidarity of physicists is such that if these conditions are not met, we shall not only fail to have the men from M.I.T with us, but that many men who have already planned to join the new Laboratory will reconsider their commitments or come with such misgivings as to reduce their usefulness." A rebellion, he concluded, would mean "a real delay in our work."

Groves had wanted the scientists commissioned as a security measure and because their work might be hazardous. He was hardly interested in the politics of the question, but delay was unthinkable. He compromised. Conant wrote a letter, co-signed by Groves, that Oppenheimer could use in recruiting; it allowed the new laboratory civilian administration and civilian staff until the time of hazardous large-scale trials. Then anyone who wanted to stay would have to accept a commission (a provision Groves chose later not to pursue). The Army would administer the community it was building around the laboratory. Laboratory security would be Oppenheimer's responsibility, and he would report to Groves.

Robert Oppenheimer thus acquired for Los Alamos what Leo Szilard had not been able to organize in Chicago: scientific freedom of speech. The price the new community paid, a social but more profoundly a political price, was a guarded barbed-wire fence around the town and a second guarded barbed-wire fence around the laboratory itself, emphasizing that the scientists and their families were walled off where knowledge of their work was concerned not only from the world but even from each other. "Several of the European-born were unhappy," Laura Fermi notes, "because living inside a fenced area reminded them of concentration camps."

The heavy-water installation at Vemork in southern Norway became a target of British sabotage operations in the winter of 1942–43. The British had been planning to send in two glider-loads of demolition experts, thirty-four trained volunteers; when Groves requested Allied action soon after his appointment to administer the Manhattan Project they moved ahead to comply. An advance party of four Norwegian commandos parachuted into the Rjukan area on October 18 to prepare the way, but bad planning and bad weather brought disaster to the gliders on the night of November 19 when they crossed the North Sea from Scotland; both crashed in Norway, one into a mountainside, and the fourteen men who survived the separate disasters were captured by German occupation forces and executed the same day.

R. V. Jones, an Oxford protégé of Cherwell who was now director of intelligence for the British Air Staff, then had "one of the most painful decisions that I had to make" —whether to send another demolition party after the first. "I reasoned that we had already decided, before the tragedy of the first raid and therefore free from sentiment, that the heavy water plant must be destroyed; casualties must be expected in war, and so if we were right in asking for the first raid we were probably right in asking that it be repeated."

This time six men, Norwegians native to the region and trained as Special Forces, parachuted onto a frozen lake thirty miles northwest of Vemork on February 16, 1943, the night of a full moon. "Here lay the Hardanger Vidda," one of them, Knut Haukelid, writes of the high plateau that surrounded the lake, "the largest, loneliest and wildest mountain area in northern Europe." The men wore white jumpsuits over British Army uniforms and parachuted with skis, supplies, a shortwave radio and eighteen sets of plastic explosives, one for each of the eighteen stainless-steel electrolysis cells of the High Concentration Plant—which happened to have been designed by a refugee physical chemist, Lief Tronstad, who was now responsible to the Norwegian High Command in London for intelligence and sabotage. Haukelid, a powerfully built mountaineer, says they weathered "one of the worst storms I have ever experienced in the mountains" to rendezvous some days later with the four Norwegians of the original advance party, who had been forced to hide out on the barren Hardanger

Vidda and were malnourished and weak. The new arrivals fattened up their compatriots while one of them skied on to Rjukan to gather the latest information about the plant. He returned to report minefields laid around the obvious approaches, guards on the suspension bridge that crossed the sheer gorge above the shelf on which the hydrochemical facility was built but only fifteen German soldiers on duty despite the forewarning of the failed glider attack. The factory itself was fitted with searchlights and guarded with machine guns.

The commandos set out mid-evening on Saturday, February 27, leaving one man behind to guard the radios. They carried cyanide capsules and agreed that if anyone was wounded he would take his own life rather than allow himself to be captured and risk betraying his comrades. They had camped high on the mountain across the gorge from the plant, which was located to take advantage of the fall of water from the lake that fed it, Tinnsjö. "Halfway down we sighted our objective for the first time, below us on the other side. The great seven-storey factory building bulked large on the landscape. . . . [The wind] was blowing fairly hard, but nevertheless the hum of the machinery came up to us through the ravine. We understood how the Germans could allow themselves to keep so small a guard there. The colossus lay like a mediaeval castle, built in the most inaccessible place, protected by precipices and rivers."

They crashed down through soft snow all the way to the bottom of the gorge, crossed the frozen river, climbed up toward the plant on the other side. Above at the elevation of the shelf was a seldom-used railroad siding leading into the compound that they hoped the Germans had chosen not to mine. "It was a dark night and there was no moon," Haukelid remembers. The searchlights were kept turned off and the high wind "drowned all the noise we made. Half an hour before midnight we came to a snow-covered building five hundred yards from Vemork, where we ate a little chocolate and waited for the change of sentries." They divided into two groups, a demolition party and a covering party. "We were well armed: five tommy-guns among nine men, and everyone had a pistol, a knife and hand grenades."

In an hour, time for the sentries to settle, they attacked. Haukelid in the covering party led the way. With bolt cutters they snipped "the thin little iron chain which barred the way to one of the most important military objectives in Europe." The covering party dispersed to its prearranged positions—Haukelid and one other man took up posts twenty yards from the Wehrmacht barracks, a flimsy wooden building they saw they could easily shoot through—and the demolition party moved ahead. The doors on the ground floor of the plant were locked, but Tronstad in London had identified for the commandos a cable intake that they could crawl along that led directly to the heavy-water facility. Two men looked for some other entrance while two disappeared into the cable intake.

After what seemed to Haukelid an interminable delay he heard an explosion, "but an astonishingly small, insignificant one. Was this what we had come over a thousand miles to do?" The guards were slow to check; only one German soldier appeared and seemed not to realize what had happened; he tried the doors to the plant, found them locked, looked to see if snow falling from the mountain above had detonated a land mine and returned to his quarters. The Norwegians moved out fast. They had descended to the river before the sirens began to sound.

The operation was successful. No one was injured on either side. All eighteen cells had been blown open, spilling nearly half a ton of heavy water into the drains. Not only would the plant require weeks to repair; because it was a cascade, pumping water of increasing deuterium concentration from one cell to the next, it would need almost a year of operation after repair simply to reach equilibrium again on its own and begin producing. General Nikolaus von Falkenhorst, the commander in chief of the occupying German Army in Norway, called the Vemork attack "the best coup I have ever seen." Whatever German physicists might be doing with heavy water, they would do it more slowly now.

In Japan both the Army Air Force and the Imperial Navy had moved separately since 1941 to promote atomic bomb research. The Riken, Yoshio Nishina's prestigious Tokyo laboratory, primarily served the Army, exploring the theoretical possibilities of U235 separation by way of the gaseous barrier diffusion, gaseous thermal diffusion, electromagnetic and centrifuge processes. In the spring of 1942 the Navy committed itself to developing nuclear power for propulsion:

> The study of nuclear physics is a national project. Research in this field is continuing on a broad scale in the United States, which has recently obtained the services of a number of Jewish scientists, and considerable progress has been made. The objective is the creation of tremendous amounts of energy through nuclear fission. Should this research prove successful, it would provide a stupendous and dependable source of power which could be used to activate ships and other large pieces of machinery. Although it is not expected that nuclear energy will be realized in the near future, the possibility of it must not be ignored. The Imperial Navy, accordingly, hereby affirms its determination to foster and assist studies in this field.

Soon after that nonviolent affirmation, however, the Naval Technological Research Institute appointed a secret committee of leading Japanese scientists—corresponding to the U.S. National Academy of Sciences committee—to meet monthly to follow research progress until it could report decisively for or against a Japanese atomic bomb. The committee included Nishina, who was forthwith elected chairman. An elderly appointee was Hantarō Nagaoka, whose Saturnian atomic model had nearly anticipated Ernest Rutherford's planetary model in the early years of the century.

The Navy committee met first on July 8 with the Navy's chief technical officers at an officers' club at Shiba Park in Tokyo. It noted that the United States was probably working on a bomb and agreed that whether and how soon Japan could produce such a weapon was as yet uncertain. To the task of answering those questions the Navy appropriated 2,000 yen, about $4,700, somewhat less than the Uranium Committee had summoned from the U.S. Treasury at Edward Teller's request at the beginning of the American program in 1939.

Nishina hardly participated in the Navy committee meetings. The fact that he was already working for the Army probably constrained him; the two services, both of which were responsible directly to the Emperor without detour through the civilian government, operated far more independently than their American counterparts and were increasingly bitter rivals. Nishina was coming to conclusions of his own, however, and at the end of 1942, when the Navy committee began to report discouragement, he met privately with a young cosmic-ray physicist in his laboratory, Tadashi Takeuchi, told his young colleague he meant to carry forward isotope separation studies and asked him to help. Takeuchi agreed.

Between December 1942 and March 1943 the Navy committee organized a ten-session physics colloquium to work through to a decision. By then it was understood that a bomb would necessitate locating, mining and processing hundreds of tons of uranium ore and that U235 separation would require a tenth of the annual Japanese electrical capacity and half the nation's copper output. The colloquium concluded that while an atomic bomb was certainly possible, Japan might need ten years to build one. The scientists believed that neither Germany nor the United States had sufficient spare industrial capacity to produce atomic bombs in time to be of use in the war.

After the final March 6 meeting the Navy representative at the colloquium reported discouragement: "The best minds of Japan, studying the subject from the point of view of their respective fields of endeavor as well as from that of national defense, came to a conclusion that can only be regarded as correct. The more they considered and discussed the problem, the more pessimistic became the atmosphere of the meeting." As a result the Navy dissolved the committee and asked its members to devote themselves to more immediately valuable research, particularly radar.

Nishina continued isotope studies for the Army, deciding on March 19 to focus on thermal diffusion as the only practical separation technology at a time of increasing national shortages. He spoke to his staff of processing several hundred tons of uranium after first building laboratory-scale diffusion apparatus. He envisioned a major program run in parallel, as the Manhattan Project was beginning to be, with weapon design and development proceeding simultaneously with U235 production.

Meanwhile a different branch of the Navy, the Fleet Administration Center, sponsored a new project in atomic bomb development at the University of Kyoto, where Tokutaro Hagiwara had made his startling early prediction of the possibility of a thermonuclear explosive. The university won support in 1943 to the extent of 600,000 yen—nearly $1.5 million—much of which it budgeted to build a cyclotron.

Robert Oppenheimer moved to Santa Fe with a small team of aides on March 15, 1943, brisk early spring. Scientists and their families arrived by automobile and train during the next four weeks. Not much was ready on the mesa, which they began to call the Hill. Groves wanted no breaches of security in the lobbies of Santa Fe hotels; the Army commandeered guest ranches in the area for quarters suitably remote and bought up Santa Fe's feeble stock of used cars and jitneys to serve as transportation through ruts and mud up and down the terrifying unbarricaded dirt switchback of the mesa access road. After flat tires and mirings, hours could be short on the Hill. Box lunches assembled in Santa Fe gave cold comfort when the delivery truck made it through.

The hardships only mattered because they slowed the work. Oppenheimer had sold it as work that would end the war to end all wars and his people believed him. The unit of measurement for wasted hours was therefore human lives. Construction crews unwilling to vary the specifications of a laboratory door or hang an unauthorized shelf initially bore the brunt of the scientists' impatience. John Manley remembers inspecting the chemistry and physics building. It needed a basement at one end for an accelerator and a solid foundation at the other end for the two Van de Graaffs—which end for which was unimportant. Rather than adjust the construction plans for terrain the contractor had drilled the basement from solid rock and used the rock debris as fill for the foundation. "This was my introduction to the Army Engineers."

Fuller Lodge, a Ranch School hall elegantly assembled of monumental hand-hewn logs, was kept to serve as a dining room and guest house. The pond south of the lodge—predictably named Ashley Pond after the Ranch School's founder—offered winter ice-skating and summer canoeing and the easeful harmonic wakes of swimming ducks. The engineers preserved the stone icehouse beside the pond that the school had used to store winter cuttings of ice and the row of tree-shaded faculty residences northeast of the lodge. Across the dirt main road that divided the mesa south of the pond the Tech Area went up in a style the Army called modified mobilization: plain one-story buildings like elongated barracks with clapboard sides and shingled roofs. T Building would house Oppenheimer and his staff and the Theoretical Physics Division; behind T, connected by a covered walkway, would be the much longer chemistry and physics building with its Van de Graaffs; behind that the laboratory shops. Farther south near the rim of the mesa above Los Alamos canyon contractors would hammer up a cryogenics laboratory and the building that would shelter Harvard's cyclotron. West and north of the Tech Area the first two-story, four-unit family apartments, painted drab green, urbanized last year's pastures and fields; more apartments, and dormitories for the unmarried, would follow.

At the beginning of April Oppenheimer assembled the scientific staff—"about thirty persons" at that point of the hundred scientists initially hired, says Emilio Segrè, who was one among them—for a series of introductory lectures. Robert Serber, thin and shy, delivered the lectures with authority despite the distraction of a lisp; they summed up the conclusions of the Berkeley summer study and incorporated the experimental fast-fission work of the past year. Edward U. Condon, the crew-cut, Alamogordo-born theoretician from Westinghouse whom Oppenheimer had chosen for associate director, revised his notes of Serber's lectures into the new laboratory's first report, a document called the *Los Alamos Primer* that was subsequently handed to all new Tech Area arrivals cleared for Secret Limited access. In twenty-four mimeographed pages the *Primer* defined the laboratory's program to build the first atomic bombs.

Serber's lectures startled the chemists and experimental physicists whom compartmentalization had kept in the dark; the scientists' euphoria at finally learning in detail what they had only previously guessed or heard hinted measures the extent to which secrecy had contorted their emotional commitment to the work. Now, following the lead of their mentors—their average age was twenty-five; Oppenheimer, Bethe, Teller, McMillan, Bacher, Segrè and Condon were older men—they could apply themselves at last with devotion. In that heady new freedom they seldom noticed the barbed wire. Similarly confined but kept uninformed because Oppenheimer and Groves decided it so, the wives served harder time.

"The object of the project," Condon summarizes what Serber told the scientists, "is to produce a *practical military weapon* in the form of a bomb in which the energy is released by a fast neutron chain reaction in one or more of the materials known to show nuclear fission." Serber said one kilogram of U235 was approximately equal to 20,000 tons of TNT and noted that nature had almost located that conversion beyond human meddling: "Since only the last few generations [of the chain reaction] will release enough energy to produce much expansion [of the critical mass], it is just possible for the reaction to occur to an interesting extent before it is stopped by the spreading of the active material." If fission had proceeded more energetically the bombs would have slept forever in the dark beds of their ores.

Serber discussed fission cross sections, the energy spectrum of secondary neutrons, the average number of secondary neutrons per fission (measured by then to be about 2.2), the neutron capture process in U238 that led to plutonium and why ordinary uranium is safe (it would have to be enriched to at least 7 percent U235, the young theoretician pointed out, "to make an explosive reaction possible"). He was already calling the bomb "the gadget," its nickname thereafter on the Hill, a bravado metonymy that Oppenheimer probably coined. The calculations Serber reported indicated a critical mass for metallic U235 tamped with a thick shell of ordinary uranium of 15 kilograms: 33 pounds. For plutonium similarly tamped the critical mass might be 5 kilograms: 11 pounds. The heart of their atomic bomb would then be a cantaloupe of U235 or an orange of Pu239 surrounded by a watermelon of ordinary uranium tamper, the combined diameter of the two nested spheres about 18 inches. Shaped of such heavy metal the tamper would weigh about a ton. The critical masses would eventually have to be determined by actual test, Serber said.

He went on to speak of damage. Out to a radius of a thousand yards around the point of explosion the area would be drenched with neutrons, enough to produce "severe pathological effects." That would render the area uninhabitable for a time. It was clear by now—it had not been clear before—that a nuclear explosion would be no less damaging than an equivalent chemical explosion. "Since the one factor that determines the damage is the energy release, our aim is simply to get as much energy from the explosion as we can. And since the materials we use are very precious, we are constrained to do this with as high an efficiency as is possible."

Efficiency appeared to be a serious problem. "The reaction will not go to completion in an actual gadget." Untamped, a bomb core even as large as twice the critical mass would completely fission less than 1 percent of its nuclear material before it expanded enough to stop the chain reaction from proceeding. An equally disadvantageous secondary effect also tended to stop the reaction: "as the pressure builds up it begins to blow off material at the outer edge of the [core]." Tamper always increased efficiency; it reflected neutrons back into the core and its inertia—not its tensile strength, which was inconsequential at the pressures a chain reaction would generate—slowed the core's expansion and helped keep the core surface from blowing away. But even with a good tamper they would need more than one critical mass per bomb for reasonable efficiency.

Detonation was equally a problem. To detonate their bombs they would have to rearrange the core material so that its effective neutron number, which corresponded to Fermi's *k*, changed from less than 1 to more than 1. But however they rearranged the material—firing one subcritical piece into another subcritical piece inside the barrel of a cannon seemed to be the simplest option—they would have no slow, smooth transition as Fermi had with CP-1. If they fired one piece into another at the high velocity of 3,000 feet per second it would take the pieces about a thousandth of a second to assemble themselves. But since more than one critical mass was necessary for an efficient explosion the pieces would be supercritical before they had completely mated. If a stray neutron then started a chain reaction, the resulting inefficient explosion would proceed from beginning to end in a few millionths of a second. "An explosion started by a premature neutron will be all finished before there is time for the pieces to move an appreciable distance." Which meant that the neutron background—spontaneous-fission neutrons from the tamper, neutrons knocked from light-element impurities, neutrons from cosmic rays—would have to be kept as low as possible and the rearrangement of the core material managed as fast as possible. On the other hand, they did not have to worry that a fizzle would drop an intact bomb into enemy hands; even a fizzle would release energy equivalent to at least sixty tons of TNT.

Predetonation would reduce the bomb's efficiency, Serber repeated; so also might postdetonation. "When the pieces reach their best position we want to be very sure that a neutron starts the reaction before the pieces have a chance to separate and break." So there might be a third basic component to their atomic bomb besides nuclear core and confining tamper: an initiator—a Ra + Be source or, better, a Po + Be source, with the radium or polonium attached perhaps to one piece of the core and the beryllium to the other, to smash together and spray neutrons when the parts mated to start the chain reaction.

Firing the pieces of core together, the Berkeley theoretician continued, "is the part of the job about which we know least at present." The summer-study group had examined several ingenious designs. The most favorable fired a cylindrical male plug of core and tamper into a mated female sphere of tamper and core, illustrated here in cross section from the *Los Alamos Primer:*

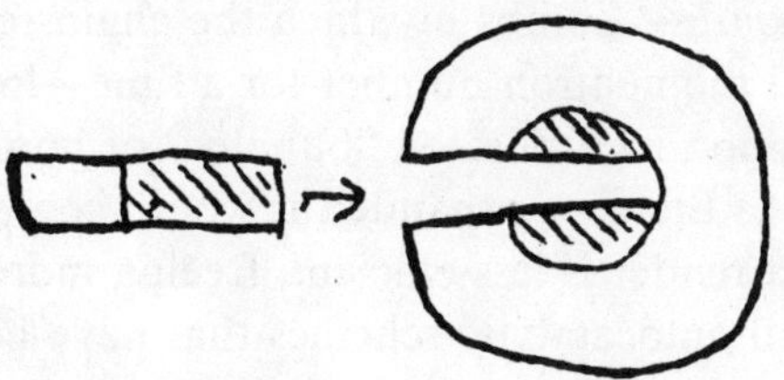

The target sphere could be simply welded to the muzzle of a cannon; then the cylinder, which might weigh about a hundred pounds, could be fired up the barrel like a shell:

> The highest muzzle velocity available in U.S. Army guns is one whose bore is 4.7 inches and whose barrel is 21 feet long. This gives a 50 lb. projectile a muzzle velocity of 3150 ft/sec. The gun weighs 5 tons. It appears that the ratio of projectile mass to gun mass is about constant for different guns so a 100 lb. projectile would require a gun weighing about 10 tons.

For a mechanism eight times lighter or with double the effective muzzle velocity they could weld two guns together at their muzzles and fire two projectiles into each other. Synchronization would be a problem with such a design and efficiency might require four critical masses instead of two, a demand which would significantly delay delivering a usable bomb.

Serber also described more speculative arrangements: sliced ellipsoidal core-tamper assemblies like halves of hard-boiled eggs that slid together; wedge-shaped quarters of core/tamper like sections of a quartered apple mounted on a ring. That was an odd and striking design, sketched in the mimeographed *Primer* as probably on a blackboard before, and it did not go unnoticed. "If explosive material were distributed around the ring and fired the pieces would be blown inward to form a sphere":

Autocatalytic bombs—bombs in which the chain reaction itself, as it proceeded, increased the neutron number for a time—looked less promising. The cleverest notion incorporated "bubbles" of boron-coated paraffin into the U235 core; as the core expanded it would compress the neutron-absorbing boron and render it less efficient, freeing more neutrons for fission chains. But: "All autocatalytic schemes that have been thought of so far require large amounts of active material, are low in efficiency unless very large amounts are used, and are dangerous to handle. Some bright ideas are needed."

Their immediate work of experiment, Serber concluded, would be measuring the neutron properties of various materials and mastering the ordnance problem—the problem, that is, of assembling a critical mass and firing the bomb. They would also have to devise a way to measure a critical mass for fast fission with subcritical amounts of U235 and Pu239. They had a deadline: workable bombs ready when enough uranium and plutonium was ready. That probably gave them two years.

The Japanese physics colloquium in Tokyo had decided in March 1943 that an atomic bomb was possible but not practically attainable by any of the belligerents in time to be of use in the present war. Robert Serber's lectures at Los Alamos in early April asserted to the contrary that for the United States an atomic bomb was both possible and probably attainable within two years. The Japanese assessment was essentially technological. Like Bohr's assessment in 1939, it overestimated the difficulty of isotope separation and underestimated U.S. industrial capacity. It also, as the Japanese government had before Pearl Harbor, underestimated American dedication. Collective dedication was a pattern of Japanese culture more than of American. But Americans could summon it when challenged, and couple it with resources of talent and capital unmatched anywhere else in the world.

The Europeans at Los Alamos complained of the barbed wire. With the exception, apparently, only of Edward Condon, who found security so oppressive he quit the project within weeks of his arrival and went back to Westinghouse, the Americans accepted the fences around their work and their lives as a necessity of war. The war was a manifestation of nationalism, not of science, and such did their duty on the Hill appear at first to be. There was "relatively little nuclear physics" at Los Alamos, Bethe says, mostly cross-section calculations. They thought they were assembled to engineer a *"practical military weapon."* That was first of all a national goal. Science—a fragile, nascent political system of limited but increasing franchise—would have to wait until the war was won. Or so it seemed. But a few among the men and women gathered at Los Alamos—certainly Robert

Oppenheimer—sniffed a paradox. They proposed in fact to win the war with an application of their science. They dreamed further that by that same application they might forestall the next war, might even end war as a means of settling differences between nations. Which must in the long run have decisive consequences, one way or the other, for nationalism.

By the time Robert Serber finished his orientation lectures at Los Alamos in mid-April most of the scientific and technical staff was on hand, many lodged temporarily in the surviving buildings of the Ranch School. Now began a second phase of the conference, to plan the laboratory's work. "If there were any ground-breaking ceremonies at Los Alamos like champagne or cutting ribbons," John Manley comments, "I was unaware of them. Most of us who were there felt that the conference in April, 1943, was really the ground-breaking ceremony." Rabi, Fermi and Samuel Allison arrived from Cambridge and Chicago to serve as senior consultants. Groves appointed a review committee—W. K. Lewis again, an engineer named E. L. Rose who was thoroughly experienced in ordnance design, Van Vleck, Tolman and one other expert—to follow planning and advise. Groves despite his formidable competence as an organizer and administrator was intellectually insecure around so many distinguished scientists, as who would not be?

They laid their plans, often during hikes into the uninhabited wild surroundings of the mesa. They had to rely heavily on theoretical anticipations of the effects they wanted to study; that was their basic constraint. Any experimental device that demonstrated a fast-neutron chain reaction to completion would use up at least one critical mass: there could be no controlled, laboratory-scale bomb tests, no squash-court demonstrations. They decided they had to analyze the explosion theoretically and work out ways to calculate the stages of its development. They needed to understand how neutrons would diffuse through the core and the tamper. They needed a theory of the explosion's hydrodynamics—the complex dynamic motions of its fluids, which the core and tamper would almost instantly become as their metals heated from solid to liquid to gas.

They needed detailed experiments to observe bomb-related nuclear phenomena and they needed integral experiments to duplicate as much as possible the full-scale operation of the bomb. They had to develop an initiator to start the chain reaction. They had to devise technology for reducing uranium and plutonium to metal, for casting and shaping that metal, possibly for alloying it to improve its properties. Particularly with plutonium, they had to discover and measure those properties in the first place and do so quickly when more than microgram quantities began to arrive.

As a sideline, because they agreed that work on the Super should continue at second priority, they wanted to construct and operate a plant for liquefying deuterium at −429°F—the cryogenics plant to be built near the south rim of the mesa.

Ordnance work was crucial. From the April discussions came immediate breakthroughs. An Oppenheimer recruit from the National Bureau of Standards who had been a protégé at Caltech, a tall, thin, thirty-six-year-old experimental physicist named Seth Neddermeyer, imagined an entirely different strategy of assembly. Neddermeyer could not quite remember after the war the complex integrations by which he came to it. An ordnance expert had been lecturing. The expert had quibbled at the physicists' use of the word "explosion" to describe firing the bomb parts together. The proper word, the expert said, was "implosion." During Serber's lectures Neddermeyer had already been thinking about what must happen when a heavy cylinder of metal is fired into a blind hole in an even heavier metal sphere. Spheres and shock waves made him think about spherically symmetrical shock waves, whatever those might be. "I remember thinking of trying to push in a shell of material against a plastic flow," Neddermeyer told an interviewer later, "and I calculated the minimum pressures that would have to be applied. Then I happened to recall a crazy thing somebody had published about firing bullets against each other. It may have had a photograph of two bullets liquefied on impact. That is what I was thinking when the ballistics man mentioned implosion."

Two bullets fired against each other recall the double-gun model of the *Los Alamos Primer.* There were other clues to Neddermeyer's new strategy placed evocatively in the *Primer* as well. That document notes that when the surface of the bomb core blows off, it "expands into the tamper material, starting a shock wave which compresses the tamper material sixteenfold." The *Primer* emphasizes more than once that the expansion of the core would be the greatest obstacle to an efficient explosion. It may have occurred to Neddermeyer that if a tamper merely by its inertia—by its tendency to stay where it is when the swelling core begins to push out against it—could resist the core's expansion and thereby increase the efficiency of the explosion, a tamper that somehow *pushed back* against the core might do even better. The compressing of the boron bubbles in the autocatalytic bomb may also have been suggestive. Finally, the *Primer* offered the interesting model of four apple-quarter wedges of core/tamper fired together by an encompassing explosive ring. "At this point," says Neddermeyer, "I raised my hand."

He proposed packing a spherical layer of high explosives around a spherical assembly of tamper and a hollow but thick-walled spherical core.

Detonated at many points simultaneously, the HE would blow inward. The shock wave from that explosion would squeeze the tamper from all sides, which in turn would squeeze the core. Squeezing the core would change its geometry from hollow shell to solid ball. What had been subcritical because of its geometry would be squeezed critical far faster and more efficiently than any mere gun could fire. "The gun will compress in one dimension," Manley remembers Neddermeyer telling them. "Two dimensions would be better. Three dimensions would be better still."

A three-dimensional squeeze inward was *implosion.* Neddermeyer had just defined a possible new way to fire an atomic bomb. The idea had been suggested previously, but no one had carried it beyond conversation. "At a meeting on ordnance problems late in April," records the Los Alamos technical history, "Neddermeyer presented the first serious theoretical analysis of the implosion. His arguments showed that the compression of a . . . sphere by detonation of a surrounding high-explosive layer was feasible, and that it would be superior to the gun method both in its high velocity and shorter path of assembly."

The response at the time was not encouraging. "Neddermeyer faced stiff opposition from Oppenheimer and, I think, Fermi and Bethe," Manley says. How do you make a shock wave spherically symmetrical? How do you keep tamper and core from squirting out in every direction as water does when squeezed between cupped hands? "Nobody . . . really took [implosion] very seriously," Manley adds. But Oppenheimer had been wrong before—even about the possibility of fission when Luis Alvarez dropped by to report it in 1939, wrong for the fifteen minutes it took him to think past the stubbornness with which he rejected any possibility he had not himself foreseen. Apparently he was learning to steer by that grudging incredulity as Bohr steered by the madness of a truly original idea. "This will have to be looked into," he told Neddermeyer in private conference after the dismissive public debate. He took his revenge for the trouble Neddermeyer was causing him by appointing that thoroughgoing loner to the newly invented post of group leader in the Ordnance Division for implosion experimentation.

The other fresh insight remembered from the April conference corrected an error that everyone wondered afterward how anyone could have overlooked. The error is perhaps a measure of how unfamiliar the physicists were with ordnance. E. L. Rose, the research engineer on Groves' review committee, woke up one day to realize that the Army cannon the physicists were basing their estimates on weighed five tons only because it had to be sturdy enough for repeated firing. A gun that wore an atomic bomb welded to its muzzle could be flimsier: it would be fired only once, after which it would vaporize and drift away. That specification cut its weight drastically and promised a practical, flyable bomb.

Fermi, superb experimentalist that he was, contributed valuably to the program of experimental studies, defining with clarity problems that needed to be examined. For him the war work was duty, however, and the eager conviction he found on the Hill puzzled him. "After he had sat in on one of his first conferences here," Oppenheimer recalls, "he turned to me and said, 'I believe your people actually *want* to make a bomb.' I remember his voice sounded surprised."

The leaders attended a party one night that April at Oppenheimer's house, the log-and-stucco former residence of the Ranch School headmaster. Edward Condon, whose father had been a builder of railroads in the West, who had worked as a newspaper reporter in tough Oakland, found occasion at Oppenheimer's party to satirize Los Alamos' Panglossian mood. He was an exceptional theoretician; he and Oppenheimer had boarded together at Göttingen; Condon thought they were fast friends. He would soon clash bitterly with Groves over compartmentalization and find that his friend the director had higher priorities than backing him up. Now, sitting in a corner at the director's house, Condon pulled from a bookshelf a copy of Shakespeare's *The Tempest* and skimmed it for speeches meant for Prospero's enchanted island that might play contrapuntally against Oppenheimer's high and dry and secret mesa where no one had a street address, where mail was censored, where drivers' licenses went nameless, where children would be born and families live and a few people die behind a post-office box in devotion to the cause of harnessing an obscure force of nature to build a bomb that might end a brutal war. There are many speeches in *The Tempest* that would have fit the occasion but one certainly that Condon would not have missed reading aloud to the assembled, Miranda's speech that Aldous Huxley borrowed for an ironic title:

> *O, wonder!*
> *How many goodly creatures are there here!*
> *How beauteous mankind is! O brave new world*
> *That has such people in't!*

The British had chosen not to bomb Vemork because Lief Tronstad, the physical chemist attached to Norwegian intelligence in London, had warned that hitting the hydrochemical facility's liquid-ammonia storage tanks would almost certainly kill large numbers of Norwegian workers. But the British had in any case long since abandoned precision bombing.

Winston Churchill had declared himself strongly in favor of strategic air attack early in the war, speaking even of extermination. In July 1940, in the desperate time after the debacle of Dunkirk and at the beginning of the Battle of Britain, Churchill had written his Minister of Aircraft Production to that effect: "But when I look round to see how we can win the war I see that there is only one sure path . . . and that is absolutely devastating, exterminating attack by very heavy bombers from this country upon the Nazi homeland. We must be able to overwhelm them by this means, without which I do not see a way through."

The slide from precision bombing attacks on industry to general attacks on cities followed less from political decisions than from inadequate technology. Bomber Command had attempted long-distance daylight precision bombing early in the war but had been unable to defend its aircraft against German fighters and flak so far from home. It therefore switched to night bombing, which reduced losses but severely impaired accuracy. If it was logical to bomb factories and other strategic targets to reduce the enemy's ability to wage war, it began to seem equally logical to bomb the blocks of workers' housing that surrounded those targets; the workers, after all, made the factories run. Sir Arthur Harris, who became chief of Bomber Command in early 1942, notes in his war memoirs of this transitional period in the summer of 1941 that "the targets chosen were in congested industrial areas and were carefully picked so that bombs which overshot or undershot the actual railway centers under attack [in this instance] should fall on these areas, thereby affecting morale. This programme amounted to a halfway stage between area and precision bombing." "Morale" is here and elsewhere in the literature of air power a euphemism for the bombing of civilians. Another sign of halfway status at this stage was permission to dump bombs before exiting Germany if crews had missed their targets.

Churchill says he authorized a study of bombing accuracy at Frederick Lindemann's suggestion which discovered in the summer of 1941 "that although Bomber Command believed they had found the target, two-thirds of crews actually failed to strike within five miles of it. . . . Unless we could improve on this there did not seem much use in continued night bombing." In November the government ordered its bomber arm to reduce operations over Germany.

To reduce strategic bombing operations was to admit failure in both theory and practice, and it was to do so at a time when the USSR was fully engaged with the German armies on the Eastern Front and Joseph Stalin was demanding the Allies open a second front in the West. Neither Britain nor the United States was nearly prepared yet to invade Europe on the ground, but both nations might offer such aid as air attack could bring.

Aiding the Soviet Union was a political justification for continuing some kind of strategic bombing campaign, though it hardly placated Stalin. Headlines proclaiming almost daily bombing raids also helped keep the home front happy when the ground war stalled.

Yet Allied politics and domestic propaganda could not have been the primary reasons for the drift from precision to area bombing, because U.S. air forces beginning to arrive in Britain in 1942 planned and carried out precision daylight bombing, though not often effectively, until much later in the war. Rather, Bomber Command switched programs in order to justify its continued existence as a service with a mission separate from Army and Navy tactical support, cutting theory to fit the facts. It found an ally in the newly ennobled Lindemann, Lord Cherwell, who calculated in March 1942 that bombing might destroy the housing of a third of the German population within a year if sufficiently pursued against industrial urban areas. Patrick Blackett and Henry Tizard thought Cherwell's estimate far too optimistic and dissented vigorously, but Cherwell had the Prime Minister's ear.

Sir Arthur Harris—"Butch," his staff came to call him, short for "the Butcher"—took over Bomber Command in February and promulgated a new approach to the air war: "It has been decided that the primary objective of your operations should now be focussed on the morale of the enemy civil population and in particular, of the industrial workers." Harris had witnessed the London Blitz; it convinced him, he writes, that "a bomber offensive of adequate weight and the right kind of bombs would, if continued for long enough, be something that no country in the world could endure." His argument was valid, of course, though what "the right kind of bombs" might be would require the work of the Manhattan Project to reveal. Hitler's terror bombing taught Britain not terror but forceful imitation. Harris certainly despised the Germans for starting and perpetuating two world wars. But he seems to have thought less about killing civilians than about solving the problem of making Bomber Command a measurably effective force. If night bombing and area bombing were the only tactics that paid a reasonable return in destruction at a reasonable price in lost aircraft and aircrew lives, then he would dedicate Bomber Command to perfecting those tactics and measure success not in factories rendered inoperative but in acres of cities flattened. Which is to say, area bombing was invented to give bombers targets they could hit.

An incendiary attack on the old Baltic port of Lübeck in March burned much of the town and produced four-figure casualties for the first time in the bombing campaign. On May 20, to demonstrate Bomber Command's effectiveness at a time of public debate, Harris mustered every aircraft he could find—hundreds of two-engine bombers of light payload and even training planes—to launch a thousand-bomber raid on Cologne. For that successful assault he organized what came to be called a bomber "stream," the aircraft flying in massed continuous formations to overwhelm defenses rather than in small and vulnerable packets as before, and destroyed some eight square miles of the ancient city on the Rhine with 1,400 tons of bombs, two-thirds of them incendiary. Finally, in August, encouraged by Cherwell, Bomber Command deployed a Pathfinder force: skilled advance crews that marked targets with colored flares so that less experienced pilots following in the lethal stream could more easily find their aiming points.

No fleet of bombers could yet accurately deliver enough high explosives to raze a city. The Lübeck bombing had been planned to test the theory that area bombing worked best by starting fires. If the bombloads were incendiary, then the massed aircraft might combine their destructiveness, wind and weather cooperating, rather than disperse it on isolated targets. The theory worked at Lübeck and again at Cologne and because it worked it won adoption. At the end of 1942 the British Chiefs of Staff called for "the progressive destruction and dislocation of the enemy's war industrial and economic system, and the undermining of his morale to a point where his capacity for armed resistance is fatally weakened." Churchill and Roosevelt affirmed the British plan for an aerial war of attrition in a directive issued at the conclusion of the Casablanca Conference in late January 1943.

On May 27, 1943, as work began at Los Alamos following the April conferences, Bomber Command ordered Hamburg attacked. Its *Most Secret Operation Order No. 173* stated its new policy of mass destruction explicitly:

> INFORMATION
>
> The importance of HAMBURG, the second largest city in Germany with a population of one and a half millions, is well known . . . . The total destruction of this city would achieve immeasurable results in reducing the industrial capacity of the enemy's war machine. This, together with the effect on German morale, which would be felt throughout the country, would play a very important part in shortening and in winning the war.
>
> 2. The "Battle of Hamburg" cannot be won in a single night. It is estimated that at least 10,000 tons of bombs will have to be dropped to complete the process of elimination . . . . This city should be subjected to sustained attack . . . .
>
> 3. . . . It is hoped that the night attacks will be preceded and/or followed by heavy daylight attacks by the United States VIIIth Bomber Command.
>
> INTENTION
>
> 4. To destroy HAMBURG.

The operation was code-named Gomorrah. Notice the significant claim that it would help shorten and win the war.

Operation Gomorrah began on the night of July 24, 1943, a hot summer Saturday in Hamburg under clear skies. Pathfinder bombers used radar to aid marking, and the initial Hamburg aiming point was chosen not for its strategic significance but for its distinctive radar reflection: a triangle of land at the junction of the Alster and North Elbe rivers, near the oldest part of the city and far from any war industry. Bomber Command had learned to adjust targeting for creep-back, the tendency of bombardiers to release their bombs as quickly as possible upon approaching the flak-infested aiming point that led to a gradual backup of impacts. From the ground the bombs seemed to unroll in the direction of the bomber stream's approach; survivors named the phenomenon "carpet bombing." Targeters incorporated creep-back into their calculations by setting the aiming point several miles forward of the intended target area. The creep-back districts behind the Hamburg aiming point to a distance of four miles were entirely residential.

To give the bombers further advantage Churchill had authorized the first use of the secret radar-jamming device known as Window: bales of 10.5-inch strips of aluminum foil to be pushed out of the bombers en route to the target to disperse on the wind and cloud German defensive radar. Window worked so well that of the 791 planes of the initial raid only twelve were lost.

Hamburg sustained heavy damage that first night but not damage even on the scale of Cologne; 1,300 tons of high explosives and almost 1,000 tons of incendiaries killed about 1,500 people and left many thousands homeless. More important for what would follow, the first raid seriously disrupted communications and overwhelmed firefighting forces.

Daylight precision bombing by American B-17's followed on July 25 and 26, attacks meant for a submarine yard and an aircraft engine factory. Smoke from the British bombing and from German defensive generators obscured the targets and they were only lightly damaged.

Harris ordered a maximum bombing effort against Hamburg again for the night of July 27. Targeters fixed the same aiming point but aligned the bomber stream to approach from the northeast rather than the north to set its creep-back over districts dense with workers' apartment buildings. Since the mix of 787 bombers for this second raid would include more Halifaxes and Stirlings, and they could carry less weight of weapons and fuel than the longer-distance Lancasters, the mix of bombs was also changed, high explosives reduced and incendiaries increased to more than 1,200 tons. More experienced pilots also came aboard, higher-ranking officers signing on to observe the effects of Window. These accidents of arrangement contributed their share to the night's catastrophe.

At 6 P.M. in Hamburg on July 27 the temperature was 86 degrees and the humidity 30 percent. Fires still burned in stores of coal and coke in the western sector of the city. Since the fires would render a blackout ineffective most of Hamburg's firefighting equipment had been moved to the area to douse them. "It was completely quiet," recalls a German woman who lived in a district targeted for creep-back, miles to the northeast. ". . . It was an enchantingly beautiful summer night."

Pathfinders started dropping yellow markers and bombs at fifty-five minutes past midnight on July 28. Five minutes later the main bomber stream arrived. Marking was good and creep-back was slow. Later arrivals began to notice a difference between this raid and others they had flown: "Most of the raids we did looked like gigantic firework displays over the target area," a flight sergeant remarks, "but this was 'the daddy of them all.' " A flight lieutenant distinguishes the difference:

> The burning of Hamburg that night was remarkable in that I saw not many fires but one. Set in the darkness was a turbulent dome of bright red fire, lighted and ignited like the glowing heart of a vast brazier. I saw no flames, no outlines of buildings, only brighter fires which flared like yellow torches against a background of bright red ash. Above the city was a misty red haze. I looked down, fascinated but aghast, satisfied yet horrified. I had never seen a fire like that before and was never to see its like again.

The summer heat and low humidity, the mix of high-explosive and incendiary bombs that made kindling and then ignited it and the absence of firefighting equipment in the bombed districts conspired to assemble a new horror. An hour after the bombing began the horror had a name, recorded first in the main log of the Hamburg Fire Department: *Feuersturm*: firestorm. A Hamburg factory worker remembers its beginning, some twenty minutes into the one-hour bombing raid:

> Then a storm started, a shrill howling in the street. It grew into a hurricane so that we had to abandon all hope of fighting the [factory] fire. It was as though we were doing no more than throwing a drop of water on to a hot stone.The whole yard, the canal, in fact as far as we could see, was just a whole, great, massive sea of fire.

Small fires had coalesced into larger fires and, greedy for oxygen, had sucked air from around the coalescing inferno and fanned further fires there. That created the wind, a thermal column above the city like an invisible chimney above a hearth; the wind heated the fury at the center of the firestorm to more than 1,400 degrees, heat sufficient to melt the windows of a streetcar, wind sufficient to uproot trees. A fifteen-year-old Hamburg girl recalls:

> Mother wrapped me in wet sheets, kissed me, and said, "Run!" I hesitated at the door. In front of me I could see only fire—everything red, like the door to a furnace. An intense heat struck me. A burning beam fell in front of my feet. I shied back but, then, when I was ready to jump over it, it was whirled away by a ghostly hand. I ran out to the street. The sheets around me acted as sails and I had the feeling that I was being carried away by the storm. I reached . . . a five-storey building in front of which we had arranged to meet again. . . . Someone came out, grabbed me by the arm, and pulled me into the doorway.

The fire filled the air with burning embers and melted the streets, a nineteen-year-old milliner reports:

> We came to the door which was burning just like a ring in a circus through which a lion has to jump. . . . The rain of large sparks, blowing down the street, were each as large as a five-mark piece. I struggled to run against the wind in the middle of the street but could only reach a house on the corner . . . .
>
> We got to the Löschplatz [park] all right but I couldn't go on across the Eiffestrasse because the asphalt had melted. There were people on the roadway, some already dead, some still lying alive but stuck in the asphalt. They must have rushed on to the roadway without thinking. Their feet had got stuck and then they had put out their hands to try to get out again. They were on their hands and knees screaming.

The firestorm completely burned out some eight square miles of the city, an area about half as large as Manhattan. The bodies of the dead cooked in pools of their own melted fat in sealed shelters like kilns or shriveled to small blackened bundles that littered the streets. Or worse, as the woman who was once the fifteen-year-old girl horribly recreates:

> Four-storey-high blocks of flats [the next day] were like glowing mounds of stone right down to the basement. Everything seemed to have melted and pressed the bodies away in front of it. Women and children were so charred as to be unrecognizable; those that had died through lack of oxygen were half-charred and recognizable. Their brains had tumbled from their burst temples and their insides from the soft parts under the ribs. How terribly these people must have died. The smallest children lay like fried eels on the pavement.

Bomber Command killed at least 45,000 Germans that night, the majority of them old people, women and children.

The bombing of Hamburg was hardly unique. It was one atrocity in a war of increasing atrocities. Between 1941 and 1943 the German Army on the Eastern Front captured and enclosed in prisoner-of-war camps without food or shelter some two million Soviet soldiers; at least one million of them died of exposure and starvation. During the same period the Final Solution to the Jewish Question—the vast Nazi program to exterminate the European Jews—began in deadly earnest after the Wannsee Conference of coordinating agencies met in suburban Berlin on January 20, 1942. Whatever moral issues such atrocities raise, they resulted from the progressive escalation of the war by all its belligerents in pursuit of victory. (Even the Final Solution: because the Nazis believed the Jews constituted a separate nation lodged subversively in their midst—nationality being defined in the Nazi canon primarily in terms of race—and as such the nation with which the Third Reich was preeminently at war. It was Hitler's particular perversity to define victory over the Jews as extermination; the Allies in their defensive war against Germany and Japan wanted only total surrender, in return for which the mass killing of combatants and civilians would stop.)

One way the belligerents could escalate was to improve their death technologies. Better bombers and better bomber defenses such as Window were hardware improvements; so were the showers at the death camps efficiently pumped with the deadly fumigant Zyklon B. The bomber-stream system and allowance for creep-back were software improvements; so were the schedules Adolf Eichmann devised that kept the trains running efficiently to the camps.

The other way the belligerents could escalate was to enlarge the range of permissible victims their death technologies might destroy. Civilians had the misfortune to be the only victims left available. Better hardware and software began to make them also accessible in increasing numbers. No great philosophical effort was required to discover acceptable rationales. War begot psychic numbing in combatants and civilians alike; psychic numbing prepared the way for increasing escalation.

Extend war by attrition to include civilians behind the lines and war becomes total. With improving technology so could death-making be. The bombing of Hamburg marked a significant step in the evolution of death technology itself, massed bombers deliberately churning conflagration. It was still too much a matter of luck, an elusive combination of weather and organization and hardware. It was still also expensive in crews and matériel. It was not yet perfect, as no technology can ever be, and therefore seemed to want perfecting.

The British and the Americans would be enraged to learn of Japanese brutality and Nazi torture, of the Bataan Death March and the fathomless horror of the death camps. By a reflex so mindlessly unimaginative it may be merely mammalian, the bombing of distant cities, out of sight and sound and smell, was generally approved, although neither the United States nor Great Britain admitted publicly that it deliberately bombed civilians. In Churchill's phrase, the enemy was to be "de-housed." The Jap and the Nazi in any case had started the war. "We must face the fact that modern warfare as conducted in the Nazi manner is a dirty business," Franklin Roosevelt told his countrymen. "We don't like it—we didn't want to get in it—but we are in it and we're going to fight it with everything we've got."

The Los Alamos review committee headed by W.K. Lewis of MIT reported its findings on May 10, 1943. It approved the laboratory's nuclear physics research program. It recommended that theoretical investigation of the thermonuclear bomb continue at second priority, subordinate to fission bomb work. It proposed a major change in the chemistry program: final purification of plutonium on the Hill, because Los Alamos would be ultimately responsible for the performance of the plutonium bomb and because the scarce new element would be used and reused for experiments during the months before a sufficient quantity accumulated to load a bomb and would have to be frequently repurified. The Lewis committee also concurred in a recommendation Robert Oppenheimer had made in March that ordnance development and engineering should begin immediately at Los Alamos rather than wait until nuclear physics studies were complete. General Groves accepted the committee's findings; they dictated an immediate doubling of Hill personnel. Thereafter until the end of the war the Los Alamos working population would double every nine months. The dust of construction never settled; housing would always be short, water scarce, electricity intermittent. Groves spent not a penny more than necessary on comforts for civilians.

The bottom pole piece of the Harvard cyclotron had been laid on April 14; by the first week in June Robert Wilson's cyclotron group saw signs of a beam. The Wisconsin long-tank Van de Graaff came on line at 4 million volts on May 15 and the 2 MV short-tank Van de Graaff on June 10. In July the first physics experiment completed at Los Alamos counted the number of secondary neutrons Pu239 emitted when it fissioned. "In this experiment," says the Los Alamos technical history, "the neutron number was measured from an almost invisible speck of plutonium and found to be somewhat greater even than for U^{235}." The experiment thus established what had not yet been confirmed despite the expensive rush of building: that plutonium emitted sufficient secondary neutrons to chain-react.

The speck of plutonium was Glenn Seaborg's 200-milligram sample of Met Lab oxide, which he had sent to Los Alamos at the beginning of the month. Seaborg had worked himself sick at the Met Lab that spring—an upper respiratory infection compounded with exhaustion and a persistent fever—and came to New Mexico with his wife during July to vacation. ("I guess I deliberately chose to be near the plutonium," he muses. "I wonder why?") Too much peace and quiet at a guest ranch threatened to exhaust him further and on July 21 he and his wife moved to the adobe-style La Fonda Hotel in Santa Fe. Compartmentalization put Los Alamos off limits. The Seaborgs were ready to return to Chicago on Friday, July 30, and Seaborg proposed to carry the Pu sample, most of the world's supply, back with him on the train. Robert Wilson and another physicist made the transfer before dawn in the restaurant where the Seaborgs were having breakfast in Santa Fe, Wilson arriving in a pickup armed Western-style with his personal Winchester .32 deer-hunting rifle to guard a highly valuable but barely visible treasure. "Then I just put it in my pocket and then into my suitcase," Seaborg remembers. He proceeded to Chicago unarmed.

To direct the expanded Ordnance Division Groves asked the Military Policy Committee in Washington to recommend a good man, preferably a military officer. Vannevar Bush knew a *naval* officer—would Groves mind? "Of course not," the general humphed. Bush proposed Captain William S. "Deke" Parsons, a 1922 Annapolis graduate then responsible under Bush for field-testing the proximity fuse.*

Parsons had also worked on early radar development and served as gunnery officer on a destroyer and experimental officer at the Naval Proving Ground in Dahlgren, Virginia. He was forty-three, cool, vigorous, trim, nearly bald, spit-and-polish but innovative; "all his life," one of the men who worked for him at Los Alamos testifies in praise, "he fought the silly regulations and the conservatism of the Navy." Groves liked him; "within a few minutes [of meeting him]," the general says, "I was sure he was the man for the job." Oppenheimer interviewed the man for the job in Washington and agreed. Parsons was married to Martha Cluverius, a Vassar graduate and the daughter of an admiral; with two blond daughters and a cocker spaniel the couple arrived at Los Alamos in an open red convertible in June.

Parsons' first order of business was the plutonium gun. Because it needed a muzzle velocity of at least 3,000 feet per second it would have to

*The proximity fuse was a miniature radar unit shaped to replace the ballistic nose of anti-aircraft shells. It sensed its proximity to a target—an enemy plane—and exploded the shell it rode at a preset range, often turning a miss into a kill. Its development was another of Bush's responsibilities and it was one of science's most important contributions to the war. Merle Tuve, Richard Roberts and most of the physics team at the Department of Terrestrial Magnetism of the Carnegie Institution had turned from fission research in August 1940 to develop it.

be 17 feet long. It should weigh no more than a ton, a fifth of the usual weight of a gun that size, which meant it would have to be machined from strong high-alloy steel. It would not require rifling but needed three independently operated primers to make sure it fired. Parsons arranged for the Navy's gun-design section to engineer it.

Norman F. Ramsey, a tall young Columbia physicist, the son of a general, served under Parsons as group leader for delivery: for devising a way to deliver the bombs to their targets and drop them. In June he contacted the U.S. Air Force to identify a combat aircraft that could carry a 17-foot bomb. "As a result of this survey," Ramsey writes, "it was apparent that the B-29 was the only United States aircraft in which such a bomb could be conveniently carried internally, and even this plane would require considerable modification so that the bomb could extend into both front and rear bomb bays. . . . Except for the British Lancaster, all other aircraft would require such a bomb to be carried externally." The Air Force was not about to allow a historic new weapon of war to be introduced to the world in a British aircraft, but the B-29 Superfortress was a new design still plagued with serious problems. The first service-test model had not yet flown when Ramsey began his aircraft survey in June; a flight-test model had crashed into a Seattle packing house in February and killed the plane's entire test crew and nineteen packing-house workers.

Ramsey did not have to wait for access to a B-29 to begin collecting data on the long bomb's ballistics, however. He mocked up a scale model and arranged to see it dropped:

> On August 13, 1943, the first drop tests of a prototype atomic bomb were made at the Dahlgren Naval Proving Ground [by a Navy TBF aircraft] to determine stability in flight. These tests were on a 14/23 scale model of a bomb shape which was then thought probably suitable for a gun assembly. Essentially, the model consisted of a long length of 14-inch pipe welded into the middle of a split standard 500-pound bomb. It was officially known at Dahlgren as the "Sewer Pipe Bomb." . . . The first test . . . was an ominous and spectacular failure. The bomb fell in a flat spin such as had rarely been seen before. However, an increase in fin area and a forward movement of the center of gravity provided stability in subsequent tests.

In the meantime Seth Neddermeyer, whose implosion experimentation group Parsons inherited, had visited a U.S. Bureau of Mines laboratory at Bruceton, Pennsylvania, to experiment with high explosives. Edwin McMillan, who was interested in implosion, went with the Caltech physicist:

> At that point it was just Seth and myself with a few helpers. The first cylindrical implosions were done at Bruceton. You take a piece of iron pipe, wrap the explosives around it, and ignite it at several points so that you get a converging wave and squash the cylinder in. That was the birth of the experimental work on implosion, long before experimental work on the gun method.

Back at Los Alamos Neddermeyer set up a small research station on South Mesa, the next mesa south of the Hill across Los Alamos canyon. He fired his first tests in an arroyo on Independence Day, 1943, using iron pipe set in cans packed with TNT. Experimenting with cylinders rather than spheres simplified calculation. Because he wanted to recover the results he packed only limited amounts of explosive. "Those tests of course could not be very sophisticated," says McMillan. ". . . They did show that you could take metal pipes and close them right in so that they became like solid bars, indicating that this was a practical method." They also showed that the squeeze was far from uniform: the pipes emerged from the arroyo dust twisted and deformed.

When Parsons, a thoroughly pragmatic engineer, had time to look over Neddermeyer's work he was openly contemptuous. He doubted if implosion could ever be made reliable enough for field use. Neddermeyer presented his initial results at one of the weekly colloquia Oppenheimer had instituted at Hans Bethe's suggestion to keep everyone with a white badge—everyone cleared for secrets—informed of Tech Area progress. Richard P. Feynman, a brilliant, outspoken New York–born graduate-student theoretician from Princeton, summarized the opinion of the assembly in a phrase: "It stinks." In the name of lightheartedness Parsons was crueler. "With everyone grinding away in such dead earnest here," he told the group, "we need a touch of relief. I question Dr. Neddermeyer's seriousness. To my mind he is gradually working up to what I shall refer to as the Beer-Can Experiment. As soon as he gets his explosives properly organized, we will see this done. The point to watch for is whether he can blow in a beer can without splattering the beer." Implosion was even harder to do than that.

John von Neumann, the Hungarian mathematician who had come to the United States in 1930 and joined the Institute for Advanced Study, had been examining for the NDRC the complex hydrodynamics of shock waves formed by shaped charges, technology which was being applied to the American tank-killing infantry weapon known as the bazooka. Like Rabi, von Neumann had agreed to serve as an occasional Oppenheimer consultant. He visited Los Alamos at the end of the summer and looked into implosion theory, another warren of hydrodynamic complexity. Neddermeyer had devised "a simple theory that worked up to a certain level of violence in the shockwave." Von Neumann, he says, "is generally credited with originating the science of large compressions. But I knew it before and had done it in a naive way. Von Neumann's was more sophisticated."

"Johnny was quite interested in high explosives," Edward Teller remembers. Teller and von Neumann renewed their youthful acquaintance during the mathematician's visit to the Hill. "In my discussions with him some crude calculations were made," Teller continues. "The calculation is indeed simple as long as you assume that the material to be accelerated is incompressible, which is the usual assumption about solid matter. . . . In materials driven by high explosives, pressures of more than 100,000 atmospheres occur." Von Neumann knew that, Teller says, as he did not. On the other hand:

> If a shell moves in one-third of the way toward the center you obtain under the assumption of an incompressible material a pressure in excess of eight million atmospheres. This is more than the pressure in the center of the earth and it was known to me (but not to Johnny), that at these pressures, iron is not incompressible. In fact I had rough figures for the relevant compressibilities. The result of all this was that in the implosion significant compressions will occur, a point which had not been previously discussed.

It had been clear from the beginning that implosion, by squeezing a hollow shell of plutonium to a solid ball, could effectively "assemble" it as a critical mass much faster than the fastest gun could fire. What von Neumann and Teller now realized, and communicated to Oppenheimer in October 1943, was that implosion at more violent compressions than Neddermeyer had yet attempted should squeeze plutonium to such unearthly densities that a solid subcritical mass could serve as a bomb core, avoiding the complex problem of compressing hollow shells. Nor would predetonation threaten from light-element impurities. Develop implosion, in other words, and they could deliver a more reliable bomb more quickly.

It was possible at that point to estimate roughly the size and shape of a bomb that worked by fast implosion. The big gun bomb would be just under 2 feet in diameter and 17 feet long. An implosion bomb—a thick shell of high explosives surrounding a thick shell of tamper surrounding a plutonium core surrounding an initiator—would be just under 5 feet in diameter and a little over 9 feet long: a man-sized egg with tail fins.

Norman Ramsey started planning full-scale drop tests that autumn as the aspens brightened to yellow at Los Alamos. He offered to practice with a Lancaster. The Air Force insisted he practice with a B-29 even though the

97

97. Beginning in 1944, U.S. Air Force B-29's systematically firebombed Japanese cities. *L. to r.,* Generals Lauris Norstad, Curtis LeMay and Thomas Power.

98

98. At the Potsdam Conference in July 1945 President Harry Truman welcomed the bomb as a substitute for Soviet entry into the Pacific war. *L. to r.,* Soviet Premier Joseph Stalin, Truman, British Prime Minister Winston Churchill.

99. Henry L. Stimson, Secretary of War, directed bomb development. 100. Jimmy Byrnes, Secretary of State, advised Truman to use the bomb to force the unconditional surrender of the Japanese.

99

100

101

101. The Hiroshima bomb, Little Boy, was a cannon with a U235 bullet and three U235 target rings fitted to its muzzle. Tinian, August 1945.

102

102. Hiroshima prestrike briefing on Tinian. *L. to r.*, first row, Joseph Buscher, unknown; second row, Norman Ramsey, Paul Tibbets; third row, Thomas Ferrell, Adm. Parnell, Deke Parsons, Luis Alvarez; fourth row, left of Parsons, Charles Sweeney, right of Parsons, Thomas Ferebee, right of Alvarez, Theodore Van Kirk; Harold Agnew.

103

103. Crew of the *Enola Gay* before Hiroshima mission: *l. to r.,* standing, John Porter (ground maintenance officer), Theodore Van Kirk (navigator), Thomas Ferebee (bombardier), Paul Tibbets (pilot), Robert Lewis (copilot), Jacob Beser (radar countermeasures officer); kneeling, Joseph Stiborik (radar operator), Robert Caron (tail gunner), Richard Nelson (radio operator), Robert Shumard (assistant engineer), Wyatt Duzenbury (flight engineer). Not shown: Deke Parsons (weaponeer), Morris Jeppson (electronics test officer).

104

104. The mushroom cloud over Hiroshima, August 6, 1945, photographed from the strike mission B-29. 105. The *Enola Gay* landing at Tinian after the Hiroshima strike.

105

106

106. A panorama of Hiroshima damage. Some roads have been cleared. Buildings left standing were earthquake-reinforced. Little Boy exploded with a yield equivalent to 12,500 tons of TNT (12.5 KT). Modern atomic artillery shells deliver equal yield; one Minuteman III missile is armed with the equivalent of 84 Hiroshimas.

107

107. Miyuki Bridge, Hiroshima, 1.4 miles from the hypocenter, 11 A.M., August 6, 1945.

108

108. The Hiroshima fireball instantly raised surface temperatures within a mile of the hypocenter well above 1,000° F.

109

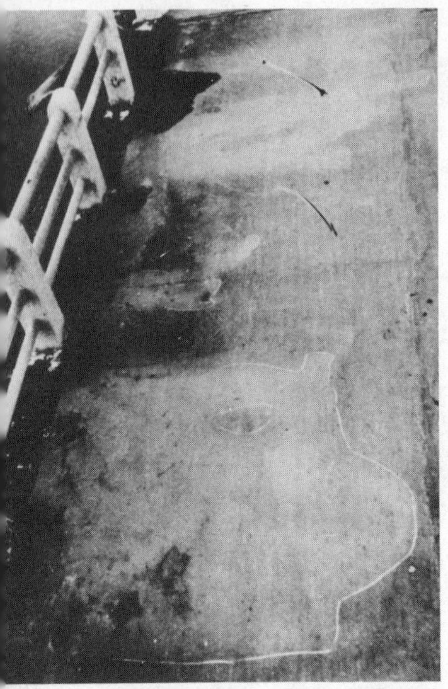

109. A man pulling a cart shadowed in unburned asphalt, Hiroshima.

110

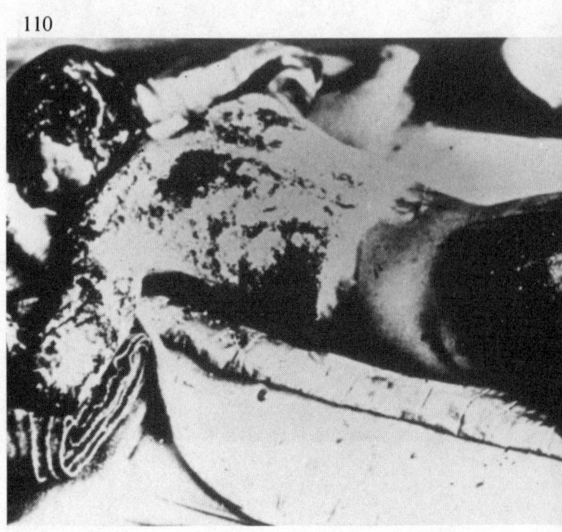

110. Thermal burns on a soldier exposed within half a mile of the Hiroshima hypocenter. His sash protected his waist.

111

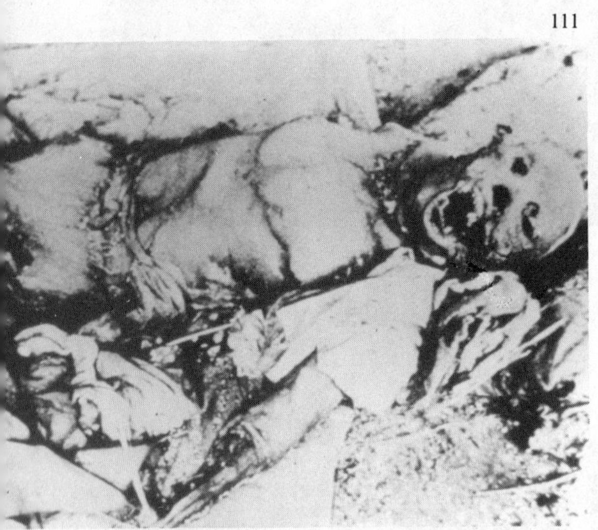

111. Unidentified corpse, Hiroshima. Deaths to the end of 1945 totaled 140,000.

112

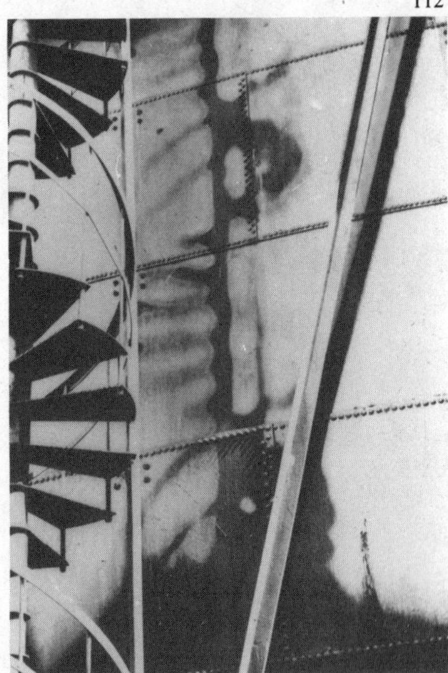

112. Staircase on a gas storage tank shadowed in uncharred paint, Hiroshima.

113

113. Fat Man was ready on Tinian on August 8, 1945, and flew the following day. Note graffiti on tail assembly.

114

114. The plutonium bomb exploded over Nagasaki near the largest Christian church in Japan at 1102 hours, August 9, 1945, with a yield estimated at 22 kilotons.

115

115. Fat Man snapped trees at Nagasaki; the less powerful Hiroshima bomb only knocked them down. 116. Collecting the dead for cremation.

116

117

117. A student exposed half a mile from the Nagasaki hypocenter.

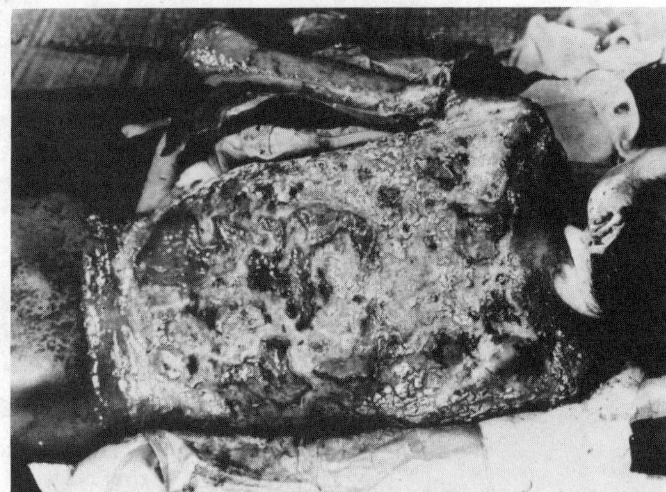

118

118. Flash burns, Nagasaki.

119. Near the Nagasaki hypocenter, noon, August 10, 1945.

119

120

121

120. Dr. Michihiko Hachiya, director of the Hiroshima Communications Hospital. His diary chronicled the disaster.

121. Emperor Hirohito decided after Nagasaki, over his ministers' objections, to end the war and cited "a new and most cruel bomb" in his August 15 surrender proclamation.

122. Los Alamos received the Army-Navy E for excellence for its work.

122

123. Mike I, the first true thermonuclear bomb, tested at Eniwetok in the Marshall Islands on November 1, 1952. Yield: 10.4 megatons (i.e., millions of tons of TNT equivalent). Pipes carried off radiation to diagnostic equipment; their arrangement confirms the linear Teller-Ulam configuration. Note man seated in foreground for scale. 124. Mike vaporized the island of Elugelab and left a crater half a mile deep and two miles wide. 125. The Mark 17 H-bomb, the first deliverable thermonuclear weapon. Yield: megaton-range. Weight: 21 tons.

123

124

125

126. The Mike shot. Its fireball expanded to a diameter of 3 miles.

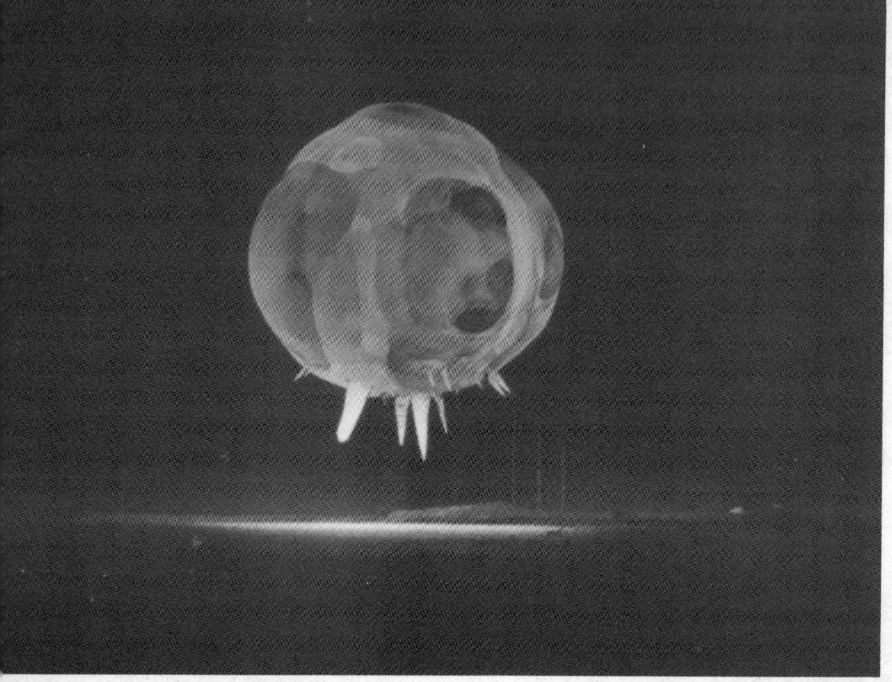

127

127. Early fireball of a postwar atomic bomb test. "I could see a great bare eyeball," Michihiko Hachiya dreamed after Hiroshima, "bigger than life, hovering over my head, staring point-blank at me."

128

128. Margrethe and Niels Bohr at their summer cottage in Tisvilde. "We are in a completely new situation that cannot be resolved by war."

new polished-aluminum intercontinental bombers were just beginning production and still scarce. "In order that the aircraft modifications could begin," Ramsey writes in his third-person report on this work, "Parsons and Ramsey selected two external shapes and weights as representative of the current plans at Site Y. . . . For security reasons, these were called by the Air Force representatives the 'Thin Man' and the 'Fat Man,' respectively; the Air Force officers tried to make their phone conversations sound as though they were modifying a plane to carry Roosevelt (the Thin Man) and Churchill (the Fat Man). . . . Modification of the first B-29 officially began November 29, 1943."

A captain of the Danish Army who was also a member of the Danish underground visited Niels Bohr at the House of Honor in Copenhagen early in 1943. After tea the two men retired to Bohr's greenhouse where hidden microphones might not overhear their conversation. The British had instructed the underground that they would soon be sending Bohr a set of keys. Blind holes had been drilled in the bows of two of the keys, identical microdots implanted and the holes sealed. A captioned diagram located the holes. "Professor Bohr should gently file the keys at the point indicated until the hole appears," the document explained. "The message can then be syringed or floated out onto a micro-slide." The captain offered to extract the microdot and have it enlarged. Bohr was no secret agent; he accepted the offer gratefully.

When the message arrived it proved to be a letter from James Chadwick. "The letter contained an invitation to my father to go to England, where he would find a very warm welcome," Aage Bohr remembers. ". . . Chadwick told my father that he would be able to work freely on scientific matters. But it was also mentioned that there were special problems in which his co-operation would be of considerable help." Bohr understood that Chadwick might be hinting about work on nuclear fission. The Danish physicist was still skeptical of its application. He would not stay in Denmark, he wrote Chadwick in return, "if I felt that I could be of real help . . . but I do not think that this is probable. Above all I have to the best of my judgment convinced myself that, in spite of all future prospects, any immediate use of the latest marvelous discoveries of atomic physics is impracticable." If an atomic bomb were a serious possibility Bohr would leave. Otherwise he had compelling reasons to stay "to help resist the threat against the freedom of our institutions and to assist in the protection of the exiled scientists who have sought refuge here."

The threat against Danish institutions that Bohr was helping to resist was peculiar to the German occupation of Denmark. Germany relied heavily on Danish agriculture, which supplied meat and butter rations to 3.6 million Germans in 1942 alone. It was a labor-intensive agriculture of small farms and it could only continue with the cooperation of the farmers and, more broadly, of the entire Danish population. Not to arouse resistance the Nazis had allowed Denmark to keep its constitutional monarchy and continue to govern itself. The Danes in turn had extracted an extraordinary price for agreeing to cooperate under foreign occupation: the security of Danish Jews. To the Danes the eight thousand Jews in Denmark, 95 percent of them in Copenhagen, were Danish citizens first of all; their security was therefore a test of German good faith. "Danish statesmen and heads of government," reports a historian, "one after the other, had made the security of the Jews a *conditio sine qua non* for the maintenance of a constitutional Danish government."

But resistance, especially strikes and sabotage, gradually increased as the Danish people felt the occupation's burden and as the tides of war began to turn against the Axis powers. The German surrender at Stalingrad on February 2, 1943, may have appeared to many Danes to be a turning point. Mussolini's resignation and arrest the following summer on July 25 and the impending surrender of Italy certainly did. On August 28 the Nazi plenipotentiary for Denmark, Dr. Karl Rudolf Werner Best, presented the Danish government with an ultimatum at Hitler's orders demanding that it declare a state of national emergency, forbid strikes and meetings and introduce a curfew, a ban on arms, press censorship at German hands and the death penalty for harboring arms and for sabotage. With the King's permission the government refused. On August 29 the Nazis reoccupied Copenhagen, disarmed the Danish Army, blockaded the royal palace and confined the King.

One reason for the takeover was Nazi determination to eliminate the Danish Jews, whose exemption from the Final Solution infuriated Hitler. The Nazis had arrested several Jewish notables on August 29 (they had planned to arrest Bohr but had decided the deed would be less obvious during a general roundup). In early September Bohr learned from the Swedish ambassador in Copenhagen that his emigré colleagues, including his collaborator Stefan Rozental, were slated for arrest. He contacted the underground, which helped the emigrés escape across the Öresund to Sweden. Rozental endured nine stormy hours crowded with other refugees in a rowboat borrowed from a city park before his exhausted party made Swedish landfall.

Bohr's turn came soon after. The Swedish ambassador took tea at the House of Honor on September 28 and hinted that Bohr would be arrested within a few days. *Even professors* were leaving Denmark, Margrethe Bohr remembers the diplomat emphasizing. The next morning word came through her brother-in-law that an anti-Nazi German woman working at Gestapo offices in Copenhagen had seen orders authorized in Berlin for the arrest and deportation of Niels and Harald Bohr.

"We had to get away the same day," Margrethe Bohr said afterward. "And the boys would have to follow later. But many were helping. Friends arranged for a boat, and we were told we could take one small bag." In the late afternoon of September 29 the Bohrs walked through Copenhagen to a seaside suburban garden and hid in a gardener's shed. They waited for night. At a prearranged time they left the shed and crossed to the beach. A motorboat ran them out to a fishing boat. Threading minefields and German patrols they crossed the Öresund by moonlight and landed at Linhamm, near Malmö.

Bohr had learned at the last minute that the Nazis planned to round up all the Danish Jews the next evening and deport them to Germany. Leaving his wife in southern Sweden to await the crossing of their sons he rushed to Stockholm to appeal to the Swedish government for aid. He discovered that the Swedes had offered to intern the Danish Jews but the Germans had denied that any roundup was planned.

In fact it proceeded on schedule while Bohr worked his way through the Swedish bureaucracy, but fell far short of success. The Danes, warned in advance, had spontaneously hidden their Jewish fellow citizens away. Only some 284 elderly rest-home residents had been seized. The more than seven thousand Jews remaining in Denmark were temporarily safe. But few of them planned at first to leave the country; it was far from certain that Sweden would accept them and there seemed nowhere else to go.

Meeting with the Swedish Undersecretary for Foreign Affairs on September 30 Bohr had urged that Sweden make public its protest note to the German Foreign Office. He saw that publicity would alert the potential victims, signal Swedish sympathy and bring pressure to bear on the Nazis to desist. The Undersecretary told him Sweden planned no further intervention beyond the confidential note. Bohr appealed to the Foreign Minister on October 2, failed to win publication of the note and determined to dispense with intermediaries. Rozental says the Danish laureate "went to see Princess Ingeborg (the sister of the Danish king Christian X) and while there expressed the desire to be received by the King of Sweden." Bohr also contacted the Danish ambassador and influential Swedish academic colleagues. Rozental describes the crucial meeting with the King:

> The audience . . . took place that afternoon. . . . King Gustaf said that the Swedish Government had tried a similar approach to the Germans once be-

> fore, when the occupying power had started deporting Jews from Norway. The . . . approach, however, had been rejected. . . . Bohr objected that in the meantime the situation had changed decisively by reason of the Allied victories, and he suggested that the offer by the Swedish government to assume responsibility for the Danish Jews should be made public. The King promised to talk to the Foreign Minister at once, but he emphasized the great difficulties of putting the plan into operation.

The difficulties were overcome. Swedish radio broadcast the Swedish protest that evening, October 2, and reported the country ready to offer asylum. The broadcast signaled a route of escape; in the next two months 7,220 Jews crossed to safety in Sweden with the active help of the Swedish coast guard. One refugee's report of what first alerted him in hiding to the idea of escape is typical: "At the pastor's house I heard on the Swedish radio that the Bohr brothers had fled to Sweden by boat and that the Danish Jews were being cordially received." With personal intervention on behalf of the principle of openness, which exposes crime as well as error to public view, Niels Bohr played a decisive part in the rescue of the Danish Jews.

Stockholm was alive with German agents and there was fear that Bohr would be assassinated. "The stay in Stockholm lasted only a short time," remembers Aage Bohr. ". . . A telegram was received from Lord Cherwell . . . with an invitation to come to England. My father immediately accepted and requested that I should be permitted to accompany him." Aage was twenty-one at the time and a promising young physicist. "It was not possible for the rest of the family to follow; my mother and brothers stayed in Sweden."

Bohr went first. The British flew their diplomatic pouch back and forth from Stockholm in an unarmed two-engine Mosquito bomber, a light, fast aircraft that could fly high enough to avoid the German anti-aircraft batteries on the west coast of Norway—flak usually topped out at 20,000 feet. The Mosquito's bomb bay was fitted for a single passenger. On October 6 Bohr donned a flight suit and strapped on a parachute. The pilot supplied him with a flight helmet with built-in earphones for communication with the cockpit and showed him the location of his oxygen hookup. Bohr also took delivery of a stick of flares. In case of attack the pilot would dump the bomb bay and Bohr would parachute into the cold North Sea; the flares would aid his rescue if he survived.

"The Royal Air Force was not used to such great heads as Bohr's," says Robert Oppenheimer wryly. Aage Bohr describes the near-disaster:

> The Mosquito flew at a great height and it was necessary to use oxygen masks; the pilot gave word on the inter-com when the supply of oxygen should be

> turned on, but as the helmet with the earphones did not fit my father's head, he did not hear the order and soon fainted because of lack of oxygen. The pilot realized that something was wrong when he received no answer to his inquiries, and as soon as they had passed over Norway he came down and flew low over the North Sea. When the plane landed in Scotland, my father was conscious again.

The vigorous fifty-eight-year-old was none the worse for wear. "Once in England and recovered," Oppenheimer continues the story, "he learned from Chadwick what had been going on." Aage arrived a week later and father and son toured Britain observing the developing activities there of the Tube Alloys project, which included a section of a pilot-scale gaseous-diffusion plant. But the center of gravity had long since shifted to the United States. The British were preparing to recover a share of the initiative by sending a mission to Los Alamos to help design the bombs; they wanted Bohr on their team to increase its influence and prestige. By then the Danish theoretician had taken what Oppenheimer calls a "good first look." At how nuclear weapons would change the world, Oppenheimer means. He emphasizes Bohr's developing understanding then with a potent simile: "It came to him as a revelation, very much as when he learned of Rutherford's discovery of the nucleus [thirty] years before."

So Niels Bohr prepared in the early winter of 1943 to travel to America once again with an important and original revelation in hand, this one in the realm not of physics but of the political organization of the world.

He was willing to be impressed by a mighty progress of industry. "The work on atomic energy in the USA and in England proved to have advanced much further than my father had expected," Aage Bohr understates. Robert Oppenheimer pitches his summary closer to the shock of surprise a refugee released from the suspended animation that had been occupied Denmark would have felt: "To Bohr the enterprises in the United States seemed completely fantastic."

They were.

15

Different Animals

The 59,000 acres of Appalachian semiwilderness along the Clinch River in eastern Tennessee that Brigadier General Leslie R. Groves acquired for the Manhattan Engineer District as one of his first official acts, in September 1942, extended from the Cumberland foothills in a series of parallel, southwestern-running ridge valleys. Groves liked the geology, which offered isolation for his several enterprises, but the new reservation was nearly as primitive as Los Alamos would be. The Clinch, a meandering tributary of the Tennessee, defined the reservation's southeastern and southwestern boundaries. Eastward twenty miles was Knoxville, a city of nearly 112,000, farther east the wall of Great Smoky Mountains National Park. Five unpaved county roads traversed the ninety-two square miles of depleted valleys and scrub-oak ridges, an area seventeen miles long and seven miles wide that supported only about a thousand families in rural poverty. In the ridge-barricaded valleys of this impoverished hill country, far from prying eyes, the United States Army intended to construct the futuristic factories that would separate U235 from U238 in quantity sufficient to make an atomic bomb.

To do so it had first to improve communications and build a town. Into the gummy red eastern-Tennessee clay in the winter of 1942 and the spring of 1943 its contractors cut fifty-five miles of rail roadbed and three hundred miles of paved roads and streets. They improved the important county roads to four-lane highways. Stone & Webster, the hard-pressed Boston engineering corporation, laid out a town plan so unimaginative that the MED rejected it and passed the assignment to the ambitious young architectural firm of Skidmore, Owings and Merrill, which produced a well-sited arrangement of housing using innovative new materials that saved enough money to allow for such amenities in the best residences as fireplaces and porches. The new town, planned initially for thirteen thousand workers, took its name from its location lining a long section of the northwesternmost valley: Oak Ridge. The entire reservation, fenced with barbed wire and controlled through seven guarded gates, was named, after a nearby Tennessee community, the Clinton Engineer Works. Its workers would come to call it Dogpatch in homage to the hillbilly comic strip "Li'l Abner." The new gates closed off public access on April 1.

Groves planned to build electromagnetic isotope separation plants and a gaseous-diffusion plant at Clinton; plutonium production, he realized during his first months on the project, would proceed at such a scale and generate so vast a quantity of potentially dangerous radioactivity that it would require a separate reservation of its own. Of the three processes, Ernest Lawrence's electromagnetic method was farthest along.

Electromagnetic isotope separation enlarged and elaborated Francis Aston's 1918 Cavendish invention, the mass spectrograph. As a 1945 report prepared by Lawrence's staff explains, the method "depends on the fact that an electrically charged atom traveling through a magnetic field moves in a circle whose radius is determined by its mass"—which was also a basic principle of Lawrence's cyclotron. The lighter the atom, the tighter the circle it made. Form ions of a vaporous uranium compound and start them moving at one side of a vacuum tank permeated by a strong magnetic field and the moving ions as they curved around would separate into two beams. Lighter U235 atoms would follow a narrower arc than heavier U238 atoms; across a four-foot semicircle the separation might be about three-tenths of an inch. Set a collecting pocket at the point where the U235 ion beam separately arrived and you could catch the ions. "When the ions strike the bottom of the collecting pocket . . . they give up their charge and are deposited as flakes of metal." Schematically, with slotted electrodes to accelerate the ions, the arrangement would look like the illustration on page 488.

Late in 1941 Lawrence had installed such a 180-degree mass spectrometer in place of the dees in the Berkeley 37-inch cyclotron. By running it continuously for a month his crews produced a partially separated 100-microgram sample of U235. That was several hundred million times less than the 100 kilograms Robert Oppenheimer had originally estimated would be

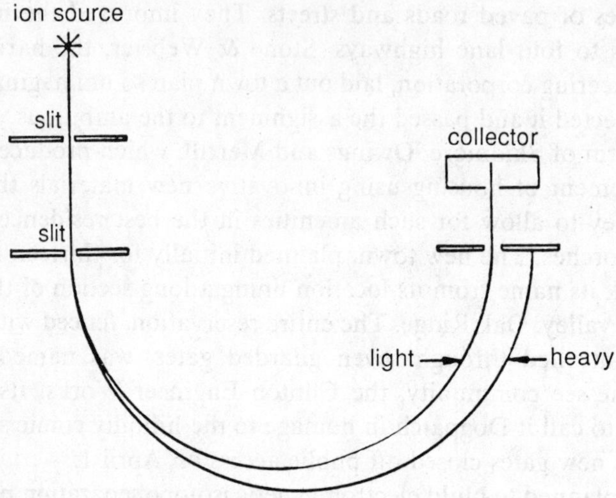

Magnetic field perpendicular to plane of drawing

necessary to make a bomb. The demonstration proved the basic principle of electromagnetic separation even as it dramatized the method's monumental prodigality: Lawrence was proposing to separate uranium atom by individual atom.

Enlarging the equipment, increasing the accelerating voltage, multiplying the number of sources and collectors set side by side between the poles of the same magnet were obvious ways to improve output and efficiency. Lawrence had committed his time to winning the war; now he committed his beautiful new 184-inch cyclotron. Instead of cyclotron dees he had D-shaped mass-spectrometer tanks installed between the pole faces of its 4,500-ton magnet. Making the new instrument work, through the spring and summer of 1942, solved the most difficult design problems. It acquired a name along the way: *calutron*, another *tron* from the *U*niversity of *Cal*ifornia.

To separate 100 grams—about 4 ounces—of U235 per day, Lawrence estimated in the autumn of 1942, would require some 2,000 4-foot calutron tanks set among thousands of tons of magnets. If a bomb needed 30 kilograms—66 pounds—of U235 for reasonable efficiency, as the Berkeley summer study group had just worked out, 2,000 such calutrons could enrich material enough for one bomb core every 300 days. That assumed the system worked reliably, which so far its laboratory predecessors had hardly done. Yet in 1942 electromagnetic separation still looked so much more promising to James Bryant Conant than either the plutonium approach or gaseous barrier diffusion that he had offered up for debate the possibility of pursuing it exclusively. Lawrence was self-confident but not foolhardy; he insisted that the two dark horses should continue to run the race alongside the favorite.

Groves was less impressed. So was the first Lewis committee that had visited Chicago and Berkeley when Fermi was building CP-1 in the winter of 1942. The Lewis committee judged gaseous diffusion the best approach because it was most like existing technology—diffusion was a phenomenon familiar to petroleum engineers and a gaseous-diffusion plant would be essentially an enormous interconnected assemblage of pipes and pumps. Electromagnetic separation by contrast was a batch process untested at such monumental scale; Berkeley planned a system of 4-foot tanks set vertically between the pole faces of large square electromagnets, two tanks to a gap and a total of 96 tanks per unit. To reduce the amount of iron needed for the magnet cores the arrangement would be not rectangular but oval, like a racetrack:

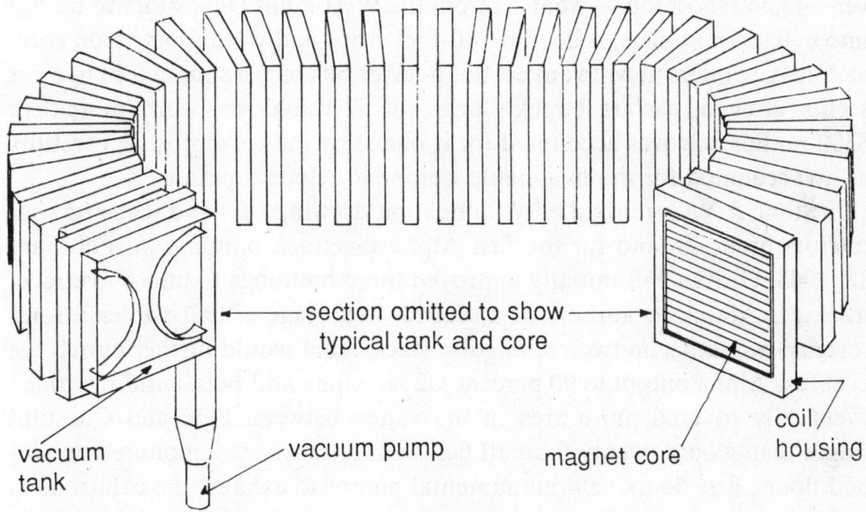

And *racetrack* it was called, though its official designation was Alpha. Berkeley could promise only 5 grams of enriched uranium per day per racetrack, but Groves thought 2,000 tanks well beyond Stone & Webster's capability and cut the number back to 500, reasoning, as Lawrence recalled later, "that the art and science of the process would go forward and that by the time the plant was built substantially higher production rates would be assured." Five grams per day per racetrack with only five racetracks would mean 1,200 days per 30-kilogram bomb even if the Alpha calutrons produced nearly pure U235, which they did not—their best production was around 15 percent. Groves counted on improvements and forged ahead.

He had to begin building before he knew precisely what to build. He worked from the general to the particular, from outline to detail. Fully six months before he decided how many calutrons to authorize, his predecessors, Colonel James Marshall and Lieutenant Colonel Kenneth Nichols, had moved to solve one serious problem of supply. The United States was critically short of copper, the best common metal for winding the coils of electromagnets. For recoverable use the Treasury offered to make silver bullion available in copper's stead. The Manhattan District put the offer to the test, Nichols negotiating the loan with Treasury Undersecretary Daniel Bell. "At one point in the negotiations," writes Groves, "Nichols . . . said that they would need between five and ten thousand tons of silver. This led to the icy reply: 'Colonel, in the Treasury we do not speak of tons of silver; our unit is the Troy ounce.' " Eventually 395 million troy ounces of silver—13,540 short tons—went off from the West Point Depository to be cast into cylindrical billets, rolled into 40-foot strips and wound onto iron cores at Allis-Chalmers in Milwaukee. Solid-silver bus bars a square foot in cross section crowned each racetrack's long oval. The silver was worth more than $300 million. Groves accounted for it ounce by ounce, almost as carefully as he accounted for the fissionable isotope it helped separate.

Stone & Webster had only foundation drawings in hand when its contractors broke ground for the first Alpha racetrack building on February 18, 1943. Groves had initially approved three buildings to house five racetracks. In March he authorized a second, Beta stage of half-size calutrons, seventy-two tanks on two rectangular tracks, that would further enrich the eventual Alpha output to 90 percent U235. Alpha and Beta buildings alone eventually covered more area in the valley between Pine and Chestnut ridges than would twenty football fields. Racetracks were mounted on second floors; first floors held monumental pumps to exhaust the calutrons to high vacuum, more cubic feet of vacuum than the combined total volume pumped down everywhere else on earth at that time. Eventually the Y-12 complex counted 268 permanent buildings large and small—the calutron structures of steel and brick and tile, chemistry laboratories, a distilled water plant, sewage treatment plants, pump houses, a shop, a service station, warehouses, cafeterias, gatehouses, change houses and locker rooms, a paymaster's office, a foundry, a generator building, eight electric substations, nineteen water-cooling towers—for an output measured in the best of times in grams per day. An inspection trip in May 1943 awed even Ernest Lawrence.

By August, twenty thousand construction workers swarmed over the area. An experimental Alpha unit saw successful operation. Lawrence was urging Groves then to double the Alpha plant. With ten Alpha racetracks instead of five he estimated he could separate half a kilogram of U235 per day at 85 percent enrichment. An Army engineer's less exuberant summary, written six days after Lawrence's, predicted 900 grams per month with existing Alpha and Beta stages beginning in November 1943, for a total of 22 kilograms of bomb-grade U235 in the first year of operation. Faced with new estimates from Los Alamos that summer that an efficient uranium gun would probably require 40 kilograms—88 pounds—of the rarer uranium isotope, Groves bought Lawrence's proposal. The doubling would add four new 96-tank tracks of advanced design designated Alpha II and a proportionate number of Beta tracks, at a cost of $150 million more than the $100 million already authorized. If everything worked at Y-12, Groves justified his proposal to the Military Policy Committee, he would then have a 40-kilogram bomb core around the beginning of 1945.

The Army had contracted with Tennessee Eastman, a manufacturing subsidiary of Eastman Kodak, to operate the electromagnetic separation plant. By late October 1943, when Stone & Webster finished installing the first Alpha racetrack, the company had assembled a work force of 4,800 men and women. They were trained to run and maintain the calutrons—without knowing why—twenty-four hours a day, seven days a week.

The big square racetrack magnets wrapped with silver windings were encased in boxes of welded steel. Oil that circulated through the boxes was supposed to insulate the windings and carry heat away. The first magnets tested at the end of October leaked electricity. If moisture in the circulating oil was shorting out the coils, the normal heat of operation would correct the problem by evaporating the water. Tennessee Eastman pushed on. Vacuum leaks in the calutron tanks were numerous and hard to find—one supervisor remembers spending most of a month looking for one leak. Inexperienced operators had trouble striking and maintaining a steady ion beam. Groves recalls that the powerful magnets unexpectedly "moved the intervening tanks, which weighed some fourteen tons each, out of position by as much as three inches. . . . The problem was solved by securely welding the tanks into place, using heavy steel tie straps. Once that was done, the tanks stayed where they belonged."

The magnets dried out but continued to short. Something was seriously wrong. Early in December Tennessee Eastman shut the entire 96-tank racetrack down. The company's engineers would have to break open one of the windings and examine it. That was major trauma; the unit must then be returned to Allis-Chalmers and rebuilt.

The inspectors found disaster: two major troubles. "The first lay in the design," writes Groves, "which placed the heavy current-carrying silver bands too close together. The other lay in the excessive amount of rust and other dirt particles in the circulating oil. These bridged the too narrow gap between the silver bands and resulted in shorting." Groves arrived seething from Washington on December 15 to view the remains. The design's inadequacy forced the general to order all forty-eight magnets hauled back to Milwaukee to be cleaned and rebuilt. The second Alpha track would not come on line until mid-January 1944. They would lose at least a month of production.

Tennessee Eastman's 4,800 employees reported for work in the shambles of gloomy halls. Rather than lose them from boredom the company scheduled classes, conferences, lectures, motion pictures, games. Serious men in double-breasted suits scouted the state for chess and checker sets. At the end of 1943 Y-12 was dead in the water with hardly a gram of U235 to show for all its enormous expense.

Gaseous-diffusion research had progressed at Columbia University since John Dunning and Eugene Booth had first demonstrated measurable U235 separation in November 1941. By the spring of 1942 Harold Urey could note in a progress report that "three methods for the separation of the uranium isotopes have now reached the engineering stage. They are the English and the American diffusion methods, and the centrifuge method." With the authorization of the full-scale plant Dunning's staff, which had grown to include about ninety people, increased in early 1943 to 225. Franz Simon's diffusion method would have operated at low gas pressures and in incremental ten-unit stages but required extremely large pumps; Columbia designed a high-pressure system with more conventional pumps, a continuous, interconnected cascade of some four thousand stages. In a postwar memoir Groves reviews the design, which was both reliably simple and expensively tedious:

> The method was completely novel. It was based on the theory that if uranium gas was pumped against a porous barrier, the lighter molecules of the gas, containing U-235, would pass through more rapidly than the heavier U-238 molecules. The heart of the process was, therefore, the barrier, a porous thin metal sheet or membrane with millions of submicroscopic openings per square inch. These sheets were formed into tubes which were enclosed in an airtight vessel, the diffuser. As the gas, uranium hexafluoride, was pumped through a long series, or cascade, of these tubes it tended to separate, the enriched gas moving up the cascade while the depleted moved down. However, there is so little difference in mass between the hexafluoride of U-238 and U-235 that it was impossible to gain much separation in a single diffusion step. That was why there had to be several thousand successive stages.

In schematic cross section the stages looked like this:

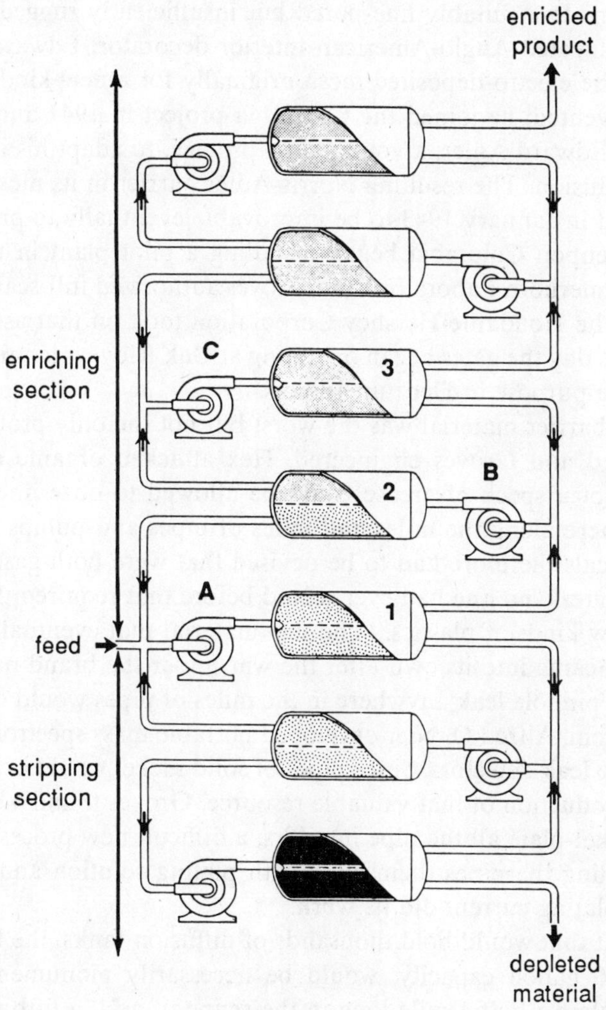

"Further development of barriers is needed," Urey had concluded in his progress report, "but we now feel confident that the problem can be solved." It had not been solved when Groves committed the Manhattan Project to a $100 million gaseous-diffusion plant, however; no practical barrier was yet in hand. The American process required finer-pored material than the British; the material also had to be rugged enough to withstand the higher pressure of the heavy, corrosive gas.

Columbia had been experimenting with copper barriers but abandoned them late in 1942 in favor of nickel, the only common metal that resisted hexafluoride corrosion. Compressed nickel powder made a suitably rugged but insufficiently fine-pored barrier material; electro-deposited nickel mesh made a suitably fine-pored but insufficiently rugged alternative. A self-educated Anglo-American interior decorator, Edward Norris, had devised the electro-deposited mesh originally for a new kind of paint sprayer he invented; he joined the Columbia project in 1941 and worked with chemist Edward Adler, a young Urey protégé, to adapt his invention to gaseous diffusion. The resulting Norris-Adler barrier in its nickel incarnation seemed in January 1943 to be improvable eventually to production quality, whereupon Columbia began installing a pilot plant in the basement of Schermerhorn Laboratory and Groves authorized full-scale barrier production. The Houdaille-Hershey Corporation took on that assignment on April 1, the day the gates began operating at Oak Ridge, planning a new factory for the purpose in Decatur, Illinois.

Suitable barrier material was the worst but not the only problem Columbia studied and Groves engineered. Hex attacked organic materials ferociously: not a speck of grease could be allowed to ooze into the gas stream anywhere along the miles and miles of pipes and pumps and barriers. Pump seals therefore had to be devised that were both gastight and greaseless, a puzzle no one had ever solved before that required the development of new kinds of plastics. (The seal material that eventually served at Oak Ridge came into its own after the war under the brand name Teflon.) A single pinhole leak anywhere in the miles of pipes would confound the entire system; Alfred O. Nier developed portable mass spectrometers to serve as subtle leak detectors. Since pipes of solid nickel would exhaust the entire U.S. production of that valuable resource, Groves found a company willing to nickel-plate all the pipe interiors, a difficult new process accomplished by filling the pipes themselves with plating solution and rotating them as the plating current did its work.

The plant that would hold thousands of diffusion tanks, the largest of them of 1,000 gallon capacity, would be necessarily monumental: four stories high, almost half a mile long in the shape of a U, a fifth of a mile wide, 42.6 acres under roof, some 2 million square feet, more than twice the total ground area of Y-12's Alpha and Beta buildings. K-25, as the gaseous-diffusion complex was designated, needed more than a narrow ridge valley. The building and operating contractors, Kellex and Union Carbide, found a relatively flat site along the Clinch River at the southwestern end of the reservation; the first surveying, for the coal-fired power plant needed to run the factory, began on May 31, 1943.

Rather than designing and setting thousands of different columns for footings the construction contractors leveled and compacted the entire K-

25 foundation area, plowing, drying and moving in the process nearly 100,000 cubic yards of red clay. That took months; the first concrete—200,000 cubic yards—was not poured until October 21. By then the continuing failure to develop an adequate barrier material had led Groves to decide to lop off the unfinished plant's upper stages and limit its enrichment potential to less than 50 percent U235—it would have been capable of taking natural uranium all the way to pure U235 with its full complement of diffusers—and to use this enriched material to feed the Beta calutrons at Y-12.

Kellex succeeded in devising a promising new barrier material in the autumn of 1943 that combined the best features of the Norris-Adler barrier and the compressed nickel-powder barrier. The problem then was what to do about the Houdaille-Hershey plant under construction in Decatur, which was designed to produce Norris-Adler. Should it be stripped and reequipped to manufacture the new barrier at the price of some delay in starting up K-25? Or should the several barrier-development teams make a final concerted effort to improve Norris-Adler to production quality? Over these significant questions Groves and Harold Urey violently clashed.

Kellex wanted to strip the Houdaille-Hershey plant and convert it, preferring delay to the risk of failure. Urey thought abandoning the Norris-Adler barrier would mean forgoing the production of U235 by gaseous diffusion in time to shorten the war. In which case he saw no reason to continue building K-25; its high priority, he argued, would even hinder the war effort by displacing more immediately useful production.

Groves decided to submit the dispute to an unusual review committee: the experts who had worked on gaseous diffusion in England. With the renewal of interchange between the British and American atomic bomb programs that autumn the British had arranged to send a delegation to work in America. Led by Wallace Akers of ICI, the group included Franz Simon and Rudolf Peierls. It met with both sides—Kellex and Columbia—on December 22 and then settled in to review American progress.

The participants reconvened early in January 1944. The new barrier, the British concluded, would probably be superior eventually to the Norris-Adler, but they thought the months of research on the Norris-Adler must count decisively in its favor if time was of the essence. The new barrier had been manufactured so far only by hand in small batches. Yet K-25 would require acres of it to fill the planned 2,892 stages of the diffusion plant's cascade.

Then Kellex set a trap: it proposed to produce the new barrier by hand by piecework—thousands of workers each duplicating the simple laboratory process Kellex had initially devised—and claimed that by doing so it could match or beat the Norris-Adler production schedule. When the British had recovered from their surprise at the novelty of the proposal they signaled their preference for the new barrier by agreeing that if production was possible it ought to be pursued. That agreement sprang the trap; with the British implicitly committed, the American engineers revealed that they could only manufacture the new barrier by stripping the Houdaille-Hershey plant and forgoing Norris-Adler production entirely.

Groves in any case had already decided, the day before the January meeting, to switch over to the new barrier; the British review then simply ratified his decision. By changing barriers rather than abandoning gaseous diffusion he confirmed what many Manhattan Project scientists had not yet realized: that the commitment of the United States to nuclear weapons development had enlarged from the seemingly urgent but narrow goal of beating the Germans to the bomb. Building a gaseous-diffusion plant that would interfere with conventional war production, would eventually cost half a billion dollars but would almost certainly not contribute significantly to shortening the war meant that nuclear weapons were thenceforth to be counted a permanent addition to the U.S. arsenal. Urey saw the point and withdrew; "from that time forward," write his colleague biographers, "his energies were directed to the control of atomic energy, not its applications."

Twelve days after Enrico Fermi proved the chain reaction in Chicago on December 2, 1942, Groves had assembled a list of criteria for a plutonium production area and definitely and finally ruled out Tennessee. "The Clinton site . . . was not far from Knoxville," he comments, "and while I felt that the possibility of serious danger was small, we could not be absolutely sure; no one knew what might happen, if anything, when a chain reaction was attempted in a large reactor. If because of some unknown and unanticipated factor a reactor were to explode and throw great quantities of highly radioactive materials into the atmosphere when the wind was blowing toward Knoxville, the loss of life and the damage to health in the area might be catastrophic." Such an accident might "wipe out all semblance of security in the project," Groves could imagine, and it might render the electromagnetic and gaseous-diffusion plants "inoperable." Better to site plutonium production somewhere far away.

The production piles needed plentiful electricity and water for blowing and cooling the helium that was planned to cool them. For safety they needed space. Those criteria suggested the great river systems of the Far West, particularly the Columbia River basin. Groves sent out an officer who would administer the plutonium reservation along with the civilian engineer who would supervise construction for Du Pont. Besides picking the site he wanted the two men to get used to working together. They did, agreeing on a promising location in south-central Washington State, and arrived back in Groves' office on New Year's Eve to report. The general received a real estate appraisal on January 21, 1943. By then he had already personally walked the ground.

Eastward of the Cascade Range, twenty air miles east of the city of Yakima, the blue, cold, fast-running Columbia River bends east, then northeast, abruptly ninety degrees southeast and finally due south through a flat, arid scrubland on its last excursion toward the continental interior before it makes its great bend below Pasco to course directly westward two hundred fifty miles to the sea. Even that far inland the river is wide and deep and veined in season with salmon, but the sandy plain surrounding wins little of the river's water and the barrier of the Cascades denies it more than six inches a year of rain.

The site Groves' representatives discovered, and Groves acquired at the end of January at a cost of about $5.1 million, was contained within the eastward excursion of the Columbia: some 500,000 acres, about 780 square miles, devoted primarily to sheep grazing but varied with a few irrigated orchards and vineyards and a farm or two thriving in wartime on irrigated crops of peppermint. Temperatures ranged from a maximum of 114° in the long, dry summers to rare −27° winter lows. Roads were sparse on the roughly circular thirty-mile tract. A Union Pacific railroad line crossed one corner; a double electric power line of 230 kilovolts traversed the northwest sector on its way from Grand Coulee Dam to Bonneville Dam. Gable Mountain, an isolated basalt outcropping that rose five hundred feet above the sedimentary plain a few miles southwest of the ninety-degree river bend, divided the riverside land at the bend from the interior. Midway down the tract where a ferry crossed the Columbia, a half-abandoned riverside village, population about 100, supplied a base of buildings and gave the Hanford Engineer Works its name.

Groves could hardly build Hanford until he knew more about the plant that would go there. It was clear that he would need enormous quantities of concrete to shield the production piles and chemical processing buildings; his Hanford engineer searched out accessible beds of gravel and aggregate to quarry. An accident might release radioactivity into the air; that called for thorough meteorological work. The river water needed study; so did the river's valuable salmon, to see how they would take to mild doses of transient radioactivity from pile discharge flow. Roads had to be paved, power sources tapped, hutments and barracks built for tens of thousands of construction workers.

What had come up once again for discussion early in 1943 was how the plutonium production piles—the Du Pont engineers were beginning to call them *reactors*—should be cooled. Crawford Greenewalt, in charge of plutonium production for Du Pont, continued to plan for helium cooling because the noble gas had no absorption cross section at all for neutrons. But it would need to be pumped through the piles under high pressure; that would require large, powerful compressors Greenewalt was not at all sure he had time to build. Enormous steel tanks would be needed to contain the gas; they would have to allow access to the pile but still remain airtight, a formidable challenge to engineer or even simply to weld.

Eugene Wigner came to the project's rescue. Fermi had found *k* for CP-1 higher than he expected. The Stagg Field pile had been assembled largely from uranium oxide. Its graphite had varied in quality, improving along the way. A production pile of pure uranium metal and high-quality graphite would find *k* higher yet—high enough, Wigner calculated, to make water cooling practical.

Wigner's team designed a 28- by 36-foot graphite cylinder lying on its side and penetrated through its entire length horizontally by more than a thousand aluminum tubes. Two hundred tons of uranium slugs the size of rolls of quarters would fill these tubes. Chain-reacting within 1,200 tons of graphite, the uranium would generate 250,000 kilowatts of heat; cooling water pumped through the aluminum tubes around the uranium slugs at the rate of 75,000 gallons per minute would dissipate that heat. The slugs would not go naked into the torrent; Wigner intended that they also should be separately sheathed in aluminum—canned. When they had burned long enough—100 days—to transmute about 1 atom in every 4,000 into plutonium the irradiated slugs could be pushed out the back of the pile simply by loading fresh slugs in at the front. The hot slugs would fall into a deep pool of pure water that would safely confine their intense but short-lived fission-product radioactivity. After 60 days they could be fished out and carted off for chemical separation.

The Wigner design was elegantly simple. Greenewalt saw engineering problems—in particular the question whether corrosion of the aluminum tubes would block the flow of cooling water—and studied helium and water side by side until the middle of February. Corrosion studies were promising. "With water of high purity," writes Arthur Compton, "the evidence indicated that no serious difficulties from this source should arise." Greenewalt opted then for water cooling. Wigner, whom Leo Szilard calls "the conscience of the Project from its early beginnings to its very end," who worried constantly about German progress, wondered angrily why it had taken Du Pont three months to see the value of a system he and his group had judged superior in the summer of 1942.

With that basic decision construction could begin at Hanford. Three production piles would go up at six-mile intervals along the Columbia River, two upstream and one downstream of its ninety-degree bend. Ten miles south, screened behind Gable Mountain, Du Pont would build four chemical-separation plants paired at two sites. The former town of Hanford would become a central construction camp serving all five construction areas.

The work proceeded slowly, dogged by recruiting problems. The nation at war had moved beyond full employment to severe labor shortages and men and women willing to camp out on godforsaken scrubland far from any major city were hard to find. Frequent sandstorms plagued the area, writes Leona Woods, now Leona Marshall after marrying fellow physicist John Marshall of Fermi's staff. "Local storms were caused by tearing up the desert floor for roads, and construction sites were suffocating. Wind-blown sand covered faces, hair, and hands and got into eyes and teeth. . . . After each storm, the number of people quitting might be as much as twice the average. When the storms were at their worst, buses and other traffic came to a stop until the roads were visible through the grey-black clouds of dust." Stoics who stayed on called the dust "termination powder."

"The most essential thing to bring with you is a padlock," a project recruiting pamphlet ominously announced. "The next important things are towels, coat hangers and a thermos bottle. Don't bring cameras or guns." Hanford, says Marshall, "was a tough town. There was nothing to do after work except fight, with the result that occasionally bodies were found in garbage cans the next morning." Du Pont built saloons with windows hinged for easy tear-gas lobbing. Eventually some 5,000 construction workers struggled in the desert dust and Du Pont built more than two hundred barracks to house them. Meat rationing stopped at the edge of the reservation; there were no meatless Tuesdays in the vast Hanford mess halls, a significant enticement for recruiting. The gray coyotes of the region fed sleek in turn on rabbits killed by cars and trucks driving the new reservation roads.

By August 1943 work had begun on the water-treatment plants for the three piles, capacity sufficient to supply a city of one million people. Du Pont released pile-design drawings in Wilmington, Delaware, on October 4 and the company's engineers staked out the first pile, 100-B, beside the Columbia on October 10. After excavating, reports an official history, "work gangs began to lay the first of 390 tons of structural steel, 17,400 cubic yards of concrete, 50,000 concrete blocks, and 71,000 concrete bricks that went into the pile buildings. Starting with the foundations for the pile and the deep water basins behind it where the irradiated slugs would be collected after discharge, the work crews were well above ground by the end of the year." The forty-foot windowless concrete monolith they were building was hollow, however: installation of B pile would not begin until February 1944.

"There was a large change of scale from the Chicago to the Hanford piles," Laura Fermi remarks. "As Fermi would have put it, they were different animals." So also were Ernest Lawrence's behemoth mass spectrometers and John Dunning's gaseous-diffusion factory with its 5 million barrier tubes. The mighty scale of the works at Clinton and Hanford is a measure of the desperation of the United States to protect itself from the most serious potential threat to its sovereignty it had yet confronted—even though that threat, of a German atomic bomb, proved to be an image in a darkened mirror. It is also a measure of the sheer recalcitrance of heavy-metal isotopes. Niels Bohr had insisted in 1939 that U235 could be separated from U238 only by turning the country into a gigantic factory. "Years later," writes Edward Teller, "when Bohr came to Los Alamos, I was prepared to say, 'You see . . .' But before I could open my mouth, he said, 'You see, I told you it couldn't be done without turning the whole country into a factory. You have done just that.' "

The monumental scale reveals another desperation as well: how ambitiously the nation was moving to claim the prize. And to deny it to others, even to the British until Winston Churchill turned Franklin Roosevelt's head at the conference in Quebec in August 1943, where Operation Overlord, the 1944 invasion of Europe across the beaches of Normandy, was planned. Before then, in June, Groves had demonstrated this last desperation at its most overweening: he proposed to the Military Policy Committee that the United States attempt to acquire total control of all the world's known supplies of uranium ore. When the Union Minière refused to reopen its flooded Shinkolobwe Mine in the Belgian Congo, Groves had to turn to the British, who owned a significant minority interest in the Belgian firm, for help; after Quebec the partnership evolved into an agreement between the two nations known as the Combined Development Trust to search out world supplies. That uranium is common in the crust of the earth to the extent of millions of tons Groves may not have known. In 1943, when the element in useful concentrations was thought to be rare, the general, acting on behalf of the nation to which he gave unquestioning devotion, exercised himself to hoard for his country's exclusive use every last pound. He might as well have tried to hoard the sea.

Work toward an atomic bomb had begun in the USSR in 1939. A thirty-six-year-old nuclear physicist, Igor Kurchatov, the head of a major laboratory since his late twenties, alerted his government then to the possible military significance of nuclear fission. Kurchatov suspected that fission research might be under way already in Nazi Germany. Soviet physicists realized in 1940 that the United States must also be pursuing a program when the names of prominent physicists, chemists, metallurgists and mathematicians disappeared from international journals: secrecy itself gave the secret away.

The German invasion of the USSR in June 1941 temporarily ended what had hardly been begun. "The advance of the enemy turned everyone's thoughts and energies to one single job," writes Academician Igor Golovin, a colleague of Kurchatov and his biographer: "to halt the invasion. Laboratories were deserted. Equipment, instruments and books were packed up, and valuable records shipped east for safety." The invasion rearranged research priorities. Radar now took first place, naval mine detection second, atomic bombs a poor third. Kurchatov moved to Kazan, four hundred miles east of Moscow beyond Gorky, to study defenses against naval mines.

In Kazan at the end of 1941 he heard from George Flerov, one of the two young physicists in his Moscow laboratory who had discovered the spontaneous fission of uranium in 1940 and reported their discovery in a cable to the *Physical Review.** Flerov had attended an international meeting of scientists in Moscow in October and heard Peter Kapitza, Ernest Rutherford's protégé, when asked what scientists could do to help the war effort, respond in part:

> In recent years a new possibility—nuclear energy—has been discovered. Theoretical calculations show that, if a contemporary bomb can for example destroy a whole city block, an atomic bomb, even of small dimensions, if it can be realized, can easily annihilate a great capital city having a few million inhabitants.

Thus recalled to their earlier work, Flerov challenged Kurchatov as he had already in a similar letter challenged the State Defense Committee that "no time must be lost in making a uranium bomb." The first requirement was fast-neutron research, he wrote. The MAUD Report had only just made that necessity clear to the United States.

Kurchatov disagreed. Research toward a uranium weapon seemed too far removed from the immediate necessities of war. But the Soviet government in the meantime had assembled an advisory committee that included Kapitza and the senior Academician Abram Joffe, Kurchatov's mentor. The

* Spontaneous fission, a relatively rare nuclear event, differs from fission caused by neutron bombardment; it occurs without outside stimulus as a natural consequence of the instability of heavy nuclei.

committee endorsed atomic bomb research and recommended Kurchatov to head it. Somewhat reluctantly he accepted.

"So it was that from early 1943 on," writes his colleague A. P. Alexandrov, "work on this difficult problem was resumed in Moscow under the leadership of Igor Kurchatov. Nuclear scientists were recalled from the front, from industry, from the research institutes which had been evacuated to the rear. Auxiliary work began in many places." Auxiliary work included building a cyclotron. Kurchatov moved his institute out of the Soviet capital to an abandoned farm near the Moscow River in the summer of 1943. An artillery range nearby offered an area for explosives testing; "Laboratory No. 2" would be the Soviet Union's Los Alamos. By January 1944 Kurchatov had assembled a staff of only about twenty scientists and thirty support personnel. "Even so," writes Herbert York, "they did experiments and made theoretical calculations concerning the reactions involved in both nuclear weapons and nuclear reactors, they began work designed to lead to the production of suitably pure uranium and graphite, and they studied various possible means for the separation of uranium isotopes." But the Soviet Bear was not yet fully aroused.

"The kind of man that any employer would have fired as a troublemaker." Thus Leslie Groves described Leo Szilard in an off-the-record postwar interview, as if the general had arrived first at fission development and Szilard had only been a hireling. Groves seems to have attributed Szilard's brashness to the fact that he was a Jew. Upon Groves' appointment to the Manhattan Project he almost immediately judged Szilard a menace. They proceeded to fight out their profound disagreements hand to hand.

The heart of the matter was compartmentalization. Alice Kimball Smith, the historian of the atomic scientists whose husband Cyril was associate division leader in charge of metallurgy at Los Alamos, defines the background of the conflict:

> If the Project could have been run on ideas alone, says Wigner, no one but Szilard would have been needed. Szilard's more staid scientific colleagues sometimes had trouble adjusting to his mercurial passage from one solution to another; his army associates were horrified, and to make matters worse, Szilard freely indulged in what he once identified as his favorite hobby—baiting brass hats. General Groves, in particular, had been outraged by Szilard's unabashed view that army compartmentalization rules, which forbade discussion of lines of research that did not immediately impinge on each other, should be ignored in the interests of completing the bomb.

The issue for Szilard was openness within the project to facilitate its work. "There is no way of telling beforehand," he wrote in a 1944 discussion of the problem, "what man is likely to discover and invent a new method which will make the old methods obsolete." The issue for Groves, to the contrary, was security.

At first Szilard bent the rules and Groves threatened him. In late October 1942, while Fermi moved toward building CP-1, Szilard apparently badgered the Du Pont engineers who arrived in Chicago to take over pile design. Arthur Compton saw this activity as obstructive but not necessarily subversive; on October 26 he wired Groves that he had given Szilard two days TO REMOVE BASE OF OPERATIONS TO NEW YORK. ACTION BASED ON EFFICIENT OPERATION OF ORGANIZATION NOT ON RELIABILITY. ANTICIPATE PROBABLE RESIGNATION. Compton did not know his man. Szilard would not resign, for the simple reason that he believed he was needed to help beat Germany to the bomb. Compton proposed surveillance: SUGGEST ARMY FOLLOW HIS MOTIONS BUT NO DRASTIC ACTION NOW. Two days later Compton hurriedly wired Groves to desist: SZILARD SITUATION STABILIZED WITH HIM REMAINING CHICAGO OUT OF CONTACT WITH ENGINEERS. SUGGEST YOU NOT ACT WITHOUT FURTHER CONSULTATION CONANT AND MYSELF.

Groves had prepared drastic action indeed. On the stationery of the Office of the Chief of Engineers, over a signature block reserved for the Secretary of War, he had drafted a letter to the U.S. Attorney General calling Leo Szilard an "enemy alien" and proposing that he "be interned for the duration of the war." Compton's telegram forestalled an ugly arrest and the letter was never signed or sent.

But the incident raised the issue of Szilard's loyalty and prejudiced Groves implacably against him. Szilard responded forthrightly; he assembled a large collection of documents from the 1939–40 period demonstrating his part in carrying the news of fission to Franklin Roosevelt and, pointedly, his efforts to enforce voluntary secrecy among physicists in the United States, Britain and France. Compton, waffling, sent the documents to Groves in mid-November with an implicit endorsement of Szilard's stand. The first Groves-Szilard confrontation thus ended in stalemate. Szilard saw how much raw power Groves commanded. Groves learned how deep were Szilard's roots in the evolution of atomic energy research and perhaps also that men he considered vital to the project—Fermi, Teller, Wigner—were Szilard colleagues of long standing and would have to be taken into account.

As political dissidents have done in the Soviet Union, Szilard embarked next on a careful campaign to negotiate changes by insisting meticulously on the enforcement of his legal rights. His opening sally came December 4, two days after Fermi proved the chain reaction. In a quiet memorandum to Arthur Compton he noted that the official responsible for handling NDRC patents had requested patent applications "for inventions relating to the chain reaction." That raised the question, Szilard wrote, of how to deal with inventions "made and disclosed before we had the benefit of the financial support of the government." He and Fermi would be glad to file a joint application, but only if they could be sure they retained their rights to their earlier separate inventions. The memorandum continues in this straightforward style until its final paragraph, which throws down the gauntlet:

> My present request clearly represents a change of [my] attitude with respect to patents on the uranium work, and I would appreciate an opportunity to explain to you and also to the government agency which may be involved, my reasons for it.

Previously Szilard had believed he would have equal voice in fission development. Since he had now been compartmentalized, his freedom of speech restrained, his loyalty challenged, he was prepared to actuate the only leverage at hand, his legal right to his inventions.

Compton sent Szilard's request to Lyman Briggs, whose responsibilities within the OSRD included patent matters; Briggs thought the Army ought to handle it. Szilard waited until the end of December, heard nothing and advanced further into the field. In a second memorandum he told Compton he wanted to apply for a patent on "the basic inventions which underlie our work on the chain reaction on unseparated uranium . . . which were made before government support for this research was forthcoming." The patent could be registered in his name alone or jointly with Fermi; he would be willing "to assign this patent at this time to the government for such financial compensation as may be deemed fair and equitable." The memorandum mentions no amount; according to Army security files Szilard asked for $750,000. But the issue was not compensation; the issue was representation:

> I wish to take this opportunity to mention that the question of patents was discussed by those who were concerned in 1939 and 1940. At that time it was proposed by the scientists that a government corporation should be formed which would look after the development of this field and . . . be the recipient of the patents. It was assumed that the scientists would have adequate representation within this government owned corporation. . . .
>
> In the absence of such a government owned corporation in which the scientists can exert their influence on the use of funds, I do not now propose to

assign to the government, without equitable compensation, patents covering the basic inventions.

Burdened by Manhattan Project security, with Du Pont taking over plutonium production and the Army moving hundreds of thousands of cubic yards of earth in unprecedented construction, Leo Szilard was advancing singlehandedly to attempt to extricate the process of decision from governmental restraints and to return it to the hands of the atomic scientists.

Compton understood the extent of the challenge. He sent Szilard's two memoranda directly to Conant, whose office received them on January 11, 1943. "Szilard's case is perhaps unique," Compton wrote the NDRC chairman, "in that for a number of years the development of this project has continuously occupied his primary attention. . . . There is no doubt that he is among the few to whom the United States Government can look for establishing basic claims for invention. The matter is thus one of real importance to our Government."

Before Washington could respond Szilard had to fight off a harassing attack from the flank. It strengthened his resolve. He discovered that a French patent filed originally by Frédéric Joliot's group had been published in Australia and he and Fermi had missed the deadline for filing challenges. Some of their claims overlapped the French work. "This is, I am afraid, an irreparable loss," he told Compton. He had now started writing down his own inventions, he said, and hoped to file a number of patents in the near future. Until he had done so he wanted to be removed from the payroll of the University of Chicago to avoid legal complications. In the meantime he would toil on once again as a free volunteer: "It would not be my intention to interrupt or slow down the work which I am doing in the laboratory at present."

Conant bumped Compton's letter up to Bush, who answered it personally and to the point with Yankee canniness. Inventions scientists made after joining the project belonged to the project, Bush told Compton; unless Szilard had disclosed his previous inventions to the University of Chicago at the time of his employment he had only a very short leg to stand on, if any at all. Genially the OSRD director outlined the proper legal procedure for secret patent filings and then kicked at the leg Szilard had left: "It is my understanding that none of this procedure has been gone through with in the case of Dr. Szilard." Bush either did not understand or chose to misunderstand Szilard's idea of an autonomous organization of scientists to guide nuclear energy development: "I gather that Dr. Szilard is particularly anxious that the proceeds arising from his early activity in invention in this field, if such eventuate, should in some way become available for the furtherance of scientific research." He thought that was admirable, but he also thought it had nothing to do with the government. Nor did he intend that it should.

By the time Bush's letter reached Compton the Met Lab director had gone another round with Szilard. Szilard asked for a raise based upon the value to the project of his inventions. Compton took the position that Szilard had signed over all his rights to his inventions to the government for as long as he was in the government's employ. Szilard would not sign a renewal contract under those terms. Trying to keep him aboard, Compton proposed raising his salary from $550 to $1,000 a month on the basis that the higher level was "comparable with the other original sponsors of this project, Messrs. Fermi and Wigner." That might have been acceptable to Szilard, since it tacitly acknowledged the special worth of the three physicists' participation, including presumably their early inventions, but Compton had to clear it with Conant. Until the arrangement was cleared and a new contract signed Szilard would remain off the payroll.

Compton reported Bush's response to Szilard in late March. There matters stood until early May, when Szilard with restrained exasperation proposed to proceed with filing patent applications. He asked that Groves designate someone to act as his legal adviser. The Army general supplied a Navy captain, Robert A. Lavender, who was attached to the OSRD in Washington, and Szilard met frequently with Lavender in the spring and early summer to discuss his claims.

Somewhere along the way Groves put Szilard under surveillance. The brigadier still harbored the incredible notion that Leo Szilard might be a German agent. The surveillance was already months old in mid-June when the MED's security office suggested discontinuing it. Groves rejected the suggestion out of hand: "The investigation of Szilard should be continued despite the barrenness of the results. One letter or phone call once in three months would be sufficient for the passing of vital information and until we know for certain that he is 100% reliable we cannot entirely disregard this person." He apparently equated disagreement with disloyalty and scaled the ratio of the two conditions directly: anyone who caused him as much pain as Leo Szilard must be a spy. It followed that he ought to be watched.

The surveillance of an innocent but eccentric man makes gumshoe comedy. Szilard traveled to Washington on June 20, 1943, and in preparation for the visit an Army counterintelligence agent reviewed his file:

> The surveillance reports indicate that Subject is of Jewish extraction, has a fondness for delicacies and frequently makes purchases in delicatessen stores, usually eats his breakfast in drug stores and other meals in restaurants, walks

> a great deal when he cannot secure a taxi, usually is shaved in a barber shop, speaks occasionally in a foreign tongue, and associates mostly with people of Jewish extraction. He is inclined to be rather absent minded and eccentric, and will start out a door, turn around and come back, go out on the street without his coat or hat and frequently looks up and down the street as if he were watching for someone or did not know for sure where he wanted to go.

Armed with these profundities a Washington agent observed the Subject arriving at the Wardman Park Hotel at 2030 hours—8:30 P.M.—on June 20 and composed a contemporary portrait:

> Age, 35 or 40 yrs; height, 5′6″; weight, 165 lbs; medium build; florid complexion; bushy brown hair combed straight back and inclined to be curly, slight limp in right leg causing droop in right shoulder and receding forehead. He was wearing brown suit, brown shoes, white shirt, red tie and no hat.

Szilard worked the next morning at the Carnegie Institution with Captain Lavender. Wigner arrived at the Wardman Park for an overnight stay ("Mr. Wigner is approximately 40 years of age, medium build, bald head, Jewish features and was conservatively dressed") and the two Hungarians, both of them presumably with justice on their minds, went off for a tour of the Supreme Court (the cabbie "said that they did not talk in a foreign tongue and there was nothing in their conversation to attract his attention. . . . He said they more or less gave him the impression that they were 'on a lark' "). In the evening they sat "on a bench by the [hotel] tennis courts where both pulled off their coats, rolled up their sleeves and talked in a foreign language for some time."

Wigner checked out early in the morning; Szilard took a cab to the Navy Building at 17th and Constitution Avenue, "entered the reception room . . . and told one of the ladies that he wished to see Commander Lewis Strauss about personal business. He stated that he had an appointment. . . . He also told the lady that he was a friend of Commander Strauss' and was interested in getting into a branch of the Navy." The Naval Research Laboratory had continued work on nuclear power for submarine propulsion independently of the Manhattan Project and that institution may have been the one Szilard had in mind. Or he may have been practicing misdirection. Strauss took him to lunch at the Metropolitan Club and apparently discouraged him from transferring; back at his hotel he wired Gertrud Weiss that he expected to arrive at the King's Crown at 8:30 P.M. and left that afternoon for New York.

Since he worked for Vannevar Bush, Lavender was hardly a disinterested consultant; when he met again with Szilard on July 14 he informed the physicist that his documents "failed to disclose an operable pile," meaning that in his opinion Szilard could not claim a patentable invention. (Ten years after the end of the war Szilard and Fermi won a joint patent for their invention of the nuclear reactor.) Szilard realized then, if not before, that he needed private counsel and asked that an attorney who could act in his behalf be cleared.

The battle was almost decided. Szilard retreated to New York. He negotiated now not only with Lavender but with Army Lieutenant Colonel John Landsdale, Jr., Groves' chief of security. In an October 9 letter to Szilard, Groves summed up the blunt exchange over which the three men bargained: "You were assured [by Lavender and Landsdale] that as soon as you were able to convey full rights [to any inventions made prior to government employment], negotiations would be entered into with a view to acquisition by the Government of any rights you may have and your reemployment on Government contracts. . . . I repeat this assurance." That is, Szilard could trade his patent rights, if any, for the privilege of working to beat the Germans to the bomb.

Groves and Szilard arranged a temporary truce—the general may have imagined it was a surrender—at a meeting in Chicago on December 3. The Army agreed to pay Szilard $15,416.60 to reimburse him for the twenty months when he worked unpaid and out-of-pocket at Columbia and for lawyers' fees.

The general had attempted several times to force Szilard to sign a document promising "not to give any information of any kind relating to the project to any unauthorized person." Szilard had consistently agreed verbally to that restriction and just as consistently refused as a matter of honor to sign. He meant to continue protesting and on January 14, 1944, he began again with a three-page letter to Vannevar Bush. He knew fifteen people, he told Bush, "who at one time or another felt so strongly about [compartmentalization] that they intended to reach the President." The central issue as always was freedom of scientific speech: "Decisions are often clearly recognized as mistakes at the time when they are made by those who are competent to judge, but . . . there is no mechanism by which their collective views would find expression or become a matter of record."

In this letter for the first time Szilard emphasized a purpose to his urgency beyond beating the Germans to the bomb: that the bomb might be used and become grimly known.

> If peace is organized before it has penetrated the public's mind that the potentialities of atomic bombs are a reality, it will be impossible to have a peace that is based on reality. . . . Making some allowances for the further development of the atomic bomb in the next few years . . . this weapon will be so pow-

> erful that there can be no peace if it is simultaneously in the possession of any two powers unless these two powers are bound by an indissoluble political union.... It will hardly be possible to get political action along that line unless high efficiency atomic bombs have actually been used in this war and the fact of their destructive power has deeply penetrated the mind of the public.

Which was the explanation Szilard now gave for challenging the Army and Du Pont: "This for me personally is perhaps the main reason for being distressed by what I see happening around me."

Bush insisted in return that all was well. "I feel that the record when this effort is over," he wrote Szilard, "will show clearly that there has never at any time been any bar to the proper expression of opinion by scientists and professional men within their appropriate sphere of activity in this whole project." But he was willing to meet with Szilard if that was what the physicist wanted. In February, preparing for that meeting, Szilard drafted forty-two pages of notes. Much in those notes is specific and local; here and there basic issues are joined.

Since invention is unpredictable, Szilard writes, "the only thing we can do in order to play safe is to encourage sufficiently large groups of scientists to think along those lines and to give them all the basic facts which they need to be encouraged to such activity. This was not done in the past [in the Manhattan Project] and it is not being done at present." He tracked the consequences of the government's policies of restriction:

> The attitude taken toward foreign born scientists in the early stages of this work had far reaching consequences affecting the attitude of the American born scientists. Once the general principle that authority and responsibility should be given to those who had the best knowledge and judgment is abandoned by discriminating against the foreign born scientists, it is not possible to uphold this principle with respect to American born scientists either. If authority is not given to the best men in the field there does not seem to be any compelling reason to give it to the second-best men and one may give it to the third- or fourth- or fifth-best men, whichever of them appears to be the most agreeable on purely subjective grounds.

Wigner's early discouragement was an "incalculable loss," Szilard thought; the fact that Fermi was excluded from centrifuge development work at Columbia "visibly affected" him "and he has from that time on shown a very marked attitude of being always ready to be of service rather than considering it his duty to take the initiative."

Finally, Szilard judged the Met Lab moribund, its services rejected and its spirit broken, and pronounced its epitaph:

> The scientists are annoyed, feel unhappy and incapable of living up to their responsibility which this unexpected turn in the development of physics has thrown into their lap. As a consequence of this, the morale has suffered to the point where it almost amounts to a loss of faith. The scientists shrug their shoulders and go through the motions of performing their duty. They no longer consider the overall success of this work as their responsibility. In the Chicago project the morale of the scientists could almost be plotted in a graph by counting the number of lights burning after dinner in the offices in Eckhart Hall. At present the lights are out.

But Leo Szilard at least was not yet done with protest.

Enrico Fermi took the initiative at least once during the war. Perhaps influenced by the enthusiasm he found at Los Alamos for weapons-making, he proposed at the time of the April 1943 conference—privately to Robert Oppenheimer, it appears—that radioactive fission products bred in a chain-reacting pile might be used to poison the German food supply.

The possibility of using radioactive material bred in a nuclear reactor as a weapon of war had been mentioned by Arthur Compton's National Academy of Sciences committee in 1941. German development of such a weapon began worrying the scientists at the Met Lab late in 1942, on the assumption that Germany might be a year or more ahead of the United States in pile development. If CP-1 went critical in December 1942, they argued, the Germans might have had time by then to run a pile long enough to create fiercely radioactive isotopes that could be mixed with dust or liquid to make radioactive (but not fissionable) bombs. Germany might then logically attempt preemptively to attack the Met Lab, if not American cities. German development of radioactive warfare, another vision in a dark mirror, seemed to the leaders of the Manhattan Project to require countering by examination into parallel U.S. development; the S-1 Committee gave such assignment to a subcommittee consisting of James Bryant Conant as chairman and Arthur Compton and Harold Urey as members. That subcommittee went to work sometime before May 1943, probably before February.

Fermi would have known of the Met Lab discussions. His proposal to Oppenheimer at the April conference was different from those essentially defensive concerns, however, and clearly offensive in intent. He may well have been motivated in part by his scientific conservatism: may have asked himself what recourse was open to the United States if a fast-fission bomb proved impossible—it could not be demonstrated by experiment for at least two years—and have found the answer in the formidable neutron flux of CP-1 and its intended successors. Oppenheimer swore Fermi to intimate secrecy within the larger secrecy of the Manhattan Project; when the Italian laureate returned to Chicago he went quietly to work.

In May Oppenheimer traveled to Washington. Among other duties he reported Fermi's ideas to Groves and learned of the Conant subcommittee. Back at Los Alamos on May 25 he wrote Fermi a warm letter reporting what he had found. He attributed the subcommittee assignment to a request from the Army Chief of Staff, George Marshall, although it seems far likelier that the study originated within the Manhattan Project. "I therefore, with Groves' knowledge and approval, discussed with [Conant] the application [i.e., poisoning German food supplies] which seemed to us so promising."

Oppenheimer had also discussed Fermi's idea with Edward Teller. The isotope the men identified that "appears to offer the highest promise" was strontium, probably strontium 90, which the human body takes up in place of calcium and deposits dangerously and irretrievably in bone. Teller thought that separating the strontium from other pile products "is not a very major problem." Oppenheimer wanted to delay the work until "the latest safe date," he told Fermi further, so that they would have "a much better chance of keeping your plan quiet." He did not even want to include Compton in any immediate discussion. Summarizing, he wrote in part:

> I should recommend delay if that is possible. (In this connection I think that we should not attempt a plan unless we can poison food sufficient to kill a half a million men, since there is no doubt that the actual number affected will, because of non-uniform distribution, be much smaller than this.)

There is no better evidence anywhere in the record of the increasing bloody-mindedness of the Second World War than that Robert Oppenheimer, a man who professed at various times in his life to be dedicated to *Ahimsa* ("the Sanscrit word that means doing no harm or hurt," he explains) could write with enthusiasm of preparations for the mass poisoning of as many as five hundred thousand human beings.

Mid-1943 was in any case a season of great apprehension among the atomic scientists, who saw Nazi Germany beginning to lose the war and sensed that country's desperation. The Manhattan Project expected to produce atomic bombs by early 1945; if Germany had begun fission research in 1939 at similar scale it should have bombs nearly in hand. Hans Bethe and Edward Teller wrote Oppenheimer in a memorandum on August 21:

> Recent reports both through the newspapers and through secret service, have given indications that the Germans may be in possession of a powerful new weapon which is expected to be ready between November and January. There

> seems to be a considerable probability that this new weapon is tubealloy [i.e., uranium]. It is not necessary to describe the probable consequences which would result if this proves to be the case.
>
> It is possible that the Germans will have, by the end of this year, enough material accumulated to make a large number of gadgets which they will release at the same time on England, Russia and this country. In this case there would be little hope for any counter-action. However, it is also possible that they will have a production, let us say, of two gadgets a month. This would place particularly Britain in an extremely serious position but there would be hope for counter-action from our side before the war is lost, provided our own tubealloy program is drastically accelerated in the next few weeks.

The memorandum goes on to criticize the handling of production "entirely by large companies" —the Hungarian threnody Szilard and Wigner also sounded—and to propose a crash program directed by Urey and Fermi to build heavy-water piles. Nothing seems to have come of the Bethe-Teller proposal—Hitler's secret weapons proved to be the V-1 and V-2 rockets then in development at Peenemünde, the first of which crossed the English coast on June 13, 1944—but it captures the mid-war mood.

Less worrisome was radioactive dusting. Conant's subcommittee considered the possibilities and concluded that they were "rather remote." Conant emphasized that he thought it *"extremely* unlikely that a radioactive weapon will be used against the U.S. and unlikely the weapon will be used at all." Groves eventually proposed to George Marshall that a handful of officers be trained in the use of Geiger counters and sent to England to observe. Preparing for the Normandy invasion, Marshall approved.

It was easier for Americans guarded by the wide moat of the Atlantic than for the British to dismiss the possibility of radioactive attack. Sir John Anderson, Chancellor of the Exchequer, a scientist and the member of Churchill's cabinet responsible for the Tube Alloys program, discussed the question with Conant at lunch at the Cosmos Club in Washington in August 1943. He was concerned particularly about German heavy-water production because British scientists believed they had found a way to separate light from heavy water at five times the efficiency of existing processes and feared their German counterparts might have made the same discovery. Heavy water would certainly work to moderate a chain-reacting pile. And such a machine might be used to breed radioactive isotopes for dusting London.

The British therefore kept closer watch on the High Concentration Plant at Vemork in Norway. It had not been damaged beyond repair. To the contrary, intelligence sources reported that summer, it had begun production again in April; German scientists had shipped heavy water from laboratory stocks in Germany to refill the various cells and speed restoration of the cascade.

When Niels Bohr escaped from Stockholm to Scotland on October 6, 1943, he carried with him Werner Heisenberg's drawing of an experimental heavy-water reactor. Bohr met more than once in London that autumn with Sir John Anderson; Anderson matched up Bohr's information with the Conant subcommittee's radioactive-warfare study and the Norwegian underground's news of Vemork's renewed production and concluded that the plant once again urgently required attack. The Nazis had significantly increased security at Vemork, which ruled out another commando raid. After British and American representatives discussed the problem in Washington George Marshall authorized precision bombing.

American Eighth Air Force B-17's climbed northeast from British bases before dawn on the morning of November 16. To minimize Norwegian casualties the aircraft were scheduled to drop their bombs during the Norsk Hydro lunch period, between 11:30 A.M. and noon. No German fighters came up from the defensive airfields of western Norway to delay them and they elected to circle over the North Sea to kill time before penetrating the Scandinavian peninsula. That alerted German flak, which took a limited toll as the bombers crossed the coast. One hundred forty got through to Vemork and released more than seven hundred 500-pound bombs. None hit the aiming point but four destroyed the power station and two damaged the electrolysis unit that supplied hydrogen to the High Concentration Plant, effectively shutting it down.

Abraham Esau of the Reich Research Council decided then to rebuild in Germany. To expedite construction the council planned to dismantle the Vemork plant and remove it to the Reich. The Norwegian underground reported that decision to London. Anderson was less concerned with the plant itself—Germany had only limited hydroelectricity to divert to its operation—than with the heavy water preserved in its cascade. British intelligence asked the Norwegians to keep watch.

Word came by way of clandestine shortwave radio from the Rjukan area on February 9, 1944, that the heavy water would be transported under guard to Germany within a week or two—not enough warning to prepare and drop in a squad of saboteurs. Knut Haukelid, who had spent the past year living on the land and organizing future military operations, was the only trained commando in the area except for the radio operator. He would have to destroy the heavy water alone with whatever amateur help he could assemble.

Haukelid slipped into Rjukan at night and met secretly with the new chief engineer at Vemork, Alf Larsen. Larsen agreed to help and they discussed possible operations. The heavy water, of enrichments varying from 97.6 down to 1.1 percent, would be transferred to some thirty-nine drums labeled potash-lye. "A one-man attack on Vemork," writes Haukelid, "I considered out of the question. . . . The only practical possibility, therefore, was to try to carry out an attack on the transport in one way or another." He and Larsen, joined later by the Vemork transport engineer, considered the various stages of the journey. The drums of water would go by train from Rjukan to the head of Lake Tinnsjö. From there the cars would be run onto a rail ferry to travel the length of the lake, proceeding beyond Tinnsjö again by train to the port where they would be loaded aboard a ship bound for Germany. Blowing up the trains would be difficult and bloody, since they would be crowded with Norwegian passengers; Haukelid finally decided to attempt to sink the ferry, which also carried passengers, into the 1,300-foot lake. The transport engineer agreed to arrange to dispatch the heavy water on a Sunday morning, when the ferry was usually least crowded.

Sabotaging the boat would almost certainly mean the deaths of some of the shipment's German guards, which would call down heavy reprisals in the Tinnsjö area against the Norwegian population. Haukelid radioed London for permission, emphasizing that his engineer compatriots had questioned if the results were worth the reprisals:

> The fact that the Germans were using heavy water for atomic experiments, and that an atomic explosion might possibly be brought about, was a thing we now talked of openly. At Rjukan they doubted very much whether the Germans had come in sight of a solution. They also doubted whether an explosion of the kind could be brought about at all.

The British begged to differ:

> The answer came from London the same day:
>
> "Matter has been considered. It is thought very important that the heavy water shall be destroyed. Hope it can be done without too disastrous results. Send our best wishes for success in the work. Greetings."

So Knut Haukelid laid his plans. He put on workman's clothes, packed his Sten gun into a violin case, identified which ferry would make the run on Sunday, February 20, 1944, the appointed day, and rode it with one eye on his watch. The *Hydro* was flat and bargelike with twin smokestacks jutting up side by side through its boxy superstructure. It reached the deepest part of the lake about thirty minutes after sailing and took twenty minutes then to cross to shallower waters. "We had therefore a margin of twenty minutes in which the explosion must take place." For even such generous leeway Haukelid needed something better than a time fuse: he needed electric detonators and a clock. He visited a Rjukan hardware-store owner at night for the detonators but was suspiciously turned away. One of his local compatriots had better luck. A handyman retired from Norsk Hydro donated one alarm clock to the cause; Alf Larsen supplied a backup. Haukelid modified them so that their hammers struck not bells but contact plates, closing a battery-powered electrical circuit that could fire the detonators.

Months earlier the British had dropped supplies to the Norwegian commando that included sticks of plastic explosive. Haukelid strung the stubby sticks together to make a circumferential loop to cut a hole in the bottom of the ferry. "As the Tinnsjö is narrow, the ferry must sink in less than five minutes, or else it would be possible to beach her. I . . . spent many hours sitting and calculating how large the hole must be for the ferry to sink quickly enough." To test his timing mechanism he hooked up a few spare detonators at his cabin on the mountain above Rjukan after a long night's work, set the alarm for evening and lay down to sleep. The detonators went off on schedule; he bolted bewildered from bed, grabbed the nearest gun and reflexively covered the door. "The timing apparatus seemed to be working properly."

On Saturday Haukelid and a local compatriot, Rolf Sörlie, slipped into Rjukan. It was crowded with German soldiers and SS police. An hour before midnight "Rolf and I went over to the bridge which crossed the river Maan and had a look at our target." The freight cars "had been run up under some lamps, and were guarded. . . . The train was to go at eight next morning, and the ferry was due to leave . . . at ten."

From the bridge the two men slipped to a back street where they met their driver in a car Haukelid had arranged with its owner to steal in the name of the King and return on Sunday morning. The owner had modified the car to run on methane and they were a long hour starting it. They picked up Larsen, who was prepared to escape Norway to avoid arrest after the work was done. He brought a suitcase of valuables and had come directly from a dinner party where he had heard a visiting concert violinist mention plans to leave on the morning ferry and had tried unsuccessfully to convince the musician to stay in the area one more day to sample its excellent skiing. Another Rjukan man also joined them. They drove to the lake well past the middle of the night:

> Armed with Sten guns, pistols and hand-grenades, we crept . . . down toward the ferry. The bitterly cold night set everything creaking and crackling; the ice on the road snapped sharply as we went over it. When we came out on the

> bridge by the ferry station, there was as much noise as if a whole company was on the march.
>
> Rolf and the other Rjukan man were told to cover me while I went on board to reconnoitre. All was quiet there. Was it possible that the Germans had omitted to place a guard at the weakest point in the whole route to the transport?
>
> Hearing voices in the crew's quarters, forward, I stole to the companion[way] and listened. There must be a party going on down there, and a game of poker. The other two followed me on to the deck of the ferry. We went down to the third-class accommodation and found a hatchway leading to the bilges. But before we had got the hatch open we heard steps, and took cover behind the nearest table or chair. The ferry watchman was standing in the doorway.

Haukelid thought fast. "The situation was awkward, but not dangerous." He told the watchman they were escaping the Gestapo and needed a place to hide:

> The watchman immediately showed us the hatchway in the deck, and told us that they had several times had illicit things with them on their trips.
>
> The Rjukan man now proved invaluable. He talked and talked with the watchman, while Rolf and I flung our sacks down under the deck and began to work.
>
> It was an anxious job, and it took time.

Haukelid and Sörlie found themselves standing on the bottom plates of the boat in a foot of cold water. They had to tape the two alarm-clock timers to one of the steel stringers that braced the ferry's hull, attach four electric detonators to the timers, attach high-speed fuses to the loop of plastic explosive, lay the charge of explosive on the bottom plates and then, most dangerously, hook up batteries to detonators and detonators to fuses. "The charge was placed in the water and concealed. It consisted of nineteen pounds of high explosive laid in the form of a sausage. We laid it forward, so that the rudder and propeller would rise above the surface when water began to come in [to prevent navigating the boat to shallower water]. . . . When the charge exploded, it would blow about eleven square feet out of the ship's side." The sausage was some twelve feet around.

Sörlie went up on deck. Haukelid set his alarms to go off at 10:45 A.M. "Making the last connection was a dangerous job; for an alarm clock is an uncertain instrument, and contact between the hammer and the alarm was avoided by not more than a third of an inch. Thus there was one third of an inch between us and disaster." Everything worked and he finished at 4 A.M.

The Rjukan man had convinced the watchman by then that the escapees he had sheltered needed to return to Rjukan to collect their possessions. Haukelid considered warning their benefactor but decided that might endanger the mission and only thanked him and shook his hand.

Ten minutes from the ferry station Haukelid and Larsen left the car to ski to Kongsberg, forty miles away around the lake, where they would catch a train for the first leg of their escape to Sweden. Sörlie carried a report for London to the clandestine radio. The driver returned the stolen car and he and the Rjukan man strolled home. At Haukelid's suggestion the Norsk Hydro transport engineer had arranged a foolproof alibi: over the weekend doctors at the local hospital operated on him for appendicitis, no questions asked.

With fifty-three people aboard including the concert violinist the *Hydro* sailed on time. Forty-five minutes into the crossing Haukelid's charge of plastic explosive blew the hull. The captain felt the explosion rather than heard it, and though Tinnsjö is landlocked he thought they might have been torpedoed. The bow swamped first as Haukelid had intended; while the passengers and crew struggled to release the lifeboats, the freight cars with their thirty-nine drums of heavy water—162 gallons mixed with 800 gallons of dross—broke loose, rolled overboard and sank like stones. Of passengers and crew twenty-six drowned. The concert violinist slipped high and dry into a lifeboat; when his violin case floated by, someone was kind enough to fish it out for him.

Kurt Diebner of German Army Ordnance counted the full effect on German fission research of the Vemork bombing and the sinking of the *Hydro* in a postwar interview:

> When one considers that right up to the end of the war, in 1945, there was virtually no increase in our heavy-water stocks in Germany . . . it will be seen that it was the elimination of German heavy-water production in Norway that was the main factor in our failure to achieve a self-sustaining atomic reactor before the war ended.

The race to the bomb, such as it was, ended for Germany on a mountain lake in Norway on a cold Sunday morning in February 1944.

Despite Pearl Harbor and the subsequent Japanese sweep across a million square miles of Southeast Asia and the western Pacific, the Pacific theater commanded less attention in the United States in the earlier years of the war than did the European. Partly that neglect was a result of the deliberate national policy that gave priority to Europe. "Europe was Washington's darling," Pacific Fleet Admiral William F. Halsey would write in a memoir, "the South Pacific was only a stepchild." But Americans also found it difficult at first to take seriously an Asian island people who were small in stature and radically different in culture. Reporting from the Solomon Islands east of New Guinea late in 1942, *Time-Life* correspondent John Hersey found the typical U.S. marine "very uneasy about what he feels is Washington's ignorance of the Pacific. Sure, he argues, Hitler has to be beaten, but that doesn't mean we have to go on thinking of the Japs as funny little ring-tailed monkeys." The U.S. Ambassador to Japan at the time of the Pearl Harbor attack, Boston-born Joseph C. Grew, confronted a similar skepticism when he returned from Japanese internment and battled it by traveling the nation lecturing:

> The other day a friend, an intelligent American, said to me: "Of course there must be ups and downs in this war; we can't expect victories every day, but it's merely a question of time before Hitler will go down to defeat before the steadily growing power of the combined air and naval and military forces of the [Allies]—and then, we'll mop up the Japs." Mark well those words, please. "And then we'll mop up the Japs."

Grew thought such bravado ill-advised. "The Japanese have known what we thought of them," he told his audiences— "that they were little fellows physically, that they were imitative, that they were not really very important in the world of men and nations." To the contrary, said Grew, they were "united," "frugal," "fanatical" and "totalitarian" :

> At this very moment, the Japanese feel themselves, man for man, superior to you and to me and to any of our peoples. They admire our technology, they may have a lurking dread of our ultimate superiority of resources, but all too many of them have contempt for us as human beings. . . . The Japanese leaders *do* think that they can and will win. They are counting on our underestimates, on our apparent disunity before—and even during—war, on our unwillingness to sacrifice, to endure, and to fight.

So far Grew's lecture might have been merely exhortation. But he went on to emphasize a phenomenon that Americans fighting in the Pacific were just then beginning to encounter. " 'Victory or death' is no mere slogan for these soldiers," Grew noted. "It is plain, matter-of-fact description of the military policy that controls their forces, from the highest generals to the newest recruits. The man who allows himself to be captured has disgraced himself and his country."

Which was exactly what Marine Major General Alexander A. Vandegrift was finding at the time, late 1942, in the Solomons at Guadalcanal.

"General," he wrote the Marine Commandant in Washington, "I have never heard or read of this kind of fighting. These people refuse to surrender. The wounded will wait until men come up to examine them . . . and blow themselves and the other fellow to death with a hand grenade."

It was frightening. It required a corresponding escalation of violence to combat. John Hersey felt the need to explain:

> A legend has grown up that this young man [i.e., the U.S. marine] is a killer; he takes no prisoners, and gives no quarter. This is partly true, but the reason is not brutality, not just vindictive remembrance of Pearl Harbor. He kills because in the jungle he must, or be killed. This enemy stalks him, and he stalks the enemy as if each were a hunter tracking a bear cat. Quite frequently you hear marines say: "I wish we were fighting against Germans. They are human beings, like us. Fighting against them must be like an athletic performance—matching your skill against someone you know is good. Germans are misled, but at least they react like men. But the Japs are like animals. Against them you have to learn a whole new set of physical reactions. You have to get used to their animal stubbornness and tenacity. They take to the jungle as if they had been bred there, and like some beasts you never see them until they are dead."

As an explanation for unfamiliar behavior, bestiality had the advantage that it made killing a formidable enemy easier emotionally. But it also, by dehumanizing him, made him seem yet more alien and dangerous. So did the other common attribution that evolved during the war to explain Japanese behavior: that the Japanese were fanatics, believers, as Grew had preached, "in the incorruptible certainty of their national cause." The historian William Manchester, a marine at Guadalcanal, argues more objectively from a longer perspective postwar:

> At the time it was impolitic to pay the slightest tribute to the enemy, and Nip determination, their refusal to say die, was commonly attributed to "fanaticism." In retrospect it is indistinguishable from heroism. To call it anything less cheapens the victory, for American valor was necessary to defeat it.

Whether bestiality, fanaticism, or heroism, the refusal of Japanese soldiers to surrender required new tactics and strong stomachs to defeat. In his best-selling 1943 book *Guadalcanal Diary* war correspondent Richard Tregaskis reported those tactics from the first land battles of the Pacific war at Guadalcanal:

> The general summarized the fighting. . . . The toughest job, he said, had been to clean out scores of dugout caves filled with Japs. Each cave, he said, had

> been a fortress in itself, filled with Japs who were determined to resist until they were all killed. The only effective way to finish off these caves, he said, had been to take a charge of dynamite and thrust it down the narrow cave entrance. After that had been done, and the cave blasted, you could go in with a submachine gun and finish off the remaining Japs. . . .
>
> "You've never seen such caves and dungeons," said the general. "There would be thirty or forty Japs in them. And they absolutely refused to come out, except in one or two isolated cases."

The statistics of the Solomons campaign told the same story: of 250 Japanese manning the garrison on Guadalcanal when the marines first landed only three allowed themselves to be taken prisoner; more than 30,000 Japanese shipped in to fight died before the island was secure, compared to 4,123 Americans. Similar patterns obtained elsewhere. The proportion of captured to dead Japanese in the North Burma campaign was 142 to 17,166, about 1:120 when a truism among Western nations is that the loss of one-fourth to one-third of an army—4:1—usually bodes surrender. Paralleling Japanese resistance, Allied losses grew.

As the slow, bloody push up the Pacific toward the Japanese home islands gained momentum through 1943, the question the behavior of Japanese soldiers raised was whether such standards applied not only to the military but to the civilians of Japan as well. Grew had sought to answer that question in his lectures the year before:

> I know Japan; I lived there for ten years. I know the Japanese intimately. The Japanese will not crack. They will not crack morally or psychologically or economically, even when eventual defeat stares them in the face. They will pull in their belts another notch, reduce their rations from a bowl to a half bowl of rice, and fight to the bitter end. Only by utter physical destruction or utter exhaustion of their men and materials can they be defeated. That is the difference between the Germans and the Japanese. That is what we are up against in fighting Japan.

In the meantime the United States manufactured flamethrowers to burn Japanese soldiers from their caves. A seasoned journalist who had traveled in Japan before the war, Henry C. Wolfe, called in *Harper's* for the firebombing of Japan's "inflammable," "matchbox" cities. "It seems brutal to be talking about burning homes," Wolfe explained. "But we are engaged in a life-and-death struggle for national survival, and we are therefore justified in taking any action that will save the lives of American soldiers and sailors. We must strike hard with everything we have at the spot where it will do the most damage to the enemy."

The month Wolfe's call to aerial battle appeared in *Harper's* —January 1943—Franklin Roosevelt met with Winston Churchill at Casablanca. In the course of the meeting the two leaders discussed what terms of surrender they would eventually insist upon; the word "unconditional" was discussed but not included in the official joint statement to be read at the final press conference. Then, on January 24, to Churchill's surprise, Roosevelt inserted the word ad lib: "Peace can come to the world," the President read out to the assembled journalists and newsreel cameras, "only by the total elimination of German and Japanese war power. . . . The elimination of German, Japanese and Italian war power means the unconditional surrender of Germany, Italy, and Japan." Roosevelt later told Harry Hopkins that the surprising and fateful insertion was a consequence of the confusion attending his effort to convince French General Henri Girard to sit down with Free French leader Charles de Gaulle:

> We had so much trouble getting those two French generals together that I thought to myself that this was as difficult as arranging the meeting of Grant and Lee—and then suddenly the Press Conference was on, and Winston and I had had no time to prepare for it, and the thought popped into my mind that they had called Grant "Old Unconditional Surrender," and the next thing I knew I had said it.

Churchill immediately concurred— "Any divergence between us, even by omission, would on such an occasion and at such a time have been damaging or even dangerous to our war effort" —and unconditional surrender became official Allied policy.

16
Revelations

"How would you like to work in America?" James Chadwick asked Otto Frisch in Liverpool one day in November 1943.

"I would like that very much," Frisch remembers responding.

"But then you would have to become a British citizen."

"I would like that even more."

Within a week the British had cleared the Austrian emigré for citizenship. Following instructions "to pack all my necessary belongings into one suitcase and to come to London by the night train" Frisch made the rounds of government offices with other emigré scientists in one crowded day—swearing allegiance to the King, picking up a passport, collecting a visa stamp at the American Embassy—and hurried back to Liverpool, where the delegation would board the converted luxury liner *Andes* the next morning. Headed by Wallace Akers of ICI, the British group included the men General Groves would ask to review barrier development as well as men going to Los Alamos: Frisch, Rudolf Peierls, William G. Penney, George Placzek, P. B. Moon, James L. Tuck, Egon Bretscher and Klaus Fuchs among others. Chadwick would join them, as would the hydrodynamicist Geoffrey Taylor.

Akers maneuvered around the transport shortage by loading them for the Liverpool pier in black mortuary limousines; a hearse for the luggage completed the cortège. On the *Andes* Frisch had an entire eight-berth cabin to himself. Unconvoyed they zigzagged west. America was luxury; traveling up from Newport News Frisch's train stopped in Richmond, Virginia:

> I wandered out into the streets. There I was greeted by a completely incredible spectacle: fruit stalls with pyramids of oranges, illuminated by bright acetylene flares! After England's blackout, and not having seen an orange for a couple of years, that sight was enough to send me into hysterical laughter.

Groves in Washington lectured them on security. A succession of trains delivered them into a fantastic landscape—Frisch and another man in December, the larger group early in 1944—and there in the bright sunlight of a pine-shouldered mesa was Robert Oppenheimer smoking a pipe and shading his close-cropped military haircut with a pork-pie hat: "Welcome to Los Alamos, and who the devil are you?"

They were Churchill's flying wedge. The bomb had been theirs to begin with as much as anybody's, but more immediate urgencies had demanded their attention and now they were couriers sent along to help build it and then to bring it home. America was giving the bomb away to another sovereign state, proliferating. Churchill had negotiated the renewed collaboration at Quebec in August:

> It is agreed between us
>
> First, that we will never use this agency against each other.
>
> Secondly, that we will not use it against third parties without each other's consent.
>
> Thirdly, that we will not either of us communicate any information about Tube Alloys to third parties except by mutual consent.

Niels Bohr and his son Aage followed next as consultant to the Tube Alloys directorate and junior scientific officer, respectively; the British were paying their salaries. Groves' security men met father and son at dockside, assigned them cover names—Nicholas and James Baker—and spirited them off to a hotel, there to discover NIELS BOHR stenciled bold and black on the Danish laureate's luggage. At Los Alamos, warmly welcomed, Nicholas and James Baker became Uncle Nick and Jim.

The first order of business was Heisenberg's drawing of a heavy-water reactor, which Bohr had previously revealed to Groves. Oppenheimer convened a conference of experts on the last day of 1943 to see if they could find any new reason to believe a pile might serve as a weapon. "It was clearly a drawing of a reactor," Bethe recalled after the war, "but when we saw it our conclusion was that these Germans were totally crazy—did they want to throw a reactor down on London?" That was not Heisenberg's purpose, but Bohr wanted to be sure. Bethe and Teller prepared the consequent report, "Explosion of an inhomogeneous uranium–heavy water pile." It found that such an explosion "will liberate energies which are probably smaller, and certainly not much larger, than those obtainable by the explosion of an equal mass of TNT."

If Heisenberg's drawing told the physicists anything it ought to have told them that the Germans were far behind; it depicted sheets of uranium rather than lumps, an inefficient arrangement Heisenberg had clung to for a time even when his colleagues had argued the advantages of a three-dimensional lattice. Samuel Goudsmit, a Dutch physicist in America who would soon lead a front-line Manhattan Project intelligence mission into Germany, remembers a more convoluted conclusion: "At that time we thought this meant simply that they had succeeded in keeping their real aims secret, even from a scientist as wise as Bohr."

Oppenheimer appreciated the salutary effect of Bohr's presence. "Bohr at Los Alamos was marvelous," he told an audience of scientists after the war. "He took a very lively technical interest. . . . But his real function, I think for almost all of us, was not the technical one." Here two texts of the postwar lecture diverge; both versions illuminate Oppenheimer's state of mind in 1944 as he remembered it. In unedited transcript he said Bohr "made the enterprise which looked so macabre seem hopeful" ; edited, that sentence became: "He made the enterprise seem hopeful, when many were not free of misgiving."

How Bohr did so Oppenheimer and even Bohr had work to explain. Oppenheimer outlines an explanation in his lecture:

> Bohr spoke with contempt of Hitler, who with a few hundred tanks and planes had tried to enslave Europe for a millennium. He said nothing like that would ever happen again; and his own high hope that the outcome would be good, and that in this the role of objectivity, the cooperation which he had experienced among scientists would play a helpful part; all this, all of us wanted very much to believe.

"He said nothing like that would ever happen again" is a key; Austrian emigré theoretician Victor Weisskopf supplies another:

> In Los Alamos we were working on something which is perhaps the most questionable, the most problematic thing a scientist can be faced with. At that time physics, our beloved science, was pushed into the most cruel part of reality and we had to live it through. We were, most of us at least, young and somewhat inexperienced in human affairs, I would say. But suddenly in the midst of it, Bohr appeared in Los Alamos.

> It was the first time we became aware of the sense in all these terrible things, because Bohr right away participated not only in the work, but in our discussions. Every great and deep difficulty bears in itself its own solution. . . . This we learned from him.

"They didn't need my help in making the atom bomb," Bohr later told a friend. He was there to another purpose. He had left his wife and children and work and traveled in loneliness to America for the same reason he had hurried to Stockholm in a dark time to see the King: to bear witness, to clarify, to win change, finally to rescue. His revelation—which was equivalent, as Oppenheimer said, to his revelation when he learned of Rutherford's discovery of the nucleus—was a vision of the complementarity of the bomb. In London and at Los Alamos Bohr was working out its revolutionary consequences. He meant now to communicate his revelation to the heads of state who might act on it: to Franklin Roosevelt and Winston Churchill first of all.

In December, before he first went out to Los Alamos, at a small reception at the Danish Embassy in Washington where he and Aage lived when they visited that city, Bohr had renewed his acquaintance with Supreme Court Associate Justice Felix Frankfurter. The justice was short, crackling, bright, Vienna-born, an agnostic Zionist Jew, an ardent patriot, a close friend of Franklin Roosevelt and one of the President's longtime advisers. Bohr had met him in England in 1933 in connection with the rescue of the emigré academics; when Bohr visited Washington in 1939, the year Frankfurter was elevated to the Court, the two men developed what Frankfurter calls a "warm friendly relation." The December tea offered no opportunity to talk privately, but on his way out Frankfurter proposed to invite Bohr to lunch in chambers at the Supreme Court. He already understood that something was up.

The justice was three years older than the physicist, born in 1882, the same year as Roosevelt. He had emigrated to the United States with his family in 1894, grown up on New York's Lower East Side, graduated at nineteen from the City College of New York and made a brilliant showing at Harvard Law. He worked for Henry Stimson when Stimson was U.S. Attorney for the Southern District of New York, before the Great War, and in Washington when Stimson served as Secretary of War the first time, under William Howard Taft. Harvard invited Frankfurter to a professorship at its law school in 1914. He held that post until Roosevelt appointed him to the Supreme Court, but he was intensely active politically across those academic years, a one-man recruiting agency for the New Deal, a loyal friend who supported Roosevelt's ill-advised 1937 scheme to pack the Court to overwhelm its conservative resistance to his innovative legislation.

After Bohr returned to Washington from Los Alamos, in mid-February, the two men kept their appointment for lunch. Both left wartime memoranda describing the meeting. "We talked about the recent events in Denmark," Frankfurter writes, "the probable course of the war, the state of England . . . our certainty of German defeat and what lay ahead. Professor Bohr never remotely hinted the purpose of his visit to this country."

Fortunately Frankfurter had heard about the project he called X. He says he heard from "some distinguished American scientists," but he certainly heard from a distraught young Met Lab scientist who had penetrated all the way to Frankfurter and Eleanor Roosevelt in 1943 with complaints about Du Pont. "I had thus become aware of X—aware, that is, that there was such a thing as X and of its significance." Since Frankfurter knew Bohr's field he assumed X was the reason for Bohr's visit:

> And so . . . I made a very oblique reference to X so that if I was right in my assumption that Professor Bohr was sharing in it, he would know that I knew something about it. . . . He likewise replied in an innocent remote way, but it soon became clear to both of us that two such persons, who had been so long and so deeply preoccupied with the menace of Hitlerism and who were so deeply engaged in the common cause, could talk about the implications of X without either of us making any disclosure to the other.

Eminent jurist and eminent physicist thus easily dispatched that modest obstacle.

"Professor Bohr then expressed to me," Frankfurter goes on, "his conviction that X might be one of the greatest boons to mankind or might become the greatest disaster . . . and he made it clear to me that there was not a soul in this country with whom he could or did talk about these things except Lord Halifax [the British ambassador] and Sir Ronald Campbell [a British representative on the Anglo-American Combined Policy Committee]." Bohr picks up the narrative in third-person voice: "On hearing this F said that, knowing President Roosevelt, he was confident that the President would be very responsive to such ideas as B outlined."

Bohr had found his go-between. "B met F again one of the last days of March," Bohr records in his wartime memorandum, "and learned that in the meantime F had had occasion to speak with the President and that the President shared the hope that the project might bring about a turning point in history." Frankfurter describes his meeting with Roosevelt:

> On this particular occasion I was with the President for about an hour and a half and practically all of it was consumed by this subject. He told me the whole thing "worried him to death" (I remember the phrase vividly), and he

> was very eager for all of the help he could have in dealing with the problem. He said he would like to see Professor Bohr and asked me whether I would arrange it. When I suggested to him that the solution of this problem might be more important than all the schemes for a world organization, he agreed and authorized me to tell Professor Bohr that he, Bohr, might tell our friends in London that the President was most eager to explore the proper safeguards in relation to X.

Much controversy surrounds this meeting, because Roosevelt later implicitly repudiated it. Why, if the President was worried to death about the postwar implications of the bomb, did he entrust a mission to the British to so informal an arrangement? He had not even met Niels Bohr. An answer to this question would answer a more substantive question: whether Roosevelt was in fact interested in exploring ideas of international control or whether he was already committed to perpetuating an Anglo-American monopoly (the Quebec Agreement implied commitment, and he had recently discussed cornering the world uranium and thorium markets with Groves and Bush).

Why did Roosevelt entrust so important a mission to Bohr? In fact, the commission worked the other way around: Bohr had come to the United States representing the British, representing at least Sir John Anderson, who had encouraged his visit as much to promote discussing the issues Bohr had raised as to bolster the British Los Alamos mission. If the commission was informal it was no more so than any number of other back-channel arrangements between the British and the Americans. Roosevelt simply responded to what he took to be a British approach. He seems to have assumed—correctly—that British statesmen around Churchill were using Bohr to communicate to the President ideas about wartime and postwar arrangements to which Churchill was not yet committed. He responded candidly with loyalty to his British counterpart, Bohr adds: "F also informed B that as soon as the question had been brought up, the President had said it was a matter for Prime Minister Churchill and himself to find the best ways of handling the project to the benefit of all mankind, and that he should heartily welcome any suggestion to this purpose from the Prime Minister." The President would be happy to discuss new ideas for postwar relations, but the British would first have to convince the P.M.; Roosevelt would not deal behind Churchill's back. Frankfurter implies this understanding: "I wrote out such a formula for Bohr to take to London—a communication to Sir John Anderson, who was apparently Bohr's connecting link with the British government."

Complicating Bohr's discussions, in March and later, was the question of what to do about the USSR. Bohr considered the question in the following perspective. Tell the Soviet Union soon, before the first bombs were nearly built, that a bomb project was under way, and the confidence might lead to negotiations on postwar arms control. Let the Soviet Union discover the information on its own, build the bombs and drop them, oppose the Soviets at the end of the war with an Anglo-American nuclear monopoly, and the likeliest outcome was a nuclear arms race.

Bohr's revelation of the complementarity of the bomb was far more fundamental than this contemporary political question. But the contemporary political question was an aspect of the larger issue and partly obscured it from view. The bomb was opportunity and threat and would always be opportunity and threat—that was the peculiar, paradoxical hopefulness. But political conditions would necessarily differ before and after it was deployed.

At the end of March 1944, Bohr seemingly had a mandate from the President of the United States to talk to the Prime Minister of Great Britain. The British in whom Bohr had been confiding were properly impressed. "Halifax considered this development to be so important," writes Aage Bohr, "that he thought my father should go to London immediately." Father and son crossed the Atlantic again, this time by military aircraft, in early April.

Anderson had been working to soften Churchill up. The tall, dark Chancellor of the Exchequer, whom Oppenheimer describes as a "conservative, dour, remarkably sweet man," sent the Prime Minister a long memorandum on March 21. He suggested opening Tube Alloys to wider discussion within the British government. Echoing Bohr, he saw the possibility of international proliferation of nuclear weapons after the war. He thought the only alternative to a vicious arms race was international agreement. He proposed "communicating to the Russians in the near future the bare fact that we expected, by a given date, to have this devastating weapon; and . . . inviting them to collaborate with us in preparing a scheme for international control."

Churchill circled "collaborate" and wrote in the margin: "on no account."

When Bohr arrived Anderson wrote the Prime Minister again, going over the same arguments but adding that he now believed Roosevelt was attending the subject and would welcome discussion. He even supplied a draft message Churchill might send to initiate an exchange. The response was equally waspish: "I do not think any such telegram is necessary nor do I wish to widen the circle who are informed."

Churchill was in no mood to see Bohr; the Danish laureate cooled his heels for weeks. While he waited he heard from the Soviets. Peter Kapitza had written Bohr shortly after the Bohrs escaped from Denmark—the letter found its way from Stockholm to the Soviet Embassy in London— "to let you know that you will be welcome to the Soviet Union where everything will be done to give you and your family a shelter and where we now have all the necessary conditions for carrying on scientific work." After alerting the Tube Alloys security officer Bohr went to the embassy in Kensington Gardens to collect the letter; on his return he reported his conversation with the embassy's counsellor. Amid much talk about the greatness of Russian science and how few friends Russia had counted before the war was the heart of the matter:

> The Counsellor then said that he knew that B had recently been to America, and B said that he had received from the journey many encouraging expressions of the wish for international cultural co-operation and that he hoped soon to come to Russia also. The Counsellor next asked what information B had received about the work of American scientists during the war, and B answered that the American scientists, just like the Russian and the British, had surely made very large contributions to the war effort which would no doubt be of great importance for an appreciation of science everywhere after the war. B thereafter told a little about the situation in Denmark during the occupation.

Quickly changing the subject. But for Bohr the blunt question and Kapitza's invitation to come to Moscow were enough to indicate that the Soviets had at least an inkling of the bomb project and might be working on their own. Which meant there was very little time left to convince them that a secret arms race had not already begun. He carried that urgency with him when he was called with Cherwell, finally, on May 16, to 10 Downing Street.

"We came to London full of hopes and expectations," Aage Bohr remembers. "It was, of course, a rather novel situation that a scientist should thus try to intervene in world politics, but it was hoped that Churchill, who possessed such imagination and who had often shown such great vision, would be inspired by the new prospects." Niels Bohr cherished that hope. His British friends had not prepared him.

"One of the blackest comedies of the war," C. P. Snow characterizes the disastrous confrontation. The definitive account is from R. V. Jones, Cherwell's protégé, who had helped make arrangements and who was surprised to find Bohr wandering a few hours later in Old Queen Street outside the Tube Alloys office:

> When I asked him how the meeting had gone he said: "It was terrible. He scolded us like two schoolboys!" From what he told me at that time and af-

> terwards, it appeared that the meeting misfired from the start. Churchill was in a bad mood, and he berated Cherwell for not having arranged the interview in a more regular manner. He then said he knew why Cherwell had done it—it was to reproach him about the Quebec Agreement. This, of course, was quite untrue, but it meant that Bohr's "set piece" talk was thrown right out of gear. Bohr, who used to say that accuracy and clarity were complementary (and so a short statement could never be precise), was not easy to hear, and all that Churchill seemed to gather was that he was worried about the likely state of the post-war world and that he wanted to tell the Russians about the progress towards the bomb. As regards the post-war world Churchill told him: "I cannot see what you are talking about. After all this new bomb is just going to be bigger than our present bombs. It involves no difference in the principles of war. And as for any post-war problems there are none that cannot be amicably settled between me and my friend, President Roosevelt."

Bohr got only the bare thirty minutes of his scheduled appointment, most of which Churchill had monopolized. "As he was leaving," Aage Bohr concludes, "my father asked for permission to write Churchill, whereupon the latter answered, 'It will be an honour for me to receive a letter from you,' adding, 'but not about politics!' "

"We did not speak the same language," Bohr said afterward. His son found him "somewhat downcast." He was angrier than that; in his seventy-second year, still stinging, he told an old friend: "It was terrible that no one over there" —England and America both— "had worked on the solution of the problems that would arise when it became possible to release nuclear energy; they were completely unprepared." And further, "It was perfectly absurd to believe that the Russians cannot do what others can. . . . There never was any secret about nuclear energy."

Churchill's obduracy was compound but straightforward. He was up to his neck in preparations for the Normandy invasion; he sniffed conspirators encroaching back-channel and instinctively swatted them down; he resented the awe his colleagues accorded this certified great man ("I did not like the man when you showed him to me, with his hair all over his head, at Downing Street," he gnawed at Cherwell afterward); he could not listen carefully enough, or was too certain of his own opinions, to be convinced that the bomb would change the rules. A year later the seventy-year-old Prime Minister had budged no further. "In all the circumstances," he wrote Anthony Eden in 1945, "our policy should be to keep the matter so far as we can control it in American and British hands and leave the French and Russians to do what they can. You can be quite sure that any power that gets hold of the secret will try to make the article and this touches the existence of human society. This matter is out of all relation to anything else that exists in the world, and I could not think of participating in any disclosure to third or fourth parties at the present time."

"He had always had a naive faith in 'secrets,' " concludes C. P. Snow. "He had been told by the best authorities that this 'secret' wasn't keepable and that the Soviets would soon have the bomb themselves. Perhaps, with one of his surges of romantic optimism, he deluded himself into not believing it. He was only too conscious that British power, and his own, was now just a vestige. So long as the Americans and British had the bomb in sole possession, he could feel that that power hadn't altogether slipped away. It is a sad story."

Bohr wrote Churchill on May 22; the letter was circumspect but political after all and conveyed what he had not been allowed to convey at the meeting: "that the President is deeply concerned in his own mind with the stupendous consequences of the project, in which he sees grave dangers, but also unique opportunities." Bohr did not spell out these opportunities. He even seemed to step back from offering advice: "The responsibility for handling the situation rests, of course, with the statesmen alone. The scientists who are brought into confidence can only offer the statesmen all such information about technical matters as may be of importance for their decisions." Those technical matters, however, Bohr made sure to note, included the probability of proliferation and of bigger bombs—he had learned of the Super at Los Alamos.

Apparently Churchill did not trouble himself to respond.

Bohr stayed on in London for several more weeks. He was thus on hand for D-Day, Tuesday, June 6, 1944. "The greatest amphibious assault ever attempted," Dwight D. Eisenhower, the Supreme Allied Commander, called that invasion of Europe across the English Channel with an initial force of 156,000 British, Canadian and American soldiers supported by 1,200 warships, 1,500 tanks and 12,000 aircraft. By the time Bohr and his son left England at the end of the week to return to the United States the Allies had secured the invasion beaches and begun advancing inland with a force bolstered now to 326,000 men. "The way home," Eisenhower instructed his armies, "is via Berlin."

For Bohr the way home was via Washington. He reported his dismal experience with Churchill to Felix Frankfurter on June 18. Frankfurter immediately carried the news to Roosevelt, who was amused to hear another tale of Churchillian pugnacity:

> About a week later F told B that this information had been heartily welcomed by the President who had said that he regarded the steps taken as a favourable development. During the talk the President had expressed the wish to see B,

and as a preliminary step F advised B to give an account of his views in a brief memorandum.

The Bohrs turned to the task as Washington steamed, the last days of June and the first days of July dawning in the high eighties and sweltering above 100° by afternoon. Aage Bohr recalls the document's preparation:

> It was worked out in the tropical heat of Washington and, like all my father's work, underwent many stages before it was ready for delivery. In the morning, my father would usually bring up new ideas for alterations that he had thought out during the night. There was no secretary to whom we could entrust such documents, and therefore I typed them; meanwhile my father darned socks and sewed buttons on for us, a job which he carried out with his usual thoroughness and manual skill.

Sewing on buttons, darning socks, suffering in the heat that seemed equatorial to a Dane of the cold North Sea, Bohr worked and reworked his memorandum to maximum generality of expression, a political analysis as reserved as any scientific paper. It says all that he had seen up to that time, which was almost everything essential.

Late in life Bohr explained the starting point of his revelation in a single phrase. *"We are in a completely new situation that cannot be resolved by war,"* he confided to a friend. He had already grasped that fundamental point when he arrived at Los Alamos in 1943 and told Oppenheimer that nothing like Hitler's attempt to enslave Europe would ever happen again. "First of all," Oppenheimer confirms, "[Bohr] was clear that if it worked, this development was going to bring an enormous change in the situation of the world, in the whole situation of war and the tolerability of war."

The weapon devised as an instrument of major war would end major war. It was hardly a weapon at all, the memorandum Bohr was writing in sweltering Washington emphasized; it was "a far deeper interference with the natural course of events than anything ever before attempted" and it would "completely change all future conditions of warfare." When nuclear weapons spread to other countries, as they certainly would, no one would be able any longer to *win.* A spasm of mutual destruction would be possible. But not war.

That was new ground, ground the nations had never walked before. It was new as Rutherford's nucleus had been new and unexplored. Bohr had searched the forbidding territory of the atom when he was young and discovered multiple structures of paradox; now he searched it again by the dark light of the energy it released and discovered profound political change.

Nations existed in a condition of international anarchy. No hierarchical authority defined their relations with one another. They negotiated voluntarily as self-interest moved them and took what they could get. War had been their final negotiation, brutally resolving their worst disputes.

Now an ultimate power had appeared. If Churchill failed to recognize it he did so because it was not a battle cry or a treaty or a committee of men. It was more like a god descending to the stage in a gilded car. It was a mechanism that nations could build and multiply that harnessed unlimited energy, a mechanism that many nations *would* build in self-defense as soon as they learned of its existence and acquired the technical means. It would seem to confer security upon its builders, but because there would be no sure protection against so powerful and portable a mechanism, in the course of time each additional unit added to the stockpiles would *decrease* security by adding to the general threat until insecurity finally revealed itself to be total at every hand.

By the necessity, commonly understood, to avoid triggering a nuclear holocaust, the *deus ex machina* would have accomplished then what men and nations had been unable to accomplish by negotiation or by conquest: the abolition of major war. Total security would be indistinguishable from total insecurity. A menacing standoff would be maintained suspiciously, precariously, at the brink of annihilation. Before the bomb, international relations had swung between war and peace. After the bomb, major war among nuclear powers would be self-defeating. No one could win. World war thus revealed itself to be historical, not universal, a manifestation of destructive technologies of limited scale. Its time would soon be past. The pendulum now would swing wider: between peace and national suicide; between peace and total death.

Bohr saw that far ahead—all the way to the present, when menacing standoff has been achieved and maintained for decades without formal agreement but at the price of smaller client wars and holocaustal nightmare and a good share of the wealth of nations—and stepped back. He wondered if such apocalyptic precariousness was necessary. He wondered if the war-weary statesmen of the day, taught the consequences of his revelation, could be induced to forestall those consequences, to adjourn the game when the stalemate revealed itself rather than illogically to play out the menacing later moves. It was clear at least that the new weapons would be appallingly dangerous. If the statesmen could be brought to understand that the danger of such weapons would be common and mutual, might they not negotiate commonly and mutually to ban them? If the end would be a warless world either way, but one way with the holocaustal machinery in place and the other way with its threat only considered and understood, what did they have to lose? Negotiating peace rather than allowing the *deus ex machina* inhumanly to impose standoff might show the common threat to contain within itself, complementarily, common promise. Much good might follow. *"It appeared to me,"* Bohr wrote in 1950 of his lonely wartime initiative, *"that the very necessity of a concerted effort to forestall such ominous threats to civilization would offer quite unique opportunities to bridge international divergencies."* That, in a single sentence, was the revelation of the complementarity of the bomb.

"Much thought has naturally been given to the question of [arms] control," Bohr flattered Franklin Roosevelt in his 1944 document, knowing that hardly any thought had yet been given, "but the further the exploration of the scientific problems concerned is proceeding" —to thermonuclear weapons, Bohr means— "the clearer it becomes that no kind of customary measures will suffice for this purpose and that especially the terrifying prospect of a future competition between nations about a weapon of such formidable character can only be avoided through a universal agreement in true confidence."

Bohr was no fool. Obviously no nation could be expected to trust another nation's bare word about something so vital to survival. Each would want to see for itself that the other was not secretly building bombs. That meant the world would have to open up. He knew very well how suspicious the Soviet Union would be of such an idea; he hoped, however, that the dangers of a nuclear arms race might appear serious enough to make evident the compensating advantages:

> The prevention of a competition prepared in secrecy will therefore demand such concessions regarding exchange of information and openness about industrial efforts including military preparations as would hardly be conceivable unless at the same time all partners were assured of a compensating guarantee of common security against dangers of unprecedented acuteness.

Nor was the urge to suspicious secrecy unique to the Soviets; the Americans and the British were even then risking an arms race by keeping their work on the atomic bomb secret from their Soviet allies. Oppenheimer elaborates:

> [Bohr] was clear that one could not have an effective control of . . . atomic energy . . . without a very open world; and he made this quite absolute. He thought that one would have to have privacy, for he needed privacy, as we all do; we have to make mistakes and be charged with them only from time to time. One would have to have respect for individual quiet, and for the quiet process of government and management; but in principle everything that

might be a threat to the security of the world would have to be open to the world.

Openness would accomplish more than forestalling an arms race. As it did in science, it would reveal error and expose abuse. Men performed in secrecy, behind closed doors and guarded borders and silenced printing presses, what they were ashamed or afraid to reveal to the world. Bohr talked to George Marshall after the war, when the Chief of Staff had advanced to Secretary of State. "What it would mean," he told him, "if the whole picture of social conditions in every country were open for judgment and comparison, need hardly be enlarged upon." The great and deep difficulty that contained within itself its own solution was not, finally, the bomb. It was the inequality of men and nations. The bomb in its ultimate manifestation, nuclear holocaust, would eliminate that inequality by destroying rich and poor, democratic and totalitarian alike in one final apocalypse. It followed complementarily that the opening up of the world necessary to prevent (or reverse) an arms race would also progressively expose and alleviate inequality, but in the direction of life, not death:

> Within any community it is only possible for the citizens to strive together for common welfare on the basis of public knowledge of the general conditions of the country. Likewise, real co-operation between nations on problems of common concern presupposes free access to all information of importance for their relations. Any argument for upholding barriers of information and intercourse, based on concern for national ideals or interests, must be weighed against the beneficial effects of common enlightenment and the relieved tension resulting from such openness.

That statement, from an open letter Bohr wrote to the United Nations in 1950, is preceded by another, a vision of a world evolved to the relative harmony of the nations of Scandinavia that once confronted each other and the rest of Europe as aggressively and menacingly as the Soviet Union and the United States had come by 1950 to do. Notice that Bohr does not propose a world government of centralized authority but a consortium: "An open world where each nation can assert itself solely by the extent to which it can contribute to the common culture and is able to help others with experience and resources must be the goal to put above everything else." And most generally and profoundly: "The very fact that knowledge is itself the basis for civilization points directly to openness as the way to overcome the present crisis."

Such an effort would begin with the United States, Bohr suggested to Roosevelt in the summer of 1944, because the United States had achieved clear advantage: "The present situation would seem to offer a most favourable opportunity for an early initiative from the side which by good fortune has achieved a lead in the efforts of mastering mighty forces of nature hitherto beyond human reach." Concessions would demonstrate goodwill; "indeed, it would appear that only when the question is taken up . . . of what concessions the various powers are prepared to make as their contribution to an adequate control arrangement, [will it] be possible for any one of the partners to assure themselves of the sincerity of the intentions of the others."

The untitled memorandum Bohr prepared for Franklin Roosevelt in Washington in 1944 went to Felix Frankfurter for review on July 5 along with a cover letter apologizing for its inadequacies. Bohr worried through the hot night and composed another apology the next day: "I have had serious anxieties," he confided, "that [the memorandum] may not correspond to your expectations and perhaps not at all be suited for the purpose." Frankfurter had the good sense to recognize the document's merit—it is still the only comprehensive and realistic charter for a postnuclear world—and about a week later told Bohr he had handed it to the President. Bohr and his son left Washington soon after, on a Friday in mid-July, to work at Los Alamos, understanding that Roosevelt would arrange a meeting in good time.

That time came in August as the President prepared to meet the Prime Minister in Quebec. Bohr returned to the U.S. capital; "on August 26th at 5 p.m.," he writes, "B was received by the President in the White House in a completely private manner." Roosevelt "was very cordial and in excellent spirits," says Aage Bohr, as well he might have been after the rapid advances of the Allied armies across Europe. He had read Bohr's memorandum; he "most kindly gave B an opportunity to explain his views and spoke in a very frank and encouraging manner about the hopes he himself entertained." FDR liked to charm; he charmed Bohr with stories, Aage Bohr recounts:

> Roosevelt agreed that an approach to the Soviet Union of the kind suggested must be tried, and said that he had the best hopes that such a step would achieve a favourable result. In his opinion Stalin was enough of a realist to understand the revolutionary importance of this scientific and technical advance and the consequences it implied. Roosevelt described in this connection the impression he had received of Stalin at the meeting in Teheran, and also related humorous anecdotes of his discussion and debates with Churchill and Stalin. He mentioned that he had heard how the negotiations with Churchill in London had gone, but added that the latter had often reacted in this way at the first instance. However, Roosevelt said, he and Churchill always managed

to reach agreement, and he thought that Churchill would eventually come around to sharing his point of view in this matter. He would discuss the problems with Churchill at their forthcoming meeting and hoped to see my father soon afterwards.

The interview lasted an hour and a half. To Robert Oppenheimer in 1948 Bohr reported a more specific commitment from the President: he "left with Professor Bohr the impression," Oppenheimer writes, "that, after discussion with the Prime Minister, he might well ask [Bohr] to undertake an exploratory mission to the Soviet Union."

"It is hardly necessary to mention the encouragement and gratitude my father felt after his talk with Roosevelt," Aage Bohr goes on; "these were days filled with the greatest optimism and expectation." Bohr saw Frankfurter in Boston and told him about the meeting. Frankfurter suggested Bohr restate his case in a thank-you note, which Bohr managed to compress into one long page by September 7. Frankfurter passed it to Roosevelt's aide. Bohr settled in eagerly to wait.

The two heads of state saved their Tube Alloy discussions for the end of the conference, late September, when they retreated to Roosevelt's estate in the Hudson Valley at Hyde Park. "This was another piece of black comedy," writes C. P. Snow. " . . .Roosevelt surrendered without struggle to Churchill's view of Bohr." The result was a secret aide-mémoire, obviously of Churchill's composition, that misrepresented Bohr's proposals, repudiated them and recorded for the first time the Anglo-American position on the new weapon's first use:

> The suggestion that the world should be informed regarding tube alloys, with a view to an international agreement regarding its control and use, is not accepted. The matter should continue to be regarded as of the utmost secrecy; but when a "bomb" is finally available, it might perhaps, after mature consideration, be used against the Japanese, who should be warned that this bombardment will be repeated until they surrender.
>
> 2. Full collaboration between the United States and the British Government in developing tube alloys for military and commercial purposes should continue after the defeat of Japan unless and until terminated by joint agreement.
>
> 3. Enquiries should be made regarding the activities of Professor Bohr and steps taken to ensure that he is responsible for no leakage of information particularly to the Russians.

The next day, September 20, Churchill wrote Cherwell in high dudgeon:

> The President and I are much worried about Professor Bohr. How did he come into this business? He is a great advocate of publicity. He made an un-

> authorized disclosure to Chief Justice [*sic*] Frankfurter who startled the President by telling him he knew all the details. He says he is in close correspondence with a Russian professor, an old friend of his in Russia to whom he has written about the matter and may be writing still. The Russian professor has urged him to go to Russia in order to discuss matters. What is all this about? It seems to me Bohr ought to be confined or at any rate made to see that he is very near the edge of mortal crimes. I had not visualized any of this before. . . . I do not like it at all.

Anderson, Halifax and Cherwell all defended Bohr to Churchill after the Hyde Park outburst, as did Bush and Conant to FDR. The Danish laureate was not confined. But neither was he invited to meet again with the President of the United States. There would be no exploratory mission to the USSR.

How much the world lost that September is immeasurable. The complementarity of the bomb, its mingled promise and threat, would not be canceled by the decisions of heads of state; their frail authority extends not nearly so far. Nuclear fission and thermonuclear fusion are not acts of Parliament; they are levers embedded deeply in the physical world, discovered because it was possible to discover them, beyond the power of men to patent or to hoard.

Edward Teller had arrived at Los Alamos in the April of its founding in 1943 prepared to participate fully in its work. He was then thirty-five years old, dark, with bushy, mobile black eyebrows and a heavy, uneven step; "youthful," Stanislaw Ulam remembers, "always intense, visibly ambitious, and harboring a smouldering passion for achievement in physics. He was a warm person and clearly desired friendship with other physicists." Teller's son Paul, his first child, had been born in February. The Tellers had shipped to the primitive New Mexico mesa two machines they considered vital to their peace of mind, a Steinway concert grand piano Mici Teller had bought for her husband for two hundred dollars at a Chicago hotel sale and a new Bendix automatic washer. They were assigned an apartment; the Steinway nearly filled the living room.

Teller had striven on behalf of nuclear energy since Bohr's first public announcement of the discovery of fission in Washington in 1939. He had helped Robert Oppenheimer organize Los Alamos and recruit its staff. He expected to contribute to the planning of the new laboratory's program and he did. "It was essential that the whole laboratory agree on one or a very few major lines of development," writes Hans Bethe, "and that all else be considered of low priority. Teller took an active part in the decision on what were to be the major lines. . . . A distribution of work among the members of the Theoretical Division was agreed upon in a meeting of all scientists of the division and Teller again had a major voice."

But Teller had received no concomitant administrative appointment that April, and the omission aggrieved him. He was qualified to lead the Theoretical Division; Oppenheimer appointed Hans Bethe instead. He was qualified to lead a division devoted to work toward a thermonuclear fusion weapon, a Super, but no such division was established. The laboratory had decided at its opening conference, and the Lewis committee had affirmed in May, that thermonuclear research should be restricted largely to theoretical studies and held to distant second priority behind fission: an atomic bomb, since it would trigger any thermonuclear arrangement, necessarily came first; there was a war on and manpower was limited.

"That I was named to head the [Theoretical] division," Bethe comments, "was a severe blow to Teller, who had worked on the bomb project almost from the day of its inception and considered himself, quite rightly, as having seniority over everyone then at Los Alamos, including Oppenheimer." Bethe believed he was chosen because his "more plodding but steadier approach to life and science would serve the project better at that stage of its development, where decisions had to be adhered to and detailed calculations had to be carried through, and where, therefore, a good deal of administrative work was inevitable." Teller saw his old friend's steadier approach differently: "Bethe was given the job to organize the effort and, in my opinion, in which I may well have been wrong, he overorganized it. It was much too much of a military organization, a line organization." On the other hand, Teller has repeatedly praised Oppenheimer's direction of Los Alamos, direction which included Bethe's appointment and ratified Bethe's decisions:

> Throughout the war years, Oppie knew in detail what was going on in every part of the Laboratory. He was incredibly quick and perceptive in analyzing human as well as technical problems. Of the more than ten thousand people who eventually came to work at Los Alamos, Oppie knew several hundred intimately, by which I mean that he knew what their relationships with one another were and what made them tick. He knew how to organize, cajole, humor, soothe feelings—how to lead powerfully without seeming to do so. He was an exemplar of dedication, a hero who never lost his humanness. Disappointing him somehow carried with it a sense of wrongdoing. Los Alamos' amazing success grew out of the brilliance, enthusiasm and charisma with which Oppenheimer led it.

"I believe maybe [Teller] resented my being placed on top of him," Bethe concludes. "He resented even more that there would be an end to free and general discussion. . . . He resented even more that he was removed [by lack of administrative contact] from Oppenheimer."

The theoretical complexity of the Super challenged Teller as the fission bomb had not; it also offered a line of work along which he might lead. "When Los Alamos was established in the spring of 1943," he writes and the technical history of the laboratory confirms, "the exploration of the Super was among its objectives." He accepted the postponement of that exploration through the summer of 1943, helping Bethe with the more immediate problem of developing means to calculate the critical mass and nuclear efficiency of various bomb designs. During the summer, experimental studies at Purdue found that the fusion reaction cross section for deuterium was much larger than expected; Teller cited that result to the Purdue Los Alamos Governing Board in September to propose renewing the Super investigation. Then John von Neumann arrived on the Hill to endorse and extend Seth Neddermeyer's implosion work and for a few months Teller was caught up in reconnoitering that new territory.

Emilio Segrè won a new workshop that 1943 autumn. At Berkeley he had measured the rate of spontaneous fission—naturally occurring fission without neutron bombardment—in uranium and plutonium. The measurements were difficult because the rates were low for the small samples Segrè had to use, but they were crucial. They determined how cleansed of light-element impurities the bomb cores would have to be—there was no point in purifying past the spontaneous background—and they determined how fast the gun assemblies would have to fire to avoid predetonation. Segrè moved off the Los Alamos mesa to protect his new and more capacious measuring instruments from the radiation other experiments generated there:

> At this time I acquired a special small laboratory for measuring spontaneous fission, the like of which I have never seen before or since. It was a log cabin that had been occupied by a ranger and it was located in a secluded valley a few miles from Los Alamos. It could be reached only by a jeep trail that passed through fields of purple and yellow asters and a canyon whose walls were marked with Indian carvings. On this trail we once found a large rattlesnake. The cabin-laboratory, in a grove shaded by huge broadleaf trees, occupied one of the most picturesque settings one could dream of.

In December at this Pajarito Canyon field station Segrè made a significant discovery. The spontaneous fission rate for natural uranium was much the same at the field station as at Berkeley, but at the field station the rate was seemingly higher for U235. Segrè deduced that cosmic-ray neutrons, which were usually too slow to fission U238 but effective to fission

U235, caused the difference. Cosmic rays batter neutrons from the upper reaches of the atmosphere and the field station was 7,300 feet nearer that region than was sea-level Berkeley. Shield out such stray neutrons and the U235 bomb core could be purified less rigorously than they had assumed. Predetonation would be less likely: the gun that assembled the U235 to critical mass would need less muzzle velocity and could be significantly shorter and lighter. Thus was Little Boy engendered, Thin Man's modest brother, a gun assembly six feet long instead of seventeen that would weigh less than 10,000 pounds, an easy load for a B-29: in a log cabin in a grove beyond fields of bright asters, up a trail visited by rattlesnakes.

Gun research was already advanced. "The first task of the gun group," Edwin McMillan remembers, "was to set up a test stand where experiments could be done. You have to have a gun emplacement, and a gun, and a sand butt, which is nothing but a huge box full of sand that you fire projectiles into so that you can find the pieces afterwards, and because there might be somebody else out there." The site they chose was Anchor Ranch, a former working ranch three miles southwest of the mesa that the Army had bought as part of the reservation; they fired the first shot on September 17, 1943.

Until the following March the group used a three-inch Navy anti-aircraft gun fitted with unrifled barrels. With it they tested propellants—eventually choosing cordite—and studied scale-model projectiles and targets. Knowing that the uranium bullet would complete a critical assembly they decided that it should not impact upon the target core but pass freely through; within microseconds of its arrival at spherical configuration it would in any case have vaporized.

From the beginning the plutonium gun with its nearly unattainable muzzle velocity of 3,000 feet per second had been a gamble. When von Neumann that autumn celebrated the advantages of implosion the Governing Board gave the novel approach its strong endorsement. Through the fall and early winter of 1943 Neddermeyer's experiments made only slow progress, however. He added few men to his group. He continued to work methodically with metal cylinders wrapped with solid slabs of high explosive. By spacing several detonators symmetrically around the wrap he could start implosion simultaneously at different points on the HE surface. From each point of detonation a detonation wave shaped like an expanding bubble would travel inward toward the metal cylinder; by varying the spacing of the detonators and the thickness of the HE Neddermeyer hoped to find a configuration that smoothed the convex, multiple shock waves to one uniform cylindrical squeeze. He was working to the same end with small metal balls, scale models of an eventual bomb core. But "the first successful HE flash photographs of imploding cylinders," notes the Los Alamos technical history, "showed that there were . . . very serious asymmetries in the form of jets which traveled ahead of the main mass. A number of interpretations of these jets were proposed, including the possibility that they were optical illusions." They were all too real. "Absolutely awful results," says Bethe. Oppenheimer decided Neddermeyer needed help. Groves agreed. Conant knew just the man.

"Everything in books [about the Manhattan Project] looks so simple, so easy, and everybody was friends with everybody," George Kistiakowsky told an audience wryly long after the war. He remembered a different Los Alamos. The tall, outspoken Ukrainian-born Harvard chemist had begun studying explosives for the National Defense Research Committee in 1940; "by 1943 I thought I knew something about them." What he knew about them was original and unorthodox: "that they could be made into precision instruments, a view which was very different from that of military ordnance." He had already won von Neumann to his view, which had prepared the Hungarian mathematician in turn to endorse the precision instrument of implosion. Conant similarly trusted Kistiakowsky's judgment. In 1941 Conant had abandoned his skepticism toward the atomic bomb because of Kistiakowsky; now the explosives expert found the Harvard president seeking his help to advance Neddermeyer's work:

> I began going to Los Alamos as a consultant in the Fall of 1943, and then pressure was put on me by Oppenheimer and General Groves and particularly Conant, which really mattered, to go there on full time. I didn't want to, partly because I didn't think the bomb would be ready in time and I was interested in helping win the war. I also had what looked like an awfully interesting overseas assignment all fixed up for myself. Well, instead, unwillingly, I went to Los Alamos. That gave me a wonderful opportunity to act as a reluctant bride throughout the life of the project, which helped at times.

Kistiakowsky arrived in late January 1944 and took up residence in a small stone cabin that had been the Ranch School's pump house, an accommodation he negotiated in preference to the men's dormitory—he was forty-four years old and divorced. He quickly discovered, as he suspected, that everything was not easy and everybody was not friends:

> After a few weeks . . . I found that my position was untenable because I was essentially in the middle trying to make sense of the efforts of two men who were at each other's throats. One was Captain [Deke] Parsons who tried to run his division the way it is done in military establishments—very conservative. The other was, of course, Seth Neddermeyer, who was the exact opposite of

> Parsons, working away in a little corner. The two never agreed about anything and they certainly didn't want me interfering.

While Kistiakowsky struggled with that dilemma the theoreticians began to glimpse how a successful implosion mechanism might be designed.

The previous spring the Polish mathematician Stanislaw Ulam, then thirty-four years old and a member of the faculty at the University of Wisconsin, had found himself unhappy merely teaching in the midst of war: "It seemed a waste of my time; I felt I could do more for the war effort." He had noticed that letters from his old friend John von Neumann often bore Washington rather than Princeton postmarks and deduced that von Neumann was involved in war work; now he wrote asking for advice. Von Neumann proposed they meet between trains in Chicago to talk and turned up impressively chaperoned by two bodyguards. Eventually Hans Bethe sent along an official invitation. In the winter of 1943 Ulam and his wife Françoise, who was then two months pregnant, rode the Sante Fe Chief to New Mexico as so many others had done before them. "The sun shone brilliantly, the air was crisp and heady, and it was warm even though there was a lot of snow on the ground—a lovely contrast to the rigors of winter in Madison."

The day of his arrival Ulam met Edward Teller for the first time—he was assigned to Teller's group—who "talked to me on that first day about a problem in mathematical physics that was part of the necessary theoretical work in preparation for developing the idea of a 'super' bomb." Teller's preemption of Ulam's first days at Los Alamos for Super calculations was symptomatic of the discord that had been widening between him and Hans Bethe, who needed every available theoretical physicist and mathematician to concentrate on the difficult problem of implosion. Teller had contributed enthusiastically and crucially to the most interesting part of the work. "However," Bethe complains, "he declined to take charge of the group which would perform the detailed calculations on the implosion. Since the theoretical division was very shorthanded, it was necessary to bring in new scientists to do the work that Teller declined to do." That was one reason the British team had been invited to Los Alamos.

Teller recalls no specific refusal. "[Bethe] wanted me to work on calculational details at which I am not particularly good," he counters, "while I wanted to continue not only on the hydrogen bomb, but on other novel subjects."

The Los Alamos Governing Board reevaluated the Super once again in February 1944, learning that despite deuterium's more favorable cross section it would still be difficult to ignite. A Super would almost certainly require tritium. The small tritium samples studied so far had been transmuted in a cyclotron by bombarding lithium with neutrons. Large-scale tritium production, like large-scale plutonium production, would require production reactors, but the piles at Hanford were unfinished and previously committed. "Both because of the theoretical problems still to be solved and because of the posssibility that the Super would have to be made with tritium," reports the Los Alamos technical history, "it appeared that the development would require much longer than originally anticipated." Work could continue—the Super was too portentous a weapon to ignore—but only to the extent that it "did not interfere with the main program."

Von Neumann soon drafted Ulam to help work out the hydrodynamics of implosion. The problem was to calculate the interactions of the several shock waves as they evolved through time, which meant trying to reduce the continuous motion of a number of moving, interacting surfaces to some workable mathematical model. "The hydrodynamical problem was simply stated," Ulam comments, "but very difficult to calculate—not only in detail, but even in order of magnitude."

He remembers in particular a long discussion early in 1944 when he questioned "all the ingenious shortcuts and theoretical simplifications which von Neumann and other . . . physicists suggested." He had argued instead for "simpleminded brute force—that is, more realistic, massive numerical work." Such work could not be done reliably by hand with desktop calculating machines. Fortunately the laboratory had already ordered IBM punchcard sorters to facilitate calculating the critical mass of odd-shaped bomb cores. The IBM equipment arrived early in April 1944 and the Theoretical Division immediately put it to good use running brute-force implosion numbers. Hydrodynamic problems, detailed and repetitious, were particularly adaptable to machine computation; the challenge apparently set von Neumann thinking about how such machines might be improved.

Then a member of the newly arrived British mission made a proposal that paid his mission's way. James L. Tuck was a tall, rumpled Cherwell protégé from Oxford who had worked in England developing shaped charges for armor-piercing shells. A shaped charge is a charge of high explosive arranged in such a way—usually hollowed out like an empty ice cream cone with the open end pointed forward—that its normally divergent, bubble-shaped shock wave converges into a high-speed jet. Such a ferocious jet can punch its way through the thick armor of a tank to spray death inside.

It had just become clear from theoretical work that the several diverging shock waves produced by multiple detonators in Neddermeyer's experiments reinforced each other where they collided and produced points of high pressure; such pressure nodes in turn caused the jets and irregularities that spoiled the implosion. Rather than continue trying to smooth out a colliding collection of divergent shock waves, Tuck sensibly proposed that the laboratory consider designing an arrangement of explosives that would produce a converging wave to begin with, fitting the shock wave to the shape it needed to squeeze. Such explosive arrangements were called lenses by analogy with optical lenses that similarly focus light.

No one wanted to tackle anything so complex so late in the war. Geoffrey Taylor, the British hydrodynamicist, arrived in May to offer further insight into the problem. He had developed an understanding of what came to be called Raleigh-Taylor instabilities, instabilities formed at the boundaries between materials. Accelerate heavy material against light material, he demonstrated mathematically, and the boundary between the two will be stable. But accelerate light material against heavy material and the boundary between the two will be unstable and turbulent, causing the two materials to mix in ways extremely difficult to predict. High explosive was light compared to tamper. All of the tamper materials under consideration except uranium were significantly lighter than plutonium. Raleigh-Taylor instabilities would constrain subsequent design. They would also make it difficult to predict bomb yield.

As the IBM results clarified shock-wave behavior the physicists began seriously to doubt if a uniform wrap of HE could ever be made to produce a symmetrical explosion. Complex though explosive lenses might be, they were apparently the only way to make implosion work. Von Neumann turned to their formulation. "You have to assume that you can control the velocity of the detonation wave in a chemical explosive very accurately," Kistiakowsky explains, "so if you start the wave at certain points by means of detonators you can predict exactly where it will be at a given time. Then you can design the charge." It was soon clear that the velocity of the converging shock waves from the several explosive lenses that would surround the bomb core could vary by no more than 5 percent. That was the demanding limit within which von Neumann designed and Kistiakowsky, Neddermeyer and their staffs began to work.

In the spring of 1944 the two difficult personal conflicts—between Teller and Bethe and between Kistiakowsky and Neddermeyer—forced Oppenheimer to intervene. First, Bethe writes, Teller withdrew from fission development:

> With the pressure of work and lack of staff, the Theoretical Division could ill afford to dispense with the services of any of its members, let alone one of

> such brilliance and high standing as Teller. Only after two failures to accomplish the expected and necessary work, and only on Teller's own request, was he, together with his group, relieved of further responsibility for work on the wartime development of the atomic bomb.

A letter from Oppenheimer to Groves on May 1, 1944, seeking to replace Teller with Rudolf Peierls, corroborates Bethe's account: "These calculations," it says in part, "were originally under the supervision of Teller who is, in my opinion and Bethe's, quite unsuited for this responsibility. Bethe feels that he needs a man under him to handle the implosion program." It was, Oppenheimer notes, a question of the "greatest urgency."

Ulam remembers that Teller threatened to leave. Oppenheimer stepped in then to save him for the project. He encouraged Teller to give himself over to the Super—encouragement, Teller wrote in 1955, perhaps disingenuously, that he needed to move him on from the immediate task at hand:

> Oppenheimer . . . continued to urge me with detailed and helpful advice to keep exploring what lay beyond the immediate aims of the laboratory. This was not easy advice to give, nor was it easy to take. It is easier to participate in the work of the scientific community, particularly when a goal of the highest interest and urgency has been clearly defined. Every one of us considered the present war and the completion of the A-bomb as the problems to which we wanted to contribute most. Nevertheless, Oppenheimer . . . and many of the most prominent men in the laboratory continued to say that the job at Los Alamos would not be complete if we should remain in doubt whether or not a thermonuclear bomb was feasible.

To that end Oppenheimer in May discussed tritium production with Groves and Du Pont's Crawford Greenewalt. The chemical company had built a pilot-scale air-cooled pile at Oak Ridge that produced neutrons to spare; Greenewalt agreed to put some of them to use bombarding lithium.

Teller departed the Theoretical Division. Rudolf Peierls took his place. Oppenheimer arranged then to meet with Teller weekly for an hour of freewheeling talk. That was a remarkable concession when the laboratory was working overtime six days a week to build a bomb before the end of the war. Oppenheimer may well have thought Teller's imaginative originality worth it. He also understood his extreme sensitivity to slight. Later that summer, when Cherwell visited Los Alamos, Oppenheimer gave a party and inadvertently failed to invite Peierls, who was deputy head of the British mission under James Chadwick. Oppenheimer sought out Peierls the next day and apologized, adding: "But there is an element of relief in this situation: it might have happened with Edward Teller."

George Kistiakowsky adjusted himself to Seth Neddermeyer until he felt that not only he but also the project was suffering; then he reviewed his alternatives and, on June 3, wrote Oppenheimer a memorandum. He and Neddermeyer had established a certain *modus vivendi*, he wrote, but it was not what he had been asked to do, which was to administer implosion work while Neddermeyer did the science, and it was "not based on mutual confidence and a friendly give-and-take."

He proposed three possible solutions. He could resign, the solution he thought best and fairest to Neddermeyer. Or Neddermeyer could resign, but that would disturb his staff and slow the work; it would also be unfair to a good physicist. Or Neddermeyer could "take over more vigorous scientific and technical direction of the project but dissociate himself completely from all administrative and personnel matters."

Oppenheimer had come to value Kistiakowsky too highly to choose any of these alternatives. He proposed a fourth. Kistiakowsky worked out the details and the men met painfully to present it to Neddermeyer on a Thursday evening in mid-June: Kistiakowsky would assume full responsibility for implosion work as an associate division leader under Parsons. Neddermeyer and Luis Alvarez, recently arrived from Chicago, would become senior technical advisers. Neddermeyer left the meeting early, as well he might. "I am asking you to accept the assignment," Oppenheimer wrote him the same evening. ". . . In behalf of the success of the whole project, as well as the peace of mind and effectiveness of the workers in the H. E. program, I am making this request of you. I hope you will be able to accept it." With enduring bitterness Neddermeyer did.

The air-cooled, pilot-scale reactor at Oak Ridge had gone critical at five o'clock in the morning on November 4, 1943; the loading crews, realizing during the night that they were nearing criticality sooner than expected, had enjoyed rousting Arthur Compton and Enrico Fermi out of bed at the Oak Ridge guest house to witness the event. The pile, which was designated X-10, was a graphite cube twenty-four feet on a side drilled with 1,248 channels that could be loaded with canned uranium-metal slugs and through which large fans blew cooling air. The channels extended for loading through the seven feet of high-density concrete that composed the pile face; at the back they opened onto a subterranean pool like the pools planned for Hanford into which irradiated slugs could be pushed to shield them until they lost their more intense short-term radioactivity. Chemists then processed the slugs in a remote-controlled pilot-scale separations plant using the chemical separation processes Glenn Seaborg and his colleagues had developed at ultramicrochemical scale in Chicago.

A few days before Compton moved to Oak Ridge to supervise the X-10 operation, at the end of November, workers discharged the first five tons of irradiated uranium from the pile. Chemical separations began the following month. By the summer of 1944 batches of plutonium nitrate containing gram quantities of plutonium had begun arriving at Los Alamos. The man-made element was quickly used and reused in extensive experiments to study its unfamiliar chemistry and metallurgy—more than two thousand separate experiments by the end of the summer.

Not chemistry or metallurgy but physics nearly condemned the plutonium bomb to failure that summer. More than a year previously Glenn Seaborg had warned that the isotope Pu240 might form along with desirable Pu239 when uranium was irradiated to make plutonium. Pu240, an even-numbered isotope, was likely to exhibit a much higher rate of spontaneous fission than Pu239. The plutonium samples Emilio Segrè had studied at his isolated log-cabin laboratory fissioned spontaneously at acceptable rates. They had been transmuted from uranium in one of the Berkeley cyclotrons. U238 needed one neutron to transmute to Pu239; for Pu240 it required two, and far more neutrons bombarded the uranium slugs cooking in the X-10 pile than a cyclotron could generate. When Segrè measured the spontaneous fission rate of the X-10 plutonium he found it much higher than the Berkeley rate. The rate for Hanford plutonium, which would be exposed to an even heavier neutron flux, was likely to be higher still. That meant they would not need to cleanse the plutonium so thoroughly of light-element impurities. But it also signaled catastrophe. They could not use a gun to assemble a critical mass of such stuff: approaching each other even at 3,000 feet per second, the plutonium bullet and target would melt down and fizzle before the two parts had time to join.

Oppenheimer alerted Conant on July 11. The two men met with Compton, Groves, Nichols and Fermi in Chicago six days later and the next day Oppenheimer wrote Groves to confirm their conclusions. Pu240 was apparently long-lived, and since the two isotopes were elementally identical it could not be removed chemically. They had not considered separating Pu239 from Pu240 electromagnetically. Such an effort with isotopes that differed by only one mass unit and were highly toxic would dwarf the vast calutron operation at Oak Ridge and could not possibly be accomplished in time to influence the outcome of the war. "It appears reasonable," Oppenheimer ended, "to discontinue the intensive effort to achieve higher purity for plutonium and to concentrate attention on methods of assembly which do not require a low neutron background for their success. At the present time the method to which an over-riding priority must be assigned is the method of implosion."

That necessity was painful, as the Los Alamos technical history makes clear: "The implosion was the only real hope, and from current evidence not a very good one." Oppenheimer agonized over the problem to the point that he considered resigning his directorship. Robert Bacher, the sturdy leader of the Experimental Physics Division, took long walks with him in those days to share his pain and eventually dissuaded him. There was no one else who could do the job, Bacher argued; without Oppenheimer there would be no bomb in time to shorten the war and save lives.

Action changed Oppenheimer's mood. "The Laboratory had at this time strong reserves of techniques, of trained manpower, and of morale," says the technical history. "It was decided to attack the problems of the implosion with every means available, 'to throw the book at it.'" Going over the prospects with Bacher and Kistiakowsky, Oppenheimer decided to carve two new divisions out of Parsons' Ordnance Division: G (for Gadget) under Bacher to master the physics of implosion and X (for eXplosives) under Kistiakowsky to perfect explosive lenses. The Navy captain howled, Kistiakowsky remembers:

> [Oppenheimer] called a big meeting of all the group heads, and there he sprang on Parsons the fact that I had plans for completely re-designing the explosives establishment. Parsons was furious—he felt that I had by-passed him and that was outrageous. I can understand perfectly how he felt but I was a civilian, so was Oppie, and I didn't have to go through him. . . . From then on Parsons and I were not on good terms. He was extremely suspicious of me.

Parsons had his hands full in any case designing the uranium gun, Little Boy, and arranging its eventual use. Oppenheimer prevailed: they would throw the book at implosion. In the months ahead the laboratory, which had swollen to 1,207 full-time employees by the previous May 1, would once again double and redouble in size.

Philip Abelson, the young Berkeley physicist to whom Luis Alvarez had run from his barber chair in January 1939 to announce the news of fission, had moved to the Naval Research Laboratory in 1941 to work on uranium enrichment for the Navy and had made valuable progress independently of the Manhattan Project in the intervening years. The Navy was interested in nuclear power as a motive force for submarines, to extend their range and to allow them to travel farther submerged. But a pile of the sort Fermi would build would be unwieldy; "it had become pretty obvious," Abelson recalls, "that a reactor fueled with natural uranium would be big as a barn." Increase the ratio of U235 to U238 in the reactor fuel—enrich the uranium—and the reactor could be correspondingly smaller; with enough enrichment, small enough to fit inside the hull of a submarine in the space previously reserved for diesel engines, batteries and fuel.

Enrichment and separation are different goals, but the same technologies achieve them. Abelson began work by looking up the record of those technologies. Gaseous barrier diffusion was under study then at Columbia, electromagnetic separation at Berkeley, centrifuge separation at the University of Virginia. Abelson decided to try a process that had been pioneered in Germany before the war: liquid thermal diffusion (using glass tubes, Otto Frisch had experimented unsuccessfully with a similar process, gaseous thermal diffusion, at Birmingham). Thermal diffusion relied on the tendency of lighter isotopes to diffuse toward a hotter region while heavier isotopes diffused toward a colder region. The mechanism for driving such diffusion could be simple: a hot pipe inside a cold pipe with liquid uranium hexafluoride flowing between the two pipe walls. Depending on the difference in temperature and the spacing between the two pipes more or less diffusion would occur. At the same time the heating and cooling of the hex would start a convection current flowing up the hot pipe wall and down the cold. That would bring the U235-enriched fluid to the top of the column where it could be tapped off. To increase the enrichment a number of columns could be connected in series to make a cascade like the cascade of barrier tanks planned for K-25.

Abelson's first technical contribution, in 1941, was inventing a relatively cheap way to make uranium hexafluoride. He processed the first hundred kilograms of hex produced in the United States. For the nominal sum of one dollar the Army contracted to borrow his patented process for Oak Ridge. He never saw the dollar.

The experimental thermal-diffusion columns Abelson built at the Naval Research Laboratory in 1941 and 1942 were 36 feet tall, each consisting of three pipes arranged one inside the other. The hot inner pipe, 1¼ inches in diameter, carried high-pressure steam at about 400° F. Surrounding that nickel pipe a copper pipe contained the liquid hex. The critical spacing between the two pipes where the hex flowed measured only about one-tenth of an inch. Surrounding both pipes a 4-inch pipe of galvanized iron carried water at about 130°, just above hex's melting point, to cool the hex.

Pumps that circulated the water were the only moving parts. "The apparatus was run continuously with no shut down or break down what so ever," Abelson reported to the Navy early in 1943. "Indeed, so constant were the various temperatures and operating characteristics that practically no attention was required to insure successful operation. Many days passed in which operating personnel did not touch any control device." To stop the flow of the hex out of a column Abelson simply dipped the bend of a U-shaped metal drain tube into a bucket of dry ice and alcohol, which froze the hex and plugged the tube. A flame to warm the tube started the flow again.

Abelson's January 4, 1943, report, submitted jointly with his NRL colleague Ross Gunn, indicated that uranium could be enriched within a single thermal-diffusion column from its natural U235 content of 0.7 percent up to 1 percent or better. With several thousand columns connected in series Abelson thought he could produce 90 percent pure U235 at the rate of 1 kilogram per day at a total construction cost of no more than $26 million. Ninety percent purity was entirely sufficient to make a bomb. (That estimate proved optimistic, however, and equilibrium time for such a cascade appeared to be as long as 600 days.)

Another choice, more in keeping with the Navy's interest in submarine propulsion, emphasized quantity enrichment rather than quality. Abelson proposed building a plant of 300 48-foot columns operating in parallel to make large amounts of slightly enriched uranium immediately. Chicago could use such slightly enriched uranium to advance its pile work, Abelson thought. He did not yet know that CP-1 had gone critical just one month before his report. "Information concerning the many experiments performed by [the Chicago] workers in the last six months has been denied to us," he complained. "It is vitally necessary that there be an exchange of technical information if proper plans are to be made for future plants." The NRL had been the first research center Groves visited when he took charge of the Manhattan Project in September 1942. Months before then, Franklin Roosevelt had specifically instructed Vannevar Bush to exclude the Navy from atomic bomb development. Groves followed the NRL's research and Bush encouraged its funding through the Military Policy Committee. But by 1943 the official flow of information on nuclear energy research ran from the Navy to the Army one-way only.

Unofficially, however, several of Groves' compartments leaked. In November 1943 the Navy authorized Abelson to build his 300-column plant. He had searched for a sufficient source of steam—thermal diffusion used volcanic magnitudes of steam, one reason the Manhattan Project had chosen not to pursue it—and located the Naval Boiler and Turbine Laboratory at the Philadelphia Navy Yard. "They were testing good-sized boilers that would go into ships," Abelson says. "They had the capability of making quantities of steam at a thousand pounds per square inch and they had Navy people standing twenty-four-hour watches to deliver the steam." The boiler laboratory's waste steam would supply his 300-column plant, but before scaling up that far he planned to test his design by building and operating only the first 100 columns. Construction began in January 1944, with completion scheduled for July. By now Abelson knew more about the

Manhattan Project. He knew that the barriers which Houdaille-Hershey had been stripped and reequipped to manufacture were not yet passing inspection and that K-25, the gaseous-diffusion plant, was therefore woefully behind schedule. He knew Los Alamos had been founded with Robert Oppenheimer as its director. He knew Berkeley was struggling to make its calutrons work. He saw that his thermal-diffusion process might come to the bomb project's rescue and he was generous enough and worried enough about the war to offer it despite the Army's several previous rebuffs.

He chose not to work through the limited official channels that the Army and the OSRD had devised to constrict the flow of information. "I wanted to let Oppenheimer know what we were doing. Someone in the Bureau of Ships knew one of the people in the [Navy] Bureau of Ordnance who was going out to Los Alamos. I remember that I met the man at the old Warner Theater here in Washington, up in the balcony—real cloak and dagger stuff." Abelson briefed the BuOrd officer about the plant he was building. He said that he expected to be producing 5 grams a day of material enriched to 5 percent U235 by July. This vital information the BuOrd man carried to Los Alamos and passed along to Edward Teller. Teller in turn briefed Oppenheimer. Oppenheimer apparently conspired then with Deke Parsons, the Hill's ranking Navy man, to concoct a cover story: that Parsons had learned of the Abelson work on a visit to the Philadelphia Navy Yard. With the Navy thus protected, Oppenheimer on April 28 alerted Groves.

Oppenheimer had seen Abelson's January 1943 report only a few months previously, a year after it was written. He was not impressed. Like his colleagues Oppenheimer had considered only those processes that enriched natural uranium all the way up to bomb grade, a requirement thermal diffusion could not efficiently meet. Now he realized that Abelson's process offered a valuable alternative, the alternative Abelson had proposed in his report to help Chicago advance its pile work: slight enrichment of larger quantities. Feeding even slightly enriched material into the Oak Ridge calutrons would greatly increase their efficiency. A thermal-diffusion plant could therefore substitute at least temporarily for the stalled lower stages of the K-25 plant and supplement the output of the Alpha calutrons. Abelson's 100-column plant with the columns operating in parallel, Oppenheimer calculated, should produce about 12 kilograms a day of uranium of 1 percent enrichment.

"Dr. Oppenheimer . . . suddenly told me that we had [made] a terrible scientific blunder," Groves testified after the war. "I think he was right. It is one of the things that I regret the most in the whole course of the operation.

We had failed to consider [thermal diffusion] as a portion of the process as a whole." From the beginning the leaders of the Manhattan Project had thought of the several enrichment and separation processes as competing horses in a race. That had blinded them to the possibility of harnessing the processes together. Groves had partly opened his eyes when barrier troubles delayed K-25; then he had decided to cancel the upper stages of the K-25 cascade and feed the lower-stage product to the Beta calutrons for final enrichment. So he was prepared to understand immediately Oppenheimer's similar point about the value of a thermal-diffusion plant: "I at once decided that the idea was well worth investigating."

Groves appointed a committee of men thoroughly experienced by now in Manhattan Project troubleshooting: W. K. Lewis, Eger Murphree and Richard Tolman. They visited the Philadelphia Navy Yard on June 1 and turned in their conclusions on June 3. They thought Oppenheimer's estimate of 12 kilograms a day of 1 percent U235 optimistic but emphasized the possibility—with 300 columns instead of 100—of producing 30 kilograms per day of 0.95 percent U235.

Groves thought bigger than that. He had a power plant with 238,000 kilowatts rated capacity coming on line within weeks in the K-25 area at Oak Ridge that K-25 would not be ready to draw on until the end of the year. It was designed to generate electricity to run the barrier diffusers but it made electricity by making steam. The steam could serve a thermal-diffusion plant that would enrich uranium for the Alpha and Beta calutrons until such a time as K-25 needed electricity. Then the permanent K-25 installation could be phased in gradually and the temporary thermal-diffusion plant phased out.

The proposal cleared the Military Policy Committee on June 12, 1944. On June 18 Groves contracted with the engineering firm of H. K. Ferguson to build a 2,100-column thermal-diffusion plant beside the power plant on the Clinch River in ninety days or less. That extraordinary deadline allowed no time for design. Ferguson would assemble the operation from twenty-one identical copies—"Chinese copies," Groves called them—of Philip Abelson's 100-column unit in the Philadelphia Navy Yard.

The general must have appreciated the fortuity of his decision when he learned the following month of the plutonium crisis at Los Alamos. But the thermal-diffusion plant was not immediately Oak Ridge's savior. Ferguson managed to build a capacious 500-foot barn of black metal siding and began operating the first rack of columns in sixty-nine days, by September 16, but steam leaked out almost as fast as it could be blown in and couplings needed extensive repair and even partial redesign. The gaseous-diffusion plant, K-25, was more than half completed but no barrier tubes shipped from Houdaille-Hershey yet met even minimum standards. The Alpha calutrons smeared uranium all over the insides of their vacuum tanks, catching no more than 4 percent of the U235; that valuable fraction, reprocessed and fed into the Beta calutrons, reached the Beta collectors in turn at only 5 percent efficiency. Five percent of 4 percent is two thousandths. A speck of U235 stuck to an operator's coveralls was well worth searching out with a Geiger counter and retrieving delicately with tweezers. No essence was ever expressed more expensively from the substance of the world with the possible exception of the human soul.

In the Pacific the island war advanced. As the Army under General Douglas MacArthur pushed up from Australia across New Guinea toward the Philippines, the Marines under Admiral Chester Nimitz island-hopped from Guadalcanal to Bougainville in the Solomons, north across the equator to Tarawa in the Gilberts, farther north to Kwajalein and Eniwetok in the Marshalls. That brought them, by the summer of 1944, within striking distance of the Japanese inner defense perimeter to the west. Its nearest bastions were the Marianas, a chain of volcanic islands at the right corner of a roughly equilateral triangle of which the Philippine main island of Luzon was the left corner and the Japanese main island of Honshu the apex. The United States wanted the Marianas as primary bases for further advance: Guam for the Navy; Saipan and Tinian for the new B-29 Superfortresses that the Army Air Force had begun deploying temporarily at great risk and expense in China's Szechwan province, ferrying aviation fuel and bombs over the Himalayas to support their mission, which was the high-altitude precision bombing of Japan. By contrast, only fifteen hundred miles of open water separated Saipan and Tinian from Tokyo and the islands could be supplied securely by sea.

Nimitz named the Marianas campaign Operation Forager; it began in mid-June with heavy bombing of the island airfields. Then 535 ships carrying 127,571 troops sailed from Eniwetok, the largest force of men and ships yet assembled for a Pacific naval operation. "We are through with flat atolls now," Holland Smith, the Marine commanding general, briefed his officers. "We learned to pulverize atolls, but now we are up against mountains and caves where Japs can dig in. A week from now there will be a lot of dead Marines."

Intelligence estimates put 15,000 to 17,000 Japanese troops on Saipan, 10,000 on smaller Tinian three miles to the south. The marines invaded Saipan first, on the morning of June 15, and won a long but shallow beachhead onto which, by afternoon, amphtracs had delivered 20,000 men. *Time* correspondent Robert Sherrod was among them dodging shells from

Japanese artillery inland; he had seen action before on the Aleutian island of Attu and on Tarawa and knew the Japanese as America had come to know them:

> Nowhere have I seen the nature of the Jap better illustrated than it was near the airstrip at dusk. I had been digging a foxhole for the night when one man shouted: "There is a Jap under those logs!" The command post security officer was dubious, but he handed concussion grenades to a man and told him to blast the Jap out. Then a sharp ping of the Jap bullet whistled out of the hole and from under the logs a skinny little fellow—not much over 5 ft. tall—jumped out waving a bayonet.
>
> An American tossed a grenade and it knocked the Jap down. He struggled up, pointed his bayonet into his stomach and tried to cut himself open in approved hara-kiri fashion. The disemboweling never came off. Someone shot the Jap with a carbine. But, like all Japs, he took a lot of killing. Even after four bullets had thudded into his body he rose to one knee. Then the American shot him through the head and the Jap was dead.

While the marines advanced into Saipan, fighting off the harrowing Japanese frontal assaults they learned to call banzai charges, 155-millimeter Long Toms brought ashore and set up in the southern sector of the island began softening up Tinian. That smaller island of thirty-eight square miles, ten miles long and shaped much like Manhattan, was far less rugged than Saipan. Its highest elevation, Mount Lasso, rose only 564 feet above sea level; its lowlands were planted in sugar cane; it had roads and a railway to recommend it to tank operations. To the disadvantage of amphibious assault the island was a raised platform protected on all sides by steep cliffs 500 to 600 feet high—The Rock, the marines would come to call it. It had two major beaches, one near Tinian Town on the southwest coast and the other, which the marines named Yellow, on the east coast at the island's waist. Navy frogmen explored both by night and found them heavily mined and defended.

Two other smaller beaches on the northwest coast hardly deserved the name; one was 60 yards long and the other 150 yards. The United States had made no division-strength landing across any beach less than twice the length of those two toeholds combined in the entire course of the war. The Japanese on Tinian accordingly defended them with nothing more than a few mines and two 25-man blockhouses. The marines coded them White 1 and White 2 and chose them for their assault.

The invasion of Tinian began on July 24, two weeks after Saipan had been secured. Because of the larger island's proximity the marines could deploy shore-to-shore rather than ship-to-shore, embarking in LST's and smaller craft directly from Saipan. A feint at Tinian Town beach decoyed the Japanese defenses and the invaders achieved complete tactical surprise, rushing ashore and pushing inland as fast as possible to escape the dangerously narrow landings. By the end of the day, when the advance halted to organize a solid defense against the Japanese troops rushing up the island from Tinian Town, most of the tanks had been brought ashore, four howitzer batteries were in place and a spare battalion was even at hand. The defenders had killed fifteen marines and wounded fewer than two hundred; the American perimeter extended inland more than two miles.

With the coming of darkness the Japanese began a mortar barrage. Near midnight their artillery arrived and they added it in. The marines answered with their howitzers. To watch for the expected Japanese counterattack they illuminated the area with flares. The attack started at 0300 hours, Japanese soldiers rushing the American lines head-on in the naked light of the flares. Against strong Marine defenses challenge quickly became slaughter.

The marines needed only four days to advance down the island. They encountered tanks and infantry and in the mild terrain easily destroyed them. They took Tinian Town on July 31, that night shattered a last banzai charge from the south and the next day, August 1, 1944, declared the island secure. More than 6,000 Japanese combatants died compared to 300 Americans. Another 1,500 marines were wounded. Soon the Seabees would arrive to begin bulldozing airfields.

Saipan before had been bloodier: 13,000 U.S. casualties, 3,000 marines killed, 30,000 Japanese defenders dead. But a more grotesque slaughter had engulfed the island's population of civilians. Believing as propaganda had prepared them that the Americans would visit upon them rape, torture, castration and murder, 22,000 Japanese civilians had made their way to two sea cliffs 80 and 1,000 feet high above jagged rocks and, despite appeals from Japanese-speaking American interpreters and even fellow islanders, had flung themselves, whole families at a time, to their deaths. The surf ran red with their blood; so many broken bodies floated in the water that Navy craft overrode them to rescue. Not all the dead had volunteered their sacrifice; many had been rallied, pushed or shot by Japanese soldiers.

The mass suicide on Saipan—a Jonestown of its day—instructed Americans further in the nature of the Jap. Not only soldiers but also civilians, ordinary men and women and children, chose death before surrender. On their home islands the Japanese were 100 million strong, and they would take a lot of killing.

"The view was stupendous, and the wind was bitter cold," Leona Marshall recalls of a day at Hanford, Washington, in September 1944 when she,

Enrico Fermi and Crawford Greenewalt climbed giddily to the top of a twelve-story tower to survey the secret reservation. They could see the Columbia River running deep and blue in both directions out of sight over the horizon; they could see the gray desert and the distant hazy mountains. By then construction was more than two-thirds completed and nearer at hand they overlooked a city of industrial buildings and barracks and three massive blockhouses, the three plutonium production reactors sited on the river's western shore. The number of construction workers had peaked at 42,400 the previous June. Marshall was working now at Hanford; Fermi and Greenewalt had traveled out to monitor the start-up of the B pile, the first one finished. The day the construction teams left it, September 13, Fermi had inserted the first aluminum-canned uranium slug to begin the loading, the Pope conferring his blessing as he had on the piles at Chicago and Oak Ridge.

Slug canning had almost come to a crisis. Two years of trial-and-error effort had not produced canning technology adequate to seal the uranium slugs, which quickly oxidized upon exposure to air or water, away from corrosion. Only in August had the crucial step been devised, by a young research chemist who had followed the problem from Du Pont in Wilmington to Chicago and then to Hanford: putting aside elaborate dips and baths he tried soaking the bare slugs in molten solder, lowering the aluminum cans into the solder with tongs and canning the slugs submerged. The melting point of the aluminum was not much higher than the melting point of the solder, but with careful temperature control the canning technique worked.

Greenewalt then pushed production around the clock. Slugs accumulated in the reactor building faster than the loading crews could use them and Marshall and Fermi observed them there on one of their inspections:

> Enrico and I went to the reactor building . . . to watch the loading.The slugs were brought to the floor in solid wooden blocks in which holes were drilled, each of a size to contain a slug, and the wooden blocks were stacked much as had been the slug-containing graphite bricks in CP-1. Idly I teased Fermi saying it looked like a chain-reacting pile. Fermi turned white, gasped, and reached for his slide rule. But after a couple of seconds he relaxed, realizing that under no circumstances could natural uranium and natural wood in any configuration cause a chain reaction.

Tuesday evening, September 26, 1944, the largest atomic pile yet assembled on earth was ready. It had reached dry criticality—the smaller loading at which it would have gone critical without cooling water if its operators had not restrained it with control rods—the previous Friday. Now the Columbia circulated through its 1,500 loaded aluminum tubes. "We arrived in the control room as the du Pont brass began to assemble," Marshall remembers. "The operators were all in place, well-rehearsed, with their start-up manuals on their desks." Some of the observers had celebrated with good whiskey; their exhalations braced the air. Marshall and Fermi strolled the room checking readings. The operators withdrew the control rods in stages just as Fermi had once directed for CP-1; once again he calculated the neutron flux on his six-inch slide rule. Gradually gauges showed the cooling water warmed, flowing in at 50°F and out at 140°. "And there it was, the first plutonium-production reactor operating smoothly and steadily and quietly. . . . Even in the control room one could hear the steady roaring sound of the high-pressure water rushing through the cooling tubes."

The pile went critical a few minutes past midnight; by 2 A.M. it was operating at a higher level of power than any previous chain reaction. For the space of an hour all was well. Then Marshall remembers the operating engineers whispering to each other, adjusting control rods, whispering more urgently. "Something was wrong. The pile reactivity was steadily decreasing with time; the control rods had to be withdrawn continuously from the pile to hold it at 100 megawatts. The time came when the rods were completely withdrawn. The reactor power began to drop, down and down."

Early Wednesday evening B pile died. Marshall and Fermi had slept by then and returned. They talked over the mystery with the engineers, who first suspected a leaking tube or boron in the river water somehow plating out on the cladding. Fermi chose to remain open-minded. The charts, which seemed to show a straight-line failure, might be hiding the shallow curve of an exponential decline in reactivity, which would mean a fission product undetected in previous piles was poisoning the reaction.

Early Thursday morning the pile came back to life. By 7 A.M. it was running well above critical again. But twelve hours later it began another decline.

Princeton theoretician John A. Wheeler had counseled Crawford Greenewalt on pile physics since Du Pont first joined the project. He was stationed at Hanford now and he followed the second failure of the pile closely. He had been "concerned for months," he writes, "about fission product poisons." B pile's heavy breathing convinced him such a poisoning had occurred. The mechanism would be compound: "A non-[neutron-]absorbing mother fission product of some hours' half-life decays to a daughter dangerous to neutrons. This poison itself decays with a half-life of some hours into a third nuclear species, non-absorbing and possibly even stable." So the pile would chain-react, making the mother product; the mother product would decay to the daughter; as the volume of daughter product increased, absorbing neutrons, the pile would decline; when sufficient daughter product was present, enough neutrons would be absorbed to starve the chain reaction and the pile would shut down. Then the daughter product would decay to a non-absorbing third element; as it decayed the pile would stir; eventually too little daughter product would remain to inhibit the chain reaction and the pile would go critical again.

Fermi had left for the night; Wheeler on watch calculated the likely half-lives based on the blooming and fading of the pile. By morning he thought he needed two radioactivities with half-lives totaling about fifteen hours:

> If this explanation made sense, then an inspection of the chart of nuclei showed that the mother had to be 6.68 hr [iodine]^{135} and the daughter 9.13 hr [xenon]^{135}. Within an hour Fermi arrived with detailed reactivity data which checked this assignment. Within three hours two additional conclusions were clear. (a) The cross section for absorption of thermal neutrons by Xe^{135} was roughly 150 times that of the most absorptive nucleus previously known, [cadmium]^{113}. (b) Almost every Xe^{135} nucleus formed in a high flux reactor would take a neutron out of circulation. Xenon had thrust itself in as an unexpected and unwanted extra control rod. To override this poison more reactivity was needed.

Greenewalt called Samuel Allison in Chicago on Friday afternoon. Allison passed the bad news to Walter Zinn at Argonne, the laboratory in the forest south of Chicago where CP-1 was meant to be housed and where several piles now operated. Zinn had just shut down CP-3, a shielded six-foot tank filled with 6.5 tons of heavy water in which 121 aluminum-clad uranium rods were suspended. Disbelieving, Zinn started the 300-kilowatt reactor up again and ran it at full power for twelve hours. It was primarily a research instrument and it had never been run so long at full power before. He found the xenon effect. Laborious calculations at Hanford over the next three days confirmed it.

Groves received the news acidly. He had ordered Compton to run CP-3 at full power full time to look for just such trouble. Ever the optimist, Compton apologized in the name of pure science: the mistake was regrettable but it had led to "a fundamentally new discovery regarding neutron properties of matter." He meant xenon's consuming appetite for neutrons. Groves would have preferred to blaze trails less flamboyantly.

If Du Pont had built the Hanford production reactors to Eugene Wigner's original specifications, which were elegantly economical, all three piles would have required complete rebuilding now. Fortunately Wheeler had fretted about fission-product poisoning. After the massive wooden shield blocks that formed the front and rear faces of the piles had been pressed, a year previously, he had advised the chemical company to increase the count of uranium channels for a margin of safety. Wigner's 1,500 channels were arranged cylindrically; the corners of the cubical graphite stacks could accommodate another 504. That necessitated drilling out the shield blocks, which delayed construction and added millions to the cost. Du Pont had accepted the delay and drilled the extra channels. They were in place now when they were needed, although not yet connected to the water supply.

D pile went critical with a full 2,004-tube loading on December 17, 1944; B pile followed on December 28. Plutonium production in quantity had finally begun. Groves was enthusiastic enough at year's end to report to George Marshall that he expected to have eighteen 5-kilogram plutonium bombs on hand in the second half of 1945. "Looks like a race," Conant noted for his history file on January 6, 1945, "to see whether a fat man or a thin man will be dropped first and whether the month will be July, August or September."

17

The Evils of This Time

The bombs James Bryant Conant speculated about early in 1945 were crude designs of uncertain yield. The previous October he had traveled out to Los Alamos to ascertain their prospects. To Vannevar Bush he reported that the gun method of detonation seemed "as nearly certain as any untried new procedure can be." The availability of a uranium gun bomb, which Los Alamos expected would explode with a force equivalent to about 10,000 tons of TNT, now depended only on the separation of sufficient U235. Implosion looked far more questionable; intensive work was just then getting under way following Oppenheimer's August reorganization of the laboratory. Conant estimated the yield of the first implosion design, whether lensed or not, "as an order of magnitude only" at about 1,000 tons TNT equivalent. That was so relatively modest a result that he invited Bush to consider the gun bomb strategic and the implosion bomb tactical.

For the past three years Bush and Conant had concentrated their efforts entirely on these first crude bombs. Now they were interested in improvements. During the summer of 1944, Conant says, on an earlier inspection trip to Los Alamos, he and Bush had found leisure and privacy to discuss "what the policy of the United States should be after the war was over." As a result they had sent Secretary of War Henry L. Stimson a joint memorandum on September 19 that independently raised some of the issues Niels Bohr had raised with Franklin Roosevelt in August, in particular that "the progress of this art and science is bound to be so rapid in the next five years in some countries that it would be extremely dangerous for this government to assume that by holding secret its present knowledge we should be secure." They did not see the bomb's complementarity, but did see that whatever control arrangement the United States and Great Britain devised—they favored a treaty—would somehow have to include the Soviet Union; if the Soviets were not informed, as Bush told Conant, the exclusion would lead "to a very undesirable relationship indeed on the subject with Russia."

Roosevelt had returned from Hyde Park troubled that Felix Frankfurter and Bohr had somehow breached Manhattan Project security, Bush and perhaps Conant had talked to Bohr and the two administrators had submitted to Stimson at his request a more detailed proposal incorporating Bohr's ideas. In doing so they had explicitly recommended that the United States sacrifice some portion of its national sovereignty in exchange for effective international control, understanding as they did so that they would have to answer vigorous opposition:

> In order to meet the unique situation created by the development of this new art we would propose that free interchange of all scientific information on this subject be established under the auspices of an international office deriving its power from whatever association of nations is developed at the close of the present war. We would propose further that as soon as practical the technical staff of this office be given free access in all countries not only to the scientific laboratories where such work is contained, but to the military establishments as well. We recognize that there will be great resistance to this measure, but believe the hazards to the future of the world are sufficiently great to warrant this attempt.

But how great in fact were the hazards? That was something else Conant traveled to Los Alamos in October to find out. If the argument for allowing the nation's military establishments to be inspected depended on the dangers of a thermonuclear explosive, it was speculative and therefore weak: the thermonuclear was still only an idea on paper that might not work. How much could fission weapons be improved? How much destructiveness of either kind might a bomber—or, as Bush and Conant briefed Stimson, "a robot plane or guided missile"—eventually visit upon the cities of the world?

What Conant learned first of all was that others had already begun to ask the same questions. The technological imperative, the urge to improvement even if the objects to be improved are weapons of mass destruction, was already operating at Los Alamos. Under intense pressure to produce a first crude weapon in time to affect the outcome of the war, people had found occasion nevertheless to think about building a better bomb. Conant reported to Bush:

> By various methods that seem quite possible of development within six months after the first bomb is perfected, it should be possible to increase the efficiency . . . in which case the same amount of material would yield something like 24,000 Tons TNT equivalent. Further developments along this same line hold a possibility of producing a single bomb with such amounts of materials and such efficiencies as to run this figure up to several hundred thousand Tons TNT equivalent, or even perhaps a million Tons TNT equivalent. . . . All these possibilities reside only in perfecting the efficiency of the use of elements "25"[U235] and "49"[Pu239]. You will thus see that a considerable "super" bomb is in the offing quite apart from the use of other nuclear reactions.

A million tons TNT equivalent was devastation indeed—the world war then raging would consume a total of about three million tons of explosives by its end—but Edward Teller, Conant found, had already dismissed such improvements as picayune:

> It seems that the possibility of inciting a thermonuclear reaction involving heavy hydrogen is somewhat less now than appeared at first sight two years ago. I heard an hour's talk on this subject by the leading theoretical man at L.A. The most hopeful procedure is to use tritium (the radioactive isotope of hydrogen made in a pile) as a sort of booster in the reaction, the fission bomb being used as the detonator and the reaction involving the atoms of liquid deuterium being the prime explosive. Such a gadget should produce an explosive equivalent to 100,000,000 Tons of TNT, which in turn should produce Class B damage over an area of 3,000 square miles!
>
> This last real super bomb is probably at least as distant now as was the fission bomb when you and I first heard of the enterprise.

The thermonuclear was something of a Rorschach test. If it could be made to work at all it was, like a fire, potentially unlimited; to build it larger you only piled on more heavy hydrogen. As Los Alamos paid less attention to Teller's Super his projection of its destructive potential grew more-grandiose.

Robert Oppenheimer also commited himself at that time to exploring the thermonuclear—after the war was won—in a letter to Richard Tolman on September 20, 1944. "I should like," he emphasized, ". . . to put in writing at an early date the recommendation that the subject of initiating violent thermonuclear reactions be pursued with vigor and diligence, and promptly." A way station on the road to a full-scale thermonuclear might be a boosted fission bomb with a small charge of heavy hydrogen confined possibly within the core of an implosion device:

> In this connection I should like to point out that [fission] gadgets of reasonable efficiency and suitable design can almost certainly induct significant thermonuclear reactions in deuterium even under conditions where these reactions are not self-sustaining. . . . It is not at all clear whether we shall actually make this development during the present project, but it is of great importance that such . . . gadgets form an experimentally possible transition from a simple gadget to the super and thus open the possibility of a not purely theoretical approach to the latter.

(In fact not deuterium but tritium proved to be the necessary ingredient of a boosted fission bomb, and such weapons were not developed until long after the end of the war.)

Alluding then to the larger consequences that Bohr had revealed, Oppenheimer emphasized once more the urgency he attached to the pursuit of an H-bomb: "In general, not only for the scientific but also for the political evaluation of the possibilities of our project, the critical, prompt, and effective exploration of the extent to which energy can be released by thermonuclear reactions is clearly of profound importance."

Working against the clock to build weapons that might end a long and bloody war strained life at Los Alamos but also heightened it. "I always pitied our Army doctors for their thankless job," comments Laura Fermi:

> They had prepared for the emergencies of the battlefields, and they were faced instead with a high-strung bunch of men, women, and children. High-strung because altitude affected us, because our men worked long hours under unrelenting pressure; high-strung because we were too many of a kind, too close to one another, too unavoidable even during relaxation hours, and we were all [as Groves had warned his officers not entirely tongue-in-cheek] crackpots; high-strung because we felt powerless under strange circumstances, irked by minor annoyances that we blamed on the Army and that drove us to unreasonable and pointless rebellion.

They made the best of it. Mici Teller waged pointed rebellion saving the backyard trees to preserve a playground for her son. "I told the soldier in his big plow to leave me please the trees here," one of her friends remembers her recounting, "so Paul could have shade but he said, 'I got orders to level off everything so we can plant it,' which made no sense as it was planted by wild nature and suits me better than dust. The soldier left, but was back next day and insisted he had more orders 'to finish this neck of the woods.' So I called all the ladies to the danger and we put chairs under the trees and sat on them. So what could he do? He shook his head and went away and has not come again." Contrariwise, to clear a ski area on the hill to the west of the mesa, George Kistiakowsky wrapped the trees with half-necklaces of plastic explosive and thus noisily but efficiently cut them down. "Then we scrounged equipment to build a rope tow and it became a nice little ski slope," he recalls.

The Fermis moved to Los Alamos in September 1944 and requested one of the less coveted fourplex apartments rather than the Ranch School faculty cottage that had been prepared for them, to make a point about social snobbery. The Peierls, Rudolf and energetic Genia—Otto Frisch's dish-drying coach in Birmingham—lived below. The mix of birthplaces and citizenships was typical of the Hill: Peierls a German Jew, his wife a Russian, both with British citizenship; Laura Fermi still nostalgic for Rome but she and her husband new American citizens as of July. "Oppie has whistled," Fermi would announce with a yawn when the morning siren sounded. "It is time to get up." The Italian laureate directed a new operation, F (for Fermi) Division, a catchall designed to take advantage of his versatility as both theoretician and experimentalist. One of the groups he caught was Teller's. "That young man has imagination," the forty-three-year-old Italian emigré told his wife drolly of the thirty-six-year-old Hungarian. "Should he take full advantage of his inventiveness, he will go a long way." Teller stayed up late at night working out ideas and playing the piano and hardly ever appeared in the Tech Area before late morning.

"Parties," remembers Fuze Development group leader Robert Brode's articulate wife Bernice, "both big and brassy and small and cheerful, were an integral part of mesa life. It was a poor Saturday night that some large affair was not scheduled, and there were usually several of them. . . . On [Saturday nights] we raised whoopie, on Sundays we took trips, the rest of the week we worked." Single men and women sponsored dorm parties fueled with tanks of punch made potent with mixed liquors and pure Tech Area grain alcohol and invited wall-to-wall crowds. The singles removed all the furniture from their dormitory common rooms to make areas for dancing and by unwritten rule kept their upstairs doors open through the night.

Square dancing evolved as a natural Saturday evening activity in that Southwestern setting. ("Everybody was wearing Western clothes—jeans, boots, parkas," Stanislaw Ulam's French wife Françoise remembers noticing with surprise when she and her husband arrived on the Hill. "There was a feeling of mountain resort, in addition to army camp.") The dances were first held in Deke Parsons' living room, then the theater, then Fuller Lodge, finally expanding to crowd the large mess hall. Eventually even the Fermis attended with their daughter Nella to learn the vigorous reels. Long after mother and daughter had been persuaded from the sidelines Fermi sat unbudging, mentally working out the steps. When he was ready he asked Bernice Brode, one of the leaders, to be his partner. "He offered to be head couple, which I thought most unwise for his first venture, but I couldn't do anything about it and the music began. He led me out on the exact beat, knew exactly each move to make and when. He never made a mistake, then or thereafter, but I wouldn't say he enjoyed himself. . . . He [danced] with his brains instead of his feet."

Theater sometimes supplied a Saturday alternative. At a performance of *Arsenic and Old Lace* Robert Oppenheimer surprised and delighted the audience by appearing powdered sepulchrally white with flour as the first of the crowd of corpses emerging from the cellar in the last act. Donald Flanders, tall and bearded, known as Moll, Computation group leader in the Theoretical Division, wrote a comic ballet, *Sacre du Mesa,* set to George Gershwin music. Despite his beard and his lack of ballet training Flanders danced the part of General Groves. Samuel Allison's son Keith appeared as Oppenheimer, dancing on a large table wearing suitably casual clothes and a pork-pie hat. "The main stage prop," Bernice Brode notes, "was a mechanical brain with flashing lights and noisy bangs and sputters, which did consistently wrong calculations, for example, 2 + 2 = 5. In the grand but hectic finale, the wrong calculations were revealed as the real sacred mystery of the mesa."

Kistiakowsky preferred less formally intellectual entertainment:

> I played a lot of poker with important people like Johnny Von Neumann, Stan Ulam, etc. . . . When I came to Los Alamos I discovered that these people didn't know how to play poker and offered to teach them. At the end of the evening they got annoyed occasionally when we added up the chips. I used to point out that if they had tried to learn violin playing, it would cost them even more per hour. Unfortunately, before the end of the war, these great theoretical minds caught on to poker and the evening's accounts became less attractive from my point of view.

And Robert Wilson, Cyclotron Program group leader, who served on the advisory Town Council, discovered even more elemental activities on the Hill despite security screening before employment and roving military police:

> Of the many problems that were presented to us during my term of office, the most memorable was when the M.P.'s who guarded the site chose to place one of our women's dorms off-limits. They recommended that we close the dorm and dismiss the occupants. A tearful group of young ladies appeared before us to argue to the contrary. Supporting them, a determined group of bachelors argued even more persuasively against closing the dorm. It seems that the girls

> had been doing a flourishing business of requiting the basic needs of our young men, and at a price. All understandable to the army until disease reared its ugly head, hence their interference. By the time we got that matter straightened out—and we did decide to continue it—I was a considerably more learned physicist than I had intended to be a few years earlier when going into physics was not all that different from taking the cloth.

Married or single, the occupants of Post Office Box 1663 were young and healthy; they produced so many babies that Groves ordered either the reservation commander or the laboratory director—both versions of the story survive—to staunch the flood. Oppenheimer, if Oppenheimer it was, refused the duty. With justification: his wife Kitty bore him a second child, a daughter, Katherine, called Toni, on December 7, 1944. So many people wanted to see the boss's baby that the hospital identified the crib with a sign and lines formed to file past the nursery window.

Crowded together behind a fence, Hill families worried about epidemic disease. A pet dog that had bitten several children turned up rabid and pet owners debated angrily with parents about which category of dependent should be kept on a leash. More frightening was the sudden death of a young chemist, a group leader's wife, from an unidentified form of paralysis. Fearing an outbreak of poliomyelitis, doctors closed the schools, put Santa Fe off limits and ordered all children indoors.

No new cases appeared, the danger abated with the continuation of cold weather and work and play resumed. "I don't think I shall ever again live in a community where so many brains were," comments Edwin McMillan's wife Elsie, Ernest Lawrence's sister-in-law, "nor shall I ever live in a community so confined that visitors expected us to fight with each other. We didn't have telephones, we didn't have the bright lights, but I don't think I shall ever live in a community that had such deep roots of cooperation and friendship."

Some reserved Sundays for church and hobbies; others devoted the day to outings. The Oppenheimers maintained magnificent riding horses and rode regularly on Sunday morning but only once in three years found time for an overnight excursion. Kistiakowsky bought one of Oppenheimer's quarter horses and refreshed himself trailing in the mountains after his late Saturday poker nights; the Army stabled the private animals along with the remuda it kept for the mounted MP's who patrolled the mesa fences. Emilio Segrè found excellent fly-fishing. "The streams are full of big trouts," he announced happily to newcomers. "All you have to do is throw in a line and they bite you, even if you are shouting." Fermi took up angling, says Segrè, "but he went about it in a peculiar way. He had tackle different from what anyone else used for trout fishing, and he developed theories about the way fish should behave. When these were not substantiated by experiment, he showed an obstinacy that would have been ruinous in science." Fermi insisted on fishing for trout with worms, arguing that the condemned creatures should be offered an authentic final meal, not the dry flies of tradition. Segrè made a point of reviewing the subtleties of trout fishing with his old friend. "Oh, I see, Emilio," Fermi eventually countered, "it is a battle of wits."

Mountain climbing had long been a Hans Bethe hobby. He and Fermi, among others, sometimes scaled Lake Peak across the Rio Grande in the Sangre de Cristos, one of Bethe's admiring group leaders remembers, to "sit there in the sunshine" at 12,500 feet "discussing physics problems. This is how many discoveries were made." Leona Marshall, who moved with Fermi to Los Alamos, recalls less Olympian hours with "nothing to do but admire the view and gasp for breath."

Equally strenuous excursions went out to area landmarks. Genia Peierls and Bernice Brode determined to find the Stone Lions, prehistoric lifesized twin effigies of crouching mountain lions carved in tuff, reported beside a ruined pueblo on a distant mesa. They gathered up a carload of Navy ensigns and another of young bachelors from the British mission and drove within ten miles of their goal, then set out walking, Genia Peierls leading the way in tennis shoes without socks: "Best for stones, best for bunions." Lunch at two in the afternoon by a cool canyon stream encouraged the weary ensigns to drop anchor, but Mrs. Peierls had cowed the young British mission men from similar protest. "OK, we proceed to Stone Lions without U. S. Navy. All aboard." More hiking, crossing desert country from mesa to mesa, the Rio Grande below. The American woman was impressed with the Stone Lions; not so the Russian. "House cats only, my dear, not well made and maybe not even old." "On the way back," Bernice Brode recalls, "the young men . . . looked out over the wide expanse of the desert region and the ribbon of water shining in the setting sun. One of them, dark and slim, wearing tortoise shell rimmed glasses, spoke in his soft voice with a slight German accent. 'I have not seen New York, nor Chicago, but I have seen the Stone Lions.' He smiled pleasantly as we walked on. His name was Klaus Fuchs." Penny-in-the-slot Fuchs, Genia Peierls nicknamed him, because the quiet, hardworking emigré theoretician only spoke when spoken to.

On a hike through Frijoles Canyon with the Fermis, Niels Bohr stopped to admire a skunk, an animal unknown to Europeans, but it chose not to instruct the vigorous Dane in the pungency of its defenses. Bears sometimes appeared on the trails, prompting warnings in the daily bulletin: "Remember that these are not tame bears like those in Yellowstone Park."

A family cat turned up with a suppurating jaw; the Hill's Army veterinarian recognized the bone necrosis as a sign of radiation poisoning from Tech Area contamination and kept the animal alive to observe its unusual symptomatology, about which not much was yet known. Its tongue swelled and its hair fell out in patches; its heartsick owner eventually asked that the animal be destroyed.

A low-power radio station began broadcasting to Hill residents on Christmas Eve, 1943, drawing on several fine collections of classical records, including Oppenheimer's; the few New Mexicans beyond the Hill who could receive the station's signals were puzzled that announcers never introduced live performers by their last names. The "Otto" who occasionally played classical piano selections was Otto Frisch. A golf course opened in June 1944. Men and women fielded baseball, softball and basketball teams. The Army divided up the old Ranch School truck garden east of Fuller Lodge into victory-gardening plots but had no water to spare for irrigation.

Life was rougher for construction workers, machinists, soldiers and WAC's: minimal barracks, jerrybuilt dormitories, muddy trailer courts. Hillbilly construction families invited once in the interest of authenticity to the square dancing at the mess hall arrived drunk and nearly caused a riot; thereafter a man in uniform guarded the door. Hans Bethe recalls that one wild machinist late in the war, when the laboratory took what help it could find, slit a fellow worker's throat "from cover to cover." The Indians from San Ildefonso and other pueblos and ranches in the area lived better for their work on the Hill as cleaning women and maintenance men. The hand-coiled black pottery of Maria Martinez soon graced many Los Alamos apartments.

In winter a pall of coal smoke hung over the mesa. The men the Army assigned to service the apartment furnaces stoked them so hot that apartment walls sometimes sizzled. Los Alamos sat high and dry surrounded by pine forests, and fire worried everyone. The main machine shop in the Tech Area caught fire one night early in 1945; Eleanor Jette remembers watching her husband Eric, Metal Reduction group leader in the Chemistry and Metallurgy Division, standing with Oppenheimer and the Hill commanding officer on the fire escape of the administration building grimly overseeing the firefighters. "Jesus," she heard someone say, "let's be thankful it isn't D building. That place is as hot as seven million dollars. Every time it gets too hot for them to work, they slap on another coat of paint." Her husband worked in D building; she did not know he worked with plutonium but understood that "hot" meant radioactive. "Damn," he told her when she asked. "You mustn't be upset. We're so careful it's fantastic." A fire in the plutonium-handling areas would be a major disaster; after the machine-shop fire Groves ordered a fireproof plutonium works built with steel walls and a steel roof and filtering systems for both incoming and outgoing air.

Robert Oppenheimer oversaw all this activity with self-evident competence and an outward composure that almost everyone came to depend upon. "Oppenheimer was probably the best lab director I have ever seen," Teller repeats, "because of the great mobility of his mind, because of his successful effort to know about practically everything important invented in the laboratory, and also because of his unusual psychological insight into other people which, in the company of physicists, was very much the exception." "He knew and understood everything that went on in the laboratory," Bethe concurs, "whether it was chemistry or theoretical physics or machine shop. He could keep it all in his head and coordinate it. It was clear also at Los Alamos that he was intellectually superior to us." The Theoretical Division leader elaborates:

> He understood immediately when he heard anything, and fitted it into the general scheme of things and drew the right conclusions. There was just nobody else in that laboratory who came even close to him. In his knowledge. There was human warmth as well. Everybody certainly had the impression that Oppenheimer cared what each particular person was doing. In talking to someone he made it clear that that person's work was important for the success of the whole project. I don't remember any occasion at Los Alamos in which he was nasty to any person, whereas before and after the war he was often that way. At Los Alamos he didn't make anybody feel inferior, not anybody.

Yet Oppenheimer felt inferior himself, had always felt for the actions of his life, as he confessed many years afterward, "a very great sense of revulsion and of wrong." At Los Alamos for the first time he seems to have found alleviation of that loathing. He may have discovered there a process of self-analysis anchored in complementarity that served him more comprehensively later in his life: "In an attempt to break out and be a reasonable man, I had to realize that my own worries about what I did were valid and were important, but that they were not the whole story, that there must be a complementary way of looking at them, because other people did not see them as I did. And I needed what they saw, and needed them." Certainly he found the more traditional alleviation of losing himself in work.

Whatever his burden of morale and work in those years, Oppenheimer also carried his full share of private pain. He was kept under constant surveillance, his movements monitored and his rooms and telephones bugged; strangers observed his most intimate hours. His home life cannot have been happy. Kitty Oppenheimer responded to the stress of living at isolated Los Alamos by drinking heavily; eventually Martha Parsons, the admiral's daughter, took over the duties of social leadership on the Hill. Army security officers hounded the director of the central laboratory of the nation's most important secret war project mercilessly; at least one of them, Peer de Silva, was convinced Oppenheimer was a Soviet spy. They interrogated him frequently, fishing for the names of people he knew or believed to be members of the Communist Party, hoping to trip him up. He invented circumstances and volunteered the names of friends to protect his own, indiscretions that would return in time to haunt him.

During the first Los Alamos summer he heard from Jean Tatlock, the unhappy woman he had loved before he met his wife. Loyally, even though she had been and still might be a Communist and he knew himself to be spied upon, he went to her; an FBI document coldly summarizes a security man's peepshow version of that meeting:

> On June 14, 1943, Oppenheimer traveled via Key Railway from Berkeley to San Francisco on the evening of June 14, 1943, where he was met by Jean Tatlock who kissed him. They dined at the Xochimilcho Cafe, 787 Broadway, San Francisco, then proceeded at 10:50 P.M. to 1405 Montgomery Street and entered a top floor apartment. Subsequently, the lights were extinguished and Oppenheimer was not observed until 8:30 A.M. next day when he and Jean Tatlock left the building together.

In January 1944 Jean Tatlock committed suicide. "I wanted to live and to give and I got paralyzed somehow," her suicide note said. It was a paralysis of the spirit Oppenheimer seemingly had to resist in himself.

Planning began in March 1944 for a full-scale test of an implosion weapon. Sometime between March and October Oppenheimer proposed a code name for that test. The first man-made nuclear explosion would be a historic event and its designation therefore a name that history might remember. Oppenheimer coded the test and the test site Trinity. Groves wrote him in 1962 to find out why, speculating that he chose the name because it is common to rivers and peaks in the American West and would be inconspicuous.

"I did suggest it," Oppenheimer responded, "but not on [that] ground. . . . Why I chose the name is not clear, but I know what thoughts were in my mind. There is a poem of John Donne, written just before his death, which I know and love. From it a quotation:

> *As West and East*
> *In all flatt Maps—and I am one—are one,*
> *So death doth touch the Resurrection.*"

The poem was Donne's "Hymne to God My God, in My Sicknesse," and among its subtleties it construes a complementarity that parallels the complementarity of the bomb that Bohr had recently revealed to Oppenheimer. ("Bohr was deeply in this," Bethe testifies, "and this was his real interest, and Bohr had long conversations with Oppenheimer which brought Oppenheimer into this at a very early stage. Oppenheimer was very much indoctrinated by Bohr's ideas of international control.") That dying leads to death but might also lead to resurrection—as the bomb for Bohr and Oppenheimer was a weapon of death that might also end war and redeem mankind—is one way the poem expresses the paradox.

"That still does not make a Trinity," Oppenheimer's letter to Groves goes on, "but in another, better known devotional poem Donne opens, 'Batter my heart, three person'd God;—.' Beyond this, I have no clues whatever." Nor must Groves have had; but the fourteenth of Donne's *Holy Sonnets* equally explores the theme of a destruction that might also redeem:

Batter my heart, three person'd God; for you
As yet but knocke, breathe, shine, and seeke to mend;
That I may rise, and stand, o'erthrow mee, and bend
Your force to breake, blowe, burn and make me new.
I, like an usurpt towne, to another due,
Labour to admit you, but Oh, to no end;
Reason, your viceroy in mee, mee should defend,
But is captiv'd, and proves weake or untrue.
Yet dearly I love you, and would be loved faine,
But am betroth'd unto your enemie:
Divorce me, untie, or breake that knot againe,
Take mee to you, imprison me, for I
Except you enthrall me, never shall be free,
Nor ever chaste, except you ravish me.

That is poetry perhaps martial enough, ardent enough and sufficiently fraught with paradox to supply a code name for the first secret test of a millennial force newly visited upon the world.

Oppenheimer did not doubt that he would be remembered to some degree, and reviled, as the man who led the work of bringing to mankind for the first time in its history the means of its own destruction. He cherished the complementary compensation of knowing that the hard riddle the bomb would pose had two answers, two outcomes, one of them transcendent. Such understanding justified the work at Los Alamos if anything did, and the work in turn healed the split between self and overweening conscience that hurt him. He had long recognized the possibility of such a convalescence and evoked it explicitly in the epistle on discipline he wrote his brother Frank in 1932 that concluded in Pauline measure: "Therefore I think that all things which evoke discipline: study, and our duties to men and to the commonwealth, war, and personal hardship, and even the need for subsistence, ought to be greeted by us with profound gratitude; for only through them can we attain to the least detachment; and only so can we know peace." At Los Alamos, if only for a time, he located that detachment in duties to men and to the commonwealth that Bohr was teaching him to believe might be worthy, not macabre. He was not the first man to find himself in war.

To develop implosion Los Alamos had to develop diagnostics, ways to see and to measure events that began and ended in considerably less time than the blink of an eye. The iron pipes Seth Neddermeyer imploded could be studied by aiming a high-speed flash camera down their bores, but how could the physicists of G Division observe the shaping of a detonation wave as it passed through solid blocks of high explosives, or the compression of the metal sphere which those explosives completely surrounded? They were competent research scientists who had been working within narrow technological constraints for a year and a half; diagnostics demanded imagination and they brought all their frustrated creativity to the task.

X-raying was a reliable approach; the Ordnance Division had already used X rays to study the behavior of small spherical arrangements of explosives. X rays reveal differences in density—dense bone casts a darker shadow than lighter flesh—and since the detonation wave of a developing implosion changed the density of the explosive material as it burned its way through, X rays could make that wave visible. But adapting X-ray diagnostics to implosion studies on an increasing scale meant protecting fragile X-ray equipment from the repeated blasts of as much as two hundred pounds of high explosives at a time. That challenge the physicists met by the unorthodox expedient of mounting their implosion tests between two closely spaced blockhouses with the X-ray unit in one building and the radiography equipment in the other, accessible to the test event through protected ports. Ultimately flash X-ray equipment—high-current X-ray tubes that pulsed as rapidly as every ten-millionth of a second—proved most useful for detonation-wave studies.

The behavior of a test unit's HE shell was easier to study with X rays and high-speed photography than was the compression of its denser metal core. For following the metal core as it squeezed to less than half its previous volume Los Alamos developed several different diagnostic methods and used them in complement.

One method set the test unit within a magnetic field and measured changes in field configuration as the metal sphere compressed. Since HE is essentially transparent to magnetism, this method allowed the physicists eventually to study full-scale assemblies. It gave reliable measure of shock waves reflected from the core and of the troublesome detonation-wave intersections that caused jets and spalling.

Carefully spaced prearranged wires contacted by the metal sphere as it imploded supplied information not only about the timing of the implosion but also about material velocities at various depths within the core. That provided direct, quantitative data which the Theoretical Division could use to check how well its hydrodynamic theory fit the facts. The Electric Method group began by measuring the high-explosive acceleration of flat metal plates. Early in 1945 it adapted its techniques to partial spheres and eventually to spheres surrounded by HE lens systems with only one lens removed to access the necessary wires.

Duplicated at another test site, the blockhouse arrangement that served to protect ordinary X-ray equipment also served to shield the most unusual diagnostic method the scientists devised: firing pulsed X rays from a betatron through scale-model implosion units into a cloud chamber and photographing the resulting ionization tracks with a stereoscopic camera.* The betatron method needed an ingenious timing circuit to trigger in quick but precise sequence the explosive charge, the betratron X-ray pulse, the expansion of the diaphragm of the cloud chamber that made the ionization tracks visible as droplets in the fog and the camera shutters that photographed them.

The fifth successful method G Division developed varied the betatron method by incorporating an intense source of gamma radiation within the core itself. The source, radioactive lanthanum extracted from among fission products of the Oak Ridge air-cooled pile, gave the method its name: RaLa. Not a cloud chamber but alignments of rugged ionization chambers served to register the changing patterns of radiation from the RaLa cores as they compressed. Since no one knew at first how extensively the radiolanthanum would contaminate the test site, Luis Alvarez, who coordinated the first experiment, borrowed two tanks from the Army's Dugway Proving Ground in Utah to use as temporary blockhouses. He recalls spectacular results:

> I was sitting in the tank when the first explosion went off. George Kistiakowsky was in one tank and I was in the other. We were looking through the periscopes and all that happened was that it blew a lot of dust in our eyes.

* A betatron accelerates electrons to high speeds in a magnetic field; such beta ray-like electrons can then be directed onto a target to produce intense beams of high-energy X rays.

> And then—we hadn't thought about this possibility at all—the whole forest around us caught on fire. These pieces of white-hot metal went flying off into the wild blue yonder setting trees on fire. We were almost surrounded.

Implosion lens development had begun the previous winter, says Bethe, when John von Neumann "very quickly designed an arrangement which was obviously correct from the theoretical point of view—I had tried and failed." Now in the fall and winter of 1944–45 Kistiakowsky had to make the theoretical arrangement work.

An optical lens takes advantage of the fact that light travels at different velocities in different media. Light traveling through air slows when it encounters glass. If the glass curves convexly, as a magnifying glass is curved, the light that encounters the thicker center must follow a longer path than the light that encounters the thinner edges. The effect of these differing path lengths is to direct the light toward a focal point.

The implosion lens system von Neumann designed was made up of truncated pyramidal blocks about the size of car batteries. The assembled lenses formed a sphere with their smaller ends pointing inward. Each lens consisted of two different explosive materials fitted together—a thick, fast-burning outer layer and a shaped slow-burning solid inclusion that extended to the surface of the face of the block that pointed toward the bomb core:

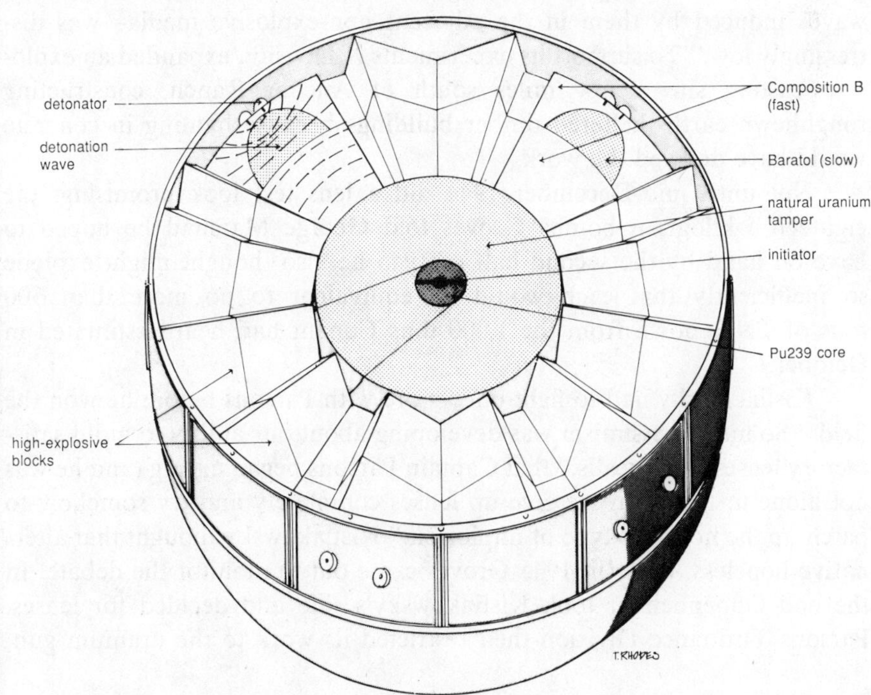

The fast-burning outer layer functioned for the detonation wave as air around an optical lens functions for light. The slower-burning shaped inclusion functioned as a magnifying glass, directing and reshaping the wave. A detonator would ignite the fast-burning explosive. That material would develop a spherical detonation wave. When the apex of the wave advanced into the apex of the inclusion, however, it would begin burning more slowly. The delay would give the rest of the wave time to catch up. As the detonation wave encountered and burned through the inclusion it thus reshaped itself from convex to concave, from a spherical wave expanding from a point to a spherical wave converging on a point, emerging fitted to the convex curve of the spherical tamper. Before the reshaped wave reached the tamper it passed through a second layer of solid blocks of fast-burning explosive to add to its force. The heavy natural-uranium tamper then served to smooth out any minor irregularities as the spherical shock wave compressed it passing through to the plutonium core.

Kistiakowsky would apologize after the war for a research program "too frequently reduced to guesswork and empirical shortcuts" because the field had been grossly neglected. "Prior to this war the subject of explosives attracted very little scientific interest," he wrote in an introduction to a technical history of X Division's work, "these materials being looked upon as blind destructive agents rather than precision instruments; the level of fundamental knowledge concerning detonation waves—and strong shock waves induced by them in the adjacent non-explosive media—was distressingly low." To support its experiments X Division expanded an explosives-casting site a few miles south of Anchor Ranch, constructing roughhewn earth-sheltered timber buildings because hauling in concrete would have delayed the work.

Not until mid-December 1944 did a lens test look promising; the eighteen 5-kilogram bombs Groves told George Marshall he hoped to have on hand by the second half of 1945 he also thought might explode so inefficiently that each would be equivalent to no more than 500 tons of TNT, down from the 1,000 tons Conant had heard estimated in October.

Kistiakowsky had to fight once more with Parsons before he won the field. "So much pessimism was developing about our ability to build satisfactory lenses," he recalls, "that Captain Parsons began urging (and he was not alone in this) that we give up lenses completely and try somehow to patch up the non-lens type of implosion." Kistiakowsky thought that alternative hopeless. Early in 1945 Groves came out to monitor the debate. In the end Oppenheimer took Kistiakowsky's side and decided for lenses. Parsons' Ordnance Division then restricted its work to the uranium gun,

Little Boy, and to engineering the weapons for the battlefield. X and G Divisions worried about implosion.

Finishing the high-explosive castings by machining them was the most dramatic innovation Kistiakowsky introduced. He wanted to shape the HE components entirely by machining from solid pre-cast blocks but lacked sufficient time to develop and build the elaborate remote-controlled machinery the innovative technology would have required. He settled instead for precision casting with machine finishing and used his limited supply of machinists primarily to turn out the necessary molds. Molds gave him "the greatest agony," he remembers; the HE components of the bomb totaled "something in the nature of a hundred or so pieces, which had to fit together to within a precision of a few thousandths of an inch on a total size of five feet and make a sphere. So we had to have very precise molds." Eventually mold procurement paced Fat Man's testing and delivery.

But even with the necessary molds on hand, casting HE was far from simple, another technology that had to be learned by trial and error. In February 1945 Kistiakowsky chose an explosive called Composition B to serve as the fast-burning component of Fat Man's lenses and a mixture he had commissioned from a Navy research laboratory, Baratol, for the slow-burning component. Composition B was poured as a hot slurry of wax, molten TNT and a non-melting crystalline powder, RDX, that was 40 percent more powerful than TNT alone. Baratol slurried barium nitrate and aluminum powder with TNT, stearoxyacetic acid and nitrocellulose:

> We learned gradually that these large castings, fifty pounds and more each, had to be cooled in just certain ways, otherwise you get air bubbles in the middle or separations of solids and liquids, all of which screwed up the implosion completely. So it was a slow process. The explosive was poured in and then people sat over that damned thing watching it as if it was an egg being hatched, changing the temperature of the water running through the various cooling tubes built into the mold.

The wilderness reverberated that winter to the sounds of explosions, gradually increasing in intensity as the chemists and physicists applied small lessons at larger scale. "We were consuming daily," says Kistiakowsky, "something like a ton of high performance explosives, made into dozens of experimental charges." The total number of castings, counting only those of quality sufficient to use, would come to more than 20,000. X Division managed more than 50,000 major machining operations on those castings in 1944 and 1945 without one explosive accident, vindication of Kistiakowsky's precision approach. A RaLa test on February 7, 1945, showed definite improvement in implosion symmetry. On March 5, after a strained round of conferences, Oppenheimer froze lens design. However scarce plutonium might be, no one doubted that Fat Man would have to be tested at full scale before a military weapon could be trusted to work.

A problem small in scale but difficult of solution was the initiator, the minuscule innermost component of the bombs. The chain reaction required a neutron or two to start it off. No one wanted to trust a billion dollars' worth of uranium or several hundred million dollars' worth of plutonium to spontaneous fission or a passing cosmic ray. Neutron sources had been familiar laboratory devices for more than a decade, ever since James Chadwick bombarded beryllium with alpha particles from polonium and broke the elusive neutral particle free in the first place. In his early lectures at Los Alamos Robert Serber had discussed using a radium-beryllium source in a gun bomb with the radium attached to one piece of core material and the beryllium to the other, arranged to smash together when the gun was fired and the two core components mated to complete a critical assembly. Radium released dangerous quantities of gamma radiation, however, and Edward Condon noted in the *Los Alamos Primer* that "some other source such as polonium . . . will probably prove more satisfactory." Polonium emitted copious quantities of alpha particles energetic enough to knock neutrons from beryllium but very little gamma radiation.

The challenge of initiator development was to design a source of sufficient neutron intensity that released those neutrons only at the precise moment they were needed to initiate the chain reaction. In the case of the uranium gun that requirement would be relatively easy to meet, since the alpha source and the beryllium could be separated with the bullet and the target core. But the implosion bomb offered no such convenient arrangement for separation and for mixing. Polonium and beryllium had to be intimately conjoined in Fat Man at the center of the plutonium core but inert as far as neutrons were concerned until the fraction of a microsecond when the imploding shock wave squeezed the plutonium to maximum density. Then the two materials needed instantaneously to mix.

Polonium, element 84 on the periodic table, was a strange metal. Marie and Pierre Curie had isolated it by hand from pitchblende residues (at backbreaking concentrations of a tenth of a milligram per ton of ore) in 1898 and named it in honor of Marie Curie's native Poland. Physically and chemically it resembled bismuth, the next element down the periodic table, except that it was a softer metal and emitted five thousand times as much alpha radiation as an equivalent mass of radium, which caused the ionized, excited air around a pure sample to glow with an unearthly blue light.

Po210, the isotope of polonium that interested Los Alamos, decayed to lead 206 with the emission of an alpha particle and a half-life of 138.4 days. The range of Po210's alphas was some 38 millimeters in air but only a few hundredths of a millimeter in solid metals; the alphas gave up their energies ionizing atoms along the way and finally came to a stop. That meant the polonium for an initiator could be safely confined within a sandwich of metal foils. Sandwiching the foils in turn might be concentric shells of light, silvery beryllium. The entire unit need be no larger than a hazelnut.

"I think I probably had the first idea [for an initiator design]," Bethe remembers, "and Fermi had a different idea, and I thought mine was better for once, and then I was the chairman of a committee of three to watch the development of the initiator." Segregating the Po210 from the beryllium was straightforward. Making sure the two elements mixed thoroughly at the right instant was not, and the primary difference between initiator designs—many were invented and tested during the winter of 1944–45—was their differing mixing mechanisms. A quantity of Po210 equivalent in alpha activity to 32 grams of radium, thoroughly mixed with beryllium, would produce some 95 million neutrons per second, but that would be no more than nine or ten neutrons in the brief ten-millionth of a second when they would be useful in an imploding Fat Man to start the chain reaction; therefore the mixing had to be certain and thorough. Initiator design has never been declassified, but irregularities machined into the beryllium outer surface that induced turbulence in the imploding shock wave probably did the job: the Fat Man initiator may have been dimpled like a golf ball.

To supply ten neutrons to initiate a chain reaction men labored for years. Bertrand Goldschmidt, a French chemist who had once been Marie Curie's personal assistant and who came to the United States after the invasion of France to work with Glenn Seaborg at the Met Lab, extracted the first half-curie of initiator polonium from old radon capsules at a New York cancer hospital (polonium is a daughter product of radium decay). Quantity production required using scarce neutrons from the Oak Ridge air-cooled pile to transmute bismuth one step up the periodic table to Po. Charles A. Thomas, research director for the Monsanto Chemical Company, a consultant on chemistry and metallurgy, took responsibility for purifying the Po, for which purpose he borrowed the indoor tennis court on his mother-in-law's large and securely isolated estate in Dayton, Ohio, and converted it to a laboratory.

Thomas shipped the Po on platinum foil in sealed containers, but another nasty characteristic of polonium caused shipping troubles: for reasons never satisfactorily explained by experiment, the metal migrates from place to place and can quickly contaminate large areas. "This isotope has been observed to migrate upstream against a current of air," notes a postwar British report on polonium, "and to translocate under conditions where it would appear to be doing so of its own accord." Chemists at Los Alamos learned to look for it embedded in the walls of shipping containers when Thomas' foils came up short.

Initiator studies proceeded in G Division at a test site established in Sandia Canyon, one mesa south of the Hill. The Initiator group drilled blind holes in large turbine ball bearings—screwballs, the experimenters called them—inserted test initiators and plugged the holes with bolts. After imploding the screwballs they recovered the remains and examined them to see how well the Po and Be had mixed. Mixing, unfortunately, could not be a conclusive measure of effectiveness. Bethe's committee selected the most promising design on May 1, 1945, but only a full-scale test culminating in a chain reaction could prove definitively that the design worked.

Progress toward a Japanese atomic bomb, never rapid, slowed to frustration and futility across the middle years of the Pacific war. After the Imperial Navy had bowed out of atomic energy research Yoshio Nishina had continued patriotically to pursue it even though he privately believed that Japan in challenging the United States had invited certain disaster. On July 2, 1943, Nishina had met with his Army liaison, a Major General Nobuuji, to report that he had "great expectations" for success. He noted that the Air Force had asked him to study uranium as a possible aircraft fuel, as an explosive and as a source of power, and he had recently received a request for assistance from another Army laboratory, which had contributed 2,000 yen to his expenses. Nobuuji promptly discouraged such consultations. "The main point," Nishina agreed, "is to complete the project as rapidly as possible." His calculations, he told Nobuuji, indicated that 10 kilograms of U235 of at least 50 percent purity should make a bomb, although cyclotron experiments would be necessary to determine "whether 10 kg. will be sufficient, or whether it will require 20 kg. or even 50 kg." He wanted help finishing his 60-inch cyclotron:

> The 250-ton, 1.5 meter accelerator is ready for operation except for certain components which are unavailable as they are being used in the construction of munitions. If this accelerator is completed we believe we can accomplish a great deal. At this moment the U.S. plans to construct an accelerator ten times as great but we are unsure as to whether they can accomplish this.

The previous March Nishina had discarded as impractical under wartime conditions in Japan all methods of isotope separation except gaseous thermal diffusion. Otto Frisch had tried gaseous thermal diffusion (differing from Philip Abelson's *liquid* thermal diffusion) at Birmingham early in 1941 and proved it inadequate for separating uranium isotopes, but Nishina had no knowledge of that secret work. The Riken team had designed a thermal column much like the laboratory-scale column Abelson had built at the Naval Research Laboratory in Washington: of concentric 17-foot pipes, the inner pipe heated to 750°F—electrically heated in the Riken configuration—and the outer pipe cooled with water.

Nishina did not meet again with Nobuuji until seven months later, in February 1944, when he reported difficulty producing uranium hexafluoride. His team had managed to develop a method for generating elemental fluorine but had not yet been able to combine the gas with uranium using an old and inefficient process that Abelson in the United States had discarded before he began his thermal-diffusion studies. Nishima also had a problem with his diffusion column that Abelson would have appreciated: it leaked. "To achieve an airtight system," Nishina told Nobuuji, "we used [sealing] wax and finally achieved our goal. Solder could not be used because of the corrosive properties of the fluorine." He was "in the middle of developing this [hexafluoride-generating] process but can see the end in sight." His 1.5-meter cyclotron was now in operation but only at low energy; his explanation for that compromise comments pointedly on the condition of the Japanese industrial economy by 1944:

> We have been unable to obtain any superior, high-frequency-generating vacuum tubes . . . for the cyclotron. . . . As a result of this constraint, the low operating voltages limit the population of neutrons we can produce. . . . In order to liberate many high-energy neutrons, a high-voltage vacuum tube is required. But, unfortunately, they are difficult to acquire.

By summer Nishina's group had manufactured some 170 grams of uranium hexafluoride—in the United States hex was now being produced by the ton—and in July attempted a first thermal separation. Gauges at the top and bottom of the column, intended to measure a difference in pressure—showing that separation was taking place—indicated no difference at all. "Well, don't worry," Nishina told his team. "Just keep on with it, just keep giving it more gas."

He reconvened with Nobuuji on November 17, 1944, to report that "since February of this year there has not been a great deal of progress." He was losing as much as half his hexafluoride to corrosion effects:

> We thought the materials we had used to make this apparatus for working with the [hexafluoride] were made of impure metals. Therefore we next used the most highly-refined metals available for the system. However, they were

> still eaten away. It was therefore necessary to reduce the pressure of the system . . . to compensate for this erosion.

The cyclotron was operating at higher but not yet full power; Nishina was using it, he told Nobuuji, "to assay the concentrated, separated material." Significantly missing from the November 17 conference report is any mention of measurable separation of U235 from U238. Nishina's staff had understood for more than a year that he did not believe his country could build an atomic bomb in time to affect the outcome of the war. Whether he continued research out of loyalty, or because he thought such knowledge would be valuable after the war, or to win support for his laboratory and deferment from military service for his young men, the bare record does not reveal. On the occasion of the November 17 conference he once again complained of the lack of sufficiently powerful vacuum tubes for his cyclotron and told Nobuuji, contrary to the evidence of experiment, that the Riken's efforts at isotope separation were "now at a midpoint in their practical solution." Nobuuji might have been more helpful if he had understood even the most basic facts of the work. An exchange between the two men late in the meeting indicates the military liaison was as innocent of nuclear physics as a stone:

> *Nobuuji:* If uranium is to be used as an explosive, 10 kg is required. Why not use 10 kg of a conventional explosive?
>
> *Nishina:* That's nonsense.

A B-29, specially modified, first dropped an atomic bomb—a dummy Thin Man—at Muroc Army Air Force Base in California on March 3, 1944. Restrained by sway-bracing, a bomb hung singly in the B-29's bomb bay from a single release, and the first series of tests ended ignominiously that season when a release cable loosened and dumped one onto closed bomb-bay doors at 24,000 feet. "The doors were then opened," a technical report notes, "and the bomb tore free, considerably damaging the doors." A second series of tests in June went better. Word that Fat Man would be heavier than previously estimated encouraged Norman Ramsey's Delivery group to replace the original bomb-release mechanism, which had been modified from a standard glider tow release, with a sturdier British Lancaster bomber design.

Lessons learned, the Air Force began modifying seventeen more B-29's at the Glenn L. Martin plant in Omaha, Nebraska, in August; that month the service prepared to train a special group to deliver the first atomic bombs. The 393rd Bombardment Squadron, then based at Fairmont, Nebraska, in training for Europe, would form the nucleus of the new organization. Late in August Henry H. ("Hap") Arnold, the commanding general of the U.S. Army Air Forces, approved the assignment of an Illinois-born lieutenant colonel, Paul W. Tibbets, twenty-nine years old, to be group commander.

Tibbets may well have been the best bomber pilot in the Air Force. He had led the first B-17 bombing mission from England into Europe, had carried Dwight Eisenhower to his Gibraltar command post before the invasion of North Africa and had led the first bomber strike of that invasion. More recently he had been test-piloting the B-29, which in 1944 was just beginning to come on line, working with the physics department of the University of New Mexico in Albuquerque to determine how well the new bomber could defend itself against fighter attack at high altitude. He was a man of medium height and stocky build with dark, wavy hair and a widow's peak, full-faced and square-jawed, a pipe smoker. His father was a candy wholesaler in Florida and a disciplinarian from whom Tibbets probably acquired his reserved perfectionism; he was closer to his mother, the former Enola Gay Haggard of Glidden, Iowa. He had chosen an Air Force career, he told a postwar interviewer, after his mother had supported him in that choice against his father's opposition:

> When I was in college, studying to be a doctor, I realized that I had always wanted to fly. In 1936, my desire to do something about it reached the point where a family showdown on the subject developed. During the discussion, a few tempers flared, but my mother never said a word. In the end, still undecided, I got her off to the side and asked her what she thought. Despite the things that had been said on the subject, and the fact that most of the people in the discussion had included the statement, "You'll kill yourself in an airplane," Mother said, quite calmly and with positive assurance, "You go ahead and fly. You will be all right."

So far he had been, and now he had won a new assignment. He flew to Second Air Force headquarters in Colorado Springs at the beginning of September 1944 to report to commanding Major General Uzal Ent. An aide installed him in the general's anteroom. An officer came out, introduced himself, took Tibbets aside and asked him if he had ever been arrested. Tibbets considered the situation and decided to answer honestly to this stranger that he had been, as a teenager in North Miami Beach, caught *in flagrante delicto* in the backseat of a car with a girl. Lieutenant Colonel John Lansdale, Jr., who was responsible to Groves for atomic bomb intelligence and security, knew about the arrest and had questioned Tibbets to test his honesty. Now he led him into Ent's office. Norman Ramsey and

Deke Parsons were waiting there. "I'm satisfied," Lansdale said. The physicist and the Navy officer briefed Tibbets on the Manhattan Project and the Muroc bombing tests. Lansdale cautioned him at length on security. After the three men left, Ent specified Tibbets' assignment. "You have to put together an outfit and deliver this weapon," the pilot remembers the Second Air Force commander saying. "We don't know anything about it yet. We don't know what it can do. . . . You've got to mate it to the airplane and determine the tactics, the training, and the ballistics—everything. These are all parts of your problem. This thing is going to be very big. I believe it has the potential and possibility of ending the war." The delivery program within the Air Force had been codenamed Silverplate, Ent told him. If Tibbets needed anything, he had only to use that magic word; Arnold had accorded it the highest priority in the service.

The Air Force chose Wendover Field, Utah, as home base for the new organization. Tibbets flew to Utah early in September, looked the base over and liked what he saw. It was sited between low mountain ranges on the desert salt flats in gritty and secure isolation 125 miles west of Salt Lake City near the Utah-Nevada border; the flat basin, the sink of an ancient and enormous freshwater lake of which the Great Salt Lake is a brackish remnant, offered miles of desolation for bombing practice. Pioneers bound for California had suffered the crossing once—their wagon ruts could still be viewed nearby. The 393rd moved to Wendover in September and with the addition of troop-carrier and other support components became the 509th Composite Group. In October it began receiving its new B-29's.

A Boeing product, the B-29 was a revolutionary aircraft, the first intercontinental bomber. It was conceived in the late 1930s by ambitious officers within what was then still the Army Air Corps as the vehicle of their vision of wars fought at great distance by strategic air power. As early as September 1939 they proposed its use from bases in the Philippines, Siberia or the Aleutians in the event of war against Japan. It was the world's first pressurized bomber and at 70,000 pounds the heaviest production bomber ever built, 135,000 pounds loaded, a weight that required an 8,000-foot runway to lumber airborne. In appearance it was a sleek, polished-aluminum tube 99 feet long intersected by huge 141-foot wings—two B-29's would fill a football field—with a classic sinusoidal tail nearly three stories tall. Four Wright 18-cylinder radial engines that each developed 2,200 horsepower propelled it at altitude at 350 miles per hour maximum speed—it cruised at 220—and it was designed to fly a 4,000-mile mission with up to 20,000 pounds of bombs, though 12,000 pounds was nearer its operational load. It could cruise above 30,000 feet, out of range of flak and of most enemy fighters. Turbosuperchargers boosted engine power; outsized 16.5-foot propellers turned more slowly than those of any other aircraft; wing flaps, the world's largest, adjusted a fifth of wing area to adapt the high-speed, long-range, low-drag wing for takeoff and landing.

On the ground the B-29 rested level on three point landing gear: retractable wheels at the nose and under each wing. The plane's eleven-man crew occupied two pressurized sections within the five joined sections of the fuselage; tandem bomb bays fore and aft of the wings separated the nose section from the waist and tail, and to pass back from the nose to the waist required crawling through a pressurized one-man tunnel. The standard B-29 crew counted pilot, copilot, bombardier, flight engineer, navigator and radio operator in the nose section, three gunners and a radar operator in the waist and another gunner in the tail. Because electrical wiring was less vulnerable to battle damage than pneumatic or hydraulic tubing, the aircraft systems with the exception of the hydraulic wheel brakes operated entirely on electric motors, more than 150 in all, with a gasoline-powered donkey engine in the rear fuselage supplying current on the ground. Analog computers ran a central gun-control system, but all the guns were stripped from 509th bombers except the 20-millimeter cannon in the tail.

If the B-29's engines were powerful they were also notoriously susceptible to fires. To improve their horsepower-to-weight ratio Wright had used magnesium for their crankcases and accessory housings. Engine cooling was inadequate and exhaust valves tended to overheat and stick; an engine would then sometimes swallow a valve and catch fire. If the fire reached the magnesium, a metal commonly used in incendiary bombs, the engine would usually burn through the main wing spar and peel off the wing. To prevent such disasters Boeing improved engine cooling but the basic design fault persisted; there was no time to develop a new power plant if the aircraft was to serve the war for which it was invented. (One Delivery group physicist remembers skimming along at Wendover for miles after takeoff, mowing sagebrush, to cool the engines before climbing to altitude.)

Once at altitude the flight crews of the 509th practiced bombing runs, bombardiers aiming from above 30,000 feet through their Norden bombsights at progressively smaller target circles limed on the ground. Crews that had flown in cloudy Europe wondered why they were training in visual bombing; an odd evasive maneuver instructed them at least in the explosive potential of the unknown weapon they would carry. Tibbets briefed no one on the atomic bomb but directed his crews to nose their aircraft over into a sharp 155-degree diving turn immediately after bomb release. Diving the huge bombers rapidly increased their airspeed; by perfecting the maneuver the crews could escape ten miles from the delayed explosion, "safe from destruction" by a bomb of 20,000 tons TNT equivalent, writes Groves, "by a factor of two." Before they practiced their diving turns they dropped bombs of concrete and bombs filled with HE. These crudely riveted Fat Man imitations, painted bright orange for visibility, they called Pumpkins. The 509th worked hard; the winter wind howled over the Wendover reservation, trapping tumbleweeds on the barbed-wire fences; crews careened into Salt Lake City on weekends to blow out. Tibbets opened their mail, bugged their telephones, had them followed and shipped off those who broke security to the secure but miserable Aleutians for the duration of the war. He held authority over 225 officers and 1,542 enlisted men. With his silverplated requisitions he commandeered from around the world the best pilots, bombardiers, navigators and flight engineers he could find.

One of them, Captain Robert Lewis of Brooklyn, New York, stocky and blond, twenty-six years old, an abrasive but gifted pilot whom Tibbets had personally trained, had spent part of the summer of 1944 at Grand Island, Nebraska, teaching a senior officer with hundreds of combat hours behind him to fly B-29's. Thus checked out, Major General Curtis LeMay rode a C-54 to India late in August to take over the 20th Bomber Command, based in India with forward airfields in China from which it was attempting with fewer than two hundred B-29's to bomb Japan. The bombers had to ferry their own fuel and ordnance from India to China over the Himalayas before each mission—seven supply flights for each bombing strike, up to twelve gallons burned for each one gallon delivered. "It didn't work," LeMay writes in his autobiography. "No one could have made it work. It was founded on an utterly absurd logistic basis. Nevertheless, our entire Nation howled like a pack of wolves for an attack on the Japanese homeland."

Curtis LeMay was a wild man, hard-driving and tough, a bomber pilot, a big-game hunter, a chewer of cigars, dark, fleshy, smart. "I'll tell you what war is about," he once said bluntly—but he said it after the war—"you've got to kill people, and when you've killed enough they stop fighting." Through most of the war he seems to have held to the preference for precision bombing over area bombing that had distinguished the U.S. Air Force from the British since Churchill's and Cherwell's intervention of 1942. Sometimes in Europe precision bombing had served, though never decisively. Over Japan, so far, it had failed. And failure was LeMay's *bête noire.*

His father had been a failure, an odd-job drifter, forever moving his family around. The LeMays lived all over Ohio, in Pennsylvania, out in the wilds in Montana, in California. Curtis Emerson LeMay, born in Columbus, Ohio, in 1906, was the first of seven children. The two memories of early childhood he chooses to offer in his autobiography are linked. Of first seeing an airplane and chasing it madly: "I wanted not only the substance of the mysterious object, not only that part I could have touched with my hands. I wished also in vague yet unforgettable fashion for the drive and speed and energy of the creature." And of compulsively running away from home: "truancy" that "bordered on mania," his mother told him. "I had to grow older," LeMay writes, "and be burdened with a lot of responsibilities, and begin to nourish ambition—I had to do these things before I could manage to control my temper and discipline my activities."

He delivered telegrams and packages and boxes of candy. He delivered newspapers, sold newspapers, wholesaled newspapers to delivery boys, supporting himself and sometimes his family: "When the grocer hesitates about putting that latest basket of groceries on the bill, then you'd better be ready to come up with cash in hand. Very early in life I was convinced bitterly of this necessity. . . . The larder was a vague mystery which Pop didn't bother to penetrate." LeMay resented the missing childhood but moved on. He paid his own way through Ohio State by working nights at a steel foundry. ROTC in college led to the Ohio National Guard because the Guard had higher priority on Army flying-school enrollments than the Army Reserve. He won his wings in 1929 and never looked back: mess officer, navigation officer, General Headquarters navigator, B-10's, B-17's. In England in 1943 and 1944 he worked night and day to improve precision bombing. He won quick promotion.

Arnold sent him to the Pacific because he needed someone who could get the job done:

> General Arnold, fully committed to the B-29 program all along, had crawled out on a dozen limbs about a thousand times, in order to achieve physical resources and sufficient funds to build those airplanes and get them into combat. . . . So he finds they're not doing too well. He has to keep juggling missions and plans and people until the B-29s *do* do well. General Arnold was absolutely determined to get results out of this weapons system.

The B-29 had to be used, that is, successfully used, or men who had staked their careers and their convictions would be shamed, resources squandered that might have aided elsewhere in the war, lives lost futilely and millions of dollars wasted. The justification recurs.

The first B-29 to arrive in the Marianas landed on Saipan on October 12, 1944. Brigadier General Haywood S. Hansell, Jr., assigned to lead the 21st Bomber Command, flew it. As Arnold's chief of staff Hansell had helped formulate the doctrine of precision bombing and believed strongly in its central premise—that wars could be won by selectively destroying the enemy's key industries of war. A stream of new bombers followed the new commander out to the Marianas; the first U.S. aircraft to fly over Tokyo since the Doolittle raid of 1942 was a B-29 on November 1 soaring high and light on a photoreconnaissance mission. A French journalist living in Tokyo at the time, Robert Guillain, remembers his sense of anticlimax:

> The city waited. Millions of lives were suspended in the silence of the radiant autumn afternoon. For a moment, antiaircraft fire shook the horizon with a noise of doors slamming in the sky. Then—nothing: the all-clear was sounded without sight of a plane. The radio announced that a single B-29 had flown over the capital without dropping any bombs.

That seemed a reprieve and for a time only reconnaissance missions disturbed the ill-defended city. "One day the visitor finally appeared, flying at 35,000 feet," Guillain continues; "he even left his signature chalked on the blue sky: a line of pure white like some living thing that seemed to nose an almost imperceptible silver fly ahead of it." Back in the Marianas Hansell was teaching his men to navigate together, to fly in formation; they had trained in the United States only as individual crews.

Hansell received his first target directive on November 11. The Joint Chiefs of Staff had approved it and it reflected their conviction that bombing and naval blockade alone could not bring the Pacific war to a timely end. In September the Combined Chiefs—British and American together—had established a planning date for the end of the war: eighteen months after the defeat of Germany. The U.S. Joint Chiefs judged an invasion of the Japanese home islands essential to achieve that goal. The target directive Hansell received therefore gave first priority to the precision bombing of the Japanese aircraft industry (to cripple Japanese air defenses before an American invasion), second priority to supporting Pacific operations (MacArthur was even then reoccupying the Philippines, returning as he had promised he would) and third priority to testing the efficacy of area incendiary attacks. These priorities, putting precision bombing first, suited Hansell's own.

His crews flew their first raid on Japan from Saipan on November 24. Their target was the Musashi aircraft engine factory north of Tokyo ten miles from the Imperial Palace. A hundred planes began the mission. Seventeen aborted; six were unable to release their bombs. Flak was heavy and the target buried in undercast. But totally unexpected at the high altitude at which the bombers flew was a 140-mile-per-hour wind. They were blown with it over the target and their ground speed was therefore nearly 450 mph, impossible for the bombardiers. As a result only twenty-four planes managed to bomb the factory area—the rest scattered their loads over the docks and warehouses around Tokyo Bay—and only sixteen bombs hit the target. "I did not anticipate the extremely high wind velocities above thirty thousand feet," Hansell said later, "and they came as a very disagreeable surprise." The Air Force had discovered the jet stream.

LeMay was then still working with his 20th Bomber Command out of India and China. Supporting the indifferent military campaigns of Chiang Kai-shek was an activity he abhorred but was sometimes forced to perform. For six months Claire Chennault, the leathery Texan who headed the U.S. air staff assigned to the Nationalist Chinese Army, had been promoting the bombing of Hankow, the riverside city on the Yangtze five hundred miles inland from Shanghai from which Japan supplied its Asian mainland armies. With a renewed Japanese drive in interior China in November Chennault pressed for a Hankow attack. LeMay resisted diverting his command from Japanese home-island targets; the Joint Chiefs had to compel his participation. B-24's and B-25's were also massing for the strike; Chennault particularly wanted LeMay to load his aircraft with incendiaries and bomb from 20,000 feet rather than from above 30,000 feet in order to sow a denser pattern. LeMay reserved one aircraft in five for high explosives. Seventy-seven B-29's took part in the raid on December 18 and burned the Hankow river district down; fires raged out of control for three days. The lesson was not lost on Washington, nor on LeMay.

At Los Alamos the same week Groves, Parsons, Conant, Oppenheimer, Kistiakowsky, Ramsey and several other leaders met in Oppenheimer's office to discuss preparing Pumpkins—they called them blockbusters—for Tibbets' 509th Composite Group. The first Fat Man design, the 1222, had already been changed because it had proved so difficult to assemble—assembly required inserting, threading nuts onto and tightening more than 1,500 bolts—and redesign meant the loss of about 80 percent of the tooling work done at the Pacific Aviation Company in Los Angeles through the autumn. The first unit of a new, simpler design, the 1291, would be ready in three days, on December 22. "Captain Parsons said that the blockbuster production for the 1291 gadget between 15 February and 15 March would require a minimum of 30 blockbusters," the minutes of the meeting report, "so that each B-29 could drop at least two. . . . An additional 20 blockbusters should be produced for H.E. testing. . . . Following that, 75 units should be produced for overseas shipment."

Groves wanted none of it. He wanted no dummy 1291's drop-tested outside the continental United States and he saw no reason to build 75

Pumpkins for overseas target practice for Tibbets' crews. It was the end of 1944 and he was feeling the pressure of accumulating Manhattan Project delays: "General Groves indicated that too much valuable time was being taken from other problems to devote time to the blockbuster program." Conant asked how long the blockbuster program would have to continue; Parsons answered combatively that it would have to continue as long as Tibbets' group operated so that 509th crews could maintain their bombing skills. He relented to reveal that "Colonel Tibbets' Group expected to reach peak combat training by 1 July."

Since Parsons had not succeeded in person in convincing Groves of the importance of bomb-assembly and bombing practice he wrote the general a forceful memorandum on the day after Christmas. There were major differences, he pointed out, between the "gun gadget" and the "implosion gadget," particularly in terms of final assembly:

> It is believed fair to compare the assembly of the gun gadget to the normal field assembly of a torpedo, as far as mechanical tests are involved. . . . The case of the implosion gadget is very different, and is believed comparable in complexity to rebuilding an airplane in the field. Even this does not fully express the difficulty, since much of the assembly involves bare blocks of high explosives and, in all probability, will end with the securing in position of at least thirty-two boosters and detonators, and then connecting these to firing circuits, including special coaxial cables and high voltage condenser circuit. . . . I believe that anyone familiar with advance base operations . . . would agree that this is the most complex and involved operation which has ever been attempted outside of a combined laboratory and ammunition depot.

Parsons' simple and compelling point: the assembly team as well as the bombardiers needed practice. Groves relented; Tibbets got his Pumpkins.

More conventional bombs were falling regularly now on Japan, if not yet to devastating effect. Robert Guillain, the French journalist, remembers the first night raid over Tokyo at the end of November:

> Suddenly there was an odd, rhythmic buzzing that filled the night with a deep, powerful pulsation and made my whole house vibrate: the marvelous sound of the B-29s passing invisibly through a nearby corner of sky, pursued by the barking of antiaircraft fire. . . . I went up on my terrace roof. . . . The B-29s caught in the sweeping searchlight beams went tranquilly on their way followed by the red flashes of ack-ack bursts which could not reach them at that altitude. A pink light spread across the horizon behind a near hill, growing bigger, bloodying the whole sky. Other red splotches lit up like nebulas else-

where on the horizon. It was soon to be a familiar sight. Feudal Tokyo was called Edo, and the people there had always been terrified by the frequent accidental fires they euphemistically called "flowers of Edo." That night, all Tokyo began to blossom.

While Parsons and Groves were debating Pumpkins, Lauris Norstad, who had succeeded Hansell in Washington as Hap Arnold's chief of staff when Hansell moved to the Marianas, passed along word to his predecessor that a trial fire raid on Nagoya, Japan's third-largest city, was an "urgent requirement." Hansell resisted. "With great difficulty," he wrote Norstad, he had "implanted the principle that our mission is the destruction of primary targets by sustained and determined attacks using precision bombing methods both visual and radar" and he was "beginning to get results." Ironically, he feared that area bombing would slacken his crews' hard-won skills. Norstad sympathized but insisted that Nagoya was only a test, "a special requirement resulting from the necessity of future planning." Nearly one hundred of Hansell's B-29's flew incendiaries to Nagoya, at the southern end of the Nobi Plain two hundred miles southwest of Tokyo, on January 3, 1945, and started numerous small fires that resisted coalescing.

In three months of hard flying, taking regular losses, Hansell had managed to destroy none of his nine high-priority targets. His determination not to rise to the bait Washington was offering—Billy Mitchell, the Air Force's earliest strategic champion, had pointed out the vulnerability of Japanese cities to fire as long ago as 1924—doomed his command. Norstad flew out to Guam to relieve Hansell of duty on January 6. Curtis LeMay arrived from China the next day. "LeMay is an operator," Norstad told Hansell, "the rest of us are planners. That's all there is to it." As if to encourage the new commander to independence, Hap Arnold suffered a major heart attack on January 15 and withdrew for a time to Miami sunshine to heal.

LeMay officially took command on January 20. He had 345 B-29's in the Marianas and more arriving. He had 5,800 officers and 46,000 enlisted men. And he had all Hansell's problems to solve: the jet stream; the terrible Japanese weather, seven days of visual bombing a month with luck and not much weather prediction because the Soviets refused to cooperate from Siberia, whence the weather came; B-29 engines that overheated and burned out while straining up the long climb to altitude; indifferent bombing:

> General Arnold needed results. Larry Norstad had made that very plain. In effect he had said: "You go ahead and get results with the B-29. If you don't get results, you'll be fired. If you don't get results, also, there'll never be any Strategic Air Forces of the Pacific. . . . If you don't get results it will mean

eventually a mass amphibious invasion of Japan, to cost probably half a million more American lives."

LeMay set his crews to intensive training. They were beginning to get radar units and he saw to it that they were able at least to identify the transition from water to land. He ordered high-altitude precision strikes but experimented with firebombing as well; 159 tons on Kobe on February 3 burned out a thousand buildings. Not good enough: "another month of indifferent operations," LeMay calls February:

> When I summed it all up, I realized that we had not accomplished very much during those six or seven weeks. We were still going in too high, still running into those big jet stream winds upstairs. Weather was almost always bad.
>
> I sat up nights, fine-tooth-combing all the pictures we had of every target which we had attacked or scouted. I examined Intelligence reports as well.
>
> Did actually very much in the way of low-altitude flak exist up there in Japan? I just couldn't find it.
>
> There was food for thought in this.

There was food for thought as well in two compelling February horrors. One occurred halfway around the world, in Europe, where LeMay had flown so often before. The other began nearby. The hardbitten general from Ohio who despised failure and was failing in Japan could not have avoided learning in detail of both.

The European event was the bombing of Dresden, the capital of the German state of Saxony, on the Elbe River 110 miles south of Berlin, famous for its art and its graceful and delicate architecture. In February 1945 the Russian front advanced to less than eighty miles to the east; refugees streamed west from that deadly harrowing and into the Saxon city. Lacking significant war industry, Dresden had not been a bombing target before and was essentially undefended. It counted in its suburbs 26,000 Allied prisoners of war.

Winston Churchill instigated the Dresden raid. The Secretary of State for Air responded to a phone call from the Prime Minister sometime in January with tactical proposals; the P.M. countered as testily as he had countered in the matter of Niels Bohr:

> I did not ask you last night about plans for harrying the German retreat from Breslau. On the contrary, I asked whether Berlin, and no doubt other large cities in East Germany should not now be considered especially attractive targets. I am glad that this is "under consideration." Pray report to me tomorrow what is going to be done.

Dresden's number thus came up. On the cold night of February 13, 1,400 Bomber Command aircraft dropped high explosives and nearly 650,000 incendiaries on the city; six planes were lost. The firestorm that ensued was visible two hundred miles away. The next day, just after noon, 1,350 American heavy bombers flew over to attack the railroad marshaling yards with high explosives but found nine-tenths cover of cloud and smoke and bombed a far larger area, encountering no flak at all.

The American novelist Kurt Vonnegut, Jr., was a young prisoner of war in Dresden at the time of the attack. He described his experience to an interviewer long after the war:

> The first fancy city I'd ever seen. A city full of statues and zoos, like Paris. We were living in a slaughterhouse, in a nice new cement-block hog barn. They put bunks and straw mattresses in the barn, and we went to work every morning as contract labor in a malt syrup factory. The syrup was for pregnant women. The damned sirens would go off and we'd hear some other city getting it—*whump a whump a whumpa whump.* We never expected to get it. There were very few air-raid shelters in town and no war industries, just cigarette factories, hospitals, clarinet factories. Then a siren went off—it was February 13, 1945—and we went down two stories under the pavement into a big meat locker. It was cool there, with cadavers hanging all around. When we came up the city was gone.... The attack didn't sound like a hell of a lot either. *Whump.* They went over with high explosives first to loosen things up, and then scattered incendiaries.... They burnt the whole damn town down....
>
> Every day [afterward] we walked into the city and dug into basements and shelters to get the corpses out, as a sanitary measure. When we went into them, a typical shelter, an ordinary basement usually, looked like a streetcar full of people who'd simultaneously had heart failure. Just people sitting there in their chairs, all dead. A fire storm is an amazing thing. It doesn't occur in nature. It's fed by the tornadoes that occur in the midst of it and there isn't a damned thing to breathe. We brought the dead out. They were loaded onto wagons and taken to parks, large, open areas in the city which weren't filled with rubble. The Germans got funeral pyres going, burning the bodies to keep them from stinking and from spreading disease. One hundred thirty thousand corpses were hidden underground.

Nearer at hand Curtis LeMay could see the intensity and ferocity of Japanese resistance increasing as American forces fought their way toward the home islands. The latest hellhole was Iwo Jima—Sulfur Island—a mass of volcanic ash and rock only seven square miles in area with a dormant volcano at one end, Mount Suribachi, that had risen from the sea within historic times. Miasmic with sulfur fumes, a steam of rotten eggs, Iwo lacked fresh water but supported two airfields from which Japanese fighter-bombers departed to attack LeMay's B-29's shining on their hardstands on Guam, Saipan and Tinian. It was nine hundred miles closer to Tokyo than the Marianas and its radar outposts gave Honshu antiaircraft batteries and defensive fighter units ample warning when B-29's dispatched for strategic assault passed overhead.

The Japanese understood the island's strategic position and had prepared for months, often under bombardment from U.S. Navy and Air Force planes, to defend it. Fifteen thousand men turned Iwo Jima into a fortress of bunkers, ditches, trenches, 13,000 yards of tunnels, 5,000 pillboxes and fortified cave entrances, vast galleys and wards built into Suribachi, blockhouses with thick concrete walls. The emplacements were armed with the largest concentration of artillery the Japanese had assembled anywhere up to that day: coastal defense guns in concrete bunkers, fieldpieces of all calibers shielded in caves, rocket launchers, tanks buried in the sand up to their turrets, 675-pound spigot mortars, long-barreled anti-aircraft guns cranked down parallel to the ground. The Japanese commander, Lieutenant General Tadamichi Kuribayashi, taught his men a new strategy: "We would all like to die quickly and easily, but that would not inflict heavy casualties. We must fight from cover as long as we possibly can." His soldiers and marines, increased in strength now to more than 21,000, would no longer throw away their lives in banzai charges. They would resist to the death. "I am sorry to end my life here, fighting the United States of America," Kuribayashi wrote his wife. "But I want to defend this island as long as I can." He expected no rescue. "They meant to make the conquest of Iwo so costly," says William Manchester, who fought not this battle but the next one, Okinawa, "that the Americans would recoil from the thought of invading their homeland."

Washington secretly considered sanitizing the island with artillery shells loaded with poison gas lobbed in by ships standing well offshore; the proposal reached the White House but Roosevelt curtly vetoed it. It might have saved thousands of lives and hastened the surrender—arguments used to justify most of the mass slaughters of the Second World War, and neither the United States nor Japan had signed the Geneva Convention prohibiting such use—but Roosevelt presumably remembered the world outcry that had followed German introduction of poison gas in the First World War and decided to leave the sanitizing of Iwo Jima to the U.S. Marines.

They began landing on Saturday, February 19, at 9 A.M., after weeks of naval barrage and bombing. A less well-defended foe would have been pulverized by that battering; the Japanese dug in on Iwo Jima were only groggy from the long disturbance of their sleep. The Navy ferried the marines to shore in amphtracs, gave them over to the deep and treacherous black pumice of the beaches and ran out to reload. The Japanese commanded Suribachi, the high ground; they had zeroed in on every point of consequence on the flat island and now stood back to fire. On the beaches, says Manchester, men were more often killed by artillery than by bullets:

> The invaders were taking heavy mortar and artillery fire. Steel sleeted down on them like the lash of a desert storm. By dusk 2,420 of the 30,000 men on the beachhead were dead or wounded. The perimeter was only four thousand yards long, seven hundred yards deep in the north and a thousand yards in the south. It resembled Doré's illustrations of the *Inferno*. Essential cargo—ammo, rations, water—was piled up in sprawling chaos. And gore, flesh, and bones were lying all about. The deaths on Iwo were extraordinarily violent. There seemed to be no clean wounds; just fragments of corpses. It reminded one battalion medical officer of a Bellevue dissecting room. Often the only way to distinguish between Japanese and marine dead was by the legs; Marines wore canvas leggings and Nips khaki puttees. Otherwise identification was completely impossible. You tripped over strings of viscera fifteen feet long, over bodies which had been cut in half at the waist. Legs and arms, and heads bearing only necks, lay fifty feet from the closest torsos. As night fell the beach reeked with the stench of burning flesh.

After that first awful night, when the Japanese might have squandered themselves in counterattacks but chose instead to hold fast to their defensive redoubts, the leaders of the invasion understood that they would pay with American lives for every foot of the island they captured. Kuribayashi's final order to his men demanded of them the same sacrifice: "We shall infiltrate into the midst of the enemy and annihilate them," he exhorted. "We shall grasp bombs, charge the enemy tanks and destroy them. With every salvo we will, without fail, kill the enemy. Each man will make it his duty to kill ten of the enemy before dying!" Slow, cruel fighting continued for most of a month. In the end, late in March, when shell and fire had changed the very landscape, victory had cost 6,821 marines killed and 21,865 wounded of some 60,000 committed, a casualty ratio of 2 to 1, the highest in Marine Corps history. Of Japanese defenders, 20,000 died on Iwo Jima; only 1,083 allowed themselves to be captured.

That so many were dying to protect his B-29 crews when their results were inconsequential to the war catalyzed LeMay to radical departure. The deaths had to be justified, the debt of death repaid.

One more incendiary test, 172 planes over Tokyo on February 23, produced the best results of any bombing so far, a full square mile of the city burned out. But LeMay had long known that fire would burn down

Japan's wooden cities if properly set. Proper setting, not firebombing itself, was the problem he struggled to solve.

He studied strike photographs. He reviewed intelligence reports. "The Japanese just didn't seem to have those 20- and 40-millimeter [antiaircraft] guns," he remembers realizing. "That's the type of defense which must be used against bombers coming in to attack at a low or medium altitude. Up at twenty-five or thirty thousand feet they have to shoot at you with 80- or 90-millimeter stuff, or they're never going to knock you down. . . . But 88-millimeter guns, *if you come in low,* are impotent. You're moving too fast."

Low-altitude firebombing had other important advantages. Flying low saved fuel coming and going from the Marianas: the B-29's could carry more bombs. Flying low put less strain on the big Wright engines: fewer aircraft would have to abort or ditch. LeMay added in another variable and proposed to bomb at night; his intelligence sources indicated that Japanese fighters lacked airborne radar units. With little or no light flak or fighter cover Tokyo would be nearly defenseless. Why not, then, LeMay reasoned, take out B-29 guns and gunners and further increase the bomb load? He decided to leave the tail gunner as an observer and pull the rest.

He discussed his plan with only a few members of his staff. They worked out a target zone, a flat, densely crowded twelve square miles of workers' houses adjacent to the northeast corner of the Imperial Palace in central Tokyo. Even two decades after the war LeMay felt the need to justify the site as in some sense industrial: "All the people living around that Hattori factory where they make shell fuses. That's the way they disperse their industry: little kids helping out [at home], working all day, little bits of kids." The U.S. Strategic Bombing Survey notes frankly that 87.4 percent of the target zone was residential, and LeMay goes on to more candid admission later in his autobiography:

> No matter how you slice it, you're going to kill an awful lot of civilians. Thousands and thousands. But, if you don't destroy the Japanese industry, we're going to have to invade Japan. And how many Americans will be killed in an invasion of Japan? Five hundred thousand seems to be the lowest estimate. Some say a million.
>
> . . . We're at war with Japan. We were attacked by Japan. Do you want to kill Japanese, or would you rather have Americans killed?

A little later in the war a spokesman for the Fifth Air Force would point out that since the Japanese government was mobilizing civilians to resist invasion, "the entire population of Japan is a proper military target."

Onto the proper military target of working-class Tokyo LeMay decided to drop two kinds of incendiaries. His lead crews would carry M47's, 100-pound oil-gel bombs, 182 per aircraft, each of which was capable of starting a major fire. Behind those crews his major force would sow M69's, 6-pound gelled-gasoline bombs, 1,520 per aircraft. He eschewed magnesium bombs because those more rigid weapons smashed all the way through the tile roofs and light wooden floors of Japanese houses and buried themselves ineffectually in the earth. LeMay also remembers including a few high explosives in the mix to demoralize the firemen.

He delayed seeking approval of his plan until the day before the raid was scheduled to go, taking responsibility for it himself and determined to risk the gamble. Norstad approved on March 8 and alerted the Air Force public relations staff to the possibility of "an outstanding strike." Arnold was informed the same afternoon. LeMay's crews were stunned to hear they would fly their sorties unarmed at staggered levels between five and seven thousand feet. "You're going to deliver the biggest firecracker the Japanese have ever seen," LeMay told them. Some of them thought he was crazy and considered mutiny. Others cheered.

From Guam first, from Saipan next and then from Tinian 334 B-29's took off for Tokyo in the late afternoon of March 9. They were loaded with more than 2,000 tons of incendiaries.

They flew toward a city that an Associated Press correspondent who knew it well had described in 1943 in a best-selling book as "grim, drab and grubby." Freed from Japanese detention in Manila and then in Shanghai, Russell Brines had brought home a message about the people he had lived among before the war and whose language he spoke:

> "We will fight," the Japanese say, "until we eat stones!" The phrase is old; now revived and ground deeply into Japanese consciousness by propagandists skilled in marshaling their sheeplike people. . . . [It] means they will continue the war until every man—perhaps every woman and child—lies face downward on the battlefield. Thousands of Japanese, maybe hundreds of thousands, accept it literally. To ignore this suicide complex would be as dangerous as our pre-war oversight of Japanese determination and cunning which made Pearl Harbor possible. . . .
>
> American fighting men back from the front have been trying to tell America this is a war of extermination. They have seen it from foxholes and barren strips of bullet-strafed sand. I have seen it from behind enemy lines. Our picture coincides. This *is* a war of extermination. The Japanese militarists have made it that way.

The fighting men of the Navy and the Air Force had seen particular evidence of Japanese doggedness that autumn and winter in the appearance of kamikazes, planes loaded with high explosives and deliberately flown to ram ships. Between October and March young Japanese pilots, most of them barely qualified university students, sacrificed themselves in some nine hundred sorties. Navy fighters and antiaircraft guns shot most of the kamikazes down. About four hundred U.S. ships were hit and only about one hundred sunk or severely damaged in a fleet of thousands, but the attacks were alien and terrifying; they served to confirm for Americans the extent of Japanese desperation even as they further depleted Japan's waning air defenses.

LeMay's pathfinders arrived first over Tokyo a little after midnight on March 10. On the district of Shitamachi on the flatlands east of the Sumida River where 750,000 people lived crowded into wood-and-paper houses they marked a diagonal of fire and then crossed it to ignite a gigantic, glowing X. At 0100 the main force of B-29's came on and began methodically bombing the flatlands. The wind was blowing at 15 miles per hour. The bombers carried their 1,520 M69's in 500-pound clusters that broke apart a few hundred feet above the ground. Main-force intervalometers—the bomb-bay mechanisms that spaced the release of the clusters—had been set for 50-foot intervals. Each planeload then covered about a third of a square mile of houses. If only a fifth of the incendiaries started fires, that was one fire for every 30,000 square feet—one fire for every fifteen or twenty closely spaced houses. Robert Guillain remembers a deadlier density:

> The inhabitants stayed heroically put as the bombs dropped, faithfully obeying the order that each family defend its own house. But how could they fight the fires with that wind blowing and when a single house might be hit by ten or even more of the bombs . . . that were raining down by the thousands? As they fell, cylinders scattered a kind of flaming dew that skittered along the roofs, setting fire to everything it splashed and spreading a wash of dancing flames everywhere.

By 0200 the wind had increased to more than 20 miles per hour. Guillain climbed to his roof to observe:

> The fire, whipped by the wind, began to scythe its way through the density of that wooden city. . . . A huge borealis grew. . . . The bright light dispelled the night and B-29's were visible here and there in the sky. For the first time, they flew low or middling high in staggered levels. Their long, glinting wings, sharp as blades, could be seen through the oblique columns of smoke rising from the city, suddenly reflecting the fire from the furnace below, black silhouettes gliding through the fiery sky to reappear farther on, shining golden against the dark roof of heaven or glittering blue, like meteors, in the searchlight beams spraying the vault from horizon to horizon. . . . All the Japanese in the gar-

> dens near mine were out of doors or peering up out of their holes, uttering cries of admiration—this was typically Japanese—at this grandiose, almost theatrical spectacle.

Something worse than a firestorm was kindled in Tokyo that night. The U.S. Strategic Bombing Survey calls it a conflagration, begun when the high wind heeled over the pillar of hot and burning gases that the fires had volatilized and convection had carried up into the air:

> The chief characteristic of the conflagration . . . was the presence of a fire front, an extended wall of fire moving to leeward, preceded by a mass of preheated, turbid, burning vapors. The pillar was in a much more turbulent state than that of [a] fire storm, and being usually closer to the ground, it produced more flame and heat, and less smoke. The progress and destructive features of the conflagration were consequently much greater than those of [a] fire storm, for the fire continued to spread until it could reach no more material. . . . The 28-mile-per-hour wind, measured a mile from the fire, increased to an estimated 55 miles at the perimeter, and probably more within. An extended fire swept over 15 square miles in 6 hours. Pilots reported that the air was so violent that B-29s at 6,000 feet were turned completely over, and that the heat was so intense, even at that altitude, that the entire crew had to don oxygen masks. The area of the fire was nearly 100 percent burned; no structure or its contents escaped damage. The fire had spread largely in the direction of the natural wind.

A bombardier who flew through the black turbulence above the conflagration remembers it as "the most terrifying thing I've ever known."

In the shallower canals of Shitamachi, where people submerged themselves to escape the fire, the water boiled.

The Sumida River stopped the conflagration from sweeping more than 15.8 square miles of the city. The Strategic Bombing Survey estimates that "probably more persons lost their lives by fire at Tokyo in a 6-hour period than at any [equivalent period of] time in the history of man." The fire storm at Dresden may have killed more people but not in so short a space of time. More than 100,000 men, women and children died in Tokyo on the night of March 9–10, 1945; a million were injured, at least 41,000 seriously; a million in all lost their homes. Two thousand tons of incendiaries delivered that punishment—in the modern notation, two kilotons. But the wind, not the weight of bombs alone, created the conflagration, and therefore the efficiency of the slaughter was in some sense still in part an act of God.

Hap Arnold sent LeMay a triumphant telex: CONGRATULATIONS. THIS MISSION SHOWS YOUR CREWS HAVE GOT THE GUTS FOR ANYTHING. Certainly

LeMay did; having gambled and succeeded, he quickly pushed on. His B-29's firebombed Nagoya on March 11; firebombed Osaka by radar on March 13; firebombed Kobe on March 16—stocks of M69's were running low and M17A1 clusters of 4-pound magnesium thermite bombs, less effective, had to be substituted; firebombed Nagoya again on March 18. "Then," says LeMay, "we ran out of bombs. Literally." In ten days and 1,600 sorties the Twentieth Air Force burned out 32 square miles of the centers of Japan's four largest cities and killed at least 150,000 people and almost certainly tens of thousands more. "I consider that for the first time," LeMay wrote Norstad privately in April, "strategic air bombardment faces a situation in which its strength is proportionate to the magnitude of its task. I feel that the destruction of Japan's ability to wage war lies within the capability of this command." He had found a method, LeMay had begun to believe, whereby the Air Force might end the Pacific war without invasion.

At Oak Ridge guests removed their shoes before entering a house. Hiring was still increasing on the muddy Tennessee reservation and construction continuing, challenges to the meager ground cover that a Tennessee Eastman employee was moved to immortalize anonymously in verse:

In order not to check in late,
I've had to lose a lot of weight,
From swimming through a fair-sized flood
And wading through the goddam mud.

I've lost my rubbers and my shoes
Perpetually I have the blues
My spirits tumble with a thud
Because of all this goddam mud.

It's in my system so that when
I cut my finger now and then
Instead of bleeding just plain blood
Out pours a stream of goddam mud.

Mud measured progress: Ernest Lawrence's calutrons, built at such great expense, had begun enriching uranium. A minimum of 100 grams per day—3.5 ounces—of 10 percent U235 came through the Alpha racetracks beginning in late September 1944. But poor planning for chemical recovery of that feed from the Beta tanks wasted some 40 percent of it, as Mark Oliphant reported to James Chadwick from Oak Ridge early in November: "This loss or hold-up . . . has resulted in a very serious delay in the production of material for the first weapon. . . . The chemistry, viewed as a whole, I believe to present an appalling example of lack of coordination, of inefficiency, and bad management."

A copy of Oliphant's complaint went to Groves, who must have acted quickly; the troubleshooting Australian physicist could report to the general two weeks later that "the output from the beta tracks has shown an abrupt and very satisfying upward trend." In his letter to Chadwick, Oliphant had noted a Beta output of only 40 grams per day; now "an output of about 90 grams per day [has] been reached and there [is] reason for believing that this level would be maintained, or even increased, during the coming months." He concluded optimistically that "there is now a definite hope that continued effort on the part of the operating company and others will lead early in the New Year to a plant output of the order of that expected."

As of January 1945 on any given day about 85 percent of some 864 Alpha calutron tanks operated to produce 258 grams—9 ounces—of 10 percent enriched product; at the same time 36 Beta tanks converted the accumulated Alpha product to 204 grams—7.2 ounces—per day of 80 percent enriched U235, sufficient enrichment to make a bomb. James Bryant Conant calculated in his handwritten history notes on January 6 that a kilogram of U235 per day would mean one gun bomb every six weeks. It follows that the gun bomb required about 42 kilograms—92.6 pounds, about 2.8 critical masses—of U235. Without further improvement the calutrons alone could produce that much material in 6.8 months, and Conant noted after conferring with Groves that "it looks as if 40–45 kg . . . will be obtained by July 1." Ernest Lawrence's monumental effort had succeeded; every gram of U235 in the one Little Boy that should be ready by mid-1945 would pass at least once through his calutrons.

Conant also contrasted his assumptions of June 1944 with his assumptions at the beginning of the new year to draw up a problematic balance sheet: while he had previously "believed a few bombs might do the trick" of ending the war, at the beginning of 1945 he was "convinced many bombs will now be required (German experience)." The German experience was probably the determined German resistance that was prolonging the war in Europe, particularly the counteroffensive through the Ardennes known as the Battle of the Bulge that had begun in mid-December and still threatened Allied lines at the time of Conant's notes. It was partly Allied frustration with such continuing resistance that would lead in another month to the atrocity of the Dresden bombing.

Houdaille-Hershey was finally delivering satisfactory barrier tubes for the K-25 gaseous-diffusion plant. Union Carbide had scheduled barrier delivery to take advantage of K-25's organization as a cascade; as individual tanks, called converters, arrived, workers hooked them into the system and tested them for leaks in atmospheres of nitrogen and helium with the portable mass spectrometers that Alfred Nier had designed. When a stage was leakproof and otherwise ready it could be operated without further delay, and the first stage of the enormous K-25 cascade was charged with uranium hexafluoride on January 20, 1945. Enrichment by gaseous barrier diffusion in the most advanced automated industrial plant in the world had begun. It would proceed efficiently with only normal maintenance for decades.

The pipes in Philip Abelson's scaled-up thermal-diffusion plant, S-50, leaked so badly they had to be welded, which delayed production, but all twenty-one racks had begun enriching uranium by March. Juggling the different enrichment processes to produce maximum output in minimum time then became a complex mathematical and organizational challenge. Lieutenant Colonel Kenneth D. Nichols, Groves' talented and long-suffering assistant, worked out the scheduling. Based on Nichols' schedule Groves decided in mid-March not to build more Alpha calutrons, as Lawrence had proposed, but to construct instead a second gaseous-diffusion plant and a fourth Beta plant. Though he certainly expected his atomic bombs to end the war, Groves seems to have justified the new construction by the Joint Chiefs' conservative estimate that the Pacific war would end eighteen months after the European; his new plants could not be completed before February 15, 1946, he explained in his proposal, but "on the assumption that the war with Japan will not be over before July, 1946, it is planned to proceed with the additions to the two plants unless instructions to the contrary are received." Perhaps he was simply being prudent.

Early in 1945 Oak Ridge began shipping bomb-grade U235 to Los Alamos. Between shipments Groves took no chances with a substance far more valuable gram for gram than diamonds. Although the Army had condemned all the land and ejected the original inhabitants from the Clinton reservation area, at the dead end of a dusty reservation back road cattle grazed in a pasture beside a white farmhouse. A concrete silo towered over the road, which was sheltered by a steep bluff. From the air the scene resembled any number of small Tennessee holdings, but the silo was a machine-gun emplacement, the farm was manned by security guards, and built into the side of the bluff a concrete bunker shielded a bank-sized vault completely encircled with guarded walkways. In this pastoral fortress Groves stored his accumulating grams of U235. Armed couriers transported it as uranium tetrafluoride in special luggage by car to Knoxville, where they boarded the overnight express to Chicago. They passed on the luggage the next morning to their Chicago counterparts, who held reserved space on the Santa Fe Chief. Twenty-six hours later, in midafternoon, the Chicago couriers debarked at Lamy, the stranded desert way station that served Santa Fe. Los Alamos security men met the train and completed the transfer to the Hill, where chemists waited eagerly to reduce the rare cargo to metal.

Plutonium production at Hanford depended as much on chemical separation as it did on chain-reacting piles. The chemistry was Glenn Seaborg's, spectacularly scaled up a billionfold directly from his team's earlier ultramicrochemical work. The plutonium in the slugs irradiated in the Hanford piles emerged mixed to the extent of only about 250 parts per million with uranium and highly radioactive fission products. Carrier chemistry—the fractional crystallization of Marie Curie and Otto Hahn—was therefore required to help the scant plutonium along. The man-made metal is extremely poisonous if ingested but only mildly radioactive. To make it safe to handle it also needed to be purified to less than 1 part in 10 million of fission products. And because the pile slugs developed such a burden of radioactivity, all but the final chemical processing had to be carried out by remote control behind thick shielding.

Seaborg's team developed two separation processes to take advantage of the different chemistries of plutonium's several different valence states. One process used bismuth phosphate as a carrier; the other used lanthanum fluoride. Bismuth phosphate, scaled up directly from Met Lab experiments, served the primary purpose of uranium and fission-product decontamination. Lanthanum fluoride, applied at pilot scale at Oak Ridge, then concentrated the plutonium from the large volume of solution in which it was suspended.

Hanford was the largest plant Du Pont had ever constructed and operated; not least among its facilities were the chemical separation buildings. "Originally eight separation plants were considered necessary," writes Groves, "then six, then four. Finally, with the benefit of the operating experience and information obtained from the Clinton semi-works, we decided to build only three, of which two would operate and one would serve as a reserve." For safety the plants went up behind Gable Mountain ten miles southwest of the riverside piles. Each building was 800 feet long, 65 feet wide and 80 feet tall, poured-concrete structures so massive the workers called them Queen Marys; the British ocean liner of that name was only a fifth again as long. The Queen Marys were essentially large concrete boxes, says Groves, containment buildings "in which there were individual cells containing the various parts involved in the process equipment. To provide protection from the intense radioactivity, the cells were surrounded by concrete walls seven feet thick and were covered by six feet of concrete."

Each Queen Mary contained forty cells, and each cell's lid, which could be removed by an overhead crane that rolled the length of the building's long canyon, weighed 35 tons. Irradiated slugs ejected from a production pile would be stored in pools of water 16.5 feet deep to remain until the most intense and therefore short-lived of their fission-product radioactivities decayed away, the water glowing blue around them with Cerenkov radiation, a sort of charged-particle sonic boom. The slugs would then move in shielded casks on special railroad cars to one of the Queen Marys, where they would first be dissolved in hot nitric acid. A standard equipment group occupied two cells: a centrifuge, a catch tank, a precipitator and a solution tank, all made of specially fabricated corrosion-resistant stainless steel. The liquid solution that the slugs had become would move through these units by steam-jet syphoning, a low-maintenance substitute for pumps. There were three necessary steps to the separation process: solution, precipitation and centrifugal removal of the precipitate. These would repeat from equipment group to equipment group down the canyon of the separation building. The end products would be radioactive wastes, stored on site in underground tanks, and small quantities of highly purified plutonium nitrate.

Once the Queen Marys were contaminated with radioactivity no repair crews could enter them. Equipment operators had to be able to maintain them entirely by remote control. The operators trained at Du Pont in Delaware, at Oak Ridge and on mockups at Hanford, but the engineer in charge, Raymond Genereaux, sought more authoritative qualification. And found it: he required his operators, one hundred of whom arrived at Hanford in October 1944, to install the process equipment into the first completed separation building by remote control, pretending the canyon was already radioactive. They did, awkwardly at first but with increasing confidence as practice improved their remote-manipulation skills.

"When the Queen Marys began to function," Leona Marshall remembers, "dissolving the irradiated slugs in concentrated nitric acid, great plumes of brown fumes blossomed above the concrete canyons, climbed thousands of feet into the air, and drifted sideways as they cooled, blown by winds aloft." B-pile slugs traveled by rail into the 221-T separation plant beginning on December 26, 1944. "The yields in the first plant runs . . . ranged between 60 and 70 per cent," Seaborg notes proudly, and "reached 90 per cent early in February 1945." Lieutenant Colonel Franklin T. Matthias, Groves' representative at Hanford, personally carried the first small batch of plutonium nitrate by train from Portland to Los Angeles, where he turned it over to a Los Alamos security courier. Thereafter shipments—small subcritical batches in metal containers in wooden boxes—

traveled in convoy by Army ambulance via Boise, Salt Lake City, Grand Junction and Pueblo to Los Alamos.

Bertrand Goldschmidt, the French chemist who worked with Glenn Seaborg, puts the Manhattan Engineer District at the height of its wartime development in perspective with a startling comparison. It was, he writes in a memoir, "the astonishing American creation in three years, at a cost of two billion dollars, of a formidable array of factories and laboratories—as large as the entire automobile industry of the United States at that date."

One of the mysteries of the Second World War was the lack of an early and dedicated American intelligence effort to discover the extent of German progress toward atomic bomb development. If, as the record repeatedly emphasizes, the United States was seriously worried that Germany might reverse the course of the war with such a surprise secret weapon, why did its intelligence organizations, or the Manhattan Project, not mount a major effort of espionage?

Vannevar Bush had raised the question of espionage with Franklin Roosevelt at their crucial meeting on October 9, 1941, when Bush apprised the President of the MAUD Report, but the OSRD director got no satisfactory answer, probably because the United States was not yet a belligerent. Groves in his memoirs passes the buck to the existing intelligence agencies—Army G-2, the Office of Naval Intelligence and the Office of Strategic Services, the forerunner of the CIA—and attributes the inadequacy of their information to "the unfortunate relationships that had grown up among [them]." Why he failed to confront the issue himself until late 1943, when George Marshall asked him directly to do so, he chooses not to say. One reason was certainly security, a Groves obsession; in order to know what to look for, intelligence agents would have to be briefed on at least isotope-separation technologies and nuclear-fission research, which would mean that any agent captured or turned might well give American secrets away. When Groves finally did take responsibility for intelligence gathering he picked scientific personnel who had not worked within the Manhattan Project and authorized paramilitary operations to advance only into areas already occupied. That at least is how he intended his intelligence unit to operate; in practice it frequently claimed its prizes in the no-man's-land between fighting fronts, by hook or by crook.

The unit Groves authorized in late 1943 somehow acquired the name Alsos, Greek for "grove" and thus obscurely revealing; the brigadier thought to have it renamed, "but I decided that to change it . . . would only draw attention to it." To head the Alsos mission he chose Lieutenant Colonel Boris T. Pash, a former high school teacher turned Army G-2 security officer, FBI trained, who had made himself notorious in domestic intelligence circles for his flamboyant investigation of Communist activities among members of the staff of Ernest Lawrence's Berkeley laboratory. Pash, trim and Slavic, with rimless glasses and light, thin hair, spoke Russian fluently and was a great hunter of Communists. His background helps explain why: his Russian emigré father was the Metropolitan—senior bishop—of the Eastern Orthodox Church in North America. It was Pash who had interrogated Robert Oppenheimer about his Communist affiliations while a clandestine recording device in the next room preserved the physicist's damaging evasions on blank sound motion picture film; he concluded without hard evidence that Oppenheimer was a Communist Party member gone underground and possibly a spy. Whatever Groves thought of Pash's Red-baiting, he chose him to head Alsos because he delivered the goods: "his thorough competence and great drive had made a lasting impression on me."

Pash set up a base in London in 1944 as the Allied armies pushed through France after the Normandy invasion. He then crossed the Channel with a squad of Alsos enlisted men and wheeled toward Paris by jeep. "The ALSOS advance party joined the 102nd U.S. Cavalry Group on Highway 188 at Orsay," a contemporary military intelligence report notes. The American force stopped outside Paris—Charles de Gaulle had persuaded Franklin Roosevelt to allow the Free French to enter the city first—but Pash decided to improvise: "Colonel Pash and party then proceeded to cut across-country to Highway 20 and joined second elements of a French armored division. The ALSOS Mission then entered the City of Paris 0855 hrs., 25 August 1944. The party proceeded to within the city in the rear of the first five French vehicles to enter, being the first American unit to enter Paris." The five French vehicles were tanks. In his unarmored jeep Pash drew repeated sniper fire. He dodged among the back streets of Paris and by the end of the day had achieved his goal, the Radium Institute on the Rue Pierre Curie. There he settled in for the evening to drink celebratory champagne with Frédéric Joliot.

Joliot knew less about German uranium research than anyone had expected. Pash moved his base to liberated Paris and began following up promising leads. One of the most significant pointed to Strasbourg, the old city on the Rhine in Alsace-Lorraine, which Allied forces began occupying in mid-November. Pash found a German physics laboratory installed there in a building on the grounds of Strasbourg Hospital. His scientific counterpart on the Alsos team was Samuel A. Goudsmit, a Dutch theoretical physicist and Paul Ehrenfest protégé who had studied criminology and had previously worked at the MIT Radiation Laboratory. Goudsmit followed

Pash to Strasbourg, began laboriously examining documents and hit the jackpot. He recalls the experience in a postwar memoir:

> It is true that no precise information was given in these documents, but there was far more than enough to get a view of the whole German uranium project. We studied the papers by candlelight for two days and nights until our eyes began to hurt. . . . The conclusions were unmistakable. The evidence at hand proved definitely that Germany had no atom bomb and was not likely to have one in any reasonable form.

But paper evidence was not good enough for Groves; as far as he was concerned, he could close the books on the German program only when he had accounted for all the Union Minière uranium ore the Germans had confiscated when they invaded Belgium in 1940, some 1,200 tons in all, the only source of untraced bomb material available to them during the war with the mines at Joachimsthal under surveillance and the Belgian Congo cut off.

Pash had already liberated part of that supply, some 31 tons, from a French arsenal in Toulouse where it had been diverted and secretly stored. Moving into Germany with the Allied armies after they crossed the Rhine late in March he acquired a larger force of men, two armored cars mounted with .50-caliber machine guns and four machine-gun-mounted jeeps and began tracking the German atomic scientists themselves. "Washington wanted absolute proof," Pash remembers, "that no atomic activity of which it did not know was being carried on by the Nazis. It also wanted to be sure that no prominent German scientist would evade capture or fall into the hands of the Soviet Union." Alsos moved through Heidelberg and picked up Walther Bothe, whose laboratory contained Germany's only functioning cyclotron. Documents there pointed to Stadtilm, near Weimar, as the location of Kurt Diebner's laboratory. The small town proved to have become the central office of the German atomic research program as well, and although Werner Heisenberg and his group from the Kaiser Wilhelm Institutes had moved to southern Germany to escape Allied bombing and the advancing Russian and Allied armies, there was a small amount of uranium oxide at Stadtilm to reward Pash's search.

Pash missed the ore rescue. Groves' liaison man with the British had been watching a factory at Stassfurt, near Magdeburg in northern Germany, since late 1944, when documents captured in Brussels indicated it might house the balance of the Belgium ore. By early April 1945 the Red Army had advanced too close to that prize to leave it uninspected any longer; Groves arranged to assemble a mixed British and American strike force led by Lieutenant Colonel John Lansdale, Jr., the security officer who had cleared Paul Tibbets, to move in. The team met with the Twelfth Army Group's G-2 in Göttingen to seek approval for the Stassfurt mission; Lansdale describes the confrontation in a report:

> We outlined to him our proposal and advised him that if we found the material we were after we proposed to remove it and that it would be necessary that we act with the utmost secrecy and greatest dispatch inasmuch as a meeting between the Russian armies and Allied armies apparently would soon take place and the area in which the material appeared to be was a part of the proposed Russian zone of occupation. [The G-2] was very perturbed at our proposal and foresaw all kinds of difficulties with the Russians and political repercussions at home. Said he must see the Commanding General.

That was calm, no-nonsense Omar Bradley:

> He went alone in to see General Bradley, who at that time was in conference with [the] Ninth Army Commander within whose area Stassfurt then was. Both of them gave unqualified approval to our project, General Bradley being reported to have remarked "to hell with the Russians."

On April 17, led by an infantry-division intelligence officer familiar with the area, Lansdale and his team struck for Stassfurt:

> The plant was a mess both from our bombings and from looting by the French workmen. After going through mountains of paper we located the lager or inventory of papers which disclosed the presence of the material we sought at the plant. . . . This ore was fortunately stored above ground. It was in barrels in open sided sheds and had obviously been there a long time, many of the barrels being broken open. Approximately 1100 tons of ore were stored there. This was in various forms, mostly the concentrates from Belgium and about eight tons of uranium oxide.

Lansdale instructed his group to take inventory and went off to Ninth Army headquarters. That organization assigned him two truck companies. He moved on to the nearest railhead within the permanent American zone of occupation but found the commanding officer there too busy evacuating some ten thousand Allied prisoners of war to be able to offer more help than half a dozen men for guard duty. Lansdale improvised, located empty airport hangars nearby where the ore could be stored awaiting shipment out of Germany and arranged to have them cleared of booby traps. Then he returned to Stassfurt:

> Many of the barrels in which the material was packed were broken open and the majority of those not broken open were in such a weakened condition that

> they could not stand transportation. [A British and an American officer] and I took a jeep and scouting around the country found in one small town a paper bag factory which had a large supply of very heavy bags. We later sent a truck and obtained 10,000 of these. We also discovered in a mill a quantity of wire and the necessary implements for closing the bags. By the evening of 19th April we had a large crew busily engaged in repacking the material and that night the movement of the material to [the railhead] started.

Boris Pash in the meantime continued to chase down the German atomic scientists. Alsos documents placed Werner Heisenberg, Otto Hahn, Carl von Weizsäcker, Max von Laue and the others in their organization in the Black Forest region of southwestern Germany in the resort town of Haigerloch. By late April the German front had broken and the French were moving ahead. Pash and his forces, which now included a battalion of combat engineers, got word in the middle of the night and raced around Stuttgart in their jeeps and trucks and armored cars to beat the French to Haigerloch. They drew German fire along the way and returned it. In the meantime Lansdale in London reassembled his British-American team and flew over to follow Pash in. The story is properly Pash's:

> Haigerloch is a small, picturesque town straddling the Eyach River. As we approached it, pillowcases, sheets, towels and other white articles attached to flagpoles, broomsticks and window shutters flew the message of surrender.
>
> . . . While our engineer friends were busy consolidating the first Alsos-directed seizure of an enemy town, [Pash's men] led teams in a rapid operation to locate Nazi research facilities. They soon found an ingenious set-up that gave almost complete protection from aerial observation and bombardment—a church atop a cliff.
>
> Hurrying to the scene, I saw a box-like concrete entrance to a cave in the side of an 80-foot cliff towering above the lower level of the town. The heavy steel door was padlocked. A paper stuck on the door indicated the manager's identity.
>
> . . . When the manager was brought to me, he tried to convince me that he was only an accountant. When he hesitated at my command to unlock the door, I said: "Beatson, shoot the lock off. If he gets in the way, shoot him."
>
> The manager opened the door.
>
> . . . In the main chamber was a concrete pit about ten feet in diameter. Within the pit hung a heavy metal shield covering the top of a thick metal cylinder. The latter contained a pot-shaped vessel, also of heavy metal, about four feet below the floor level. Atop the vessel was a metal frame. . . . [A] German prisoner . . . confirmed the fact that we had captured the Nazi uranium "machine" as the Germans called it—actually an atomic pile.

Pash left Goudsmit and his several colleagues behind at Haigerloch on April 23 and rushed to nearby Hechingen. There he found the German scientists, all except Otto Hahn, whom he picked up in Tailfingen two days later, and Werner Heisenberg, whom he located with his family at a lake cottage in Bavaria.

The pile at Haigerloch had served for the KWI's final round of neutron-multiplication studies. One and a half tons of carefully husbanded Norsk-Hydro heavy water moderated it; its fuel consisted of 664 cubes of metallic uranium attached to 78 chains that hung down into the water from the metal "shield" Pash describes. With this elegant arrangement and a central neutron source the KWI team in March had achieved nearly sevenfold neutron multiplication; Heisenberg had calculated at the time that a 50 percent increase in the size of the reactor would produce a sustained chain reaction.

"The fact that the German atom bomb was not an immediate threat," Boris Pash writes with justifiable pride, "was probably the most significant single piece of military intelligence developed throughout the war. Alone, that information was enough to justify Alsos." But Alsos managed more: it prevented the Soviet Union from capturing the leading German atomic scientists and acquiring a significant volume of high-quality uranium ore. The Belgian ore confiscated at Toulouse was already being processed through the Oak Ridge calutrons for Little Boy.

At Los Alamos in late 1944 Otto Frisch, always resourceful at invention, proposed a daring program of experiments. Enriched uranium had begun arriving on the Hill from Oak Ridge. By compounding the metal with hydrogen-rich material to make uranium hydride it had become possible to approach an assembly of critical mass responsive to fast as well as slow neutrons. Frisch was leader of the Critical Assemblies group in G Division. Making a critical assembly involved stacking several dozen 1½-inch bars of hydride one at a time and measuring the increased neutron activity as the cubical stack approached critical mass. Usually the small bars were stacked within a boxlike framework of larger machined bricks of beryllium tamper to reflect back neutrons and reduce the amount of uranium required. Dozens of these critical-assembly experiments had gone forward during 1944. "By successively lowering the hydrogen content of the material as more U^{235} became available," the Los Alamos technical history points out, "experience was gained with faster and faster reactions."

But it was impossible to assemble a complete critical mass by stacking bars; such an assembly would run away, kill its sponsors with radiation and melt down. Frisch nearly caused a runaway reaction one day by leaning too close to a naked assembly—he called it a Lady Godiva—that was just subcritical, allowing the hydrogen in his body to reflect back neutrons. "At that moment," he remembers, "out of the corner of my eye I saw that the little red [monitoring] lamps had stopped flickering. They appeared to be glowing continuously. The flicker had speeded up so much that it could no longer be perceived." Instantly Frisch swept his hand across the top of the assembly and knocked away some of the hydride bars. "The lamps slowed down again to a visible flicker." In two seconds he had received by the generous standards of the wartime era a full day's permissible dose of radiation.

Despite that frightening experience, Frisch wanted to work with full critical masses to determine by experiment what Los Alamos had so far been able to determine only theoretically: how much uranium Little Boy would need. Hence his daring proposal:

> The idea was that the compound of uranium-235, which by then had arrived on the site, enough to make an explosive device, should indeed be assembled to make one, but leaving a big hole so that the central portion was missing; that would allow enough neutrons to escape so that no chain reaction could develop. But the missing portion was to be made, ready to be dropped through the hole so that for a split second there was the condition for an atomic explosion, although only barely so.

Brilliant young Richard Feynman laughed when he heard Frisch's plan and named it: he said it would be like tickling the tail of a sleeping dragon. The Dragon experiment it became.

At a remote laboratory site in Omega Canyon that Fermi also used, Frisch's group built a ten-foot iron frame, the "guillotine," that supported upright aluminum guides. The experimenters surrounded the guides at table level with blocks of uranium hydride. To the top of the guillotine they raised a hydride core slug about two by six inches in size. It would fall under the influence of gravity, accelerating at 32 feet per second/per second. When it passed between the blocks it would momentarily form a critical mass. Mixed with hydride, the U235 would react much more slowly than pure metal would react later in Little Boy. But the Dragon would stir, and its dangerous stirring would give Frisch a measure of the fit between theory and experiment:

> It was as near as we could possibly go towards starting an atomic explosion without actually being blown up, and the results were most satisfactory. Everything happened exactly as it should. When the core was dropped through the hole we got a large burst of neutrons and a temperature rise of several degrees in that very short split second during which the chain reaction proceeded as a sort of stifled explosion. We worked under great pressure be-

cause the material had to be returned by a certain date to be made into metal. . . . During those hectic weeks I worked about seventeen hours a day and slept from dawn till mid-morning.

The official Los Alamos history measures the significance of Frisch's Dragon-tickling:

> These experiments gave direct evidence of an explosive chain reaction. They gave an energy production of up to twenty million watts, with a temperature rise in the hydride up to 2°C per millisecond. The strongest burst obtained produced 10^{15} neutrons. The dragon is of historical importance. It was the first controlled nuclear reaction which was supercritical with prompt neutrons alone.

By April 1945 Oak Ridge had produced enough U235 to allow a near-critical assembly of pure metal without hydride dilution. The little bars arrived at the Omega site packed in small, heavy boxes everyone took pains to set well apart; unpacked and unwrapped, the metal shone silver in Frisch's workbench light. Gradually it oxidized, to blue and then to rich plum. Frisch had walked in the snow at Kungälv puzzling out the meaning of Otto Hahn's letters to his aunt; in the basement at Bohr's institute in Copenhagen he had borrowed a name from biology for the process that made these small exotic bars deadly beyond measure; at Birmingham with Rudolf Peierls he had toyed with a formula and had first seen clearly that no more plum-colored metal than now lay scattered on his workbench would make a bomb that would change the world. At Los Alamos in Southwestern spring, dénouement: he would assemble as near a critical mass of U235 as anyone might ever assemble by hand and not be destroyed.

April 12, Thursday, was the day Frisch completed his critical assembly experiments with metallic U235. The previous day Robert Oppenheimer had written Groves the cheering news that Kistiakowsky had managed to produce implosive compressions so smoothly symmetrical that their numbers agreed with theoretical prediction. April 12 in America was Friday, April 13, in Japan, and on the night of that unlucky day B-29's bombing Tokyo bombed the Riken. The wooden building housing Yoshio Nishina's unsuccessful gaseous thermal diffusion experiment did not immediately burn; firemen and staff managed to extinguish the fires that threatened it. But after the other fires were out the building suddenly burst into flame. It burned to the ground and took the Japanese atomic bomb project with it. In Europe John Lansdale was preparing to rush to Stassfurt to confiscate what remained of the Belgian uranium ore; when Groves heard of the success of that adventure later in April he wrote a memorandum to George Marshall that closed the German book:

> In 1940 the German Army in Belgium confiscated and removed to Germany about 1200 tons of uranium ore. So long as this material remained hidden under the control of the enemy we could not be sure but that he might be preparing to use atomic weapons.
>
> Yesterday I was notified by cable that personnel of my office had located this material near Stassfurt, Germany and that it was now being removed to a safe place outside of Germany where it will be under the complete control of American and British authorities.
>
> The capture of this material, which was the bulk of uranium supplies available in Europe, would seem to remove definitely any possibility of the Germans making use of an atomic bomb in this war.

The day these events cluster around, April 12, saw another book closed: at midday, in Warm Springs, Georgia, while sitting for a portrait, Franklin Delano Roosevelt in the sixty-third year of his life was shattered by a massive cerebral hemorrhage. He lingered comatose through the afternoon and died at 3:35 P.M. He had served his nation as President for thirteen years.

When the news of Roosevelt's death reached Los Alamos, Oppenheimer came out from his office onto the steps of the administration building and spoke to the men and women who had spontaneously gathered there. They grieved as Americans everywhere grieved for the loss of a national leader. Some also worried about whether the Manhattan Project would continue. Oppenheimer scheduled a Sunday morning memorial service that everyone in and out of the Tech Area might attend.

"Sunday morning found the mesa deep in snow," Philip Morrison recalls of that day, April 15. "A night's fall had covered the rude textures of the town, silenced its business, and unified the view in a soft whiteness, over which the bright sun shone, casting deep blue shadows behind every wall. It was no costume for mourning, but it seemed recognition of something we needed, a gesture of consolation. Everybody came to the theater, where Oppie spoke very quietly for two or three minutes out of his heart and ours." It was Robert Oppenheimer at his best:

> When, three days ago, the world had word of the death of President Roosevelt, many wept who are unaccustomed to tears, many men and women, little enough accustomed to prayer, prayed to God. Many of us looked with deep trouble to the future; many of us felt less certain that our works would be to a good end; all of us were reminded of how precious a thing human greatness is.

> We have been living through years of great evil, and of great terror. Roosevelt has been our President, our Commander-in-Chief and, in an old and unperverted sense, our leader. All over the world men have looked to him for guidance, and have seen symbolized in him their hope that the evils of this time would not be repeated; that the terrible sacrifices which have been made, and those that are still to be made, would lead to a world more fit for human habitation. . . .
>
> In the Hindu scripture, in the Bhagavad-Gita, it says, "Man is a creature whose substance is faith. What his faith is, he is." The faith of Roosevelt is one that is shared by millions of men and women in every country of the world. For this reason it is possible to maintain the hope, for this reason it is right that we should dedicate ourselves to the hope, that his good works will not have ended with his death.

Vice President Harry S. Truman of Independence, Missouri, who knew only the bare fact of the Manhattan Project's existence, said later that when he heard from Eleanor Roosevelt that he must assume the Presidency in Franklin Roosevelt's place, "I kept thinking, 'The lightning has struck. The lightning has struck!' " Between the Thursday of Roosevelt's death and the Sunday of the memorial service on the Hill, Otto Frisch delivered to Robert Oppenheimer his report on the first experimental determination of the critical mass of pure U235. Little Boy needed more than one critical mass, but the fulfillment of that requirement was now only a matter of time. The lightning had struck at Los Alamos as well.

PART THREE

LIFE AND DEATH

What will people of the future think of us? Will they say, as Roger Williams said of some of the Massachusetts Indians, that we were wolves with the minds of men? Will they think that we resigned our humanity? They will have the right.

C. P. Snow

I see that as human beings we have two great ecstatic impulses in us. One is to participate in life, which ends in the giving of life. The other is to avoid death, which ends tragically in the giving of death. Life and death are in our gift, we can activate life and activate death.

Gil Elliot

18

Trinity

Within twenty-four hours of Franklin Roosevelt's death two men told Harry Truman about the atomic bomb. The first was Henry Lewis Stimson, the upright, white-haired, distinguished Secretary of War. He spoke to the newly sworn President following the brief cabinet meeting Truman called after taking the oath of office on the evening of the day Roosevelt died. "Stimson told me," Truman reports in his memoirs, "that he wanted me to know about an immense project that was under way—a project looking to the development of a new explosive of almost unbelievable destructive power. That was all he felt free to say at the time, and his statement left me puzzled. It was the first bit of information that had come to me about the atomic bomb, but he gave me no details."

Truman had known of the Manhattan Project's existence since his wartime Senate work as chairman of the Committee to Investigate the National Defense Program, when he had attempted to explore the expensive secret project's purpose and had been rebuffed by the Secretary of War himself. That a senator of watchdog responsibility and bulldog tenacity would call off an investigation into unaccounted millions of dollars in defense-plant construction on Stimson's word alone gives some measure of the quality of the Secretary's reputation.

Stimson was seventy-seven years old when Truman assumed the Presidency. He could remember stories his great-grandmother told him of her childhood talks with George Washington. He had attended Phillips Andover when the tuition at that distinguished New England preparatory school was sixty dollars a year and students cut their own firewood. He had graduated from Yale College and Harvard Law School, had served as Secretary of War under William Howard Taft, as Governor General of the Philippines under Calvin Coolidge, as Secretary of State under Herbert Hoover. Roosevelt had called him back to active service in 1940 and with able assistance especially from George Marshall and despite insomnia and migraines that frequently laid him low he had built and administered the most powerful military organization in the history of the world. He was a man of duty and of rectitude. "The chief lesson I have learned in a long life," he wrote at the end of his career, "is that the only way you can make a man trustworthy is to trust him; and the surest way to make him untrustworthy is to distrust him and show your distrust." Stimson sought to apply the lesson impartially to men and to nations. In the spring of 1945 he was greatly worried about the use and consequences of the atomic bomb.

The other man who spoke to Truman, on the following day, April 13, was James Francis Byrnes, known as Jimmy, sixty-six years old, a private citizen of South Carolina since the beginning of April but before then for three years what Franklin Roosevelt had styled "assistant President": Director of Economic Stabilization and then Director of War Mobilization, with offices in the White House. While FDR ran the war and foreign affairs, that is, Byrnes had run the country. "Jimmy Byrnes . . . came to see me," writes Truman of his second briefing on the atomic bomb, "and even he told me few details, though with great solemnity he said that we were perfecting an explosive great enough to destroy the whole world." Then or soon afterward, before Truman met with Stimson again, Byrnes added a significant twist to his tale: "that in his belief the bomb might well put us in a position to dictate our own terms at the end of the war."

At that first Friday meeting Truman asked Byrnes to transcribe his shorthand notes on the Yalta Conference, three months past, which Byrnes had attended as one of Roosevelt's advisers and about which Truman, merely the Vice President then, knew little. Yalta represented nearly all Byrnes' direct experience of foreign affairs. It was more than Truman had. Under the circumstances the new President found it sufficient and informed his colleague that he meant to make him Secretary of State. Byrnes did not object. He insisted that he be given a free hand, however, as Roosevelt had given him in domestic affairs, and Truman agreed.

"A small, wiry, neatly made man," a team of contemporary observers describes Jimmy Byrnes, "with an odd, sharply angular face from which his sharp eyes peer out with an expression of quizzical geniality." Dean Acheson, then an Assistant Secretary of State, thought Byrnes overconfident and insensitive, "a vigorous extrovert, accustomed to the lusty exchange of South Carolina politics." Truman assayed the South Carolinian most shrewdly a few months after their April discussion in a private diary he intermittently kept:

> Had a long talk with my able and conniving Secretary of State. My but he has a keen mind! And he is an honest man. But all country politicians are alike. They are sure all other politicians are circuitous in their dealings. When they are told the straight truth, unvarnished, it is never believed—an asset *sometimes.*

A politician's politician, Byrnes had managed in his thirty-two years of public life to serve with distinction in all three branches of the federal government. He was self-made from the ground up. His father died before he was born. His mother learned dressmaking to survive. Young Jimmy found work at fourteen, his last year of formal education, in a law office, but in lieu of classroom study one of the law partners kindly guided him through a comprehensive reading list. His mother in the meantime taught him shorthand and in 1900, at twenty-one, he earned appointment as a court reporter. He read for the law under the judge whose circuit he reported and passed the bar in 1904. He ran first, in 1908, for solicitor, the South Carolina equivalent of district attorney, and made himself known prosecuting murderers. More than forty-six stump debates won him election to Congress in 1910; in 1930, after fourteen years in the House and five years out of office, he was elected to the Senate. By then he was already actively promoting Franklin Roosevelt's approaching presidential bid. Byrnes served as one of the candidate's speechwriters during the 1932 campaign and afterward worked hard as Roosevelt's man in the Senate to push through the New Deal. His reward, in 1941, was a seat on the United States Supreme Court, which he resigned in 1942 to move to the White House to take over operating the complicated wartime emergency program of wage and price controls, the assistant Presidency of which Roosevelt spoke.

In 1944 everyone understood that Roosevelt's fourth term would be his last. The man he selected for Vice President would therefore almost certainly take the Democratic Party presidential nomination in 1948. Byrnes expected to be that man and Roosevelt encouraged him. But the assistant President was a conservative Democrat from the Deep South, and at the last minute Roosevelt compromised instead on the man from Missouri, Harry S. Truman. "I freely admit that I was disappointed," Byrnes writes with understatement approaching lockjaw, "and felt hurt by President Roosevelt's action." He made a point of visiting the European front with George Marshall in September 1944, in the midst of the presidential campaign; when he returned FDR had to appeal to him formally by letter—a document Byrnes could show around—to endorse the ticket with a speech.

Byrnes undoubtedly regarded Truman as a usurper: if not Truman but he had been Roosevelt's choice he would be President of the United States now. Truman knew Byrnes' attitude but needed the old pro badly to help him run the country and face the world. Hence the prize of State. The Secretary of State was the highest-ranking member of the cabinet and under the rules of succession then obtaining was the officer next in line for the Presidency as well when the Vice Presidency was vacant. Short of the Presidency itself, State was the most powerful office Truman had to give.

Vannevar Bush and James Bryant Conant had needed months to convince Henry Stimson to take up consideration of the bomb's challenge in the postwar era. He had not been ready in late October 1944 when Bush pressed him for action and he had not been ready in early December when Bush pressed him again. By then Bush knew what he thought the problem needed, however:

> We proposed that the Secretary of War suggest to the President the establishment of a committee or commission with the duty of preparing plans. These would include the drafting of legislation and the drafting of appropriate releases to be made public at the proper time. . . . We were all in agreement that the State Department should now be brought in.

Stimson allowed one of his trusted aides, Harvey H. Bundy, a Boston lawyer, father of William P. and McGeorge, at least to begin formulating a membership roster and list of duties for such a committee. But he did not yet know even in broad outline what basic policy to recommend.

Bohr's ideas, variously diluted, floated by that time in the Washington air. Bohr had sought to convince the American government that only early discussion with the Soviet Union of the mutual dangers of a nuclear arms race could forestall such an arms race once the bomb became known. (He would try again in April to see Roosevelt; Felix Frankfurter and Lord Halifax, the British ambassador, would be strolling in a Washington park discussing Bohr's best avenue of approach when the bells of the city's churches began tolling the news of the President's death.) Apparently no one within the executive branch was sufficiently convinced of the *inevitability* of Bohr's vision. Stimson was as wise as any man in government, but late in December he cautioned Roosevelt that the Russians should earn the right to hear the baleful news:

> I told him of my views as to the future of S-1 [Stimson's code for the bomb] in connection with Russia: that I knew they were spying on our work but had not yet gotten any real knowledge of it and that, while I was troubled about the possible effect of keeping from them even now that work, I believed that it was essential not to take them into our confidence until we were sure to get a real quid pro quo from our frankness. I said I had no illusions as to the possibility of keeping permanently such a secret but that I did not think it was yet time to share it with Russia. He said he thought he agreed with me.

In mid-February, after talking again to Bush, Stimson confided to his diary what he wanted in exchange for news of the bomb. Bohr's conviction that only an open world modeled in some sense on the republic of science could answer the challenge of the bomb had drifted, in Bush's mind, to a proposal for an international pool of scientific research. Of such an arrangement Stimson wrote that "it would be inadvisable to put it into full force yet, until we had gotten all we could in Russia in the way of liberalization in exchange for S-1." That is, the quid pro quo Stimson thought the United States should demand from the Soviet Union was the democratization of its government. What for Bohr was the inevitable outcome of a solution to the problem of the bomb—an open world where differences in social and political conditions would be visible to everyone and therefore under pressure to improve—Stimson imagined should be a precondition to any initial exchange.

Finally in mid-March Stimson talked to Roosevelt, their last meeting. That talk came to no useful end. In April, with a new President in the White House, he prepared to repeat the performance.

In the meantime the men who had advised Franklin Roosevelt were working to convince Harry Truman of the increasing perfidy of the Soviet Union. Averell Harriman, the shrewd multimillionaire Ambassador to Moscow, had rushed to Washington to brief the new President. Truman says Harriman told him the visit was based on "the fear that you did not understand, as I had seen Roosevelt understand, that Stalin is breaking his agreements." To soften that condescension Harriman added that he feared Truman "could not have had time to catch up with all the recent cables." The self-educated Missourian prided himself on how many pages of documents he could chew through per day—he was a champion reader—and undercut Harriman's condescension breezily by instructing the ambassador to "keep on sending me long messages."

Harriman told Truman they were faced with a "barbarian invasion of Europe." The Soviet Union, he said, meant to take over its neighbors and install the Soviet system of secret police and state control. "He added that he was not pessimistic," the President writes, "for he felt that it was possible for us to arrive at a workable basis with the Russians. He believed that this would require a reconsideration of our policy and the abandonment of any illusion that the Soviet government was likely soon to act in accordance with the principles to which the rest of the world held in international affairs."

Truman was concerned to convince Roosevelt's advisers that he meant to be decisive. "I ended the meeting by saying, 'I intend to be firm in my dealings with the Soviet government.' " Delegates were arriving in San Francisco that April, for example, to formulate a charter for a new United Nations to replace the old and defunct League. Harriman asked Truman if he would "go ahead with the world organization plans even if Russia dropped out." Truman remembers responding realistically that "without Russia there would not be a world organization." Three days later, having heard from Stalin in the meantime and met the arriving Soviet Foreign Minister, Vyacheslav Molotov, he retreated from realism to bluster. "He felt that our agreements with the Soviet Union had so far been a one-way street," an eyewitness recalls, "and that he could not continue; it was now or never. He intended to go on with the plans for San Francisco and if the Russians did not wish to join us they could go to Hell."

Stimson argued for patience. "In the big military matters," Truman reports him saying, "the Soviet government had kept its word and the military authorities of the United States had come to count on it. In fact . . . they had often done better than they had promised." Although George Marshall seconded Stimson's argument and Truman could not have had two more reliable witnesses, it was not counsel the new and untried President wanted to hear. Marshall added a crucial justification that Truman took to heart:

> He said from the military point of view the situation in Europe was secure but that we hoped for Soviet participation in the war against Japan at a time when it would be useful to us. The Russians had it within their power to delay their entry into the Far Eastern war until we had done all the dirty work. He was inclined to agree with Mr. Stimson that the possibility of a break with Russia was very serious.

Truman could hardly tell the Russians to go to hell if he needed them to finish the Pacific war. Marshall's justification for patience meant Stalin had the President over a barrel. It was not an arrangement Harry Truman intended to perpetuate.

He let Molotov know. They had sparred diplomatically at their first meeting; now the President attacked. The issue was the composition of the postwar government of Poland. Molotov discussed various formulas, all favoring Soviet dominance. Truman demanded the free elections that he understood had been agreed upon at Yalta: "I replied sharply that an agreement had been reached on Poland and that there was only one thing to do, and that was for Marshal Stalin to carry out that agreement in accordance with his word." Molotov tried again. Truman replied sharply again, repeating his previous demand. Molotov hedged once more. Truman proceeded to lay him low: "I expressed once more the desire of the United States for friendship with Russia, but I wanted it clearly understood that this could be only on a basis of the mutual observation of agreements and not on the basis of a one-way street." Those are hardly fighting words; Molotov's reaction suggests that the President spoke more pungently at the time:

> "I have never been talked to like that in my life," Molotov said.
>
> I told him, "Carry out your agreements and you won't get talked to like that."

If Truman felt better for the exchange, it disturbed Stimson. The new President had acted without knowledge of the bomb and its potentially fateful consequences. It was time and past time for a full briefing.

Truman agreed to meet with Stimson at noon on Wednesday, April 25. The President was scheduled to address the opening session of the United Nations conference in San Francisco by radio that evening. One more conditioning incident intervened; on Tuesday he received a communication from Joseph Stalin, "one of the most revealing and disquieting messages to reach me during my first days in the White House." Molotov had reported Truman's tough talk to the Soviet Premier. Stalin replied in kind. Poland bordered on the Soviet Union, he wrote, not on Great Britain or the United States. "The question [of] Poland had the same meaning for the security of the Soviet Union as the question [of] Belgium and Greece for the security of Great Britain"—but "the Soviet Union was not consulted when those governments were being established there" following the Allied liberation. The "blood of the Soviet people abundantly shed on the fields of Poland in the name of the liberation of Poland" demanded a Polish government friendly to Russia. And finally:

> I am ready to fulfill your request and do everything possible to reach a harmonious solution. But you demand too much of me. In other words, you demand that I renounce the interests of security of the Soviet Union, but I cannot turn against my country.

With this blunt challenge on his mind Truman received his Secretary of War.

Stimson had brought Groves along for technical backup but left him waiting in an outer office while he discussed issues of general policy. He began dramatically, reading from a memorandum:

> Within four months we shall in all probability have completed the most terrible weapon ever known in human history, one bomb of which could destroy a whole city.

We had shared the development with the British, Stimson continued, but we controlled the factories that made the explosive material "and no other nation could reach this position for some years." It was certain that we would not enjoy a monopoly forever, and "probably the only nation which could enter into production within the next few years is Russia." The world "in its present state of moral advancement compared with its technical development," the Secretary of War continued quaintly, "would be eventually at the mercy of such a weapon. In other words, modern civilization might be completely destroyed."

Stimson emphasized what John Anderson had emphasized to Churchill the year before: that founding a "world peace organization" while the bomb was still a secret "would seem to be unrealistic":

> No system of control heretofore considered would be adequate to control this menace. Both inside any particular country and between the nations of the world, the control of this weapon will undoubtedly be a matter of the greatest difficulty and would involve such thorough-going rights of inspection and internal controls as we have never heretofore contemplated.

That brought Stimson to the crucial point:

> Furthermore, in the light of our present position with reference to this weapon, the question of sharing it with other nations and, if so shared, upon what terms, becomes a primary question of our foreign relations.

Bohr had proposed to inform other nations of the common dangers of a nuclear arms race. At the hands of Stimson and his advisers that sensible proposal had drifted to the notion that the issue was sharing the weapon itself. As Commander in Chief, as a veteran of the First World War, as a man of common sense, Truman must have wondered what on earth his Secretary of War was talking about, especially when Stimson added that "a certain moral responsibility" followed from American leadership in nuclear technology which the nation could not shirk "without very serious responsibility for any disaster to civilization which it would further." Was the United States morally obligated to give away a devastating new weapon of war?

Now Stimson called in Groves. The general brought with him a report on the status of the Manhattan Project that he had presented to the Secretary of War two days earlier. Both Stimson and Groves insisted Truman read the document while they waited. The President was restive. He had a threatening note from Stalin to deal with. He had to prepare to open the United Nations conference even though Stimson had just informed him that allowing the conference to proceed in ignorance of the bomb was a sham. A scene of darkening comedy followed as the proud man who had challenged Averell Harriman to keep sending him long messages tried to avoid public instruction in the minutiae of a secret project he had fought doggedly as a senator to investigate. Groves misunderstood completely:

> Mr. Truman did not like to read long reports. This report was not long, considering the size of the project. It was about twenty-four pages and he would constantly interrupt his reading to say, "Why, I don't like to read papers." And Mr. Stimson and I would reply: "Well we can't tell you this in any more concise language. This is a big project." For example, we discussed our relations with the British in about four or five lines. It was that much condensed. We had to explain all the processes and we might just say what they were and that was about all.

After the reading of the lesson, Groves notes, "a great deal of emphasis was placed on foreign relations and particularly on the Russian situation"—Truman reverting to his immediate problems. He "made it very definite," Groves adds for the record, "that he was in entire agreement with the necessity for the project."

The final point in Stimson's memorandum was the proposal Bush and Conant had initiated to establish what Stimson called "a select committee . . . for recommending action to the Executive and legislative branches of our government." Truman approved.

In his memoirs the President describes his meeting with Stimson and Groves with tact and perhaps even a measure of private humor: "I listened with absorbed interest, for Stimson was a man of great wisdom and foresight. He went into considerable detail in describing the nature and the power of the projected weapon. . . . Byrnes had already told me that the weapon might be so powerful as to be capable of wiping out entire cities and killing people on an unprecedented scale." That was when Byrnes had crowed that the new bombs might allow the United States to dictate its own terms at the end of the war. "Stimson, on the other hand, seemed at least as much concerned with the role of the atomic bomb in the shaping of history as in its capacity to shorten this war. . . . I thanked him for his enlightening presentation of this awesome subject, and as I saw him to the door I felt how fortunate the country was to have so able and so wise a man in its service." High praise, but the President was not sufficiently impressed at the outset with Stimson and Harriman to invite either man to accompany him to the next conference of the Big Three. Both found it necessary, when the time came, to invite themselves. Jimmy Byrnes went at the President's invitation and sat at the President's right hand.

Discussion between Truman and his various advisers was one level of discourse in the spring of 1945 on the uses of the atomic bomb. Another was joined two days after Stimson and Groves briefed the President when a Target Committee under Groves' authority met for the first time in Lauris Norstad's conference room at the Pentagon. Brigadier General Thomas F. Farrell, who would represent the Manhattan Project as Groves' deputy to the Pacific Command, chaired the committee; besides Farrell it counted two other Air Force officers—a colonel and a major—and five scientists, including John von Neumann and British physicist William G. Penney. Groves opened the meeting with a variant of his usual speech to Manhattan Project working groups: how important their duty was, how secret it must be kept. He had already discussed targets with the Military Policy Committee and now informed his Target Committee that it should propose no more than four.

Farrell laid down the basics: B-29 range for such important missions no more than 1,500 miles; visual bombing essential so that these untried and valuable bombs could be aimed with certainty and their effects photographed; probable targets "urban or industrial Japanese areas" in July, August or September; each mission to be given one primary and two alternate targets with spotter planes sent ahead to confirm visibility.

Most of the first meeting was devoted to worrying about the Japanese weather. After lunch the committee brought in the Twentieth Air Force's top meteorologist, who told them that June was the worst weather month in Japan; "a little improvement is present in July; a little bit better weather is present in August; September weather is bad." January was the best month, but no one intended to wait that long. The meteorologist said he could forecast a good day for bombing operations only twenty-four hours ahead, but he could give two days' notice of bad weather. He suggested they station submarines near the target areas to radio back weather readings.

Later in the afternoon they began considering targets. Groves had extended Farrell's guidelines:

> I had set as the governing factor that the targets chosen should be places the bombing of which would most adversely affect the will of the Japanese people to continue the war. Beyond that, they should be military in nature, consisting either of important headquarters or troop concentrations, or centers of production of military equipment and supplies. To enable us to assess accurately the effects of the bomb, the targets should not have been previously damaged by air raids. It was also desirable that the first target be of such size that the damage would be confined within it, so that we could more definitely determine the power of the bomb.

But such pristine targets had already become scarce in Japan. If the first choice the Target Committee identified at its first meeting was hardly big enough to confine the potential damage, it was the best the enemy had left to offer:

> Hiroshima is the largest untouched target not on the 21st Bomber Command priority list. Consideration should be given to this city.

"Tokyo," the committee notes continue, "is a possibility but it is now practically all bombed and burned out and is practically rubble with only the palace grounds left standing. Consideration is only possible here."

The Target Committee did not yet fully understand the level of authority it commanded. With a few words to Groves it could exempt a Japanese city from Curtis LeMay's relentless firebombing, preserving it through spring mornings of cherry blossoms and summer nights of wild monsoons for a more historic fate. The committee thought it took second priority behind LeMay rather than first priority ahead, and in emphasizing these mistaken priorities the colonel who reviewed the Twentieth Air Force's bombing directive for the committee revealed what the United States' policy in Japan in all its deadly ambiguity had become:

> It should be remembered that in our selection of any target, the 20th Air Force is operating primarily to laying waste all the main Japanese cities, and that they do not propose to save some important primary target for us if it interferes with the operation of the war from their point of view. Their existing procedure has been to bomb the hell out of Tokyo, bomb the aircraft, manufacturing and assembly plants, engine plants and in general paralyze the aircraft industry so as to eliminate opposition to the 20th Air Force operations. The 20th Air Force is systematically bombing out the following cities with the prime purpose in mind of not leaving one stone lying on another:
>
> Tokyo, Yokohama, Nagoya, Osaka, Kyoto,
> Kobe, Yawata & Nagasaki.

If the Japanese were prepared to eat stones, the Americans were prepared to supply them.

The colonel also advised that the Twentieth Air Force planned to increase its delivery of conventional bombs steadily until it was dropping 100,000 tons a month by the end of 1945.

The group decided to study seventeen targets including Tokyo Bay, Yokohama, Nagoya, Osaka, Kobe, Hiroshima, Kokura, Fukuoka, Nagasaki and Sasebo. Targets already destroyed would be culled from the list. The weather people would review weather reports. Penney would consider "the size of the bomb burst, the amount of damage expected, and the ultimate distance at which people would be killed." Von Neumann would be responsible for computations. Adjourning its initial meeting the Target Committee planned to meet again in mid-May in Robert Oppenheimer's office at Los Alamos.

A third level of discourse on the uses of the bomb revealed itself as Henry Stimson assembled the committee that Bush and Conant had proposed to him and he had proposed in turn to the President. On May 1, the day German radio announced the suicide of Adolf Hitler in the ruins of Berlin, George L. Harrison, a special Stimson consultant and the president of the New York Life Insurance Company, prepared for the Secretary of War an entirely civilian committee roster consisting of Stimson as chairman, Bush, Conant, MIT president Karl Compton, Assistant Secretary of State William L. Clayton, Undersecretary of the Navy Ralph A. Bard and a special representative of the President whom the President might choose. Stimson modified the list to include Harrison as his alternate and carried it to Truman for approval on May 2. Truman agreed and Stimson apparently assumed his interest in the project, but the President significantly did not even bother to name his own man to the list. Stimson wrote in his diary that night:

> The President accepted the present members of the committee and said that they would be sufficient even without a personal representative of himself. I said I should prefer to have such a representative and suggested that he should be a man (a) with whom the President had close personal relations and (b) who was able to keep his mouth shut.

Truman had not yet announced his intention to appoint Byrnes Secretary of State because the holdover Secretary, Edward R. Stettinius, Jr., was heading the United States delegation to the United Nations in San Francisco and the President did not want to undercut his authority there. But word of the forthcoming appointment had diffused through Washington.

Acting on it, Harrison suggested that Stimson propose Byrnes. On May 3 Stimson did, "and late in the day the President called me up himself and said that he had heard of my suggestion and it was fine. He had already called up Byrnes down in South Carolina and Byrnes had accepted." Bundy and Harrison, Stimson told his diary, "were tickled to death." They thought their committee had acquired a second powerful sponsor. In fact they had just welcomed a cowbird into their nest.

Stimson sent out invitations the next day. He proposed calling his new group the Interim Committee to avoid appearing to usurp congressional prerogatives: "when secrecy is no longer required," he explained to the prospective members, "Congress might wish to appoint a permanent Post War Commission." He set the first informal meeting of the Interim Committee for May 9.

The membership would assemble in the wake of momentous change. The war in Europe had finally ground to an end. Supreme Allied Commander Dwight D. Eisenhower celebrated the victory on national radio the evening of Tuesday, May 8, 1945, V-E Day:

> I have the rare privilege of speaking for a victorious army of almost five million fighting men. They, and the women who have so ably assisted them, constitute the Allied Expeditionary Force that has liberated western Europe. They have destroyed or captured enemy armies totalling more than their own strength, and swept triumphantly forward over the hundreds of miles separating Cherbourg from Lübeck, Leipzig and Munich. . . .
>
> These startling successes have not been bought without sorrow and suffering. In this Theater alone 80,000 Americans and comparable numbers among their Allies, have had their lives cut short that the rest of us might live in the sunlight of freedom. . . .
>
> But, at last, *this* part of the job is done. No more will there flow from this Theater to the United States those doleful lists of death and loss that have brought so much sorrow to American homes. The sounds of battle have faded from the European scene.

Eisenhower had watched Colonel General Alfried Jodl sign the act of military surrender in a schoolroom in Rheims—the temporary war room of the Supreme Headquarters Allied Expeditionary Force—in the early morning hours of May 7. Eisenhower's aides had attempted then to draft a suitably eloquent message to the Combined Chiefs reporting the official surrender. "I tried one myself," Eisenhower's chief of staff Walter Bedell Smith remembers, "and like all my associates, groped for resounding phrases as fitting accolades to the Great Crusade and indicative of our dedication to the great task just accomplished." The Supreme Commander listened quietly for a time, thanked everyone for trying and dictated his own unadorned report:

> The mission of this Allied force was fulfilled at 0241, local time, May 7th, 1945.

Better to be brief, better than resounding phrases. Twenty million Soviet soldiers and civilians died of privation or in battle in the Second World War. Eight million British and Europeans died or were killed and another five million Germans. The Nazis murdered six million Jews in ghettos and concentration camps. Manmade death had ended thirty-nine million human lives prematurely; for the second time in half a century Europe had become a charnel house.

There remained the brutal conflict Japan had begun in the Pacific and refused despite her increasing destruction to end by unconditional surrender.

Officially Byrnes was retired to South Carolina. In fact he was visiting Washington surreptitiously, absorbing detailed evening briefings by State Department division chiefs at his apartment at the Shoreham Hotel. On the afternoon of V-E Day he spent two hours closeted alone with Stimson. Then Harrison, Bundy and Groves joined them. "We all discussed the function of the proposed Interim Committee," Stimson records. "During the meeting it became very evident what a tremendous help Byrnes would be as a member of the committee."

The next morning the Interim Committee met for the first time in Stimson's office. The gathering was preliminary, to fill in Byrnes, State's Clayton and the Navy's Bard on the basic facts, but Stimson made a point of introducing the former assistant President as Truman's personal representative. The membership was thus put on notice that Byrnes enjoyed special status and that his words carried extra weight.

The committee recognized that the scientists working on the atomic bomb might have useful advice to offer and created a Scientific Panel adjunct. Bush and Conant put their heads together and recommended Arthur Compton, Ernest Lawrence, Robert Oppenheimer and Enrico Fermi for appointment.

Between the first and second meetings of the Interim Committee its *Doppelgänger,* the Target Committee, met again for two days, May 10 and 11, at Los Alamos. Added to the full committee as advisers were Oppenheimer, Parsons, Tolman and Norman Ramsey and for part of the deliberations Hans Bethe and Robert Brode. Oppenheimer took control by devising and presenting a thorough agenda:

A. Height of Detonation
B. Report on Weather and Operations
C. Gadget Jettisoning and Landing
D. Status of Targets
E. Psychological Factors in Target Selection
F. Use Against Military Objectives
G. Radiological Effects
H. Coordinated Air Operations
I. Rehearsals
J. Operating Requirements for Safety of Airplanes
K. Coordination with 21st [Bomber Command] Program

Detonation height determined how large an area would be damaged by blast and depended crucially on yield. A bomb detonated too high would expend its energy blasting thin air; a bomb detonated too low would expend its energy excavating a crater. It was better to be low than high, the committee minutes explain: "The bomb can be detonated as much as 40% below the optimum with a reduction of 24% in area of damage whereas a detonation [only] 14% above the optimum will cause the same loss in area." The discussion demonstrates how uncertain Los Alamos still was of bomb yield. Bethe estimated a yield range for Little Boy of 5,000 to 15,000 tons TNT equivalent. Fat Man, the implosion bomb, was anybody's guess: 700, 2,000, 5,000 tons? "With the present information the fuse would be set at 2,000 tons equivalent but fusing for the other values should be available at the time of final delivery. . . . Trinity data will be used for this gadget."

The scientists reported and the committee agreed that in an emergency a B-29 in good condition could return to base with a bomb. "It should make a normal landing with the greatest possible care. . . . The chances of [a] crash initiating a high order [i.e., nuclear] explosion are . . . sufficiently small [as to be] a justifiable risk." Fat Man could even survive jettisoning into shallow water. Little Boy was less forgiving. Since the gun bomb contained more than two critical masses of U235, seawater leaking into its casing could moderate stray neutrons sufficiently to initiate a destructive slow-neutron chain reaction. The alternative, jettisoning Little Boy onto land, might loose the U235 bullet down the barrel into the target core and set off a nuclear explosion. For temperamental Little Boy, the minutes note, unluckily for the aircrew, "the best emergency procedure that has so far been proposed is . . . the removal of the gun powder from the gun and the execution of a crash landing."

Target selection had advanced. The committee had refined its qualifications to three: "important targets in a large urban area of more than three miles diameter" that were "capable of being damaged effectively by blast" and were "likely to be unattacked by next August." The Air Force had agreed to reserve five such targets for atomic bombing. These included:

(1) *Kyoto*—This target is an urban industrial area with a population of 1,-000,000. It is the former capital of Japan and many people and industries are now being moved there as other areas are being destroyed. From the psychological point of view there is the advantage that Kyoto is an intellectual center for Japan and the people there are more apt to appreciate the significance of such a weapon as the gadget. . . .

(2) *Hiroshima*—This is an important army depot and port of embarkation in the middle of an urban industrial area. It is a good radar target and it is such a size that a large part of the city could be extensively damaged. There are adjacent hills which are likely to produce a focusing effect which would considerably increase the blast damage. Due to rivers it is not a good incendiary target.

The other three targets proposed were Yokohama, Kokura Arsenal and Niigata. An unsung enthusiast on the committee suggested a spectacular sixth target for consideration, but wiser heads prevailed: "The possibility of bombing the Emperor's palace was discussed. It was agreed that we should not recommend it but that any action for this bombing should come from authorities on military policy."

So the Target Committee sitting in Oppenheimer's office at Los Alamos under the modified Lincoln quotation that Oppenheimer had posted on the wall—THIS WORLD CANNOT ENDURE HALF SLAVE AND HALF FREE—remanded four targets to further study: Kyoto, Hiroshima, Yokohama and Kokura Arsenal.

The committee and its Los Alamos consultants were not unmindful of the radiation effects of the atomic bomb—its most significant difference in effect from conventional high explosives—but worried more about radiation danger to American aircrews than to the Japanese. "Dr. Oppenheimer presented a memo he had prepared on the radiological effect of the gadget. . . . The basic recommendations of this memo are (1) for radiological reasons no aircraft should be closer than 2½ miles to the point of detonation (for blast reasons the distance should be greater) and (2) aircraft must avoid the cloud of radio-active materials."

Since the expected yields of the bombs under discussion made them something less than city-busters, the Target Committee considered following Little Boy and Fat Man with conventional incendiary raids. Radioactive clouds that might endanger LeMay's follow-up crews worried the targeters, though they thought an incendiary raid delayed one day after an atomic bombing might be safe and "quite effective."

With a better sense for having visited Los Alamos of the weapons it was targeting, the Target Committee scheduled its next meeting for May 28 at the Pentagon.

Vannevar Bush thought the second Interim Committee meeting on May 14 produced "very frank discussions." The group, he decided, was "an excellent one." These judgments he passed along to Conant, who had been unable to attend. Stimson won approval of the Scientific Panel as constituted and discussed the possibility of assembling a similar group of industrialists. As his agenda noted, such a group would "advise of [the] likelihood of other nations repeating what our industry has done"—that is, whether other nations could build the vast, innovative industrial plant necessary to produce atomic bombs.

That May Monday morning the committee received copies of Bush's and Conant's September 30, 1944, memorandum to Stimson, the discussion framed on Bohr's ideas of the free exchange of scientific information and inspection not only of laboratories throughout the world but also of military installations. Bush promptly hedged his commitment to so open a world:

> I . . . said that while we made the memorandum very explicit, that it certainly did not indicate that we were irrevocably committed to any definite line of action but rather felt that we ought to express our ideas early in order that there might be discussion as [a] result of which we might indeed change our thoughts as we studied into the subject further, and I said also that we would undoubtedly write the memorandum a little differently today due to the lapse of time since last September.

At the end of the meeting Byrnes took his copy along and studied it with interest.

The Secretary of State–designate was learning fast. When the Interim Committee met again on Friday, May 18, with Groves sitting in, Byrnes brought up the Bush-Conant memorandum as soon as draft press releases announcing the dropping of the first atomic bomb on Japan had been reviewed. It was Bush's turn to be absent; Conant passed along the news:

> Mr. Byrnes spent considerable time discussing our memorandum of last fall, which he had read carefully and with which he was much impressed. It apparently stimulated his thinking (which was all that we had originally desired I imagine). He was particularly impressed with our statement that the Russians might catch up in three to four years. This premise was violently opposed by the General [i.e., Groves], who felt that twenty years was a much better figure. . . . The General is basing his long estimate on a very poor view of Russian ability, which I think is a highly unsafe assumption. . . .

> There was some discussion about the implications of a time interval as short as four years and various international problems were discussed, particularly the question of whether or not the President should tell the Russians of the existence of the weapon after the July test.

Bohr's proposal to enlist the Soviet Union in discussions before the atomic bomb became a reality here slips to the question of whether or not to tell the Soviets the bare facts after the first bomb had been tested but before the second was dropped on Japan. Byrnes thought the answer to that question might depend on how quickly the USSR could duplicate the American accomplishment. The Interim Committee's recording secretary, 2nd Lieutenant R. Gordon Arneson, remembered after the war of this confrontation that "Mr. Byrnes felt that this point was a very important one." The veteran of House and Senate cloakrooms was at least as concerned as Henry Stimson to extract a quid pro quo for any exchange of information, as Conant's next comment to Bush demonstrates:

> This question [i.e., whether or not to tell the Russians about the atomic bomb before using it on Japan] led to the review of the Quebec Agreement which was shown once more to Mr. Byrnes. He asked the General what we had got in exchange, and the General replied only the arrangements controlling the Belgium-Congo [*sic*]. . . . Mr. Byrnes made short work of this line of argument.

The Quebec Agreement of 1943 renewed the partnership of the United States and Great Britain in the nuclear enterprise; Groves was justifying it as an exchange for British help in securing the Union Minière's agreement to sell the two nations all its uranium ore. The British-American relation was built on deeper foundations than that, and Conant moved quickly to limit the damage of Groves' blunder:

> Some of us then pointed out the historic background and [that] our connection with England flowed from the original agreement as to the complete exchange of scientific information. . . . I can foresee a great deal of trouble on this front. It was interesting that Mr. Byrnes felt that Congress would be most curious about this phase of the matter.

If Byrnes had begun his service on the Interim Committee respecting the men who had carried the Manhattan Project forward, he must have conceived less respect for them now. Both Stimson and Bush, Conant told Byrnes, had talked to Churchill in Quebec. If, as it seemed, they could be conned by the British into giving away the secrets of the bomb—whatever

Byrnes imagined those might be—for the price of a few tons of uranium ore, how much was their judgment worth? Why give away something so stupendous as the bomb unless you got something equally stupendous in return? Byrnes believed international relations worked like domestic politics. The bomb was power, newly minted, and power was to politics as money was to banking, a medium of enriching exchange. Only naïfs and fools gave it away.

Enter Leo Szilard.

As the man who had thought longer and harder than anyone else about the consequences of the chain reaction, Szilard had chafed at his continuing exile from the high councils of government. Another politically active Met Lab scientist, Eugene Rabinowitch, a younger man, confirms "the feeling which was certainly shared . . . by others that we were surrounded by a kind of soundproof wall so that you could write to Washington or go to Washington and talk to somebody but you never got any reaction back." With the successful operation of the production reactors and separation plants at Hanford the work of the Met Lab had slowed; Compton's people, Szilard particularly, found time to think about the future. Szilard says he began to examine "the wisdom of testing bombs and using bombs." Rabinowitch remembers "many hours spent walking up and down the Midway [the wide World's Fair sward south of the University of Chicago main campus] with Leo Szilard and arguing about these questions and about what can be done. I remember sleepless nights."

There was no point in talking to Groves, Szilard reasoned in March 1945, nor to Bush or Conant for that matter. Secrecy barred discussion with middle-level authorities. "The only man with whom we were sure we would be entitled to communicate," Szilard recalls, "was the President." He prepared a memorandum for Franklin Roosevelt and traveled to Princeton to enlist once again the durable services of Albert Einstein.

Except for some minor theoretical calculations for the Navy, Einstein had been excluded from wartime nuclear development. Bush explained why to the director of the Institute for Advanced Study early in the war:

> I am not at all sure that if I place Einstein in entire contact with his subject he would not discuss it in a way that it should not be discussed. . . . I wish very much that I could place the whole thing before him . . . but this is utterly impossible in view of the attitude of people here in Washington who have studied into his whole history.

The great theoretician whose letter to Roosevelt helped alert the United States government to the possibility of an atomic bomb was thus spared by concern for security and by hostility to his earlier outspoken politics—his pacifism and probably also his Zionism—from contributing to that weapon's development. Szilard could not show Einstein his memorandum. He told his old friend simply that there was trouble ahead and asked for a letter of introduction to the President. Einstein complied.

From Chicago Szilard approached Roosevelt through his wife. Eleanor Roosevelt agreed to see him on May 8 to pursue the matter. Thus fortified, he wandered to Arthur Compton's office to confess his out-of-channel sins. Compton surprised him by cheering him on. "Elated by finding no resistance where I expected resistance," Szilard reports, "I went back to my office. I hadn't been in my office for five minutes when there was a knock on the door and Compton's assistant came in, telling me that he had just heard over the radio that President Roosevelt had died. . . .

"So for a number of days I was at a complete loss for what to do," Szilard goes on. He needed a new avenue of approach. Eventually it occurred to him that a project as large as the Met Lab probably employed someone from Kansas City, Missouri, Harry Truman's original political base. He found a young mathematician named Albert Cahn who had worked for Kansas City boss Tom Pendergast's political machine to earn money for graduate school. Cahn and Szilard traveled to Kansas City later that month, dazzled Pendergast's hoodlum elite with who knows what Szilardian tale "and three days later we had an appointment at the White House."

Truman's appointments secretary, Matthew Connelly, barred the door. After he read the Einstein letter and the memorandum he relaxed. "I see now," Szilard remembers him saying, "this is a serious matter. At first I was a little suspicious, because the appointment came through Kansas City." Truman had guessed the subject of Szilard's concern. At the President's direction Connelly sent the wandering Hungarian to Spartansburg, South Carolina, to talk to a private citizen named Jimmy Byrnes.

A University of Chicago dean, a scientist named Walter Bartky, had accompanied Szilard to Washington. For added authority Szilard enlisted Nobel laureate Harold Urey and the three men boarded the overnight train south. Compartmentalization was working: "We did not quite understand why we were sent by the President to see James Byrnes. . . . Was he to . . . be the man in charge of the uranium work after the war, or what? We did not know." Truman had alerted Byrnes that the delegation was on its way. The South Carolinian received it warily at his home. He read the letter from Einstein first—"I have much confidence in [Szilard's] judgment," the theoretician of relativity testified—then turned to the memorandum.

It was a prescient document. It argued that in preparing to test and then use atomic bombs the United States was "moving along a road leading to the destruction of the strong position [the nation] hitherto occupied in the world." Szilard was referring not to a moral advantage but to an industrial: as he wrote elsewhere that spring, U.S. military strength was "essentially due to the fact that the United States could outproduce every other country in heavy armaments." When other countries acquired nuclear weapons, as they would in "just a few years," that advantage would be lost: "Perhaps the greatest immediate danger which faces us is the probability that our 'demonstration' of atomic bombs will precipitate a race in the production of these devices between the United States and Russia."

Much of the rest of the memorandum asked the sort of questions the Interim Committee was also asking about international controls versus attempting to maintain an American monopoly. But Szilard echoed Bohr in pleading for what no one among the national leaders concerned with the problem seemed able to grasp, that "these decisions ought to be based not on the *present* evidence relating to atomic bombs, but rather on the situation which can be expected to confront us in this respect a few years from now." By present evidence the bombs were modest and the United States held them in monopoly; the difficulty was deciding what the future would bring. Szilard first offended Byrnes in his memorandum by concluding that "this situation can be evaluated only by men who have first-hand knowledge of the facts involved, that is, by the small group of scientists who are actively engaged in this work." Having thus informed Byrnes that he thought him unqualified, Szilard then proceeded to tell him how his inadequacies might be corrected:

> If there were in existence a small subcommittee of the Cabinet (having as its members the Secretary of War, either the Secretary of Commerce or the Secretary of the Interior, a representative of the State Department, and a representative of the President, acting as the secretary of the Committee), the scientists could then submit to such a committee their recommendations.

It was H. G. Wells' Open Conspiracy emerging again into the light; it amused Byrnes, a man who had climbed to the top across forty-five years of hard political service, not at all:

> Szilard complained that he and some of his associates did not know enough about the policy of the government with regard to the use of the bomb. He felt that scientists, including himself, should discuss the matter with the Cabinet, which I did not feel desirable. His general demeanor and his desire to participate in policy making made an unfavorable impression on me.

Byrnes proceeded to demonstrate the dangers of a lack of firsthand knowledge, Szilard remembers:

> When I spoke of my concern that Russia might become an atomic power, and might become an atomic power soon, if we demonstrated the power of the bomb and if we used it against Japan, his reply was, "General Groves tells me there is no uranium in Russia."

So Szilard explained to Byrnes what Groves, busy buying up the world supply of high-grade ore, apparently did not understand: that high-grade deposits are necessary for the extraction of so rare an element as radium but that low-grade ores, which undoubtedly existed in the Soviet Union, were entirely satisfactory where so abundant an element as uranium was concerned.

To Szilard's argument that using the atomic bomb, even testing the atomic bomb, would be unwise because it would disclose that the weapon existed, Byrnes took a turn at teaching the physicist a lesson in domestic politics:

> He said we had spent two billion dollars on developing the bomb, and Congress would want to know what we had got for the money spent. He said, "How would you get Congress to appropriate money for atomic energy research if you do not show results for the money which has been spent already?"

But Byrnes' most dangerous misunderstanding from Szilard's point of view was his reading of the Soviet Union:

> Byrnes thought that the war would be over in about six months. . . . He was concerned about Russia's postwar behavior. Russian troops had moved into Hungary and Rumania, and Byrnes thought it would be very difficult to persuade Russia to withdraw her troops from these countries, that Russia might be more manageable if impressed by American military might, and that a demonstration of the bomb might impress Russia. I shared Byrnes' concern about Russia's throwing around her weight in the postwar period, but I was completely flabbergasted by the assumption that rattling the bomb might make Russia more manageable.

Shadowed by one of Groves' ubiquitous security agents, the three discouraged men caught the next train back to Washington.

There on the same day the Target Committee was meeting, this time with Paul Tibbets as well as Tolman and Parsons on hand. Much of the discussion concerned Tibbets' training program for the 509th Composite Group. He had sent his best crews to Cuba for six weeks to give them radar experience and flying time over water. "On load and distance tests," the committee minutes report, "Col. Tibbets stated crews had taken off at 135,-000 lbs. gross load, flown 4300 miles with 10,000 lb. bomb load, bombed from 32,000 ft. and returned to base with 900 gallons of fuel. This is in excess of the expected target run and further tests will reduce the loading to reach the S.O.P. [standard operating procedure] of 500 gallons of fuel on return." The 509th was in the process of staging out to Tinian. Pumpkin production was increasing; nineteen had been shipped to Wendover and some of them dropped.

LeMay was also keeping busy. "The 3 reserved targets for the first unit of this project were announced. With current and prospective rate of [Twentieth Air Force] H.E. bombing, it is expected to complete strategic bombing of Japan by 1 Jan 46 so availability of future targets will be a problem." If the Manhattan Project did not hurry, that is, there would be no cities left in Japan to bomb.

Kyoto, Hiroshima and Niigata were the three targets reserved. The committee completed its review by abandoning any pretension that its objectives there were military:

> The following conclusions were reached:
>
> (1) not to specify aiming points, this is to be left to later determination at base when weather conditions are known.
> (2) to neglect location of industrial areas as pin point target, since on these three targets such areas are small, spread on fringes of cities and quite dispersed.
> (3) to endeavor to place first gadget in center of selected city; that is, not to allow for later 1 or 2 gadgets for complete destruction.

And that was that; the Target Committee would schedule no more meetings but would remain on call.

Stimson abhorred bombing cities. As he wrote in his third-person memoir after the war, "for thirty years Stimson had been a champion of international law and morality. As soldier and Cabinet officer he had repeatedly argued that war itself must be restrained within the bounds of humanity. . . . Perhaps, as he later said, he was misled by the constant talk of 'precision bombing,' but he had believed that even air power could be limited in its use by the old concept of 'legitimate military targets.' " Fire-bombing was "a kind of total war he had always hated." He seems to have conceived the idea that even the atomic bomb could be somehow humanely applied, as he discussed with Truman on May 16:

> I am anxious to hold our Air Force, so far as possible, to the "precision" bombing which it has done so well in Europe. I am told that it is possible and adequate. The reputation of the United States for fair play and humanitarianism is the world's biggest asset for peace in the coming decades. I believe the same rule of sparing the civilian population should be applied, as far as possible, to the use of any new weapons.

But the Secretary of War had less control over the military forces he was delegated to administer than he would have liked, and nine days later, on May 25, 464 of LeMay's B-29's—nearly twice as many as flew the first low-level March 9 incendiary raid—once again successfully burned out nearly sixteen square miles of Tokyo, although the Strategic Bombing Survey asserts that only a few thousand Japanese were killed compared to the 86,000 it totals for the earlier conflagration. The newspapers made much of the late-May fire raid; Stimson was appalled.

On May 30 Groves crossed the river from his Virginia Avenue offices and hove into view. Stimson's frustration at the bombing of Japanese cities ignited a fateful exchange, as the general later told an interviewer:

> I was over in Mr. Stimson's office talking to him about some matter in connection with the bomb when he asked me if I had selected the targets yet. I replied that I had that report all ready and I expected to take it over to General Marshall the following morning for his approval. Mr. Stimson then said: "Well, your report is all finished, isn't it?" I said: "I haven't gone over it yet, Mr. Stimson. I want to be sure that I've got it just right." He said: "Well, I would like to see it" and I said: "Well, it's across the river and it would take a long time to get it." He said: "I have all day and I know how fast your office operates. Here's a phone on this desk. You pick it up and you call your office and have them bring that report over." Well, it took about fifteen or twenty minutes to get that report there and all the time I was stewing and fretting internally over the fact that I was shortcutting General Marshall. . . . But there was nothing I could do and when I protested slightly that I thought it was something that General Marshall should pass on first, Mr. Stimson said: "This is one time I'm going to be the final deciding authority. Nobody's going to tell me what to do on this. On this matter I am the kingpin and you might just as well get that report over here." Well in the meantime he asked me what cities I was planning to bomb, or what targets. I informed him and told him that Kyoto was the preferred target. It was the first one because it was of such size that we would have no question about the effects of the bomb. . . . He immediately said: "I don't want Kyoto bombed." And he went on to tell me about its long history as a cultural center of Japan, the former ancient capital, and a great many reasons why he did not want to see it bombed. When the report came over and I handed it to him, his mind was made up. There's no question

> about that. He read it over and he walked to the door separating his office from General Marshall's, opened it and said: "General Marshall, if you're not busy I wish you'd come in." And then the Secretary really double-crossed me because without any explanation he said to General Marshall: "Marshall, Groves has just brought me his report on the proposed targets." He said: "I don't like it. I don't like the use of Kyoto."

So Kyoto at least, the Rome of Japan, founded in 793, famous for silk and cloisonné, a center of the Buddhist and Shinto religions with hundreds of historic temples and shrines, would be spared, though Groves would continue to test his superior's resolve in the weeks to come. The Imperial Palace in Tokyo had been similarly spared even as Tokyo was laid waste around it. There were still limits to the destructiveness of war: the weapons were still modest enough to allow such fine discriminations.

The Interim Committee was to meet in full dress with its Scientific Panel on Thursday, May 31, and on Friday, June 1, with its industrial advisers. The Joint Chiefs of Staff prepared the ground for those meetings on May 25 when they issued a formal directive to the Pacific commanders and to Hap Arnold defining U.S. military policy toward Japan in the months to come:

> The Joint Chiefs of Staff direct the invasion of Kyushu (operation OLYMPIC) target date 1 November 1945, in order to:
>
> (1) Intensify the blockade and aerial bombardment of Japan.
> (2) Contain and destroy major enemy forces.
> (3) Support further advances for the purpose of establishing the conditions favorable to the decisive invasion of the industrial heart of Japan.

Truman had not yet signed on for the Japanese invasion. One of his advisers favored a naval blockade to starve the Japanese to surrender. The President would soon tell the Joint Chiefs that he would judge among his options "with the purpose of economizing to the maximum extent possible the loss of American lives." Marshall, with MacArthur concurring from the field, estimated that casualties—killed, wounded and missing—in the first thirty days following an invasion of the southernmost Japanese home island would not exceed 31,000. An invasion of the main island of Honshu across the plain of Tokyo would be proportionately more violent.

When Szilard returned to Washington from South Carolina he looked up Oppenheimer, just arrived in town for the Interim Committee meeting, to lobby him. So hard was the Los Alamos director working to complete the first atomic bombs that Groves had doubted two weeks earlier if he could break free for the May 31 meeting. Oppenheimer would not for the world have missed the chance to advise at so high a level. But his candid vision of the future of the weapon he was building was as unromantic as his understanding of its immediate necessity was, in Szilard's view, misinformed:

> I told Oppenheimer that I thought it would be a very serious mistake to use the bomb against the cities of Japan. Oppenheimer didn't share my view. He surprised me by starting the conversation by saying, "The atomic bomb is shit." "What do you mean by that?" I asked him. He said, "Well, this is a weapon which has no military significance. It will make a big bang—a very big bang—but it is not a weapon which is useful in war." He thought that it would be important, however, to inform the Russians that we had an atomic bomb and that we intended to use it against the cities of Japan, rather than taking them by surprise. This seemed reasonable to me.... However, while this was necessary it was certainly not sufficient. "Well," Oppenheimer said, "don't you think that if we tell the Russians what we intend to do and then use the bomb in Japan, the Russians will understand it?" And I remember that I said, "They'll understand it only too well."

Stimson's insomnia troubled him on the night of May 30 and he arrived at the Pentagon the next morning feeling miserable. His committee assembled at 10 A.M. Marshall, Groves, Harvey Bundy and another aide attended by invitation, but Stimson's attention was focused on the four scientists, three of them Nobel laureates. The elderly Secretary of War welcomed them warmly, congratulated them on their accomplishments and was concerned to convince them that he and Marshall understood that the product of their labor would be more than simply an enlarged specimen of ordnance. The handwritten notes he prepared emphasize the awe in which he held the bomb; he was not normally a histrionic man:

> S.1
>
> Its *size* and *character*
> We don't think it *mere* new *weapon*
> *Revolutionary Discovery* of Relation of man to universe
> Great History Landmark like
>   *Gravitation*
>   *Copernican* Theory
> But,
> Bids fair [to be] *infinitely greater,* in *respect* to its *Effect*
>   —on the ordinary affairs of man's life.
> May *destroy* or *perfect* International *Civilization*
> May [be] *Frankenstein or* means for World Peace

Oppenheimer was surprised and impressed. When Roosevelt died, he told an audience late in life, he had felt "a terrible bereavement . . . partly because we were not sure that anyone in Washington would be thinking of what needed to be done in the future." Now he saw that "Colonel Stimson was thinking hard and seriously about the implications for mankind of the thing we had created and the wall into the future that we had breached." And though Oppenheimer knew Stimson had never sat down to talk with Niels Bohr, the Secretary seemed to be speaking in terms derived at some near remove from Bohr's understanding of the complementarity of the bomb.

After Stimson's introduction Arthur Compton offered a technical review of the nuclear business, concluding that a competitor would need perhaps six years to catch up with the United States. Conant mentioned the thermonuclear and asked Oppenheimer what gestation period that much more violent mechanism would require; Oppenheimer estimated a minimum of three years. The Los Alamos director took the floor then to review the explosive forces involved. First-stage bombs, he said, meaning crude bombs like Fat Man and Little Boy, might explode with blasts equivalent to 2,000 to 20,000 tons of TNT. That was an upward revision of the estimate Bethe had supplied the Target Committee at Los Alamos in mid-May. Second-stage weapons, Oppenheimer went on—meaning presumably advanced fission weapons with improved implosion systems—might be equal to 50,000 to 100,000 tons of TNT. Thermonuclear weapons might range from 10 million to 100 million tons TNT equivalent.

These were numbers most of the men in the room had seen before and were inured to. Apparently Byrnes had not; they worried him gravely: "As I heard these scientists . . . predict the destructive power of the weapon, I was thoroughly frightened. I had sufficient imagination to visualize the danger to our country when some other country possessed such a weapon." For now the President's personal representative bided his time.

Entirely in energetic character, Ernest Lawrence spoke up for staying ahead of the rest of the world by knowing more and doing more than any other country. He made explicit a future course for the nation about which the previous record of all the meetings and deliberations is oddly silent, a course based on assumptions diametrically opposite to Oppenheimer's profound insight that the atomic bomb was shit:

> Dr. Lawrence *recommended* that a program of plant expansion be vigorously pursued and at the same time a sizable stock pile of bombs and material should be built up. . . . Only by vigorously pursuing the necessary plant expansion and fundamental research . . . could this nation stay out in front.

That was a prescription for an arms race as soon as the Soviet Union took up the challenge. Arthur Compton immediately signed on. So did his brother Karl. Oppenheimer contented himself with a footnote about materials allocation. Stimson eventually summarized the discussion:

1. *Keep our industrial plant intact.*
2. *Build up sizeable stock piles of material for military use and for industrial and technical use.*
3. *Open the door to industrial development.*

Oppenheimer demurred that the scientists should be released to return to their universities and get back to basic science; during the war, he said, they had been plucking the fruits of earlier research. Bush emphatically agreed.

The committee turned to the question of international control and Oppenheimer took the lead. His exact words do not survive, only their summary in the meeting notes kept by the young recording secretary, Gordon Arneson, but if that summary is accurate, then Oppenheimer's emphasis was different from Bohr's and misleading:

> *Dr. Oppenheimer* pointed out that the immediate concern had been to shorten the war. The research that had led to this development had only opened the door to future discoveries. Fundamental knowledge of this subject was so widespread throughout the world that early steps should be taken to make our developments known to the world. He thought it might be wise for the United States to offer to the world free interchange of information with particular emphasis on the development of peace-time uses. The basic goal of all endeavors in the field should be the enlargement of human welfare. If we were to offer to exchange information before the bomb was actually used, our moral position would be greatly strengthened.

Where was Bohr's understanding that the bomb was a source of terror but for that very reason also a source of hope, a means of welding together nations by their common dread of a menacing nuclear standoff? The problem was not exchanging information to improve America's moral standing; the problem was leaders sitting down and negotiating a way beyond the mutual danger the new weapons would otherwise install. The opening up would emerge *out of* those negotiations, necessarily, to guarantee safety; it could not in the real world of secrecy and suspicion realistically *precede* them. In 1963, lecturing on Bohr, Oppenheimer understood well enough the fundamental weakness of his proposal:

> Bush and Compton and Conant were clear that the only future they could envisage with hope was one in which the whole development would be interna-

> tionally controlled. Stimson understood this; he understood that it meant a very great change in human life; and he understood that the central problem at that moment lay in our relations with Russia. . . . But there were differences: Bohr was for action, for timely and responsible action. He realized that it had to be taken by those who had the power to commit and to act. He wanted to change the whole framework in which this problem would appear, early enough so that the problem would be altered by it. He believed in statesmen; he used the word over and over again; he was not very much for committees. The Interim Committee was a committee, and proved itself by appointing another committee, the scientific panel.

No one should presume to judge these men as they struggled with a future that even a mind as fundamental as Niels Bohr's could only barely imagine. But if Robert Oppenheimer ever had a chance to present Bohr's case to those who had the power to commit and to act he had it that morning. He did not speak the Dane's hard plain truths. He spoke instead as Aaron to Bohr's Moses. And Bohr, though he waited nearby in Washington, had not been invited to appear in the star chamber of that darkly paneled room.

Even Stimson thought Oppenheimer's proposals misguided. He asked immediately "what would be the position of democratic governments as against totalitarian regimes under such a program of international control coupled with scientific freedom"—as if opening up the world would leave either democratic or totalitarian nations unchanged, a confusion that Oppenheimer's confusion inspired. Which led to further confusion: "The Secretary said . . . it was his own feeling that the democratic countries had fared pretty well in this war. *Dr. Bush* endorsed this view vigorously." Bush then unwittingly outlined a domestic model of what Bohr's larger open world might be: "He said that our tremendous advantage stemmed in large measure from our system of team work and free interchange of information." And promptly lapsed back into Stimson's extended status quo: "He expressed some doubt, however, of our ability to remain ahead permanently if we were to turn over completely to the Russians the results of our research under free competition with no reciprocal exchange."

Odder and odder, and Byrnes sitting among them trying to imagine a weapon equivalent to 100 million tons of TNT, trying to imagine what it would mean to possess such a weapon and listening to these highly educated men, men almost entirely of the Eastern establishment, of Harvard and MIT and Princeton and Yale, blithely proposing, it seemed, to give away the knowledge of how to make such a weapon.

Stimson left to attend a White House ceremony and they went on to speak of Russia, which Byrnes knew as an advancing brutality currently devouring Poland, and Oppenheimer again took the lead:

> *Dr. Oppenheimer* pointed out that Russia had always been very friendly to science and suggested that we might open up this subject with them in a tentative fashion and in the most general terms without giving them any details of our productive effort. He thought we might say that a great national effort had been put into this project and express hope for cooperation with them in this field. He felt strongly that we should not prejudge the Russian attitude in this matter.

Oppenheimer found an ally then in George Marshall, who "discussed at some length the story of charges and counter-charges that have been typical of our relations with the Russians, pointing out that most of these allegations have proven unfounded." Marshall thought Russia's reputation for being uncooperative "stemmed from the necessity of maintaining security." He believed a way to begin was to forge "a combination among like-minded powers, thereby forcing Russia to fall in line by the very force of this coalition." Such bulldozing had worked in the gunpowder days now almost past but it would not work in the days of the bomb; that power would be big enough, as Oppenheimer's estimates clarified, to make one nation alone a match for the world.

The surprise of the morning was perhaps Marshall's idea for an opening to Moscow: "He raised the question whether it might be desirable to invite two prominent Russian scientists to witness the [Trinity] test." Groves must have winced; after the years of secrecy, after the thousands of numb man-hours of security work, that would be a renunciation worthy of Bohr himself.

Byrnes had heard enough. He had sat behind Franklin Roosevelt at Yalta making notes. In all but the formalities he outranked even Henry Stimson. He put his foot down and the seasoned committeemen moved smoothly into line:

> *Mr. Byrnes* expressed a fear that if information were given to the Russians, even in general terms, Stalin would ask to be brought into the partnership. He felt this to be particularly likely in view of our commitments and pledges of cooperation with the British. In this connection *Dr. Bush* pointed out that even the British did not have any of our blue prints on plants. *Mr. Byrnes* expressed the view, *which was generally agreed to by all present,* that the most desirable program would be to push ahead as fast as possible in production and research to make certain that we stay ahead and at the same time make every effort to better our political relations with Russia.

When Stimson returned, Compton summed up the sense of the crucial discussion the Secretary of War had missed—"the need for maintaining ourselves in a position of superiority while at the same time working toward adequate political agreements." Marshall left them for duty and the rest of the committee trooped off to lunch.

They sat at adjoining tables in a Pentagon dining room. They were a civilian committee; separate conversations converged on the same question, only briefly mentioned during the morning and not taken up: was there no way to let this cup pass from them? Must Little Boy be dropped on the Japanese in surprise? Could their stubborn enemy not be warned in advance or a demonstration arranged?

Stimson, at the focus of one conversation (Byrnes the center of the other), may have spoken then of his outrage at the mass murder of civilians and his complicity; Oppenheimer remembered such a statement at some time during the day and lunch was the only unstructured occasion:

> [Stimson emphasized] the appalling lack of conscience and compassion that the war had brought about . . . the complacency, the indifference, and the silence with which we greeted the mass bombings in Europe and, above all, Japan. He was not exultant about the bombings of Hamburg, of Dresden, of Tokyo. . . . Colonel Stimson felt that, as far as degradation went, we had had it; that it would take a new life and a new breath to heal the harm.

The only recorded response to Stimson's *mea culpa* is Oppenheimer's admiration for it, but there were a number of responses to the question of warning the Japanese or demonstrating the atomic bomb. Oppenheimer could not think of a suitably convincing demonstration:

> You ask yourself would the Japanese government as then constituted and with divisions between the peace party and the war party, would it have been influenced by an enormous nuclear firecracker detonated at a great height doing little damage and your answer is as good as mine. I don't know.

Since the Secretary of State–designate had power to commit and to act, the significant responses to the question are Byrnes'. In a 1947 memoir he recalled several:

> We feared that, if the Japanese were told that the bomb would be used on a given locality, they might bring our boys who were prisoners of war to that area. Also, the experts had warned us that the static test which was to take place in New Mexico, even if successful, would not be conclusive proof that a bomb would explode when dropped from an airplane. If we were to warn the

> Japanese of the new highly destructive weapon in the hope of impressing them and if the bomb then failed to explode, certainly we would have given aid and comfort to the Japanese militarists. Thereafter, the Japanese people probably would not be impressed by any statement we might make in the hope of inducing them to surrender.

In a later television interview he emphasized a more political concern: "The President would have had to take the responsibility of telling the world that we had this atomic bomb and how terrific it was . . . and if it didn't prove out what would have happened to the way the war went God only knows."

Someone among the assembled, Ernest Lawrence remembers, concluded that the "number of people that would be killed by the bomb would not be greater in general magnitude than the number already killed in fire raids," making those slaughters a baseline, as indeed before the awful potential of the new weapon they were.

These troubled men returned to Stimson's office and spent most of the afternoon considering the effect of the bombing on the Japanese and their will to fight. Someone unnamed chose to discredit the atomic bomb's destructiveness, asserting it "would not be much different from the effect caused by any Air Corps strike of present dimensions." Oppenheimer defended his creation's pyrotechnics, citing the electromagnetic and nuclear radiation it would expel:

> *Dr. Oppenheimer* stated that the visual effect of an atomic bombing would be tremendous. It would be accompanied by a brilliant luminescence which would rise to a height of 10,000 to 20,000 feet. The neutron effect of the explosion would be dangerous to life for a radius of at least two-thirds of a mile.

It was probably during this afternoon discussion that Oppenheimer reported an estimate prepared at Los Alamos of how many deaths an atomic bomb exploded over a city might cause. Arthur Compton remembers the number as 20,000, an estimate based on the assumption, he says, that the city's occupants would seek shelter when the air raid began and before the bomb went off. He recalls Stimson bringing up Kyoto then, "a city that must not be bombed." The Secretary still insisted passionately that "the objective was military damage . . . not civilian lives."

The contradiction in Stimson's caveat persisted into his summary of the afternoon's findings, which he offered before he left the meeting at three thirty:

> After much discussion concerning various types of targets and the effects to be produced, *the Secretary expressed the conclusion, on which there was general*

> *agreement, that we could not give the Japanese any warning; that we could not concentrate on a civilian area; but that we should seek to make a profound psychological impression on as many of the inhabitants as possible. At the suggestion of Dr. Conant the Secretary agreed that the most desirable target would be a vital war plant employing a large number of workers and closely surrounded by workers' houses.*

Which had been the general formula in Europe, but according to Curtis LeMay the Japanese worked at home, as families:

> We were going after military targets. No point in slaughtering civilians for the mere sake of slaughter. Of course there is a pretty thin veneer in Japan, but the veneer was there. It was their system of dispersal of industry. All you had to do was visit one of those targets after we'd roasted it, and see the ruins of a multitude of tiny houses, with a drill press sticking up through the wreckage of every home. The entire population got into the act and worked to make those airplanes or munitions of war . . . men, women, children. We knew we were going to kill a lot of women and kids when we burned [a] town. Had to be done.

Stimson had now left the meeting. Arthur Compton wanted to talk about problems at the Met Lab. Before that final discussion the spirit of Leo Szilard bustled through the room. Groves had just learned of another round of Szilardian conspiracy. The general was wrathful: "*General Groves* stated that the program has been plagued since its inception by the presence of certain scientists of doubtful discretion and uncertain loyalty." Szilard had traveled on to New York after talking to Oppenheimer and that very morning had looked up Boris Pregel, the Russian-born French metals speculator and bon vivant who had helped out in the early Columbia days and whose mine on Great Bear Lake supplied the Manhattan Project with uranium ore. On May 16 Szilard had sent Pregel a version of his Truman memorandum. (Groves knew all this from what he calls "secret intelligence sources.") Meeting with Pregel fresh from the May 28 meeting with Byrnes, Szilard had "expressed the opinion," says Groves, "that someone high in the Government [i.e., Byrnes] had been completely misinformed as to [Russian] sources of ore by the [U.S.] Army. He claimed that the misinformation was given intentionally." Two could play at sniffing conspiracy, and even in the midst of debate on the necessity of total death in total war, they did.

The next morning, June 1, the Interim Committee met with four industrialists. Walter S. Carpenter, the president of Du Pont, estimated that the Soviet Union would need "at least four or five years" to construct a plutonium production facility like Hanford. James White, president of Tennessee Eastman, "doubted whether Russia would be able to secure sufficient precision in its equipment to make [an electromagnetic separation plant] possible" at all. George Bucher, the president of Westinghouse, thought that if the Soviets acquired the services of German technicians and scientists they might build an electromagnetic operation in three years. A vice president of Union Carbide, James Rafferty, offered the longest odds: ten years to build a gaseous-diffusion plant from the ground up—but only three years if the Soviets ferreted out barrier technology by espionage.

Mentally Byrnes added processing time to plant construction: "I concluded that any other government would need from seven to ten years, at least, to produce a bomb." From a political point of view seven years was a millennium.

Stimson still quailed at destroying entire cities with atomic bombs. In the afternoon, absenting himself from the Interim Committee discussions, he distanced that horror by pursuing the precision-bombing question further with Hap Arnold, whom he says he "sternly questioned." "I told him of my promise from [War Department Undersecretary for Air Robert] Lovett that there would be only precision bombing in Japan. . . . I wanted to know what the facts were." Arnold told Stimson the one about dispersed Japanese industry. Area bombing was the only way to get at all those drill presses. "He told me, however, that they were trying to keep it down as far as possible." Stimson was willing a few days later to pass that tale along to Truman, with a brace of ambivalent motives thrown in for good measure:

> I told him how I was trying to hold the Air Force down to precision bombing but that with the Japanese method of scattering its manufacture it was rather difficult to prevent area bombing. I told him I was anxious about this feature of the war for two reasons: First, because I did not want to have the United States get the reputation for outdoing Hitler in atrocities; and second, I was a little fearful that before we could get ready, the Air Force might have Japan so thoroughly bombed out that the new weapon would not have a fair background to show its strength. He said he understood.

While Stimson was away Byrnes swiftly and decisively co-opted the committee. "Mr. Byrnes felt that it was important there be a final decision on the question of the use of the weapon," recording secretary Arneson recalled after the war. He described the decision-making process in the minutes he took on June 1:

> *Mr. Byrnes recommended,* and the Committee *agreed,* that the Secretary of War should be advised that, while recognizing that the final selection of the target was essentially a military decision, the present view of the Committee

> was that the bomb should be used against Japan as soon as possible; that it be used on a war plant surrounded by workers' homes; and that it be used without prior warning.

It remained to carry the decision to the President for endorsement. Byrnes headed straight for the White House as soon as the Interim Committee adjourned:

> I told the President of the final decision of his Interim Committee. Mr. Truman told me he had been giving serious thought to the subject for many days, having been informed as to the investigation of the committee and the consideration of alternative plans, and that with reluctance he had to agree that he could think of no alternative and found himself in accord with what I told him the Committee was going to recommend.

Truman saw his Secretary of War five days later. The President, Stimson noted in his diary, "said that Byrnes had reported to him already about [the Interim Committee's decision] and that Byrnes seemed to be highly pleased with what had been done."

Harry Truman did not give the order to drop the atomic bomb on June 1. But he appears to have made the decision then, with a little help from Jimmy Byrnes.

After the Interim Committee meeting on May 31 Robert Oppenheimer had sought out Niels Bohr. "I was very deeply impressed with General Marshall's wisdom," he remembered in 1963, "and also that of Secretary Stimson; and I went over to the British mission and met Bohr and tried to comfort him; but he was too wise and too worldly to be comforted, and he left for England very soon after that, quite uncertain about what, if anything, would happen."

Before Bohr left, late in June, he attempted one last time to see a high official of the United States government—Stimson—Harvey Bundy sending in a message on June 18 to the Secretary: "Do you want to try and work in a meeting with Professor Bohr, the Dane, before you get away this week?"

At the side of the memorandum, in bold script, whether from exhaustion or impatience or because he understood that the matter had been taken out of his hands, Henry Stimson struck finally: "No."

No one doubted that Little Boy would work if any design would. Otto Frisch's Dragon experiments had proven the efficacy of the fast-neutron chain reaction in uranium. The gun mechanism was wasteful and inefficient but U235 was forgiving. It remained to test implosion. While doing so the physicists could also compare their theory of the progress of such an exotic release of energy with the huge blinding fact. Trinity would be the largest physics experiment ever attempted up to that time.

The hard work of finding a proving ground sufficiently barren and remote and organizing it fell to a compact, close-cropped Harvard experimental physicist named Kenneth T. Bainbridge. His task, the Los Alamos technical history notes, "was one of establishing under conditions of extreme secrecy and great pressure a complex scientific laboratory in a barren desert." Bainbridge was well qualified. From Cooperstown, New York, the son of a wholesale stationer, he had worked under Ernest Rutherford at the Cavendish and had designed and built the Harvard cyclotron that now served the Manhattan Project's purposes on the Hill. He had brought back word of the MAUD Committee report to Vannevar Bush in the summer of 1941 and had worked at MIT and in Great Britain on radar. Robert Bacher had recruited him for Los Alamos in the summer of 1943. Beginning in March 1944 he took charge of Trinity.

He needed a flat, desolate site with good weather, near enough to Los Alamos to make travel convenient but far enough away to obscure obvious connection. From map data he chose eight sites, including a desert training area in southern California, the Texas Gulf sandbar region now known as Padre Island and several barren dry valleys in southern New Mexico. Riding three-quarter-ton weapons carriers with Robert Oppenheimer and a team of Army officers in May 1944, Bainbridge led an exploration of the New Mexico sites through late snow; carrying along food and water and sleeping bags, he remembers, they "followed unmapped ranch trails past deserted areas of dry farming lands beaten by too many years of drought and high winds." For Oppenheimer it was a rare escape from the daily burdens of directing Los Alamos, one he was not able to repeat. Several explorations later Bainbridge chose a flat scrub region some sixty miles northwest of Alamogordo between the Rio Grande and the Sierra Oscura, known ominously from Spanish times as the Jornada del Muerto—the dry and therefore dangerous Dead Man's Trail, the Journey of Death. Two hundred ten miles south of Los Alamos, the Jornada formed the northwest sector of the Alamogordo Bombing Range; with the permission of Second Air Force Commander Uzal Ent, Bainbridge staked out an eighteen-by-twenty-four-mile claim.

The demands of the implosion crisis in the autumn of 1944 reduced Trinity's priority, says Bainbridge, "almost to zero . . . until the end of February 1945." With bomb physics well in hand by then Oppenheimer set the test shot's target date at July 4 and Bainbridge got busy. His staff of twenty-five increased across the next five months to more than 250. Herbert Anderson, P. B. Moon, Emilio Segrè and Robert Wilson carried major responsibilities; William G. Penney, Enrico Fermi and especially Victor Weisskopf served as consultants.

The Army leased the David McDonald ranch in the middle of the Jornada site and renovated it for a field laboratory and Military Police station. About 3,400 yards northwest of McDonald Ranch Bainbridge marked out Ground Zero. From that center, at compass points roughly north, west and south at 10,000-yard distances, Corps of Engineers contractors built earth-sheltered bunkers with concrete slab roofs supported by oak beams thicker than railroad ties. N-10000, 5.7 miles from Zero, would house recording instruments and searchlights; W-10000 would house searchlights and banks of high-speed cameras; S-10000 would serve as the control bunker for the test. Another five miles south beyond S-10000 a Base Camp of tents and barracks took shape.

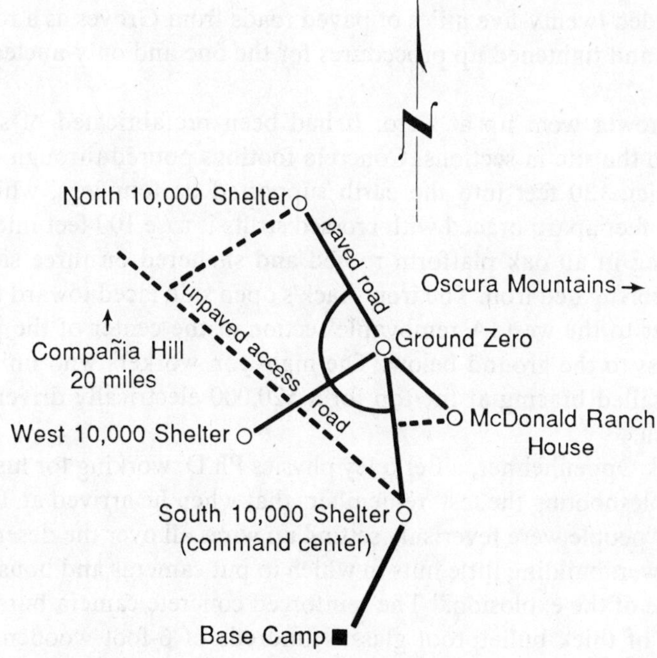

A hill named Compañia twenty miles northwest of Zero on the edge of the Jornada would serve as a VIP scenic overlook. The Oscuras to the east rose more than 4,000 feet above the high alkaline plain.

The Jornada was host to gray hard mesquite, to yucca sharp as the swords of samurai, to scorpions and centipedes men shook in the morning from their boots, to rattlesnakes and fire ants and tarantulas. The MP's hunted antelope with machine guns for fresh meat and for sport. Groves authorized only cold showers for his troops; their isolated duty would win them eventual award for the lowest VD rate in the entire U.S. Army. The well water, fouled with gypsum, made a sovereign purgative. It also stiffened the hair.

Contractors built two towers. One, 800 yards south of Zero, they bolted together 20 feet high in trestles of heavy beams like those that framed the bunkers. It supported a wide platform like an outdoor dance floor and one day in early May the builders returned from a mandated layoff to find it had vanished. Bainbridge had seen it stacked with 100 tons of high explosives in wooden boxes, had packed canisters of dissolved hot Hanford slugs at the center and before dawn on May 7 had blown the entire stack, the largest chemical explosion ever deliberately set off, merely to practice routines and try out instruments. The dirt roads had caused delays; he demanded twenty-five miles of paved roads from Groves as a result and got them, and tightened up procedures for the one and only nuclear test to come.

The tower went up at Zero. It had been prefabricated of steel and shipped to the site in sections. Concrete footings poured through the hard desert caliche 20 feet into the earth supported its four legs, which were spaced 35 feet apart; braced with crossed struts it rose 100 feet into the air, culminating in an oak platform roofed and sheltered on three sides with sheets of corrugated iron. The iron shack's open side faced toward the camera bunker to the west. A removable section at the center of the platform gave access to the ground below. The high-iron workers who finished the tower installed bracing at the top for a $20,000 electrically driven heavy-duty winch.

Frank Oppenheimer, a Berkeley physics Ph.D. working for his brother now troubleshooting the test, remembers that when he arrived at Trinity in late May "people were feverishly setting up wires all over the desert, building the tower, building little huts in which to put cameras and house people at the time of the explosion." The reinforced concrete camera bunkers had portholes of thick bulletproof glass. Hundreds of 6-foot wooden T-poles strung thick as a loom frame with 500 miles of wire walked away from Zero to the instrument bunkers safe miles beyond; other wires buried underground ran protected inside miles of premium garden hose.

Besides photographic studies three kinds of experiments concerned Bainbridge and his team. One set, by far the most extensive, would measure blast, optical and nuclear effects with seismographs, geophones, ionization chambers, spectrographs, films and a variety of gauges. A second would study the implosion in detail and check the operation of the new exploding-wire detonators Luis Alvarez had invented. Experiments planned by Herbert Anderson to reveal the explosive yield radiochemically made up the third category. Harvard physicist David Anderson (no relation) arranged to acquire two Army tanks for that work and to pressurize them and line them with lead; Herbert Anderson and Fermi meant to ride them close to the crater at Zero immediately after the shot, scoop up some of the radioactive debris with a tethered cup hitched to a rocket fired into the crater and retrieve the material for laboratory measurement. Its ratio of fission products to unfissioned plutonium would reveal the yield.

By May 31 enough plutonium had arrived at Los Alamos from Hanford to begin critical-mass experiments. Seth Neddermeyer's shell-configured core had been abandoned even though thin-walled shells give the highest compressions in implosion. Designing out their hydrodynamic instabilities required calculations too dificult to accomplish by hand. Berkeley theoretician Robert Christy designed a more conservative solid core, two mated hemispheres totaling less than one critical mass that implosion would squeeze to at least double their previous density, shortening the distance that fission neutrons would have to travel between nuclei and rendering the mass supercritical. Frisch's group confirmed the core configuration experimentally on June 24. For the high-density form of Pu the critical mass within a heavy tamper is eleven pounds; even with a nut-sized central hollow to encapsulate an initiator the Trinity core cannot have been larger than a small orange.

Delivery of full-sized molds for the implosion lens segments paced the test; they began arriving in quantity only in June, and on June 30 the committee responsible for deciding the test date moved it back to July 16 at the earliest. Kistiakowsky's group worked night and day at S-Site to make enough lenses. "Most troublesome were the air cavities in the interior of the large castings," he recalled after the war, "which we detected by x-ray inspection techniques but could not repair. More rejects than acceptable castings were usually our unfortunate lot."

Groves met with Oppenheimer and Parsons on June 27 to lay plans for shipping the first atomic bombs to the Pacific. They agreed to send the Little Boy U235 projectile by water and the several U235 target pieces later by air; the shipping program acquired the code name Bronx because of that New York borough's adjacency to Manhattan. The metallurgists at Los Alamos cast one target piece before the end of June and the U235 bullet on July 3. The next day, Independence Day, the Combined Policy Committee met in Washington and the British officially gave their approval, as the Quebec Agreement provided, for the use of atomic bombs on Japan.

Truman had agreed to meet with Stalin and Churchill in the Berlin suburb of Potsdam sometime during the summer; he told Stimson on June 6 that he had succeeded in postponing the conference until July 15 "on purpose," Stimson wrote in his diary, "to give us more time." Though Truman and Byrnes had not yet decided to tell Stalin about the atomic bomb, a successful test would change the Pacific equation; they might not need a Soviet invasion of Manchuria to challenge the Japanese and might therefore have to trade away less in Europe. To make sure the President had news of the test at Potsdam, Groves decided during the first week in July to fix the test date at July 16, subject to the vagaries of the weather. He had learned late in June of the possibility of dangerous radioactive fallout over populated areas of New Mexico—"What are you," he berated the Los Alamos physician who gave him the news, "some kind of Hearst propagandist?"—or he would not have waited even on the weather.

So the shot was set for sometime in mid-July, in the heat of the desert summer when the temperature on the Jornada often burned above 100° late in the day. Oppenheimer wired Arthur Compton and Ernest Lawrence: ANY TIME AFTER THE 15TH WOULD BE A GOOD TIME FOR OUR FISHING TRIP. BECAUSE WE ARE NOT CERTAIN OF THE WEATHER WE MAY BE DELAYED SEVERAL DAYS.

The senior men arranged a betting pool with a one-dollar entry fee, wagering on the explosive yield. Edward Teller optimistically picked 45,000 tons TNT equivalent. Hans Bethe picked 8,000 tons, Kistiakowsky 1,400. Oppenheimer chose a modest 300 tons. Norman Ramsey took a cynical zero. When I. I. Rabi arrived a few days before the test the only bet left was for 18,000 tons; whether or not he believed that might be the Trinity yield, he bought it.

As of July 9 Kistiakowsky did not yet have enough quality lens castings on hand to assemble a complete charge. Oppenheimer further compounded his troubles by insisting on firing a Chinese copy of the gadget a few days before the Trinity shot to test its high-explosive design at full scale with a nonfissionable core. Each unit would require ninety-six blocks of explosive. Kistiakowsky resorted to heroic measures:

> In some desperation, I got hold of a dental drill and, not wishing to ask others to do an untried job, spent most of one night, the week before the Trinity test, drilling holes in some faulty castings so as to reach the air cavities indicated on our x-ray inspection films. That done, I filled the cavities by pouring molten explosive slurry into them, and thus made the castings acceptable. Overnight, enough castings were added to our stores by my labors to make more than two spheres.

"You don't worry about it," he adds fatalistically. "I mean, if fifty pounds of explosives goes in your lap, you won't know it."

Navy Lieutenant Commander Norris E. Bradbury, a brisk, energetic Berkeley physics Ph.D., took charge of assembling the high explosives. On Wednesday, July 11, he met with Kistiakowsky to sort the charges according to their quality. "The castings were personally inspected by Kistiakowsky and Bradbury for chipped corners, cracks, and other imperfections," writes Bainbridge. ". . . Only first-quality castings which were not chipped or which could be easily repaired were used for the Trinity assembly. The remainder of the castings were diverted for the Creutz charge"—so named for Edward Creutz, the physicist who was running the Chinese copy test. The castings were waxy, mottled, brown with varnish. They weighed in total, for each device, about 5,000 pounds.

Everyone felt the pressure of the approaching test. It took its toll. "That last week in many ways dragged," Elsie McMillan remembers; "in many ways it flew on wings. It was hard to behave normally. It was hard not to think. It was hard not to let off steam. We also found it hard not to overindulge in all the natural activities of life." In a letter to Eleanor Roosevelt in 1950 Oppenheimer recalled an odd group delusion:

> Very shortly before the test of the first atomic bomb, people at Los Alamos were naturally in a state of some tension. I remember one morning when almost the whole project was out of doors staring at a bright object in the sky through glasses, binoculars and whatever else they could find; and nearby Kirtland Field reported to us that they had no interceptors which had enabled them to come within range of the object. Our director of personnel was an astronomer and a man of some human wisdom; and he finally came to my office and asked whether we would stop trying to shoot down Venus. I tell this story only to indicate that even a group of scientists is not proof against the errors of suggestion and hysteria.

By then the two small plutonium hemispheres had been cast, and plated against corrosion and to absorb alpha particles with nickel, which made the assembly, as metallurgist Cyril Smith would write, "beautiful to gaze upon." But "an unscheduled change began to be evident three or four days before the scheduled date." Plating solution trapped beneath the plating on the flat faces of the hemispheres began to blister the nickel, spoiling the fit. "For a time," says Smith, "postponement of the whole event was threatened." Completely filing off the blisters would expose the plutonium. The metallurgists salvaged the castings by grinding only partway through the blisters and smoothing the bumpy fit with sheets of gold foil. The core of the first atomic bomb would go to its glory dressed in improvised offerings of nickel and gold.

A tropical air mass moved north over Trinity on July 10, just as the test meteorologist, Caltech-trained Jack M. Hubbard, thirty-nine years old, had predicted. Hubbard had resisted the July 16 date, a Monday, since he first heard of it; he expected bad weather that weekend. The Gulf air suspended salt crystals that diffused a slight haze. On July 12, worrying about Potsdam, Groves confirmed the test for the morning of July 16. Bainbridge passed the word to Hubbard. "Right in the middle of a period of thunderstorms," the meteorologist stormed to his journal, "what son-of-a-bitch could have done this?" Groves had been awarded such scurrilous genealogy before.

The general's decision started Norris Bradbury and his crews of Special Engineering Detachment GI's—SED's, the science-trained recruits were called—assembling the Trinity and Creutz high-explosive charges at two separate canyon sites near Los Alamos mesa that Thursday. They debated filling the small air spaces between the castings with grease. Kistiakowsky decided against such filler, writes Bainbridge, "on the basis that the castings assembled were much better than any previously made and that the air spaces left by the spacer materials were insignificant." The charges, each of which had been X-rayed one last time and numbered, were papered into snugness instead with facial tissue and Scotch tape. The simplified and improved casing of the unit to be tested, which was designated model 1561, differed from the earlier 1222 casing of bolted pentagons; it featured an equatorial band of five segments machined from dural castings to which were bolted large upper and lower domed polar caps. When the explosives that lined the lower hemisphere had been papered into place Bradbury's SED's winched down the heavy tamper sphere of natural uranium, which filled the cavity like the pit in an avocado. The tamper was missing a cylindrical plug; the resulting hole would receive the core assembly. The explosive blocks that formed the upper shell followed next.

For transport to Trinity one set of castings was temporarily left out, replaced by a trapdoor plug through which the core assembly could be positioned in the tamper. The reserved castings—an inner of solid Composition B, an outer lensed—were boxed separately with one spare of each type. The men completed preparing the HE assembly for the slow drive down to Trinity by bagging it in waterproof Butvar plastic, boxing it in a braced shipping crate of knotty pine and lashing the resulting package securely to the bed of a five-ton Army truck. A tarpaulin then muffled its secrets in inconclusive drape.

The plutonium core left the Hill first, at three that Thursday afternoon, shock-mounted in a field carrying case studded with rubber bumpers with a strong wire bail. It rode with Philip Morrison in the backseat of an Army sedan like a distinguished visitor, a carload of armed guards clearing the way ahead and another of pit-assembly specialists bringing up the rear. Morrison also delivered a real and a simulated initiator. At about six o'clock a sunburned young sergeant in a white T-shirt and summer uniform pants carried the plutonium core in its field case into the room at McDonald Ranch where it would spend the night. Guards surrounded the ranchhouse to keep vigil.

For security and to encounter less road traffic the HE assembly would make the trip by night; Kistiakowsky deliberately scheduled that more conspicuous convoy to leave at one minute after midnight on Friday, July 13, to put reverse English on the day's unlucky reputation. He rode in the lead car with the security guards. He soon dozed off and was then startled awake by the scream of the car's siren as the convoy ran through Santa Fe; the Army wanted no late-night drunken drivers rolling out of sidestreets to collide with its truckload of handmade high explosives. Beyond Santa Fe the convoy slowed again to below thirty miles an hour; the haul to Trinity took eight hours and Kistiakowsky got some sleep.

On Friday morning at nine the pit-assembly team gathered in white lab coats at McDonald Ranch to begin the final phase of its work. Brigadier General Thomas Farrell was on hand as Groves' deputy, Robert Bacher as the team's senior adviser. Bainbridge looked in; so did Oppenheimer. The ranchhouse room where the core had spent the night had been thoroughly vacuumed in preparation and its windows sealed against dust with black electrical tape to convert it to a makeshift clean room. On a table there the assemblers spread crisp brown wrapping paper and laid out the pieces of their puzzle: two gold-faced, nickel-plated hemispheres of plutonium, a shiny beryllium initiator hot with polonium alphas and, to confine these crucial elements, the several pieces of plum-colored natural uranium that formed the cylindrical 80-pound plug of tamper. Before assembly began Bacher asked for a receipt from the Army for the material it would soon explode. Los Alamos was officially an extension of the University of California working for the Army under contract and Bacher wanted to document the university's release from responsibility for some millions of dollars' worth of plutonium that would soon be vaporized. Bainbridge thought the ceremony a waste of time but Farrell saw its point and agreed. To relieve the tension Farrell insisted on hefting the hemispheres first to confirm that he was getting good weight. Like polonium but much less intensely, plutonium is an alpha emitter; "when you hold a lump of it in your hand," says Leona Marshall, "it feels warm, like a live rabbit." That gave Farrell pause; he set the hemispheres down and signed the receipt.

The parts were few but the men worked carefully. They nested the initiator between the two plutonium hemispheres; they nested the nickel ball in turn in its hollowed plug of tamper. That required the morning and half the afternoon. Two men lugged the heavy boxed assembly on a barrow out to the car. It arrived in its lethal dignity at Zero at 3:18 P.M.

There Norris Bradbury's crew had been busy with the five-foot sphere of high explosives Kistiakowsky had delivered that morning. At 1 P.M. the truck driver had backed his load under the tower. The men had used a jib winch to lift off the wooden packing crate, had swung it aside and lowered around the sphere a massive set of steel tongs suspended from the main winch anchored one hundred feet up at the top of the tower. With the tongs securing the sphere its two tons were winched up off the truck bed; the driver pulled the truck away and the winch lowered the preassembled unit to a skid set on the asphalt-paved ground. "We were scared to death that we would drop it," Bradbury recalls, "because we didn't trust the hoist and it was the only bomb immediately available. It wasn't that we were afraid of setting it off, but we might damage it in some way." Before they opened the upper polar cap to expose the trapdoor plug they erected a white tent over the assembly area; thereafter a diffused glow of sunlight illuminated their work.

Inserting the plug courted disaster, team member Boyce McDaniel remembers:

> The [high-explosive] shell was incomplete, one of the lenses was missing. It was through this opening that the cylindrical plug containing the plutonium and initiator was to be inserted. . . . In order to maximize the density of the uranium in the total assembly, the clearance between the plug and the spherical shell had been reduced to a few thousandths of an inch. Back at Los Alamos, three sets of these plugs and [tamper spheres] had been made. However, in the haste of last minute production, the various units had not been made interchangeable, so not all of the plugs would fit into all [holes]. Great care had been exercised to make sure, however, that mating pieces had been shipped to [Trinity].
>
> Imagine our consternation when, as we started to assemble the plug in the hole, deep down in the center of the high explosive shell, it would not enter! Dismayed, we halted our efforts in order not to damage the pieces, and stopped to think about it. Could we have made a mistake. . . ?

Bacher saw the cause and calmed them: the plug had warmed and expanded in the hot ranchhouse but the tamper, set deep within the insulation of its shell of high explosives, was still cool from Los Alamos. The men left the two pieces of heavy metal in contact and took a break. When they checked the assembly again the temperatures had equalized. The plug slid smoothly into place.

Then it was the turn of the explosives crew. Oppenheimer watched over them, conspicuous in his pork-pie hat, wasted to 116 pounds by a recent bout of chicken pox and the stress of months of late nights and seven-day weeks. In the motion picture that documents this historic assembly he darts in and out of the frame like a foraging water bird, pecking at the open well of the bomb. Someone hands Bradbury a strip of Scotch tape and his arms disappear into the well to secure a block of explosive. He finished the work in late evening under lights. The detonators were not yet installed. That would be the next day's challenge after the unit had been hauled to the top of the tower.

The following morning, Saturday, around eight, Bradbury supervised raising the test device to its high platform. The openings into the casing where the detonators would be inserted had been covered and taped to keep out dust; as the bulky sphere rose into the air it revealed itself generously bandaged as if against multiple wounds. It stopped at fifteen feet long enough to allow a crew of GI's to stack depths of striped ticking-covered Army mattresses up nearly to its skid, a prayer in cotton batting against a damaging fall. Then it started up again, twisting slowly, seeming on its thin, braided steel cable to levitate, rising the full height of the tower and diminishing slightly with distance as it rose. Two sergeants received it into the tower shack through the open floor, replaced the floor panel and lowered the unit onto its skid, positioning it with its north and south polar caps at left and right rather than above and below as they had been positioned during assembly, the same posture in which its militant armored twin, Fat Man, would ride to war in the bomb bay of a B-29. The delicate work of inserting the detonators then began.

Disaster loomed again that day. The Creutz group at Los Alamos had fired the Chinese copy, measured the simultaneity of its implosion by the magnetic method and called Oppenheimer to report the dismaying news that the Trinity bomb was likely to fail. "So of course," says Kistiakowsky, "I immediately became the chief villain and everybody lectured me." Groves flew in to Albuquerque in his official plane with Bush and Conant at noon; they were appalled at the news and added their complaints to Kistiakowsky's full burden:

> Everybody at headquarters became terribly upset and focused on my presumed guilt. Oppenheimer, General Groves, Vannevar Bush—all had much to say about that incompetent wretch who forever after would be known to the world as the cause of the tragic failure of the Manhattan Project. Jim

> Conant, a close personal friend, had me on the carpet it seemed for hours, coldly quizzing me about the causes of the impending failure.
>
> Sometime later that day Bacher and I were walking in the desert and as I timidly questioned the results of the magnetic test Bob accused me of challenging no less than Maxwell's equations themselves! At another point Oppenheimer became so emotional that I offered him a month's salary against ten dollars that our implosion charge would work.

In the midst of this contretemps all of Little Boy but its U235 target pieces slipped away. With two Army officers in escort, a closed black truck and seven carloads of security guards left Los Alamos Saturday morning for Kirtland Air Force Base in Albuquerque. A manifest describes the truck's expensive cargo:

> a. 1 box, wt. about 300 lbs, containing projectile assembly of active material for the gun type bomb.
> b. 1 box, wt. about 300 lbs, containing special tools and scientific instruments.
> c. 1 box, wt. about 10,000 lbs, containing the inert parts for a complete gun type bomb.

Two DC-3's waiting at Kirtland flew the crates and their officer escorts to Hamilton Field, near San Francisco, from which another security convoy escorted them to Hunter's Point Naval Shipyard to await the sailing of the U.S.S. *Indianapolis,* the heavy cruiser that would deliver them to Tinian.

At Trinity gloom was everywhere. A physical chemist from Los Alamos, Joseph O. Hirschfelder, remembers Oppenheimer's discomfiture that Saturday evening at the hotel where the guests invited to view the test had begun to assemble: "We drove to the Hilton Hotel in Albuquerque, where Robert Oppenheimer was meeting with a large group of generals, Nobel laureates, and other VIP's. Robert was very nervous. He told [us] about some experimental results which Ed Creutz had obtained earlier in the day which indicated that the [Trinity] atom bomb would be a dud."

Oppenheimer searched for calm in the midst of this latest evidence of the physical world's relentlessness and found a breath of it in the *Bhagavad-Gita,* the seven-hundred-stanza devotional poem interpolated into the great Aryan epic *Mahabharata* at about the same time that Greece was declining from its golden age. He had discovered the *Gita* at Harvard; at Berkeley he had learned Sanskrit from the scholar Arthur Ryder to set himself closer to the original text and thereafter a worn pink copy occupied an honored place on the bookshelf closest to his desk. There are meanings enough for a lifetime in the *Gita,* dramatized as a dialogue between a warrior prince named Arjuna and Krishna, the principal avatar of Vishnu (and

Vishnu the third member of the Hindu godhead with Brahma and Shiva—a Trinity again). Vannevar Bush records the particular meaning Oppenheimer clutched that desperate Saturday in July:

> His was a profoundly complex character. . . . So my comment will be brief. I simply record a poem, which he translated from the Sanscrit, and which he recited to me two nights before [Trinity]:
>
> *In battle, in forest, at the precipice in the mountains,*
> *On the dark great sea, in the midst of javelins and arrows,*
> *In sleep, in confusion, in the depths of shame,*
> *The good deeds a man has done before defend him.*

Back at Base Camp Oppenheimer slept no more than four hours that night; Farrell heard him stirring restlessly on his bunk in the next room of the quarters they shared, racked with coughing. Chain-smoking as much as meditative poetry drove him through his days.

Sturdy Hans Bethe found a way back from the precipice, Kistiakowsky remembers:

> Sunday morning another phone call came with wonderful news. Hans Bethe spent the whole night of Saturday analyzing the electromagnetic theory of this experiment and discovered that the instrumental design was such that even a perfect implosion could not have produced oscilloscope records different from what was observed. So I became again acceptable to local high society.

When Groves called, Oppenheimer chatted happily about the Bethe results. The general interrupted: "What about the weather?" "The weather is whimsical," the whimsical physicist said. The Gulf air mass had stagnated over the test site. But change was coming. Jack Hubbard, the meteorologist, predicted light and variable winds the next day.

Stagnation exacerbated the July heat. Camera crews replacing battery packs damaged by a blown circuit burned their hands on metal camera housings. Frank Oppenheimer, thin enough not to suffer the heat unduly, hurried to construct a last-minute experiment less aloof than readings of light and radiation: he set out boxes filled with excelsior and posts nailed with corrugated iron strips to simulate the fragile Japanese houses where LeMay's ubiquitous drill presses lurked. Groves had forbidden the construction of full-scale housing for the test, more scientific tomfoolery, a waste of money and time. Norris Bradbury's instructions for bomb assembly as of Saturday listed *"Gadget complete"*; for "Sunday, 15 July, all day," he advised his crews to "look for rabbits' feet and four-leaved clovers.

Should we have the Chaplain down there?" Rabbits' feet would turn up, but even chaplains would have had trouble finding a stem of clover on the Jornada.

Oppenheimer, Groves, Bainbridge, Farrell, Tolman and an Army meteorologist met with Hubbard at McDonald Ranch at four that afternoon to consider the weather. Hubbard reminded them that he had never liked the July 16 date. He thought the shot could go as scheduled, he noted in his journal, "in less than optimum conditions, which would require sacrifices." Groves and Oppenheimer repaired to another room to confer. They decided to wait and see. They had scheduled a last weather conference for the next morning at 0200 hours; they would make up their minds then. The shot was set for 0400 and they let that time stand.

Sometime early that evening Oppenheimer climbed the tower to perform a final ritual inspection. There before him crouched his handiwork. Its bandages had been removed and it was hung now with insulated wires that looped from junction boxes to the detonator plugs that studded its dark bulk, an exterior ugly as Caliban's. His duty was almost done.

At dusk the tired laboratory director was calm. He stood with Cyril Smith beside the reservoir at McDonald Ranch where cattle had watered and spoke of families and home, even of philosophy, and Smith found himself soothed. A storm was blowing up. Oppenheimer looked beneath it to anchorage, to the darkening Oscuras. "Funny how the mountains always inspire our work," the metallurgist heard him say.

With the weather changing from stagnant to violent and with everyone short of sleep, moods swung at Base Camp. The occasion of Fermi's satire that evening made Bainbridge furious. It merely irritated Groves:

> I had become a bit annoyed with Fermi . . . when he suddenly offered to take wagers from his fellow scientists on whether or not the bomb would ignite the atmosphere, and if so, whether it would merely destroy New Mexico or destroy the world. He had also said that after all it wouldn't make any difference whether the bomb went off or not because it would still have been a well worth-while scientific experiment. For if it did fail to go off, we would have proved that an atomic explosion was not possible.

On the realistic grounds, the Italian laureate explained with his usual candor, that the best physicists in the world would have tried and failed.

Bainbridge was furious because Fermi's "thoughtless bravado" might scare the soldiers, who did not have the benefit of a knowledge of thermonuclear ignition temperatures and fireball cooling effects. But a new force was about to be loosed on the world; no one could be absolutely certain—

Fermi's point—of the outcome of its debut. Oppenheimer had assigned Edward Teller the deliciously Tellerian task of trying to think of any imaginable trick or turn by which the explosion might escape its apparent bounds. Teller at Los Alamos that evening raised the same question Fermi had, but questioned Robert Serber, no mere uninformed GI:

> Trying to find my way home in the darkness, I bumped into an acquaintance, Bob Serber. That day we had received a memo from our director . . . saying that we would have to be [at Trinity] well before dawn, and that we should be careful not to step on a rattlesnake. I asked Serber, "What will you do tomorrow about the rattlesnakes?" He said, "I'll take a bottle of whiskey." I then went into my usual speech, telling him how one could imagine that things might get out of control in this, that, or a third manner. But we had discussed these things repeatedly, and we could not see how, in actual fact, we could get into trouble. Then I asked him, "And what do you think about it?" There in the dark Bob thought for a moment, then said, "I'll take a second bottle of whiskey."

Rabi, the real mystic among them, spent the evening playing poker.

Bainbridge managed a little sleep. He headed the Arming Party charged with arming the bomb. He was due at Zero by 11 P.M. to prepare the shot. An MP sergeant woke him at ten; he picked up Kistiakowsky and Joseph McKibben, the tall, lanky Missouri-born physicist responsible for running the countdown, and assembled with Hubbard and his crew and two security men. "On the way in," Bainbridge remembers, "I stopped at S 10,000 and locked the main sequence timing switches. Pocketing the key I returned to the car and continued to Point Zero." A young Harvard physicist, Donald Hornig, was busy in the tower. He had designed the 500-pound X-unit of high-voltage capacitors that fired Fat Man's multiple detonators with microsecond simultaneity, a crucial Luis Alvarez invention, and now was disconnecting the unit Bainbridge's crews had used for practice runs and connecting the new unit reserved for the shot. In static test this Fat Man would be fired through cables from the S-10000 control bunker; the one to be shipped to Tinian, self-contained, would carry onboard batteries. Cables or batteries would charge the X-unit and on command it would discharge its capacitors to the detonators, vaporizing wires imbedded in the explosive blocks to start shock waves to set off the HE. "Soon after our arrival," says Bainbridge, "Hornig completed his work and returned to S 10,000. Hornig was the last man to leave the top of the tower."

Hubbard operated a portable weather station at the tower; to measure wind speed and direction the two sergeants who worked with him inflated and released helium balloons. At eleven o'clock he found the wind blowing across Zero toward N-10000. At midnight the Gulf air mass had thickened to 17,000 feet and arranged two inversions—cooler air above warmer—within its layered depths that might loop the radioactive Trinity column back down to the ground directly below.

To an observer traveling toward the desert from Los Alamos "the night was dark with black clouds, and not a star could be seen."

Thunderstorms began lashing the Jornada at about 0200 hours on July 16, drenching Base Camp and S-10000. "It was raining cats and dogs, lightning and thunder," Rabi remembers. "[We were] really scared [that] this object there in the tower might be set off accidentally. So you can imagine the strain on Oppenheimer." Winds gusted to thirty miles an hour. Hubbard hung on at Zero for last-minute readings—only misting drizzle had yet reached the tower area—and arrived eight minutes late for the 0200 weather conference at Base Camp, to find Oppenheimer waiting for him outside the weather center there. Hubbard told him they would have to scrub 0400 but should be able to shoot between 0500 and 0600. Oppenheimer looked relieved.

Inside they found an agitated Groves waiting with his advisers. "What the hell is wrong with the weather?" the general greeted his forecaster. Hubbard took the opportunity to repeat that he had never liked July 16. Groves demanded to know when the storm would pass. Hubbard explained its dynamics: a tropical air mass, night rain. Afternoon thunderstorms took their energy from the heating of the earth and collapsed at sunset; this one, contrariwise, would collapse at dawn. Groves growled that he wanted a specific time, not an explanation. I'm giving you both, Hubbard rejoined. He thought Groves was ready to cancel the shot, which seems unlikely given the pressure from Potsdam. He told Groves he could postpone if he wanted but the weather would relent at dawn.

Oppenheimer applied himself to soothe his bulky comrade. Hubbard was the best man around, he insisted, and they ought to trust his forecast. The others at the meeting—Tolman and two Army meteorologists, one more than before—agreed. Groves relented. "You'd better be right on this," he threatened Hubbard, "or I will hang you." He ordered the meteorologist to sign his forecast and set the shot for 0530. Then he went off to roust the governor of New Mexico out of bed to the telephone to warn him he might have to declare martial law.

Bainbridge at Zero was less concerned with local effects than with distant, even though he had personally locked open the circuits that communicated with the shelters. "Sporadic rain was a disturbing factor," he recalls. ". . . We had none of the lightning reported by those at the Base Camp about 16,000 yards away or at S 10,000, but it made interesting conversation as many of the wires from N, S, W 10,000 ended at the tower." About 0330 a gust of wind at Base Camp collapsed Vannevar Bush's tent; he found his way to the mess hall, where from 0345 the cooks began serving a breakfast of powdered eggs, coffee and French toast.

The gods sent Emilio Segrè happier amusement. He had distracted himself through the evening with André Gide's *The Counterfeiters* and slept through the worst of the Base Camp storm. "But my attention was attracted by an unbelievable noise whose nature escaped me completely. As the noise persisted, Sam Allison and I went out with a flashlight and, much to our surprise, found hundreds of frogs in the act of making love in a big hole that had filled with water."

Hubbard departed Base Camp at 0315 for S-10000. The rain had moved on. He telephoned Zero; one of his men there said the clouds were opening and a few stars shone. By 0400 the wind was shifting toward the southwest, away from the shelters. The meteorologist prepared his final forecast at S-10000. He called Bainbridge at 0440. "Hubbard gave me a complete weather report," the Trinity director recalls, "and a prediction that at 5:30 a.m. the weather at Point Zero would be possible but not ideal. We would have preferred no inversion layer at 17,000 feet but not at the expense of waiting over half a day. I called Oppenheimer and General Farrell to get their agreement that 5:30 a.m. would be T = 0." Hubbard, Bainbridge, Oppenheimer and Farrell each had veto over the shot. They all agreed. Trinity would fire at 0530 hours July 16, 1945—just before dawn.

Bainbridge had arranged to report each step of the final arming process to S-10000 in case anything went wrong. "I drove McKibben to W 900 so that he could throw the timing and sequence switches there while I checked off his list." Back at Zero Bainbridge called in the next step "and threw the special arming switch which was not on McKibben's lines. Until this switch was closed the bomb could not be detonated from S 10,000. The final task was to switch on a string of lights on the ground which were to serve as an 'aiming point' for a B-29 practice bombing run. The Air Force wanted to know what the blast effects would be like on a plane 30,000 feet up and some miles away. . . . After turning on the lights, I returned to my car and drove to S 10,000." Kistiakowsky, McKibben and the security guards rode with him. They were the last to leave the site. Behind them searchlight beams converged on the tower.

The Arming Party arrived at S-10000, the earth-sheltered concrete control bunker, at about 0508. Hubbard gave Bainbridge his signed forecast. "I unlocked the master switches," Bainbridge concludes, "and McKibben started the timing sequence at −20 minutes, 5:09:45 a.m." Oppenheimer would watch the shot from S-10000, as would Farrell, Donald

Hornig and Samuel Allison. With the beginning of the final countdown Groves left by jeep for Base Camp. For protection against common disaster he wanted to be physically separated from Farrell and Oppenheimer.

Busloads of visitors from Los Alamos and beyond had begun arriving at Compañia Hill, the viewing site twenty miles northwest of Zero, at 0200. Ernest Lawrence was there, Hans Bethe, Teller, Serber, Edwin McMillan, James Chadwick come to see what his neutron was capable of and a crowd of other men, including Trinity staff no longer needed down on the plain. "With the darkness and the waiting in the chill of the desert the tension became almost unendurable," one of them remembers. The shortwave radio requisitioned to advise them of the schedule refused to work until after Allison began broadcasting the countdown. Richard Feynman, a future Nobel laureate who had entered physics as an adolescent via radio tinkering, tinkered the radio to life. Men began moving into position. "We were told to lie down on the sand," Teller protests, "turn our faces away from the blast, and bury our heads in our arms. No one complied. We were determined to look the beast in the eye." The radio went dead again and they were left to watch for the warning rockets to be fired from S-10000. "I wouldn't turn away . . . but having made all those calculations, I thought the blast might be rather bigger than expected. So I put on some suntan lotion." Teller passed the lotion around and the strange prophylaxis disturbed one observer: "It was an eerie sight to see a number of our highest-ranking scientists seriously rubbing sunburn lotion on their faces and hands in the pitch-blackness of the night, twenty miles from the expected flash."

The countdown continued at S-10000. At 0525 a green Very rocket went up. That signaled a short wail of the siren at Base Camp. Shallow trenches had been bulldozed below the south rim of the Base Camp reservoir for protection and since these men watched ten miles closer to Zero than the crowd on Compañia Hill they planned to use them. Rabi lay down next to Kenneth Greisen, a Cornell physicist, facing south away from Zero. Greisen remembers that he was "personally nervous, for my group had prepared and installed the detonators, and if the shot turned out to be a dud, it might possibly be our fault." Groves found refuge between Bush and Conant, thinking "only of what I would do if, when the countdown got to zero, nothing happened." Victor Weisskopf remembers that "groups of observers had arranged small wooden sticks at a distance of 10 yds from our observation place in order to estimate the size of the explosion." The sticks were posted on the rim of the reservoir. "They were arranged so that their [height] corresponded to 1000 ft. at zero point." Philip Morrison relayed the countdown to the Base Camp observers by loudspeaker.

The two-minute-warning rocket fizzled. A long wail of the Base Camp siren signaled the time. The one-minute warning rocket fired at 0529. Morrison also meant to look the beast in the eye and lay down on the slope of the reservoir facing Zero. He wore sunglasses and held a stopwatch in one hand and a piece of welder's glass in the other. The welder's glass was stockroom issue: Lincoln Super-visibility Lens, Shade #10.

At S-10000 someone heard Oppenheimer say, "Lord, these affairs are hard on the heart." McKibben had been marking off the minutes and Allison broadcasting them. At 45 seconds McKibben turned on a more precise automatic timer. "The control post was rather crowded," Kistiakowsky notes, "and, having now nothing to do, I left as soon as the automatic timer was thrown in . . . and went to stand on the earth mound covering the concrete dugout. (My own guess was that the yield would be about 1 kt [i.e., 1,000 tons, 1 kiloton], and so five miles seemed very safe.)"

Teller prepared himself further at Compañia Hill: "I put on a pair of dark glasses. I pulled on a pair of heavy gloves. With both hands I pressed the welder's glass to my face, making sure no stray light could penetrate around it. I then looked straight at the aim point."

Donald Hornig at S-10000 monitored a switch that could cut the connection between his X-unit in the tower and the bomb, the last point of interruption if anything went wrong. At thirty seconds before T = 0 four red lights flashed on the console in front of him and a voltmeter needle flipped from left to right under its round glass cover to register the full charging of the X-unit. Farrell noticed that "Dr. Oppenheimer, on whom had rested a very heavy burden, grew tenser as the last seconds ticked off. He scarcely breathed. He held on to a post to steady himself. For the last few seconds, he stared directly ahead."

At ten seconds a gong sounded in the control bunker. The men lying in their shallow trenches at Base Camp might have been laid out for death. Conant told Groves he never imagined seconds could be so long. Morrison studied his stopwatch. "I watched the second-hand until T = −5 seconds," he wrote the day of the shot, "when I lowered my head onto the sand bank in such a way that a slight rise in the ground completely shielded me from Zero. I placed the welding glass over the right lens of my sun glasses, the left lens of which was covered by an opaque cardboard shield. I counted seconds and at zero began to raise my head just over the protecting rise." Ernest Lawrence on Compañia Hill had planned to watch the shot through the windshield of a car, allowing the glass to filter out damaging ultraviolet, "but at the last minute decided to get out . . . (evidence indeed I was excited!)." Robert Serber, his bottles of whiskey to succor him, stared twenty miles toward distant Zero with unprotected eyes. The last decisive inaction was Hornig's:

> Now the sequence of events was all controlled by the automatic timer except that I had the knife switch which could stop the test at any moment up until the actual firing . . . I don't think I have ever been keyed up as I was during those final seconds . . . I kept telling myself "the least flicker of that needle and you have to act." It kept on coming down to zero. I kept saying, "Your reaction time is about half a second and you can't relax for even a fraction of a second." . . . My eyes were glued on the dial and my hand was on the switch. I could hear the timer counting . . . three . . . two . . . one. The needle fell to zero. . . .

Time: 0529:45. The firing circuit closed; the X-unit discharged; the detonators at thirty-two detonation points simultaneously fired; they ignited the outer lens shells of Composition B; the detonation waves separately bulged, encountered inclusions of Baratol, slowed, curved, turned inside out, merged to a common inward-driving sphere; the spherical detonation wave crossed into the second shell of solid fast Composition B and accelerated; hit the wall of dense uranium tamper and became a shock wave and squeezed, liquefying, moving through; hit the nickel plating of the plutonium core and squeezed, the small sphere shrinking, collapsing into itself, becoming an eyeball; the shock wave reaching the tiny initiator at the center and swirling through its designed irregularities to mix its beryllium and polonium; polonium alphas kicking neutrons free from scant atoms of beryllium: one, two, seven, nine, hardly more neutrons drilling into the surrounding plutonium to start the chain reaction. Then fission multiplying its prodigious energy release through eighty generations in millionths of a second, tens of millions of degrees, millions of pounds of pressure. Before the radiation leaked away, conditions within the eyeball briefly resembled the state of the universe moments after its first primordial explosion.

Then expansion, radiation leaking away. The radiant energy loosed by the chain reaction is hot enough to take the form of soft X rays; these leave the physical bomb and its physical casing first, at the speed of light, far in front of any mere explosion. Cool air is opaque to X rays and absorbs them, heating; "the very hot air," Hans Bethe writes, "is therefore surrounded by a cooler envelope, and only this envelope"—hot enough at that—"is visible to observers at a distance." The central sphere of air, heated by the X rays it absorbs, reemits lower-energy X rays which are absorbed in turn at its boundaries and reemitted beyond. By this process of downhill leapfrogging, which is known as radiation transport, the hot sphere begins to cool itself. When it has cooled to half a million degrees—

in about one ten-thousandth of a second—a shock wave forms that moves out faster than radiation transport can keep up. "The shock therefore separates from the very hot, nearly isothermal [i.e., uniformly heated] sphere at the center," Bethe explains. Simple hydrodynamics describes the shock front: like a wave in water, like a sonic boom in air. It moves on, leaving behind the isothermal sphere confined within its shell of opacity, isolated from the outside world, growing only slowly by radiation transport on this millisecond scale of events.

What the world sees is the shock front and it *cools* into visibility, the first flash, milliseconds long, of a nuclear weapon's double flash of light, the flashes too closely spaced to distinguish with the eye. Further cooling renders the front transparent; the world if it still has eyes to see looks *through* the shock wave into the hotter interior of the fireball and "because higher temperatures are now revealed," Bethe continues, "the total radiation increases toward a second maximum": the second, longer flash. The isothermal sphere at the center of the expanding fireball continues opaque and invisible, but it also continues to give up its energy to the air beyond its boundaries by radiation transport. That is, as the shock wave cools, the air behind it heats. A cooling wave moves in reverse of the shock wave, eating into the isothermal sphere. Instead of one simple thing the fireball is thus several things at once: an isothermal sphere invisible to the world; a cooling wave moving inward toward that sphere, eating away its radiation; a shock front propagating into undisturbed air, air that has not yet heard the news. Between each of these parts lay further intervening regions of buffering air.

Eventually the cooling wave eats the isothermal sphere completely away and the entire fireball becomes transparent to its own radiation. Now it cools more slowly. Below about 9000°F it can cool no more. Then, concludes Bethe, "any further cooling can only be achieved by the rise of the fireball due to its buoyancy, and the turbulent mixing associated with this rise. This is a slow process, taking tens of seconds."

The high-speed cameras at W-10000 recorded the later stages of the fireball's development, Bainbridge reports, tracking its huge swelling from the eyeball it had been:

> The expansion of the ball of fire before striking the ground was almost symmetric . . . except for the extra brightness and retardation of a part of the sphere near the bottom, a number of blisters, and several spikes that shot radially ahead of the ball below the equator. Contact with the ground was made at 0.65 ms [i.e., thousandths of a second]. Thereafter the ball became rapidly smoother. . . . Shortly after the spikes struck the ground (about 2 ms) there appeared on the ground ahead of the shock wave a wide skirt of lumpy mat-

> ter. . . . At about 32 ms [when the fireball had expanded to 945 feet in diameter] there appeared immediately behind the shock wave a dark front of absorbing matter, which traveled slowly out until it became invisible at 0.85 s [the expanding front about 2,500 feet across]. The shock wave itself became invisible [earlier] at about 0.10 s. . . .
>
> The ball of fire grew even more slowly to a [diameter] of about [2,000 feet], until the dust cloud growing out of the skirt almost enveloped it. The top of the ball started to rise again at 2 s. At 3.5 s a minimum horizontal diameter, or neck, appeared one-third of the way up the skirt, and the portion of the skirt above the neck formed a vortex ring. The neck narrowed, and the ring and fast-growing pile of matter above it rose as a new cloud of smoke, carrying a convection stem of dust behind it. . . . The stem appeared twisted like a left-handed screw.

But men saw what theoretical physics cannot notice and what cameras cannot record, saw pity and terror. Rabi at Base Camp felt menaced:

> We were lying there, very tense, in the early dawn, and there were just a few streaks of gold in the east; you could see your neighbor very dimly. Those ten seconds were the longest ten seconds that I ever experienced. Suddenly, there was an enormous flash of light, the brightest light I have ever seen or that I think anyone has ever seen. It blasted; it pounced; it bored its way right through you. It was a vision which was seen with more than the eye. It was seen to last forever. You would wish it would stop; altogether it lasted about two seconds. Finally it was over, diminishing, and we looked toward the place where the bomb had been; there was an enormous ball of fire which grew and grew and it rolled as it grew; it went up into the air, in yellow flashes and into scarlet and green. It looked menacing. It seemed to come toward one.
>
> A new thing had just been born; a new control; a new understanding of man, which man had acquired over nature.

To Teller at Compañia Hill the burst "was like opening the heavy curtains of a darkened room to a flood of sunlight." Had astronomers been watching they could have seen it reflected from the moon, literal moonshine.

Joseph McKibben made a comparison at S-10000: "We had a lot of flood lights on for taking movies of the control panel. When the bomb went off, the lights were drowned out by the big light coming in through the open door in the back."

It caught Ernest Lawrence at Compañia Hill in the act of stepping from his car: "Just as I put my foot on the ground I was enveloped with a warm brilliant yellow white light—from darkness to brilliant sunshine in an instant and as I remember I momentarily was stunned by the surprise."

To Hans Bethe at Compañia Hill "it looked like a giant magnesium flare which kept on for what seemed a whole minute but was actually one or two seconds."

Serber at Compañia Hill risked blindness but glimpsed an earlier stage of the fireball:

> At the instant of the explosion I was looking directly at it, with no eye protection of any kind. I saw first a yellow glow, which grew almost instantly to an overwhelming white flash, so intense that I was completely blinded. . . . By twenty or thirty seconds after the explosion I was regaining normal vision. . . . The grandeur and magnitude of the phenomenon were completely breathtaking.

Segrè at Base Camp imagined apocalypse:

> The most striking impression was that of an overwhelmingly bright light. . . . I was flabbergasted by the new spectacle. We saw the whole sky flash with unbelievable brightness in spite of the very dark glasses we wore. . . . I believe that for a moment I thought the explosion might set fire to the atmosphere and thus finish the earth, even though I knew that this was not possible.

Not light but heat disturbed Morrison at Base Camp:

> From ten miles away, we saw the unbelievably brilliant flash. That was not the most impressive thing. We knew it was going to be blinding. We wore welder's glasses. The thing that got me was not the flash but the blinding heat of a bright day on your face in the cold desert morning. It was like opening a hot oven with the sun coming out like a sunrise.

It unfolded in silence, a ballistics expert watching from Compañia Hill realized with awe:

> The flash of light was so bright at first as to seem to have no definite shape, but after perhaps half a second it looked bright yellow and hemispherical with the flat side down, like a half-risen sun but about twice as large. Almost immediately a turgid rising of this luminous mass began, great swirls of flame seeming to ascend within a rather rectangular outline which expanded rapidly in height. . . . Suddenly out of the center of it there seemed to rise a narrower column to a considerably greater height. Then as a climax, which was exceedingly impressive in spite of the fact that the blinding brightness had subsided, the top of the slenderer column seemed to mushroom out into a thick parasol of a rather bright but spectral blue. . . . All this seemed very fast . . .

> and was followed by a feeling of letdown that it was all over so soon. Then came the awe-inspiring realization that it was twenty miles away, that what had flared up and died so brilliantly and quickly was really a couple of miles high. The feeling of the remoteness of this thing which had seemed so near was emphasized by the long silence while we watched the grey smoke grow into a taller and taller twisting column, a silence broken after a minute or so that seemed much longer by a quite impressive bang, about like the crack of a five-inch anti-aircraft gun at a hundred yards.

"Most experiences in life can be comprehended by prior experiences," Norris Bradbury comments, "but the atom bomb did not fit into any preconceptions possessed by anybody."

As the fireball rose into the air, Joseph W. Kennedy reports, "the overcast of strato-cumulus clouds directly overhead [became] pink on the underside and well illuminated, as at a sunrise." Weisskopf noticed that "the path of the shock wave through the clouds was plainly visible as an expanding circle all over the sky where it was covered by clouds." "When the red glow faded out," writes Edwin McMillan, "a most remarkable effect made its appearance. The whole surface of the ball was covered with a purple luminescence, like that produced by the electrical excitation of the air, and caused undoubtedly by the radioactivity of the material in the ball."

Fermi had prepared an order-of-magnitude experiment to determine roughly the bomb's yield:

> About 40 seconds after the explosion the air blast reached me. I tried to estimate its strength by dropping from about six feet small pieces of paper before, during and after the passage of the blast wave. Since, at the time, there was no wind, I could observe very distinctly and actually measure the displacement of the pieces of paper that were in the process of falling while the blast was passing. The shift was about 2½ meters, which, at the time, I estimated to correspond to the blast that would be produced by ten thousand tons of T.N.T.

"From the distance of the source and from the displacement of the air due to the shock wave," Segrè explains, "he could calculate the energy of the explosion. This Fermi had done in advance having prepared himself a table of numbers, so that he could tell immediately the energy liberated from this crude but simple measurement." "He was so profoundly and totally absorbed in his bits of paper," adds Laura Fermi, "that he was not aware of the tremendous noise."

Frank Oppenheimer found his brother watching beside him outside the control bunker at S-10000:

> And so there was this sense of this ominous cloud hanging over us. It was so brilliant purple, with all the radioactive glowing. And it just seemed to hang there forever. Of course it didn't. It must have been just a very short time until it went up. It was very terrifying.
>
> And the thunder from the blast. It bounced on the rocks, and then it went—I don't know where else it bounced. But it never seemed to stop. Not like an ordinary echo with thunder. It just kept echoing back and forth in that Jornada del Muerto. It was a very scary time when it went off.
>
> And I wish I would remember what my brother said, but I can't—but I think we just said, "It worked." I think that's what we said, both of us. "It worked."

Trinity director Bainbridge appropriately pronounced its benediction: "No one who saw it could forget it, a foul and awesome display."

At Base Camp Groves "personally thought of Blondin crossing Niagara Falls on his tightrope, only to me the tightrope had lasted for almost three years, and of my repeated, confident-appearing assurances that such a thing was possible and that we would do it." Sitting up in their trenches before the blast wave arrived, he and Conant and Bush ceremoniously shook hands.

The blast had knocked Kistiakowsky down at S-10000. He scrambled up to watch the fireball rise and darken and mushroom purple auras, then moved to claim his bet. "I slapped Oppenheimer on the back and said, 'Oppie, you owe me ten dollars.' " The distracted Los Alamos director searched his wallet. "It's empty," he told Kistiakowsky, "you'll have to wait." Bainbridge went around congratulating the S-10000 leaders on the success of the implosion method. "I finished by saying to Robert, 'Now we are all sons of bitches.' . . . [He] told my younger daughter later that it was the best thing anyone said after the test."

"Our first feeling was one of elation," Weisskopf remembers, "then we realized we were tired, and then we were worried." Rabi elaborates:

> Naturally, we were very jubilant over the outcome of the experiment. While this tremendous ball of flame was there before us, and we watched it, and it rolled along, it became in time diffused with the clouds. . . . Then it was washed out with the wind. We turned to one another and offered congratulations, for the first few minutes. Then, there was a chill, which was not the morning cold; it was a chill that came to one when one thought, as for instance when I thought of my wooden house in Cambridge, and my laboratory in New York, and of the millions of people living around there, and this power of nature which we had first understood it to be—well, there it was.

Oppenheimer looked again into the *Gita* for a model sufficiently scaled:

> We waited until the blast had passed, walked out of the shelter and then it was extremely solemn. We knew the world would not be the same. A few people laughed, a few people cried. Most people were silent. I remembered the line from the Hindu scripture, the *Bhagavad-Gita:* Vishnu is trying to persuade the Prince that he should do his duty and to impress him he takes on his multi-armed form and says, "Now I am become Death, the destroyer of worlds." I suppose we all thought that, one way or another.

Other models also came to mind, Oppenheimer told an audience shortly after the war:

> When it went off, in the New Mexico dawn, that first atomic bomb, we thought of Alfred Nobel, and his hope, his vain hope, that dynamite would put an end to wars. We thought of the legend of Prometheus, of that deep sense of guilt in man's new powers, that reflects his recognition of evil, and his long knowledge of it. We knew that it was a new world, but even more we knew that novelty itself was a very old thing in human life, that all our ways are rooted in it.

The successful director of the Los Alamos bomb laboratory left with Farrell in a jeep. Rabi watched him arrive at Base Camp and saw a change:

> He was in the forward bunker. When he came back, there he was, you know, with his hat. You've seen pictures of Robert's hat. And he came to where we were in the headquarters, so to speak. And his walk was like "High Noon"—I think it's the best I could describe it—this kind of strut. He'd done it.

"When Farrell came up to me," Groves continues the story, "his first words were, 'The war is over.' My reply was, 'Yes, after we drop two bombs on Japan.' I congratulated Oppenheimer quietly with 'I am proud of you,' and he replied with a simple 'Thank you.' " The theoretical physicist who was also a poet, who found physics, as Bethe says, "the best way to do philosophy," had staked his claim on history. It was a larger claim, but more ambivalent, than any Nobel Prize.

The horses in the MP stable still whinnied in fright; the paddles of the dusty Aermotor windmill at Base Camp still spun away the energy of the blast; the frogs had ceased to make love in the puddles. Rabi broke out a bottle of whiskey and passed it around. Everyone took a swig. Oppenheimer went to work with Groves on a report for Stimson at Potsdam. "My faith in the human mind has been somewhat restored," Hubbard overheard him say. He estimated the blast at 21,000 tons—21 kilotons. Fermi knew from his paper experiment that it was at least 10 KT. Rabi had wagered 18. Later that morning Fermi and Herbert Anderson would don white surgical scrub suits and board the two lead-lined tanks to drive near Zero. Fermi's tank broke down after only a mile of approach and he had to walk back. Anderson clanked on. Through the periscope the young physicist studied the crater the bomb had made. The tower—the $20,000 winch, the shack, the wooden platform, the hundred feet of steel girders—was gone, vaporized down to the stubby twisted wreckage of its footings. What had been asphalt paving was now fused sand, green and translucent as jade. The cup strung to Anderson's rocket scooped up debris. His later radiochemical measurements confirmed 18.6 KT. That was nearly four times what Los Alamos had expected. Rabi won the pot.

Fermi experienced a delayed reaction, he told his wife: "For the first time in his life on coming back from Trinity he had felt it was not safe for him to drive. It had seemed to him as if the car were jumping from curve to curve, skipping the straight stretches in between. He had asked a friend to drive, despite his strong aversion to being driven." Stanislaw Ulam, who chose not to attend the shot, watched the buses returning: "You could tell at once they had had a strange experience. You could see it on their faces. I saw that something very grave and strong had happened to their whole outlook on the future."

A bomb exploded in a desert damages not much besides sand and cactus and the purity of the air. Stafford Warren, the physician responsible for radiological safety at Trinity, had to search to discover more lethal effects:

> Partially eviscerated dead wild jack rabbits were found more than 800 yards from zero, presumably killed by the blast. A farm house three miles away had doors torn loose and suffered other extensive damage. . . .
>
> The light intensity was sufficient at nine miles to have caused temporary blindness and this would be longer lasting at shorter distances. . . . The light together with the heat and ultraviolet radiation would probably cause severe damage to the unprotected eye at 5–6 miles; damage sufficient to put personnel out of action several days if not permanently.

The boxes of excelsior Frank Oppenheimer had set out, and the pine boards, also recorded the coming of the light: they were charred beyond 1,000 yards, slightly scorched up to 2,000 yards. At 1,520 yards—nine-tenths of a mile—exposed surfaces had heated almost instantly to 750°F.

William Penney, the British physicist who had studied blast effects for the Target Committee, held a seminar at Los Alamos five days after Trinity. "He applied his calculations," Philip Morrison remembers. "He predicted that this [weapon] would reduce a city of three or four hundred thousand people to nothing but a sink for disaster relief, bandages, and hospitals. He made it absolutely clear in numbers. It was reality."

Around the time of the Trinity shot, in the predawn dark at Hunter's Point in San Francisco Bay, a floodlit crane had loaded onto the deck of the *Indianapolis* the fifteen-foot crate that carried the Little Boy gun assembly. Two sailors carried aboard the Little Boy bullet in a lead bucket shouldered between them on a crowbar. They followed the two Los Alamos Army officers to the cabin of the ship's flag lieutenant, who had vacated it for the voyage. Eyebolts had been welded to its deck. The sailors strapped the lead bucket to the eyebolts. One of the officers padlocked it into place. They would take turns guarding it around the clock for the ten-day voyage to Tinian.

At 0836 Pacific War Time, four hours after the light flung from the Jornada del Muerto blanched the face of the moon, the *Indianapolis* sailed with its cargo under the Golden Gate and out to sea.

19
Tongues of Fire

At the end of March 1945, as Curtis LeMay's bombers shuttled back and forth burning cities, Colonel Elmer E. Kirkpatrick, a plainspoken Army engineer, arrived in the Marianas to locate a small corner where he could lodge Paul Tibbets' 509th Composite Group. Kirkpatrick met with LeMay and then with Pacific Fleet commander Chester Nimitz on Guam on the day he arrived, March 30, and found the commanding officers cooperative. LeMay personally flew Kirkpatrick to Tinian on April 3. The next day, he reported to Groves, he "covered most of the island [and] decided on our sites and the planning forces went to work on layouts." Though there was no shortage of B-29's, he found that cement and buildings were scarce; "housing and life here is a little rugged for everyone except [general] officers & the Navy. Tents or open barracks." Kirkpatrick flew back to Guam on April 5 "to dig up some materials some place" and "to get authority for the work I required," threaded his way through the Air Force and Navy chains of command with his letters of authority from Washington and by the end of the day had seen a telex sent to Saipan "directing them to give me enough material to get the essential things done." A Navy construction battalion—the SeaBees—would build the buildings and hardstands and dig the pits from which the bombs, too large for ground-level clearance, would be lifted up into the bomb bays of Tibbets' B-29's.

By early June, when Tibbets arrived to inspect the accommodations and confer with LeMay, Kirkpatrick could report that "progress has been very satisfactory and I have the feeling now that we can't miss." He sat in on an evening meeting between Tibbets and LeMay and heard evidence that the Twentieth Air Force commander did not yet appreciate the power of an atomic bomb:

> LeMay does not favor high altitude bombing. Work is not as accurate but, more important, visibility at such altitudes is extremely poor especially during the period June to November. Tibbets advised him that the weapon would destroy a plane using it at an altitude of less than 25,000 feet.

Kirkpatrick demonstrated his progress to Groves with an impressive list: five warehouses, an administration building, roads and parking areas and nine magazines completed; pits completed except for lifts; hardstands for parking the 509th aircraft completed except for asphalt paving; generator buildings and compressor shed completed; one air-conditioned building where the bombs would be assembled to be completed by July 1; two more assembly buildings to be completed by August 1 and August 15. Of the 509th's men more than 1,100 had already staged out by ship "and more [are] coming in every week."

The first of Tibbets' combat crews arrived June 10, flying themselves to Tinian in advanced, specially modified new B-29's. The early-model aircraft delivered to the group the previous autumn had become obsolete, Tibbets explained to readers of the *Saturday Evening Post* after the war:

> Tests showed us that the B-29's we had weren't good enough for atom bombing. They were heavy, older types. Top cylinders were overheating and causing valve failures in the long climb to 30,000 feet at 80 per cent of full power. . . .
>
> I asked for new, light-weight B-29's and fuel-injection systems to replace carburetors.

He got those improvements and more: quick-closing pneumatic bomb doors, fuel flow meters, reversible electric propellers.

The new aircraft had been modified to accommodate the special bombs they would carry and the added crew. The cylindrical tunnel that connected the pressurized forward and waist sections of the plane had to be partly cut away and reworked so that the larger bomb, Fat Man, would fit in the forward bomb bay. Guide rails were installed to prevent the tail assemblies from hanging up during fallout. An extra table, chair, oxygen outlet and interphone station for the weaponeers responsible for monitoring a bomb during flight went in forward of the radio operator's station in the forward section. "The performance of these special B-29's was exceptional," writes the engineer in charge of their procurement. "They were without doubt the finest B-29's in the theater." By the end of June, eleven of the new bombers shone on their hardstands in the Pacific sun.

To men used to the blizzards and dust of Wendover, Utah, the 509th's historian claims, Tinian "looked like the Garden of Paradise." The surrounding blue ocean and the palm groves may have occasioned that vision. Philip Morrison, who came out after Trinity to help assemble Fat Man, saw more reverberantly what the island had become, as he told a committee of U.S. Senators later in 1945:

> Tinian is a miracle. Here, 6,000 miles from San Francisco, the United States armed forces have built the largest airport in the world. A great coral ridge was half-leveled to fill a rough plain, and to build six runways, each an excellent 10-lane highway, each almost two miles long. Beside these runways stood in long rows the great silvery airplanes. They were there not by the dozen but by the hundred. From the air this island, smaller than Manhattan, looked like a giant aircraft carrier, its deck loaded with bombers. . . .
>
> And all these gigantic preparations had a grand and terrible outcome. At sunset some day the field would be loud with the roar of motors. Down the great runways would roll the huge planes, seeming to move slowly because of their size, but far outspeeding the occasional racing jeep. One after another each runway would launch its planes. Once every 15 seconds another B-29 would become air-borne. For an hour and a half this would continue with precision and order. The sun would go below the sea, and the last planes could still be seen in the distance, with running lights still on. Often a plane would fail to make the take-off, and go skimming horribly into the sea, or into the beach to burn like a huge torch. We came often to sit on the top of the coral ridge and watch the combat strike of the 313th wing in real awe. Most of the planes would return the next morning, standing in a long single line, like beads on a chain, from just overhead to the horizon. You could see 10 or 12 planes at a time, spaced a couple of miles apart. As fast as the near plane would land, another would appear on the edge of the sky. There were always the same number of planes in sight. The empty field would fill up, and in an hour or two all the planes would have landed.

A resemblance in shape between Tinian and Manhattan had inspired the SeaBees to name the island's roads for New York City streets. The 509th happened to be lodged immediately west of North Field at 125th Street and Eighth Avenue, near Riverside Drive, in Manhattan, the environs of Columbia University where Enrico Fermi and Leo Szilard had identified secondary neutrons from fission: the wheel had come full circle.

"The first half of July," Norman Ramsey writes of 509th activity, "was occupied with establishing and installing all of the technical facilities needed for assembly and test work at Tinian." In the meantime the group's flight crews practiced navigating to Iwo Jima and back and bombing with standard general-purpose bombs and then with Pumpkins such bypassed islands still nominally in Japanese hands as Rota and Truk.

Harry Truman and Jimmy Byrnes left suburban Potsdam in an open car to tour ravaged Berlin at about the same time on July 16, 1945, that Groves and Oppenheimer at Trinity were preparing their first report of the tower shot's success. The Potsdam Conference, appropriately coded TERMINAL, was supposed to have begun that afternoon, but Joseph Stalin was late arriving by armored train from Moscow. (He apparently suffered a mild heart attack the previous day.) The Berlin tour gave Truman an opportunity to view at close hand the damage Allied bombing and Red Army shelling had done.

Byrnes was officially Secretary of State now, invested in a sweltering ceremony in the White House Rose Garden on July 3 attended by a crowd of his former House, Senate and Supreme Court colleagues. After Byrnes swore the oath of office Truman had kidded him: "Jimmy, kiss the Bible." Byrnes complied, then gave as good as he got: passed the Bible to the President and bade him kiss it as well. Truman did so; understanding the byplay between the former Vice President and the man who had missed his turn, the crowd laughed. Four days later the two leaders boarded the cruiser *Augusta* for the Atlantic crossing to Antwerp and now they rode side by side into Berlin, conquerors in snap-brim hats and natty worsteds.

Though he had arrived before them in Potsdam, Henry Stimson did not accompany the President and his favorite adviser on their tour. The Secretary of War had consulted with Truman the day before Byrnes' swearing-in—proposing to give the Japanese "a warning of what is to come and definite opportunity to capitulate"—and as he was leaving had asked the President plaintively if he had not invited his Secretary of War to attend the forthcoming conference out of solicitude for his health. That was it, Truman had said quickly, and Stimson had replied that he could manage the trip and would like to go, that Truman ought to have advice "from the top civilians in our Department." The next day, the day of Byrnes' investiture, Truman accorded the elderly statesman permission. But Stimson had traveled separately on the military transport *Brazil* via Marseilles, was lodged separately in Potdam from the President and his Secretary of State and would not be included in their daily private discussions. One of Stimson's aides felt that "Secretary Byrnes was a little resentful of Mr. Stimson's presence there. . . . The Secretary of the Navy wasn't there so why should Mr. Stimson be there?" Byrnes in his 1947 account of his career, *Speaking Frankly,* narrates an entire chapter about Potsdam without once mentioning Stimson's name, relegating his rival to a brief separate discussion of the decision to use the atomic bomb on Japan and awarding him there the dubious honor of having chosen the targets. In fact, Stimson at Potsdam would be reduced to serving Truman and Byrnes as not much more than a messenger boy. But the messages he brought were fateful.

"We reviewed the Second Armored Division," Truman reports his Berlin tour in his impromptu diary, ". . . Gen. [J. H.] Collier, who seemed to know his stuff, put us in a reconnaissance car built with side seats and no top, just like a hoodlum wagon minus the top, or a fire truck with seats and no hose, and we drove slowly down a mile and a half of good soldiers and some millions of dollars worth of equipment—which had amply paid its way to Berlin." The destroyed city fired an uneasy burst of associations:

> Then we went on to Berlin and saw absolute ruin. Hitler's folly. He overreached himself by trying to take in too much territory. He had no morals and his people backed him up. Never did I see a more sorrowful sight, nor witness retribution to the nth degree. . . .
>
> I thought of Carthage, Baalbec, Jerusalem, Rome, Atlantis; Peking, Babylon, Nineveh; Scipio, Rameses II, Titus, Hermann, Sherman, Jenghis Khan, Alexander, Darius the Great. But Hitler only destroyed Stalingrad—and Berlin. I hope for some sort of peace—but I fear that machines are ahead of morals by some centuries and when morals catch up perhaps there'll be no reason for any of it.
>
> I hope not. But we are only termites on a planet and maybe when we bore too deeply into the planet there'll be a reckoning—who knows?

The "Proposed Program for Japan" that Stimson had offered to Truman on July 2 had reckoned up that country's situation—which included the possible entry of the Soviet Union, at present neutral, into the Pacific war—and judged it desperate:

> Japan has no allies.
>
> Her navy is nearly destroyed and she is vulnerable to a surface and underwater blockade which can deprive her of sufficient food and supplies for her population.
>
> She is terribly vulnerable to our concentrated air attack upon her crowded cities, industrial and food resources.
>
> She has against her not only the Anglo-American forces but the rising forces of China and the ominous threat of Russia.

> We have inexhaustible and untouched industrial resources to bring to bear against her diminishing potential.
>
> We have great moral superiority through being the victim of her first sneak attack.

On the other hand, Stimson had argued, because of the mountainous Japanese terrain and because "the Japanese are highly patriotic and certainly susceptible to calls for fanatical resistance to repel an invasion," America would probably "have to go through with an even more bitter finish fight than in Germany" if it attempted to invade. Was there, then, any alternative? Stimson thought there might be:

> I believe Japan *is* susceptible to reason in such a crisis to a much greater extent than is indicated by our current press and other current comment. Japan is not a nation composed wholly of mad fanatics of an entirely different mentality from ours. On the contrary, she has within the past century shown herself to possess extremely intelligent people, capable in an unprecedentedly short time of adopting not only the complicated technique of Occidental civilization but to a substantial extent their culture and their political and social ideas. Her advance in these respects . . . has been one of the most astounding feats of national progress in history. . . .
>
> It is therefore my conclusion that a carefully timed warning be given to Japan. . . .
>
> I personally think that if in [giving such a warning] we should add that we do not exclude a constitutional monarchy under her present dynasty, it would substantially add to the chances of acceptance.

Within the text of his proposal the Secretary of War several times characterized it as "the equivalent of an unconditional surrender," but others did not see it so. Before Byrnes left for Potsdam he had carried the document to ailing Cordell Hull, a fellow Southerner and Franklin Roosevelt's Secretary of State from 1933 to 1944, and Hull had immediately plucked out the concession to the "present dynasty"—the Emperor Hirohito, in whose mild myopic figure many Americans had personified Japanese militarism—and told Byrnes that "the statement seemed too much like appeasement of Japan."

It may have been, but by the time they arrived in Potsdam, Stimson, Truman and Byrnes had learned that it was also the minimum condition of surrender the Japanese were prepared to countenance, whatever their desperate situation. U.S. intelligence had intercepted and decoded messages passing between Tokyo and Moscow instructing Japanese ambassador Naotake Sato to attempt to interest the Soviets in mediating a Japanese surrender. "The foreign and domestic situation for the Empire is very serious," Foreign Minister Shigenori Togo had cabled Sato on July 11, "and even the termination of the war is now being considered privately. . . . We are also sounding out the extent to which we might employ the USSR in connection with the termination of the war. . . . [This is] a matter with which the Imperial Court is . . . greatly concerned." And pointedly on July 12:

> It is His Majesty's heart's desire to see the swift termination of the war. . . . However, as long as America and England insist on unconditional surrender our country has no alternative but to see it through in an all-out effort for the sake of survival and the honor of the homeland.

Unconditional surrender seemed to the Japanese leadership a demand to give up its essential and historic polity, a demand that under similar circumstances Americans also might hesitate to meet even at the price of their lives: hence Stimson's careful qualification of his proposed terms of surrender. But to the extent that the imperial institution was tainted with militarism, an offer to preserve it might also seem an offer to preserve the militaristic government that ran the country and that had started and pursued the war. Certainly many Americans might think so and might conclude in consequence that their wartime sacrifices were being callously betrayed.

Hull considered these difficulties while Byrnes sailed the Atlantic and sent along a cable of further advice on July 16. The Japanese might reject a challenge to surrender, the former Secretary of State argued, even if it allowed the Emperor to remain on the throne. In that case not only would the militarists among them be encouraged by what they would take to be a sign of weakening Allied will, but also "terrible political repercussions would follow in the U.S. . . . Would it be well *first* to await the climax of Allied bombing and Russia's entry into the war?"

The point of warning the Japanese was to encourage an early surrender in the hope of avoiding a bloody invasion; the trouble with waiting until the Soviet Union entered the war was that it left Truman where he had dangled uncomfortably for months: over Stalin's barrel, dependent on the USSR for military intervention in Manchuria to tie up the Japanese armies there. Hull's delaying tactic might improve the first prospect; but it might also secure the second.

Another message arrived in Potsdam that evening, however, that changed the terms of the equation, a message for Stimson from George Harrison in Washington announcing the success of the Trinity shot:

> Operated on this morning. Diagnosis not yet complete but results seem satisfactory and already exceed expectations. Local press release necessary as in-

terest extends great distance. Dr. Groves pleased. He returns tomorrow. I will keep you posted.

"Well," Stimson remarked to Harvey Bundy with relief, "I have been responsible for spending two billions of dollars on this atomic venture. Now that it is successful I shall not be sent to prison in Fort Leavenworth." Happily the Secretary of War carried the cable to Truman and Byrnes, just returned to Potsdam from Berlin.

In Stimson's welcome news Byrnes saw a more general reprieve. It informed his overnight response to Hull. "The following day" Hull says, "I received a message from Secretary Byrnes agreeing that the statement [warning the Japanese] should be delayed and that, when it was issued, it should not contain this commitment with regard to the Emperor." Byrnes had good reason to delay a warning now: to await the readying of the first combat atomic bombs. Those weapons would answer Hull's first objection; if the Japanese ignored a warning, then the United States could deliver a brutally retributive response. With such weapons in the U.S. arsenal unconditional surrender need not be compromised. And America no longer required the Soviet Union's aid in the Pacific; the problem now would be not dealing the Soviets in but stalling to keep them out. "Neither the President nor I," Byrnes affirms, "were anxious to have them enter the war after we had learned of this successful test."

Byrnes and others within the American delegation came to realize that preserving the Emperor might be sensible policy if Hirohito alone could persuade the far-flung Japanese armies, undefeated and with a year's supply of ammunition on hand, to lay down their arms. The new Secretary of State, who was drafting a suitable declaration, sought a formula that would not arouse the American people but might reassure the Japanese. The Joint Chiefs produced its first version: "Subject to suitable guarantees against further acts of aggression, the Japanese people will be free to choose their own form of government." The Japanese polity resided in the Imperial House, not in the people, but provision for popular government was as conditional an unconditional surrender as the enemy would be allowed.

George Harrison cabled Stimson on July 21 that "all your local military advisors engaged in preparation definitely favor your pet city": Groves still coveted Kyoto. Stimson quickly returned that he was "aware of no factors to change my decision. On the contrary new factors here tend to confirm it."

Harrison also asked Stimson to alert him by July 25 "if [there is] any change in plans" because "[the] patient [is] progressing rapidly." At the same time Groves requested permission from George Marshall to brief Douglas MacArthur, who had not yet been told about the new weapon, in view of "the imminence of the use of the atomic fission bomb in operations against Japan, 5 to 10 August." The 509th had begun flying Pumpkin missions over Japan the previous day for combat experience and to accustom the enemy to small, unescorted flights of B-29's at high altitude.

Groves' eyewitness narrative of the Trinity test had arrived that Saturday just before noon. Stimson sought out Truman and Byrnes and had the satisfaction of riveting them to their chairs by reading it aloud. Groves estimated "the energy generated to be in excess of the equivalent of 15,000 to 20,000 tons of TNT" and allowed his deputy, Thomas F. Farrell, to call the visual effects "unprecedented, magnificent, beautiful, stupendous and terrifying." Kenneth Bainbridge's "foul and awesome display" became at Farrell's hand "that beauty the great poets dream about but describe most poorly and inadequately," which Farrell presumably meant for a superlative. "As to the present war," Farrell opined, "there was a feeling that no matter what else might happen, we now had the means to insure its speedy conclusion and save thousands of American lives." Stimson saw that Truman was "tremendously pepped up" by the report. "[He] said it gave him an entirely new feeling of confidence."

The President met the next day to discuss Groves' results with Byrnes, Stimson and the Joint Chiefs, including Marshall and Hap Arnold. Arnold had long maintained that conventional strategic bombing by itself could compel the Japanese to surrender. In late June, when invasion was being decided, he had rushed LeMay to Washington to work the numbers. LeMay figured he could complete the destruction of the Japanese war machine by October 1. "In order to do this," writes Arnold, "he had to take care of some 30 to 60 large and small cities." Between May and August LeMay took care of fifty-eight. But Marshall disagreed with the Air Force assessment. The situation in the Pacific, he had told Truman in June, was "practically identical" to the situation in Europe after Normandy. "Airpower alone was not sufficient to put the Japanese out of the war. It was unable alone to put the Germans out." He explained his reasoning at Potsdam to an interviewer after the war:

> We regarded the matter of dropping the [atomic] bomb as exceedingly important. We had just gone through a bitter experience at Okinawa [the last major island campaign, when the Americans lost more than 12,500 men killed and missing and the Japanese more than 100,000 killed in eighty-two days of fighting]. This had been preceded by a number of similar experiences in other Pacific islands, north of Australia. The Japanese had demonstrated in each case they would not surrender and they would fight to the death. . . . It was

> expected that resistance in Japan, with their home ties, would be even more severe. We had had the one hundred thousand people killed in Tokyo in one night of [conventional] bombs, and it had had seemingly no effect whatsoever. It destroyed the Japanese cities, yes, but their morale was not affected as far as we could tell, not at all. So it seemed quite necessary, if we could, to shock them into action. . . . We had to end the war; we had to save American lives.

Before Groves' report arrived, Dwight Eisenhower, a hard and pragmatic commander, had angered Stimson with a significantly different assessment. "We'd had a nice evening together at headquarters in Germany," the Supreme Allied Commander remembers, "nice dinner, everything was fine. Then Stimson got this cable saying the bomb had been perfected and was ready to be dropped." The cable was the second Harrison had sent, the day after the Trinity test when Groves arrived back in Washington:

> Doctor has just returned most enthusiastic and confident that the little boy is as husky as his big brother. The light in his eyes discernible from here to Highhold and I could have heard his screams from here to my farm.

Highhold was Stimson's Long Island estate, 250 miles from Washington—the Trinity flash had been visible even farther from Zero than that. Harrison's farm was 50 miles outside the capital. Eisenhower found the allegorical code less than amusing and the subject baleful:

> The cable was in code, you know the way they do it. "The lamb is born" or some damn thing like that. So then he told me they were going to drop it on the Japanese. Well, I listened, and I didn't volunteer anything because, after all, my war was over in Europe and it wasn't up to me. But I was getting more and more depressed just thinking about it. Then he asked for my opinion, so I told him I was against it on two counts. First, the Japanese were ready to surrender and it wasn't necessary to hit them with that awful thing. Second, I hated to see our country be the first to use such a weapon. Well . . . the old gentleman got furious. And I can see how he would. After all, it had been his responsibility to push for all the huge expenditure to develop the bomb, which of course he had a right to do, and *was* right to do. Still, it was an awful problem.

Eisenhower also spoke to Truman, but the President concurred in Marshall's judgment, having already formed his own. "Believe Japs will fold up before Russia comes in," he confided to his diary almost as soon as he heard of the Trinity success. "I am sure they will when Manhattan appears over their homeland."

When to issue the Potsdam Declaration now became essentially a question of when the first atomic bombs would be ready to be dropped. Stimson queried Harrison, who responded on July 23:

> Operation may be possible any time from August 1 depending on state of preparation of patient and condition of atmosphere. From point of view of patient only, some chance August 1 to 3, good chance August 4 to 5 and barring unexpected relapse almost certain before August 10.

Stimson had also asked for a target list, "always excluding the particular place against which I have decided. My decision has been confirmed by highest authority." Harrison complied: "Hiroshima, Kokura, Niigata in order of choice here."

Which meant that Nagasaki had not yet, as of the last full week in July, been added to the list. Within days it would be. Official Air Force historians speculate that LeMay's staff proposed it. The requirement for visual bombing was probably the reason. Hiroshima was 440 miles southwest of Niigata. Nagasaki, over the mountains from Kokura on Kyushu, was a further 220 miles southwest of Hiroshima. If one city was socked in, another might be clear. Nagasaki was certainly also added because it was one of the few major cities left in Japan that had not yet been burned out.

A revealing third cable completed the day's communications from Harrison (the metallurgists at Los Alamos had finished the Pu core for Fat Man that day). It concerned possible future deliveries of atomic bombs and hinted at a forthcoming change in design, probably to the so-called "mixed" implosion bomb with a core of U235 and plutonium alloyed together. Such a core could draw on the resources of both Oak Ridge and Hanford:

> First one of tested type [i.e., Fat Man] should be ready at Pacific base about 6 August. Second one ready about 24 August. Additional ones ready at accelerating rate from possibly three in September to we hope seven or more in December. The increased rate above three per month entails changes in design which Groves believes thoroughly sound.

Stimson reported Harrison's several estimates to Truman on Tuesday morning, July 24. The President was pleased and said he would use them to time the release of the Potsdam Declaration. The Secretary took advantage of the moment to appeal to Truman to consider assuring the Japanese privately that they could keep their Emperor if they persisted in making that concession a condition of surrender. Deliberately noncommittal, the President said he had the point in mind and would take care of it.

Stimson left and Byrnes joined Truman for lunch. They discussed how to tell Stalin as little as possible about the atomic bomb. Truman wanted protective cover when Stalin learned that his wartime allies had developed an epochal new weapon behind his back but wanted to give as little as possible away. Byrnes also devised a more immediate reason for circumspection, he told the historian Herbert Feis in 1958:

> As a result of his experience with the Russians during the first week of the Conference he had come to the conclusion that it would be regrettable if the Soviet Union entered the [Pacific] war, and . . . he was afraid that if Stalin were made fully aware of the power of the new weapon, he might order the Soviet Army to plunge forward at once.

But in fact Stalin already knew about the Trinity test. His agents in the United States had reported it to him. It appears he was not immediately impressed. There is gallows humor in Truman's elaborately offhand approach to the Soviet Premier at the end of that day's plenary session at the Cecilienhof Palace, stripped and shabby, where pale German mosquitoes homing through unscreened windows dined on the sanguinary conquerors. Truman left behind his translator, rounded the baize-covered conference table and sidled up to his Soviet counterpart, both men dissimulating. "I casually mentioned to Stalin that we had a new weapon of unusual destructive force. The Russian Premier showed no special interest. All he said was that he was glad to hear it and hoped we would make 'good use of it against the Japanese.' " "That," concludes Robert Oppenheimer dryly, knowing how much at that moment the world lost, "was carrying casualness rather far."

If Stalin was not yet impressed with the potential of the bomb, Truman in his private diary was waxing apocalyptic, biblical visions mingling in his autodidact's mind with doubt that the atom could be decomposed and denial that the new weapon would be used to slaughter civilians:

> We have discovered the most terrible bomb in the history of the world. It may be the fire destruction prophesied in the Euphrates Valley Era, after Noah and his fabulous Ark.
>
> Anyway we "think" we have found a way to cause a disintegration of the atom. An experiment in the New Mexican desert was startling—to put it mildly. . . .
>
> This weapon is to be used against Japan between now and August 10th. I have told the Sec. of War, Mr. Stimson, to use it so that military objectives and soldiers and sailors are the target and not women and children. Even if the Japs are savages, ruthless, merciless and fanatic, we as the leader of the world for the common welfare cannot drop this terrible bomb on the old Capital or the new.

> He & I are in accord. The target will be a purely military one and we will issue a warning statement asking the Japs to surrender and save lives. I'm sure they will not do that, but we will have given them the chance. It is certainly a good thing for the world that Hitler's crowd or Stalin's did not discover this atomic bomb. It seems to be the most terrible thing ever discovered, but it can be made the most useful.

The Tuesday Truman mentioned the new weapon to Stalin the Combined Chiefs met with their Soviet counterparts; Red Army chief of staff General Alexei E. Antonov announced that Soviet troops were assembling on the Manchurian border and would be ready to attack in the second half of August. Stalin had said August 15 before. Byrnes was anxious that the Soviets might prove uncharacteristically punctual.

That afternoon in Washington Groves drafted the historic directive releasing the atomic bomb to use. It passed up through Harrison for transmission by radio EYES ONLY to Marshall "in order that your approval and the Secretary of War's approval might be obtained as soon as possible." (A small map of Japan cut from a large National Geographic Society map and a one-page description of the chosen targets, which now included Nagasaki, followed by courier.) Marshall and Stimson approved the directive at Potsdam and presumably showed it to Truman, though it does not record his formal authorization; it went out the next morning to the new commander of the Strategic Air Force in the Pacific:

> To General Carl Spaatz, CG, USASTAF:
>
> 1. The 509 Composite Group, 20th Air Force will deliver its first special bomb as soon as weather will permit visual bombing after about 3 August 1945 on one of the targets: Hiroshima, Kokura, Niigata and Nagasaki. . . .
>
> 2. Additional bombs will be delivered on the above targets as soon as made ready by the project staff. . . .
>
> 3. Dissemination of any and all information concerning the use of the weapon against Japan is reserved to the Secretary of War and the President of the United States. . . .
>
> 4. The foregoing directive is issued to you by direction and with the approval of the Secretary of War and of the Chief of Staff, USA.

As Groves drafted the directive the metallurgists at Los Alamos finished casting the rings of U235 that fitted together to form the gun bomb's target assembly, the last components needed to complete Little Boy.

Strategy and delivery intersected on July 26 and synchronized. The *Indianapolis* arrived at Tinian. Three Air Transport Command C-54 cargo planes departed Kirtland Air Force Base with the three separate pieces of the Little Boy target assembly; two more ATC C-54's departed with Fat Man's initiator and plutonium core. Meanwhile Truman's staff released the Potsdam Declaration to the press at 7 P.M. for dispatch from Occupied Germany at 9:20. It offered on behalf of the President of the United States, the President of Nationalist China and the Prime Minister of Great Britain to give Japan "an opportunity to end this war":

> Following are our terms. We will not deviate from them. There are no alternatives. We shall brook no delay.
>
> There must be eliminated for all time the authority and influence of those who have deceived and misled the people of Japan into embarking on world conquest. . . .
>
> Until such a new order is established . . . points in Japanese territory . . . shall be occupied.
>
> . . . Japanese sovereignty shall be limited to the islands of Honshu, Hokkaido, Kyushu, Shikoku and such minor islands as we determine.
>
> The Japanese military forces, after being completely disarmed, shall be permitted to return to their homes with the opportunity to lead peaceful and productive lives.
>
> We do not intend that the Japanese shall be enslaved as a race or destroyed as a nation, but stern justice shall be meted out to all war criminals. . . . Freedom of speech, of religion, and of thought, as well as respect for the fundamental human rights shall be established.
>
> Japan shall be permitted to maintain such industries as will sustain her economy. . . .
>
> The occupying forces of the Allies shall be withdrawn from Japan as soon as these objectives have been accomplished and there has been established in accordance with the freely expressed will of the Japanese people a peacefully inclined and responsible government.
>
> We call upon the government of Japan to proclaim now the unconditional surrender of all Japanese armed forces. . . . The alternative for Japan is prompt and utter destruction.

"We faced a terrible decision," Byrnes wrote in 1947. "We could not rely on Japan's inquiries to the Soviet Union about a negotiated peace as proof that Japan would surrender unconditionally without the use of the bomb. In fact, Stalin stated the last message to him had said that Japan would 'fight to the death rather than accept unconditional surrender.' Under the circumstances, agreement to negotiate could only arouse false hopes. Instead, we relied upon the Potsdam Declaration."

The text of that somber document went out by radio to the Japanese from San Francisco; Japanese monitors picked it up at 0700 hours Tokyo time July 27. The Japanese leaders debated its mysteries all day. A quick Foreign Office analysis noted for the ministers that the Soviet Union had preserved its neutrality by not sponsoring the declaration, that it specified what the Allies meant by unconditional surrender and that the term itself had been applied specifically only to the nation's armed forces. Foreign Minister Togo disliked the demand for occupation and the stripping away of Japan's foreign possessions; he recommended waiting for a Soviet response to Ambassador Sato's representations before responding.

The Prime Minister, Baron Kantaro Suzuki, came during the day to the same position. The military leaders disagreed. They recommended immediate rejection. Anything less, they argued, might impair morale.

The next day Japanese newspapers published a censored version of the Potsdam text, leaving out in particular the provision allowing disarmed military forces to return peacefully to their homes and the assurance that the Japanese would not be enslaved or destroyed. In the afternoon Suzuki held a press conference. "I believe the Joint Proclamation by the three countries," he told reporters, "is nothing but a rehash of the Cairo Declaration. As for the Government, it does not find any important value in it, and there is no other recourse but to ignore it entirely and resolutely fight for the successful conclusion of the war." In Japanese Suzuki said there was no other recourse but to *mokusatsu* the declaration, which could also mean "treat it with silent contempt." Historians have debated for years which meaning Suzuki had in mind, but there can hardly be any doubt about the rest of his statement: Japan intended to fight on.

"In the face of this rejection," Stimson explained in *Harper's* in 1947, "we could only proceed to demonstrate that the ultimatum had meant exactly what it said when it stated that if the Japanese continued the war, 'the full application of our military power, backed by our resolve, will mean the inevitable and complete destruction of the Japanese armed forces and just as inevitably the utter devastation of the Japanese homeland.' For such a purpose the atomic bomb was an eminently suitable weapon."

The night of Suzuki's press conference the five C-54's from Albuquerque arrived at Tinian, six thousand miles nearer Japan, while three B-29's departed Kirtland each carrying a Fat Man high-explosive preassembly.

The U.S. Senate in the meantime ratified the United Nations Charter.

The *Indianapolis* had sailed on to Guam after unloading the Little Boy gun and bullet at Tinian on July 26; from Guam it continued unescorted toward Leyte in the Philippines, where two weeks of training would ready the crew, 1,196 men, to join Task Force 95 at Okinawa preparing for the

November 1 Kyushu invasion. With the destruction of the Japanese surface fleet and air force, unescorted sailing had become commonplace on courses through rear areas, but the *Indianapolis,* an older vessel, lacked sonar gear for submarine detection and was top-heavy. Japanese submarine I-58 discovered the heavy cruiser in the Philippine Sea a little before midnight on Sunday, July 29, and mistook it for a battleship. Easily avoiding detection while submerging to periscope depth, I-58 fired a fanwise salvo of six torpedoes from 1,500 yards. Lieutenant Commander Mochitsura Hashimoto, I-58's commanding officer, remembers the result:

> I took a quick look through the periscope, but there was nothing else in sight. Bringing the boat on to a course parallel with the enemy, we waited anxiously. Every minute seemed an age. Then on the starboard side of the enemy by the forward turret, and then by the after turret there rose columns of water, to be followed immediately by flashes of bright red flame. Then another column of water rose from alongside Number 1 turret and seemed to envelop the whole ship—"A hit, a hit!" I shouted as each torpedo struck home, and the crew danced round with joy. . . . Soon came the sound of a heavy explosion, far greater than that of the actual hits. Three more heavy explosions followed in quick succession, then six more.

The torpedoes and following explosions of ammunition and aviation fuel ripped away the cruiser's bow and destroyed its power center. Without power the radio officer was unable to send a distress signal—he went through the motions anyway—or the bridge to communicate with the engine room. The engines pushed the ship forward unchecked, scooping up water through the holes in the hull and leaving behind the sailors thrown overboard who had been sleeping on deck in the tropical heat. The order to abandon ship, when it came, had to be passed by word of mouth.

With the ship listing to 45 degrees frightened and injured men struggled to follow disaster drill. Fires lit the darkness and smoke sickened. The ship's medical officer found some thirty seriously burned men in the port hangar where the aviation fuel had exploded; at best they got morphine for their screams and rough kapok lifejackets strapped on over their burns. They went overboard with the others into salt water scummed with nauseating fuel oil. It was possible to walk down the hull to the keel and jump into the water but the spinning number three screw with its lethal blades chopped to death the unwary.

Some 850 men escaped. The stern rose up a hundred feet straight into the air and the ship plunged. The survivors heard screams from within the disappearing hull. Then they were left to the night and the darkness in twelve-foot swells.

Most had kapok lifejackets. Few had found their way to life rafts. They floated instead in clusters, linked together, stronger men swimming the circumferences to catch sleepers before they drifted away; one group numbered between three and four hundred souls. They pushed the wounded to the center where the water was calmer and prayed the distress call had gone out.

The captain had found two empty life rafts and later that night encountered one more occupied. He ordered the rafts lashed together. They sheltered ten men and he thought them the only survivors. Through the night a current carried the swimmers southwest while wind blew the rafts northeast; by the light of morning rafts and swimmers had separated beyond discovery.

More than fifty injured swimmers died during the night. Their comrades freed them from their jackets in the morning and let them go. The wind abated and the sun glared from the oil slick, blinding them with painful photophobia. And then the sharks came. A seaman swimming for a floating crate of potatoes thrashed in the water and was gone. Elemental terror: the men pressed together in their groups, some clusters deciding to beat the water, some to hang immotile as flotsam. A shark snapped away both a sailor's legs and his unbalanced torso, suspended in its lifejacket, flipped upside down. One survivor remembered counting twenty-five deadly attacks; the ship's doctor in his larger group counted eighty-eight.

They won no rescue. They passed through Monday and Monday night and Tuesday and Tuesday night without water, sinking lower and lower in the sea as the kapok in their lifejackets waterlogged. Eventually the thirst-crazed drank seawater. "Those who drank became maniacal and thrashed violently," the doctor testifies, "until the victims became comatose and drowned." The living were blinded by the sun; their lifejackets abraded their ulcerating skin; they burned with fever; they hallucinated.

Wednesday and Wednesday night. The sharks circled and darted in to foray after flesh. Men in the grip of group delusions followed one swimmer to an island he thought he saw, another to the ghost of the ship, another down into the ocean depths where fountains of fresh water seemed to promise to slake their thirst; all were lost. Fights broke out and men slashed each other with knives. Saturated lifejackets with waterlogged knots dragged other victims to their deaths. "We became a mass of delirious, screaming men," says the doctor grimly.

Thursday morning, August 2, a Navy plane spotted the survivors. Because of negligence at Leyte the *Indianapolis* had not yet even been missed. A major rescue effort began, ships steaming to the area, PBY's and PBM's dropping food and water and survival gear. The rescuers found 318 naked and emaciated men. The fresh water they drank, one of them remembers, tasted "so sweet [it was] the sweetest thing in your life." Through the 84-hour ordeal more than 500 men had died, their bodies feeding sharks or lost to the depths of the sea.

After making good his escape, submarine commander Hashimoto reminisces, "at length, on the 30th, we celebrated our haul of the previous day with our favorite rice with beans, boiled eels, and corned beef (all of it tinned)."

The day of the I-58's feast of canned goods Carl Spaatz telexed Washington with news:

> HIROSHIMA ACCORDING TO PRISONER OF WAR REPORTS IS THE ONLY ONE OF FOUR TARGET CITIES . . . THAT DOES NOT HAVE ALLIED PRISONER OF WAR CAMPS.

It was too late to reconsider targets, prisoners of war or not. Washington telexed back the next day:

> TARGETS ASSIGNED . . . REMAIN UNCHANGED. HOWEVER IF YOU CONSIDER YOUR INFORMATION RELIABLE HIROSHIMA SHOULD BE GIVEN FIRST PRIORITY AMONG THEM.

The die was cast.

Once Trinity proved that the atomic bomb worked, men discovered reasons to use it. The most compelling reason Stimson stated in his *Harper's* apologia in 1947:

> My chief purpose was to end the war in victory with the least possible cost in the lives of the men in the armies which I had helped to raise. In the light of the alternatives which, on a fair estimate, were open to us I believe that no man, in our position and subject to our responsibilities, holding in his hands a weapon of such possibilities for accomplishing this purpose and saving those lives, could have failed to use it and afterwards looked his countrymen in the face.

The Scientific Panel of the Interim Committee—Lawrence, Compton, Fermi, Oppenheimer—had been asked to conjure a demonstration of sufficient credibility to end the war. Meeting at Los Alamos on the weekend of June 16–17, debating long into the night, it found in the negative. Even Fermi's ingenuity was not sufficient to the task of devising a demonstration persuasive enough to decide the outcome of a long and bitter conflict. Recognizing "our obligation to our nation to use the weapons to help save American lives in the Japanese war," the panel first surveyed the opinions of scientific colleagues and then stated its own:

> Those who advocate a purely technical demonstration would wish to outlaw the use of atomic weapons, and have feared that if we use the weapons now our position in future negotiations will be prejudiced. Others emphasize the opportunity of saving American lives by immediate military use, and believe that such use will improve the international prospects, in that they are more concerned with the prevention of war than with the elimination of this specific weapon. We find ourselves closer to these latter views; we can propose no technical demonstration likely to bring an end to the war; we see no acceptable alternative to direct military use.

The bomb was to prove to the Japanese that the Potsdam Declaration meant business. It was to shock them to surrender. It was to put the Russians on notice and serve, in Stimson's words, as a "badly needed equalizer." It was to let the world know what was coming: Leo Szilard had dallied with that rationale in 1944 before concluding in 1945 on moral grounds that the bomb should not be used and on political grounds that it should be kept secret. Teller revived a variant rationale in early July 1945, in replying to Szilard about a petition Szilard was then circulating among Manhattan Project scientists protesting the bomb's impending use:

> First of all let me say that I have no hope of clearing my conscience. The things we are working on are so terrible that no amount of protesting or fiddling with politics will save our souls. . . .
>
> But I am not really convinced of your objections. I do not feel that there is any chance to outlaw any one weapon. If we have a slim chance of survival, it lies in the possibility to get rid of wars. The more decisive the weapon is the more surely it will be used in any real conflicts and no agreements will help.
>
> Our only hope is in getting the facts of our results before the people. This might help to convince everybody that the next war would be fatal. For this purpose actual combat-use might even be the best thing.

The bomb was also to be used to pay for itself, to justify to Congress the investment of $2 billion, to keep Groves and Stimson out of Leavenworth prison.

"To avert a vast, indefinite butchery," Winston Churchill summarizes in his history of the Second World War, "to bring the war to an end, to give peace to the world, to lay healing hands upon its tortured peoples by a manifestation of overwhelming power at the cost of a few explosions, seemed, after all our toils and perils, a miracle of deliverance."

The few explosions did not seem a miracle of deliverance to the civilians of the enemy cities upon whom the bombs would be dropped. In their behalf—surely they have claim—something more might be said about reasons. The bombs were authorized not because the Japanese refused to surrender but because they refused to surrender unconditionally. The debacle of conditional peace following the First World War led to the demand for unconditional surrender in the Second, the earlier conflict casting its dark shadow down the years. "It was the insistence on unconditional surrender that was the root of all evil," writes the Oxford moralist G. E. M. Anscombe in a 1957 pamphlet opposing the awarding of an honorary degree to Harry Truman. "The connection between such a demand and the need to use the most ferocious methods of warfare will be obvious. And in itself the proposal of an unlimited objective in war is stupid and barbarous."

As before in the Great War for every belligerent, that was what the Second World War had become: stupid and barbarous. "For men to choose to kill the innocent as a means to their ends," Anscombe adds bluntly, "is always murder, and murder is one of the worst of human actions. . . . In the bombing of [Japanese] cities it was certainly decided to kill the innocent as a means to an end." In the decision of the Japanese militarists to arm the Japanese people with bamboo spears and set them against a major invasion force to fight to the death to preserve the homeland it was certainly decided to kill the innocent as a means to an end as well.

The barbarism was not confined to the combatants or the general staffs. It came to permeate civilian life in every country: in Germany and Japan, in Britain, in Russia, certainly in the United States. It was perhaps the ultimate reason Jimmy Byrnes, the politician's politician, and Harry Truman, the man of the people, felt free to use and compelled to use a new weapon of mass destruction on civilians in undefended cities. "It was the psychology of the American people," I. I. Rabi eventually decided. "I'm not justifying it on military grounds but on the existence of this mood of the military with the backing of the American people." The mood, suggests the historian Herbert Feis, encompassed "impatience to end the strain of war blended with a zest for victory. They longed to be done with smashing, burning, killing, dying—and were angry at the defiant, crazed, useless prolongation of the ordeal."

In 1945 *Life* magazine was the preeminent general-circulation magazine in the United States. It served millions of American families for news and entertainment much as television a decade later began to do. Children read it avidly and reported on its contents in school. In the last issue of *Life* before the United States used the atomic bomb a one-page picture story appeared, titled, in 48-point capitals, A JAP BURNS. Its brief text, for those who could tear their eyes away from the six postcard-sized black-and-white photographs showing a man being burned alive long enough to read the words, savored horror while complaining of ugly necessity:

> When the 7th Australian Division landed near Balikpapan on the island of Borneo last month they found a town strongly defended by Japanese. As usual, the enemy fought from caves, from pillboxes, from every available hiding place. And, as usual, there was only one way to advance against them: burn them out. Men of the 7th, who had fought the Japs before, quickly applied their flamethrowers, soon convinced some Japs that it was time to quit. Others, like the one shown here, refused. So they had to be burned out.
>
> Although men have fought one another with fire from time immemorial, the flamethrower is easily the most cruel, the most terrifying weapon ever developed. If it does not suffocate the enemy in his hiding place, its quickly licking tongues of flame sear his body to a black crisp. But so long as the Jap refuses to come out of his holes and keeps killing, this is the only way.

In a single tabloid page *Life* had assembled a brutal allegory of the later course of the Pacific war.

Little Boy was ready on July 31. It lacked only its four sections of cordite charge, a precaution prepared when the weapon was designed at Los Alamos but decided upon at Tinian, for safety on takeoff and in the event visual bombing proved impossible, in which case Tibbets had orders to bring the bomb back. Three of Tibbets' full complement of fifteen B-29's flew a last test that last day of July with a dummy Little Boy. They took off from Tinian, rendezvoused over Iwo Jima, returned to Tinian, dropped unit L6 into the sea and practiced their daredevil diving turn. "With the completion of this test," writes Norman Ramsey, "all tests preliminary to combat delivery of a Little Boy with active material were completed." That unit would be number L11, and the sturdy tungsten-steel target holder screwed to its muzzle, the best in stock, was the first one Los Alamos had received; it had served four times for firing tests at Anchor Ranch late in 1944 before being packed in cosmoline for the voyage out to Tinian.

Since everything was ready, Farrell telexed Groves to report that the mission could be flown on August 1; he would assume that the Spaatz directive of July 25 authorized such initiative unless Groves replied to the contrary. The commanding general of the Manhattan Project let his deputy's interpretation stand. Little Boy would have flown on August 1 if a typhoon had not approached Japan that day to intervene.

So the mission waited on the weather. On August 2, Thursday, the three B-29's that carried Fat Man preassemblies arrived from New Mexico.

The assembly team of Los Alamos scientists and military ordnance technicians went to work immediately to prepare one Fat Man for a drop test and a second with higher-quality HE castings for combat. The third preassembly would be held in reserve for the plutonium core scheduled to be shipped from Los Alamos in mid-August. "By August 3," recalls Paul Tibbets, "we were watching the weather and comparing it to the [long-range] forecast. The actual and forecast weather were almost identical, so we got busy."

Among other necessities, getting busy involved briefing the crews of the seven 509th B-29's that would fly the first mission for weather reporting, observation and bombing. Tibbets scheduled the briefing for 1500 hours on August 4. The crews arrived between 1400 and 1500 to find the briefing hut completely surrounded by MP's armed with carbines. Tibbets walked in promptly at 1500; he had just returned from checking out the aircraft he intended to use to deliver Little Boy, usually piloted by Robert Lewis: B-29 number 82, as yet unnamed. Deke Parsons joined him on the briefing platform. A radio operator, Sergeant Abe Spitzer, kept an illegal diary of his experiences at Tinian that describes the briefing.

The moment had arrived, Tibbets told the assembled crews. The weapon they were about to deliver had recently been tested successfully in the United States; now they were going to drop it on the enemy.

Two intelligence officers undraped the blackboards behind the 509th commander to reveal aerial photographs of the targets: Hiroshima, Kokura, Nagasaki. (Niigata was excluded, apparently because of weather.) Tibbets named them and assigned three crews—"finger crews"—to fly ahead the day of the drop to assess their cloud cover. Two more aircraft would accompany him to photograph and observe; the seventh would wait beside a loading pit on Iwo Jima as a spare in case Tibbets' plane malfunctioned.

The 509th commander introduced Parsons, who wasted no words. He told the crews the bomb they were going to drop was something new in the history of warfare, the most destructive weapon ever made: it would probably almost totally destroy an area three miles across.

They were stunned. "It is like some weird dream," Spitzer mused, "conceived by one with too vivid an imagination."

Parsons prepared to show a motion picture of the Trinity test. The projector refused to start. Then it started abruptly and began chewing up leader. Parsons told the projectionist to shut the machine off and improvised. He described the shot in the Jornada del Muerto: how far away the light had been seen, how far away the explosion had been heard, the effects of the blast wave, the formation of the mushroom cloud. He did not identify the source of the weapon's energy, but with details—a man knocked down at 10,000 yards, men 10 and 20 miles away temporarily blinded—he won their rapt attention.

Tibbets took over again. They were now the hottest crews in the Air Force, he warned them. He forbade them to write letters home or to discuss the mission even among themselves. He briefed them on the flight. It would probably go, he said, early on the morning of August 6. An air-sea rescue officer described rescue operations. Tibbets closed with a challenge, a final word Spitzer paraphrases in his diary:

> The colonel began by saying that whatever any of us, including himself, had done before was small potatoes compared to what we were going to do now. Then he said the usual things, but he said them well, as if he meant them, about how proud he was to have been associated with us, about how high our morale had been, and how difficult it was not knowing what we were doing, thinking maybe we were wasting our time and that the "gimmick" was just somebody's wild dream. He was personally honored and he was sure all of us were, to have been chosen to take part in this raid, which, he said—and all the other big-wigs nodded when he said it—would shorten the war by at least six months. And you got the feeling that he really thought this bomb would end the war, period.

The following morning, Sunday, Guam reported that weather over the target cities should improve the next day. "At 1400 on August 5," Norman Ramsey records, "General LeMay officially confirmed that the mission would take place on August 6."

That afternoon the loading crew winched Little Boy onto its sturdy transport dolly, draped it with a tarpaulin to protect it from prying eyes—there were still Japanese soldiers hiding out on the island, hunted at night by security forces like raccoons—and wheeled it to one of the 13 by 16-foot loading pits Kirkpatrick had prepared. A battery of photographers followed along to record the proceedings. The dolly was wheeled over the nine-foot pit on tracks; the hydraulic lift came up to relieve it of its bomb and detachable cradle; the crew wheeled the dolly away, removed the tracks, rotated the bomb 90 degrees and lowered it into the pit.

The world's first combat atomic bomb looked like "an elongated trash can with fins," one of Tibbets' crew members thought. With its tapered tail assembly that culminated in a boxed frame of stabilizing baffle plates it was 10½ feet long and 29 inches in diameter. It weighed 9,700 pounds, an armored cylinder jacketed in blackened dull steel with a flat, rounded nose. A triple fusing system armed it. The main fusing component was a radar unit adapted from a tail-warning mechanism developed to alert combat pilots

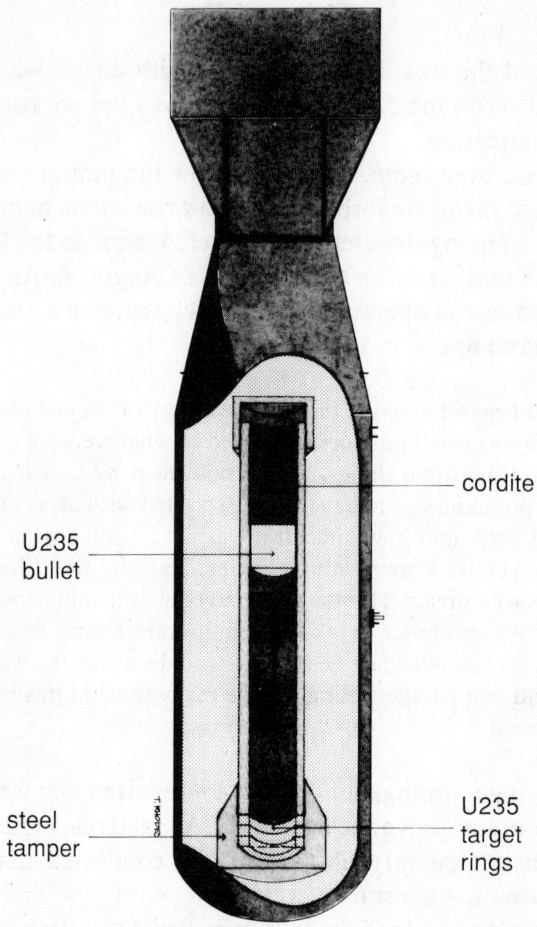

when enemy aircraft approached from behind. "This radar device," notes the Los Alamos technical history, "would close a relay [i.e., a switch] at a predetermined altitude above the target." For reliability Little Boy and Fat Man each carried four such radar units, called Archies. Rather than an approaching enemy aircraft, the bomb Archies would bounce their signals off the approaching enemy ground. An agreed reading by any two of the units would send a firing signal into the next stage of the fusing system, the technical history explains:

> This stage consisted of a bank of clock-operated switches, started by arming wires which were pulled out of the clocks when the bomb dropped from the plane's bomb bay. These clock switches were not closed until 15 seconds after the bomb was released. Their purpose was to prevent detonation in case the A[rchie] units were fired by signals reflected from the plane. A second arming

> device was a [barometric] pressure switch, which did not close until subject to a pressure corresponding to 7000 feet altitude.

Once it passed through the clock and barometric arming devices the Little Boy firing signal went directly to the primers that lit the cordite charges to fire the gun. Externally the fusing system revealed itself in trailing whips of radar antennae, clock wires threaded into holes in the weapon's upper waist and holes in its tapered tail assembly that admitted external air to guarantee accurate barometry.

Loading the bomb was delicate: the fit was tight. A ground crew towed the B-29 to a position beside the loading pit, running onto a turntable the main landing gear on the wing nearer the pit. Towing the aircraft around on the turntable through 180 degrees positioned it over the pit. The hydraulic lift raised Little Boy to a point directly below the open bomb doors. A plumb bob hung from the single bomb shackle for a point of reference and jacks built into the bomb cradle allowed the crew to line up the bomb eye.

"The operation can be accomplished in 20 to 25 minutes," a Boeing engineer commented in an August report, "but is a rather ticklish procedure, as there is very little clearance with the catwalks and, once installed, nothing holds the bomb but the single shackle and adjustable sway braces bearing on it."

Though he flew it as his own, Robert Lewis had never named B-29 number 82. The day of the loading Tibbets consulted the officers in Lewis' crew—but not Lewis—and did so. The 509th commander chose not pinups or puns but his mother's given names, Enola Gay, because she had assured him he would not be killed flying when he fought out with his father his decision to become a pilot. "Through the years," Tibbets told an interviewer once, "whenever I got in a tight spot in a plane I always remembered her calm assurance. It helped. In getting ready for the big one I rarely thought of what might happen, but when I did, those words of Mom's put an end to it." He "wrote a note on a slip of paper," located a sign painter among the service personnel—the man had to be dragged away from a softball game—and told him to "paint that on the strike ship, nice and big." Foot-high, squared brushstrokes went on at a 30-degree angle beneath the pilot's window of the bullet-nosed plane, the middle name flush-right below the first.

Lewis, a sturdy, combative two-hundred-pounder, had known for a day or two that Tibbets would pilot the mission, a disappointment, but still considered the special B-29 his own. When he dropped by late in the afternoon to inspect it and found ENOLA GAY painted on its fuselage he was furious. "What the hell is *that* doing on *my* plane?" one of his crew mates remembers him yelling. He found out that Tibbets had authorized the christening and marched off to confront him. The 509th commander told him coolly, rank having its privileges, that he didn't think the junior officer would mind. Lewis minded, but he could do no more than stow away his resentment for the war stories he would tell.

"By dinnertime on the fifth," Tibbets narrates, "all [preparations were] completed. The atom bomb was ready, the planes were gassed and checked. Takeoff was set for [2:45] a.m. I tried to nap, but visitors kept me up. [Captain Theodore J.] Dutch [Van Kirk, the *Enola Gay*'s navigator,] swallowed two sleeping tablets, then sat up wide awake all night playing poker." The weapon waiting in the bomb bay took its toll on nerves.

"Final briefing was at 0000 of August 6," Ramsey notes—midnight. Tibbets emphasized the power of the bomb, reminded the men to wear the polarized goggles they had been issued, cautioned them to obey orders and follow their protocols. A weather officer predicted moderate winds with clouds over the targets clearing at dawn. Tibbets called forward a Protestant chaplain who delivered a prayer composed for the occasion on the back of an envelope; it asked the Almighty Father "to be with those who brave the heights of Thy heaven and who carry the battle to our enemies."

After the midnight briefing the crews ate an early breakfast of ham and eggs and Tibbets' favorite pineapple fritters. Trucks delivered them to their hardstands. At the *Enola Gay*'s hardstand, writes Ramsey, "amid brilliant floodlights, pictures were taken and retaken by still and motion picture photographers (as though for a Hollywood premiere)." A photograph shows ten of the twelve members of the strike plane's crew posed in flight coveralls under the forward fuselage by the nose wheel: boyish Van Kirk in overseas cap with his coveralls unzipped down his chest to expose a white T-shirt; Major Thomas Ferebee, the bombardier, a handsome Errol Flynn copy with an Errol Flynn mustache, resting a friendly hand on Van Kirk's shoulder; Tibbets standing at the center of it all easily smiling, belted and trim, his hands in his pockets; at Tibbets' left Robert Lewis, the only crew member wearing a weapon; small, wiry Lieutenant Jacob Beser beside Lewis awkwardly smiling, a Jewish technician from Baltimore added for the flight, responsible for electronic countermeasures to screen the Archie units from Japanese radar. In front of the officers kneel the slimmer, mostly younger enlisted men (though the entire flight crew was young, Tibbets now all of thirty years old): radar operator Sergeant Joseph Stiborik; tail gunner Staff Sergeant Robert Caron, Brooklyn-born, wearing a Dodgers baseball cap; radio operator Private Richard R. Nelson; assistant engineer Sergeant Robert H. Shumard; flight engineer Staff Sergeant

Wyatt Duzenbury, thirty-two, a former Michigan tree surgeon who thought the bomb looked like a tree trunk. An eleventh member of the crew, 2nd Lieutenant Morris Jeppson, an ordnance expert, would assist Deke Parsons in arming and monitoring Little Boy. Parsons, the twelfth man, resisted photographing but was flying the mission as weaponeer.

The three weather planes and the Iwo Jima standby had already left. Tibbets ordered Wyatt Duzenbury to start engines at 0227 hours. Pilot and copilot sat side by side just back of the point where the cylindrical fuselage began to curve inward to form the bullet-shaped nose; Ferebee, the bombardier, sat a step down ahead of them within the nose itself, an exposed position but a good view. Almost everything inside the aircraft was painted a dull lime green. "It was just another mission," Tibbets says, "if you didn't let imagination run away with your wits." As Dimples Eight Two, the *Enola Gay*'s unlikely designation that day, he reconstructs his dialogue with the Tinian control tower:

> I forgot the atom bomb and concentrated on the cockpit check.
>
> I called the tower. "Dimples Eight Two to North Tinian Tower. Taxi-out and take-off instructions."
>
> "Dimples Eight Two from North Tinian Tower. Take off to the east on Runway A for Able."
>
> At the end of the runway, another call to the tower and a quick response: "Dimples Eight Two cleared for take-off."
>
> Bob Lewis called off the time. Fifteen seconds to go. Ten seconds. Five seconds. Get ready.

At that moment the *Enola Gay* weighed 65 tons. It carried 7,000 gallons of fuel and a four-ton bomb. It was 15,000 pounds overweight. Confident the aircraft was maintained too well to falter, Tibbets decided to use as much of the two-mile runway as he needed to build RPM's and manifold pressure before roll-up.

He eased the brakes at 0245, the four fuel-injected Wright Cyclone engines pounding. "The B-29 has lots of torque in take-off," he notes. "It wants to swerve off the runway to the left. The average mass-production pilot offsets torque by braking his right wheels. It's a rough ride, you lose ten miles an hour and you delay the take-off." Nothing so crude for Tibbets. "Pilots of the 509th Group were taught to cancel torque by leading in with the left engines, advancing throttles ahead of the right engines. At eighty miles an hour, you get full rudder control, advance the right-hand engines to full power and, in a moment, you're airborne." Takeoff needed longer than a moment for the *Enola Gay*'s overloaded flight. As the runway disappeared beneath the big bomber Lewis fought the urge to pull back the yoke. At the last possible takeoff point he thought he did. Not he but Tibbets did and abruptly they were flying, an old dream of men, climbing above a black sea.

Ten minutes later they crossed the northern tip of Saipan on a course northwest by north at 4,700 feet. The air temperature was a balmy 72°. They were flying low not to burn fuel lifting fuel and for the comfort of the two weaponeers, Parsons and Jeppson, who had to enter the unpressurized, unheated bomb bay to finish assembling the bomb.

That work began at 0300. It was demanding in the cramped confines of the loaded bomb bay but not dangerous; there was only minimal risk of explosion. The green plugs that blocked the firing signal and prevented accidental detonation were plugged into the weapon; Parsons confirmed that fact first of all. Next he removed a rear plate; removed an armor plate beneath, exposing the cannon breech; inserted a wrench into the breech plug and rotated the wrench about sixteen times to unscrew the plug; removed it and placed it carefully on a rubber pad. He inserted the four sections of cordite one at a time, red ends to breech. He replaced the breech plug and tightened it home, connected the firing line, reinstalled the two metal plates and with Jeppson's help removed and secured the tools and the catwalk. Little Boy was complete but not yet armed. The charge loading took fifteen minutes. They spent another fifteen minutes checking monitoring circuitry at the panel installed at the weaponeer's position in the forward section. Then, except for monitoring, their work was done until time to arm the bomb.

Robert Lewis kept a journal of the flight. William L. Lawrence, the *New York Times* science editor attached to the Manhattan Project, had traveled out to Tinian expecting to go along. When he learned to his bitter disappointment that his participation had been deleted he asked Lewis to take notes. The copilot imagined himself writing a letter to his mother and father but appears to have sensed that the world would be looking over his shoulder and styled his entries with regulation Air Force bonhomie. "At forty-five minutes out of our base," he began self-consciously, "everyone is at work. Colonel Tibbets has been hard at work with the usual tasks that belong to the pilot of a B-29. Captain Van Kirk, navigator, and Sergeant Stiborik, radio operator, are in continuous conversation, as they are shooting bearings on the northern Marianas and making radar wind runs." No mention of Parsons or Jeppson, oddly enough, though Lewis could have seen the bomb hanging in its bay through the round port below the tunnel opening straight back from his copilot's seat.

The automatic pilot, personified as George, was flying the plane, which Tibbets stationed below 5,000 feet. The commander realized he was tired, Lewis records: "The colonel, better known as 'the Old Bull,' shows signs of a tough day. With all he's had to do to get this mission off, he is deserving of a few winks, so I'll have a bite to eat and look after 'George.' "

Rather than sleep Tibbets crawled through the thirty-foot tunnel to chat with the waist crew, wondering if they knew what they were carrying. "A chemist's nightmare," the tail gunner, Robert Caron, guessed, then "a physicist's nightmare." "Not exactly," Tibbets hedged. Tibbets was leaving by the time Caron put two and two together:

> [Tibbets] stayed . . . a little longer, and then started to crawl forward up the tunnel. I remembered something else, and just as the last of the Old Man was disappearing, I sort of tugged at his foot, which was still showing. He came sliding back in a hurry, thinking maybe something was wrong. "What's the matter?"
>
> I looked at him and said, "Colonel, are we splitting atoms today?"
>
> This time he gave me a really funny look, and said, "That's about it."

Caron's third try, which he styles "a lucky guess," apparently decided Tibbets to complete the crew's briefing; back in his seat he switched on the interphone, called "Attention!" and remembers saying something like "Well, boys, here's the last piece of the puzzle." They carried an atomic bomb, he told them, the first to be dropped from an airplane. They were not physicists; they understood at least that the weapon was different from any other ever used in war.

Lewis took control from George to weave his way through a mass of towering cumuli, clouds black in the darkness that swept aside to reveal a sky shot with stars. "At 4:30," he jotted, "we saw signs of a late moon in the east. I think everyone will feel relieved when we have left our bomb with the Japs and get half way home. Or, better still, all the way home." Ferebee in the nose was quiet; Lewis suspected he was thinking of home, "in the midwest part of old U.S.A." The bombardier was in fact from Mocksville, North Carolina, close enough to the Midwest for a native of New York. Dawn lightening a little past 0500 cheered them; "it looks at this time," Lewis wrote coming out of the clouds, "that we will have clear sailing for a long spell."

At 0552 they approached Iwo Jima and Tibbets began climbing to 9,-300 feet to rendezvous with the observation and photography planes. The *Enola Gay* circled left over Iwo, found its two escorts and moved on, its course continuing northwest by north toward the archipelago of green islands the men called the Empire.

"After leaving Iwo we began to pick up some low stratus," Lewis resumes his narrative, "and before long we were flying on top of an undercast. At 07:10 the undercast began to break up a little bit. Outside of a high thin cirrus and the low stuff it's a very beautiful day. We are now about two hours from Bombs Away." They flew into history through a middle world, suspended between sky and sea, drinking coffee and eating ham sandwiches, engines droning, the smell of hot electronics in the air.

At 0730 Parsons visited the bomb bay for the last time to arm Little Boy, exchanging its green plugs for red and activating its internal batteries. Tibbets was about to begin the 45-minute climb to altitude. Jeppson worked his console. Parsons told Tibbets that Little Boy was "final." Lewis overheard:

> The bomb was now independent of the plane. It was a peculiar sensation. I had a feeling the bomb had a life of its own now that had nothing to do with us. I wished it were over and we were at this same position on the way back to Tinian.

"Well, folks, it won't be long now," the copilot added as Tibbets increased power to climb.

The weather plane at Hiroshima reported in at 0815 (0715 Hiroshima time). It found two-tenths cloud cover lower and middle and two-tenths at 15,000 feet. The other two target weather reports followed. "Our primary is the best target," Lewis wrote enthusiastically, "so, with everything going well so far, we will make a bomb run on Hiroshima." "It's Hiroshima," Tibbets announced to the crew.

They leveled at 31,000 feet at 0840. They had pressurized the aircraft and heated it against an outside temperature of −10° F. Ten minutes later they achieved landfall over Shikoku, the smaller home island east of Hiroshima, a city which looks southeastward from the coast of Honshu into the Inland Sea. "As we are approaching our target, Ferebee, Van Kirk and Stiborik are coming into their own, while the colonel and I are standing by and giving the boys what they need." Correcting course, Lewis means, aligning the plane. He got excited then or busy: "There will be a short intermission while we bomb our target." But bombing the target was the main event.

The crew pulled on heavy flak suits, cumbersome protection the pilots disdained. No Japanese fighters came up to meet them, nor were they bothered by flak.

The two escort planes dropped back to give the *Enola Gay* room. Tibbets reminded his men to wear their protective goggles.

They carried no maps. They had studied aerial photographs and knew the target city well. It was distinctive in any case, sited on a delta divided by the channels of seven distributaries. "Twelve miles from the target," Tibbets remembers, "Ferebee called, 'I see it!' He clutched in his bombsight and took control of the plane from me for a visual run. Dutch [Van Kirk] kept giving me radar course corrections. He was working with the radar operator. . . . I couldn't raise them on the interphone to tell them Ferebee had the plane." The bombardier flew the plane through his bombsight, the knurled knobs he adjusted instructing the automatic pilot to make minor corrections in course. They crossed the Inland Sea on a heading only five degrees south of due west. Van Kirk noticed eight large ships south of them in Hiroshima harbor. The *Enola Gay*'s ground speed then was 285 knots, about 328 miles per hour.

Above a fork in the Ōta River in central Hiroshima a T-shaped bridge spanned the river and connected to the island formed by the two distributaries. The Aioi Bridge, not a war plant surrounded by workers' houses, was Ferebee's chosen aiming point. Second Army headquarters was based nearby. Tibbets had called the bridge the most perfect AP he'd seen in the whole damn war:

> Ferebee had the drift well killed but the rate was off a little. He made two slight corrections. A loud "blip" on the radio notified the escort B-29's that the bomb would drop in two minutes. After that, Tom looked up from his bombsight and nodded to me; it was going to be okay.
>
> He motioned to the radio operator to give the final warning. A continuous tone signal went out, telling [the escorts]: "In fifteen seconds she goes."

The distant weather planes also heard the radio signal. So did the spare B-29 parked on Iwo Jima. It alerted Luis Alvarez in the observation plane to prepare to film the oscilloscopes he had installed there; the radio-linked parachute gauges he had designed to measure Little Boy's explosive yield hung in the bomb bay waiting to drop with the bomb and float down toward the city.

Hiroshima unrolled east to west in the cross hairs of Thomas Ferebee's Norden bombsight. The bomb-bay doors were open. Ferebee had flown sixty-three combat missions in Europe before returning to the United States to instruct and then to join the 509th. Before the war he had wanted to be a baseball player and had got as far as spring tryouts with a major-league team. He was twenty-four years old.

"The radio tone ended," Tibbets says tersely, "the bomb dropped, Ferebee unclutched his sight." The arming wires pulled out to start Little Boy's clocks. The first combat atomic bomb fell away from the plane, then nosed down. It was inscribed with autographs and messages, some of them obscene. "Greetings to the Emperor from the men of the *Indianapolis,*" one challenged.

Four tons lighter, the B-29 jumped. Tibbets dove away:

> I threw off the automatic pilot and hauled *Enola Gay* into the turn.
>
> I pulled antiglare goggles over my eyes. I couldn't see through them; I was blind. I threw them to the floor.
>
> A bright light filled the plane. The first shock wave hit us.
>
> We were eleven and a half miles slant range from the atomic explosion, but the whole airplane cracked and crinkled from the blast. I yelled "Flak!" thinking a heavy gun battery had found us.
>
> The tail gunner had seen the first wave coming, a visible shimmer in the atmosphere, but he didn't know what it was until it hit. When the second wave came, he called out a warning.
>
> We turned back to look at Hiroshima. The city was hidden by that awful cloud . . . boiling up, mushrooming, terrible and incredibly tall.
>
> No one spoke for a moment; then everyone was talking. I remember Lewis pounding my shoulder, saying, "Look at that! Look at that! Look at that!" Tom Ferebee wondered about whether radioactivity would make us all sterile. Lewis said he could taste atomic fission. He said it tasted like lead.

"Fellows," Tibbets announced on the interphone, "you have just dropped the first atomic bomb in history."

Van Kirk remembers the two shock waves—one direct, one reflected from the ground—vividly:

> [It was] very much as if you've ever sat on an ash can and had somebody hit it with a baseball bat. . . . The plane bounced, it jumped and there was a noise like a piece of sheet metal snapping. Those of us who had flown quite a bit over Europe thought that it was anti-aircraft fire that had exploded very close to the plane.

The apparent proximity of the explosion would be one of its trademarks, much as its heat had seemed intimate to Philip Morrison and his colleagues at Trinity.

Turning, diving, circling back to watch, the crew of the *Enola Gay* missed the early fireball; when they looked again Hiroshima smothered under a pall. Lewis in a postwar interview:

> I don't believe anyone ever expected to look at a sight quite like that. Where we had seen a clear city two minutes before, we could now no longer see the city. We could see smoke and fires creeping up the sides of the mountains.

Van Kirk:

> If you want to describe it as something you are familiar with, a pot of boiling black oil. . . . I thought: Thank God the war is over and I don't have to get shot at any more. I can go home.

It was a sentiment hundreds of thousands of American soldiers and sailors would soon express, and it was hard-earned.

Leaving the scene the tail gunner, Robert Caron, had a long view:

> I kept shooting pictures and trying to get the mess down over the city. All the while I was describing this on the intercom. . . . The mushroom itself was a spectacular sight, a bubbling mass of purple-gray smoke and you could see it had a red core in it and everything was burning inside. As we got farther away, we could see the base of the mushroom and below we could see what looked like a few-hundred-foot layer of debris and smoke and what have you.
>
> I was trying to describe the mushroom, this turbulent mass. I saw fires springing up in different places, like flames shooting up on a bed of coals. I was asked to count them. I said, "Count them?" Hell, I gave up when there were about fifteen, they were coming too fast to count. I can still see it—that mushroom and that turbulent mass—it looked like lava or molasses covering the whole city, and it seemed to flow outward up into the foothills where the little valleys would come into the plain, with fires starting up all over, so pretty soon it was hard to see anything because of the smoke.

Jacob Beser, the electronic countermeasures officer, an engineering student at Johns Hopkins before he enlisted, found an image from the seashore for the turmoil he saw:

> That city was burning for all she was worth. It looked like . . . well, did you ever go to the beach and stir up the sand in shallow water and see it all billow up? That's what it looked like to me.

Little Boy exploded at 8:16:02 Hiroshima time, 43 seconds after it left the *Enola Gay,* 1,900 feet above the courtyard of Shima Hospital, 550 feet southeast of Thomas Ferebee's aiming point, Aioi Bridge, with a yield equivalent to 12,500 tons of TNT.

"It was all impersonal," Paul Tibbets would come to say. It was not impersonal for Robert Lewis. "If I live a hundred years," he wrote in his journal, "I'll never quite get these few minutes out of my mind." Nor would the people of Hiroshima.

* * *

In my mind's eye, like a waking dream, I could still see the tongues of fire at work on the bodies of men.

Masuji Ibuse, *Black Rain*

The settlement on the delta islands of the Ōta River in southwestern Honshu was named Ashihara, "reed field," or Gokaura, "five villages," before the feudal lord Terumoto Mōri built a fortress there between 1589 and 1591 to secure an outlet for his family holdings on the Inland Sea. Mōri called his fortress Hiro-shima-jō, "broad-island castle," and gradually the town of merchants and artisans that grew up around it acquired its name. It was an 800-foot rectangle of massive stone walls protected within a wide rectangular moat, one corner graced by a high white pagoda-like tower with five progressively inset roofs. The Mōri family soon lost its holdings to the stronger Fukushima family, which lost them in turn to the Asano family in 1619. The Asanos had the good sense to have allied themselves closely with the Tokugawa Shogunate and ruled Hiroshima fief within that alliance for the next two and a half centuries. Across those centuries the town prospered. The Asanos saw to its progressive enlargement by filling in the estuarial shallows to connect its islands. Divided then into long, narrow districts by the Ōta's seven distributaries, Hiroshima assumed the form of an open, extended hand.

The restoration of the Meiji emperor in 1868 and the abolition of the feudal clan system transformed Hiroshima fief into Hiroshima Prefecture and the town, like the country, began vigorously to modernize. A physician was appointed its first mayor in 1889 when it officially became a city; the population that celebrated the change numbered 83,387. Five years of expensive landfill and construction culminated that year in the opening of Ujina harbor, a reclamation project that established Hiroshima as a major commercial port. Railroads came through at the turn of the century.

By then Hiroshima and its castle had found further service as an army base and the Imperial Army Fifth Division was quartered in barracks within and around the castle grounds. The Fifth Division was the first to be shipped to battle when Japan and China initiated hostilities in 1894; Ujina harbor served as a major point of embarkation and would continue in that role for the next fifty years. The Meiji emperor moved his headquarters to the castle in Hiroshima in September, the better to direct the war, and the Diet met in extraordinary session in a provisional Diet building there. Until the following April, when the limited mainland war ended with a Japanese victory that included the acquisition of Formosa and the southern part of Manchuria, Hiroshima was *de facto* the capital of Japan. Then the emperor returned to Tokyo and the city consolidated its gains.

It acquired further military and industrial investments in the first three decades of the twentieth century as Japan turned to increasing international adventure. By the Second World War, an American study noted in the autumn of 1945, "Hiroshima was a city of considerable military importance. It contained the 2nd Army headquarters, which commanded the defense of all of southern Japan. The city was a communication center, a storage point, and an assembly area for troops. To quote a Japanese report, 'Probably more than a thousand times since the beginning of the war did the Hiroshima citizens see off with cries of "Banzai" the troops leaving from the harbor.' " From Hiroshima in 1945 the Japanese Army general staff prepared to direct the defense of Kyushu against the impending American invasion.

Earlier in the war the city's population had approached 400,000, but the threat of strategic bombing, so ominously delayed, had led the authorities to order a series of evacuations; on August 6 the resident population numbered some 280,000 to 290,000 civilians plus about 43,000 soldiers. Given that proportion of civilian to military—more than six to one—Hiroshima was not, as Truman had promised in his Potsdam diary, a "purely military" target. It was not without responsibility, however, in serving the ends of war.

"The hour was early, the morning still, warm, and beautiful," a Hiroshima physician, Michihiko Hachiya, the director of the Hiroshima Communications Hospital, begins a diary of the events Little Boy entrained on August 6. "Shimmering leaves, reflecting sunlight from a cloudless sky, made a pleasant contrast with shadows in my garden." The temperature at eight o'clock was 80 degrees, the humidity 80 percent, the wind calm. The seven branches of the Ōta flowed past crowds of citizens walking and bicycling to work. The streetcars that clanged outside Fukuya department store two blocks north of Aioi Bridge were packed. Thousands of soldiers, bare to the waist, exercised at morning calesthenics on the east and west parade grounds that flanked Hiroshima Castle a long block west of the T-shaped bridge. More than eight thousand schoolgirls, ordered to duty the day before, worked outdoors in the central city helping to raze houses to clear firebreaks against the possibility of an incendiary attack. An air raid alert at 7:09—the 509th weather plane—had been called off at 7:31 when the B-29 left the area. Three more *B-sans* approaching just before 8:15 sent hardly anyone to cover, though many raised their eyes to the high silver instruments to watch.

"Just as I looked up at the sky," remembers a girl who was five years old at the time and safely at home in the suburbs, "there was a flash of white light and the green in the plants looked in that light like the color of dry leaves."

Closer was more brutal illumination. A young woman helping to clear firebreaks, a junior-college student at the time, recalls: "Shortly after the voice of our teacher, saying 'Oh, there's a B!' made us look up at the sky, we felt a tremendous flash of lightning. In an instant we were blinded and everything was just a frenzy of delirium."

Closer still, in the heart of the city, no one survived to report the coming of the light; the constrained witness of investigative groups must serve instead for testimony. A Yale Medical School pathologist working with a joint American-Japanese study commission a few months after the war, Averill A. Liebow, observes:

> Accompanying the flash of light was an instantaneous flash of heat. . . Its duration was probably less than one tenth of a second and its intensity was sufficient to cause nearby flammable objects . . . to burst into flame and to char poles as far as 4,000 yards away from the hypocenter [i.e., the point on the ground directly below the fireball]. . . . At 600–700 yards it was sufficient to chip and roughen granite. . . . The heat also produced bubbling of tile to about 1,300 yards. It has been found by experiment that to produce this effect a temperature of [3,000° F] acting for four seconds is necessary, but under these conditions the effect is deeper, which indicates that the temperature was higher and the duration less during the Hiroshima explosion.

"Because the heat in [the] flash comes in such a short time," adds a Manhattan Project study, "there is no time for any cooling to take place, and the temperature of a person's skin can be raised [120° F] . . . in the first millisecond at a distance of [2.3 miles]."

The most authoritative study of the Hiroshima bombing, begun in 1976 in consultation with thirty-four Japanese scientists and physicians, reviews the consequences of this infernal insolation, which at half a mile from the hypocenter was more than three thousand times as energetic as the sunlight that had shimmered on Dr. Hachiya's leaves:

> The temperature at the site of the explosion . . . reached [5,400° F] . . . and primary atomic bomb thermal injury . . . was found in those exposed within [2 miles] of the hypocenter. . . . Primary burns are injuries of a special nature and not ordinarily experienced in everyday life.

This Japanese study distinguishes five grades of primary thermal burns ranging from grade one, red burn, through grade three, white burn, to grade five, carbonized skin with charring. It finds that "severe thermal burns of over grade 5 occurred within [0.6 to 1 mile] of the hypocenter . . . and those of grades 1 to 4 [occurred as far as 2 to 2.5 miles] from the hypocenter. . . . Extremely intense thermal energy leads not only to carbonization but also to evaporation of the viscerae." People exposed within half a mile of the Little Boy fireball, that is, were seared to bundles of smoking black char in a fraction of a second as their internal organs boiled away. "Doctor," a patient commented to Michihiko Hachiya a few days later, "a human being who has been roasted becomes quite small, doesn't he?" The small black bundles now stuck to the streets and bridges and sidewalks of Hiroshima numbered in the thousands.

At the same instant birds ignited in midair. Mosquitoes and flies, squirrels, family pets crackled and were gone. The fireball flashed an enormous photograph of the city at the instant of its immolation fixed on the mineral, vegetable and animal surfaces of the city itself. A spiral ladder left its shadow in unburned paint on the surface of a steel storage tank. Leaves shielded reverse silhouettes on charred telephone poles. The black-brushed calligraphy burned out of a rice-paper name card posted on a school building door; the dark flowers burned out of a schoolgirl's light blouse. A human being left the memorial of his outline in unspalled granite on the steps of a bank. Another, pulling a handcart, protected a handcart- and human-shaped surface of asphalt from boiling. Farther away, in the suburbs, the flash induced dark, sunburn-like pigmentation sharply shadowed deep in human skin, streaking the shape of an exposed nose or ear or hand raised in gesture onto the faces and bodies of startled citizens: the mask of Hiroshima, Liebow and his colleagues came to call that pigmentation. They found it persisting unfaded five months after the event.

The world of the dead is a different place from the world of the living and it is hardly possible to visit there. That day in Hiroshima the two worlds nearly converged. "The inundation with death of the area closest to the hypocenter," writes the American psychiatrist Robert Jay Lifton, who interviewed survivors at length, "was such that if a man survived within a thousand meters (.6 miles) and was out of doors . . . more than nine tenths of the people around him were fatalities." Only the living, however inundated, can describe the dead; but where death claimed nine out of ten or, closer to the hypocenter, ten out of ten, a living voice describing necessarily distorts. Survivors are like us; but the dead are radically changed, without voice or civil rights or recourse. Along with their lives they have been deprived of participation in the human world. "There was a fearful silence which made one feel that all people and all trees and vegetation were dead," remembers Yōko Ōta, a Hiroshima writer who survived. The silence was the only sound the dead could make. In what follows among the living, remember them. They were nearer the center of the event; they died because they were members of a different polity and their killing did not therefore count officially as murder; their experience most accurately models the worst case of our common future. They numbered in the majority in Hiroshima that day.

Still only light, not yet blast: Hachiya:

> I asked Dr. Koyama what his findings had been in patients with eye injuries.
>
> "Those who watched the plane had their eye grounds burned," he replied. "The flash of light apparently went through the pupils and left them with a blind area in the central portion of their visual fields.
>
> "Most of the eye-ground burns are third degree, so cure is impossible."

And a German Jesuit priest reporting on one of his brothers in Christ:

> Father Kopp . . . was standing in front of the nunnery ready to go home. All of a sudden he became aware of the light, felt that wave of heat, and a large blister formed on his hand.

A white burn with the formation of a bleb is a grade-four burn.

Now light and blast together; they seemed simultaneous to those close in. A junior-college girl:

> Ah, that instant! I felt as though I had been struck on the back with something like a big hammer, and thrown into boiling oil. . . . I seem to have been blown a good way to the north, and I felt as though the directions were all changed around.

The first junior-college girl, the one whose teacher called everyone to look up:

> The vicinity was in pitch darkness; from the depths of the gloom, bright red flames rise crackling, and spread moment by moment. The faces of my friends who just before were working energetically are now burned and blistered, their clothes torn to rags; to what shall I liken their trembling appearance as they stagger about? Our teacher is holding her students close to her like a mother hen protecting her chicks, and like baby chicks paralyzed with terror, the students were thrusting their heads under her arms.

The light did not burn those who were protected inside buildings, but the blast found them out:

> That boy had been in a room at the edge of the river, looking out at the river when the explosion came, and in that instant as the house fell apart he was blown from the end room across the road on the river embankment and landed on the street below it. In that distance he passed through a couple of windows inside the house and his body was stuck full of all the glass it could hold. That is why he was completely covered with blood like that.

The blast wave, rocketing several hundred yards from the hypocenter at 2 miles per second and then slowing to the speed of sound, 1,100 feet per second, threw up a vast cloud of smoke and dust. "My body seemed all black," a Hiroshima physicist told Lifton, "everything seemed dark, dark all over. . . . Then I thought, 'The world is ending.' " Yōko Ōta, the writer, felt the same chill:

> I just could not understand why our surroundings had changed so greatly in one instant. . . . I thought it might have been something which had nothing to do with the war, the collapse of the earth which it was said would take place at the end of the world.

"Within the city," notes Hachiya, who was severely injured, "the sky looked as though it had been painted with light *sumi* [i.e., calligraphy ink], and the people had seen only a sharp, blinding flash of light; while outside the city, the sky was a beautiful, golden yellow and there had been a deafening roar of sound." Those who experienced the explosion within the city named it *pika,* flash, and those who experienced it farther away named it *pika-don,* flash-boom.

The houses fell as if they had been scythed. A fourth-grade boy:

> When I opened my eyes after being blown at least eight yards, it was as dark as though I had come up against a black-painted fence. After that, as if thin paper was being peeled off one piece at a time, it gradually began to grow brighter. The first thing that my eyes lighted upon then was the flat stretch of land with only dust clouds rising from it. Everything had crumbled away in that one moment, and changed into streets of rubble, street after street of ruins.

Hachiya and his wife ran from their house just before it collapsed and terror opened out into horror:

> The shortest path to the street lay through the house next door so through the house we went—running, stumbling, falling, and then running again until in headlong flight we tripped over something and fell sprawling into the street. Getting to my feet, I discovered that I had tripped over a man's head.
>
> "Excuse me! Excuse me, please!" I cried hysterically.

A grocer escaped into the street:

> The appearance of people was . . . well, they all had skin blackened by burns. . . . They had no hair because their hair was burned, and at a glance you couldn't tell whether you were looking at them from in front or in

> back. . . . They held their arms [in front of them] . . . and their skin—not only on their hands, but on their faces and bodies too—hung down. . . . If there had been only one or two such people . . . perhaps I would not have had such a strong impression. But wherever I walked I met these people. . . . Many of them died along the road—I can still picture them in my mind—like walking ghosts. . . . They didn't look like people of this world. . . . They had a very special way of walking—very slowly. . . . I myself was one of them.

The peeled skin that hung from the faces and bodies of these severely injured survivors was skin that the thermal flash had instantly blistered and the blast wave had torn loose. A young woman:

> I heard a girl's voice clearly from behind a tree. "Help me, please." Her back was completely burned and the skin peeled off and was hanging down from her hips. . . .
>
> The rescue party . . . brought [my mother] home. Her face was larger than usual, her lips were badly swollen, and her eyes remained closed. The skin of both her hands was hanging loose as if it were rubber gloves. The upper part of her body was badly burned.

A junior-college girl:

> On both sides of the road, bedding and pieces of cloth had been carried out and on these were lying people who had been burned to a reddish-black color and whose entire bodies were frightfully swollen. Making their way among them are three high school girls who looked as though they were from our school; their faces and everything were completely burned and they held their arms out in front of their chests like kangaroos with only their hands pointed downward; from their whole bodies something like thin paper is dangling—it is their peeled-off skin which hangs there, and trailing behind them the unburned remnants of their puttees, they stagger exactly like sleepwalkers.

A young sociologist:

> Everything I saw made a deep impression—a park nearby covered with dead bodies waiting to be cremated . . . very badly injured people evacuated in my direction. . . . The most impressive thing I saw was some girls, very young girls, not only with their clothes torn off but with their skin peeled off as well. . . . My immediate thought was that this was like the hell I had always read about.

A five-year-old boy:

> That day after we escaped and came to Hijiyama Bridge, there were lots of naked people who were so badly burned that the skin of their whole body was hanging from them like rags.

A fourth-grade girl:

> The people passing along the street are covered with blood and trailing the rags of their torn clothes after them. The skin of their arms is peeled off and dangling from their finger tips, and they go walking silently, hanging their arms before them.

A five-year-old girl:

> People came fleeing from the nearby streets. One after another they were almost unrecognizable. The skin was burned off some of them and was hanging from their hands and from their chins; their faces were red and so swollen that you could hardly tell where their eyes and mouths were. From the houses smoke black enough to scorch the heavens was covering the sky. It was a horrible sight.

A fifth-grade boy compiling a list:

> The flames which blaze up here and there from the collapsed houses as though to illuminate the darkness. The child making a suffering, groaning sound, his burned face swollen up balloon-like and jerking as he wanders among the fires. The old man, the skin of his face and body peeling off like a potato skin, mumbling prayers while he flees with faltering steps. Another man pressing with both hands the wound from which blood is steadily dripping, rushing around as though he has gone mad and calling the names of his wife and child—ah—my hair seems to stand on end just to remember. This is the way war really looks.

But skin peeled by a flash of light and a gust of air was only a novelty among the miseries of that day, something unusual the survivors could remember to remember. The common lot was random, indiscriminate and universal violence inflicting terrible pain, the physics of hydraulics and leverage and heat run riot. A junior-college girl:

> Screaming children who have lost sight of their mothers; voices of mothers searching for their little ones; people who can no longer bear the heat, cooling their bodies in cisterns; every one among the fleeing people is dyed red with blood.

The thermal flash and the blast started fires and very quickly the fires became a firestorm from which those who could ambulate ran away and those who sustained fractures or were pinned under houses could not; two months later Liebow's group found the incidence of fractures among Hiroshima survivors to be less than 4.5 percent. "It was not that injuries were few," the American physicians note; "rather, almost none who had lost the capacity to move escaped the flames." A five-year-old girl:

> The whole city . . . was burning. Black smoke was billowing up and we could hear the sound of big things exploding. . . . Those dreadful streets. The fires were burning. There was a strange smell all over. Blue-green balls of fire were drifting around. I had a terrible lonely feeling that everybody else in the world was dead and only we were still alive.

Another girl the same age:

> I really have to shudder when I think of that atom bomb which licked away the city of Hiroshima in one or two minutes on the 6th of August, 1945. . . .
>
> We were running for our lives. On the way we saw a soldier floating in the river with his stomach all swollen. In desperation he must have jumped into the river to escape from the sea of fire. A little farther on dead people were lined up in a long row. Al little farther on there was a woman lying with a big log fallen across her legs so that she couldn't get away.
>
> When Father saw that he shouted, "Please come and help!"
>
> But not a single person came to help. They were all too intent on saving themselves.
>
> Finally Father lost his patience, and shouting, "Are you people Japanese or not?" he took a rusty saw and cut off her leg and rescued her.
>
> A little farther on we saw a man who had been burned black as he was walking.

A first-grade girl whose mother was pinned under the wreckage of their house:

> I was determined not to escape without my mother. But the flames were steadily spreading and my clothes were already on fire and I couldn't stand it any longer. So screaming, "Mommy, Mommy!" I ran wildly into the middle of the flames. No matter how far I went it was a sea of fire all around and there was no way to escape. So beside myself I jumped into our [civil defense] water tank. The sparks were falling everywhere so I put a piece of tin over my head to keep out the fire. The water in the tank was hot like a bath. Beside me there were four or five other people who were all calling someone's name. While I was in the water tank everything became like a dream and sometime or other I became unconscious. . . . Five days after that [I learned that] Mother had finally died just as I had left her.

Similarly a woman who was thirteen at the time who was still haunted by guilt when Lifton interviewed her two decades later:

> I left my mother there and went off. . . . I was later told by a neighbor that my mother had been found dead, face down in a water tank . . . very close to the

> spot where I left her. . . . If I had been a little older or stronger I could have rescued her. . . . Even now I still hear my mother's voice calling me to help her.

"Beneath the wreckage of the houses along the way," recounts the Jesuit priest, "many have been trapped and they scream to be rescued from the oncoming flames."

"I was completely amazed," a third-grade boy remembers of the destruction:

> While I had been thinking it was only my house that had fallen down, I found that every house in the neighborhood was either completely or half-collapsed. The sky was like twilight. Pieces of paper and cloth were caught on the electric wires. . . . On that street crowds were fleeing toward the west. Among them were many people whose hair was burned, whose clothes were torn and who had burns and injuries. . . . Along the way the road was full to overflowing with victims, some with great wounds, some burned, and some who had lost the strength to move farther. . . . While we were going along the embankment, a muddy rain that was dark and chilly began to fall. Around the houses I noticed automobiles and footballs, and all sorts of household stuff that had been tossed out, but there was no one who stopped to pick up a thing.

But against the background of horror the eye of the survivor persisted in isolating the exceptional. A thirty-five-year-old man:

> A woman with her jaw missing and her tongue hanging out of her mouth was wandering around the area of Shinsho-machi in the heavy, black rain. She was heading toward the north crying for help.

A four-year-old boy:

> There were a lot of people who were burned to death and among them were some who were burned to a cinder while they were standing up.

A sixth-grade boy:

> Nearby, as if he were guarding these people, a policeman was standing, all covered with burns and stark naked except for some scraps of his trousers.

A seventeen-year-old girl:

> I walked past Hiroshima Station . . . and saw people with their bowels and brains coming out. . . . I saw an old lady carrying a suckling infant in her arms. . . . I saw many children . . . with dead mothers. . . . I just cannot put into words the horror I felt.

At Aioi Bridge:

> I was walking among dead people. . . . It was like hell. The sight of a living horse burning was very striking.

A schoolgirl saw "a man without feet, walking on his ankles." A woman remembers:

> A man with his eyes sticking out about two inches called me by name and I felt sick. . . . People's bodies were tremendously swollen—you can't imagine how big a human body can swell up.

A businessman whose son was killed:

> In front of the First Middle School there were . . . many young boys the same age as my son . . . and what moved me most to pity was that there was one dead child lying there and another who seemed to be crawling over him in order to run away, both of them burned to blackness.

A thirty-year-old woman:

> The corpse lying on its back on the road had been killed immediately. . . . Its hand was lifted to the sky and the fingers were burning with blue flames. The fingers were shortened to one-third and distorted. A dark liquid was running to the ground along the hand.

A third-grade girl:

> There was also a person who had a big splinter of wood stuck in his eye—I suppose maybe he couldn't see—and he was running around blindly.

A nineteen-year-old Ujina girl:

> I saw for the first time a pile of burned bodies in a water tank by the entrance to the broadcasting station. Then I was suddenly frightened by a terrible sight on the street 40 to 50 meters from Shukkeien Garden. There was a charred body of a woman standing frozen in a running posture with one leg lifted and her baby tightly clutched in her arms. Who on earth could she be?

A first-grade girl:

> A streetcar was all burned and just the skeleton of it was left, and inside it all the passengers were burned to a cinder. When I saw that I shuddered all over and started to tremble.

"The more you hear the sadder the stories get," writes a girl who was five years old at Hiroshima. "Since just in my family there is so much sadness from it," deduces a boy who was also five, "I wonder how much sadness other people must also be having."

Eyes watched as well from the other side. A history professor Lifton interviewed:

> I went to look for my family. Somehow I became a pitiless person, because if I had pity, I would not have been able to walk through the city, to walk over those dead bodies. The most impressive thing was the expression in people's eyes—bodies badly injured which had turned black—their eyes looking for someone to come and help them. They looked at me and knew that I was stronger than they. . . . I saw disappointment in their eyes. They looked at me with great expectation, staring right through me. It was very hard to be stared at by those eyes.

Massive pain and suffering and horror everywhere the survivors turned was their common lot. A fifth-grade boy:

> I and Mother crawled out from under the house. There we found a world such as I had never seen before, a world I'd never even heard of before. I saw human bodies in such a state that you couldn't tell whether they were humans or what. . . . There is already a pile of bodies in the road and people are writhing in death agonies.

A junior-college girl:

> At the base of the bridge, inside a big cistern that had been dug out there, was a mother weeping and holding above her head a naked baby that was burned bright red all over its body, and another mother was crying and sobbing as she gave her burned breast to her baby. In the cistern the students stood with only their heads above the water and their two hands, which they clasped as they imploringly cried and screamed, calling their parents. But every single person who passed was wounded, all of them, and there was no one to turn to for help.

A six-year-old boy:

> Near the bridge there were a whole lot of dead people. There were some who were burned black and died, and there were others with huge burns who died with their skins bursting, and some others who died all stuck full of broken glass. There were all kinds. Sometimes there were ones who came to us asking

> for a drink of water. They were bleeding from their faces and from their mouths and they had glass sticking in their bodies. And the bridge itself was burning furiously. . . . The details and the scenes were just like Hell.

Two first-grade girls:

> We came out to the Miyuki Bridge. Both sides of the street were piled with burned and injured people. And when we looked back it was a sea of bright red flame.
>
> *
>
> The fire was spreading furiously from one place to the next and the sky was dark with smoke. . . .
>
> The [emergency aid station] was jammed with people who had terrible wounds, some whose whole body was one big burn. . . . The flames were spreading in all directions and finally the whole city was one sea of fire and sparks came flying over our heads.

A fifth-grade boy:

> I had the feeling that all the human beings on the face of the earth had been killed off, and only the five of us [i.e., his family] were left behind in an uncanny world of the dead. . . . I saw several people plunging their heads into a half-broken water tank and drinking the water. . . . When I was close enough to see inside the tank I said "Oh!" out loud and instinctively drew back. What I had seen in the tank were the faces of monsters reflected from the water dyed red with blood. They had clung to the side of the tank and plunged their heads in to drink and there in that position they had died. From their burned and tattered middy blouses I could tell that they were high school girls, but there was not a hair left on their heads; the broken skin of their burned faces was stained bright red with blood. I could hardly believe that these were human faces.

A physician sharing his horror with Hachiya:

> Between the [heavily damaged] Red Cross Hospital and the center of the city I saw nothing that wasn't burned to a crisp. Streetcars were standing at Kawaya-cho and Kamiya-cho and inside were dozens of bodies, blackened beyond recognition. I saw fire reservoirs filled to the brim with dead people who looked as though they had been boiled alive. In one reservoir I saw a man, horribly burned, crouching beside another man who was dead. He was drinking blood-stained water out of the reservoir. . . . In one reservoir there were so many dead people there wasn't enough room for them to fall over. They must have died sitting in the water.

A husband helping his wife escape the city:

> While taking my severely-wounded wife out to the riverbank by the side of the hill of Nakahiro-machi, I was horrified, indeed, at the sight of a stark naked man standing in the rain with his eyeball in his palm. He looked to be in great pain but there was nothing that I could do for him.

The naked man may have been the same victim one of Hachiya's later visitors remembered noticing, or he may have been another:

> There were so many burned [at a first-aid station] that the odor was like drying squid. They looked like boiled octopuses. . . . I saw a man whose eye had been torn out by an injury, and there he stood with his eye resting in the palm of his hand. What made my blood run cold was that it looked like the eye was staring at me.

The people ran to the rivers to escape the firestorm; in the testimony of the survivors there is an entire subliterature of the rivers. A third-grade boy:

> Men whose whole bodies were covered with blood, and women whose skin hung from them like a kimono, plunged shrieking into the river. All these become corpses and their bodies are carried by the current toward the sea.

A first-grade girl:

> We were still in the river by evening and it got cold. No matter where you looked there was nothing but burned people all around.

A sixth-grade girl:

> Bloated corpses were drifting in those seven formerly beautiful rivers; smashing cruelly into bits the childish pleasure of the little girl, the peculiar odor of burning human flesh rose everywhere in the Delta City, which had changed to a waste of scorched earth.

A young ship designer whose response to the bombing was to rush home immediately to Nagasaki:

> I had to cross the river to reach the station. As I came to the river and went down the bank to the water, I found that the stream was filled with dead bodies. I started to cross by crawling over the corpses, on my hands and knees. As I got about a third of the way across, a dead body began to sink under my weight and I went into the water, wetting my burned skin. It pained severely. I

> could go no further, as there was a break in the bridge of corpses, so I turned back to the shore.

A third-grade boy:

> I got terribly thirsty so I went to the river to drink. From upstream a great many black and burned corpses came floating down the river. I pushed them away and drank the water. At the margin of the river there were corpses lying all over the place.

A fifth-grade boy:

> The river became not a stream of flowing water but rather a stream of drifting dead bodies. No matter how much I might exaggerate the stories of the burned people who died shrieking and of how the city of Hiroshima was burned to the ground, the facts would still be clearly more terrible.

Terrible was what a Hachiya patient found beyond the river:

> There was a man, stone dead, sitting on his bicycle as it leaned against a bridge railing. . . . You could tell that many had gone down to the river to get a drink of water and had died where they lay. I saw a few live people still in the water, knocking against the dead as they floated down the river. There must have been hundreds and thousands who fled to the river to escape the fire and then drowned.
>
> The sight of the soldiers, though, was more dreadful than the dead people floating down the river. I came onto I don't know how many, burned from the hips up; and where the skin had peeled, their flesh was wet and mushy. . . .
>
> And they had no faces! Their eyes, noses and mouths had been burned away, and it looked like their ears had melted off. It was hard to tell front from back.

The suffering in the crowded private park of the Asano family was doubled when survivors faced death a second time, another Hachiya confidant saw:

> Hundreds of people sought refuge in the Asano Sentei Park. They had refuge from the approaching flames for a little while, but gradually, the fire forced them nearer and nearer the river, until at length everyone was crowded onto the steep bank overlooking the river. . . .
>
> Even though the river is more than one hundred meters wide along the border of the park, balls of fire were being carried through the air from the opposite shore and soon the pine trees in the park were afire. The poor people faced a fiery death if they stayed in the park and a watery grave if they

> jumped in the river. I could hear shouting and crying, and in a few minutes they began to fall like toppling dominoes into the river. Hundreds upon hundreds jumped or were pushed in the river at this deep, treacherous point and most were drowned.

"Along the streetcar line circling the western border of the park," adds Hachiya, "they found so many dead and wounded they could hardly walk."

The setting of the sun brought no relief. A fourteen-year-old boy:

> Night came and I could hear many voices crying and groaning with pain and begging for water. Someone cried, "Damn it! War tortures so many people who are innocent!" Another said, "I hurt! Give me water!" This person was so burned that we couldn't tell if it was a man or a woman.
>
> The sky was red with flames. It was burning as if scorching heaven.

A fifth-grade girl:

> Everybody in the shelter was crying out loud. Those voices.... They aren't cries, they are moans that penetrate to the marrow of your bones and make your hair stand on end....
>
> I do not know how many times I called begging that they would cut off my burned arms and legs.

A six-year-old boy:

> If you think of Brother's body divided into left and right halves, he was burned on the right side, and on the inside of the left side....
>
> That night Brother's body swelled up terribly badly. He looked just like a bronze Buddha....
>
> [At Danbara High School field hospital] every classroom ... was full of dreadfully burned people who were lying about or getting up restlessly. They were all painted with mercurochrome and white salve and they looked like red devils and they were waving their arms around like ghosts and groaning and shrieking. Soldiers were dressing their burns.

The next morning, remembers a boy who was five years old at the time, "Hiroshima was all a wasted land." The Jesuit, coming in from a suburb to aid his brothers, testifies to the extent of the destruction:

> The bright day now reveals the frightful picture which last night's darkness had partly concealed. Where the city stood, everything as far as the eye could reach is a waste of ashes and ruin. Only several skeletons of buildings completely burned out in the interior remain. The banks of the rivers are covered with dead and wounded, and the rising waters have here and there covered

some of the corpses. On the broad street in the Hakushima district, naked, burned cadavers are particularly numerous. Among them are the wounded who are still alive. A few have crawled under the burned-out autos and trams. Frightfully injured forms beckon to us and then collapse.

Hachiya corroborates the priest's report:

> The streets were deserted except for the dead. Some looked as if they had been frozen by death while still in the full action of flight; others lay sprawled as though some giant had flung them to their death from a great height. . . .
>
> Nothing remained except a few buildings of reinforced concrete. . . . For acres and acres the city was like a desert except for scattered piles of brick and roof tile. I had to revise my meaning of the word destruction or choose some other word to describe what I saw. Devastation may be a better word, but really, I know of no word or words to describe the view.

The history professor Lifton interviewed is similarly at a loss:

> I climbed Hikiyama Hill and looked down. I saw that Hiroshima had disappeared. . . . I was shocked by the sight. . . . What I felt then and still feel now I just can't explain with words. Of course I saw many dreadful scenes after that—but that experience, looking down and finding nothing left of Hiroshima—was so shocking that I simply can't express what I felt. . . . Hiroshima didn't exist—that was mainly what I saw—Hiroshima just didn't exist.

Without familiar landmarks, the streets filled with rubble, many had difficulty finding their way. For Yōko Ōta the city's history itself had been demolished:

> I reached a bridge and saw that the Hiroshima Castle had been completely leveled to the ground, and my heart shook like a great wave. . . . The city of Hiroshima, entirely on flat land, was made three-dimensional by the existence of the white castle, and because of this it could retain a classical flavor. Hiroshima had a history of its own. And when I thought about these things, the grief of stepping over the corpses of history pressed upon my heart.

Of 76,000 buildings in Hiroshima 70,000 were damaged or destroyed, 48,000 totally. "It is no exaggeration to say," reports the Japanese study, "that the whole city was ruined instantaneously." Material losses alone equaled the annual incomes of more than 1.1 million people. "In Hiroshima many major facilities—prefectural office, city hall, fire departments, police stations, national railroad stations, post offices, telegram and telephone offices, broadcasting station, and schools—were totally demolished

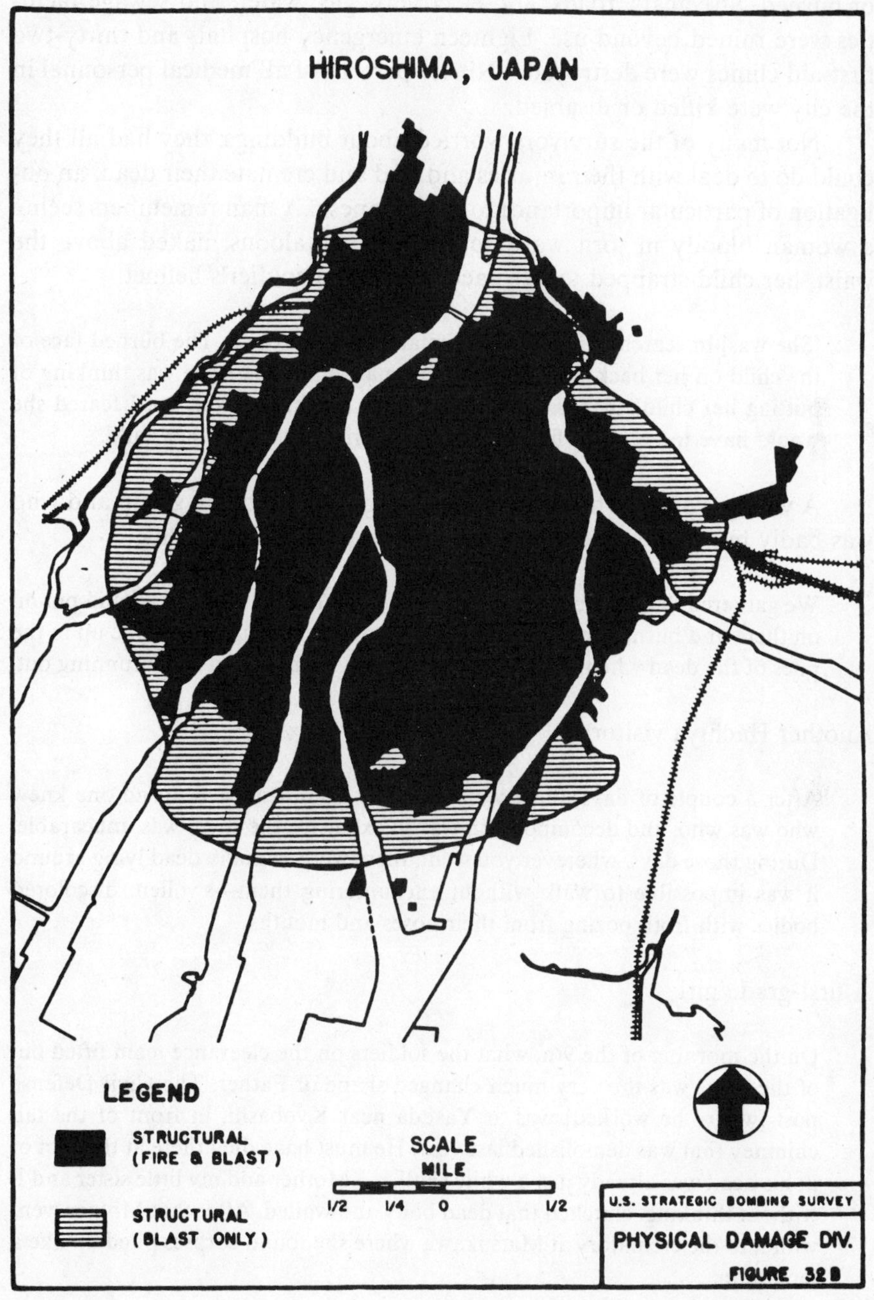
HIROSHIMA, JAPAN
LEGEND
STRUCTURAL (FIRE & BLAST)
STRUCTURAL (BLAST ONLY)
SCALE
MILE
1/2 1/4 0 1/2
U.S. STRATEGIC BOMBING SURVEY
PHYSICAL DAMAGE DIV.
FIGURE 32 B

or burned. Streetcars, roads, and electricity, gas, water, and sewage facilities were ruined beyond use. Eighteen emergency hospitals and thirty-two first-aid clinics were destroyed." Ninety percent of all medical personnel in the city were killed or disabled.

Not many of the survivors worried about buildings; they had all they could do to deal with their injuries and find and cremate their dead, an obligation of particular importance to the Japanese. A man remembers seeing a woman bloody in torn wartime *mompei* pantaloons, naked above the waist, her child strapped to her back, carrying a soldier's helmet:

> [She was] in search of a place to cremate her dead child. The burned face of the child on her back was infested with maggots. I guess she was thinking of putting her child's bones in a battle helmet she had picked up. I feared she would have to go far to find burnable material to cremate her child.

A young woman who had been in charge of a firebreak group and who was badly burned on one shoulder recalls the mass cremations:

> We gathered the dead bodies and made big mountains of the dead and put oil on them and burned them. And people who were unconscious woke up in the piles of the dead when they found themselves burning and came running out.

Another Hachiya visitor:

> After a couple of days, there were so many bodies stacked up no one knew who was who, and decomposition was so extensive the smell was unbearable. During those days, wherever you went, there were so many dead lying around it was impossible to walk without encountering them—swollen, discolored bodies with froth oozing from their noses and mouths.

A first-grade girl:

> On the morning of the 9th, what the soldiers on the clearance team lifted out of the ruins was the very much changed shape of Father. The Civil Defense post [where he worked] was at Yasuda near Kyobashi, in front of the tall chimney that was demolished last year. He must have died there at the foot of it; his head was already just a white skull. . . . Mother and my little sister and I, without thinking, clutched that dead body and wailed. After that Mother went with it to the crematory at Matsukawa where she found corpses piled up like a mountain.

Having moved his hospital sickbed to a second-floor room with blown-out windows that fire had sterilized, Hachiya himself could view and smell the ruins:

> Towards evening, a light southerly wind blowing across the city wafted to us an odor suggestive of burning sardines. . . . Towards Nigitsu was an especially large fire where the dead were being burned by the hundreds. . . . These glowing ruins and the blazing funeral pyres set me to wondering if Pompeii had not looked like this during its last days. But I think there were not so many dead in Pompeii as there were in Hiroshima.

Those who did not die seemed for a time to improve. But then, explains Lifton, they sickened:

> Survivors began to notice in themselves and others a strange form of illness. It consisted of nausea, vomiting, and loss of appetite; diarrhea with large amounts of blood in the stools; fever and weakness; purple spots on various parts of the body from bleeding into the skin . . . inflammation and ulceration of the mouth, throat and gums . . . bleeding from the mouth, gums, throat, rectum, and urinary tract . . . loss of hair from the scalp and other parts of the body . . . extremely low white blood cell counts when those were taken . . . and in many cases a progressive course until death.

Only gradually did the few surviving and overworked Japanese doctors realize that they were seeing radiation sickness; "atomic bomb illness," explains the authoritative Japanese study, "is the first and only example of heavy lethal and momentary doses of whole body irradiation" in the history of medicine. A few human beings had been accidentally overexposed to X rays and laboratory animals had been exposed and sacrificed for study but no large population had ever experienced so extensive and deadly an assault of ionizing radiation before.

The radiation brought further suffering, Hachiya reports in his diary:

> Following the *pika,* we thought that by giving treatment to those who were burned or injured recovery would follow. But now it was obvious that this was not true. People who appeared to be recovering developed other symptoms that caused them to die. So many patients died without our understanding the cause of death that we were all in despair. . . .
>
> Hundreds of patients died during the first few days; then the death rate declined. Now, it was increasing again. . . . As time passed, anorexia [i.e., loss of appetite] and diarrhea proved to be the most persistent symptoms in patients who failed to recover.

Direct gamma radiation from the bomb had damaged tissue throughout the bodies of the exposed. The destruction required cell division to manifest itself, but radiation temporarily suppresses cell division; hence the delayed onset of symptoms. The blood-forming tissues were damaged worst, particularly those that produce the white blood cells that fight infection. Large doses of radiation also stimulate the production of an anti-clotting factor. The outcome of these assaults was massive tissue death, massive hemorrhage and massive infection. "Hemorrhage was the cause of death in all our cases," writes Hachiya, but he also notes that the pathologist at his hospital "found changes in every organ of the body in the cases he . . . autopsied." Liebow reports "evidence of generalization of infection with masses of bacteria in . . . organs as remote from the surface [of the body] as the brain, bone marrow and eye." The operator of a crematorium in the Hiroshima suburbs, a connoisseur of mortality, told Lifton "the bodies were black in color . . . most of them had a peculiar smell, and everyone thought this was from the bomb. . . . The smell when they burned was caused by the fact that these bodies were decayed, many of them even before being cremated—some of them having their internal organs decay even while the person was living." Yōko Ōta raged:

> We were being killed against our will by something completely unknown to us. . . . It is the misery of being thrown into a world of new terror and fear, a world more unknown than that of people sick with cancer.

In the depths of his loss a boy who was a fourth-grader at Hiroshima found words for the unspeakable:

> Mother was completely bedridden. The hair of her head had almost all fallen out, her chest was festering, and from the two-inch hole in her back a lot of maggots were crawling in and out. The place was full of flies and mosquitoes and fleas, and an awfully bad smell hung over everything. Everywhere I looked there were many people like this who couldn't move. From the evening when we arrived Mother's condition got worse and we seemed to see her weakening before our eyes. Because all night long she was having trouble breathing, we did everything we could to relieve her. The next morning Grandmother and I fixed some gruel. As we took it to Mother, she breathed her last breath. When we thought she had stopped breathing altogether, she took one deep breath and did not breathe any more after that. This was nine o'clock in the morning of the 19th of August. At the site of the Japan Red Cross Hospital, the smell of the bodies being cremated is overpowering. Too much sorrow makes me like a stranger to myself, and yet despite my grief I cannot cry.

Not human beings alone died at Hiroshima. Something else was destroyed as well, the Japanese study explains—that shared life Hannah Arendt calls the common world:

> In the case of an atomic bombing . . . a community does not merely receive an impact; the community itself is destroyed. Within 2 kilometers of the atomic

bomb's hypocenter all life and property were shattered, burned, and buried under ashes. The visible forms of the city where people once carried on their daily lives vanished without a trace. The destruction was sudden and thorough; there was virtually no chance to escape. . . . Citizens who had lost no family members in the holocaust were as rare as stars at sunrise. . . .

The atomic bomb had blasted and burned hospitals, schools, city offices, police stations, and every other kind of human organization. . . . Family, relatives, neighbors, and friends relied on a broad range of interdependent organizations for everything from birth, marriage, and funerals to firefighting, productive work, and daily living. These traditional communities were completely demolished in an instant.

Destroyed, that is, were not only men, women and thousands of children but also restaurants and inns, laundries, theater groups, sports clubs, sewing clubs, boys' clubs, girls' clubs, love affairs, trees and gardens, grass, gates, gravestones, temples and shrines, family heirlooms, radios, classmates, books, courts of law, clothes, pets, groceries and markets, telephones, personal letters, automobiles, bicycles, horses—120 war-horses—musical instruments, medicines and medical equipment, life savings, eyeglasses, city records, sidewalks, family scrapbooks, monuments, engagements, marriages, employees, clocks and watches, public transportation, street signs, parents, works of art. "The whole of society," concludes the Japanese study, "was laid waste to its very foundations." Lifton's history professor saw not even foundations left. "Such a weapon," he told the American psychiatrist, "has the power to make everything into nothing."

There remains the question of how many died. The U.S. Army Medical Corps officer who proposed the joint American-Japanese study to Douglas MacArthur thought as late as August 28 that "the total number of casualties reported at Hiroshima is approximately 160,000 of which 8,000 are dead." The Jesuit priest's contemporary reckoning approaches the appalling reality and illuminates further the destruction of the common world:

How many people were a sacrifice to this bomb? Those who had lived through the catastrophe placed the number of dead at at least 100,000. Hiroshima had a population of 400,000. Official statistics place the number who had died at 70,000 up to September 1st, not counting the missing—and 130,000 wounded, among them 43,500 severely wounded. Estimates made by ourselves on the basis of groups known to us show that the number of 100,000 dead is not too high. Near us there are two barracks, in each of which forty Korean workers lived. On the day of the explosion they were laboring on the streets of Hiroshima. Four returned alive to one barracks and sixteen to the other. Six hun-

> dred students of the Protestant girls' school worked in a factory, from which only thirty or forty returned. Most of the peasant families in the neighborhood lost one or more of their members who had worked at factories in the city. Our next door neighbor, Tamura, lost two children and himself suffered a large wound since, as it happened, he had been in the city on that day. The family of our reader suffered two dead, father and son; thus a family of five members suffered at least two losses, counting only the dead and severely wounded. There died the mayor, the president of the central Japan district, the commander of the city, a Korean prince who had been stationed in Hiroshima in the capacity of an officer, and many other high-ranking officers. Of the professors of the University thirty-two were killed or severely wounded. Especially hard-hit were the soldiers. The Pioneer Regiment was almost entirely wiped out. The barracks were near the center of the explosion.

More recent estimates place the number of deaths up to the end of 1945 at 140,000. The dying continued; five-year deaths related to the bombing reached 200,000. The death rate for deaths up to the end of 1945 was 54 percent, an extraordinary density of killing; by contrast, the death rate for the March 9 firebombing of Tokyo, 100,000 deaths among 1 million casualties, was only 10 percent. Back at the U.S. Army Institute of Pathology in Washington in early 1946 Liebow used a British invention, the Standardized Casualty Rate, to compute that Little Boy produced casualties, including dead, 6,500 times more efficiently than an ordinary HE bomb. "Those scientists who invented the . . . atomic bomb," writes a young woman who was a fourth-grade student at Hiroshima—"what did they think would happen if they dropped it?"

Harry Truman learned of the atomic bombing of Hiroshima at lunch on board the *Augusta* en route home from Potsdam. "This is the greatest thing in history," he told a group of sailors dining at his table. "It's time for us to get home."

Groves called Oppenheimer from Washington on August 6 at two in the afternoon to pass along the news:

> Gen. G: I'm very proud of you and all of your people.
> Dr. O: It went all right?
> Gen. G: Apparently it went with a tremendous bang.
> Dr. O: When was this, was it after sundown?
> Gen. G: No, unfortunately, it had to be in the daytime on account of security of the plane and that was left in the hands of the Commanding General over there. . . .
> Dr. O: Right. Everybody is feeling reasonably good about it and I extend my heartiest congratulations. It's been a long road.

Gen. G: Yes, it has been a long road and I think one of the wisest things I ever did was when I selected the director of Los Alamos.

Dr. O: Well, I have my doubts, General Groves.

Gen. G: Well, you know I've never concurred with those doubts at any time.

If Oppenheimer, who knew nothing yet of the extent of the destruction, was only feeling "reasonably good" about his handiwork, Leo Szilard felt terrible when the story broke. The press release issued from the White House that day called the atomic bomb "the greatest achievement of organized science in history" and threatened the Japanese with "a rain of ruin from the air, the like of which has never been seen on this earth." In Chicago on Quadrangle Club stationery Szilard scribbled a hasty letter to Gertrud Weiss:

> I suppose you have seen today's newspapers. Using atomic bombs against Japan is one of the greatest blunders of history. Both from a practical point of view on a 10-year scale and from the point of view of our moral position. I went out of my way and very much so in order to prevent it but as today's papers show without success. It is very difficult to see what wise course of action is possible from here on.

Otto Hahn, interned with the German atomic scientists on a rural estate in England, was shattered:

> At first I refused to believe that this could be true, but in the end I had to face the fact that it was officially confirmed by the President of the United States. I was shocked and depressed beyond measure. The thought of the unspeakable misery of countless innocent women and children was something that I could scarcely bear.
>
> After I had been given some gin to quiet my nerves, my fellow-prisoners were also told the news. . . . By the end of a long evening of discussion, attempts at explanation, and self-reproaches I was so agitated that Max von Laue and the others became seriously concerned on my behalf. They ceased worrying only at two o'clock in the morning, when they saw that I was asleep.

But if some were disturbed by the news, others were elated, Otto Frisch found at Los Alamos:

> Then one day, some three weeks after [Trinity], there was a sudden noise in the laboratory, of running footsteps and yelling voices. Somebody opened my door and shouted, "Hiroshima has been destroyed!"; about a hundred thousand people were thought to have been killed. I still remember the feeling of

> unease, indeed nausea, when I saw how many of my friends were rushing to the telephone to book tables at the La Fonda Hotel in Santa Fe, in order to celebrate. Of course they were exalted by the success of their work, but it seemed rather ghoulish to celebrate the sudden death of a hundred thousand people, even if they were "enemies."

The American writer Paul Fussell, an Army veteran, emphasizes "the importance of experience, sheer vulgar experience, in influencing one's views about the first use of the bomb." The experience Fussell means is "that of having come to grips, face to face, with an enemy who designs your death":

> I was a 21-year-old second lieutenant leading a rifle platoon. Although still officially in one piece, in the German war I had been wounded in the leg and back severely enough to be adjudged, after the war, 40 percent disabled. But even if my leg buckled whenever I jumped out of the back of the truck, my condition was held to be satisfactory for whatever lay ahead. When the bombs dropped and news began to circulate that [the invasion of Japan] would not, after all, take place, that we would not be obliged to run up the beaches near Tokyo assault-firing while being mortared and shelled, for all the fake manliness of our facades we cried with relief and joy. We were going to live. We were going to grow up to adulthood after all.

In Japan the impasse persisted between civilian and military leaders. To the civilians the atomic bomb looked like a golden opportunity to surrender without shame, but the admirals and the generals still despised unconditional surrender and refused to concur. Foreign Minister Togo continued to pursue Soviet mediation as late as August 8. Ambassador Sato asked for a meeting with Molotov that day; Molotov set the meeting for eight in the evening, then moved it up to five o'clock. Despite earlier notice of the power of the new weapon, news of the devastation of a Japanese city by an American atomic bomb had surprised and shocked Stalin and prompted him to accelerate his war plans; Molotov announced that afternoon to the Japanese ambassador that the Soviet Union would consider itself at war with Japan as of the next day, August 9. Well-armed Soviet troops, 1.6 million strong, waited in readiness on the Manchurian border and attacked the ragged Japanese an hour after midnight.

In the meantime a progaganda effort that originated in the U.S. War Department was developing in the Marianas. Hap Arnold cabled Spaatz and Farrell on August 7 ordering a crash program to impress the facts of atomic warfare on the Japanese people. The impetus probably came from George Marshall, who was surprised and shocked that the Japanese had not immediately sued for peace. "What we did not take into account," he said long afterward, ". . . was that the destruction would be so complete that it would be an appreciable time before the actual facts of the case would get to Tokyo. The destruction of Hiroshima was so complete that there was no communication at least for a day, I think, and maybe longer."

The Navy and the Air Force both lent staff and facilities, including Radio Saipan and a printing press previously used to publish a Japanese-language newspaper distributed weekly over the Empire by B-29s. The working group that assembled on August 7 in the Marianas decided to attempt to distribute 6 million leaflets to forty-seven Japanese cities with populations exceeding 100,000. Writing the leaflet occupied the group through the night. A historical memorandum prepared for Groves in 1946 notes that the working group discovered in a midnight conference with Air Force commanders "a certain reluctance to fly single B-29's over the Empire, reluctance arising from the fact that enemy opposition to single flights was expected to be increased as the result of the total damage to Hiroshima by one airplane."

The proposed text of the leaflet was ready by morning and was flown from Saipan to Tinian at dawn for Farrell's approval. Groves' deputy edited it and ordered the revised text called to Radio Saipan by inter-island telephone for broadcast to the Japanese every fifteen minutes; radio transmission probably began the same day. The text described the atomic bomb as "the equivalent in explosive power to what 2,000 of our giant B-29's can carry on a single mission," suggested skeptics "make inquiry as to what happened to Hiroshima" and asked the Japanese people to "petition the Emperor to end the war." Otherwise, it threatened, "we shall resolutely employ this bomb and all our other superior weapons." Printing millions of copies of a leaflet took time, and distribution was delayed some hours further by a local shortage of T-3 leaflet bombs. Such was the general confusion that Nagasaki did not receive its quota of warning leaflets until August 10.

Assembly of Fat Man unit F31 was progressing at Tinian in the air-conditioned assembly building designed for that purpose. F31 was the second Fat Man with real high explosives that the Tinian team had assembled; the first, with lower-quality HE castings and a non-nuclear core, unit F33, had been ready since August 5 for a test drop but would not be dropped until August 8 because the key 509th crews were busy delivering Little Boy and being debriefed. The F31 Fat Man, Norman Ramsey writes,

> was originally scheduled for dropping on August 11 local time. . . . However, by August 7 it became apparent that the schedule could be advanced to August 10. When Parsons and Ramsey proposed this change to Tibbets, he expressed regret that the schedule could not be advanced two days instead of

> only one since good weather was forecast for August 9 and the five succeeding days were expected to be bad. It was finally agreed that [we] would try to be ready for August 9 provided all concerned understood that the advancement of the date by two full days introduced a large measure of uncertainty into the probability of meeting such a drastically revised schedule.

One member of the Fat Man assembly team, a young Navy ensign named Bernard J. O'Keefe, remembers the mood of urgency in the Marianas, where the war was still a daily threat:

> With the success of the Hiroshima weapon, the pressure to be ready with the much more complex implosion device became excruciating. We sliced off another day, scheduling it for August 10. Everyone felt that the sooner we could get off another mission, the more likely it was that the Japanese would feel that we had large quantities of the devices and would surrender sooner. We were certain that one day saved would mean that the war would be over one day sooner. Living on that island, with planes going out every night and people dying not only in B-29s shot down, but in naval engagements all over the Pacific, we knew the importance of one day; the *Indianapolis* sinking also had a strong effect on us.

Despite that urgency, O'Keefe adds, August 9 sat less well; "the scientific staff, dog-tired, met and warned Parsons that cutting two full days would prevent us from completing a number of important checkout procedures, but orders were orders."

The young Providence, Rhode Island, native had been a student at George Washington University in 1939 and had attended the conference there on January 25 at which Niels Bohr announced the discovery of fission. Now on Tinian more than six years later, on the night of August 7, it became O'Keefe's task to check out Fat Man for the last time before its working parts were encased beyond easy access in armor. In particular, he was required to connect the firing unit mounted on the front of the implosion sphere with the four radar units mounted in the tail by plugging in a cable inaccessibly threaded around the sphere inside its dural casing:

> When I returned at midnight, the others in my group left to get some sleep; I was alone in the assembly room with a single Army technician to make the final connection. . . .
>
> I did my final checkout and reached for the cable to plug it into the firing unit. It wouldn't fit!
>
> "I must be doing something wrong," I thought. "Go slowly; you're tired and not thinking straight."
>
> I looked again. To my horror, there was a female plug on the firing set

> and a female plug on the cable. I walked around the weapon and looked at the radars and the other end of the cable. Two male plugs. . . . I checked and double-checked. I had the technician check; he verified my findings. I felt a chill and started to sweat in the air-conditioned room.
>
> What had happened was obvious. In the rush to take advantage of good weather, someone had gotten careless and put the cable in backward.

Removing the cable and reversing it would mean partly disassembling the implosion sphere. It had taken most of a day to assemble it. They would miss the window of good weather and slip into the five days of bad weather that had worried Paul Tibbets. The second atomic bomb might be delayed as long as a week. The war would go on, O'Keefe thought. He decided to improvise. Although "nothing that could generate heat was ever allowed in an explosive assembly room," he determined to "unsolder the connectors from the two ends of the cable, reverse them, and resolder them":

> My mind was made up. I was going to change the plugs without talking to anyone, rules or no rules. I called in the technician. There were no electrical outlets in the assembly room. We went out to the electronics lab and found two long extension cords and a soldering iron. We . . . propped the door open so it wouldn't pinch the extension cords (another safety violation). I carefully removed the backs of the connectors and unsoldered the wires. I resoldered the plugs onto the other ends of the cable, keeping as much distance between the soldering iron and the detonators as I could as I walked around the weapon. . . . We must have checked the cable continuity five times before plugging the connectors into the radars and the firing set and tightening up the joints. I was finished.

So, the next day, was Fat Man, the two armored steel ellipsoids of its ballistic casing bolted together through bathtub fittings to lugs cast into the equatorial segments of the implosion sphere, its boxed tail sprouting radar antennae just as Little Boy's had done. By 2200 on August 8 it had been loaded into the forward bomb bay of a B-29 named *Bock's Car* after its usual commander, Frederick Bock, but piloted on this occasion by Major Charles W. Sweeney. Sweeney's primary target was Kokura Arsenal on the north coast of Kyushu; his secondary was the old Portuguese- and Dutch-influenced port city of Nagasaki, the San Francisco of Japan, home of that country's largest colony of Christians, where the Mitsubishi torpedoes used at Pearl Harbor had been made.

Bock's Car flew off Tinian at 0347 on August 9. The Fat Man weaponeer, Navy Commander Frederick L. Ashworth, remembers the flight to rendezvous:

> The night of our takeoff was one of tropical rain squalls, and flashes of lightning stabbed into the darkness with disconcerting regularity. The weather forecast told us of storms all the way from the Marianas to the Empire. Our rendezvous was to be off the southeast coast of Kyushu, some fifteen hundred miles away. There we were to join with our two companion observation B-29s that took off a few minutes behind us.

Fat Man was fully armed at takeoff except for its green plugs, which Ashworth changed to red only ten minutes into the mission so that Sweeney could cruise above the squalls at 17,000 feet, St. Elmo's fire glowing on the propellers of his plane. The pilot soon discovered he would enjoy no reserve of fuel; the fuel selector that would allow him to feed his engines from a 600-gallon tank of gasoline in his aft bomb bay refused to work. He circled over Yakoshima between 0800 and 0850 Japanese time waiting for his escorts, one of which never did catch up. The finger plane at Kokura reported three-tenths low clouds, no intermediate or high clouds and improving conditions, but when *Bock's Car* arrived there at 1044 heavy ground haze and smoke obscured the target. "Two additional runs were made," Ashworth notes in his flight log, "hoping that the target might be picked up after closer observation. However, at no time was the aiming point seen."

Jacob Beser controlled electronic countermeasures on the Fat Man mission as he had done on the Little Boy mission before. He remembers of Kokura that "the Japs started to get curious and began sending fighters up after us. We had some flak bursts and things were getting a little hairy, so Ashworth and Sweeney decided to make a run down to Nagasaki, as there was no sense dragging the bomb home or dropping it in the ocean."

Sweeney had enough fuel left for only one pass over the target before nursing his aircraft to an emergency landing on Okinawa. When he approached Nagasaki he found the city covered with cloud; with his fuel low he could either bomb by radar or jettison a bomb worth several hundred million dollars into the sea. It was Ashworth's call and rather than waste the bomb he authorized a radar approach. At the last minute a hole opened in the cloud cover long enough to give the bombardier a twenty-second visual run on a stadium several miles upriver from the original aiming point nearer the bay. Fat Man dropped from the B-29, fell through the hole and exploded 1,650 feet above the steep slopes of the city at 11:02 A.M., August 9, 1945, with a force later estimated at 22 kilotons. The steep hills confined the larger explosion; it caused less damage and less loss of life than Little Boy.

But 70,000 died in Nagasaki by the end of 1945 and 140,000 altogether

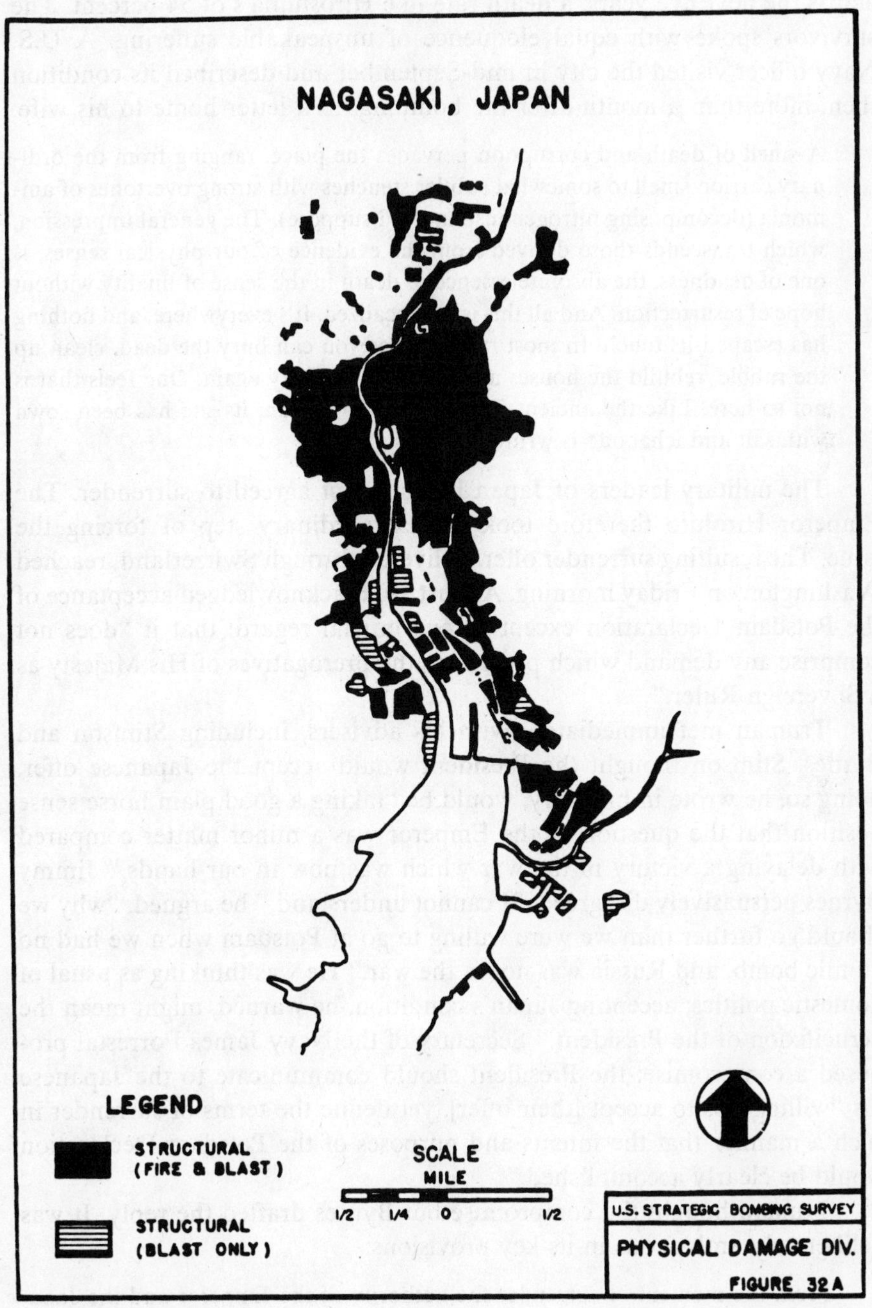
NAGASAKI, JAPAN
LEGEND
STRUCTURAL
(FIRE & BLAST)
STRUCTURAL
(BLAST ONLY)
SCALE
MILE
1/2
1/4
0
1/2
U.S. STRATEGIC BOMBING SURVEY
PHYSICAL DAMAGE DIV.
FIGURE 32A

across the next five years, a death rate like Hiroshima's of 54 percent. The survivors spoke with equal eloquence of unspeakable suffering. A U.S. Navy officer visited the city in mid-September and described its condition then, more than a month after the bombing, in a letter home to his wife:

> A smell of death and corruption pervades the place, ranging from the ordinary carrion smell to somewhat subtler stenches with strong overtones of ammonia (decomposing nitrogenous matter, I suppose). The general impression, which transcends those derived from the evidence of our physical senses, is one of deadness, the absolute essence of death in the sense of finality without hope of resurrection. And all this is not localized. It's everywhere, and nothing has escaped its touch. In most ruined cities you can bury the dead, clean up the rubble, rebuild the houses and have a living city again. One feels that is not so here. Like the ancient Sodom and Gomorrah, its site has been sown with salt and ichabod* is written over its gates.

The military leaders of Japan had still not agreed to surrender. The Emperor Hirohito therefore took the extraordinary step of forcing the issue. The resulting surrender offer, delivered through Switzerland, reached Washington on Friday morning, August 10. It acknowledged acceptance of the Potsdam Declaration except in one crucial regard: that it "does not comprise any demand which prejudices the prerogatives of His Majesty as a Sovereign Ruler."

Truman met immediately with his advisers, including Stimson and Byrnes. Stimson thought the President would accept the Japanese offer; doing so, he wrote in his diary, would be "taking a good plain horse sense position that the question of the Emperor was a minor matter compared with delaying a victory in the war which was now in our hands." Jimmy Byrnes persuasively disagreed. "I cannot understand," he argued, "why we should go further than we were willing to go at Potsdam when we had no atomic bomb, and Russia was not in the war." He was thinking as usual of domestic politics; accepting Japan's condition, he warned, might mean the "crucifixion of the President." Secretary of the Navy James Forrestal proposed a compromise: the President should communicate to the Japanese his "willingness to accept [their offer], yet define the terms of surrender in such a manner that the intents and purposes of the Potsdam Declaration would be clearly accomplished."

Truman bought the compromise but Byrnes drafted the reply. It was deliberately ambiguous in its key provisions:

> From the moment of surrender the authority of the Emperor and the Japanese Government to rule the state shall be subject to the Supreme Commander of the Allied Powers. . . .

* "The glory is departed."

> The Emperor and the Japanese High Command will be required to sign the surrender terms. . . .
>
> The ultimate form of government shall, in accordance with the Potsdam Declaration, be established by the freely expressed will of the Japanese people.

Nor did Byrnes hurry the message along; he kept it in hand overnight and only released it for broadcast by radio and delivery through Switzerland the following morning.

Stimson, still trying to bring his Air Force under control, had argued at the Friday morning meeting that the United States should suspend bombing, including atomic bombing. Truman thought otherwise, but when he met with the cabinet that afternoon he had partly reconsidered. "We would keep up the war at its present intensity," Forrestal paraphrases the President, "until the Japanese agreed to these terms, with the limitation however that there will be no further dropping of the atomic bomb." Henry Wallace, the former Vice President who was now Secretary of Commerce, recorded in his diary the reason for the President's change of mind:

> Truman said he had given orders to stop the atomic bombing. He said the thought of wiping out another 100,000 people was too horrible. He didn't like the idea of killing, as he said, "all those kids."

The restriction came none too soon. Groves had reported to Marshall that morning that he had gained four days in manufacture and expected to ship a second Fat Man plutonium core and initiator from New Mexico to Tinian on August 12 or 13. "Provided there are no unforeseen difficulties in manufacture, in transportation to the theatre or after arrival in the theatre," he concluded cautiously, "the bomb should be ready for delivery on the first suitable weather after 17 or 18 August." Marshall told Groves the President wanted no further atomic bombing except by his express order and Groves decided to hold up shipment, a decision in which Marshall concurred.

The Japanese government learned of Byrnes' reply to its offer of conditional surrender not long after midnight on Sunday, August 12, but civilian and military leaders continued to struggle in deadlocked debate. Hirohito resisted efforts to persuade him to reverse his earlier commitment to surrender and called a council of the imperial family to collect pledges of support from the princes of the blood. The Japanese people were not yet told of the Byrnes reply but knew of the peace negotiations and waited in suspense. The young writer Yukio Mishima found the suspense surreal:

> It was our last chance. People were saying that Tokyo would be [atomic-bombed] next. Wearing white shirts and shorts, I walked about the streets. The people had reached the limits of desperation and were now going about their affairs with cheerful faces. From one moment to the next, nothing happened. Everywhere there was an air of cheerful excitement. It was just as though one was continuing to blow up an already bulging toy balloon, wondering: "Will it burst now? Will it burst now?"

Strategic Air Forces commander Carl Spaatz cabled Lauris Norstad on August 10 proposing "placing [the] third atomic bomb . . . on Tokyo," where he thought it would have a salutary "psychological effect on government officials." On the other hand, continuing area incendiary bombing disturbed him; "I have never favored the destruction of cities as such with all inhabitants being killed," he confided to his diary on August 11. He had sent off 114 B-29's on August 10; because of bad weather and misgivings he canceled a mission scheduled for August 11 and restricted operations thereafter to "attacks on military targets visually or under very favorable blind bombing conditions." American weather planes over Tokyo were no longer drawing anti-aircraft fire; Spaatz thought that fact "unusual."

The vice chief of the Japanese Navy's general staff, the man who had conceived and promoted the kamikaze attacks of the past year that had added to American bewilderment and embitterment at Japanese ways, crashed a meeting of government leaders on the evening of August 13 with tears in his eyes to offer "a plan for certain victory": "sacrifice 20,000,000 Japanese lives in a special [kamikaze] attack." Whether he meant the 20 million to attack the assembled might of the Allies with rocks or bamboo spears the record does not reveal.

A B-29 leaflet barrage forced the issue the next morning. Leaflet bombs showered what remained of Tokyo's streets with a translation of Byrnes' reply. The Lord Keeper of the Privy Seal knew such public revelation would harden the military against surrender. He carried the leaflet immediately to the Emperor and just before eleven that morning, August 14, Hirohito assembled his ministers and counselors in the imperial air raid shelter. He told them he found the Allied reply "evidence of the peaceful and friendly intentions of the enemy" and considered it "acceptable." He did not specifically mention the atomic bomb; even that terrific leviathan submerged in the general misery:

> I cannot endure the thought of letting my people suffer any longer. A continuation of the war would bring death to tens, perhaps even hundreds, of thousands of persons. The whole nation would be reduced to ashes. How then could I carry on the wishes of my imperial ancestors?

He asked his ministers to prepare an imperial rescript—a formal edict—that he might broadcast personally to the nation. The officials were not legally bound to do so—the Emperor's authority lay outside the legal structure of the government—but by older and deeper bonds than law they were bound, and they set to work.

In the meantime Washington had grown impatient. Groves was asked on August 13 about "the availability of your patients together with the time estimate that they could be moved and placed." Stimson recommended proceeding to ship the nuclear materials for the third bomb to Tinian. Marshall and Groves decided to wait another day or two. Truman ordered Arnold to resume area incendiary attacks. Arnold still hoped to prove that his Air Force could win the war; he called for an all-out attack with every available B-29 and any other bombers in the Pacific theater and mustered more than a thousand aircraft. Twelve million pounds of high-explosive and incendiary bombs destroyed half of Kumagaya and a sixth of Isezaki, killing several thousand more Japanese, even as word of the Japanese surrender passed through Switzerland to Washington.

The first hint of surrender reached American bases in the Pacific by radio in the form of a news bulletin from the Japanese news agency Dōmei at 2:49 P.M. on August 14—1:49 A.M. in Washington:

> Flash! Flash! Tokyo, Aug. 14—It is learned an imperial message accepting the Potsdam Proclamation is forthcoming soon.

The bombers droned on even after that, but eventually that day the bombs stopped falling. Truman announced the Japanese acceptance in the afternoon. There were last-minute acts of military rebellion in Tokyo—a high officer assassinated, an unsuccessful attempt to steal the phonograph recording of the imperial rescript, a brief takeover of a division of Imperial Guards, wild plans for a coup. But loyalty prevailed. The Emperor broadcast to a weeping nation on August 15; his 100 million subjects had never heard the high, antique Voice of the Crane before:

> Despite the best that has been done by everyone . . . the war situation has developed not necessarily to Japan's advantage, while the general trends of the world have all turned against her interest. Moreover, the enemy has begun to employ a new and most cruel bomb, the power of which to do damage is indeed incalculable, taking the toll of many innocent lives. . . . This is the reason why We have ordered the acceptance of the provisions of the Joint declaration of the Powers. . . .
>
> The hardships and sufferings to which Our nation is to be subjected hereafter will be certainly great. We are keenly aware of the inmost feelings of all

ye, Our subjects. However, it is according to the dictate of time and fate that We have resolved to pave the way for a grand peace for all generations to come by enduring the unendurable and suffering what is insufferable. . . .

Let the entire nation continue as one family from generation to generation.

"If it had gone on any longer," writes Yukio Mishima, "there would have been nothing to do but go mad."

"An atomic bomb," the Japanese study of Hiroshima and Nagasaki emphasizes, ". . . is a weapon of mass slaughter." A nuclear weapon is in fact a total-death machine, compact and efficient, as a simple graph prepared from Hiroshima statistics demonstrates:

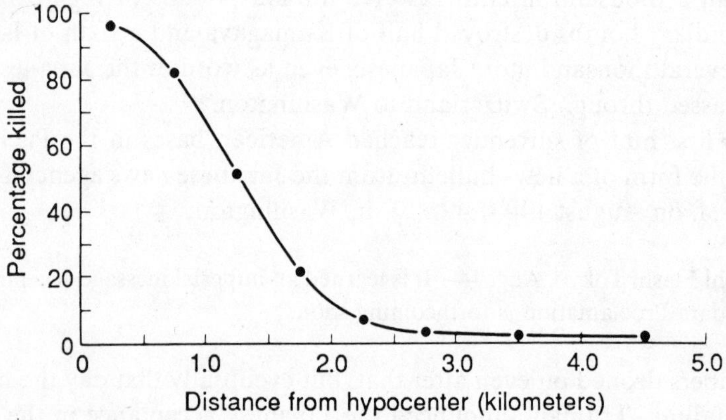

The percentage of people killed depends simply on distance from the hypocenter; the relation between death percentage and distance is inversely proportional and the killing, as Gil Elliot emphasizes, is no longer selective:

By the time we reach the atom bomb, Hiroshima and Nagasaki, the ease of access to target and the instant nature of macro-impact mean that both the choice of city and the identity of the victim has become completely randomized, and human technology has reached the final platform of self-destructiveness. The great cities of the dead, in numbers, remain Verdun, Leningrad and Auschwitz. But at Hiroshima and Nagasaki the "city of the dead" is finally transformed from a metaphor into a literal reality. The city of the dead of the future is our city and its victims are—not French and German soldiers, nor Russian citizens, nor Jews—but all of us without reference to specific identity.

"The experience of these two cities," the Japanese study emphasizes, "was the opening chapter to the possible annihilation of mankind."

On August 24, having recently heard about the man holding an eyeball, Dr. Michihiko Hachiya suffered a nightmare. Like the myth of the Sphinx—destruction to those who cannot answer its riddle, whom ignorance or inattention or arrogance misleads—the dream of this Japanese doctor who was wounded in the world's first atomic bombing and who ministered to hundreds of victims must be counted one of the millennial visions of mankind:

> The night had been close with many mosquitoes. Consequently, I slept poorly and had a frightful dream.
>
> It seems I was in Tokyo after the great earthquake and around me were decomposing bodies heaped in piles, all of whom were looking right at me. I saw an eye sitting on the palm of a girl's hand. Suddenly it turned and leaped into the sky and then came flying back towards me, so that, looking up, I could see a great bare eyeball, bigger than life, hovering over my head, staring point blank at me. I was powerless to move.

"I awakened short of breath and with my heart pounding," Michihiko Hachiya remembers.

So do we all.

Epilogue

The atomic bombing of Hiroshima and Nagasaki horrified Leo Szilard. He felt a full measure of guilt for the development of such terrible weapons of war; the shape of things to come that he had first glimpsed as he crossed Southampton Row in Bloomsbury in 1933 had found ominous residence in the world partly at his invitation. In the petition to the President that he had circulated among the atomic scientists in July 1945—the petition Edward Teller in consultation with Robert Oppenheimer had decided not to sign, writing Szilard he felt "that I should do the wrong thing if I tried to say how to tie the little toe of the ghost to the bottle from which we just helped it to escape"—Szilard had argued that large moral responsibilities devolved upon the United States in consequence of its possession of the bomb:

> The development of atomic power will provide the nations with new means of destruction. The atomic bombs at our disposal represent only the first step in this direction, and there is almost no limit to the destructive power which will become available in the course of their future development. Thus a nation which sets the precedent of using these newly liberated forces of nature for purposes of destruction may have to bear the responsibility of opening the door to an era of devastation on an unimaginable scale.

The United States set that precedent in Japan; Szilard wrote Gertrud Weiss his despairing August 6 letter saying it was difficult to see what wise course of action was possible after that; but within days he was moving to protest and debate. Upon hearing of the Nagasaki bombing he immediately asked the chaplain of the University of Chicago to include a special prayer for the dead and a collection for the survivors of the two Japanese cities in any service commemorating the end of the war. He drafted a second petition to the President calling the atomic bombings "a flagrant violation of our own moral standards" and asking that they be stopped. The Japanese surrender mooted the issue and the petition was never sent.

Besides White House and War Department press releases the United States government immediately published a detailed report on the scientific aspects of atomic bomb development, in preparation during the preceding year by Princeton physicist Henry DeWolf Smyth. *Atomic Energy for Military Purposes* was another faded echo of Niels Bohr's appeal for openness. It appalled the British, enlightened the Soviets on which approaches to isotope separation not to pursue and—Groves' intention in releasing it— defined what might be public and what secret about the atomic bomb program, thereby forestalling information leaks.

With the atomic secret, such as it was, made public Szilard went to see the Chicago chancellor, Robert Maynard Hutchins, "and told him that something needed to be done to get thoughtful and influential people to think about what the bomb might mean to the world, and how the world and America could adjust to its existence. I proposed that the University of Chicago call a three-day meeting and assemble about twenty-five of the best men to discuss the subject." Hutchins liked the idea and began contacting twice that many participants, including Henry Wallace, Tennessee Valley Authority chairman David E. Lilienthal, the ubiquitous Charles Lindbergh and a number of academics and scientists. The meeting took shape for late September.

The day after the Nagasaki bombing Ernest Lawrence had flown to New Mexico, partly to escape newspaper reporters clamoring for interviews, partly to work with Oppenheimer on a report on postwar planning that the Interim Committee had solicited from its Scientific Panel. The inventor of the cyclotron, who approved of the use of the bombs to avoid invasion and force a Japanese surrender, found his Los Alamos colleague weary, guilty and depressed. Oppenheimer wondered if the dead at Hiroshima and Nagasaki were not luckier than the survivors, whose exposure to the bombs would have lifetime effects. His mood that weekend found expression during the next weeks in letters. "You will believe that this undertaking has not been without its misgivings," he wrote his former Ethical

Culture School teacher Herbert Smith, the confessor of his youth; "they are heavy on us today, when the future, which has so many elements of high promise, is yet only a stone's throw from despair." To Haakon Chevalier, his friend at Berkeley in Depression days, Oppenheimer repeated that "the circumstances are heavy with misgiving, and far, far more difficult than they should be, had we power to remake the world to be as we think it."

Lawrence could muster only limited patience for Oppenheimer's remorse. He thought the atomic bomb a "terrible swift sword" that would end the war and might succeed in "ending all wars." He also seems to have claimed it as his own. "In one newspaper interview out of many published the day after Hiroshima," notes Stanislaw Ulam mischievously, "E. O. Lawrence 'modestly admitted,' according to the interviewer, 'that he more than anyone else was responsible for the atomic bomb.' "

From secret participants in a top-secret project the two men and their colleagues had emerged as public heroes, the artificers of a military revolution. "With the discovery of fission," C. P. Snow comments, ". . . physicists became, almost overnight, the most important military resource a nation-state could call upon." The letter on postwar planning that the Berkeley and Los Alamos directors polished that last weekend of the war tried out their new authority; in it the members of the Interim Committee Scientific Panel—Lawrence, Oppenheimer, Compton, Fermi—set aside merely technical advice to propose a radical rethinking of national policy. In doing so they began to outline the nuclear dilemma as they understood it.

They were convinced, they wrote, "that weapons quantitatively and qualitatively far more effective than now available will result from further work on these problems." (Among such weapons they thought the "technical prospects of the realization of the super bomb" to be "quite favorable.") They could not, however, "devise or propose effective military countermeasures for atomic weapons" and it was their "firm opinion that no military countermeasures will be found." They were not only "unable to outline a program that would assure this nation for the next decades hegemony in the field of atomic weapons," but were "equally unable to insure that such hegemony, if achieved, could protect us from the most terrible destruction." What followed, they thought, was the necessity of *political* change:

> The development, in the years to come, of more effective atomic weapons, would appear to be a most natural element in any national policy of maintaining our military forces at great strength; nevertheless we have grave doubts that this further development can contribute essentially or permanently to the prevention of war. We believe that the safety of this nation—as opposed to its ability to inflict damage on an enemy power—cannot lie

> wholly or even primarily in its scientific or technical prowess. It can be based only on making future wars impossible. It is our unanimous and urgent recommendation to you that, despite the present incomplete exploitation of technical possibilities in this field, all steps be taken, all necessary international arrangements be made, to this one end.

Oppenheimer had discovered such convictions in his discussions with Bohr, but Lawrence had advised the Interim Committee only two months earlier to build up stockpiles. For a time at the end of the war the reality of the bomb seems to have moved the Berkeley laureate to at least limited internationalism. "There is no doubt in my mind," he wrote during this period, "that the best channel of information about what is going on in Russia would be developed by encouraging free interchange of science and scientists. In fact it is the only avenue I can think of that has a reasonable chance of working."

Oppenheimer carried the Scientific Panel's letter to Washington a few days after the surrender and found Henry Stimson out of town. He talked instead to Stimson's aide George L. Harrison and to Vannevar Bush. "I emphasized of course that all of us would earnestly do whatever was really in the national interest, no matter how desperate and disagreeable," he wrote Lawrence after he returned to New Mexico; "but that we felt reluctant to promise that much real good could come of continuing the atomic bomb work—just like poison gases after the last war." But no more than Leo Szilard before him was Oppenheimer successful at influencing policy from outside the political process, whatever his newfound authority as a consultant:

> I had the fairly clear impression from the talks [with Harrison and Bush] that things had gone badly at Potsdam, and that little or no progress had been made in interesting the Russians in collaboration or control. I don't know how seriously an effort was made: apparently neither Churchill nor Attlee nor Stalin was any help at all, but this is only my conjecture. While I was in Washington two things happened, both rather gloomy: the President issued an absolute Ukase, forbidding any disclosures on the atomic bomb—and the terms were broad—without his personal approval. The other was that Harrison took our letter to [Jimmy] Byrnes, who sent back word just as I was leaving that "in the present critical international situation there was no alternative to pushing the [Manhattan Project] program full steam ahead." . . .I do not come away from [i.e., I still feel] a profound grief, and a profound perplexity about the course we should be following.

The Conference on Atomic Energy Control at the University of Chicago convened on a Thursday and Friday in late September. David Lilienthal kept shorthand notes; their highlights reveal remarkable prescience among the participants, as does Szilard's memory of the event. Jacob Viner, an influential University of Chicago economist, told the conference that the atomic bomb was the cheapest way yet devised of killing human beings. With two giants, he argued—the Soviet Union and the United States—world government would be impossible. "Degree of peace we have had in the past two or three centuries," Lilienthal noted of Viner's comments, "has been due to uncertainty as to who was your natural enemy.... Now no ambiguity as to target, where there are only two giants. Already having psychological effects in this country." Viner thought "atomic bomb warfare more largely a war of nerves.... Psychological warfare begins when two countries have atomic bombs... Believes atomic bomb will be peacemaking in effect; deterrent effect—the cost when it is used against you." Only five weeks after the first use of nuclear weapons in war Viner had ferreted out the essential principle of deterrence, of the balance of terror in a nuclear-armed world.

Szilard presented his ideas to the group on Friday. He emphasized that the bombs would get bigger, bypassing secrecy by quoting a public statement of Mark Oliphant's to the effect that weapons "corresponding to one million and to ten million tons of TNT ... are entirely possible." Of Szilard's talk Lilienthal noted:

> We are in an armament race.
>
> If Russia starts making atomic bombs in two or three years—perhaps five or six years—then we have an armed peace, and it will be a durable peace.
>
> But we will not have permanent peace at lesser cost than world government. But this cannot come without changed loyalty of people. If we can't have that, all we can have is a durable peace [i.e., deterrence]. Only purpose of a durable peace would be to create conditions 20–30 years from now [that] can bring about world peace. That requires shift of loyalties.
>
> If we are *sure* to get a Third World War, the later it comes the worse for us.
>
> Victor of next war will *make* a world government, even if that victor should be the United States, having lost 25 million people dead.

Szilard himself came to believe his predictions unimpressive. He thought Viner did better:

> The wisest remarks that were made at this meeting were made by Jake Viner, and what he said was this: "None of these things will happen. There will be no preventive war, and there will be no international agreement involving inspection. America will be in [sole] possession for a number of years, and the

> bomb will exert a certain subtle influence; it will be present at every diplomatic conference in the consciousness of the participants and will exert its effect. Then, sooner or later, Russia will also have the bomb, and then a new equilibrium will establish itself."

No longer shackled to Army secrecy, Leo Szilard would continue and increase his participation in the political life of his adopted country. But whether from guilt, or because nuclear physics no longer seemed to him a frontier, or because he had come to understand that the liberation of atomic energy was far more likely to enable man to destroy the earth than to leave it behind for the stars—the cause for which he had taken up studying the nucleus in the first place—he closed the loop he opened in 1932, went off in 1947 to Cold Spring Harbor Laboratory on Long Island to audit the phage course offered there and turned away from physics to biology.

H. G. Wells lived to know of Hiroshima and Nagasaki. Deeply pessimistic in his final years, he died at eighty on August 13, 1946.

Immediately after Trinity Edward Teller and Enrico Fermi had renewed theoretical work on the problem of thermonuclear ignition. "The end sought," explains the Los Alamos technical history, "was a bomb burning about a cubic meter of liquid deuterium. For such a bomb the energy release will be about ten million tons of TNT." But with the Japanese surrender hydrogen bomb studies at Los Alamos temporarily slowed and stopped. General Groves, Oppenheimer would testify later, "was unclear whether his mandate and therefore mine extended to fiddling with this next project. I so reported to the people in the laboratory, who were thinking about it." Teller, frustrated, suspected his colleagues had "[lost] their appetites for weapons work." Some had. Most simply wanted to go home. "We all felt," Hans Bethe remembers, "that, like the soldiers, we had done our duty and that we deserved to return to the type of work that we had chosen as our life's career, the pursuit of pure science and teaching. . . . Moreover, it was not obvious in [1945 and] 1946 that there was any need for a large effort on atomic weapons in peacetime."

Teller passionately disagreed. "He expressed himself as terribly pessimistic about relations with Russia," Bethe remembers of a conversation the two theoreticians had that winter. "He was terribly anti-communist, terribly anti-Russian. Now I knew that he had been anti-communist during the communist takeover in Hungary when he was about eleven, but now it came out in a much more forceful way. Teller said we had to continue research on nuclear weapons . . . it was really wrong of all of us to want to leave. The war was not over and Russia was just as dangerous an enemy as

Germany had been. I just couldn't go along with that. I thought it was more important to go home and get the universities restarted, to train young physicists again."

Bethe was returning to Cornell. Oppenheimer had offers from every direction; he turned down Harvard in late September, believing, he wrote James Bryant Conant, "that I would like to go back to California for the rest of my days." Fermi had accepted appointment to the faculty of the University of Chicago. Teller had been invited to work at Fermi's side. Leaving Los Alamos would mean leaving the Super to others, but staying at Los Alamos would mean becoming part of what Oppenheimer, free once more to indulge in casual cruelty, was calling the second team.

Norris Bradbury, the vigorous Berkeley-trained Navy physicist who had organized the assembly of the Trinity bomb, was replacing Oppenheimer as director. "In the months immediately following the war," Bradbury recalled in 1948, "the Laboratory struggled for existence and there is no better way to put it":

> Here was Los Alamos in September, 1945. The senior civilian scientists, weary of living under wartime conditions, under wartime security, on a wartime Army post, and under conditions of wartime urgency, thought longingly of their academic laboratories and classrooms. The more junior civilians thought of the academic degrees they did not have and the further education they ought to have. . . .
>
> There was even no agreement as to what sort of future should be planned for Los Alamos. There was one school of thought which held that Los Alamos should become a monument, a ghost laboratory, and that all work on the military use of atomic energy should cease. Another group looked with increasing pessimism on the deterioration of our international relations and contended that Los Alamos should become a factory for atomic weapons. The majority agreed that, for the present at least, the United States required a research laboratory devoted to the study of fundamental nuclear physics and chemistry and their possible application to military use.

Bradbury asked Teller to continue at Los Alamos as head of the Theoretical Division, the position Teller had believed he deserved when the laboratory was founded and that Oppenheimer had given to Bethe. Teller would sign on only if Bradbury promised major commitment in return. "I said we either should make a great effort to build a hydrogen bomb in the shortest possible time or develop new models of fission explosives and speed progress by at least a dozen [weapons] tests a year. Bradbury said he would like to see either program, but that neither was realistic. There no longer was governmental support for weapons work. No one was interested." Either the new director was misinformed or Teller misrepresents his position; Jimmy Byrnes had charged Oppenheimer only a few weeks earlier to maintain "full steam ahead." The immediate postwar problem at Los Alamos was not lack of support but lack of authority. The Army had run the Manhattan Project in wartime. Now the work needed congressional authorization and funding, and that was slow in coming because it depended on legislation dealing with atomic energy, a revolutionary new field. "To demand, as Teller did as a condition of his staying," writes Bethe, "that Los Alamos tackle the super-bomb on a large scale, or plan for twelve tests a year on fission bombs, was plainly unrealistic to say the least."

Teller went looking for Oppenheimer, "seeking his advice and support":

> I told him about my conversation with Bradbury, and then said: "This has been your laboratory, and its future depends on you. I will stay if you will tell me that you will use your influence to help me accomplish either of my goals, if you will help enlist support for work toward a hydrogen bomb or further development of the atomic bomb."
>
> Oppenheimer's reply was quick: "I neither can nor will do so."
>
> It was obvious and clear to me that Oppenheimer did not want to support further weapons work in any way. It was equally obvious that only a man of Oppenheimer's stature could arouse governmental interest in either program. I was not willing to work without backing, and told Oppenheimer that I would go to Chicago. He smiled: "You are doing the right thing."

Deke Parsons gave a party that night. Teller says Oppenheimer sought him out and asked him, "Now that you have decided to go to Chicago, don't you feel better?" Teller complained that he did not feel better; he felt that their work had been only a beginning. "We have done a wonderful job here," Oppenheimer countered, "and it will be many years before anyone can improve on our work in any way." The insensitivity of the remark rankled Teller as its ambiguity confused him. He would quote it frequently in the years after 1945, always to demonstrate its self-deception. It might have meant: the Soviets will not soon build a bomb. Or it might have meant: the Oppenheimer team had accomplished in fission development what a Teller team could not soon improve in thermonuclear development. Teller would read it both ways and like neither reading.

His immediate response was to take his problem to Fermi. Fermi apparently argued with him, consistent with the Interim Committee Scientific Panel letter of August 17, that the solution to the problem of nuclear weapons must be a political solution. Fermi thought Teller was overly optimistic as well about the early prospects for a successful thermonuclear. Not only was thermonuclear burning itself a hard problem; the atomic bomb would also have to be better understood and considerably improved before it could be made efficient enough to serve as a thermonuclear trigger. But the two men were good friends, and Fermi encouraged Teller to write him a letter expressing his dissent; he would be happy to pass it along to the Secretary of War for the Interim Committee file.

A year earlier, in the midst of war, James Bryant Conant had visited Los Alamos and talked to Teller about the Super. Teller had predicted then, as Conant reported to Vannevar Bush, that the Super was "probably at least as distant now as was the fission bomb when . . . I first heard of the enterprise." That estimate—between four and five years—was already optimistic compared to Fermi's. Now, in October 1945, Teller set it as an upper limit. He also first stated for the record many of the arguments for pursuing technological security that he would elaborate in the decades to come.

"When," he asked in the question-and-answer format he adopted, "could the first super bomb be tried out?" He answered with two numbers, the second an early example of what has come to be called threat inflation:

> It is my belief that five years is a conservative estimate of this time. This assumes that the development will be pursued with some vigor. The job, however, may be much easier than expected and may take no more than two years. In considering future dangers it is important not to disregard this eventuality.

How soon could another country produce such a superbomb? Faster than the United States, Teller apparently thought, despite his adopted nation's commanding technological and industrial lead: "The time needed . . . may not be much longer than the time needed by them to produce an atomic bomb."

What about moral objections? They were meaningless before the onrush of technology:

> There is among my scientific colleagues some hesitancy as to the advisability of this development on the grounds that it might make the international problems even more difficult than they are now. My opinion is that this is a fallacy. If the development is possible, it is out of our powers to prevent it.

Teller thought that civil defense measures such as the dispersal of cities might prove effective against atomic bombs but "very much less so against super bombs." He could not yet offer detailed plans for the peaceful use of thermonuclear explosives. "But I consider it a certainty that the super bomb will allow us to extend our power over natural phenomena far beyond anything we can at present imagine."

He filed his dissent as he prepared to depart Los Alamos. Teller was not one to fight for lost causes. He might have stayed, but the first team was leaving. His wife was expecting their second child. He packed his grand piano and moved to a professorship at the University of Chicago to do physics with Enrico Fermi. For a few years he would find security in family and teaching and research.

General Leslie R. Groves traveled to Los Alamos in mid-October to present a certificate of appreciation to the laboratory from the Secretary of War. "Under a brilliant New Mexico sky," Alice Kimball Smith remembers, "virtually the entire population of the mesa assembled for the outdoor ceremony" on October 16, Oppenheimer's last day as director. He still thought he would return to California and to teaching, but accepting the certificate he struck the theme that would occupy the next full decade of his life:

> It is our hope that in years to come we may look at this scroll, and all that it signifies, with pride.
>
> Today that pride must be tempered with a profound concern. If atomic bombs are to be added as new weapons to the arsenals of a warring world, or to the arsenals of nations preparing for war, then the time will come when mankind will curse the names of Los Alamos and Hiroshima.
>
> The peoples of the world must unite, or they will perish. This war, that has ravaged so much of the earth, has written these words. The atomic bomb has spelled them out for all men to understand. Other men have spoken them, in other times, of other wars, of other weapons. They have not prevailed. There are some, misled by a false sense of human history, who hold that they will not prevail today. It is not for us to believe that. By our works we are committed, committed to a world united, before the common peril, in law, and in humanity.

Besides the certificate the men and women of Los Alamos each received a memento that day: a sterling-silver pin the size of a dime stamped with a large letter *A* framing the small word BOMB. Before Oppenheimer rushed off to Washington to testify on atomic energy to House and Senate committees a newspaper reporter asked him if the atomic bomb had any significant limitations. "The limitations lie in the fact that you don't want to be on the receiving end of one," he quipped. Then he ventured prophecy: "If you ask: 'Can we make them more terrible?' the answer is yes. If you ask: 'Can we make a lot of them?' the answer is yes. If you ask: 'Can we make them terribly more terrible?' the answer is probably." *Time* featured the remarks in its International section at the end of the month with a photograph of Oppenheimer holding a pipe and looking persuasive. He was "the smartest of the lot," the newsmagazine quoted an unnamed colleague on his behalf. The public romance had begun.

I. I. Rabi returned to Columbia University, Eugene Wigner to Princeton, Luis Alvarez, Glenn Seaborg and Emilio Segrè to Berkeley, George Kistiakowsky to Harvard. Victor Weisskopf went to MIT. Stanislaw Ulam briefly and unhappily tried UCLA, then came back to Los Alamos. James Chadwick and most of the British Mission returned to Great Britain with pockets full of secrets. In September the British had given a formal farewell party for their friends on the Hill, the Brobdingnagian log cabin of Fuller Lodge jammed with men in black tie and even white tie and tails and women in long gowns not completely aired of mothballs. Genia Peierls had cooked buckets of thick soup; steak and kidney pie was served on paper plates; Winifred Moon supplied several hundred paper cartons of trifle, a dessert which she swore she would never look upon again without nausea. The Oppenheimers and the Peierlses sat at high table above the fray (the Chadwicks had not come out from Washington) and the convivial and sometimes bibulous James Tuck served as toastmaster. After dinner, Bernice Brode notes, the British staged an original pantomime based on *Babes in the Woods:*

> Good Uncle Winnie had sent his Babes to join forces with Good Uncle Franklin, to out-wit Bad Uncles Adolf and Benito. All that befell the children on their hazardous journey to the Unknown Desert was acted out by the entire Mission. . . . The end, the grand finale was a re-enacting of the [Trinity] test, with [a stepladder for] a high tower from which a pail of stuff was overturned making flashes and bangs and clatters for several minutes. This was not entirely comprehensible to many of the women, but made a tremendous hit with the men, particularly some of the details of the bangs. It was indeed a real smash hit.

Later they cleared the floor and danced, as they had danced so many Saturday nights in that strange wilderness retreat through the long years of war.

Niels Bohr, resettled in the Carlsberg House of Honor in Copenhagen, wrote Oppenheimer on November 9:

> I was very sorry that I should not see you again before my return to Denmark, but, due to difficulties in arranging passage for Margrethe and me, we could not, as we had intended to, return to U.S.A. before the secret of the project was lifted, and then it was thought advisable that I no longer postponed my return to Denmark.

> I need not say how often Aage and I think of all the kindness you and Kitty showed us in these last eventful years, where your understanding and sympathy have meant so much to me, and how closely I feel connected with you in the hope that the great accomplishment may contribute decisively to bringing about harmonious relationships between nations. . . .
>
> I trust that the whole matter is developing in a favorable way.

It was not, as Bohr knew; there was loud talk in the United States of an atomic "secret" that America would keep and protect. How little might be secret was revealed that autumn and early winter in a series of reports of Soviet stirrings in the field. The War Department learned in the middle of September, writes the historian Herbert Feis, "that the Soviet authorities were compelling the commanders of the Czechoslovak Army to give orders that all German plans, parts, models and formulas regarding the use of atomic energy, rocket weapons, and radar be turned over to them. Russian infantry and technical troops occupied Jachimov . . . and St. Joachimstal, the town and the factory—the only place in central Europe where at this time uranium was being produced." The old mine from the residues of which Martin Klaproth had first isolated the heavy gray metal he named uranium, the mine young Robert Oppenheimer had explored on a walking tour in 1921, had fallen into Soviet hands.

An attaché at the U.S. Embassy in Moscow warned on December 24 that "the U.S.S.R. is out to get the atomic bomb. This has been officially stated. The meager evidence available indicates that great efforts are being made and that super-priority will be given to the enterprise." At home was incomprehension, Herbert York remembers: "To most . . . of us, Russia was as mysterious and remote as the other side of the moon and not much more productive when it came to really new ideas or inventions. A common joke of the time said that the Russians could not surreptitiously introduce nuclear bombs in suitcases into the United States because they had not yet been able to perfect a suitcase." But if American leaders did not believe the Soviet Union could soon achieve an atomic bomb, what it would do otherwise and what were its motives had become a matter of intense debate within the U.S. government.

Tragically, that debate obscured the deeper issue then confronting the world for the first time in history. Robert Oppenheimer had testified before Congress; he had begun to work his way into the corridors of power; now, on a stormy Friday night at the beginning of November 1945, he stepped forward to examine the nuclear dilemma publicly. Freed from the constraints of the Los Alamos directorship he spoke to five hundred members of the Association of Los Alamos Scientists, a new political organization, crowded into the larger movie theater on the Hill. An unrevised transcript preserves his words much as his listeners heard them; thunder above the mesa orchestrated his bare reconnaissance. It framed anew the prospects Bohr had revealed and defined limitations and opportunities that have persisted into the present.

"I should like to talk tonight—if some of you have long memories perhaps you will regard it as justified—as a fellow scientist," Oppenheimer began with humor, "and at least as a fellow worrier about the fix we are in." Involved, he thought, were "issues which are quite simple and quite deep." One of those issues for him was *why* scientists had built the atomic bomb. He listed a number of motives: fear that Nazi Germany would build it first, hope that it would shorten the war, curiosity, "a sense of adventure," or so that the world might know "what can be done . . . and deal with it." But he thought the basic motivation was moral and political:

> When you come right down to it the reason that we did this job is because it was an organic necessity. If you are a scientist you cannot stop such a thing. If you are a scientist you believe that it is good to find out how the world works; that it is good to find out what the realities are; that it is good to turn over to mankind at large the greatest possible power to control the world and to deal with it according to its lights and its values. . . .
>
> It is not possible to be a scientist unless you believe that the knowledge of the world, and the power which this gives, is a thing which is of intrinsic value to humanity, and that you are using it to help in the spread of knowledge, and are willing to take the consequences.

The defining trust in the value of knowledge that Oppenheimer ascribes here to science echoes Bohr's succinct formulation of the value of openness: "The very fact that knowledge is itself the basis of civilization points directly to openness as the way to overcome the present crisis." Long before them Thomas Jefferson, secure in his understanding of the core principles of democracy, professed a similar conviction. "I know no safe depository of the ultimate powers of the society but the people themselves," he wrote late in life; "and if we think them not enlightened enough to exercise that control with a wholesome discretion, the remedy is not to take it from them, but to inform their discretion."

Oppenheimer went on to examine the political changes he believed the new weapons challenged mankind to explore:

> But I think the advent of the atomic bomb and the facts which will get around that they are not too hard to make, that they will be universal if people wish to make them universal, that they will not constitute a real drain on the economy

of any strong nation, and that their power of destruction will grow and is already incomparably greater than that of any other weapon—I think these things create a new situation, so new that there is some danger, even some danger in believing, that what we have is a new argument for arrangements, for hopes, that existed before this development took place. By that I mean that much as I like to hear advocates of a world federation, or advocates of a United Nations organization, who have been talking of these things for years—much as I like to hear them say that here is a new argument, I think that they are in part missing the point, because the point is not that atomic weapons constitute a new argument. There have always been good arguments. The point is that atomic weapons constitute also a field, a new field, and a new opportunity for realizing preconditions. I think when people talk of the fact that this is not only a great peril, but a great hope, this is what they should mean . . .[:] the simple fact that in this field, because it is a threat, because it is a peril . . . there exists a possibility of realizing, of beginning to realize, those changes which are needed if there is to be any peace.

Those are very far-reaching changes. They are changes in the relations between nations, not only in spirit, not only in law, but also in conception and feeling. I don't know which of these is prior; they must all work together, and only the gradual interaction of one or the other can make a reality. I don't agree with those who say the first step is to have a structure of international law. I don't agree with those who say the only thing is to have friendly feelings. All of these things will be involved. I think it is true to say that atomic weapons are a peril which affects everyone in the world, and in that sense a completely common problem, as common a problem as it was for the Allies to defeat the Nazis.

Solving that common problem, he continued, could serve as "a pilot plant for a new type of international collaboration":

I speak of it as a pilot plant because it is quite clear that the control of atomic weapons cannot be in itself the unique end of such operation. The only unique end can be a world that is united, and a world in which war will not occur.

Half a century of circumscribed and often cynical negotiations for arms control has not altered Oppenheimer's essential point, which was Bohr's hopeful vision first of the complementarity of the bomb.

Next he discussed what Bohr would have called the necessity of renunciation. Oppenheimer offered an American analogy:

The one point I want to hammer home is what an enormous change in spirit is involved. There are things which we hold very dear, and I think rightly hold

> very dear; I would say that the word democracy perhaps stood for some of them as well as any other word. There are many parts of the world in which there is no democracy. There are other things which we hold dear, and which we rightly should. And when I speak of a new spirit in international affairs I mean that even to these deepest of things which we cherish, and for which Americans have been willing to die—and certainly most of us would be willing to die—even in those deepest things, we realize that there is something more profound than that; namely, the common bond with other men everywhere. It is only if you do that that this makes sense; because if you approach the problem and say, "We know what is right and we would like to use the atomic bomb to persuade you to agree with us," then you are in a very weak position. . . .
>
> I want to express the utmost sympathy with the people who have to grapple with this problem and in the strongest terms to urge you not to underestimate its difficulty. I can think of an analogy. . . : in the days in the first half of the nineteenth century there were many people, mostly in the North, but some in the South, who thought that there was no evil on earth more degrading than human slavery, and nothing that they would more willingly devote their lives to than its eradication. Always when I was young I wondered why it was that when Lincoln was President he did not declare that the war against the South, when it broke out, was a war that slavery should be abolished, that this was the central point, the rallying point, of that war. Lincoln was severely criticized by many of the Abolitionists as you know, by many then called radicals, because he seemed to be waging a war which did not hit the thing that was most important. But Lincoln realized, and I have only in the last months come to appreciate the depth and wisdom of it, that beyond the issue of slavery was the issue of the community of the people of this country, and the issue of the Union. . . . In order to preserve the Union Lincoln had to subordinate the immediate problem of the eradication of slavery—and trust—and I think if he had had his way it would have gone so—to the conflict of these ideas in a united people to eradicate it.

For such understanding Oppenheimer celebrated Bohr, "who was here so much during the difficult days, who had many discussions with us, and who helped us reach the conclusion [that a universal renunciation of the use of force] was not only a desirable solution, but that it was the unique solution, that there were no other alternatives."

Little more in Oppenheimer's talk that stormy night carries weight down the years: practical matters of legislation, counsel to his fellow scientists to accept responsibility for the consequences of their work. In closing he delivered a final burst of realism about a timetable for change:

> I'm not sure that the greatest opportunities for progress do not lie somewhat further in the future than I had for a long time thought. . . .

> The plain fact is that in the actual world, and with the actual people in it, it has taken time, and it may take longer, to understand what this is all about. And I'm not sure, as I have said before, that in other lands it won't take longer than it does in this country.

These basic questions engaged men in 1945; they engage us still, as if the clock had stopped while only the machinery of armament with terrible and terribly more terrible weapons has kept on running.

Edward Teller returned to Los Alamos in April 1946 to chair a secret conference. Its purpose, according to a subsequent report, was "to review work that has been done on the Super for completeness and accuracy and to make suggestions concerning further work that would be needed in this field if actual construction and test of the Super were planned." John von Neumann, Stanislaw Ulam and Norris Bradbury attended the conference, as did Emil Konopinski, John Manley, Philip Morrison, Canadian theoretician J. Carson Mark and a crowd of other participants. One whose presence would vitally affect U.S. nuclear weapons policy later was Klaus Fuchs.

The Super conference examined only one design for a thermonuclear weapon, the design Teller and his group had developed during the war, the so-called classical Super, with an estimated explosive force of 10 million tons TNT equivalent—10 megatons. The ingredients for the classical Super would be an atomic bomb, a cubic meter of liquid deuterium and an indefinite amount of the rare second isotope of hydrogen, tritium, which because of its short 12.26-year half-life does not normally exist in nature but can be created in a nuclear reactor by bombarding lithium with neutrons. How these components would have been arranged in the classical Super is still secret: probably spherically, the fission trigger and hydrogen isotopes physically contiguous and contained within a heavy tamper.

"It is likely," the conference decided on the basis of the Teller group's calculations, "that a super-bomb can be constructed and will work. Definite proof of this can hardly ever be expected and a final decision can be made only by a test of the completely assembled super-bomb." The conference called for "a detailed calculation" to study mathematically the probable progress of the explosion (the hand calculations that Teller and his group had made to arrive at the classical Super were necessarily, given the complexity of the problem, rough and incomplete). The conference also found that Teller's design was "on the whole workable." Some participants had doubts; "should the doubts prove well-founded, simple modifications of the design will render the model feasible." In conclusion:

> The undertaking of the new and important Super Bomb project would necessarily involve a considerable fraction of the resources which are likely to be devoted to work on atomic developments in the next years. . . . We feel it appropriate to point out that further decision in a matter so filled with the most serious implications as is this one can properly be taken only as part of the highest national policy.

In June 1946, three months after the Super conference, the U.S. nuclear weapons stockpile consisted of only nine Fat Man bombs, of which no more than seven could be made operational for lack of initiators. The stockpile held only thirteen bombs a year later, two years after the end of the war. Plutonium production was the crucial bottleneck. The high neutron flux of the Hanford production piles had proven damaging. One had been unloaded in May to prevent further damage and the other damped back to 80 percent of its full capacity. Anything Los Alamos could do, therefore, to improve fission bomb design would significantly bolster the U.S. nuclear arsenal at a time of increasing conflict with the Soviet Union. Nevertheless, about half the Theoretical Division's time between 1946 and 1950 went to the Super. Atomic bombs by then were a matter more of engineering than of theoretical physics, to be sure, but the complexities of the thermonuclear problem also tantalized.

There ensued a curious period of optimism in Teller's life. His wife Mici bore him a second child, a daughter, in the summer of 1946 and he found more time for his family. He was caught up again in the grandeur and the deeply satisfying creativity of basic science. "The years after Los Alamos," writes Eugene Wigner, "and until the renewal of his preoccupation with national security, were perhaps Teller's most fruitful years scientifically." Teller taught, co-authored thirteen scientific papers, regularly visited Los Alamos to consult, wrote articles for the new *Bulletin of the Atomic Scientists.* In the *Bulletin* he called for an end to secrecy where "purely scientific data" are concerned. He praised as "ingenious, daring and basically sound" the Acheson-Lilienthal Report that became the basis of the Baruch Plan for international control of nuclear weapons that the United States offered to the United Nations in 1946. He recognized the absolutes of the absolute weapon in April 1946, the same month as the Super conference, in a surprising profession of faith: "Nothing that we can plan as a defense for the next generation is likely to be satisfactory; that is, nothing but world-union."

A year later he still saw no defense against atomic weapons. He described with compassionate horror the terrible devastation of Hiroshima: "One is struck by the picture of fires raging unopposed, wounds remaining unattended, sick men killing themselves with the exertions of helping their fellows." It was even possible to imagine, he wrote, "that the effects of an atomic war will endanger the survival of man." He thought in December 1947, in the wake of Soviet rejection of the Baruch Plan, that "agreement with the Russians still seems possible"; the Danes, he noted waggishly, were once similarly imperialistic and ambitious. "We must now work for world law and world government. . . . Even if Russia should not join immediately, a successful, powerful, and patient world government may secure their cooperation in the long run. . . . We [scientists] have two clear-cut duties: to work on atomic energy and to work for world government which alone can give us freedom and peace."

The extreme swings in Edward Teller's outlook remain somewhat mysterious. The British theoretician Freeman Dyson, then one of Teller's students, wrote his family from postwar Chicago that his teacher, whom he liked and admired, was nevertheless "a good example of the saying that no man is so dangerous as an idealist." The emotional timbre of Teller's Chicago-period writing differs notably from his later work. It is less choleric and more optimistic, of course, but a deeper difference is that it is informed by a much greater degree of trust in his fellow man and in the possibility that human institutions might serve to restrain the conflicts of nations. Even Russia became, for a time, "this fabulous monster," not the threatening presence Teller described to Bethe in 1945 and would invoke with increasing urgency in the years after 1949: "World government," he wrote in the *Bulletin* as late as July 1948, "is our only hope for survival. . . . I believe that we should cease to be infatuated with the menace of this fabulous monster, Russia. Our present necessary task of opposing Russia should not cause us to forget that in the long run we cannot win by working against something. We must work for something. We must work for World Government." The reasons for this difference of attitude must be partly personal and inaccessible, perhaps even to the man himself. But it would appear from subsequent events that a crucial reason for Teller's sense of security in the immediate postwar years was America's sole possession of the atomic bomb.

But hardly anyone was listening. The Cold War had begun in earnest. Oppenheimer had found his way into the high councils of government; now director of the Institute for Advanced Study at Princeton and chairman of the scientific General Advisory Committee of the newly established U.S. Atomic Energy Commission, he was internationally famous, a household name. A plaintive footnote in a 1948 Teller *Bulletin* review of the first year's work of the Atomic Energy Commission reveals the Hungarian physicist's isolation from power in those days: "Due to the limited experience of the author the account is necessarily incomplete."

In the early summer of 1949 Teller returned to Los Alamos on leave of absence from Chicago. He did not easily wrench himself away. Oppenheimer had sought him out and encouraged him; Bradbury had then sent a delegation that included Stanislaw Ulam to invite him back. He said later he went because he had decided that his writing, public speaking and political action were less than productive and that "the best contribution I could make would be to go back to Los Alamos to help develop weapons—something I knew about and that could yield concrete results." He was undoubtedly influenced by the Soviet coup in Czechoslovakia in the winter of 1948, by the blockade of Berlin that began the following summer and by the impending Communist victory over Nationalist forces in China. A more personal challenge was the fate of Hungary, which had briefly experienced democratic government once more as a republic under the protection of the Allied Control Commission. But the Red Army remained in occupation and by 1948 the Communist Party had maneuvered itself into power. A one-slate election on May 15, 1949, finished the job. Teller's father, mother, sister and nephew had survived the destruction of Hungarian Jewry and still lived in Budapest. Now they were cut off from him.

Teller was thus back at weapons work when Harry Truman announced, on September 23, 1949, the explosion of Joe I, the first Soviet atomic bomb. Like most Americans, Teller had not expected the Soviet success so soon. He called Oppenheimer on the day the Soviet test was announced in a state of arousal sufficient to cause Oppenheimer to advise him sharply, "Keep your shirt on." He testified later that his mind "did not immediately turn in the direction of working on the thermonuclear bomb," but in fact he discussed that prospect intensely at Los Alamos early in October with Ernest Lawrence and Luis Alvarez, who encouraged him. The American nuclear monopoly had ended. The fabulous monster had real claws. If the Soviet Union had tested an atomic bomb, could a Soviet hydrogen bomb be far behind? Teller decided that the only possible hope for continued national security was an all-out American effort to build the Super.

The first concrete act of the United States government in response to the Soviet Union's demonstrated mastery of explosive fission was to approve, in October 1949, a program to expand the production of uranium and plutonium. In the meantime a secret debate raged within the government: what should the United States do? Herbert York, at the time a Teller protégé, describes the debate's constituents:

> Especially considering the enormity of the issue—and most of those involved were fully aware of its enormity—the participants in the secret debate were

> very few: the members of the GAC [the AEC's General Advisory Committee], the members of the AEC and a few of their staff, the members of the [Senate and House Joint Committee on Atomic Energy] and a few of their staff, a very few top officials in the Defense Department, and a very small group of concerned scientists. . . . Altogether, there were less than one hundred people, most of whom thought of themselves—probably correctly—as being involved in making one of the most fateful decisions of all time.

Another crucial figure in the debate was Dean Acheson, now Secretary of State. Truman listened to Acheson, Secretary of Defense Louis Johnson and the Joint Chiefs of Staff more than he listened to the scientists. The Joint Chiefs told Truman, without much staff evaluation, that a Soviet H-bomb would be "intolerable." So, in as many words, did Acheson. None of the groups that participated in the debate studied the possibility that the result of the American initiative might be an arms race. The H-bomb appeared to its supporters to offer a return to nuclear superiority. On January 31, 1950, Truman announced in favor of proceeding to develop it.

Teller took the President's decision as a personal victory. Since at least the September afternoon of Oppenheimer's blunt advice to keep his shirt on he seems to have felt that his fellow physicist was personally attempting to impede him. Oppenheimer was chairman of the GAC, which had been asked in October to advise the AEC commissioners in the matter. Teller went to Washington to lobby past that committee. Among others he talked to Kenneth D. Nichols, now a general and with Groves' retirement the Army's resident nuclear weapons expert. One autumn Sunday morning on Nichols' front porch Teller argued so emotionally that Nichols finally challenged him: "Edward, why are you worrying about the situation so much?" "I'm not really worrying about the situation," Nichols remembers Teller responding; "I'm worrying about the people who should be worrying about it."

The GAC met on October 29 and 30 (Oppenheimer, Conant, Fermi, Rabi, Caltech president Lee DuBridge, metallurgist Cyril Smith, Bell Laboratories president Oliver E. Buckley, engineer Hartley Rowe, Glenn Seaborg absent). It recommended in response to the Soviet achievement that the AEC look further into increasing the production of fissionable materials. It called for "an intensification of efforts to make atomic weapons available for tactical purposes." It proposed building facilities to produce more neutrons for weapons research and development. And it strongly recommended pursuing an existing Los Alamos program to use small quantities of tritium in atomic bombs to "boost" those bombs to more efficient and more powerful explosions (an invention that was successfully tested in May 1951).

But the committee also recommended against pursuing "with high priority the development of the super bomb." It based its recommendation on essentially two arguments: that at ten megatons a Super would be a weapon of mass destruction only, with no other apparent military use; and that it would not obviously improve the security of the United States. It would not do so, in particular, because the design then under consideration—the design reviewed at the Super conference in 1946, Teller's design, the classical Super—looked as if it would require large amounts of tritium, and plutonium and tritium, both created in nuclear reactors, would compete for existing plant capacity. Tritium was as well eighty times as expensive to make, gram for gram, as plutonium. The U.S. nuclear stockpile in late 1949 consisted of about two hundred atomic bombs. Slowing production of bombs that worked for a new weapon that might not work made no sense to the scientists and engineers of the GAC. As Oppenheimer summarized in later testimony, "The [H-bomb] program we had in 1949 was a tortured thing that you could well argue did not make a great deal of technical sense. It was therefore possible to argue also that you did not want it even if you could have it."

The GAC members divided themselves into a majority and a minority to write explanatory annexes to their October 30 report. Conant drafted the majority annex, which Oppenheimer, DuBridge, Rowe, Smith and Buckley signed. It said that a Super "might become a weapon of genocide." It said such a bomb "should never be produced. . . . To the argument that the Russians may succeed in developing this weapon, we would reply that our undertaking it will not prove a deterrent to them. Should they use the weapon against us, reprisals by our large stock of atomic bombs would be comparably effective to the use of a super." Rabi drafted the minority annex, which he and Fermi signed. It argued that the H-bomb question might serve as a springboard for new arms-control efforts. It described the Super as "a weapon which in practical effect is almost one of genocide," then for good measure went even further in condemnation: "It is an evil thing considered in any light."

Teller had been the leading proponent of this enormously powerful weapon since before Pearl Harbor. His Manhattan Project colleagues on the GAC—Oppenheimer, Conant, Fermi, Rabi, Smith—had encouraged him and even worked alongside him. Nothing was said then of evil, of genocide. They who had won high position building weapons used in the mass destruction of two Japanese cities now condemned another, more ingenious weapon. They who were scientists and understood that kindling a sustained thermonuclear reaction on the earth was a historic experiment in fundamental physics proposed the indefinite postponement of that experiment, thus probably handing the triumph gratuitously to the Russians. A courier delivered a copy of the committee's report to Los Alamos in advance of a delegation of interested congressmen who had already seen it. With AEC chairman David Lilienthal's approval, John Manley, associate director of the laboratory and executive secretary of the GAC, showed it to division heads, including Teller. "Edward was of course just completely aghast," Manley recalls, "and his reaction was to offer me a bet that if we did not proceed immediately with a crash program on the Super, he would be a prisoner of war of the Russians in the United States."

Teller explained his reaction to a *Time-Life* interviewer in 1954, using an ironic variation of the argument he accepted from Oppenheimer in 1945 when he rejected Szilard's petition:

> The reasons they gave just made me mad. . . . The important thing in any science is to do the things that can be done. Scientists naturally have a right and a duty to have opinions. But their science gives them no special insight into public affairs. There is a time for scientists and movie stars and people who have flown the Atlantic to restrain their opinions lest they be taken more seriously than they should be.

It is sometimes difficult even for other scientists to remember that the atomic and hydrogen bombs were developed not only as weapons of terrible destruction. They were also, as Fermi once said, "superb physics."

At the time of the Truman announcement, J. Carson Mark headed the Theoretical Division at Los Alamos. "Truman's words," Mark comments, "didn't necessarily mean we did anything much different from what we had been doing because we really didn't know how to make a gadget that would work as a hydrogen bomb." In February 1950, when Washington learned that for seven crucial years, from 1942 to 1949, Klaus Fuchs had passed along secret information to the Soviet Union, Truman turned to a special committee of the National Security Council for advice. The committee recommended that the President clarify his somewhat vague January 31 directive to put energetic development of a hydrogen bomb clearly on record. It emphasized at the same time that there was no obvious way to speed up the weapons test schedule and no guarantee of success. Truman promulgated the committee's report as official policy.

The necessary next step toward a workable thermonuclear weapon was elaborate mathematical simulation. Without a mathematical model of the evolution of thermonuclear burning from a fission trigger event, Mark explains, "no conclusive evidence was possible short of a successful stab in the dark, since a [test] failure would not necessarily establish unfeasibility, but possibly only that the system chosen [Teller's classical Super, for example] was unsuitable." The Super problem, as the simulation came to be called, was the largest mathematical effort ever undertaken up to that time, "vastly larger," writes Stanislaw Ulam, "than any astronomical calculation done to that date on hand computers." It involved calculating the blooming of a thermonuclear explosion—its heating, its extremely complex hydrodynamics, its evolving physical reactions—in progressive increments of less than one ten-millionths of a second. The 1946 Super conference had called for such calculations, but until the further development of the electronic computer they simply could not be accomplished in any reasonable period of time. Teller said as much in a September 1947 report, discussing choosing between his classical Super and an alternative design: "I think that the decision whether considerable effort is to be put on the development of the TX-14 or the Super should be postponed for approximately two years; namely, until such time as these experiments, tests, and calculations have been carried out."

In late 1949, before the H-bomb decision, Los Alamos began detailed work preparing a machine calculation for the first primitive electronic computer, the ENIAC at the Aberdeen Proving Grounds in Maryland. After the H-bomb decision Ulam and the Theoretical Division's Cornelius J. Everett decided to go ahead by hand with a simplified version of the calculation. "We started to work each day for four to six hours with slide rule, pencil, and paper," Ulam remembers, "making frequent quantitative guesses. . . . Much of our work was done by guessing values of geometrical factors, imagining intersections of solids, estimating volumes, and estimating chances of points escaping. We did this repeatedly for hours, liberally sprinkling the guesses with constant slide-rule calculations. It was long and arduous work." They were calculating part 1 of a two-part problem, trying to see if a fission trigger heating a specified amount of deuterium and tritium would start a thermonuclear reaction. Teller's group had calculated a cruder version of part 1 between 1944 and 1946. By February 1950 Ulam saw that the amount of tritium Teller had estimated earlier was not nearly sufficient. "The result of the calculations," Ulam reported, "seems to be that the model considered is a fizzle." He increased the estimated tritium volume and began again. Even with more tritium Teller's classical Super looked distinctly unpromising. Late in April Ulam went off to Princeton to discuss his pessimistic results with von Neumann and Fermi. The three men talked with Oppenheimer as well; Ulam noticed that he "seemed rather glad to learn of the difficulties."

Ulam returned to Los Alamos and broke the news to Teller. "He was pale with fury yesterday literally," Ulam reported to von Neumann—"but I think is calmed down today." Teller at first refused to believe the calculations. He also questioned Ulam's motives for performing them; according to the official AEC history it was necessary for von Neumann to offer Teller "reassurances that the motives behind the changes [in tritium estimates] were constructive." Ironically, Ulam had favored building the Super from the beginning.

The Super problem went on the ENIAC on schedule in June. The evolving results confirmed Ulam's and Everett's findings. "In the course of the calculation," Ulam recalls, "in spite of an initial, hopeful-looking 'flare up,' the whole assembly started slowly to cool down. Every few days Johnny [von Neumann] would call in some results. 'Icicles are forming,' he would say dejectedly." At the same time Ulam and Fermi, who was visiting Los Alamos for the summer, began hand-calculating the next phase of the Super problem, which concerned the propagation of the initial thermonuclear reaction. That work, says Ulam, "turned out to be basic to the technology of thermonuclear explosions." It also predicted that Teller's Super would fizzle. The Super was simply a bad design, Hans Bethe explains, a dead end:

> That Ulam's calculations had to be done at all was proof that the H-bomb project was not ready for a "crash" program when Teller first advocated such a program in the Fall of 1949. Nobody will blame Teller because the calculations of 1946 were wrong, especially because adequate computing machines were not then available. But he was blamed at Los Alamos for leading the Laboratory, and indeed the whole country, into an adventurous program on the basis of calculations which he himself must have known to have been very incomplete. The technical skepticism of the GAC on the other hand had turned out to be far more justified than the GAC itself had dreamed in October 1949.

Between October 1950 and January 1951, Bethe goes on, Teller "was desperate. . . . He proposed a number of complicated schemes to save [the classical Super], none of which seemed to show much promise. It was evident that he did not know of any solution." He was nevertheless unwilling to retrench. He wanted most of the laboratory's time for at least another year and a half. He did not know how to make a thermonuclear, he told the GAC at an October meeting, but he was convinced it could be done. He insisted that the bottleneck was a lack of theoreticians at Los Alamos and a lack of imagination. If the Greenhouse tests of thermonuclear feasibility scheduled for the spring of 1951 at Eniwetok atoll in the Marshall Islands proved a hydrogen bomb impossible, he concluded, Los Alamos might be strong enough to continue; if the tests proved a bomb possible, the laboratory might not be strong enough to follow through. Teller's assessment won him few friends at Los Alamos.

Severe stress can be creative. So can long familiarity with a problem. By February 1951 Ulam was angry with Teller and Teller was angry with everyone. The result was a novel, entirely unexpected invention. Not even Teller had anticipated it. Bethe supplies a context for lay assessment: "The new concept was to me, who had been rather closely associated with the program, about as surprising as the discovery of fission had been to physicists in 1939." The concept has come to be called the Teller-Ulam configuration.

Afterward Teller would variously deny, acknowledge and claim credit for Ulam's contribution. Ulam would consistently acknowledge Teller's part but quietly insist upon his own. Others—Lothar Nordheim of the Theoretical Division, Herbert York—confirm, as Nordheim wrote in 1954 to the *New York Times,* that "a general principle was formulated by Dr. Stanislaw Ulam in collaboration with Teller, who shortly afterward gave it its technically practical form." Teller's most nearly generous acknowledgment appears in his 1955 essay "The Work of Many People": "Two signs of hope came within a few weeks: one sign was an imaginative suggestion by Ulam; the other sign was a fine calculation by [physicist Frederick] de Hoffmann." His unwillingness consistently to acknowledge Ulam's contribution, in contradiction of scientific ethics, suggests the importance he attached to historic rank in the matter. He came to dislike being called "the father of the H-bomb," but asserted his paternity in 1954 with a curiously explicit allegory of his on-again, off-again relationship with Los Alamos:

> It is true that I am the father in [the] biological sense that I performed a necessary function and let nature take its course. After that a child had to be born. It might be robust or it might be stillborn, but *something* had to be born. The process of conception was by no means a pleasure; it was filled with difficulty and anxiety for both parties. My act . . . aroused the emotions associated with such behavior.

Bethe sifts the evidence the other way, drolly tongue-in-cheek: "I used to say that Ulam was the father of the hydrogen bomb and Edward was the mother, because he carried the baby for quite a while."

The mechanism of the Teller-Ulam H-bomb was revealed in general terms in an official Los Alamos publication in 1983, on the occasion of the fortieth anniversary of the laboratory's founding:

> The first megaton-yield explosives (hydrogen bombs) were based on the application of x-rays produced by a primary nuclear device to compress and ignite a physically distinct secondary nuclear assembly. The process by which the time-varying radiation source is coupled to the secondary is referred to as radiation transport.

Stanislaw Ulam's basic contribution appears to have emerged from a closer look at the early development of the fission fireball, which initially radiates most of its energy as X rays. These, traveling at the speed of light, advance outward ahead of any shock wave. The classical Super and other previous designs presumably tried to pack the entire mass of thermonuclear material into the evolving fission explosion to heat it hydrodynamically—more spheres within spheres, a fatter and unworkable Fat Man. Those designs always promised to blow apart before thermonuclear burning could make much headway. Ulam suddenly realized, it seems, that if the thermonuclear materials were *physically separated* from the fission primary, the enormous flux of X rays coming off the primary might be applied somehow to start thermonuclear burning in the brief fraction of a second before the slower shock wave caught up and blew everything apart.

Ulam and Teller proceeded to develop Ulam's idea. The X rays from the primary might heat the thermonuclear secondary directly (as microwaves heat food in a microwave oven) but they could not squeeze it efficiently to the greater density that would promote fusion. Some other material would need to intervene. It turned out that ordinary plastic would serve. Dump so large a flux of X rays into a layer of dense plastic foam wrapped around a cylindrical stick of thermonuclear materials and the plastic would heat instantaneously to a plasma—a hot, ionized gas—expanding explosively at pressures thousands of times more intense than the pressures high explosives can generate. So a fission primary—a little Fat Man, no larger in today's efficient weapons than a soccer ball—might occupy one end of an evacuated cylindrical casing. Farther along the casing a layer of plastic might wrap a cylindrical arrangement of thermonuclear material. Fire the primary and the X-ray flux would radiate into the plastic at the speed of light, much faster than the expanding fission shock wave coming up behind. Configuring the plastic would be much simpler than configuring high-explosive lenses; the light-swift X rays would irradiate it simultaneously along its entire length and the resulting implosion would be beautifully symmetrical.

That, to the extent that continuing secrecy allows its reconstruction, is probably what Ulam first conceived and Teller made practical. Though it was the necessary breakthrough, it was not the end of invention. Even with

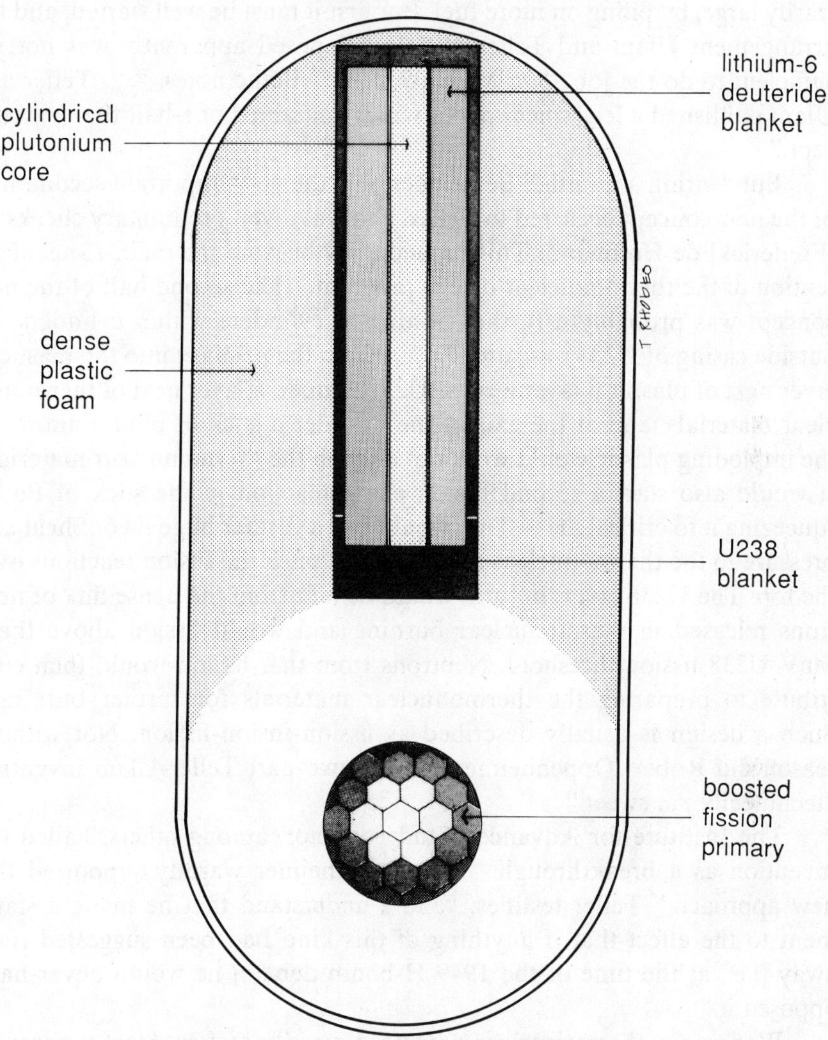

the greater heat and pressure of irradiated plastic implosion the design apparently would not evolve sufficient heat long enough to kindle a full-scale thermonuclear reaction. Such reactions depend on heating light atoms such as deuterium and tritium—increasing their velocity of motion—sufficiently to force them through the electrical barrier of the nucleus so that they can fuse to helium. The process requires heat and pressure but no critical mass. Once fusion has begun, the binding energy released in the reaction (for deuterium and tritium, 17.6 MeV) promotes further burning. A fusion weapon can therefore be made arbitrarily large, as a fire can be made arbitrarily large, by piling on more fuel. But first it must be well started, and the arrangement Ulam and Teller initially proposed apparently was not yet sufficient to do the job. "On March 9, 1951," Bethe notes, ". . . Teller and Ulam published a [classified] paper which contained one-half the new concept."

But "within a month," Bethe goes on, "the very important second half of the new concept occurred to Teller, and was given preliminary checks by [Frederick] de Hoffmann. This immediately became the main focus of attention of the thermonuclear design program." The second half of the new concept was probably a further nesting of cylinders within cylinders: an outside casing of U238 to scatter X rays from the primary into the plastic; a layer next of plastic; a layer next of U238 tamper; a layer next of thermonuclear materials; and at the axis of the cylinder a stick of plutonium. Now the imploding plastic would work not only on the thermonuclear materials. It would also start a second fission chain reaction in the stick of Pu by squeezing it to critical mass. That would add a further huge flux of heat and pressure to the thermonuclear materials and push the fusion reactions over the top. The U238 layer, in turn, would benefit from the dense flux of neutrons released in thermonuclear burning and would fission above the 1 MeV U238 fission threshold. Neutrons from that fission would then contribute to preparing the thermonuclear materials for further burning.* Such a design is usually described as fission-fusion-fission. Not without reason did Robert Oppenheimer call the two-part Teller-Ulam invention "technically . . . sweet."

The Institute for Advanced Study director, among others, hailed the invention as a breakthrough. "Dr. Oppenheimer warmly supported this new approach," Teller testifies, "and I understand that he made a statement to the effect that if anything of this kind had been suggested right away [i.e., at the time of the 1949 H-bomb debate] he would never have opposed it."

Work on a thermonuclear advanced rapidly at Los Alamos through 1951, but by then Teller's relationships had deteriorated to the point of no return. (Three years later, in a time of trial, Oppenheimer would note on a yellow pad a damning remark of Teller's that summarized the Hungarian physicist's turnabout from proponent of world government to aggressive weaponeer, a remark that echoes all the way back to his traumatic experience of revolution and counterrevolution in the Hungary of his youth: " 'Since I cannot work w[ith] the appeasers, I will work with the Fascists.' . . . Someone heard E. T. say this. Who?") In 1952, with Ernest Lawrence's

* In modern "dry" bombs, by making tritium from lithium[6] deuteride.

support and Defense Department backing, Teller won a second weapons laboratory in the Livermore Valley, fifty miles inland from Berkeley. Los Alamos was left to build the first experimental thermonuclear device, banally coded Mike.

Teller chose not to attend the Mike shot at Eniwetok on November 1, 1952. He was busy starting up his new weapons laboratory and could hardly spare the time; he certainly also felt unwelcome. The Mike device was fueled with liquid tritium and deuterium; the liquids required a cryogenic refrigeration plant to maintain their low temperature. The complex assembly weighed some 65 tons and occupied an entire laboratory building on the small island of Elugelab. Shrouded in black tar paper, shimmering in the heat, the cubic building looked in the distance like a diabolic twin to the Kaaba of Mecca.

Teller contrived nevertheless to follow the progress of the test. He stationed himself at a seismograph in a basement room of the Berkeley geology building. Herbert York, acting director at Livermore, tuned a shortwave radio to the frequency of the Mike shot telemetry. When the shot was fired he called Teller in Berkeley. The two physicists had calculated the time a seismic wave from a successful shot would require to travel under the Pacific basin to northern California—about fifteen minutes, Teller remembers:

> I watched with little patience, the seismograph making at each minute a clearly visible vibration which served as a time signal. At last the time signal came that had to be followed by the shock from the explosion and there it seemed to be: the luminous point appeared to dance wildly and irregularly. Was it only that the pencil which I held as a marker trembled in my hand?

Mike was expected to explode with an energy of a few megatons. But its designers had engineered every component to maximize its yield. It yielded 10.4 million tons TNT equivalent, a thousand times more violent than Little Boy. "This thing is the plague of Thebes," Oppenheimer once complained of the H-bomb. Now the plague found incarnation.

The test was secret. No report would reach Los Alamos until security officers on Eniwetok had time to examine and encode it. Teller knew Mike had worked before its builders did. He dictated a telegram to York to send on to Los Alamos. The message was brief but barbed: "It's a boy."

"The fireball," writes Leona Marshall Libby, "expanded to 3 miles in diameter. Observers, all evacuated to 40 miles or more away, saw millions of gallons of [atoll] lagoon water, turned to steam, appear as a giant bubble.

When the steam had evaporated, they saw that the island of Elugelab, where the bomb [building] had been, had vanished, vaporized also. In its place, a crater ½ mile deep and 2 miles wide had been torn in the reef."

The Soviet Union exploded a device with a small hydrogen component in August 1953. Its yield was probably several hundred kilotons, about half the yield of the largest fission weapon the United States had tested up to that time. "This was not a true H-bomb," Hans Bethe comments, "as I know very well because I was chairman of the committee analyzing the Russian [fallout]."

At 65 tons Mike was too large and complex a mechanism to serve as a deliverable bomb. Its designers had fueled it with liquid deuterium and tritium for simplicity in measuring the thermonuclear reactions it would test. For a deliverable bomb the thermonuclear material of choice would be lithium deuteride, a stable powder, the lithium in the form of the isotope Li6, which constitutes 7.4 percent of natural lithium but can be separated from it relatively easily. Neutrons from the fission components of a lithium-fueled bomb would produce tritium almost instantly from Li6, which would then fuse with the deuteride to develop thermonuclear burning just as the wet and bulky liquid hydrogen isotopes had done in Mike. The dry design was tested during Operation Castle in the spring of 1954; "the very first test of the series," writes Herbert York, "the *Bravo* test, was of a device using LiD as its fuel and yielding 15 megatons. It was in a form readily adaptable for delivery by aircraft, and thus was the first large American hydrogen bomb." A true Soviet thermonuclear, dropped from an aircraft in test, followed on November 23, 1955.

When Niels Bohr arrived at Los Alamos in 1943, writes Robert Oppenheimer, "his first serious question was, 'Is it really big enough?' " The bomb, Bohr meant: big enough to end world war, big enough to challenge mankind to find its way beyond man-made death to a world more open and more humane. "I do not know whether it was," Oppenheimer adds; "it did finally get to be." By 1955, if not before, the bomb had worked an essential change upon the world. Oppenheimer had already found succinct metaphoric expression of that change in a commencement address he delivered early in 1946. "It did not take atomic weapons to make war terrible," he said then. ". . . It did not take atomic weapons to make man want peace, a peace that would last. But the atomic bomb was the turn of the screw. It has made the prospect of future war unendurable. It has led us up those last few steps to the mountain pass; and beyond there is a different country."

Gil Elliot's *Twentieth Century Book of the Dead* is a useful guide to that progress. Elliot is a Scottish writer of original mind who lives in London. It occurred to him to look into the question of how many human beings have died by man-made violence in this most bloody of centuries. He discovered that few historians or statisticians have bothered to count past the men in uniform. He worked out order-of-magnitude estimates and arrived at a total (including combatants) of about 100 million dead. He calls this uncelebrated multitude a nation of the dead:

> We know as much about the nation of the dead as we might have known about any living nation fifty years ago when the techniques of social measurement were still at an early stage. The population is around one hundred million. A proper census has not yet been possible but the latest estimate based on samples of the population suggests a figure of a hundred and ten million. That's about the size of it. A large modern nation. It's very much a twentieth-century nation, as cosmopolitan in its origins as the United States. The people have always been mixed, but the real growth began in 1914. Between then and the early 1920s the population reached twenty million, and steady growth over the next twenty years brought it to almost forty million by the outbreak of the Second World War. In the early 1940s the population more than doubled, with annual increases reaching peaks of 10/12 million. Since 1945 the growth-rate has declined below any previous levels since the late 1920s. This has been accompanied by a gigantic increase in the *capacity* for expansion.

Elliot went on from counting to examine how this silent nation assembled itself, the manner of its murder. He found the weapons to have been hardware and privation: big guns, small arms in combat, small arms in massacre, aerial bombs; ghettos, camps, sieges, occupations, dislocations, famines, blockades. Behind the weapons Elliot encountered a phenomenon more basic and more malign: the war machine evolving across the decades into a total-war machine and the total-war machine managing variously and intermittently to create areas of total death: Verdun, Leningrad, Auschwitz, Hiroshima.

The most compact, efficient, inexpensive, inexorable mechanisms of total death are nuclear weapons. Since 1945 they have therefore come to dominate the field. "The lesson we should learn from all this," I. I. Rabi remarks, "and the frightening thing which we did learn in the course of the war, was . . . how easy it is to kill people when you turn your mind to it. When you turn the resources of modern science to the problem of killing people, you realize how vulnerable they really are."

The change from a total-war machine capable of gouging out pockets of total death in a living landscape to a total-death machine capable of burning and blasting and poisoning and chilling the human world is Oppenheimer's turn of the screw. Elliot elaborates:

> The hundred million or so man-made deaths of the twentieth century . . . are more directly comparable with the scale of death from disease and plague which was the accepted norm before this century. Indeed, man-made death has largely replaced these as a source of untimely death. This is the kind of change that Hegel meant when he said that a quantitative change, if large enough, could bring about a qualitative change. The quality of this particular change becomes clear if we connect the present total of deaths with the scale of death inherent in the weapons now possessed by the large powers. Nuclear strategists talk in terms of hundreds of millions of deaths, of the destruction of whole nations and even of the entire human race.

Less efficient machines required two-thirds of a century to assemble a nation of the dead; the nuclear death machine could manage it in half an hour. The nuclear death machine has become capable of creating not merely cities of the dead or nations of the dead but a world of the dead. (Even before detailed studies appeared of the potentially widespread disaster known as nuclear winter, the World Health Organization had estimated—in 1982—that a major nuclear war would kill half the population of the earth: two billion people.) Therefore, Elliot deduces:

> The moral significance is inescapable. If morality refers to relations between individuals, or between the individual and society, then there can be no more fundamental moral issue than the continuing survival of individuals and societies. The scale of man-made death is the central moral as well as material fact of our time.

This identifies what we are talking about—the modern phenomenon of total death, that is, not capitalism versus communism or democracy versus the police state—but does not explain how we arrived at the brink of so absolute an abyss. Elliot supplies a clue to the answer in his discussion of the First World War. "The one thing that stands out overall," he observes, "is that at no time, before, during or after the war, was there a living organic structure in society [e.g., church, political party, custom, body of law] with sufficient strength to resist the new man-made and machine-made creation: [organized] death."

War is ancient. That it traditionally exposes to maximum danger a biologically surplus and relatively powerless subset of the population—young males—suggests that in some circumstances of traditional intersocial conflict it confers reproductive advantage. Nor were mass slaughters ever rare. The Old Testament regularly celebrates their carnage. The histories of empires bulk thick with them.

World war differed not only in scale but also in essential organization from such more limited earlier conflicts. Total death differs from the surround of mass slaughter in its time-dependent, assembly-line linearity. Both kinds of violence emerge from a distinctly modern process: the nation-state parasitizing applied science and industrial technology to protect itself and to further its ambitions.

Though it dominates the world, the nation-state owns no long history of legitimacy. It developed in the eighteenth and nineteenth centuries, its nationalism "a doctrine invented in Europe," writes the political scientist Elie Kedourie, that "pretends to supply a criterion for the determination of the unit of population proper to enjoy a government exclusively its own. . . . Briefly, the doctrine holds that humanity is naturally divided into nations, that nations are known by certain characteristics which can be ascertained, and that the only legitimate type of government is national self-government."

> Not the least triumph of this doctrine is that such propositions have become accepted and are thought to be self-evident, that the very word nation has been endowed by nationalism with a meaning and a resonance which until the end of the eighteenth century it was far from having. These ideas have become firmly naturalized in the political rhetoric of the West which has been taken over for the use of the whole world. But what now seems natural once was unfamiliar, needing argument, persuasion, evidences of many kinds; what seems simple and transparent is really obscure and contrived, the outcome of circumstances now forgotten and preoccupations now academic, the residue of metaphysical systems sometimes incompatible and even contradictory.

Nationalism differed radically from the hierarchical feudal organization that preceded it in the West. It offered every member of society who was included within its definition the security, invested with powerful emotion, of merger into a welcoming crowd. Not the king or the nobility but the people would be its essential polity: *L'état c'est moi et moi et moi.* That was the increase in political freedom its invention installed. But complementarily, notes the economist Barbara Ward, "its essential nature is [as well] to leave other people out. . . . It can even divorce from all community of brotherhood and goodwill fellowmen who simply happen to live on the other side of a river."

The power of the state, when nationalism succeeded in acquiring it—enlarging such power in the process—amplified that essential tension. Whole populations discovered political and emotional investment in their national causes. But outlanders became certainly more alien; the Other was confirmed in his Otherness; and between nation-states so radically divided—divided, as they believed, by nature itself—opened gulfs of threatening anarchy. Bridging them was difficult in the best of circumstances and no hierarchical authority survived to mediate as the Church had once done. In international affairs the worst case came to be counted the most reliable.

Then industrial technology and applied science enormously amplified the nation-state's power and when the smoke cleared the cities of the dead and gradually the nation of the dead revealed themselves to view. "Once men lose all grip on reality," observes Barbara Ward, "there seems to be no limit to the horrors of hatred and passion and rage they can dredge up from their psychological depths, horrors which normally we use all our social institutions to check. Unleashed nationalism on the contrary removes the checks."

Which suggests, reverting to Elliot's clue, that no living organic structure could be found sufficiently strong to resist the new death organization because the entire nation was implicated: the death organization was the nation-state itself. It followed that once mechanisms could be devised with which to attack civilian populations, civilian populations would be attacked. The enemy was the enemy nation, which was no more than the corporate body of all the enemy citizens, each of whom, in uniform or not, regardless of age or sex, was individually the enemy.

But the nation-state was not the only new political system invented in early modern times. Through the two centuries of the nation-state's evolution the republic of science had been evolving in parallel. Founded on openness, international in scope, science survived in the nation-state's midst by limiting its sovereignty to a part of the world which interested the larger system hardly at all: observable natural phenomena. Within that limited compass it proved spectacularly successful, lighting up the darkness, healing the sick, feeding the multitudes. And finally with the release of nuclear energy its success brought it into direct confrontation with the political system within which it operated. In 1945 science became the first living organic structure strong enough to challenge the nation-state itself.

The conflict between science and the nation-state that has continued and enlarged since 1945 is different from traditional forms of political conflict. Bohr visited the statesmen of his day to explain it but chose to be diplomatic rather than blunt. He explained that with the coming of nuclear weapons the world would arrive at an entirely new situation that could not be resolved by war. The situation might be resolved by statesmen sitting down together and negotiating for mutual security. If they did so, the inevitable outcome of such negotiations, given the understandable suspicion on every side, must be an open world. To Winston Churchill and apparently also to Franklin Roosevelt Bohr's scenario appeared dangerously naïve. In his role as spokesman for the republic of science Bohr certainly carried news of danger, but he was never naïve. He was warning the statesmen that science was about to hand them control over a force of nature that would destroy their political system. Considering the slaughter that political system had perpetrated upon the twentieth century, he was polite enough not to add, the mechanism of its dismantling had turned up none too soon.

The bomb that science found hidden in the world and made manifest would destroy the nation-state paradoxically by rendering it defenseless. Against such small and cheap and holocaustal weapons no defense could ever be certain. The thickest shields, from fighter aircraft to Star Wars, could be penetrated merely by multiplying weapons, decoys and delivery systems. The only security from the bomb would be political: negotiation toward an open world, which would increase security by decreasing national sovereignty and damping out the violence that attended it.

The consequence of refusing to negotiate would be a temporary monopoly followed by an arms race. That road to nowhere looked so much more familiar than Bohr's open world, which even Oppenheimer sometimes confused with World Government, that the nations preferred it and took it. The bomb might be a wall, but until the wall was tested escalation by escalation, new weapons system by new weapons system, who could prove that clever men—or threatening enemies—might not find a way under its fastness or around? Nuclear weapons might also be enterprise and profit and steady work. They might secure the citadel. They might permit the nation not to send its favored sons to war. More significantly, they would deter major war and freeze into permanence the political status quo. The nation-state could roll on into perpetuity with its sovereignty intact.

So it seemed along the way. So it still seems to many. But rather than a guarantor of sovereignty the arms race has proven a *reductio ad absurdum* of sovereignty. Though they bristle with holocaustal weapons, the superpowers confront each other today totally vulnerable, totally dependent for their continued survival on mutual and reasonable restraint, their sovereignties so thoroughly compromised that they can exercise their military ambitions only through third-world skirmishes that seldom find conclusive end. The bomb, the final word on the accumulation of power—that matter properly arranged is *all* power—has saturated national sovereignty and shorted it out.

Bohr would surely emphasize that the result of either course—negotiation or arms race—must be the foreclosure of the nation-state. Negotiation to an open world would replace the nation-state with some more tolerant and peaceful and international arrangement that recognized the reality of the bomb. Alternatively the death machine that we have installed in our midst will destroy the nation-state, ours and our rival's, along with most of the rest of the human world. The weapons with which the superpowers have armed themselves—collectively the equivalent of more than one million Hiroshimas—are linked together through their warning systems into a hair-trigger, feedback-looped contrivance, and no human contrivance has ever worked perfectly nor ever will. Each side is hostage to the other side's errors. The clock ticks. Accidents happen. Nuclear war would abolish the nation-state as certainly as negotiation but instead of a living, open world would replace it with a world of the dead, a world completely closed.

Science is sometimes blamed for the nuclear dilemma. Such blame confuses the messenger with the message. Otto Hahn and Fritz Strassmann did not invent nuclear fission; they discovered it. It was there all along waiting for us, the turn of the screw. If the bomb seems brutal and scientists criminal for assisting at its birth, consider: would anything less absolute have convinced institutions capable of perpetrating the First and Second World Wars, of destroying with hardware and callous privation 100 million human beings, to cease and desist? Nor was escalation inevitable. To the contrary, it resulted from a series of deliberate choices the superpowers made in pursuit of national interests.

But if the arms race is not a creation of science (however much men trained as scientists and applying the discoveries of science may have helped it along), what constitutes that republic's armament in its continuing conflict with the nation-state?

Oddly from the perspective of previous conflicts, science's highly effective armament is the basic scientific principle of openness. Science fights the exclusivity of the nation-state, an exclusivity that has revealed itself capable of preparing to convert the living world into a dead world of corpses, by sharing its discoveries freely—in Oppenheimer's words, by "turn[ing] over to mankind at large the greatest possible power to control the world and to deal with it according to its lights and its values." That deep trust in the promise of openness to remake the world must inspire even at the brink of the abyss. Science in conflict with the nation-state demonstrates how an open world could function without chartered violence. The effectiveness of such profound civility is obscured at present because it necessarily operates from within the nation-state itself. Turning around and looking back across the half-century since 1945 demonstrates its power: it forced an end to world war, in itself an enormous deliverance.

If the arms race now makes that deliverance seem a leap out of the frying pan into the fire, science's response has been to continue to confront the nation-state with the facts and probabilities it discovers in the course of its daily work. Nuclear winter, whatever its level of severity, is one of those probabilities. Damage to the ozone layer is another. The likelihood of widespread epidemics after a nuclear war and of mass starvation because of disruptions in food transport are two more. The nation-states may have understood that nuclear weapons spoil war. The continuing arms race unfortunately demonstrates they have not yet understood that the nationalist system of exclusion and international confrontation has now become suicidal. Each new contribution to understanding—more knowledge turned over to mankind—must further erode that stubborn and potentially genocidal ignorance. Additional knowledge will certainly continue to emerge. It is not likely to prove massive armaments a blessing.

Change is possible. Americans who want the Soviet Union to change first, as Henry Stimson did, should realize that they can only pursue that cause peacefully; the Soviet Union controls a deterrent fully as dangerous as the United States' deterrent. And patriots may need reminding that the national security state is not where holy democracy began. The American Revolution foresaw a future much like Bohr's open world, in part because the framers of that revolution and the founders of the republic of science drew from a common body of Enlightenment ideas. The national security state that the United States has evolved toward since 1945 is significantly a denial of the American democratic vision: suspicious of diversity, secret, martial, exclusive, monolithic, paranoid. "Nationalism conquered both the American thesis and the Russian antithesis of the universalist faith," writes Barbara Ward. "The two great federated experiments, based upon a revolutionary concept of the destiny of all mankind, have ended, in counterpoint, as the two most powerful nation-states in history." But other nations have moderated their belligerence and tempered their ambitions without losing their souls. Sweden was once the scourge of Europe. It gave way; the empty fortress at Kungälv testifies to that. Now it abides honorably and peacefully among the nations.

Change is possible because the choice is bare: change is the only alternative to total death. The conditions have already been established, irrevocably, for the destruction of the human world or its modification into some more collegial commonality. The necessity now is to begin to dismantle the death machine. The energies rich and intelligent peoples have squandered on the elaboration of death need to be turned to the elaboration of life.

Bohr's great vision of the complementarity of the bomb can bring hope to the prospect of change. The death machine's suicidal destructiveness is ample reason to work for its dismantling. But although that road is now necessarily longer, the promise still holds, as it has held from the beginning, that negotiating away from chartered violence will be identical to negotiating toward an open world. Democracy has nothing to fear from such a world.

Negotiation is in fact already ongoing, partly by necessity, partly by inadvertence. It began when the United States and Great Britain decided to build nuclear weapons secretly and spring them on the world in surprise, thereby precipitating an arms race with the Soviet Union that eventually came to stalemate. It continued when the United States resisted preemptive war in the brief years of its nuclear monopoly; when new delivery systems made defense impossible and thereby further undermined national sovereignty; when nations tolerated overflights and then satellite reconnaissance of their previously sacrosanct territories. It elaborates in custom and tradition every time confrontation leads to prudent stand-down or front- or back-channel resolution. It progresses as commoners in every country slowly come to understand that in a nuclear world their national leaders cannot, no matter how much tribute and control they exact, protect even their citizens' bare lives, the minimum demand the commons have made in exchange for the political authority that is ultimately theirs alone to award.

It can be useful to categorize nuclear weapons as a more virulent strain of plague, to consider man-made death as a phenomenon that parallels the older phenomenon of biological death that the people of all the nations, working in peaceful concert, have brought under a measure of control. Elliot draws this comparison productively:

> Our societies are dedicated to the preservation and care of life. . . . Public death was first recognized as a matter of civilized concern in the nineteenth century, when some health workers decided that untimely death was a question between men and society, not between men and God. Infant mortality and endemic disease became matters of social responsibility. Since then, and for that reason, millions of lives have been saved. They are not saved by accident or goodwill. Human life is daily deliberately protected from nature by accepted practices of hygiene and medical care, by the control of living conditions and the guidance of human relationships. Mortality statistics are constantly examined to see if the causes of death reveal any areas needing special attention. Because of the success of these practices, the area of public death has, in advanced societies, been taken over by man-made death—once an insignificant or "merged" part of the spectrum, now almost the whole.
>
> When politicians, in tones of grave wonder, characterize our age as one of vast effort in saving human life, and enormous vigor in destroying it, they seem to feel they are indicating some mysterious paradox of the human spirit.

> There is no paradox and no mystery. The difference is that one area of public death has been tackled and secured by the forces of reason; the other has not. The pioneers of public health did not change nature, or men, but adjusted the active relationship of men to certain aspects of nature so that the relationship became one of watchful and healthy respect. In doing so they had to contend with and struggle against the suspicious opposition of those who believed that to interfere with nature was sinful, and even that disease and plague were the result of something sinful in the nature of man himself.

The pioneers of public health who proposed to secure the biological death machine for the forces of reason must have felt, at the beginning, a despair at the magnitude of the task like the despair many thoughtful citizens today feel at the magnitude of the task of similarly securing the man-made death machine. They persisted and triumphed.

Bohr's open world has already been negotiated and installed against the biological death machine. No one any longer considers disease a political issue and only modern primitives consider it a judgment of God. When the World Health Organization worked through the 1960s and 1970s to eradicate smallpox from the earth—a program the Soviet Union initiated—both the Soviet Union and the United States shared the cost of the campaign with the third-world countries involved. The Soviets were not charged with expansionism nor the Americans with imperialism. WHO health workers of diverse national origin usually found welcome; once they demonstrated progress and overcame skepticism that containment and then eradication of so pervasive a disease was possible they won enthusiastic local support. "The eradication of smallpox will represent a major milestone in the history of medicine," the director of the campaign, the American physician Donald A. Henderson, wrote in its final phase. "It will have demonstrated what can be achieved when governments throughout the world join an international organization in a common purpose." It did: the most devastating and feared natural pestilence in human history is gone, a great victory for mankind.

Man-made death is evidently more intractable than biological death. Whether the unarmed republic of science, dedicated to human felicity rather than to the accumulation of power, can force nation-states armed to the teeth to change before they destroy themselves remains to be seen. That no world wars have engulfed us since 1945 is our interim guarantee that the opening up of the world is well begun, though at any time accident or miscalculation could close it forever. That nuclear weapons proliferate and the superpowers exhaust their economies attempting to outmaneuver each other to unattainable dominance demonstrates how irrationally tenacious is our hold on traditional forms of control.

In the spring of 1957 former AEC chairman Gordon Dean asked Robert Oppenheimer to comment on Henry Kissinger's forthcoming book *Nuclear Weapons and Foreign Policy.* Oppenheimer responded:

> Of course Kissinger is right in conceiving the problems of policy planning and strategy in terms of national power, in rough analogy to the national struggles of the 19th century; yet I have the impression that there are deep things abroad in the world, which in time are going to turn the flank of all struggles so conceived. This will not happen today, nor easily as long as Soviet power continues great and unaltered; but nevertheless I think in time the transnational communities in our culture will begin to play a prominent part in the political structure of the world, and even affect the exercise of power by the states.

The preeminent transnational community in our culture is science. With the release of nuclear energy in the first half of the twentieth century that model commonwealth decisively challenged the power of the nation-state. The confrontation is ongoing and inextricably embedded in mortal risk, but it offers at least a distant prospect of felicity.

The different country that still opens before us is Bohr's open world.

Kansas City, Missouri
1981–1986

Acknowledgments

These men and women who participated in the events of this book generously made time for interviews and correspondence: Philip Abelson, Luis W. Alvarez, David L. Anderson, William A. Arnold, Hans Bethe, Rose Bethe, Eugene T. Booth, Sakae Itoh, Shigetoshi Iwamatsu, George Kistiakowsky, Willis E. Lamb, Jr., Leon Love, Alfred O. C. Nier, I. I. Rabi, Stefan Rozental, Glenn Seaborg, Emilio Segrè, Edward Teller, Stanislaw Ulam, Eugene Wigner and Herbert York.

Michael Korda took the chance of sponsorship. David Halberstam, Geoffrey Ward and Edward O. Wilson vouched for me to the Ford Foundation. Arthur L. Singer, Jr., saved the day. The Cockefair Chair in Continuing Education at the University of Missouri–Kansas City and its director, Michael Mardikes, lent support. Louis Brown offered physics coaching and wise counsel far beyond the call of any duty and is not responsible for lapses in either regard. Egon Weiss went out of his way to arrange access to the Szilard Papers. The Linda Hall Library of Science and its former director, Larry X. Besant, and the UMKC Library and its former director, Kenneth LaBudde, never failed.

I visited or corresponded with a number of institutions; their staffs guided me with competence and courtesy: American Institute of Physics Niels Bohr Library; Argonne National Laboratory; Bibliothek und Archiv für Geschichte der Max-Planck-Gesellschaft, Dahlem; Columbia University; Department of Terrestrial Magnetism, Carnegie Institution of Washington; Hiroshima Peace Culture Foundation; J. Robert Oppenheimer Memorial Committee; Lawrence Berkeley Laboratory; Library of Congress; Los Alamos National Laboratory; National Archives; Niels Bohr Institute, Copenhagen; The Readers Digest of Japan; United States Air Force Museum, Wright-Patterson AFB; United States Military Academy Library; University of California—San Diego; University of Chicago Library.

Friends and colleagues helped with research, advice, encouragement, aid: Millicent Abell, Hans and Elisabeth Archenhold, John Aubrey, Dan Baca, Roy and Sandra Beatty, David Butler, Margaret Conyngham, Gil Elliot, Jon Else, Susie Evans, Peter Francis, Kimball Higgs, Jack Holl, Ulla Holm, Joan and Frank Hood, Jim and Reiko Ishikawa, Sigurd Johansson, Tadao Kaizuka, Edda and Rainer König, Barbro Lucas, Thomas Lyons, Karen McCarthy, Donald and Britta McNemar, Yasuo Miyazaki, Hiroyuki Nakagawa, Kimiko Nakai, Rolf Neuhaus, Issei Nishimori, Fredrik Nordenham, Patricia O'Connell, Gena Peyton, Edward Quattlebaum, P. Wayne Reagan, Edward Reese, Katherine Rhodes, Timothy Rhodes, Bill Jack Rodgers, Siegfried Ruschin, Robert G. Sachs, Silva Sandow, Sabine Schaffner, Ko Shioya, R. Jeffrey Smith, Robert Stewart, Lewis H. Strauss, Linda Talbot, Sharon Gibbs Thibodeau, Josiah Thompson, Kosta Tsipsis, Erma Valenti, Joan Warnow, Spencer Weart, Paul Williams, Edward Wolowiec, Mike Yoshida.

Luis Alvarez and Emilio Segrè were kind enough to read the galleys and offered invaluable suggestions.

Mary saw it through.

Notes

ABBREVIATIONS AND SOURCES:

OHI: Oral history interview.
AIP: Center for the History of Physics, American Institute of Physics, New York, N.Y.
AHQP: Archives for the History of Quantum Physics, available at the AIP and several other repositories.
Bush-Conant File: Vannevar Bush–James B. Conant files, Office of Scientific Research and Development, S-1 (Record Group 227), National Archives.
MED: Manhattan Engineer District Records (Record Group 77), National Archives.
JRO Papers: J. Robert Oppenheimer Papers, Library of Congress.
Strauss Papers: Lewis L. Strauss Papers, Herbert Hoover Library, West Branch, Iowa.
Szilard Papers: Leo Szilard Papers, University of California at San Diego.

Chapter 1: Moonshine

PAGE

13. September 12, 1933: I derive this date from Leo Szilard's statement at Szilard (1972), p. 529, that he read about Ernest Rutherford's speech to the British Association "one morning . . . in the newspapers" and "that day . . . was walking down Southampton Row." The British Association story appeared prominently on p. 7 of *The Times* on Sept. 12.
13. "short fat man . . . wives": Szilard (1972), p. xv.
14. "Mr. Wells . . . justification": *The Times,* p. 6.
14. He knew Wells personally: Shils (1964), p. 38.
14. Szilard read Wells' tract: Weart and Szilard (1978), p. 22n.
14. he traveled to London in 1929: ibid.
14. Szilard bid: Shils (1964), p. 38.
14. "I knew languages. . . .mascot": Weart and Szilard (1978), p. 4.
14. "When I was young . . . politics": Szilard (1972), p. xix.
14. "I said to them . . . this statement": Weart and Szilard (1978), pp. 4–5.

PAGE

14. his clarity of judgment: ibid., p. 5.
15. the Eötvös Prize: von Kármán and Edson (1967), p. 22.
15. "no career in physics": Weart and Szilard (1978), p. 5.
15. "felt that his skill . . . colleagues": Wigner (1964), p. 338.
15. saved his life: Weart and Szilard (1978), p. 8.
15. "family connections": ibid., p. 7.
15. "Not long afterward . . . disappeared": ibid., p. 8.
16. around Christmastime: ibid.
16. "In the end . . . '21": ibid., p. 9.
16. "As soon . . . Einstein": Wigner (1964), p. 338.
16. Wigner remembers: ibid.
17. Szilard won his attention: Segrè (1970), p. 106.
17. von Laue . . . accepted Szilard: Weart and Szilard (1978), Fig. 1, p. 10.
17. "There was snow . . . strange": de Jonge (1978), p. 125.
17. "the air . . . counted out": ibid., p. 130.
17. Press Ball: ibid., p. 132.

PAGE

17. Mies van der Rohe: Friedrich (1972), p. 163.
17. Yehudi Menuhin: ibid., p. 219.
17. George Grosz: Grosz (1923); Friedrich (1972), p. 152.
17. "an elderly . . . shoelaces": quoted in ibid., p. 90.
17. Fyodor Vinberg: ibid., pp. 95–96.
18. "a dark . . . enigma": quoted in ibid., p. 190.
18. "No, one . . . gods": de Jonge (1978), p. 99.
18. "In order . . . of inflation": Elsasser (1978), pp. 31–32.
18–19. "During a . . . and opinion": Wigner (1964), p. 337.
19. "Berlin . . . of physics": Weart and Szilard (1978), p. 8.
19. "In order . . . original work": ibid., p. 9.
19. "I couldn't . . . to my mind": ibid.
19. Einstein, for example: Cf. Einstein's own evaluation: "Because of the understanding of the essence of Brownian motion, suddenly all doubts vanished about the correctness of Boltzmann's [statistical] interpretation of the thermodynamic laws." Cited in Pais (1982), p. 100. Cf. also Szilard (1972), p. 31ff.
19–20. "and I saw . . . to do": Weart and Szilard (1978), p. 9.
20. "Well . . . very much": ibid., p. 9ff.
20. "and next . . . degree": ibid., p. 11.
20. Six months later: ibid.
20. accepted as *Habilitationsschrift:* Szilard (1972), p. 6.
20. Szilard patents: ibid., pp. 697–706.
20–21. "A sad . . . valve": Feld (1984), p. 676.
21. pumping refrigerant: Weart and Szilard (1978), p. 12.
21. January 5, 1929: Szilard (1972), p. 528.
21. April 1, 1929: Childs (1968), p. 138ff.
21. "the mid-twenties in Germany": Weart and Szilard (1978), p. 22.
21. "had a . . . scale": Snow (1981), p. 44.
21. *Der Bund:* Weart and Szilard (1978), p. 23ff.
21. "a closely . . . spirit": ibid., p. 23.
21–22. "If we . . . own": ibid., p. 24.
22. "take over . . . parliament": ibid., p. 25.
22. "The Order . . . state": ibid., p. 28n.
22. "The Voice of the Dolphins": Szilard (1961).

PAGE

22. banding together: Weart and Szilard (1978), p. 22.
22. "the parliamentary . . . generations": ibid.
22. "I reached . . . Switzerland": ibid., p. 13.
23. Chadwick *Nature* letter: Chadwick (1932a).
23. Chadwick *Proc. Roy. Soc.* paper: Chadwick (1932b).
24. Szilard found orphan: Weart and Szilard (1978), p. 16.
24. "the liberation . . . bombs": ibid.
24. "oppressed . . . application": Wells (1914), p. 46.
24. "This book . . . time": Weart and Szilard (1978), p. 16.
24–25. "I met . . . system": ibid.
25. Such must . . . nuclear physics: ibid., pp. 12–13.
25. "All I . . . too bad": ibid., p. 13.
25. Things got . . . Meitner: ibid.
25. "They all . . . happening": ibid.
25. "He looked . . . eyes": ibid., p. 14.
25–26. "I took . . . day earlier": ibid.
26. £1595: Bank receipt dated 6 September 1933 in Szilard Papers.
26. £854: Letter to "Béla" dated 31 August 1933 in Szilard Papers.
26. "I was . . . Association": Weart and Szilard (1978), p. 17.
26. "attempted . . . developments": *The Times,* p. 6.
27–28. "Lord Rutherford . . . irritated me": Szilard (1972), p. 529.
28. "This sort of . . . Row": Weart and Szilard (1978), p. 17.
28. "I was . . . be wrong": Szilard (1972), p. 530.
28. "It occurred . . . may react": ibid., p. 183.
28. Polanyi: Semenoff (1935), p. 5.
28. "As the light . . . the street": Szilard (1972), p. 530.
28. "it suddenly . . . atomic bombs": Weart and Szilard (1978), p. 17.

Chapter 2: Atoms and Void

29. "For by convention . . . void": quoted in *Scientific American* (1949), p. 49.
29–30. "It seems . . . formed them": in *Optics,* quoted in Guillemin (1968), p. 15.
30. "Though in . . . and weight": quoted in Pais (1982), p. 82.
30. "It is . . . pursuit in life": Planck (1949), p. 13.
30. "the process . . . by any means": ibid., p. 17.

PAGE

31. "Thus . . . is settled": W. Ostwald, at a meeting of the Deutsche Gesellschaft für Naturforscher und Ärzte in 1895, quoted in Pais (1982), p. 83.
31. "The consistent . . . finite atoms": in 1883, quoted in ibid., p. 82.
31. "What the atom . . . as ever": quoted in Chadwick (1954), p. 436.
31. "republic of science": Polanyi (1962).
31. "a highly . . . free society": ibid., p. 5.
32. "Millions are . . . committed": Polanyi (1974), p. 63.
32. "the established . . . letter": Polanyi (1946), p. 43.
32. "uncertainties . . . nature": ibid.
32. "a full initiation": ibid.
32. "close personal . . . master": ibid.
32–33. "what do we . . . the world": Feynman (1963), p. 2-1.
33. "no one . . . accepted": Polanyi (1946), p. 45.
33. "Any account . . . is untrue": Polanyi (1974), p. 51.
33. an analogy: Polanyi (1962), p. 6ff.
34. "Let them . . . consequence": ibid., p. 7.
34. "growing points": ibid., p. 15.
34. "The authority . . . *above* them": ibid., p. 14. His emphasis.
34–35. "*This network* . . . neighborhoods": ibid.
35. "Physics . . . those events": Wigner (1981), p. 8.
35. three broad criteria: cf. discussion in Polanyi (1962), p. 10ff.
36. one thousand . . . physicists: cf. Segrè (1980), p. 9.
36. "his genius . . . astonished": Chadwick (1954), p. 440. Details of Rutherford's childhood selected from Eve (1939), Feather (1940) and Crowther (1974).
36. sickening insecurity: the phrase is C. P. Snow's in Snow (1967), p. 11.
37. "That's the last potato": Eve (1939), p. 11.
37. "Now Lord . . . mine": ibid., p. 342.
37. "Magnetization of . . . discharges": 1894, in Rutherford (1962), pp. 25–57.
37. the world record: Marsden (1962), p. 3.
37. "like an . . . lion's skin": Eve (1939), p. 24.
37. "A magnetic detector . . . applications": Rutherford (1962), pp. 80–104.
37. Marconi . . . in September: cf. Eve (1939), p. 35.
38. "The reason . . . the future": ibid., p. 23.

PAGE

38. "You cannot serve . . . time": quoted in Kapitza (1980), p. 267.
38. "one curious . . . mistakes": Snow (1967), p. 7.
38. "I believe . . . commercially": Oliphant (1972), p. 140ff.
38. "in his . . . cultured man": Marsden (1962), p. 16.
39. "before tempting . . . more": ibid., p. 3.
39. "I hope . . . by myself": Eve (1939), p. 34.
39. Bank of England sealing wax: Blackett (1933), p. 72: "It is curious that the most universally successful vacuum cement available for many years should have been a material of common use for quite other purposes. At one time it might have been hard to find in an English laboratory an apparatus which did not use red Bank of England sealing-wax as a vacuum cement."
40. "the almost . . . passes": J. J. Thompson in Conn and Turner (1965), p. 53.
40. "the corpuscle . . . cathode ray": Crowther (1974), p. 123.
40. "a number . . . electrification": J. J. Thompson in Conn and Turner (1965), p. 97.
41. Thompson . . . discovering X rays: cf. ibid., p. 33.
41. Frederick Smith: cf. Andrade (1957), p. 444.
41. "at a . . . discharge-tube": J. J. Thompson in Conn and Turner (1965), p. 33.
41. Röntgen, Becquerel: these details from Segrè (1980), p. 19ff.
41. "exposed . . . on the negative": quoted in ibid., p. 28.
42. "expecting . . . in the dark": quoted in ibid., p. 29.
42. "There are . . . [beta] radiation": Rutherford (1962), p. 175.
42. P. V. Villard: Segrè (1980), p. 50.
42. "The McGill . . . cannot complain": Eve (1939), p. 57.
42. In 1900 . . . radioactive gas: "A radioactive substance emitted from thorium compounds." Rutherford (1962), pp. 220–231.
42. "At the beginning . . . be examined": Soddy (1953), p. 124ff.
43. "conveyed the . . . gas!": ibid., p. 126.
43. an "isotope": Soddy (1913), p. 400.
43. "for more than . . . an institution": Soddy (1953), p. 127.

PAGE

43. "similar . . . cathode rays": Rutherford (1962), p. 549.
43. "It may . . . molecular charge": ibid., p. 606ff.
44. "playful suggestion . . . in smoke": Eve (1939), p. 102.
44. "some fool . . . unawares": ibid.
44. "It is . . . secret": Soddy (1953), p. 95.
44. "My idea . . . romances": quoted in Dickson (1969), p. 228.
44–45. "After a very . . . radium rays": Eve (1939), p. 93.
45. "I may . . . keep going": ibid., p. 123.
45. "they are . . . moving": ibid., p. 127.
45. "it remained . . . true physicist": R. H. Fowler, quoted in ibid., p. 429.
45. An eyewitness: ibid., p. 183.
45. reported the month before: "The Nature of the α Particle," Nov. 3, 1908; Rutherford (1963), pp. 134–135.
45. "After some days . . . vessel": ibid., p. 145.
45. "In this . . . style": Russell (1950), p. 91.
46. "I see . . . his apparatus": quoted in Eve (1939), p. 239.
46. "supper in . . . the motor": Russell (1950), p. 88.
46. his handshake: Oliphant (1972), p. 22.
46. "he gave . . . physical contact": ibid.
46. He could still be mortified: cf. his response to the bishop in gaiters who presumed to compare the South Island to Stoke-on-Trent in Russell (1950), p. 96.
46. "He was . . . tricks": ibid., p. 89.
46–47. "Youthful . . . fools gladly": Weizmann (1949), p. 118.
47. "Scattering of . . . rays": Feather (1940), p. 117.
47. "I was . . . to taste": Eve (1939), p. 384.
47. Philipp Lenard: cf. Andrade (1957), p. 441.
47. 100 million volts: Rutherford's calculation in 1906 cited in Feather (1940), p. 131.
47. "Such results . . . electrical forces": ibid.
48. rang a bell: Blackett (1933), p. 77.
48. But the experiment was troubled: details from Marsden (1962), p. 8ff.
48. "See if . . . surface": ibid., p. 8.
49. "I remember . . . told him": ibid.
49. "If the . . . be required": H. Geiger and E. Marsden, "On a diffuse reflection of α-particles" in Conn and Turner (1965), p. 135ff.

PAGE

49. a first quick intuition: cf. Norman Feather in Rutherford (1963), p. 22.
49–50. "It was . . . minute nucleus": quoted in Conn and Turner (1965), p. 136ff.
50. sheets of good paper: cf. photographs of these historic notes in Rutherford (1963), following p. 240.
50. a model . . . pendulum: cf. Eve (1939), p. 197.
50. "largely . . . people": Chadwick OHI, AIP, p. 11.
50. a rare snake: ibid. Cf. also Chadwick (1954), p. 442n.
50. "a most . . . it": Chadwick OHI, AIP, p. 12.
50. "a central . . . in amount": Rutherford (1963), p. 212.
50. Nagaoka had postulated: cf. Conn and Turner (1965), p. 112ff, for partial text.
50. "Campbell tells . . . optical effects": quoted in Feather (1940), p. 136.
51. "supposed to . . . rotating electrons": Rutherford (1963), p. 254.
51. "for the . . . in Manchester": Nagaoka refers in his letter to "your paper on the calculation of alpha particles which was in progress when I visited Manchester." That paper, "The number of ***a*** particles emitted by uranium and thorium and by uranium minerals," was written with Hans Geiger, appeared in the *Philosophical Magazine* in Oct. 1910 and was sent July 1910. For the text of Nagaoka's letter cf. Eve (1939), p. 200.
51. the same theoretical defect: cf. discussion in Heilbron and Kuhn (1969), p. 241ff.
52. "Bohr . . . radioactive work": Eve (1939), p. 218.

Chapter 3: Tvi

53. *Tvi:* conversations with Josiah Thompson greatly enlightened this discussion.
53. "There came . . . Niels Bohr!": Eve (1939), p. 218.
53. "an enormous . . . head": Snow (1981), p. 19.
53. "much more . . . later years": ibid.
54. "he took . . . the matter": quoted in Rozental (1967), p. 78.
54. "uttering his . . . truth": quoted in Pais (1982), p. 417.
54. "his assurance . . . vivid images": Frisch (1979), p. 94.

PAGE

54. "he would . . . as criticism": quoted in Rozental (1967), p. 79.
54. "Not often . . . of trance": quoted in Pais (1982), pp. 416–417.
54. "gloomy . . . smile": quoted in Rozental (1967), p. 215.
55. "keen worshipper": Harald Høffding, quoted in ibid., p. 13.
55. "lovable personality": the surgeon Ole Chievitz, quoted in ibid., p. 15.
56. great interrelationships: Petersen (1963), p. 9: "Bohr has said that as far back as he could remember he liked to dream of great interrelationships."
56. speaking in paradoxes: according to Høffding in his *Memoirs,* quoted in Rozental (1967), p. 13.
56. "I was . . . different character": Bohr OHI, AIP, p. 1.
56. "At a . . . his imagination": Rozental (1967), p. 15.
56. "the special . . . the family": quoted in ibid.
56. "Even as . . . fundamental problems": Petersen (1963), p. 9.
56. trouble learning to write: cf. Segrè (1980), p. 119.
56. "There runs . . . two brothers": Rozental (1967), p. 23.
57. "*à deux*": Vilhelm Slomann, quoted in ibid., p. 25.
57. "In my . . . than I": Bohr OHI, AIP, p. 1.
57. Harald . . . told whoever asked: cf. for example Richard Courant in Rozental (1967), p. 301.
57. a stick used as a probe: e.g., Rozental (1967), p. 306.
57. "believed literally . . . of faith": quoted in ibid., p. 74.
57. "I see . . . his heart": quoted without citation in Moore (1966), p. 35. Moore was allowed access to some of Bohr's unpublished private correspondence.
57. Bohr drafted . . . private letters: cf. Rozental (1967), p. 30.
57. "If the . . . will come": quoted in Cline (1965), p. 214.
58. Bohr's anxiety: this discussion is based on Lewis S. Feuer's excellent analysis in Feuer (1982) but differs in emphasis and to some extent in conclusions. Holton (1973) is also an essential source.
58. "a young . . . unusual resolution": quoted in Rozental (1967), p. 74.

PAGE

58. "an unfinished . . . in [Denmark]": Bohr (1963), p. 13.
58. "a remarkably . . . position [as human beings]": ibid.
58. "Every one . . . his initiation": quoted in Rozental (1967), p. 121.
58. "very soberly . . . social activities": Bohr (1963), p. 13.
58. "[I start] . . . bottomless abyss": ibid.
59. "Bohr kept . . . studies itself": Oppenheimer (1963), II, pp. 25–26.
59. "Certainly I . . . to madness": quoted in Rosenfeld (1963), p. 48.
59. "Thus on . . . becomes actor": quoted in ibid., p. 49.
59. "Bohr would . . . emphasized": Rozental (1967), p. 121.
60. the image that recurred: cf. ibid., pp. 77, 327–328. Examples abound in the written record.
60. "suspended in language": quoted in Petersen (1963), p. 10.
60. *"Nur die . . . die Wahrheit":* quoted in Holton (1973), p. 148.
60. "I took . . . with Høffding": Bohr OHI, AIP, p. 1.
60. Harald Høffding: cf. biographical note at Bohr (1972), p. xx.
60. "despair": Holton (1973) notes this confession on p. 144.
61. Møller taught Kierkegaard: cf. Thompson (1973), p. 88.
61. "my youth's . . . departed friend": quoted in ibid.
61. The Danish word . . . "ambiguity": paraphrased from ibid., p. 155.
61. "His leading . . . *the individual*": quoted in Holton (1973), p. 146.
61. "Only in . . . of continuity": quoted in ibid., p. 147.
61. "At that . . . multivalued functions": Bohr OHI, AIP, p. 1.
62. the solid work: cf. Bohr (1972), p. 4.
62. "took such . . . the flame": Rosenfeld (1979), p. 325.
62–63. "the experiments . . . the paper": quoted in Rosenfeld (1963), p. 39.
63. "not a professor": Bohr (1972), p. 10.
63. "This is . . . ever read": ibid., p. 501.
63. "envy would . . . the rooftops": ibid., p. 95, adjusting the idiom.
63. "four months . . . rough drafts": ibid.
64. "unpractical": Bohr OHI, AIP, p. 2.
64. "He made . . . something good": quoted in Nielsen (1963), pp. 27–28.
64. "in deepest . . . my father": Bohr (1972), p. 295.

PAGE

64. "Dr. Bohr . . . a record": quoted in ibid., pp. 98–99.
64–65. "Oh Harald! . . . little fireplace": ibid., p. 519.
65. "under threat . . . stand it": ibid., p. 523.
65. "for an . . . blustering wind": quoted in Rozental (1967), p. 44, adjusting the idiom.
65. "absolute geniuses . . . you out": quoted in ibid., p. 40.
65. "I wonder . . . his ideas": quoted in Moore (1966), p. 32.
65. "I'm longing . . . silly talk": quoted in ibid., p. 33.
65–66. "It takes . . . was different": Bohr OHI, AIP, pp. 13–14.
66. "came down . . . his name": Bohr (1963), p. 31. My chronology of Bohr's visits to Manchester and his arrangements to work there generally follows the plausible conjectures of Heilbron and Kuhn (1969), p. 233, n. 57.
66. "just then . . . atomic nucleus": Bohr (1963), p. 31.
66. Bohr had matters on his mind: cf. his letter to C. W. Oseen on Dec. 1, 1911: "I am at the moment very enthusiastic about the quantum theory (I mean its experimental side), but I am still not sure this is not due to my ignorance." Quoted in Heilbron and Kuhn (1969), p. 230, with a following discussion.
66. "the patience . . . his mind": Bohr (1963), p. 32.
66. "one of . . . of Rutherford": ibid., p. 31.
66. "Bohr's different . . . football player!": quoted in Rozental (1967), p. 46.
66. eleven Nobel Prize winners: cf. Zuckerman (1977), p. 103.
67. Manchester is always here: cited by A. S. Russell in Birks (1962), p. 93ff.
67. "an introductory . . . research": Bohr (1963), p. 32.
67. Bohr learned about radiochemistry: cf. ibid., pp. 32–33.
68–69. "Rutherford . . . thought . . . his atom": Bohr OHI, AIP, p. 13.
69. "Ask Bohr!": quoted in Rozental (1967), p. 46.
69. "It could . . . quickly": Heilbron and Kuhn (1969), p. 238; Bohr (1972), p. 559; selecting the most idiomatic phrases from each translation.

PAGE

69. "getting along . . . to you!": Bohr (1972), p. 561, adjusting the idiom.
69. July 22: Bohr to Harald Bohr, ibid.
69. "to be . . . few weeks": quoted in Heilbron and Kuhn (1969), p. 256.
69. "a very . . . these things": quoted in ibid.
70. "One must . . . mechanical sort": quoted in ibid., p. 214. My discussion here generally follows this excellent monograph.
70. "Later measurements . . . to be": Planck (1949), p. 41.
70. "a universal . . . *of action*": ibid., p. 43.
72. "The spectra . . . a butterfly": quoted in Heilbron and Kuhn (1969), p. 257, n. 117.
74. "As soon . . . to me": quoted in ibid., p. 265.
75. "There is . . . its advent": Darrow (1952), p. 53.
75–76. "There appears . . . to stop": quoted in Bohr (1963), p. 41.
76. "Every change . . . possible transitions": quoted in Feuer (1982), p. 137.
76. "principal assumptions": cf. Shamos (1959), p. 338.
76–77. "Bohr characteristically . . . phenomena": Rosenfeld (1979), p. 318.
77. "asking questions of Nature": Rosenfeld (1963), p. 51.
77. "I try . . . I think": Oppenheimer (1963), I, p. 7.
77. "He points . . . their validity": Rosenfeld (1979), p. 318.
77. "the fairyland of the imagination": quoted in Thompson (1973), p. 176.
77. "It is . . . about nature": Petersen (1963), p. 12.
77. "It was . . . so seriously": Bohr OHI, AIP, p. 13.

Chapter 4: The Long Grave Already Dug

78. October 23, 1912: Hahn (1966), p. 70. Hahn (1970), p. 102, says Oct. 12. The official program confirms the later date.
78. a wet day, etc.: cf. photo "The dedication of the Kaiser Wilhelm Institute for Chemistry" in Hahn (1966), following p. 72.
78–79. the Kaiser Wilhelm Society: details from Haber (1971), pp. 49–50.
79. an embarrassment: cf. Hahn (1966), p. 50.
79. Hahn admired women: cf. Hahn (1970), passim.

PAGE

80. For details of Meitner's early life: cf. Frisch (1979), p. 3.
80. "There was . . . close friends": Hahn (1970), p. 88.
80–81. "an emanating . . . screen": Hahn (1966), p. 71.
81. "If I . . . in prison": Hahn (1970), p. 110.
81. "worthy of . . . noble brows": quoted in ibid., p. 102.
81. Moseley: this discussion relies on Heilbron (1974).
82. "so reserved . . . like him": quoted in ibid., p. 57.
82. "Hindoos, Burmese . . . 'scented dirtiness' ": Moseley to his mother, ibid., p. 176.
82. "Some Germans . . . photographing them": ibid., p. 193.
83. "We find . . . the atom": ibid., p. 205.
83. "unbearably hot . . . start measurements": ibid., p. 206.
83. "I want . . . a thousand": ibid., pp. 207–208.
84. a billiard table: Bohr OHI, AIP, p. 7.
84. "And that . . . were away": ibid.
84. "the only . . . of success": quoted in Rozental (1967), p. 58.
84. "a definite . . . chemical properties": quoted in Eve (1939), p. 224.
84. "shy . . . and noble": ibid., p. 223.
84. "great developments . . . on mankind": quoted in ibid., p. 224.
84. "do not . . . and 'fantastic' ": Bohr (1972), p. 567.
84. "Speaking with . . . saying so": quoted in Eve (1939), p. 226.
85. "During the . . . on Physics": Heilbron (1974), pp. 211–213.
85. "Because you . . . from Moseley": Bohr OHI, AIP, p. 4.
85. Bayer Dye Works: cf. Haber (1971), p. 128.
85–86. "In the . . . thermos vessels": Hahn (1970), p. 107.
86. "It is . . . whole valley": quoted in Rozental (1967), p. 64.
87. "The eleven- . . . of Palestine": quoted in Weisgal and Carmichael (1963), p. 20.
87. "a deliberate . . . 'eternal students' ": Weizmann (1949), p. 93.
87. "inviting every . . . remuneration": ibid., p. 171.
87–88. "In the . . . thrown away": ibid., p. 134.
88. January 1915: Stein (1961), p. 140.

PAGE

88. "Really messianic . . . upon us": quoted in ibid., p. 137n.
88. "You know . . . your hands": quoted in Weizmann (1949), p. 171.
88. "So it . . . British Admiralty": ibid., p. 172. Weizmann writes 1916, but this is clearly a slip of memory. Cf. Stein (1961), p. 118. Churchill was no longer First Lord in 1916.
89. "brisk, fascinating . . . two years": Weizmann (1949), p. 173.
90. "Horse-chestnuts . . . for maize": Lloyd George (1933), pp. 49–50.
90. "When our . . . in Palestine": ibid., p. 50. Vera Weizmann affirmed the authenticity of this conversation; cf. Stein (1961), p. 120n.
90. "view with . . . this object": cf. frontispiece facsimile, Stein (1961).
90. "outstanding war . . . National Home": quoted in ibid., p. 120n.
90. "it being . . . in Palestine": ibid., frontispiece facsimile.
91. a German rocket signal, etc.: these details at Lefebure (1923), pp. 36–37; Goran (1967), p. 68; and Hahn (1970), pp. 119–120.
91. "to abstain . . . deleterious gases": Carnegie Endowment for International Peace (1915), p. 1.
92. 300,000 pads: Pound (1964), p. 131.
92. Otto Hahn helped: cf. Hahn (1970), p. 118ff.
92. 5,730 of them: Prentiss (1937), p. 148.
92. "Haber informed . . . gas-warfare": Hahn (1970), p. 118.
93. James Franck: according to ibid., p. 119ff.
93. I. G. Farben: cf. Lefebure (1923), p. 86; Haber (1971), pp. 279–280.
93. mid-June 1915: as Hahn remembers it in Hahn (1970), p. 120. Prentiss (1937) says phosgene was first used by the Germans in a cloud-gas attack against the British at Nieltje on Dec. 19, 1915. Hahn may have meant 1916.
93. "the wind . . . success": Hahn (1970), p. 120.
93. buying German dyestuffs: cf. Haber (1971), p. 189.
93. phosgene: cf. Prentiss (1937), p. 154ff.
94. chlorpicrin: cf. ibid., p. 161ff.
94. mustard gas: cf. ibid., p. 177.
95. a typical artillery barrage: estimated from the figures given at Lefebure (1923), pp. 77–80.

PAGE

95. "She began . . . of life": Goran (1967), p. 71.
95. a scientist belongs: according to ibid.
96. "Our destination . . . in doubt": Heilbron (1974), p. 271.
96. "to be . . . or systematizing": ibid., p. 271ff.
96. "full of . . . and Australians": ibid., p. 272.
96. "The one . . . food": ibid., p. 274.
97. "over ghastly . . . slippery inclines": G. E. Chadwick, quoted in ibid., p. 122.
97. "They came . . . The Farm": Masefield (1916), p. 206.
97. "one of . . . in history": quoted in Kevles (1979), p. 113.
97. Folkestone: this section relies primarily on Fredette (1976).
98. "I saw . . . the sight": quoted in ibid., pp. 20–21.
98–99. "You must . . . in war": quoted in ibid., p. 30.
100. "a basis . . . to fight": ibid., p. 39.
100. "The day . . . and subordinate": quoted in ibid., p. 111.
100. research contracts: Prentiss (1937), p. 84.
100. "the great . . . of warfare": Lefebure (1923), p. 173.
100. a vast war-gas arsenal: cf. Prentiss (1937), p. 85, for these details and statistics.
101. "Had the . . . war": Lefebure (1923), p. 176.
101. 500,000 . . . 300,000: cf. Ellis (1976), p. 62.
101. 170 million rounds: ibid.
101. "Concentrated essence of infantry": J. F. C. Fuller, quoted in Keegan (1976), p. 228.
101. "I go . . . others": Edmund Blunden, quoted in Ellis (1975), pp. 137–138.
101. 21,000 men: cf. Keegan (1976), p. 255.
101–102. "It bears . . . common needle": quoted in Ellis (1975), p. 16.
102. "For the . . . working shift": Keegan (1976), pp. 229–230.
102. a software package: this discussion benefits from Elliot (1972), p. 20ff.
102. "The basic . . . trenches": ibid., p. 20.
102. "The War . . . of victims": Sassoon (1937), II, p. 143.
103. the long grave already dug: Masefield (1916), p. 104.
103. "The war . . . human variation": Elliot (1972), p. 23.
103. Elliot stresses: ibid., p. 25.

Chapter 5: Men from Mars

PAGE

104. "Horse-drawn . . . social currents": von Kármán (1967), p. 14.
105. "the fountain . . . to oppression": Paul Ignotus, quoted in Fermi (1971), pp. 38–39.
105. 33 percent . . . illiterate: Jászi (1924), p. 7.
105. 37.5 percent of . . . arable land: McCagg (1970), p. 186.
105. 1910 statistics: cf. Nagy-Talavera (1970), p. 41n.
105. S. V. Schossberger: cf. McCagg (1970), p. 132. My discussion of this phenomenon generally follows McCagg.
106. "one day . . . almost unpronounceable": von Kármán (1967), p. 17.
106. Jewish family ennoblements: cf. McCagg (1970), p. 63.
106. "galaxy of . . . lived elsewhere": Frisch (1979), pp. 173–174.
107. Von Kármán at six: von Kármán (1967), pp. 15–16.
107. Von Neumann at six: cf. Goldstine (1972), pp. 166–167.
107. Edward Teller . . . late . . . to talk: cf. Blumberg and Owens (1976), p. 6.
107. "Johnny used . . . improved": Ulam (1976), p. 111.
107. "As a . . . were good": Teller (1962), p. 81.
107. "addiction to . . . *of Man*": Weart and Szilard (1978), p. 4.
107. "the most . . . 19th century": E. F. Kunz, quoted in Madach (1956), p. 7.
108. "In [Madach's] . . . is pessimistic": *New York Post,* Nov. 24, 1945, quoted in Weart and Szilard (1978), p. 3n.
108. "a society . . . achievement": Smith (1960), p. 78.
108. "My father . . . them ourselves": von Kármán (1967), p. 21.
108. "We had . . . mathematician": interview with Eugene Wigner, Princeton, N.J., Jan. 21, 1983.
108. Teller recalls . . . syllogism: cf. Blumberg and Owens (1976), p. 137.
109. At Princeton: cf. Goldstine (1972), p. 176.
109. even Wigner thought: cf. Heims (1980), p. 43.
109. the only authentic genius: cf. Fermi (1971), pp. 53–54.
109. "So you . . . like geniuses": Blumberg and Owens (1976), pp. 15–16.

PAGE

109. "I think . . . impressed me": ibid., p. 23.
109. "The Revolution . . . irresistible momentum": Ferenc Göndör, quoted in Völgyes (1971), p. 31.
110. "*Es ist passiert*": quoted in ibid., p. 12.
110. "the rousing . . . melodious flood": Koestler (1952), p. 63.
110. "So far . . . sadistic excesses": von Kármán (1967), p. 93.
111. "because it . . . to come": Koestler (1952), p. 67.
111. "We left . . . put down": USAEC (1954), p. 654.
111. the group of Hungarian financiers: McCagg (1970), p. 16.
111. Teller heard of corpses: Blumberg and Owens (1976), p. 18.
111. The Tellers acquired two soldiers: cf. ibid.
111–112. "I shiver . . . terrible revenge": quoted in ibid., p. 19.
112. five hundred deaths: I use Koestler's figure ("under five hundred"), the larger of the two I have found. Koestler (1952), p. 67.
112. at least five thousand deaths: Heims (1980), p. 47, citing Rudolf L. Tökes, *Béla Kun and the Hungarian Soviet Republic* (Praeger, 1967), p. 214.
112. "no desire . . . all question": Jászi (1923), p. 160, the atrocities in detail ff.
112. "that the . . . nationalities": quoted in ibid., p. 186.
112–113. "It will . . . face extinction": Ulam (1976), p. 111.
113. "dinned into . . . stay even": *Time,* Nov. 19, 1957, p. 22.
113. "I loved . . . doomed society": Coughlan (1963), p. 89.
113. "I was . . . is lasting": von Kármán (1967), p. 95.
113. "once commented . . . the 'it' ": Pais (1982), p. 39.
113. "the acquisition . . . hostile world": Weizmann (1949), p. 18.
113. "every division . . . a watershed": ibid., p. 29.
113–114. "In the . . . little understanding": Born (1981), p. 39.
114. "Only a . . . the maze": Segrè (1980), p. 124.
114. "Bohr remembered . . . of Maxwell": Oppenheimer (1963), I, p. 21.
114. "His reactions . . . as well": Rozental (1967), p. 138.

PAGE

115. "That . . . of thought": quoted in Segrè (1980), p. 124.
115. "a unique . . . unforgettable experience": Bohr (1963), p. 54.
115–116. "I shall . . . highly exciting": Heisenberg (1971), pp. 37–38.
116. "At the . . . that afternoon": ibid., p. 38.
116. "Suddenly, the . . . of hope": ibid., p. 42.
116. "You are . . . small children!": Gamow (1966), p. 51.
116. "radiant . . . walking shorts": quoted in Jungk (1958), p. 26.
116. "But now . . . very seriously": Heisenberg (1971), p. 55.
116. "It saddened . . . such fancies": ibid., p. 8.
117. "a few . . . to rise": ibid., p. 61.
117. "a coherent . . . atomic physics": ibid., p. 62.
119. "This was . . . scholarship": the interviewer was Thomas Kuhn, in 1963, quoted in Smith and Weiner (1980), p. 3.
119. one of Robert's friends: Francis Fergusson, cited in ibid., p. 2.
119. "desperately amiable . . . be agreeable": Paul Horgan, quoted in ibid.
119. "an unctuous . . . a bastard": quoted in Royal (1969), pp. 15–16.
120. "tortured him": quoted in ibid., p. 23.
120. "Still a . . . about him": Jane Didisheim Kayser, quoted in Smith and Weiner (1980), p. 6.
120. "came down . . . the time": ibid., p. 7.
121. a Goth coming into Rome: Royal (1969), p. 27.
121. "He intellectually . . . the place": quoted in ibid. Michelmore (1969), p. 11, however, has Oppenheimer himself saying: "I . . . just raided the place intellectually."
121. a typical year: cf. Smith and Weiner (1980), p. 45.
121. "although I . . . with murder": ibid., p. 46.
122. "the most . . . came alive": Michelmore (1969), p. 11.
122. "Up to . . . of wrong": Seven Springs Farm transcript, p. 5, in JRO Papers, Box 66.
122. "Generously, you . . . dead. Voila": Smith and Weiner (1980), p. 54.
122. Both of Oppenheimer's . . . friends: they are quoted in this regard in ibid., p. 32.

PAGE

123. "It came . . . in physics": ibid., pp. 45–46.
123. "a man . . . an apprentice": ibid. , p. 69.
123. "But Rutherford . . . the center": ibid., p. 75.
123-124. "perfectly prodigious . . . success": quoted in ibid., p. 77.
124. "I am . . . Harvard overnight": ibid., p. 87.
124. "The business . . . interested in": ibid, p. 88ff.
124. "The melancholy . . . been snubbed": ibid., p. 128.
124. "How is . . . worth living": ibid., p. 86.
124. "making a . . . a career": ibid., p. 90.
124. "on the . . . was chronic": quoted in Royal (1969), p. 35.
125. "noisy . . . me crazy": Smith and Weiner (1980), p. 92.
125. "doing a . . . and alarm": John Edsall, quoted in ibid.
125. "When Rutherford . . . That's bad": ibid., p. 96.
125. "sweetness": Snow (1981), p. 60.
125. "At that . . . theoretical physicist": Smith and Weiner (1980), p. 96.
125. "He said . . . probably true": quoted in ibid., p. 94.
125-126. "a great . . . to brigantines": ibid., p. 95.
126. "The [Cambridge] . . . but love": quoted in Davis (1968), p. 22.
126. "a great . . . it": quoted in ibid., p. 21.
127. "Although this . . . very much": Smith and Weiner (1980), p. 103.
127. "Not only . . . for generations": Teller (1980), p. 137.
127. "a desert": second AHQP interview, p. 18.
127. "largeness and . . . fixed up": Smith and Weiner (1980), p. 121.
128. "house and . . . and stream": ibid., p. 126.
128. Everyone went . . . except Einstein: Segrè (1980) gives Einstein's reason at p. 168.
128. "In other . . . same structure": Heisenberg (1971), p. 71.
128-129. "This hypothesis . . . be true": ibid., p. 72.
129. "with . . . liberation": ibid., p. 71.
129. "Wilhelm Wien . . . by Schrödinger": Heisenberg in Rozental (1967), p. 103.
129. "For though . . . laborious discussions": ibid.
129. "While Mrs. . . . admit that": Heisenberg (1971), pp. 75–76.

PAGE

129. "If one . . . step forward": quoted in Rozental (1967), pp. 103–104.
129-130. "utterly . . .all along": Heisenberg (1971), p. 77.
130. "It is . . . can observe": quoted in ibid.
130. "On this . . . quantum mechanics": Heisenberg in Rozental (1967), p. 105.
130. Bohr ought to have liked: cf. Heisenberg's discussion in ibid., p. 106.
131. "the great . . . scientists": Bohr (1961), p. 52.
131. "renunciation": e.g., ibid., pp. 77, 80.
132. "Two magnitudes . . . the other": Segrè (1980), p. 167.
132. "bears a . . . and object": Bohr (1961), p. 91.
132. "quantum mechanics . . . play dice": quoted in Holton (1973), p. 120.
132. "We all . . . it all": Heisenberg (1971), p. 79.
133. " 'God does . . . the last": ibid., p. 80.
133. "Nor is . . . the world": quoted in ibid., p. 81.

Chapter 6: Machines

134. "I shall . . . world": Snow (1958), p. 88.
134. "uncarpeted floor . . . volcano": Oliphant (1972), p. 19.
135. "An anomalous effect in nitrogen": Rutherford (1963), p. 585ff.
135. "gave rise . . . itself": ibid., p. 547.
136. "I occasionally . . . this method": quoted in Bohr (1963), p. 50.
137. "appeared to . . . H scintillations": Rutherford (1963), p. 585.
137. "must be . . . in air": ibid., p. 587.
137. "From the . . . is disintegrated": ibid., p. 589.
137. one . . . in 300,000: Rutherford (1965), p. 24.
138. Francis William Aston: biographical details from de Hevesy (1947).
139. "In this . . . discharge tube": ibid., p. 637.
139. building the precision instrument: cf. Aston (1927, 1933).
139. "In letters . . . atomic model": Bohr (1963), p. 52.
139-140. "that neon . . . to 1": Aston (1938), p. 105.
140. "High packing . . . the reverse": Aston (1927), p. 958.
140. "If we . . . full speed": Aston (1938), p. 106.
140-141. "the nuclear . . . door neighbor": ibid., pp. 113–114.

PAGE

141. "Stockholm . . . ever since": quoted in de Hevesy (1947), p. 645.
141. "particularly detested . . . barking kind": quoted in ibid., p. 644.
141. "What is . . . not answer": quoted in Kevles (1977), p. 96. Numbers of American physicists given here and ff.
142. Psychometricians: e.g., Eiduson (1962), Goodrich et al. (1951), Roe (1952) and Terman (1955).
142. IQ scores: cf. Roe (1952), p. 24.
142. "He is . . . his nature": ibid. , p. 22.
143. A psychological examination: Eiduson (1962).
143. "their fathers . . . knew them": ibid., p. 65.
143. "rigid . . . reserved": ibid., p. 22.
143. "shy, lonely . . . or politics": Terman (1955), p. 29.
143. a fatherly science teacher: cf. Goodrich et al. (1951), p. 17.
143. "masterfulness . . . dignity": ibid.
143. "It would . . . their students": ibid.
143. Ernest Orlando Lawrence: biographical details from Alvarez (1970), Childs (1968) and Davis (1968).
143. "almost . . . mathematical thought": Alvarez (1970), p. 253.
144. "it seemed . . . atomic nucleus": Lawrence (1951), p. 430.
144. "the tedious . . . electron volts": Alvarez (1970), p. 260.
145. "In his . . . after night": ibid., p. 261.
145–146. "This new . . . arrangement": Lawrence (1951), p. 431.
146. "It struck . . . magnetic field": Alvarez (1970), p. 261.
146. "Oh, that . . . your own": quoted in Davis (1968), p. 19.
146. "I'm going . . . famous!": quoted in Childs (1968), p. 140.
146. a battered gray Chrysler: cf. Smith and Weiner (1980), p. 135.
146. "unbelievable vitality . . . the opposite": quoted in Childs (1968), p. 143.
148. "The intensity . . . was before": quoted in Davis (1968), p. 38.
148. "having a . . . *about 300*": Lawrence and Livingston (1932), p. 32.
148. "Assuming then . . . do this": ibid., p. 34.
149. Oppenheimer told a friend: cf. Davis (1968), p. 23.
149. "men of . . . broad intuition": Rabi (1969), p. 7.
149. the tunnel effect: cf. Bethe (1968), p. 393: "This work led him on to a treatment of the ionization of the hydrogen atom by electric fields, probably the first paper describing the penetration of a potential barrier."
150. dying suns: e.g. Oppenheimer and Snyder (1939).
150–151. "You put . . . know peace": Smith and Weiner (1980), pp. 155–156.
151. "Interested to . . . the circumstances": quoted in Childs (1968), p. 174.
151. "uncommon sensitivity . . . of thinking": Eiduson (1962), p. 105–106.
151. "Were this . . . such fantasying": ibid., p. 106.
152. "This discovery . . . in him": Pais (1982), p. 253.
152. "But if . . . about them": quoted in Rozental (1967), p. 139.
152. "And Bohr . . . not quite' ": Oppenheimer (1963), I, p. 3.
153. The Bakerian Lecture: "Nuclear constitution of atoms" in Rutherford (1965), p. 14ff.
153. "the possible . . . intense field": ibid., p. 34.
153. James Chadwick: biographical details in Massey and Feather (1976). Cf. also Chadwick's various recollections.
153. "was hardly . . . of Moseley": Massey and Feather (1976), p. 50.
153–154. "But also . . . Soldiers' ": Chadwick (1954), p. 443.
154. "Before the . . . neutral particle": Chadwick OHI, AIP, pp. 35–36.
155. "And so . . . only occasionally": ibid., p. 36.
155. "It was . . . disposal": Oliphant (1972), p. 67.
155. "dour and . . . became apparent": ibid., p. 68.
155. "to conceal . . . gruff façade": quoted in Wilson (1975), p. 57.
155. "He had . . . chuckle": Massey and Feather (1976), p. 66.
155. "the physics . . . noisy": ibid., p. 12.
156. he said later: paraphrased in ibid., p. 15.
156. "were so . . . of alchemy": Chadwick (1964), p. 159.
157. "passed through . . . be found": Chadwick (1954), p. 445.
157. "the problem . . . to believe": ibid., p. 444.
157. "was a . . . really important": James Chadwick OHI, AIP, p. 49.
157. "We are . . . physics!": Snow (1967), p. 3.
157. "We are . . . complex atoms": Rutherford (1965), p. 181.

PAGE

157. the scintillation method: this discussion follows Feather (1964), esp. p. 136ff.
157. "He found . . . while counting!": Massey and Feather (1976), p. 19.
159. "The loss . . . of them": Eve (1939), p. 341.
159. his armorial bearings: cf. illustration and description in ibid., p. 342.
159. "a real . . . physicist": Segrè (1980), p. 180.
159. "I don't . . . he did": James Chadwick OHI, AIP, p. 70.
159. "Indeed . . . element investigated": Feather (1964), p. 138.
160. "that the . . . backward direction": James Chadwick OHI, AIP, p. 161.
160. "And that . . . the neutron": James Chadwick OHI, AIP, p. 71.
161. "Of course . . . very much": ibid.
161. "together . . . in Paris": Feather (1964), p. 142.
161. "They fitted . . . hydrogenous material": ibid., p. 140.
162. "Not many . . . strange": Chadwick (1964), p. 161.
162. The radiation source: cf. photograph at Crowther (1974), p. 196.
163. "For the . . . oscillograph record": Feather (1964), p. 141.
163. "the number . . . protons": Chadwick (1932b), p. 695.
163. "In this . . . were tested": ibid.
163. "Hydrogen . . . this way": ibid., p. 696.
163–164. "In general . . . of 1920": ibid., p. 697.
164. "It was . . . time": James Chadwick OHI, AIP, p. 71.
164. "But there . . . the letter": ibid., p. 72.
164. "To [Chadwick's] . . . physicist": Segrè (1980), p. 184.
165. That Wednesday: the Kapitza Club traditionally met on Tuesday, but I take it that Chadwick finished the first intense phase of his work with the writing of his letter to *Nature* dated this day, Feb. 17, 1932. His remark about wanting to be chloroformed (see below) indicates he had not yet rested from his ten-day marathon.
165. "a very . . . us all": Oliphant (1972), p. 76.
165. "one of . . . a fortnight": Snow (1981), p. 35.
165. "A beam . . . times faster": Morrison (1951), p. 48.

PAGE

165. "the prehistory . . . nuclear physics": Bethe OHI, AIP, p. 3.
166. *"the personification . . . experimentalist":* Gamow (1966), p. 213. The complete *Faust* text is translated here by Barbara Gamow.
166. "The *Neutron* . . . you agree?: ibid., p. 213.
167. "That which . . . heart in": ibid., p. 214.
167. "Now a . . . along!": ibid.

Chapter 7: Exodus

168. "Antisemitism . . . is violent": Nathan and Norden (1960), p. 37
168. "A new . . . Newton": Pais (1982), Plate II.
168. Nobel Prize nominations: cf. ibid., p. 502ff.
168. "made . . . beyond Newton": quoted in ibid., p. 508.
168. "a massive . . . muscled": Snow (1967a), p. 52.
168–169. "A powerful . . . slipped off": ibid., p. 49.
169. "to look . . . his image": Erikson, "Psychoanalytic Reflections on Einstein's Centenary," p. 157, in Holton and Elkana (1982).
169. "dressed in . . . his eyes": Infeld (1941), p. 92.
169. "finally the . . . structure": quoted in Pais (1982), p. 239. The paper is A. Einstein, *PAW* (1915), p. 844.
169. "One of . . . scientific ideas": quoted in Clark (1971), p. 290.
170. popular lectures: cf. Feuer (1982), p. 82.
170. disrupted lecture: cf. Pais (1982), p. 315ff.
170. "said that . . . German spirit": quoted in Clark (1971), p. 318.
170. Einstein mistakenly thought: cf. Einstein to Arnold Sommerfeld, Sept. 6, 1920: "I attached too much importance to that attack on me, in that I believed that a great part of our physicists took part in it. So I really thought for two days that I would 'desert' as you call it. But soon there came reflection." Quoted in ibid., p. 323ff.
170. " 'My Answer . . . disposition, then": quoted in ibid., p. 319.
170. "Everyone . . . my article": quoted in Pais (1982), p. 316.
171. "miracle . . . deeply hidden": quoted in ibid.; p. 37.

PAGE

171. "If you . . . the things": quoted in Clark (1971), p. 469.
171. "Through the . . . social environment": quoted in ibid., p. 36.
171. His father stumbled: for a careful reconstruction of this period in Einstein's life cf. ibid., p. 39ff.
171. "Politically . . . my youth": quoted in ibid., p. 315.
171. medically unfit: Pais (1982), p. 45n.
172. "victorious child": Holton and Elkana (1982), p. 151.
172. "I sometimes . . . grown up": quoted in Clark (1971), p. 27.
172. $E = mc^2$: the paper is A. Einstein, *Jahrb. Rad. Elektr.* 4, 411 (1907).
172. "It is . . . for radium": quoted in Pais (1982), p. 149.
172. "The line . . . the nose": revised from ibid., p. 148ff.
173. "like men . . . postage stamp": quoted in Holton and Elkana (1982), p. 326.
173. "great work . . . were slender": quoted in Clark (1971), p. 252.
173. "I begin . . . younger years": c. 1915, revised from Pais (1982), p. 243.
173. "were . . . ambivalent": quoted in ibid., p. 315.
173. "a new . . . eternity": quoted in Clark (1971), p. 473.
174. "first discovered . . . and dispersion": quoted in ibid., p. 475.
174. "undignified . . . annoyed": quoted in Pais (1982), p. 314.
174. "in a . . . anti-Semitism also": quoted in Young-Bruehl (1982), p. 92.
174. "I am . . . in Germany": quoted in Feuer (1982), p. xxvi.
174. 54,000 marks: deJonge (1978), p. 240.
174–175. "I was . . . years ago": Roberts (1938), p. 265.
175. "These points . . . Wittenberg!": quoted in Toland (1976), p. 96.
175. refers to Jewry more frequently: cf. Hitler (1971), index.
175–176. "no lovers . . . of decomposition": ibid., passim.
176. The sun shines in: cf. photograph of Hitler's cell in Toland (1976), between pp. 172–173.
176. lederhosen: cf. photograph of Hitler at Landsberg, ibid.
176. "I often . . . magic formula!": quoted in ibid., p. 64.
176. "If at . . . in vain": Hitler (1971), p. 679.
177. The Jewish people: sources for this discussion include Arendt (1973), Bauer (1982), Cohn (1967), Dawidowicz (1967, 1975), Laqueur (1965), Litvinoff (1976), Mendelsohn (1970), Mendes-Flohr and Reinharz (1980), Parkes (1964), Patai (1977), *The Protocols of the Meetings of the Learned Elders of Zion* (1934), Rosenberg (1970), Veblen (1919), Weizmann (1949).
177. The fantasy of Jews: cf. Cohn (1967), p. 254.
178. "the enemies of Christ": Parkes (1964) attributes this canard to Catherine II in 1762. The *Encyclopedia Judaica,* however, ascribes it to the Czarina Elizabeth Petrovna in 1742. Whether mother- or daughter-in-law made the statement, it clearly reflects imperial opinion of the Jews at the time of the Polish partition.
179. Edward Teller's grandmother: interview with Herbert York, La Jolla, Calif., June 27, 1983.
179. "The Jews . . . a citizen": Mendes-Flohr and Reinharz (1980), p. 104.
180. "Jewish disorders": quoted in Levin (1977), p. 18.
181. Jews to the U.S.: for annual numbers cf. Mendes-Flohr and Reinharz (1980), p. 374.
181. "I have . . . of course": reported by Herman Rauschning, quoted in Cohn (1967), p. 60.
181. "We owe . . . by heart": quoted in Arendt (1973), p. 360.
181–182. "At eleven . . . the accursed": Cohn (1967), p. 34.
182. "What I . . . the *goyim*": *Protocols* (1934), p. 142.
182. "produced . . . our slaves": ibid., p. 175ff.
182. "The principal . . . the Papacy": ibid., p. 193.
183. "It will . . . a merit": ibid., p. 205.
183. *Protocols* plagiarized: the best discussions of the bizarre history of the *Protocols* are Cohn (1967) and Laqueur (1965).
183. "gave them . . . state itself": Arendt (1973), p. 39.
184. "Thus the . . . larger scale": ibid., p. 360.
184. "Do you . . . need them": Richard Breitling was the journalist. Quoted in Beyerchen (1977), p. 10.
184. meeting with Goebbels: cf. Goebbels' diary entry quoted in Dawidowicz (1975), p. 68.

PAGE

185. "I have . . . couldn't happen": quoted in Blumberg and Owens (1976), p. 51.
185. "I didn't . . . to change": Otto Frisch OHI, AIP, p. 12.
185. "Civil . . . must retire": quoted in Dawidowicz (1975), p. 77.
185. "descended from . . . grandparents": quoted in ibid., p. 78.
185. a quarter of the physicists: Beyerchen (1977), p. 44.
185. Some 1,600 scholars: ibid.
186. "I decided . . . my life": quoted in Clark (1971), p. 539.
186. "We sat . . . to America": quoted in ibid., p. 543.
186. *"Ich bin . . . dafür"*: quoted in ibid., p. 544.
186. fifteen thousand: according to Pais (1982), p. 450. Clark (1971), p. 544, has $16,000.
186. "Turn around . . . again": quoted in Pais (1982), p. 318.
187. "recommended . . . me also": Eugene Wigner OHI, AIP, p. 2.
187. "There was . . . it well": ibid., p. 6.
187. Leo Szilard to Eugene Wigner: Oct. 8, 1932, Egon Weiss personal papers, USMA Library, West Point, N.Y. Trans. Edda König.
187. "close . . . 1933": Weart and Szilard (1978), p. 14.
188. "On entering . . . up here": Elsasser (1978), p. 161.
188. numbers of dismissals: Beyerchen (1977), p. 44.
188. Bethe's dismissal: cf. Bernstein (1980), p. 34.
188. "Geiger . . . personal level": ibid., p. 33.
188. "He wrote . . . nothing": ibid., p. 35.
188. Bethe at 27: telephone interview with Rose Bethe, Jan. 18, 1984.
188–189. "I was . . . whatever": interview with Hans Bethe, Ithaca, N.Y., Sept. 12, 1982.
189. "Sommerfeld . . . come back": Bernstein (1980), p. 35.
189. "His early . . . mechanics": Wigner (1969), p. 2.
189–190. "It was . . . I could": interview with Edward Teller, Stanford, Calif., June 19, 1982.
190. "an old German nationalist": quoted in Blumberg and Owens (1976), p. 49.
190. "I really . . . in Germany": ibid.
190. "quite shocked . . . the line": Frisch (1979), p. 52.

PAGE

190. "very disappointed . . . back to": Otto Frisch OHI, AIP, p. 14.
190. "To me . . . I felt": Rozental (1967), p. 137.
191. "Stern . . . Blackett had": Frisch OHI, AIP, p. 14.
191. "If physics . . . dictator": quoted in "Patrick Maynard Stuart Blackett," *Biog. Mem. F.R.S.* 21, p. 22.
191. "Lise Meitner . . . to you": Frisch OHI, AIP, p. 12ff.
192. Bohr persuaded him: so Franck told Alice Kimball Smith: Smith, "The Politics of Control—The Role of the Chicago Scientists." Symposium on the 40th Anniversary of the First Chain Reaction, University of Chicago, Dec. 2, 1982. Franck's daughters have emphasized to the contrary that his decision to resign in protest was made "for himself and by himself and nobody else had any part in making it": Beyerchen (1977), p. 16; p. 215, n. 8.
192. Max Born's reinstatement: according to Beyerchen (1977), p. 21.
192. "We decided . . . of May": Born (1971), p. 113.
192. "Ehrenfest . . . young ones": ibid., p. 113ff.
192. "the old . . . for it": Shils (1964). Shils tells the Vienna story here at length and notes that he heard it not from Szilard but from "other persons." It was, he writes, "absolutely characteristic of Szilard to launch a campaign of aid and claim no credit later for himself."
193. "that as . . . do something": ibid., p. 38.
193. the major U.S. effort: cf. Duggan and Drury (1948) and Weiner (1969).
193. "a long . . . interview": Weart and Szilard (1978), p. 32.
193. "university of exiles": Born (1971), p. 114.
193. In Switzerland: Szilard reports these activities in a letter to Beveridge dated May 23, 1933. Leo Szilard Papers.
193. "sympathetic . . . German scientists": Szilard to Beveridge, ibid.
193. "he proposed . . . them out": Frisch (1979), p. 53.
194. Benjamin Liebowitz: for biographical data cf. "A memorial service for BENJAMIN LIEBOWITZ," Egon Weiss personal papers, West Point.

PAGE

194. "It is . . . Germany": Liebowitz to Ernest P. Boas, May 5, 1933. Szilard Papers.
194. "dismissed . . . stronger": Weart and Szilard (1978), p. 36.
194–195. "rather tired . . . the parties": ibid., p. 35.
195. Locker-Lampson: cf. Clark (1971), p. 566ff.
195. "talking . . . goats": quoted in ibid., p. 603.
195. "He did . . . forward": Moon (1974), p. 23.
195. British appointments: Bentwich (1953), p. 13, puts this number at 155.
195. American contributions: ibid., p. 19: "The total American financial contribution by 1935 equalled that of the rest of the world."
196. Emergency Committee arrivals: cf. Duggan and Drury (1948), p. 25.
196. one hundred physicists: Weiner (1969), p. 217.
196. "is a . . . distraction": Nathan and Norden (1960), p. 245.
196. "fell in . . . to him": Wigner OHI, AIP, p. 5.
196. "large and . . . atmosphere": Ulam (1976), p. 69ff.
196. "I used . . . the climate": ibid., p. 158.
196. "was astonished . . . and sang": Infeld (1941), p. 245.
197. Hans Bethe walked: Hans Bethe OHI, AIP.
197. "When I . . . me away": Mendelssohn (1973), p. 164.

Chapter 8: Stirring and Digging

198. Stirring and Digging: cf. Francis Bacon, *The Advancement of Learning:* "Surely to alchemy this right is due, that it may be compared to the husbandman whereof Aesop makes the fable: that, when he died, told his sons that he had left unto them gold buried underground in his vineyard; and they digged all over the ground, and gold they found none; but by reason of their stirring and digging the mould about the roots of their vines, they had a great vintage the year following: so assuredly the search and stir to make gold hath brought to light a great number of good and fruitful inventions and experiments." Quoted in Seaborg (1958), p. xxi.

PAGE

198. the Gamows' escape: cf. Gamow (1970), p. 108ff.
198–199. "I . . . my eyes": ibid., p. 120.
199. "You see . . . to arrange": ibid., p. 122.
199. "the voice . . . they were!": ibid., p. 123.
200. "unable . . . neutron": quoted in Weart (1979), p. 44.
200. "In the . . . encouragement": revised from ibid., p. 44, and Biquard (1962), p. 36.
200–201. "the emission . . . element": Joliot, quoted in Biquard (1962), p. 36.
201. "I irradiate . . . it continues": quoted in ibid., p. 32.
201. "The following . . . working order": ibid., p. 37.
201. "The yield . . . million atoms": Joliot (1935), p. 370.
202. "never . . . of view": quoted in Weart (1979), p. 46.
202. "Marie Curie . . . her life": quoted in Biquard (1962), p. 33.
202. "one of . . . the century": Segrè (1980), p. 197ff.
202. "These . . . transmutation": ibid., p. 198, where the letter to *Nature* is reproduced as Fig. 9.15.
202. "I congratulate . . . any success": quoted in Biquard (1962), p. 39.
202–203. "we are . . . necessary precautions": Joliot (1935), p. 373.
203. "spending much . . . very long": Weart and Szilard (1978), p. 36.
203. "became . . . chain reaction": ibid., p. 17.
203. "a little . . . a job": ibid.
203. "I remember . . . memoranda": ibid., p. 19ff.
203. a patent application: cf. Szilard (1972), p. 622ff.
203. March 12, 1934: LS completed the application on Saturday, March 10. He had to wait until Monday to file.
203. books on microfilm: cf. Szilard (1972), p. 722.
203. "In accordance . . . substances": ibid., p. 622. The balance of the application seems to concern a rough early conception of a thermonuclear fusion reactor of the Shiva type with a blanket for breeding heavy-element transmutations!
204. "that the . . . the other": Weart and Szilard (1978), p. 18.
204. "None of . . . England": ibid.

PAGE

204. Fermi was prepared: cf. Holton (1974), for evidence and a discussion.
204. "I remember . . . effective": Frisch (1976b), p. 46.
205. Both Fermi's biographers: L. Fermi (1954) and Segrè (1970).
205. "Fermi must . . . handwriting": Segrè (1970), p. 8.
205. "I studied . . . physics": quoted in ibid., p. 10.
205. "a very . . . death": quoted in ibid., p. 11.
206. "the partial . . . examination": ibid., p. 12.
206. "In the . . . propagandist": Fermi to Enrico Persico, Jan. 30, 1920, in ibid., p. 194. Segrè translates the extant Fermi-Persico correspondence in an appendix, p. 189ff.
206. "shy . . . solitude": ibid., p. 33.
206. "could not . . . nebulous": ibid., p. 23.
206. "he . . . in Rome": ibid., p. 33.
206. "Fermi remembered . . . recognize him": interview with Emilio Segrè, Lafayette, Calif., June 29, 1983.
206. "toward . . . experiment": Segrè (1970), p. 23.
206. "disliked . . . possible": quoted in ibid., p. 55.
206. "enlightening simplicity": ibid.
206. "quantum engineer": quoted by Weisskopf in Weiner (1972), p. 188.
206–207. "Not a . . . pretty active": quoted in Davis (1968), p. 266.
207. "cold . . . nature": quoted in ibid., p. 265.
207. "Fermi's thumb . . . flying": L. Fermi (1954), p. 7ff.
207. "was . . . sex appeal": ibid., p. 10.
207. "perhaps . . . than against": ibid., p. 15. LF's emphasis.
208. the time was ripe: sources for this discussion include Holton (1974) and Amaldi (1977) as well as L. Fermi (1954) and Segrè (1970).
208. "A fantastic . . . intuition": quoted in Holton (1974), p. 172.
209. too remote: according to Segrè (1970), p. 72.
209. Fermi found amusing: Segrè interview, June 29, 1983.
209. Fermi skiing: L. Fermi in Badash (1980), p. 89.
209. "We had . . . radioactivity": quoted in Holton (1974), p. 173, n. 81.
209. "Since the . . . interesting": Rabi (1970), p. 16.

PAGE

210. "The location . . . of study": Segrè (1970), p. 53.
210. crude Geiger counters: cf. Amaldi (1977), p. 301, Fig. 3, and Libby (1979), p. 41.
211. 100,000 neutrons: cf. Fermi paper (hereafter FP) 84*b*, Fermi (1962), p. 674.
211. "Small cylindrical . . . seconds": ibid.
211–212. "We organized . . . our stuff": Segrè (1955), p. 258ff.
212. The next letter: FP 85*b*, Fermi (1962), p. 676.
212. "Amaldi . . . good loser": L. Fermi (1954), p. 89.
213. 800 millicuries: cf. FP 99, Fermi (1962), p. 748.
213. "a very . . . determined": FP 86*b*, ibid., p. 678.
213. "This negative . . . than 92": FP 99, ibid., p. 750.
213. "a new element": cf. partial text of Corbino's address in Segrè (1970), p. 76.
214. "The discoveries . . . wars": Weart and Szilard (1978), p. 37.
214. "Of course . . . the point": ibid., p. 39.
214. "the liberation . . . the chain": Szilard (1972), p. 639.
214. critical mass: cf. ibid., p. 642.
214. "some cheap . . . an explosion": ibid.
215. "Marie Curie . . . not corrupted": quoted in Eve (1939), p. 388.
215. "which was . . . experiments": Weart and Szilard (1978), p. 20.
215. Szilard applied to Rutherford: cf. LS to Ernest Rutherford, June 7, 1934, Szilard Papers.
215. "These experiments . . . confirmed": Weart and Szilard (1978), p. 20.
215. the problem with helium: cf. Brown (n.d.), p. 53ff.
216. early in July: Amaldi (1977), p. 305.
216. "going on . . . laboratory": ibid.
216. an unanswered question: cf. Amaldi (1977), p. 310, and FP 98 (p. 744), FP 103 (p. 755) and p. 641 of Fermi (1962).
217. "We also . . . 3 minutes": Amaldi (1977), p. 310.
217. radiative-capture problem: cf. FP 103, Fermi (1962), p. 754ff.
217. "and those . . . important": ibid., p. 756.
217. "Shortly afterwards . . . results": ibid., p. 641.
217. "In particular . . . same room": Amaldi (1977), p. 311ff.

PAGE

218. "I will . . . have been": quoted in Segrè (1970), p. 80. The colleague was Subrahmanyan Chandrasekhar.
218. "About noon . . . radioactivity": ibid.
218. "the halls . . . magic!": L. Fermi (1954), p. 98.
219. water worked: cf. FP 105*b*, Fermi (1962), p. 761ff.
219. "Fermi dictated . . . time": Segrè (1970), p. 81.
219. "They shouted . . . drunk": L. Fermi (1954), p. 100.
219. "Influence of . . . neutrons—I": FP 105*b*, Fermi (1962).
219. "The case . . . same element": ibid., p. 761.
219–220. "might never . . . found out": Hans Bethe OHI, AIP, p. 30.
220. *Physical Review* paper: A. von Grosse, *Phys. Rev.* 46:241 (1934).
220. "It was . . . characteristics": Hahn (1966), p. 141.
220. "I began . . . chamber": Amaldi (1977), p. 317.
221. "The experiments . . . results": ibid.
221. "Through these . . . weight 239": FP 107, Fermi (1962), p. 791.
221. "Other examples . . . bromine": Szilard (1972), p. 646.
221. "So I . . . a chemist": Weart and Szilard (1978), p. 18.
221–222. "understood . . . done": ibid., p. 19.
222. Frederick Alexander Lindemann: cf. especially Mendelssohn (1973), p. 168ff.
222. "If your . . . physicist": quoted in ibid., p. 168.
223. "he became . . . for arrogance": ibid., p. 169.
223. "unbending . . . gentlemen": ibid., p. 168.
223. "gracious living . . . friendship": ibid., p. 171.
223. "saw a . . . modern war": Churchill (1948), p. 79ff.
223. "the question . . . as possible": Weart and Szilard (1978), p. 41.
224. "there is . . . taking patents": ibid., p. 40.
224. "Early in . . . proper use": ibid., p. 42.
224. "from private . . . at Oxford": ibid.
224. "there appears . . . concerned": quoted in ibid., p. 18, n. 28.
224. "I daresay . . . Government anything": quoted in Szilard (1972), p. 733.
224–225. "contains . . . this country": ibid., p. 734.

PAGE

225. "Bohr in . . . that game": Rozental (1967), p. 138.
226. "The lid . . . past year": ibid., p. 153.
226. "was drowned . . . for him": Oppenheimer (1963), II, p. 30.
227. "On that . . . understand it": Frisch (1979), p. 102.
227. "Neutron capture and nuclear constitution": Bohr (1936).
227. "For still . . . to become": ibid., p. 348.
228. "the consequences . . . developed": ibid.
228. "This 1937 . . . cleared up": Wheeler (1963b), p. 40.
228–229. Rutherford's death: cf. Eve (1939), p. 424ff, and Oliphant (1972), p. 153ff.
229. "seedy": quoted in Oliphant (1972), p. 154.
229. "a wonderful . . . of hope": quoted in ibid., p. 155.
229. "I want . . . Nelson College": quoted in Eve (1939), p. 425.
229. "When the . . . life": Oliphant (1972), p. 155.
229–230. "Life is . . . encouragement": Smith and Weiner (1980), p. 204.
230. "to me . . . father": Bohr (1958), p. 73.
230. "Voltaire . . . atomic physics": quoted in Eve (1939), p. 430ff.
230. "I have . . . attractive": ibid., p. 424.
230. "On element 93": *Zeitschrift für Angewandte Chemie* 47: 653. Cf. translation in Graetzer and Anderson (1971), p. 16ff.
231. Segrè remembers: cf. Emilio Segrè OHI, AIP, p. 24, and my Segrè interview.
231. "I think . . . elements": Otto Frisch OHI, AIP, p. 38.
231. "It is . . . element": FP 98, Fermi (1962), p. 734.
231. He later told Teller, Segrè, Woods: e.g., the three citations ff.
231. "Fermi refused . . . of nuclei": Teller (1979), p. 140.
231–232. "You know . . . to himself": Segrè interview, June 29, 1983.
232. "Why was . . . allowed": Libby (1979), p. 43.

Chapter 9: An Extensive Burst

233. "I believe . . . been granted": Meitner (1964), p. 2.
233. "quite convinced . . . had done": James Chadwick OHI, AIP, p. 76.
233. "Slight . . . by nature": Frisch (1968), p. 414.

PAGE

233. "there . . . X-rays": Frisch (1978), p. 427.
234. "persuaded . . . collaboration": Meitner (1962), p. 6.
234. "not only . . . Professor Hahn": Hahn (1966), p. 66.
234. "she could . . . story-teller": Frisch (1968), p. 414.
234. "totally . . . vanity": Frisch (1978), p. 426.
234. "though . . . could play": Frisch (1968), p. 414.
234. "It . . . alert": Axelsson (1946), p. 31.
234. "the vision . . . final truth": Frisch (1978), p. 426.
234. "For Hahn . . . to explain": ibid., p. 428.
235. Hahn met Joliot: Weart (1979), p. 57.
235. "It seems . . . its interpretation": quoted in Graetzer and Anderson (1971), p. 37.
235. "Who knows . . . storm": quoted in Churchill (1948), p. 262.
235. "The years . . . conditions": Meitner (1959), p. 12.
235. Meitner feared: cf. Frisch (1968), p. 410ff.
236. "I gave . . . an emergency": an important detail; after the war Meitner bitterly accused Hahn of railroading her out of Germany so that he would not have to share the discovery of fission with her—as if he foresaw it in July. Cf. Hahn (1970), p. 199.
236. "I took . . Holland colleagues": Axelsson (1946), p. 31.
236. Physical Institute: such is LM's address in Meitner and Frisch (1939). In his postwar recollections Frisch consistently places her at the "newly-built" Nobel Institute.
236. she photographed him: cf. Szilard (1972), p. 18.
237. "He told . . . the day": quoted in Leigh Fenly, "The Agony of the Bomb, and Ecstasy of Life with Leo Szilard." *San Diego Union,* Nov. 19, 1978, p. D-8.
237. "a *very* . . . a 'stranger' ": LS to Gertrud Weiss, March 26, 1936. Trans. Edda König. Szilard Papers.
237. "stay in . . . the war?": Weart and Szilard (1978), p. 20ff.
237. Lewis L. Strauss: details of his life from Pfau (1984), which Dr. Pfau was kind enough to allow me to read in MS.

PAGE

238. "My boy . . . you down": quoted in ibid.
238. "I became . . . hospitals": Strauss (1962), p. 163.
238. the report to *Nature:* cf. Szilard (1972), pp. 140, 147ff.
238. "An isotope . . . my parents": Strauss (1962), p. 164.
238–239. "August 30 . . . Leo Szilard": Szilard Papers.
239. owned patent jointly: "Patents which have been taken out by Dr. Brasch and Dr. Szilard were to be brought into this foundation." File memorandum, Szilard Papers.
239. "asked me . . . 'surge generator' ": Strauss (1962), p. 164.
239. "In the . . . fresh fruit": Shils (1964), p. 39.
239. debates among lawyers: cf. file memorandum, Szilard Papers.
239. "On April . . . taste unchanged": M. Lenz to LS, April 15, 1938. Szilard Papers.
240. "I left . . . wife here": Emilio Segrè OHI, AIP, p. 31.
240. a map of Ethiopia: Segrè (1970), p. 87.
241. "He was . . . to fascism": ibid., p. 63.
241. "We worked . . . Civil War": quoted in ibid., p. 90.
241. "That was . . . Italy": ibid., p. 91.
241. "America . . . of Europe": ibid., p. 92.
242. "I have . . . through anything": quoted in Shirer (1960), p. 343.
242. "Rome of . . . and master": revised from Segrè (1970), p. 95.
242. Fermi told Segrè: ibid., p. 96.
242. "Jews . . . Italian race": quoted in L. Fermi (1954), p. 119.
243. "Why should . . . can't he?": quoted in Frisch (1979), p. 108.
243. "the dangers . . . other societies": Bohr (1958), p. 23.
243. "we may . . . and variety": ibid., p. 30.
243. the German delegates: according to Moore (1966), p. 218.
243. "the . . . prejudices": Bohr (1958), p. 31.
243. "destroying . . . each other": Arendt (1951), p. 478.
243. Bohr and Fermi's Nobel: cf. L. Fermi (1954), p. 120ff.
244. Goldhaber: cf. Szilard (1972), p. 141ff.
244. "the whole . . . be delayed": Churchill (1948), p. 292.

PAGE

244. "I just . . . and see": Weart and Szilard (1978), p. 21.
245. "was . . . mood": quoted in Churchill (1948), p. 301.
245. "The British . . . worse treatment": ibid., p. 301ff.
245. "conditions . . . security": quoted in ibid., p. 302.
245. "He told . . . was accepted": quoted in ibid., p. 306.
245. "that this . . . than Germans": quoted in ibid., p. 309.
245. "How horrible . . . my soul": quoted in ibid., p. 315.
246. "regard the . . . another again": quoted in ibid., p. 318.
246. invasion of British Isles: cf. ibid.
246. "This is . . . our time": quoted in ibid.
246. Lindemann drove up: this story and Lindemann's remark appear in Mendelssohn (1973), p. 172.
246. "the complete . . . of force": Churchill (1948), p. 303.
246. HAVE ON . . . DECISIONS: Weart and Szilard (1978), p. 48.
246. "As my . . . decreased": Szilard (1972), p. 185.
247. University of Rochester: Goldhaber, Hill and Szilard, *Phys. Rev.* 55:47, refers to these experiments "to be reported in the following paper." They are therefore not reported in *Phys. Rev.* 55:47 as Weart and Szilard (1978) assert (p. 53, n.1). Szilard in Weart and Szilard (1978), p. 53, says to the point: "I went up to Rochester and stayed there for two weeks and made some experiments on indium which finally cleared up the mystery." Since he wrote the Admiralty on Dec. 21, 1938 (cf. Weart and Szilard, p. 60), the Rochester work probably occurred in late November–early December.
247. "Taken . . . by fractionation": quoted in Graetzer and Anderson (1971), p. 38.
247. "You can . . . muddled up": quoted in ibid., p. 39ff.
247. Strassmann speculated: cf. Irving (1967), p. 21. Irving interviewed both Strassmann and Hahn.
248. "must be . . . alpha particles": quoted in Graetzer and Anderson (1971), p. 42.
248. Meitner wrote in warning: according to Frisch (1979), p. 115. Frisch says elsewhere that this letter has been lost. It is not included among the Hahn-Meitner correspondence in Hahn (1975). All translations from the Hahn-Meitner correspondence by Edda König.
248. "Bohr was . . . elements": Hahn (1970), p. 150.
249. "Hard . . . be withdrawn": L. Fermi (1954), p. 123.
249. "especially rich ones": quoted in Dawidowicz (1975), p. 135.
250. "your discovery . . . slow neutrons": quoted in Segrè (1970), p. 98.
250. "Most of . . . very interesting": Hahn (1975), p. 75ff.
250. Meitner's living conditions: cf. ibid., pp. 91, 93, 103.
250. Eva von Bahr-Bergius: Johansson (n.d.), p. 1, and Hahn (1975), p. 103.
250. "Of course . . . important apparatus": Hahn (1975), p. 99.
250. "Concerning . . . care of": ibid., p. 76.
250. "a little . . . somewhat better": ibid., p. 77.
251. "As much . . . like *barium*": ibid., p. 77ff.
251. KWI layout: cf. floor plan, Max Planck Society Library and Archive, Berlin-Dahlem, and illustration accompanying "Die Kernspaltung," *Bild der Wissenschaft,* Dec. 1978, pp. 68–69.
251. KWI tables: the composite worktable preserved at the Deutsches Museum in Munich would appear to be the measurement-room table with a paraffin block, flasks and filters added to represent the other work areas.
252. "forms . . . crystals": Hahn (1966), p. 154ff.
252. "The attempts . . . being perceptible": Hahn (1946), p. 58.
252. "Exciting . . . mesothorium": quoted in Irving (1967), p. 23.
253. "Perhaps you . . . somewhat bearable": Hahn (1975), p. 78ff.
253. "very warm . . . wishes": ibid., p. 79.
253. Hahn had little joy: cf. ibid., p. 78: "How much I am looking forward to it—after such a long time without you—you can imagine."
253. *Naturwissenschaften:* cf. ibid.: "But before the institute closes we still want to write something . . . for Naturwiss." (Dec. 19, 1938); p. 81: "Since yesterday we have been putting together our Ra-Ba proofs. . . . On Friday the work is supposed to be turned

PAGE

in to Naturwiss.... The whole thing is not very well suited for [them] but they will publish it quickly" (Dec. 21, 1938). Cf. also Irving (1967), p. 27. Irving has it nearly right.

253. "Your radium . . . impossible": Hahn (1975), p. 79.
253. "if you . . . New Year": ibid., p. 79ff.
254. "Our radium . . . it quickly": ibid., p. 81.
254. "Further experiments . . . altogether": Weart and Szilard (1978), p. 80.
254. "just about . . . point": Szilard (1972), p. 185.
254. Hahn's and Strassmann's paper: all quotations from Hahn and Strassmann (1939a), trans. Hans G. Graetzer.
255. "especially . . . any more": Hahn (1966), p. 157.
256. "that I . . . box": Jungk (1958), p. 68. The Rosebaud pickup version is in Irving (1967), p. 27.
256. Kungälv: for much of this history cf. Claesson (1959).
257. Frisch and Meitner in Kungälv: sources for this episode, one of the most confused in the entire story, are Frisch (1967b, 1968, 1978, 1979); Frisch OHI, AIP; Rozental (1967); Clark (1980); Meitner (1962, 1964). A close reading of Hahn (1975) is extremely important for straightening out the accumulated errors of memory.
257. a quiet inn: the building, at No. 9, had become in 1982 a veterans' hall.
257. met in the evening: Frisch OHI, AIP, p. 33.
257. a large magnet: ibid.
257. insisted Frisch read it: Rozental (1967), p. 144: "But she wouldn't listen; I had to read that letter."
257. "Barium . . . mistake": Frisch OHI, AIP, p. 33.
257. "Finally . . . my problem": Meitner (1962), p. 7.
257-258. "But it's . . . that": Frisch OHI, AIP, p. 34.
258. "But how . . . drop": Rozental (1967), p. 144.
258. "Couldn't . . . of thing": Frisch OHI, AIP, p. 34.
258. "Now . . . opposite points": ibid.
258. "Well . . . I mean": ibid.
258. "I remember . . . instability": ibid.
259. "Then . . . energy": ibid.

PAGE

259-260. "gave a . . . well": Meitner (1964), p. 4.
260. "had . . . her head": Frisch OHI, AIP, p. 34.
260. "One fifth . . . fitted": Frisch (1979), p. 116.
260. "Lise . . . much lighter": Frisch OHI, AIP, p. 37.
260. Hahn's letter of Dec. 21: cf. LM to OH, Dec. 29, 1938: "Furthermore, how about the so-called actinium? Can they be separated from lanthanum or not?" Hahn (1975), p. 83. Hahn reported that result in his Dec. 21 letter; if Meitner had received it she would have known.
260. "barium fantasy . . . something": Hahn (1975), p. 82.
261. "very exciting . . . it": ibid., p. 83.
261. "Today . . . amazing": ibid.
261. "against . . . experience": quoted in Weart (1979), p. 59.
261. "We have . . . start": Hahn (1975), p. 84.
261. "If your . . . results": ibid.
261. "keen . . . Bohr": Rozental (1967), p. 145.
261. OF-NB meeting on Jan. 3: cf. OF to LM, Jan. 13, 1939: "Only today was I able to speak with Bohr about the bursting uranium." Stuewer (1985), p. 50.
261. "I had . . . be!": Rozental (1967), p. 145. Frisch misplaces this conversation to a later time.
261. "since Bohr . . . tomorrow": quoted in Stuewer (1985), p. 51.
261-262. "I am . . . findings": Hahn (1975), p. 85ff.
262. chronology of paper development and meeting with Bohr: Stuewer (1985), p. 51, quoting a contemporary letter from OF to LM.
262. Frisch mentioned experiment to Bohr: cf. Bohr's letter to his wife quoted at Moore (1966), p. 233: "I emphasized that Frisch had also spoken of an experiment in his notes." Note that according to OF this discussion occurred before he talked to Placzek. Placzek probably did not, therefore, as OF later remembered, suggest the experiment.
262. chronology of Placzek discussion: OF to LM, Jan. 8, 1939, quoted in Stuewer (1985), p. 53.
262. "was like . . . cancer": Stuewer (1979), p. 72.

PAGE

262. "One would . . . high": Frisch (1939), p. 276.
263. Jan. 13 until 6 A.M.: Frisch confirms date and time from his original laboratory notes in Stuewer (1979), p. 72.
263. "pulses at . . . two": Frisch (1939), p. 276.
263. "At seven . . . camp": Frisch OHI, AIP, p. 35.
263. "a state . . . confusion": ibid.
263. William A. Arnold: personal communication.
263. a dividing living cell: "Bohr had always urged that a nucleus behaved like a small droplet; a uranium nucleus . . . might divide itself into two smaller nuclei . . . much as a living cell becomes two smaller cells by fission." Frisch (1978), p. 428.
264. "I wrote . . . tail": Frisch OHI, AIP, p. 36.
264. two papers for *Nature:* Meitner and Frisch (1939); Frisch (1939).
264. papers posted: Stuewer (1985), p. 53.
264. "As we . . . seasickness": Rosenfeld (1979), p. 342.
264. "We . . . Fermi family": L. Fermi (1954), p. 139.
265. "During the . . . none": ibid., p. 154.
265. Rosenfeld thought paper sent: cf. Rosenfeld (1979), p. 343.
265. "In those . . . train": Stuewer (1979), p. 77.
265. "The effect . . . directions": Rosenfeld (1979), p. 343.
265. "I was . . . off": quoted in Moore (1966), p. 231.
265. Bohr letter to *Nature:* Bohr (1939a).
265. "Can you . . . weeks": Eugene Wigner OHI, AIP, p. 28.
265–266. "they said . . . doomed": interview with Eugene Wigner, Princeton N.J., Jan. 21, 1983.
266. "Wigner told . . . me": Weart and Szilard (1978), p. 53.
266. "if, as . . . proceeding": quoted in Stuewer (1985), p. 52.
267. Rabi from Bohr himself: as he remembers it. Telephone interview Feb. 27, 1984.
267. "probably . . . night": telephone interview with Willis E. Lamb, Jr., Feb. 24, 1984.
267. Rabi told Fermi: but remembers doing so as early as Jan. 17, 1939, which is difficult to reconcile with Fermi's proposal to Dunning of a confirming experiment on Jan. 25.
267. "I remember . . . news": Fermi (1962), p. 996.
267. "spreading . . . around": Lamb interview, Feb. 24, 1984.
267. "The discovery . . . uranium": quoted in Segrè (1970), p. 217.
267. "I thought . . . Fermi": Weart and Szilard (1978), p. 53.
267–268. "I feel . . . time": ibid., p. 62.
268. Fermi/Dunning/Anderson experiment: cf. Wilson (1975), p. 69ff, and Sachs (1984), p. 18ff.
268. "He came . . . say": Wilson (1975), p. 69ff.
268. "Before I . . . Bohr's": ibid., p. 71.
269. "All we . . . thought": ibid., p. 72.
269. "in general . . . me": quoted in Blumberg and Owens (1976), p. 70.
269. "Bohr has . . . obvious": Teller (1962), p. 8ff.
269. "Fermi . . . at Princeton": quoted in Moore (1966), p. 233.
269. Anderson returned to Pupin: cf. Anderson's account in Sachs (1984), p. 24ff, which includes photostats of Anderson's entries that night in his laboratory notebook, quoted here.
270. thought Dunning would telegraph: Evidence that Dunning had not wired Fermi as of Saturday night is Fermi's unusual response to the Roberts experiment at the DTM. Cf. Bolton (n.d.), p. 18. Frisch's explanation to Bohr is quoted in Stuewer (1985), p. 53.
270. Bohr chiding Frisch: quoted in Stuewer (1985), p. 53.
270. "that no . . . results": quoted in ibid., p. 55.
270. 51 participants: cf. group portrait, Carnegie Institution archives, Washington, D.C.
270. Gamow introduced Bohr: Roberts, et al. (1939). Roberts says Tuve wrote the introductory paragraph to this contemporary paper; the conference, it says, "began . . . with a discussion by Professor Bohr and Professor Fermi." Cf. also R. B. Roberts to E. T. Roberts, Jan. 30, 1939: "The annual theoretical physics conference started Thursday with an announcement by Bohr that Hahn in Germany had discovered a radioactive isotope of barium as a product

PAGE

of bombarding uranium with neutrons." DTM archives, Carnegie Institution.

271. "The Theo . . . implications": Roberts (1979), p. 29.
271. "Fermi . . . atomic power": RBR to ETR, Jan. 30, 1939.
271. KINDLY . . . WRITING: Weart and Szilard (1978), p. 60.
271. "in a . . . anybody": quoted in Weart (1979), p. 63.
271. Joliot's response: cf. ibid., p. 63ff.
272. APO: Dr. Louis Brown, DTM, personal communication.
272. "Sat 4:30 . . . Kr?)": R. B. Roberts laboratory notes (n.p.), DTM archives.
273. "tremendous . . . energy release": RBR to ETR, Jan. 30, 1939.
273. "We promptly . . . thorium": Roberts (1979), p. 29.
273. "I told . . . night": RBR to ETR, Jan. 30, 1939.
273. all except Teller: RBR's laboratory notes.
273. Fermi amazed: Bolton (n.d.), p. 18. Bolton talked to both Roberts and Meyer; both agreed on Fermi's response.
273. "I had . . . *Nature*": quoted in Moore (1966), p. 236.
273. "There . . . phone calls": Roberts (1979), p. 30.
273. "Fermi . . . 1939": Wilson (1975), p. 73.
273. "I would . . . physics": ibid., p. 72.
273–274. "So . . . almost pitiful": Luis Alvarez OHI, AIP.
274. "About 9:30 . . . proceed": Wilson (1975), p. 28ff.
274. "I remember . . . conclusions": Alvarez OHI, AIP.
274. "The U . . . way": Smith and Weiner (1980), p. 270ff, conjecture this letter to have been written on Jan. 28, 1939. The "papers" JRO refers to must be Henry's AP story, which reached Berkeley via the *Chronicle* on Sunday, Jan. 29 (on the evidence of Abelson's "About 9:30 a.m."). JRO dated the letter "Saturday"; probably therefore Feb. 4, 1939.
274. "might . . . to hell": Smith and Weiner (1980), p. 209.
274–275. "when fission . . . bomb": quoted in Weiner (1972), p. 90.
275. "A little . . . disappear": quoted in Kevles (1977), p. 324.

Chapter 10: Neutrons

PAGE

279. I. I. Rabi: cf. Bernstein (1975).
279. "infinite": ibid., p. 64.
280. "the mystery . . . nature is": ibid., p. 50.
280. Szilard learned: cf. Weart and Szilard (1978), p. 54.
280. "and say . . . about it": ibid.
280. "Nothing known . . . quantitatively": Wilson (1975), p. 76.
281. "From the . . . precautions": Weart and Szilard (1978), p. 54.
281. at Strauss' request: Strauss (1962), p. 172.
281. "the performance . . . been completed": ibid., p. 171.
281. "No! . . . on it": Teller (1962), p. 9ff.
282. "It is . . . than ever": Rosenfeld (1979), p. 343.
282. "For example . . . square foot": quoted in Clark (1980), p. 86.
284. "From these . . . slow neutrons": Roberts et al. (1939a), p. 417.
284. "Taking a . . . blackboard": Rosenfeld (1979), p. 343.
285. "He wrote . . . the process": ibid., p. 344.
285. "It was . . . also present": Dempster (1935), p. 765.
285. Nier measured the ratio: Nier (1939).
286. "changing from . . . two MeV": Fermi (1949), p. 166.
287. "Resonance in . . . nuclear fission": Bohr (1939b).
287. "were . . . inseparable": Fermi (1962), p. 999.
287. "slow neutrons . . . in uranium": Weart and Szilard (1978), p. 64.
287. "Bohr . . . to U238": Roberts et al. (1940), second page of introduction (unnumbered).
287–288. "For fast . . . abundant isotope": Bohr (1939b), p. 419.
288. a tank of water: cf. Fermi (1962), p. 5ff.
288. "Szilard watched . . . get them' ": Booth (1969), p. 11.
288–289. "All we . . . no radium": Weart and Szilard (1978), p. 55.
289. "to see . . . uranium": ibid., p. 64.
289. "just . . . from England": Weart and Szilard (1978), p. 55.
289. "a strange . . . object": Booth (1969), p. 11.
289*n*. "Fermi and . . . was U-238": Booth (1969), p. 20.

PAGE

289*n*. "outraged": quoted in Moore (1966), p. 248.
289*n*. "It was . . . Bohr's argument": Rosenfeld (1979), p. 345.
290. Roberts' and Meyer's *Phys. Rev.* letter: Roberts et al. (1939b).
290. "As soon . . . their thoughts": Weart and Szilard (1978), p. 66.
291. Szilard-Zinn experiment: cf. Szilard (1972), p. 158ff.
291. "Everything was . . . went home": Weart and Szilard (1978), p. 55.
291. "We find . . . about two": Szilard (1972), p. 158.
291. "more than . . . absorbed": Joliot et al. (1939a), p. 471.
291. "a yield . . . captured": Fermi (1962), p. 6.
291. "I was . . . the neutrons": Teller (1962), p. 10.
292. PERFORMED . . . 50%: Strauss (1962), p. 174.
292. "That night . . . for grief": Weart and Szilard (1978), p. 55.
292. "strongly appealed . . . discoveries": LS to A. H. Compton, Nov. 12, 1942, p. 3. MED 201.
292. "such a . . . handling it": Weart and Szilard (1978), p. 56.
292. G. B. Pegram: cf. Embrey (1970).
293. "probably the . . . world's work": quoted in ibid., p. 378.
293. "Experiments in . . . be disregarded": quoted in L. Fermi (1954), p. 162.
294. "Szilard . . . certainly possible?": Stuewer (1979), p. 282.
294. "We tried . . . into physics": Teller (1979), p. 143.
294. "the enormous . . . of U235": Stuewer (1979), p. 282.
294. "it was . . . taken seriously": Fermi (1962), p. 999.
294. "it can . . . huge factory": Blumberg and Owens (1976), p. 89.
294. "two months . . . one idea": L. Fermi (1954), p. 155.
295. "There's a wop": Hans Bethe interview, Sept. 12, 1982.
295. "a . . . board room": Strauss (1962), p. 236.
295. officer taking notes: these details in ibid., p. 238.
295. "Enrico . . . predictions": L. Fermi (1954), p. 165.
295. "to discuss . . . the majority": Weart and Szilard (1978), p. 56.
295. Joliot et al. paper: Joliot et al. (1939a).

PAGE

295–296. "From that . . . no sense": Weart and Szilard (1978), p. 57.
296. a second Joliot paper: Joliot et al. (1939b).
296. "The interest . . . satisfied": ibid.
296. "I began . . . absurd": quoted in Clark (1981), p. 58ff.
296. German initiatives: cf. especially Irving (1967), the basic reference to this subject.
296. "We take . . . others": quoted in ibid., p. 34.
297. "Tempers and . . . atoms": *New York Times,* April 30, 1939, p. 35.
297–298. "There is . . . whole matter": quoted in Groueff (1967), p. 191.
298. "By separating . . . very great": Wilson (1975), p. 75.
298. "went back . . . to do": Booth (1969), p. 27.
298. "He was . . . Fermi": Wilson (1975), p. 76.
299. "The [radio] . . . neutrons present": Fermi (1962), p. 12.
299. "He liked . . . time thinking": Wilson (1975), p. 78.
299. "Szilard made . . . assistant": Emilio Segrè interview, June 29, 1983.
299. "very competent": Wilson (1975), p. 78.
300. "about ten . . . by uranium": Fermi (1962), p. 12.
300. "an average . . . perhaps 1.5": ibid., p. 13.
300. "We were . . . Placzek's helium": Weart and Szilard (1978), p. 81.
300. the resulting paper: Fermi (1962), p. 11ff.
300. "by an . . . rays": ibid., p. 15.
301. "I was . . . to think": Weart and Szilard (1978), p. 81.
301. "is an . . . possibility": ibid., p. 88.
301–302. "It seems . . . reasonable price": Szilard (1972), p. 195.
302. "Thank you . . . *of uranium*": ibid., p. 197. Fermi's emphasis.
302. "the carbon . . . canned form": ibid., p. 196.
302. "even more . . . considered": ibid., p. 213.
302. "perhaps 50 . . . uranium": ibid., p. 196.
302. about $35,000: LS to "Bill Richards," July 9, 1939. Szilard Papers.
302. "He took . . . fall": Weart and Szilard (1978), p. 82.
303. "I knew . . . really seriously": ibid.
303. "it seems . . . no escape": ibid., p. 90.

PAGE

303. "Dr. Wigner . . . and me": ibid., p. 98.
303. "He was . . . concerned": ibid., p. 82.
303. "shared the . . . be advised": Szilard (1972), p. 214.
303. "worry about . . . to Germany?": Weart and Szilard (1978), p. 82.
304. Gustav Stolper: LS implies in ibid., p. 84, that he first contacted Stolper after his first visit to Long Island. His letter to Einstein of July 19, 1939 (p. 90), however, makes it clear that he talked to Stolper before his first visit but that Stolper did not connect him to Alexander Sachs until *after* that visit. In 1945 (Hellman [1945], p. 70) Sachs implied that he had been in touch with Einstein, Wigner and Szilard before this introduction. The contemporary record cited here indicates otherwise.
304. Sunday, July 16: the letter that resulted from the first meeting was transcribed by Wigner's secretary on Monday morning; July 16, 1939, is the only Sunday between LS's July 9 letter to Fermi and his post-meeting July 19 letter to AE.
304. "We were . . . us there": Weart and Szilard (1978), p. 83.
304–305. "He came . . . soda water": Snow (1967), p. 52ff.
305. "*Daran . . . gedacht!*": Nathan and Norden (1960), p. 291; Clark (1971), p. 669ff.
305. "very quick . . . to object": Weart and Szilard (1978), p. 83.
305. Einstein dictated a letter: cf. ibid. for an English paraphrase of this first Einstein draft. The letter to Roosevelt that eventually resulted is often erroneously attributed to LS. As will become apparent, that letter grew directly from this first draft.
305. "He reported . . . this matter": ibid., p. 90.
306. Alexander Sachs: cf. Hellman (1945). Sachs' book title appears on the cover page of Notes on imminence world war in perspective accrued errors and cultural crisis of the inter-war decades, March 10, 1939, MED 319.7.
306. "took the . . . in person": Weart and Szilard (1978), p. 91.
306. "Although I . . . his promise": ibid.
306. Teller midweek: cf. LS to AE, July 19, 1939, ibid., p. 90.
306–307. "Perhaps you . . . particularly nice": Weart and Szilard (1978), p. 91.

PAGE

307. July 30: I find no reference to this date except the garbled account in Blumberg and Owens (1976), p. 94, which gives it for the earlier LS-Wigner visit. It fell somewhere between July 20, 1939, when LS called AE to confirm his proposal by letter of July 19, and August 2, 1939, when LS again wrote AE. July 30 looks possible.
307. "I entered . . . chauffeur": NOVA (1980), p. 2.
307. a third text: cf. LS to AE, July 2, 1939: "I am enclosing the German text which we drafted together in Peconic." Weart and Szilard (1978), p. 92.
307. "Yes, yes . . . than indirectly": quoted in Teller (1979), p. 144.
307. "at long . . . middle man": Weart and Szilard (1978), p. 92.
307. "that you . . . too cleverly": AE to LS (n.d.). Szilard Papers. Trans. Edda König.
307. "We will . . . too stupid": Weart and Szilard (1978), p. 96. Translation revised.
307. Lindbergh letter: cf. ibid., p. 99.
308. "the Administration . . . in America": ibid., p. 95. This is the letter Sachs ultimately delivered to Roosevelt for AE. Szilard's accompanying memorandum is in Szilard (1972), p. 201ff.
308. "If a . . . the case": Weart and Szilard (1978), p. 97ff.
308. "a horrible . . . atomic bombings": Wigner (1945), p. 28.
308. "the Hungarian conspiracy": E. P. Wigner, memorandum to LS, April 16, 1941. Szilard Papers.
308–309. "Our social . . . the eye": U.S. Senate (1945), p. 7.
309. "a perfect . . . of armour": Churchill (1948), p. 447.
309. "Adam and . . . ever since": Ulam (1976), p. 116.
309. revulsion against bombings: this discussion follows Hopkins (1966).
310. "No theory . . . people": quoted in ibid., p. 454.
310. "inhuman . . . populations": quoted in ibid., p. 455.
310. "one of . . . reprisals": quoted in ibid., p. 457.
310. "Although . . . to come": ibid.
310. "The ruthless . . . immediate reply": Roosevelt (1939), p. 454.

PAGE

311. a secret conference: cf. Irving (1967), p. 40ff.
311. Bohr-Wheeler paper: Bohr and Wheeler (1939c).
311. "Preparatory . . . Fission": Irving (1967), p. 46n.
312. "felt that . . . is abolished": von Weizsäcker (1978), p. 199ff.
312. "is . . . our man": Weart and Szilard (1978), p. 100.
313. "He says . . . matters stand": ibid., p. 101.
313. late afternoon: on the evidence of the brandy and of Sachs' evening meeting with Briggs.
313. Watson meeting: according to AS to E. Wigner, Oct.17, 1939, MED 319.7. Hewlett and Anderson (1962) identify the two participants besides Watson as Adamson and Hoover, the ordnance specialists subsequently appointed to the Uranium Committee, citing a 1947 statement filed by Adamson. Sachs' contemporary letter is more authoritative.
313. "Alex . . . up to?": Moore (1966), p. 268. I have found no other source for this quotation or the Napoleon story but take it Moore interviewed Sachs.
313. Napoleon story: ibid. Moore places this story near the end of the meeting, but it was clearly designed to catch FDR's attention. Cf. also Hellman (1945), p. 71: "The October 11th White House interview was one of a considerable series, during which Sachs, according to friends, would ease the President into the discussion with a few learned jokes."
313. "Bah! . . . visionists!": A. C. Sutcliffe, *Robert Fulton* (Macmillan, 1915), p. 98.
313. "I am . . . to him": quoted in Hellman (1945), p. 70.
314. Sachs did not read the Einstein letter: there is considerable evidence in the record to this point; cf. especially Sachs' almost- explicit admission at U.S. Senate (1945), p. 10: "The Einstein letter of August 2, from which I quoted in part in my own letter, was left with the President, along with my own letter." Hewlett and Anderson (1962), p. 17, confirm the omission: "Sachs read aloud his covering letter, which emphasized the same ideas as the Einstein communication but was more pointed on the need for funds." The scientific authority behind the meeting was nevertheless AE's, as FDR wrote AE on Oct. 19, 1939: "I found this data of such import that I have convened a board . . . to thoroughly investigate the possibilities of your suggestion." Nathan and Norden (1960), p. 297. Some have questioned the effect of the Einstein/Szilard/Sachs contact. Its effect was to convince FDR to appoint the Advisory Committee on Uranium. The emigrés were hardly to blame for the inadequacies of that committee.
314. Sachs summation: Sachs (1945).
314. Sachs intentionally: U.S. Senate (1945), p. 7.
314. "ambivalence . . . and evil": ibid., p. 9.
314. "the more . . . door neighbor": Aston (1938), p. 113ff. Also quoted in ibid.
314. "Alex . . . requires action": U.S. Senate (1945), p. 9.
315. "Don't let . . . me again": ibid.
315. Tuve deputized Roberts: Roberts (1979), p. 37.
315. Sachs breakfast: AS to E. Wigner, Oct. 17, 1939. MED 319.7.
315. Szilard began: cf. his Oct. 26, 1939, memorandum to L. Briggs (Szilard [1972], p. 204ff), which embodies "the statements and recommendations made by me at the meeting of October 21st": LS to LB, Oct. 26, 1939. Weart and Szilard (1978), p.110ff.
315. "too heavy . . . airplane": Szilard (1972), p. 202.
315. "In Aberdeen . . . prize yet": quoted in Teller (1979), p. 144.
315. ordnance depot: cf. Blumberg and Owens (1976), p. 98.
315. Roberts raised objection: Sachs notes "a strong objection" (Sachs [1945] p. 7) from "scientists who were not as much concerned as these refugee scientists"—U.S. Senate (1945), p. 11. The only other American scientist at the meeting besides Briggs was Mohler. He may have concurred with Roberts, but Roberts had the necessary fast-neutron measurements.
316. "there are . . . possibility": Roberts (1939c), p. 613.
316. the DTM had begun assessing: Roberts writes: "After Florida [i.e., March 1939] I continued work . . . on neutron scattering but my main efforts went into measuring cross-section for

PAGE

fission for neutrons of various energies. These were essential in calculating whether a chain reaction would run." Roberts (1979), p. 37. Roberts "made rough measurements of the fission cross-section for neutrons in the energy range 500–2000 kv." Roberts (1940), p. 2.

316. "very unlikely . . . reaction": Roberts (1939c), p. 613.

316. Briggs spoke up: Sachs (1945), p. 11.

316. "astonished . . . enthusiastic": Weart and Szilard (1978), p. 110.

316. "The issue . . . ahead": U.S. Senate (1945), p. 11.

316. "I said . . . is expensive": Blumberg and Owens (1976), p. 98.

316. "How much . . . need": Eugene Wigner interview, Jan. 21, 1983.

316. "The diversion . . . such recommendation": Weart and Szilard (1978), p. 110.

316–317. "For the . . . me yet": Teller (1979), p. 145.

317. $33,000: Szilard (1972), p. 205.

317. "At this . . . be cut": Weart and Szilard (1978), p. 85.

317. "All right . . . your money": Hewlett and Anderson (1962), p. 20.

317. Uranium Committee report: excerpts at Sachs (1945), p. 7ff.

317. Fermi letter: EF to AOCN, Oct. 28, 1939. A.O.C. Nier, personal communication.

317. Nier finally began preparing: A.O.C. Nier, personal communication.

Chapter 11: Cross Sections

318. "I regularly . . . every night": Otto Frisch OHI, AIP, p. 12.

318. "in a . . . any good": ibid., p. 40.

318–319. "I first . . . concentration camp": ibid., p. 39ff.

319. "So I . . . tourist": Frisch (1979), p. 120.

319. "a great . . . sobriety": ibid., p. 121.

319. "a sample . . . changed": ibid., p. 123ff.

319. "material enriched . . . bottom": ibid., p. 124.

319. "the most . . . Hitler war": Snow (1981), p. 105.

319–320. "I managed . . . on time": Frisch (1979), p. 125.

320. "That process . . . the trouble": Frisch (1971), p. 22.

320. "new explosives . . . by them": Churchill (1948), p. 386ff.

PAGE

321. when Oliphant consulted Peierls: cf. Frisch (1971), p. 123.

321. Perrin's formula: Perrin (1939).

321. Peierls' formula: Peierls (1939).

321. "of the . . . practical significance": Clark (1981), p. 85.

322. "ran her . . . times since": Frisch (1979), p. 130.

322. "Is that . . . written?": Frisch OHI, AIP, p. 39.

322. "I wondered . . . be needed?": Frisch (1979), p. 126.

323. "we had . . . to happen": Frisch (1977), p. 23.

323. 10^{-23} cm^2: ibid.

323. "Just . . . playfully": ibid., p. 22

323. "To my . . . or two": Frisch (1979), p. 126.

323. four millionths/second: Gowing (1964), p. 391.

323. "I worked . . . by them": quoted in Clark (1981), p. 88.

323. "I had . . . be possible": Frisch (1979), p. 126.

323. "The cost . . . the war": Wilson (1975), p. 55.

324. "Look . . . about that?": Frisch OHI, AIP, p. 39.

324. "They . . . me": Oliphant (1982), p. 17.

324. "I remember . . . were doing": Frisch (1977), p. 25.

324. "On the . . . in uranium": the full text appears at Gowing (1964), p. 389ff.

324. "to point . . . discussions": ibid., p. 389.

324. "the energy . . . or less": ibid., p. 391.

324. "Memorandum . . . 'super-bomb' ": Ronald M. Clark found this document among the papers of Henry Tizard and published it in Clark (1965), p. 214ff.

325. "I have . . . present time": quoted in ibid., p. 218.

325. "I have . . . on it": Frisch (1979), p. 126.

326. "heavy water . . . yet known": Irving (1967), p. 49.

326. Norsk Hydro: cf. ibid., pp. 49ff, 56ff.

327. Allier and heavy water: cf. Weart (1979), p. 130ff.

327. "The complete . . . United States": York (1976), p. 30. For York on Soviet research cf. p. 29ff.

327. Japanese studies: cf. Pacific War Research Society (1972) (hereafter PWRS) and Shapley (1978).

PAGE

327. Takeo Yasuda: PWRS (1972), p. 18ff.
328. "We are . . . any day": quoted in Moore (1966), p. 267.
328. "this . . . *coup*": Churchill (1948), p. 600.
329. "It was . . . to persecute": Rozental (1967), p. 160ff.
329. Nobel Prize medals: cf. de Hevesy (1962), p. 27.
329. 1.5 tons heavy water: Irving (1967), p. 61.
329. "What I . . . a committee": quoted in Clark (1965), p. 218.
330. "the possibility . . . the Germans": quoted in Clark (1981), p. 92ff.
330. "We entered . . . be investigated": Gowing (1964), p. 394.
330. "unnecessarily excited": quoted in Clark (1981), p. 94.
330. "I still . . . very low": quoted in Clark (1965), p. 219.
330. "Dr. Frisch . . . was feasible": quoted in Clark (1981), p. 95.
330–331. "The Committee . . . separation": Oliphant (1982), p. 17.
331. "the most . . . was wrong": Weart and Szilard (1978), p. 115.
331. Watson decided: Hewlett and Anderson (1962), p. 21.
331. "a crucial . . . application": quoted in ibid.
331. "Divergent chain . . . and carbon": Szilard (1972), p. 216ff.
331. "seemed to . . . went home": Weart and Szilard (1978), p. 115.
332. "the most . . . this research": ibid., p. 122.
332. "I worked . . . of uranium": Booth et al. (1969), p. 28.
332. "very doubtful . . . uranium": quoted in Hewlett and Anderson (1962), p. 20.
333. "These experiments . . . in uranium": Nier et al. (1940a).
333. "Furthermore . . . unseparated U": Nier et al. (1940b).
333. 400 to 500 $\times 10^{-24}$ cm^2: Nier et al. (1940a).
333. "Cartons of . . . make measurements": Wilson (1975), p. 83ff. Anderson recalls 1.5 tons of graphite here; but Fermi (1962), FP 136, p. 34, the report of this experiment, confirms the larger figure.
333. "So physicists . . . happening": Fermi (1962), p. 1000.
334. "A precise . . . delight him": Wilson (1975), p. 84.

PAGE

334. 3×10^{-27} cm^2: Fermi (1962), p. 32.
334. "scientists . . . Institution": Gowing (1964), p. 43.
334. "It is . . . goose chase": quoted in Clark (1965), p. 220.
335. Teller calculation: Hewlett and Anderson (1962), p. 32.
335. "the cross-section . . . pure uranium": Roberts et al. (1940), Introduction, second page.
335. "I came . . . miracle happened": Teller (1977), p. 11.
335. "To deflect . . . my mind": Blumberg and Owens (1976), p. 100.
335. "In the . . . to go": Teller (1979), p. 145.
335. Teller had never bothered: cf. ibid.
335. "We had . . . to me": quoted in *Forbes,* Feb. 18, 1980, p. 62.
335. "the continuance . . . mystic immunity": Roosevelt (1941), p. 184.
335–336. "Then he . . . be lost": Teller (1979), p. 145ff.
336. "but something . . . be lost": Blumberg and Owens (1976), p. 101.
336. "conquest and . . . different cause": Roosevelt (1941), pp. 184–187.
336. "My mind . . . changed since": Blumberg and Owens (1976), p. 101.
336. "That experience . . . lack meant": Bush (1970), p. 74.
337. "It was . . . certainly need": ibid., p. 33.
337. "something meshed . . . language": ibid., p. 35.
337–338. "Each of . . . to turn": ibid., p. 36.
338. "the threat . . . our minds": ibid., p. 34.
338. Bush and Conant proving impossibility: this insightful assessment comes from Dupree (1972), p. 456.
338. "I remember . . . and voice": Snow (1967b), p. 149ff.
339. Franz Simon: cf. Arms (1966).
339. "use my . . . this country": ibid., p. 111.
339. Simon joked: ibid., p. 109.
339. "It was . . . the streets": quoted in Clark (1981), p. 108.
339. "Within a . . . the matter": Moon (1977), p. 544.
339. "I do . . . taken seriously": quoted in Gowing (1964), p. 47.
340. hammered kitchen strainer: Arms (1966), p. 109, says this occurred in "late spring." Fitted against other events June is a reasonable surmise.
340. "Arms . . . separate isotopes": ibid.

PAGE

340. "The first . . . soda-water": quoted in Clark (1981), p. 110.
340. MET . . . KENT: quoted in ibid., p. 95.
341. "an anagram . . . they can": quoted in ibid., p. 96.
341. strategic bombing: cf. Burns (1967); Kennett (1982); Saundby (1961).
341. "short . . . air raid": quoted in Kennett (1982), p. 112.
341. "to undertake . . . are available": quoted in ibid., p. 113.
341-342. Hitler reserved London: ibid., p. 118.
342. "And if . . . cities out!": quoted in ibid., p. 119.
342. "will-to-resist": quoted in ibid., p. 118.
342. "a systematic . . . Lion unnecessary": quoted in ibid., p. 120.
342. HE tonnage: Harrisson (1976), p. 128.
343. deaths: ibid., p. 265.
343. Simon report: reproduced, probably in rewritten form, under a different title as part of the MAUD Report and given in this form in Gowing (1964), p. 416ff. I quote from the MAUD version, p. 416.
343. Simon delivered report: Arms (1966), p. 111.
343. Auer ordered sixty tons: Irving (1967), p. 65.
344. Joliot: the cyclotron episode appears at Weart (1979), p. 156ff.
344. Bothe graphite measurements: Bothe (1944).
345. "When . . . was on": Weart and Szilard (1978), p. 116.
345. "These galling . . . scientific fraud": Bothe (1951), p. 1ff. Trans. Louis Brown.
345. "uranium . . . not work": Frisch (1979), p. 138.
346. "only for . . . consideration": Irving (1967), p. 80. Irving's report of Harteck's meaning is here and on p. 277; the heavy-water recommendation is also here.
346. Suzuki report/Nishina: PWRS (1972), p. 19ff; Shapley (1978), p. 153.
346. Turner letter to *Phys. Rev.:* Turner (1946).
346. "It seems . . . to say": Weart and Szilard (1978), p. 126ff.
346. Turner review article: Turner (1940).
347. "a little . . . of isotopes": Weart and Szilard (1978), p. 126.
347. "it is . . . be used": Turner (1946).
347. "In $_{94}$ EkaOs^{240} . . . be expected": Turner (1946).

PAGE

347*n*. Bohr had speculated: cf. Nobel Committee presentation speech preceding McMillan (1951), p. 310ff.
348. "When a . . . a book": McMillan (1951), p. 314.
348. "Nothing very . . . very interesting": ibid., p. 315.
348. "a uranium . . . neutron capture": McMillan (1939).
348. "the two-day . . . explanation": McMillan (1951), p. 316.
349. "Segrè . . . the story": ibid., p. 317.
349. "As time . . . vacation": ibid., p. 318.
350. "When he . . . work together": ibid., p. 319.
350. "Within a . . . like uranium": Wilson (1975), p. 33.
350. "Radioactive element 93": McMillan and Abelson (1940).
350. "it might . . . contribution": Weart and Szilard (1978), p.127.
350. idea occurred to von Weizsäcker: cf. Irving (1967), p. 68.
351. "finding that . . . neptunium": McMillan (1951), p. 321.
351. "I left . . . national defense": ibid., p. 322.
352. "excellent public . . . children have": L. Fermi (1954), p. 145.
352. "an . . . annual": ibid., p. 148.
352. " 'D'you know . . . crab grass": quoted in ibid., p. 147.
352. "purposely studied . . . Americanization": Segrè (1970), p. 104.
352. Segrè at Purdue: Segrè discusses this episode, including Lawrence's attitude, in Emilio Segrè OHI, AIP, p. 33.
352. "the machine . . . I do": ibid.
352. "we had . . . scary problem": Segrè (1981), p. 11.
352. "Fermi . . . 94]": ibid.
353. "I suggested . . . his collaborators": Seaborg (1976), p. 5.
353. Two searches: both of which may be followed day by day in ibid.
353. 0.6 microgram: ibid., p. 13.
354. "key step . . . discovery": Seaborg (1958), p. 4.
354. Seaborg remembers: cf. Bickel (1980), p. 188.
354. "With this . . . *94":* Seaborg (1976), p. 25.
355. "This morning . . . *neutrons":* ibid., p. 34.
355. larger critical mass: Gowing (1964), p. 68.

PAGE

355. "This first . . . is manageable": quoted in ibid., p. 67ff.
356. "I remember . . . 28 years": James Chadwick OHI, AIP, p. 105.

Chapter Twelve: A Communication from Britain

357. Conant: cf. Conant (1970), Kistiakowsky and Westheimer (1979).
357. "the most . . . race": Conant (1970), p. 252.
357. "What shall . . . formality": quoted in ibid., p. 253.
358. "I said . . . the Interior": ibid., p. 52.
358. "I did . . . or weapon": ibid., p. 49.
358. "Conant achieved . . . chemistry": Kistiakowsky and Westheimer (1979), p. 212.
359–360. "strong belief . . . involving Briggs": Conant (1970), p. 276ff.
360–361. "Light a . . . possibilities?": quoted in Childs (1968), p. 311.
361. Compton follow-up letter: K. Compton to V. Bush, March 17, 1941. OSRD S-1, Bush-Conant File, folder 19.
361. "by nature . . . solution": ibid.
361–362. "I told . . . trail behind": VB to F. Jewett, June 7, 1941. Bush-Conant File, f. 4.
362. "a very . . . atomic weapon": Wilson (1975), p. 205.
362. Bainbridge contacted Briggs: on the evidence of V. Bush to F. Jewett, April 15, 1941: "The immediate reason being a suggestion from Bainbridge that we send a member of our group to London on the uranium problem." Bush-Conant File, f. 19.
362. "I am . . . my head": Bush (1970), p. 60.
362. "it would . . . present time": VB to FJ, April 15, 1941.
362. "It was . . . scientific problems": Compton (1956), p. 45.
362–363. "disturbed . . . bloodedly evaluate": VB to FJ, April 15, 1941.
363. "fitness . . . task": Compton (1956), p. 46.
363. "Arthur Compton . . . and strong": Libby (1979), p. 91ff.
363. "tallness . . . enormously": ibid., p. 16.
363. "a small . . . place": Compton (1967), p. 31.
364. "probably the . . . of physics": quoted in Pais (1982), p. 414.
364. "Bohr spoke . . . different manner' ": Nielson (1963), p. 27.

PAGE

364. "In 1940 . . . time later": Compton (1967), p. 44.
365. "There followed . . . interested": Compton (1956), p. 46.
365. first NAS report (May 17, 1941): Bush-Conant File, f. 3.
365. "the matter . . . applications multiply": ibid.
365–366. "And only . . . negative": Conant (1970), p. 278.
366. "authoritative and impressive": discussed in FJ to VB, June 6, 1941. Bush-Conant File, f. 4.
366. "a lurking . . . well balanced": ibid.
366. "This uranium . . . doubt": VB to FJ, June 7, 1941. Bush-Conant File, f. 4.
366–367. "We told . . . at Columbia": Seaborg (1976), p. 42.
367. "*to crush* . . . against England": Hitler directive #21, "Operation Barbarossa," Dec. 18, 1940, quoted in Churchill (1949), p. 589.
367. "What worried . . . to priorities": Conant (1970), p. 278ff. Conant (1943), p. 5, confirms this recollection.
368. Briggs learned from Lawrence: a letter dated July 10, 1941, according to Conant (1943), p. 13.
368. "In the . . . no money": Eugene T. Booth, personal communication.
368. "The government's . . . war program": Compton (1956), p. 49.
368. "More significant . . . entirely feasible": Conant (1970), p. 280.
368. eight of twenty-four physicists: Conant (1943), p. 20.
368. "In essence . . . 'draft report' ": ibid.
368. MAUD Report: given in full in Gowing (1946), p. 394ff.
369. "With the . . . in order": Conant (1943), p. 21.
369. "During July . . . uranium program": Conant (1970), p. 279.
369–370. "If each . . . efficiently": Weart and Szilard (1978), p. 138.
370. "Fritz Houtermans . . . brilliant ideas": Frisch (1979), p. 71ff.
370. "that at . . . energy": Bethe (1967), p. 216.
370. "but fell . . . the Nazis": Frisch (1979), p. 72ff.
371. Houtermans report: cf. Irving (1967), p. 84.
371. "Every neutron . . . thermal neutrons": quoted in ibid., p. 85.
371–372. "at worst . . . been defeated": quoted in Clark (1981), p. 126.

PAGE

372. "Although personally . . . Lord Cherwell": Churchill (1950), p. 814.
372. "If Congress . . . receive one": Weart and Szilard (1978), p. 146.
372. "most important . . . and determined": Conant (1943), p. 19.
372. "The minutes . . . and distressed": Oliphant (1982), p. 17.
373. "came to . . . for submarines": quoted in Davis (1968), p. 112.
373. "I'll even . . . in Berkeley": quoted in Childs (1968), p. 315. Childs attributes this wire to Lawrence. Since he was in Berkeley and Oliphant in Washington, I take it to be Oliphant's.
373. "How much . . . complete consideration": quoted in ibid., p. 316ff.
373. Oliphant sees Conant and Bush: cf. Bickel (1980), p. 166. Bickel interviewed Oliphant at length.
373. "gossip . . . subjects": Conant (1943), p. 19.
373. "non-committal . . . of fission": quoted in Gowing (1964), p. 84n.
373–374. "that the . . . serious consideration": quoted in Conant (1943), p. 20.
374. "Certain developments . . . its development": Compton (1956), p. 6.
374. "out of the blue": interview with Edward Teller, Stanford, Calif., June 19, 1982.
374. "I decided . . . bombs": Blumberg and Owens (1976), p. 110.
374–375. "Next Sunday . . . believed me": NOVA (1980), p. 3.
375. Hagiwara lecture: quoted in "Concerning uranium, Tonizo Laboratory, April 43." Document copy and translation in the private collection of P. Wayne Reagan, Kansas City, Mo.
375. Chicago meeting: Conant lists Pegram as a fourth participant. Compton, who believed the meeting to be crucial to his future and who describes it in detail, does not.
375. "It was . . . talk freely": Compton (1956), p. 7.
375. "very vigorous . . . whole field": Conant (1943), p. 21.
375. "Conant was . . . be convinced": Compton (1956), p. 7ff.
376. "I could . . . research programs": Conant (1970), p. 280.
376. "If this . . . do it": Compton (1956), p. 8.
376. "the results . . . been exposed": Conant (1943), p. 22.

PAGE

376–377. "I grew . . . a Russian' ": interview with George Kistiakowsky, Cambridge, Mass., Jan. 15, 1982.
377. "When I . . . have reservations?": Conant (1970), p. 279.
377. "counted . . . significant": ibid., p. 280.
377. Bush memorandum: VB to J. B. Conant, Oct. 9, 1941. Bush-Conant File, f. 4. Quotations describing Bush's meeting with FDR come from this memo.
378. "emphasized to . . . are over": VB to F. Jewett, Nov. 4, 1941. Bush-Conant File, f. 4.
379. Bush to expedite research: cf. VB to FDR, March 9, 1942: "In accordance with your instructions [on October 9] I have since expedited this work in every way possible." Bush-Conant File, f. 13.
379–380. "called . . . hundred pounds": Compton (1956), p. 53ff.
380. Dunning and Booth choosing gaseous diffusion: Booth et al. (1975), p. 1ff.
380. "Our . . . enriched uranium": Eugene T. Booth, personal communication.
380. barrier materials: cf. Cohen et al. (1983), p. 636ff.
380*n*. "One cannot . . . more tons": FP 143, Fermi (1962), p. 99. Herbert Anderson's headnote here confirms the chronology of this incident.
381. "He urged . . . so well": Compton (1956), p. 55.
381. October 21 in Schenectady: Compton (1956), p. 56, says Cambridge, but Hewlett and Anderson (1962), p. 46, referring to the minutes of the meeting, locate it here.
381. "I have . . . deliberation": quoted in Childs (1968), p. 321.
381. Conant scolded Lawrence: ibid., p. 319.
381. "leftwandering activities": quoted in ibid.
381. "Many of . . . of them": USAEC (1954), p. 11.
381–382. "causes and . . . with us": quoted in Childs (1968), p. 319.
382. Oppenheimer debated Lawrence: cf. ibid.
382. "It was . . . direct use": USAEC (1954), p. 11.
382. "I . . . forget it": Smith and Weiner (1980), p. 220.
382. "Kistiakowsky . . . members objected": Compton (1956), p. 56ff.

PAGE

383. "In our . . . grave responsibility": quoted in Childs (1968), p. 321.
383. "the destructiveness . . . of inertia": Compton (1956), p. 57.
383. "No help . . . helpful suggestions": ibid., p. 58.
383. "It was . . . atomic bomb": quoted in Irving (1967), p. 93.
383–384. "he was . . . his trip": E. Heisenberg (1984), p. 77ff.
384. "with . . . hospitality": ibid., p. 78.
384. "Being aware . . . this conversation": quoted in Jungk (1958), p. 103ff.
385. "The impression . . . actual events": Rozental (1967), p. 193.
385. "Heisenberg and . . . a standoff": Oppenheimer (1963), III, p. 7.
385. Heisenberg reactor drawing: reported by Hans Bethe in Bernstein (1979), p. 77.
385. "Bohr . . . all else": E. Heisenberg (1984), p. 81.
385. "bond to . . . Bohr's reply": ibid., p. 80.
386. "a state . . . despair": ibid., p. 81.
386. "he had . . . enough spoon": Mott and Peierls (1977), p. 230.
386. third NAS report: Report to the President of the National Academy of Sciences by the Academy Committee on Uranium, Nov. 6, 1941. Bush-Conant File, f. 18.
386. "The special . . . *with U235"*: ibid., p. 1.
386. *"a fission* . . . can be": ibid.
386. *"The mass* . . . fast neutrons": ibid., p. 2.
386. "may be . . . itself": ibid., p. 3.
386. "approaching . . . test": ibid., p. 4.
386. "in . . . four years": ibid.
386–387. "The possibility . . . this program": ibid., p. 6.
387. "more . . . of accomplishment": Compton (1956), p. 61.
387. "leaving Briggs . . . Ernest Lawrence": VB to FJ, Nov. 4, 1941. Bush-Conant File, f. 4.
388. "Jan 19 . . . FDR": Bush-Conant File, f. 13.
388. "The meeting . . . more firmly": Compton (1956), p. 70.
389. "a worthy . . . said Conant": ibid., p. 70ff.
389. "the construction . . . secret project": Conant (1970), p. 282.
389. December 7, 1941: I rely primarily on Prange (1982) for this summary reconstruction, but cf. also Murukami (1982), Coffey (1970) and Toland (1970).
389. "must be . . . ever seen": quoted in Prange (1982), p. 500.
390. "Negotiations with . . . disclose intent": quoted in ibid., p. 402.
390. "a defense . . . Philippines": quoted in ibid., p. 403.
390. "This dispatch . . . tasks assigned": quoted in ibid., p. 406.
390. "more . . . phrasing": quoted in ibid., p. 409.
391. "Well . . . it": quoted in ibid., p. 501.
393. Nagasaki torpedoes: cf. ibid., p. 323.

Chapter 13: The New World

395. "Szilard at . . . customers": Fermi (1962), p. 1003.
395. thirty tons of graphite: ibid., p. 546.
395. "Much of . . . with *heap":* Segrè (1970), p. 116.
396. "We . . . oxide": Fermi (1962), p. 1002.
396. The cans: cf. FP 150, ibid., p. 128.
396. "This structure . . . as possible": ibid.
396. "We were . . . exhausting work": Wilson (1975), p. 86.
396–397. "We . . . four pounds": Fermi (1962), p. 1002.
397. "Fermi tried . . . and precision": Wilson (1975), p. 87.
397. *k:* cf. FP 149, Fermi (1962), p. 120.
397. "Now that . . . Pearl Harbor": ibid., p. 1002ff.
398. "the atmosphere . . . optimism reigned": Conant (1943), II, p. 2.
398. the next day: i.e., Dec. 19, 1941. Compton gives Dec. 20 but cf. Hewlett and Anderson (1962), p. 53.
398. "On the . . . 18 months": AHC to VB et al., Dec. 20, 1941, p. 2. Bush-Conant File, folder 5.
398–399. "This figure . . . per year": Compton (1956), p. 72.
399. Anderson scouting locations: cf. his letter to Szilard, Jan. 21, 1942, Szilard Papers.
399. "egg-boiling": ibid.
399. "Each was . . . for Chicago": Compton (1956), p. 80.
399. "We will . . . war": quoted in Compton, "Operation of the Metallurgical Project," memorandum, July 28, 1944. Bush-Conant File, f. 20a.
399–400. "Finally, wearied . . . to Chicago": Compton (1956), p. 81.
400. "had come . . . moving again": L. Fermi (1954), p. 174.

PAGE

400. "was unhappy . . . quite efficiently": ibid., p. 169.
400. THANK YOU . . . ORGANIZATION: AHC to LS, Jan. 25, 1942. Szilard Papers.
401. uranium press: Libby (1979), p. 70. Libby's chronology here is garbled, however.
401. "There are . . . ordered one": L. Fermi (1954), p. 186. LF believed the pile was canned to exclude the air, but cf. FP 151, Fermi (1962), p. 137: "Particular care was taken to eliminate as much as possible the moisture."
401. "required soldering . . . the job": Wattenberg (1982), p. 23.
401. "To insure . . . their heads": L. Fermi (1954), p. 186.
401–402. "Like the . . . winter meant": Churchill (1950), p. 536.
402. "well-fed . . . fighting": Guderian, quoted in Shirer (1960), p. 862.
402. "The winter . . . certain": Churchill (1950), p. 537.
402. "The work . . . near future": quoted in Irving (1967), p. 94.
402. "Experimental Luncheon": cf. Goudsmit (1947), p. 170.
403. "too busy . . . moment": quoted in ibid., p. 171.
403. "Pure uranium-235 . . . colossal force": quoted in Irving (1967), p. 99.
403. "it would . . . to detonate": quoted in ibid., p. 100.
403. "The first . . . be done": quoted in Groves (1962), p. 335. Note that this is testimony obtained surreptitiously by bugging while its subjects, who have claimed it was mistranslated and misinterpreted, were prisoners of war. To the extent that it is reliable it is far more candid than published statements, however.
403. "In the . . . its importance": Speer (1970), p. 225.
404. "The word . . . reaction": quoted in Irving (1967), p. 108.
404. "As . . . a pineapple": quoted in ibid., p. 109.
404. "His answer . . . the war": Speer (1970), p. 226.
404–405. "Hitler had . . . see it": ibid., p. 227.
405. "on the . . . propelling machinery": ibid.
405. "In the . . . prime movers": Heisenberg (1947), p. 214.
405. "We may . . . finding out": VB to FDR, March 9, 1942. Bush-Conant File, f. 13.
405. March 9 report: "Report to the President, status of tubealloy development" (n.d.). Bush-Conant File, f. 13.
406. "I think . . . the essence": FDR to VB, March 11, 1942. Bush-Conant File, f. 13.
406. "While all . . . the other": JBC to VB, May 14, 1942. Bush-Conant File, f. 5.
406. Conant reviewed the evidence: cf. ibid.
407. "if the . . . of machinery": ibid.
407. Seaborg to Chicago: chronology and details of this section follow Seaborg (1977).
407. "This day . . . Project": ibid., p. 2.
407. 250 ppm: Seaborg (1958), p. 16.
408. "We conceived . . . were employed": ibid., p. 8.
408. "I . . . precipitation": Seaborg (1977), p. 9.
408. "Sometimes I . . . right now": ibid., p. 42.
409. "We looked . . . concrete bench": ibid., p. 56.
409. "I always . . . fit": ibid., p. 112.
410. "It was . . . balance": Seaborg (1958), p. 38.
410. "the fellows . . . Hall": Seaborg (1977), p. 66.
410. "But to . . . the north": ibid., p. 68.
410. "Our witnesses . . . stay": ibid., p. 70.
410. "it was . . . call": ibid., p. 75.
410. "I do . . . a whole": G. Breit to L. Briggs, May 18, 1942. Bush-Conant File, f. 5.
411. "a prerequisite . . . bomb": Seaborg (1977), p. 75.
411. "because . . . use helium": ibid., p. 91.
411. "a water . . . time": Weart and Szilard (1978), p. 157.
412. "Compton repeated . . . of 1944": Seaborg (1977), p. 86ff.
412–413. "Compton opened . . . people present": ibid., p. 93ff.
413. "Stated in . . . lines": Weart and Szilard (1978), p. 156.
413. "In 1939 . . . were eliminated": ibid., p. 152.
413. "The UNH . . . a minimum": Seaborg (1977), p. 148.
414. "facetiously . . . attention": interview with Glenn Seaborg, Berkeley, Calif., June 22, 1982.
414–415. "Perhaps today . . . of man": Seaborg (1977), p. 192ff.
415. "a holiday . . . hue": ibid., p. 193.

PAGE

415. "luminaries": Smith and Weiner (1980), p. 227.
415. "I considered . . . practical way": quoted in Bernstein (1980), p. 70.
415. "After the . . . effort": quoted in ibid., p. 61.
416. "The essential . . . hands full": Smith and Weiner (1980), p. 226.
416. Kiddygram: cf. ibid.
416. "our best . . . country": quoted in Bernstein (1980), p. 55.
416. early July: on July 9, 1942, Teller told Seaborg he was leaving for a month in Berkeley. Seaborg (1977), p. 111.
416. "He had . . . probably work": quoted in Bernstein (1980), p. 71.
416–417. "We were . . . do it": Teller (1962), p. 38.
417. "We had . . . hydrogen bomb": quoted in Bernstein (1980), p. 72ff.
417. "in the . . . been tragic": "Remarks by Raymond T. Birge," May 5, 1964, p. 5ff. JRO Papers, Box 248.
417. "The theory . . . do much": interview with Hans Bethe, Ithaca, N.Y., Sept. 12, 1982.
417. "My wife . . . doing it": quoted in Bernstein (1980), p. 73.
418. 35,000 eV/400 million degrees: Hawkins (1947), p. 14.
418. 85,000 tons: ibid., p. 15.
418. 500 atomic bombs: based on Bush's March estimate of 2 KT per "unit."
418. "I didn't . . . know more": Bethe interview, Sept. 12, 1982.
419. "I'll never . . . their calculations": Compton (1956), p. 127ff.
419. "I very . . . my arguments": Bethe interview, Sept. 12, 1982.
419. "It was . . . common sense": Hawkins (1947), p.15.
419. "My theories . . . conclusion": Teller (1962), p. 39.
420. "Konopinski . . . guess": ibid.
420. lithium deuteride: cf. Teller to Oppenheimer, Sept. 5, 1942, JRO Papers, Box 71: "In connection with these reactions it occurred to me that our Lithium Deuterite estimate which we made at Berkeley might be wrong . . . But even so I agree that Lithium Hydride will probably not be possible without some change in isotopic composition."
420. "We were . . . unforgettable": Bethe (1968), p. 398.
420. a major effort: cf. Hawkins (1947), p. 2.

PAGE

420. "Fast neutron . . . work": Seaborg (1977), pp. 269–271.
420. Cincinnati/Tennessee: ibid.
420. Conant notes: handwritten on yellow legal pad paper, headed "August 26, 1942. Status of the Bomb." Bush-Conant File, f. 14a.
421. Executive Committee report: "Status of Atomic Fission Project," (n.d.), Bush-Conant File, f. 12.
421. "the physicists . . . report": via Harvey Bundy. Cf. VB, "Memorandum for Mr. Bundy," July 29, 1942. OSRD S-1 Bush Report March 1942 #58.
422. "Classical engineers . . . everyone": Libby (1979), p. 90ff.
422. Wilson's meeting: Leona (Woods) Libby's is the more detailed recollection and squares with the timing of the Stone & Webster appointment in June, following which the engineering firm conducted preliminary studies during the summer. Compton apparently confuses the meeting Wilson called with the June meeting Seaborg describes when the decision to turn Pu production over to industry was first announced—also a rowdy meeting. Cf. Libby (1979), p. 90ff; Compton (1956), p. 108ff; Seaborg (1977), p. 93ff.
422. "We (some . . . & Webster": Libby (1979), p. 91.
422–423. "When Compton . . . and disbanded": ibid., p. 91ff.
423. Szilard memorandum: Weart and Szilard (1978), p. 153ff, and draft "Memorandum" dated Sept. 19, 1942, Szilard Papers.
423. "In talking . . . instrument": LS, draft "Memorandum," p. 5.
423. "I have . . . OSRD": ibid., p. 4.
423. "The situation . . . our work": Weart and Szilard (1978), p. 155.
423. "There is . . . be functional": ibid., p. 156.
423. "If we . . . it up": quoted in ibid., p. 147.
423–424. "We may . . . responsibility lies": ibid., p. 159ff.
424. "From my . . . war efforts": VB to Harvey Bundy, Aug. 29, 1942. OSRD S-1 Bush Report March 1942 #58, p. 4.
424–425. "On the . . . Oh": Groves (1948), p. 15.
425. "to take . . . project": "Memorandum

PAGE

for the Chief of Engineers," Sept. 17, 1942. MD I/I/f.25b.

425. "I thought . . . colonel": Groves (1962), p. 5.

426. Groves' father: cf. Groves (n.d.), "The Army As I Saw It."

426. "Entering West . . . I knew": ibid., p. 103.

426. "A . . . wolf": quoted in Davis (1968), p. 244.

426. "the biggest . . . of understanding": quoted in Goodchild (1980), p. 56ff.

426. "I was . . . horrified": Groves (1962), p. 19.

427. "I told . . . the soup": quoted in ibid., p. 20.

427. "His reaction . . . his wishes": ibid., p. 22.

427. "We had . . . a year": ibid., p. 23.

428. "You made . . . start moving": quoted in Groueff (1967), p. 15n. Groueff interviewed Groves at length.

428. $k = 0.995$: Hewlett and Anderson (1962), p. 70.

428. "I remember . . . beach": L. Fermi (1954), p. 191.

428. "in frigid . . . promontory": Libby (1979), p. 2.

428. "One evening . . . little kids": ibid., p. 4.

428. "As he . . . he was": ibid., p. 1.

429. "I was . . . we wanted": R. Sachs (1984), p. 33.

429. "At each . . . 'metallurgists' ": L. Fermi (1954), p. 176.

429. "one of . . . his wife": Compton (1956), p. 207.

429. "close to 1.04": Fermi (1962), p. 207.

429. "For the . . . questions asked": Wilson (1975), p. 91.

429. 1 percent improvement in k: Fermi (1962), p. 212.

430. Zinn preparations: cf. FP 181 ibid.; Wattenberg (1982); Wilson (1975), p. 108ff.

431. "The War . . . build both": Seaborg (177), p. 284ff.

431. "I left . . . be impossible": Groves (1962), p. 41.

431. "Its reasons . . . design data": ibid., p. 48.

431–432. "if we . . . casualties": ibid., p. 49.

432. "dreadful decision": Seaborg (1977), p. 343.

432. "We did . . . be intolerable": Compton (1956), p. 137.

432. "most significant . . . fission occurs": ibid., p. 136ff.

PAGE

432. delayed neutrons: Roberts et al. (1939b).

432. "The only . . . myself": Compton (1956), p. 138.

433. building CP-1: cf. Allardice and Trapnell (1955); Compton (1956), p. 132ff; FP 181, Fermi (1962); L. Fermi (1954), p. 176ff; Groueff (1967), p. 54ff; Libby (1979), p. 118ff and passim; R. Sachs (1984), pp. 32ff and 281ff; Seaborg (1977), p. 388ff; Segrè (1970), p. 120ff; Wigner (1967), p. 228ff; Wilson (1975), pp. 91ff and 108ff.

433. "Gus Knuth . . . on hand": Wilson (1975), p. 92.

433. number of layers: 57 layers/17 days/2 shifts = 1.7 per shift.

433–434. "A simple . . . and myself": Wilson (1975), p. 93.

434. "Each day . . . following shifts": Fermi (1962), p. 268.

434. pile countdown: these numbers charted in ibid., FP 181, p. 275.

434–435. "We tried . . . the business": quoted in Wilson (1975), p. 94.

436. "That night . . . in place": Fermi (1962), p. 269.

436. "I will . . . leisurely": ibid., p. 270.

436–437. "The next . . . the batter": Libby (1979), p. 120.

437. "Back we . . . received heat": ibid.

438. "several of . . . the material": Wattenberg (1982), p. 30.

438. "Fermi instructed . . . supposed to": ibid., p. 31.

438. "Again the . . . control rod": ibid.

439. "After the . . . and locked": ibid.

439. "This time . . . level off": quoted in ibid., p. 32.

440. "at first . . . about it": Wilson (1975), p. 95.

440. Fermi told tech council: cf. Seaborg (1977), p. 394. Seaborg gives $k = 1.006$, presumably a typographical error; cf. FP 181, Fermi (1962), p. 276.

440. "Then everyone . . . ZIP in!": Wilson (1975), p. 95.

440. "Nothing very . . . cannot foresee": Wigner (1967), p. 240.

442. "We each . . . except Wigner": Wattenberg (1982), p. 32.

442. "bursting . . . news": Seaborg (1977), p. 390.

442. "in my . . . University": Conant (1970), p. 290.

PAGE

442. "Jim . . . and happy": Compton (1956), p. 144.
442. "There was . . . of mankind": Weart and Szilard (1978), p. 146.

Chapter 14: Physics and Desert Country

443. "massive . . . work": Bethe (1968), p. 396.
443. "[Oppenheimer] . . . choir boy": Chevalier (1965), p. 11.
444. "He was . . . loved him": Dorothy McKibben, quoted in Else (1980), p. 9.
444. "He was . . . unspoken wishes": Chevalier (1965), p. 21.
444. "painful but . . . my voice": quoted in Davis (1968), p. 129.
444. "sometimes appeared . . . most sensitive": Segrè (1970), p. 134.
444–445. "Robert could . . . like it": quoted in Davis (1968), p. 103.
445. "But it . . . or something": Smith and Weiner (1980), p. 135.
445. "the loneliest . . . world": quoted in ibid., p. 145.
445. "My friends . . . to change": USAEC (1954), p. 8.
445. "I had . . . my students": ibid.
445. "very grave . . . those days": interview with Philip Morrison, Cambridge, Mass., Jan. 1982.
445–446. "And through . . . the community": USAEC (1954), p. 8.
446. "In the . . . and country": ibid.
446. "I never . . . to me": ibid., p. 10.
446. "Dr. [Stewart] . . . it was": ibid.
446. "I went . . . the world": ibid., p. 9.
447. "created the . . . nuclear physics": Bethe (1968), p. 396.
447. "He began . . . to try": quoted in Davis (1968), p. 79.
447. JRO meeting Groves: cf. LRG to JRO, Sept. 27, 1960. JRO Papers, Box 36.
447–448. "I became . . . no consideration": USAEC (1954), p. 12.
448. "a military . . . as officers": ibid.
448. "original . . . in Berkeley": LLG to JRO, Sept. 27, 1960.
448. "the work . . . pace": Groves (1962), p. 60.
448. "Outside the . . . Oppenheimer": ibid., p. 62.
448. "It was . . . a theorist": Bethe (1968), p. 399.
448. "snag . . . any means": Groves (1962), p. 63.

PAGE

448–449. "He's a . . . about sports": interview, March 8, 1946. Szilard Papers.
449. "After much . . . the task": Groves (1962), p. 62ff.
449. "by default . . . bad name": quoted in Davis (1968), p. 159.
449. "it was . . . astonished": Else (1980), p. 11.
449. October 15 and 19: cf. LLG to JRO, Sept. 27, 1960.
449. "For this . . . hands on": Smith and Weiner (1980), p. 231.
449. Bethe, Segrè et al.: Kunetka (1982), p. 48.
449–450. "so that . . . conditions": Groves (1962), p. 64.
450. Groves' criteria: cf. Badash (1980), p. 3ff.
450. "a delightful . . . Utah": ibid., p. 4.
450. "a lovely . . . satisfactory": Smith and Weiner (1980), p. 236.
450. "considerable . . . usable site": Badash (1980), p. 14ff.
450. "as though . . . directly there": ibid., p. 5.
450. "The school . . . its stream": Church (1960), p. 4.
451. "beautiful . . . country": Segrè (1970), p. 135.
451. "hot and . . . or moisture": L. Fermi (1954), p. 204.
451. "I remember . . . the place' ": Badash (1980), p. 15.
451. "My two . . . be combined": quoted in Royal (1969), p. 49; cf. also Brode (1960), first page of Introduction. I merge these two versions of JRO's statement; the sense is the same and the exact remark is variously attested.
451. "Nobody could . . . go crazy": quoted in Davis (1968), p. 163.
451. Corps of Engineers' appraisal: MED 319.1.
451–452. "What we . . . accelerators": Badash (1980), p. 30.
452. "The prospect . . . Los Alamos": USAEC (1954), p. 12ff.
452. "Oppenheimer . . . radar": Moyers (1984).
452. "the culmination . . . physics": quoted by JRO in Smith and Weiner (1980), p. 250.
452. "To me . . . consequence": ibid.
453. "Laboratory . . . our hi-jinks": ibid., p. 243ff.
453. "Oppenheimer's . . . a teletype": Badash (1980), p. 10.

PAGE

453. "a very . . . to come": ET to JRO, March 6, 1943. JRO Papers, Box 71.
453. Teller's prospectus: referred to in ET to JRO, Jan. 4, 1943 (misdated 1942). JRO Papers, Box 71.
453. "mental love . . . in conversation": quoted in Coughlan (1963), p. 90.
453. "fundamentally . . . somewhat introverted": quoted in Blumberg and Owens (1976), p. 77.
453–454. Alvarez disagrees: personal communication.
454. "scientific autonomy . . . our work": Smith and Weiner (1980), p. 247ff.
455. "Several of . . . camps": L. Fermi (1954), p. 201.
455. Vemork raid: cf. Haukelid (1954); Irving (1967); Jones (1967).
455. "one of . . . be repeated": Jones (1967), p. 1422.
455. "Here lay . . . Europe": Haukelid (1954), p. 71.
455. "one of . . . mountains": ibid., p. 73.
456. "Halfway down . . . and rivers": ibid., p. 92ff.
456. "It was . . . of sentries": ibid., p. 94.
456. "We were . . . grenades": ibid., p. 95.
456. "the thin . . . Europe": ibid.
457. "but an . . . to do?": ibid., p. 98.
457. "the best . . . seen": quoted in Irving (1967), p. 149.
457. Japan: cf. Pacific War Research Society (1972), p. 27ff, and Shapley (1978).
457. "The study . . . field": quoted in PWRS (1972), p. 26.
458. "The best . . . the meeting": ibid., p. 35.
459. "This was . . . Engineers": Badash (1980), p. 31.
460. "about thirty persons": Segrè (1970), p. 135.
460. *Los Alamos Primer:* Condon (1943). Designated LA-1.
460–461. "The object . . . nuclear fission": ibid., p. 1.
461. "Since only . . . active material": ibid., p. 2.
461. 7 percent U235: Condon says at least tenfold; 1/140th × 10 = 7%. Condon (1943), p. 5.
461. "to make . . . possible": ibid.
461. "the gadget": ibid., p. 7.
461. "severe . . . effects": ibid., p. 9.
461. "Since . . . is possible": ibid., p. 10.
461. "The reaction . . . gadget": ibid., p. 11.
461. "as the . . . [core]": ibid., p. 13.

PAGE

462. "An explosion . . . distance": ibid., p. 16.
462. "When the . . . break": ibid., p. 18.
462. "is . . . at present": ibid., p. 21.
463. illustration: Condon's drawing, ibid.
463. "The highest . . . 10 tons": ibid.
463. "If explosive . . . sphere": ibid., p. 22.
463. illustration: Condon's drawing, ibid.
464. "All autocatalytic . . . needed": ibid., p. 24.
464. "relatively . . . physics": Hans Bethe OHI, AIP, p. 59.
465. "If there . . . ceremony": Badash (1980), p. 31ff.
465. April conference plans: cf. Hawkins (1947), p. 16ff.
466. Neddermeyer's thoughts: as reported to Davis (1968), p. 170ff.
466. "I remember . . . implosion": quoted in ibid., p. 171.
466. "expands . . . sixteenfold": Condon (1943), p. 15.
466. "At this . . . hand": quoted in Davis (1968), p. 171.
467. "The gun . . . better still": quoted in ibid., p. 172.
467. "At a . . . of assembly": Hawkins (1947), p. 23.
467. "Neddermeyer . . . and Bethe": quoted in Davis (1968), p. 173.
467. "Nobody . . . seriously": Badash (1980), p. 34.
467. "This will . . . into": quoted in Davis (1968), p. 173.
468. "After he . . . surprised": quoted in ibid., p. 182.
468. Condon and *The Tempest:* cf. Smith and Weiner (1980), p. 252.
468. the bombing of Hamburg: cf. Kennett (1982), Middlebrook (1980), Overy (1980).
469. "But when . . . way through": quoted in Jones (1966), p. 80ff.
469. "the targets . . . bombing": quoted in Kennett (1982), p. 128.
469. "that although . . . night bombing": Churchill (1950), p. 279.
470. Headlines proclaiming raids: for a discussion of this point cf. Hopkins (1966), p. 461ff.
470. "It has . . . workers": quoted in Kennett (1982), p. 129.
470. "a bomber . . . endure": quoted in ibid., p. 130.
471. "INFORMATION . . . HAMBURG": quoted in Middlebrook (1980), p. 95.
472. Operation Gomorrah: I rely here on Middlebrook (1980).

PAGE

473. "It was . . . night": quoted in ibid., p. 253.
473. "Most of . . . them all": quoted in ibid., p. 244.
473. "The burning . . . like again": quoted in ibid.
473. "Then a . . . of fire": quoted in ibid., p. 259.
474. "Mother wrapped . . . the doorway": quoted in ibid., p. 264.
474. "We came . . . knees screaming": quoted in ibid., p. 266ff.
474. "Four-storey . . . the pavement": quoted in ibid., p. 276.
475. two million Soviet soldiers: Elliot (1972), p. 48. Elliot puts total Soviet military POWs at 5 million and POW deaths at 3 million; I use his number here of those enclosed in occupied Russia, of which he writes: "Total deprivation of entire enclosed populations . . . does not exist elsewhere in human history." The other 3 million were treated with more customary brutality.
475–476. mammalian reflex: cf. Kruuk (1972).
476. "We must . . . we've got": quoted in Hopkins (1966), p. 464.
476. Lewis committee findings: cf. Hawkins (1947), p. 24.
476. "In this . . . for U^{235}": ibid., p. 71.
477. "I guess . . . why?": interview with Glenn Seaborg, Berkeley, Calif., June 22, 1982.
477. "Then I . . . suitcase": ibid.
477. "Of course not": Groves (1962), p. 160.
477. "all his . . . the Navy": Joseph Hirschfelder, quoted in Badash (1980), p. 82.
477. "within a . . . the job": Groves (1962), p. 160.
477*n*. Tuve reassignment: cf. V. Bush to MT, Aug.14, 1941. Bush-Conant File, f. 4.
478. "As a . . . externally": Ramsey (1946), p. 6ff.
478. B-29: cf. Birdsall (1980). The first service-test model flew June 27, 1943 (ibid., p. 18).
478. "On August . . . subsequent tests": Ramsey (1946), p. 7.
479. "At that . . . gun method": Badash (1980), p. 17.
479. "Those tests . . . practical method": ibid.
479. "It stinks": quoted in Davis (1968), p. 216.

PAGE

479. "With everyone . . . the beer": quoted in ibid.
480. "a simple . . . sophisticated": quoted in ibid., p. 217.
480. "Johnny was . . . previously discussed": quoted in Blumberg and Owens (1976), p. 455.
480. JvN and ET to JRO: the official record says "autumn." I conjecture October because the governing board met Oct. 28, 1943. Hawkins (1947), p. 76.
481. "In order . . . 1943": Ramsey (1946), p. 8ff.
481. "Professor Bohr . . . micro-slide": Rozental (1967), p. 192 plate.
481. "The letter . . . help": ibid., p. 193ff.
481. "if I . . . refuge here": quoted in ibid., p. 194.
482. 3.6 million Germans: Yahil (1969), p. 118.
482. "Danish statesmen . . . government": ibid., p. 200ff.
482–483. Margrethe Bohr remembers: cf. Moore (1966), p. 302.
483. "We had . . . small bag": quoted in ibid., p. 303.
483. Bohr appeals to Swedish government: Flender (1963), p. 76. Flender interviewed Bohr at length; his account is garbled, however.
483. 284 people: Yahil (1969), p. 187.
483. Sept. 30, 1943: Yahil (1969), p. 328, puts this meeting "the day after [Bohr's] arrival in Stockholm," i.e., Oct. 1, 1943. But cf. Rozental (1967), p. 168: "on the same evening. . . ."
483. "went to . . . of Sweden": Rozental (1967), p. 169.
483. contacted the Danish ambassador: Yahil (1969), p. 330.
483–484. "The audience . . . operation": Rozental (1967), p. 169.
484. "At the . . . received": quoted in Yahil (1969), p. 219.
484. "The stay . . . in Sweden": Rozental (1967), p. 195.
484. "The Royal . . . as Bohr's": Oppenheimer (1963), III (Los Alamos version), p. 7.
484–485. "The Mosquito . . . conscious again": Rozental (1967), p. 196.
485. "Once in . . . going on": Oppenheimer (1963), III, p. 7.
485. "good first . . . years before": ibid., p. 8.
485. "The work . . . expected": Rozental (1967), p. 196.

PAGE

485. "To Bohr . . . fantastic": Oppenheimer (1963), III, p. 7.

Chapter 15: Different Animals

487. "depends on . . . its mass": Brobeck and Reynolds (1945), p. 4.
487. "When the . . . of metal": ibid., p. 5.
487. 100-microgram sample: ibid., p. 7.
489. "that . . . be assured": EOL to LRG, Aug. 3, 1943. Bush-Conant file, f. 19.
490. "At one . . . Troy ounce": Groves (1962), p. 107.
490. electromagnetic separation buildings: "Pertinent reference data, CEW." Dec. 1, 1944. MED 319.1, p. 3ff.
491. 20,000 workers: W. E. Kelley to E. H. Marsden, Aug. 9, 1943. MED misc., f. 4.
491. 40 kg U235: JRO to LRG, Sept. 25, 1943, p. 3 MED 337.
491. Army engineer's summary: W. E. Kelley to E. H. Marsden, Aug. 9, 1943. MED misc., f.4.
491. one supervisor remembers: interview with Leon Love, Oak Ridge, Tenn., 1975.
491. "moved the . . . they belonged": Groves (1962), p. 105ff.
491–492. "The first . . . shorting": ibid., p. 104ff.
492. Dunning's staff: Cohen (1983), p. 641.
492. "three methods . . . method": quoted in ibid., p. 637ff.
492. "The method . . . stages": Groves (1962), p. 111.
493. "Further . . . be solved": quoted in Cohen (1983), p. 637ff.
495. 2,892 stages: Cave Brown (1977), p. 311.
496. "from that . . . applications": Cohen (1983), p. 643.
496. "The Clinton . . . inoperable": Groves (1962), p. 69ff.
497. real estate appraisal: "Gross Appraisal, Gable Project." Jan. 21, 1943. MED 319.1.
497. Hanford description: ibid.
498. dimensions: Cave Brown (1977), p. 322.
498. 1:4000: Seaborg (1977), p. 548.
498. "With water . . . arise": Compton (1956), p. 170.
498. "the conscience . . . very end": Weart and Szilard (1978), p. 148.
499. "Local storms . . . dust": Libby (1979), p. 167.
499. "The most . . . guns": quoted in Groueff (1967), p. 141.

PAGE

499. "was a . . . morning": Libby (1979), p. 167.
499–500. "work gangs . . . the year": Hewlett and Anderson (1962), p. 216ff.
500. Forty-foot pile building: Hewlett and Anderson (1962), p. 217, give 120 feet; that measurement includes the detached exhaust stack, however. Cf. Libby (1979), p.167, and Hewlett and Anderson (1962), photo following p. 224.
500. "There was . . . animals": Badash (1980), p. 91.
500. "Years later . . . that": Teller (1962), p. 211.
500–501. Soviet research: cf. York (1976), p. 29ff; Alexandrov (1967); Golovin (1967); Szulc (1984).
501. "The advance . . . safety": Golovin (1967), p. 14.
501. "In recent . . . inhabitants": quoted in York (1976), p. 30.
501. "no time . . . bomb": ibid.
502. "So it . . . places": Alexandrov (1967), p. 12.
502. "Even so . . . isotopes": York (1976), p. 31.
502. Groves' anti-Semitism: cf. transcript of Groves interview of March 8, 1946, Szilard Papers: "Only a man with [Szilard's] brass would have pushed through to the President. Take Wigner or Fermi—they're not Jewish—they're quiet, shy, modest, just interested in learning . . . Of course, most of [Szilard's] ideas are bad, but he has so many. . . . And I'm not prejudiced. I don't like certain Jews and I don't like certain well-known characteristics of theirs but I'm not prejudiced."
502. "If the . . . bomb": Smith (1965), p. 27.
503. "There is . . . obsolete": Weart and Szilard (1978), p. 165.
503. TO REMOVE . . . NOW: AHC to LRG, Oct. 26, 1942. MED 201, Leo Szilard.
503. SZILARD . . . MYSELF: AHC to LRG, Oct. 28, 1942. MED 201, Leo Szilard.
503. "enemy alien . . . war": draft Sec. of War to Atty. Gen., Oct. 28, 1942. MED 201, Leo Szilard.
503. Compton to Groves mid-November: Nov. 13, 1942. MED 201, Leo Szilard.
504. "for inventions . . . it": LS to AHC, Dec. 4, 1942. Bush-Conant File, f. 13.
504. Compton to Briggs: cf. AHC to JBC, Jan. 7, 1943. Bush-Conant File, f. 13.

PAGE

504–505. "the basic . . . inventions": LS to AHC, Dec. 29, 1942. MED 072, Szilard patents.
504. $750,000: undated memorandum, "Leo Szilard," p. 3. MED 12, Intelligence and security.
505. "Szilard's case . . . Government": AHC to JBC, Jan. 7, 1943. Bush-Conant File, f. 13.
505. "This is . . . present": LS to AHC, Jan. 13, 1943. MED 072, Szilard patents.
505–506. "It is . . . research": VB to AHC, Jan. 29, 1943. Bush-Conant File, f. 13.
506. "comparable . . . Wigner": AHC to JBC, Feb. 3, 1943. MED 072. Compton notes at the head of this letter that it was never sent but was communicated orally.
506. "The investigation . . . person": LRG to Capt. Calvert, June 12, 1943. MED 201.
506–507. "The surveillance . . . to go": "Memorandum for the officer in charge," June 24, 1943. MED 201.
507. "Age, 35 . . . no hat": ibid.
507. "(Mr. Wigner . . . the Navy": ibid.
508. "failed to . . . pile": RAL to LRG (n.d.), "copy made for Maj. Peterson 8-2-43." MED 072.
508. "You were . . . assurance": LRG to LS, Oct. 9, 1943. MED 201.
508. Dec. 3 Chicago meeting: H. E. Metcalfe, "A memorandum of a conference held at the Chicago Area Office, U.S. Engineers, on 3 December 1943." MED 072.
508. "not to . . . person": LRG to LS, Oct. 8, 1943. MED 072.
508–509. "who at . . . around me": LS to VB, Jan. 14, 1944. Bush-Conant File, f. 13. Reprinted in Weart and Szilard (1978), p. 161ff.
509. "I feel . . . project": VB to LS, Jan. 18, 1944. Bush-Conant File, f. 13.
509. "the only . . . at present": Weart and Szilard (1978), p. 165.
509. "The attitude . . . initiative": ibid., p. 177.
510. "The scientists . . . are out": ibid., p. 178.
510. Fermi and poisoning food: cf. JRO to EF, May 25, 1943. JRO Papers, Box 33.
510. Met Lab worries: cf. A. H. Compton, J. B. Conant, H. Urey, "Radioactive material as a military weapon." MED 319.1, Literature. Appendix IV, p. 7.

PAGE

510. May 1943/before February: May is the month of JRO's letter to Fermi, which mentions the subcommittee; February is the date of a table of biological effects given in ibid., Appendix I, p. 4ff.
511. "I therefore . . . than this": JRO to EF, May 25, 1943. JRO Papers, Box 33.
511. "the Sanscrit . . . or hurt": quoted in Davis (1968), p. 330.
511–512. "Recent reports . . . large companies": HAB and ET to JRO, July 21, 1943. JRO Papers, Box 20.
512. Conant subcommittee report: MED 319.1.
512. Cosmos Club: Irving (1967), p. 166.
512–513. Vemork bombing and ferry sinking: cf. ibid., p. 174ff; Haukelid (1954), p. 149ff.
514. 97.6 to 1.1 percent: Irving (1967), p. 188.
514. "a one-man . . . another": Haukelid (1954), p. 156.
514. "The fact . . . at all": ibid., p. 160.
514. "The answer . . . Greetings": ibid., p. 161.
514–515. "We had . . . place": ibid., p. 163.
515. "As the . . . enough": ibid., p. 167ff.
515. "The timing . . . properly": ibid., p. 163.
515. "Rolf . . . at ten": ibid., p. 165.
515–516. "Armed with . . . took time": ibid., p. 166ff.
516. "The charge . . . side": ibid., p. 167.
516. "Making the . . . disaster": ibid., p. 168.
517. "When one . . . war ended": quoted in Irving (1967), p. 191.
517–518. "Europe was . . . stepchild": quoted in Costello (1981), p. 354.
518. "very uneasy . . . monkeys": Hersey (1942), p. 36.
518. "The other . . . the Japs": Grew (1942), p. 81.
518. "The Japanese . . . and nations": ibid., p. 79.
518. "united . . . totalitarian": ibid., p. 80.
518. "At this . . . to fight": ibid., p. 80ff.
518. " 'Victory or . . . his country": ibid., p. 82.
519. "General . . . hand grenade": quoted in Manchester (1980), p. 183.
519. "A legend . . . or dead": Hersey (1942), p. 36.
519. "in the . . . cause": Grew (1942), p. 80.
519. "At the . . . defeat it": Manchester (1980), p. 240.

PAGE

519–520. "The general . . . isolated cases": Tregaskis (1943), p. 79.
520. "I know . . . fighting Japan": Grew (1942), p. 82.
520. "It seems . . . the enemy": Wolfe (1943), p. 190.
521. unconditional surrender: cf. Churchill (1950), p. 695ff.
521. "We had . . . war effort": quoted in ibid., p. 687.

Chapter 16: Revelations

522. "How would . . . night train": Frisch (1979), p. 145ff.
522. hearse: Clark (1980), p. 154.
523. "I wandered . . . laughter": Frisch (1979), p. 148.
523. "Welcome . . . you?": quoted in ibid., p. 150.
523. Quebec Agreement: for complete text cf. Gowing (1964), p. 439ff.
523. Bohr's luggage: cf. Frisch (1979), p. 169.
523–524. "It was . . . London": quoted in Bernstein (1980), p. 77.
524. "Explosion . . . of pile": MED 337. Cf. also JRO to LRG, Jan. 1, 1944, same file.
524. "At that . . . Bohr": Goudsmit (1947), p. 177.
524. "Bohr at . . . one": Oppenheimer (1963), III (Los Alamos version), p. 10ff.
524. "made . . . hopeful": Oppenheimer (1963), III, p. 11.
524. "He made . . . misgiving": Oppenheimer (1963), III (Los Alamos version), p. 11.
524. "Bohr spoke . . . believe": ibid.
524–525. "In Los . . . from him": quoted in Moore (1966), p. 330.
525. "They . . . bomb": quoted in Nielson (1963), p. 29.
525. "warm . . . relation": FF memorandum headed "Private," April 18, 1945. JRO Papers, Box 34.
526. "We talked . . . Campbell": ibid.
526. "On hearing . . . B outlined": unsigned Bohr memorandum, May 6, 1945. JRO Papers, Box 34.
526. "B met . . . in history": ibid.
526–527. "On this . . . to X": FF memorandum, April 18, 1945.
527. "F also . . . Minister": NB memorandum, May 6, 1945.
527. "I wrote . . . government": FF memorandum, April 18, 1945.

PAGE

528. "Halifax . . . immediately": Rozental (1967), p. 203.
528. "conservative . . . man": Oppenheimer (1963), III (Los Alamos version), p. 8.
528. Anderson memorandum, March 21, 1944: cf. Clark (1980), p. 169—where a portion is quoted confusingly *after* a later memorandum—and Gowing (1964), p. 350ff.
528. "communicating . . . account": quoted in Clark (1980), p. 169.
528. "I do . . . informed": quoted in Gowing (1964), p. 352.
529. "to let . . . work": PK to NB, Oct. 23, 1943. JRO Papers, Box 34.
529. "The Counsellor . . . the occupation": "Conversation between B and Counsellor Zinchenko at the Soviet Embassy in London on April 20th, 1944, at 5 p.m." JRO Papers, Box 34.
529. "We came . . . new prospects": Rozental (1967), p. 203.
529. "One of . . . the war": Snow (1981), p. 112.
529. R. V. Jones: cf. Jones (1966), p. 88ff.
529–530. "When I . . . Roosevelt": ibid., p. 88.
530. "As he . . . politics!": Rozental (1967), p. 204.
530. "We . . . language": quoted in Gowing (1964), p. 355.
530. "downcast": Rozental (1967), p. 204.
530. "It was . . . nuclear energy": quoted in Nielson (1963), p. 29.
530. "I did . . . Street": quoted in Clark (1980), p. 177.
530–531. "In all . . . present time": quoted in Sherwin (1975), p. 108.
531. "He had . . . sad story": Snow (1981), p. 116.
531. "that the . . . their decisions": NB to WC, May 22, 1944. JRO Papers, Box 34.
531. "The way . . . Berlin": Chandler (1970), III, p. 1865.
531–532. "About a . . . memorandum": NB memorandum, May 6, 1945.
532. "It was . . . manual skill": Rozental (1967), p. 205ff.
532. Bohr FDR memorandum: July 3, 1944. JRO Papers, Box 21. Relevant portions of the text of this unpublished document are quoted in NB's "Open Letter to the United Nations" reprinted in Rozental (1967), p. 341.
532. "We are . . . *by war*": quoted in Nielson (1963), p. 29ff. My italics.

PAGE

532. "First of . . . of war": Oppenheimer (1963), III (Los Alamos version) p. 8.
532. "a far . . . warfare": NB memorandum, July 3, 1944.
534. *"It appeared . . . divergencies"*: Rozental (1967), p. 341. My italics.
534. "Much thought . . . confidence": NB memorandum, July 3, 1944.
534. "The prevention . . . acuteness": ibid.
534–535. "[Bohr] . . . the world": Oppenheimer (1963), III (Los Alamos version) p. 9.
535. "What it . . . enlarged upon": quoted in ibid.
535. "Within any . . . openness": Rozental (1967), p. 350.
535. "An open . . . else": ibid.
535. "The very . . . crisis": ibid., p. 351.
536. "The present . . . the others": NB memorandum, July 3, 1944.
536. "I have . . . purpose": NB to FF, July 6, 1944. JRO Papers, Box 34.
536. "on August . . . manner": NB memorandum, May 6, 1945.
536. "was very . . . spirits": Rozental (1967), p. 205.
536. "most kindly . . . entertained": NB memorandum, May 6, 1945.
536–537. "Roosevelt . . . afterwards": Rozental (1967), p. 206ff.
537. "left with . . . Union": unsigned memorandum "Notes on Bohr" dated May 20, 1948, on the stationery of the office of the Director of the Institute for Advanced Study. JRO Papers, Box 21.
537. "It is . . . expectation": Rozental (1967), p. 207.
537. "This was . . . of Bohr": Snow (1981), p. 116.
537. "The suggestion . . . Russians": quoted in Gowing (1964), p. 447.
537–538. "The President . . . at all": quoted in Clark (1981), p. 177.
538. "youthful . . . physicists": Ulam (1976), p. 151.
538–539. "It was . . . major voice": Bethe (1982). Communicated in manuscript; Ms. p. 2.
539. thermonuclear research: cf. Hawkins (1947), p. 24.
539. "That I . . . inevitable": quoted in Bernstein (1980), p. 81.
539. "Bethe was . . . organization": quoted in Blumberg and Owens (1976), p. 129ff.
539. "Throughout the . . . led it": Teller (1983), p. 190ff.

PAGE

539–540. "I believe . . . Oppenheimer": quoted in Blumberg and Owens (1976), p. 129ff.
540. "When Los . . . objectives": Teller (1955), p. 269.
540. the laboratory history confirms: cf. Hawkins (1947), p. 96.
540. "At this . . . dream of": Segrè (1970), p. 137.
541. "The first . . . out there": Badash (1980), p. 17.
541. bullet passing through: cf. Hawkins (1947), p. 131.
541–542. "the first . . . illusions": ibid., p. 77.
542. "Absolutely . . . results": quoted in Bernstein (1980), p. 85.
542. "Everything in . . . everybody": Badash (1980), p. 49.
542. "by 1943 . . . ordnance": ibid.
542. GK won JvN to his view: interview with George Kistiakowsky, Cambridge, Mass., Jan. 15, 1982.
542. "I began . . . at times": Badash (1980), p. 49ff.
542–543. "After a . . . interfering": quoted a Goodchild (1980), p. 112ff.
543. "It seemed . . . effort": Ulam (1976), p. 141.
543. "The sun . . . Madison": ibid., p. 145.
543. "talked to . . . bomb": ibid., p. 148ff.
543. "However . . . to do": Bethe (1982), Ms. p. 2.
543. "[Bethe] . . . novel subjects": quoted in Blumberg and Owens (1976), p. 131.
544. "Both . . . main program": Hawkins (1947), p. 97.
544. "the hydrodynamical . . . magnitude": Ulam (1976), p. 154.
544. "all the . . . work": ibid., p. 154ff.
545. "You have . . . the charge": Kistiakowsky interview, Jan. 15, 1982.
545. 5 percent variation: Hawkins (1947), p. 91.
545–546. "With the . . . bomb": Bethe (1982), Ms. p. 3.
546. "These calculations . . . urgency": JRO to LRG, May 1, 1944. MED 201, Peierls, R.
546. "Oppenheimer . . . was feasible": Teller (1955), p. 269.
546. "But there . . . Teller": quoted in Smith and Weiner (1980), p. 273.
547. Kistiakowsky memorandum: GBK to JRO, June 3, 1944. JRO Papers, Box 43.
547. "I am . . . accept it": quoted in Kunetka (1979), p. 88.

PAGE

548. 2000+ experiments: 2,500; cf. Smith and Weiner (1980), p. 282.
548. "It appears . . . implosion": JRO to LRG, July 18, 1944. Bush-Conant File, f. 3.
549. "The implosion . . . one": Hawkins (1947), p. 82.
549. "The Laboratory . . . at it' ": ibid.
549. "[Oppenheimer] . . . of me": quoted in Goodchild (1980), p. 118.
549. 1,207 employees: "Personnel employed at 'Y' technical area, May 1, 1944." MED 201, Personnel.
549. "it had . . . a barn": interview with Philip Abelson, Washington, D.C., Sept. 17, 1982.
550. the dollar: ibid.
550. thermal diffusion experiments: cf. Abelson (1943).
550. "The apparatus . . . device": ibid., p. 5.
551. "Information . . . future plants": ibid., p. 20.
551. "They were . . . steam": Abelson interview, Sept. 17, 1982.
551–552. Abelson knew Manhattan Project: ibid.
552. "I wanted . . . dagger stuff": telephone interview with Philip Abelson, Oct.16, 1984.
552. BuOrd man: Hewlett and Anderson (1962), p. 168, cite a visit by Deke Parsons to the Navy Yard as the origin of this contact. Abelson remembers no such visit. The official AEC historians apparently found the Parsons version in JRO's memorandum to LRG. Groves was concerned after the war to discredit a Szilard recounting of this story similar to Abelson's version. Abelson remembers quite clearly that he initiated the contact; asked if he deliberately breached compartmentalization, he answers, "I sure as hell did!" Telephone interview, Oct. 16, 1984.
552–553. "Dr. Oppenheimer . . . a whole": USAEC (1954), p. 164ff.
553. "I at . . . investigating": Groves (1962), p. 120.
553. Lewis/Murphree/Tolman conclusions: cf. "Possible utilization of Navy pilot thermal diffusion plant," dated June 3, 1944. Bush-Conant File, f. 3.
553. "Chinese copies": Groves (1962), p. 120.
554. "We are . . . Marines": quoted in Costello (1981), p. 476.

PAGE

555. "Nowhere have . . . was dead": Sherrod (1944), p. 32.
555. Tinian: cf. esp. Hough (1947).
556. "The view . . . cold": Libby (1979), p. 177.
557. "Enrico and . . . reaction": ibid., p. 178ff.
557–558. "We arrived . . . cooling tubes": ibid., p. 179ff.
558. "Something was . . . and down": ibid., p. 180ff.
558. Fermi open-minded: cf. ibid., p. 181.
558. "concerned for . . . even stable": Wheeler (1962), p. 34.
559. "If this . . . was needed": ibid., p. 34ff.
559. "a fundamentally . . . matter": quoted in Hewlett and Anderson (1962), p. 307.
560. Groves' report to Marshall: cf. J. B. Conant handwritten "Notes on history of S-1" dated Jan. 6, 1945. Bush-Conant File, f. 19.
560. "Looks like . . . September": ibid.

Chapter 17: The Evils of This Time

561. Conant to Bush: "Report on visit to Los Alamos—October 18, 1944." Bush-Conant File, f. 3.
561. "what the . . . over": Conant (1970), p. 300.
562. Niels Bohr: NB's influence on VB and JBC can be traced by careful reading. The two administrators knew little or nothing of NB's ideas on Sept. 19, 1944, when they sent their own to HLS: when VB met with Cherwell and FDR on Sept. 22, VB was disturbed that FDR was discussing postwar arrangements without benefit of briefing and gathered, apparently from FDR, that NB wanted the British and the Americans to maintain peace via bilateral postwar monopoly. Between Sept. 22 and 30, however, at least VB must have talked to NB: the memorandum he and JBC sent HLS on that later date contains and endorses all NB's basic ideas. Since Bohr was in the doghouse with FDR at the time, VB and JBC were probably politic not to credit him as their source. Cf. VB/JBC to HLS, Sept. 19, 1944, MED 76, S-1 interim committee scientific panel; VB to JBC, "Memorandum of conference," Sept. 22, 1944, Bush-Conant File, f. 20a; VB to JBC, Sept. 23, 1944, ibid.; VB/JBC to HLS,

PAGE

Sept. 30, 1944, Bush-Conant File, f. 20a.

562. "the progress . . . be secure": VB/JBC to HLS, Sept. 19, 1944.

562. "to a . . . Russia": VB to JBC, Sept. 22, 1944.

562. "In order . . . attempt": VB/JBC to HLS, Sept. 30, 1944.

562. "a robot . . . missile": ibid.

563. "By various . . . enterprise": JBC to VB, Oct. 20, 1944. Bush-Conant File, f. 3.

563–564. "I should . . . importance": USAEC (1954), p. 954ff.

564. Los Alamos: This discussion draws especially on Badash (1980), Brode (1960), L. Fermi (1954), Jette (1977), Libby (1979), Lyon and Evans (1984) and Segrè (1970).

564. "I always . . . rebellion": L. Fermi (1954), p. 231ff.

564–565. "I told . . . come again": quoted in Brode (1960), I, 7.

565. "Then we . . . slope": Badash (1980), p. 61.

565. "Oppie . . . get up": quoted in L. Fermi (1954), p. 227.

565. "That young . . . long way": quoted in ibid., p. 219.

565. "Parties . . . we worked": Brode (1960), X, 5.

565. "Everybody . . . army camp": Else (1980), p. 9.

566. "He offered . . . feet": Brode (1960), VIII, 5.

566. "The main . . . mesa": ibid., X, 7.

566. "I played . . . of view": Badash (1980), p. 61.

566–567. "Of the . . . cloth": Wilson (1975), p. 160.

567. "I don't . . . friendship": Badash (1980), p. 43.

567. "The streams . . . shouting": quoted in Brode (1960), IX, 7.

567–568. "but he . . . science": Segrè (1970), p. 140.

568. "Oh, I . . . wits": quoted in Ulam (1976), p. 165.

568. "sit there . . . made": Badash (1980), p. 81.

568. "nothing . . . breath": Libby (1979), p. 204ff.

568. "Best for . . . Fuchs": Brode (1960), IX, 7.

568. "Remember . . . Park": quoted in Lyon and Evans (1984), p. 31.

569. "from . . . cover": Hans Bethe OHI, AIP, p. 159.

PAGE

569–570. "Jesus . . . fantastic": Jette (1977), p. 84.

570. "Oppenheimer . . . exception": interview with Edward Teller, Stanford, Calif., June 19, 1982.

570. "He knew . . . us": Else (1980), p. 10.

570. "He understood . . . anybody": interview with Hans Bethe, Ithaca, N.Y., Sept. 12, 1982.

570. "a very . . . them": "Seven Springs meeting, 5/63," p. 5. JRO Papers, Box 66.

571. volunteered names to protect his own: cf. Stern and Green (1969), p. 48ff.

571. "On June . . . together": D. M. Ladd to Director FBI, Dec. 17, 1953, p. 9. JRO FBI file, doc. 65.

571. "I wanted . . . somehow": quoted in Goodchild (1980), p. 128.

571. between March and October: between the beginning of planning and the first mention of Trinity I find in the record, JBC to VB, Oct. 18, 1944.

571. "I did . . . Resurrection": JRO to LRG, Oct. 20, 1962. JRO Papers, Box 36.

572. "Bohr was . . . control": Hans Bethe OHI, AIP, p. 62.

572. "That still . . . whatever": JRO to LRG, Oct. 20, 1962.

572. Oppenheimer did not doubt: cf. his famous remark to Truman that he had blood on his hands.

572. healed the split: cf. Dyson (1979), p. 81ff, esp. Kitty Oppenheimer's choice of George Herbert's "The Collar" as "a poem . . . that she found particularly appropriate to describe how Robert had appeared to himself." "The Collar" works complementarities similar to Donne's.

573. "Therefore I . . . peace": Smith and Weiner (1980), p. 156.

574–575. "I was . . . surrounded": interview with Luis Alvarez, Berkeley, Calif., June 22, 1982.

575. "very quickly . . . failed": Bethe interview, Sept. 12, 1982.

576. "too frequently . . . shortcuts": Kistiakowsky (1949a), I-1.

576. "Prior to . . . low": ibid., I-2.

576. "So much . . . implosion": Badash (1980), p. 54.

577. "the greatest . . . molds": interview with George Kistiakowsky, Cambridge, Mass., Jan. 15, 1982.

577. Composition B/Baratol: cf. Kistiakowsky (1949b).

PAGE

577. "We learned . . . mold": Kistiakowsky interview, Jan. 15, 1982.
577. "We were . . . charges": Kistiakowsky (1980), p. 19.
578. initiator: Dr. Louis Brown, DTM, Carnegie Institution of Washington, contributed valuably to this discussion.
578. "some other . . . satisfactory": Condon (1943), p. 19.
579. "I think . . . initiator": Bethe interview, Sept. 12, 1982.
579–580. "This isotope . . . accord": quoted in Trenn (1980), p. 98.
580. screwballs: cf. Groueff (1967), p. 327.
580. Nishina's private belief: cf. Pacific War Research Society (1972) (hereafter PWRS), p. 23.
580. first Nishina/Nobuuji meeting: "Uranium project research meeting," July 2, 1943. Copies of original documents and translations in the private collection of P. Wayne Reagan, Kansas City, Mo.
581. second Nishina/Nobuuji meeting: "Uranium project research meeting," Feb. 2, 1944. P. Wayne Reagan collection.
581. 170 grams: PWRS (1972), p. 48.
581. "Well, don't . . . gas": quoted in ibid., p. 49.
581. third Nishina/Nobuuji meeting: "Uranium project research meeting," Nov. 17, 1944. P Wayne Reagan collection.
582. Nishina's staff had understood: cf. PWRS (1972), p. 41.
582. "The doors . . . doors": Ramsey (1946), p. 126.
583. "When I . . . right": quoted in Marx (1967), p. 98.
584. "I'm satisfied": quoted in Tibbets (1973), p. 51.
584. "You have . . . war": ibid.
584. Air Force chose Wendover: Tibbets has remembered making the choice, but it was determined before his appointment; no doubt he confirmed it. Cf. Capt. Derry to LRG, Aug. 29, 1944. MED 5c, Preparation and movement of personnel and equipment to Tinian.
584. B-29: cf. Birdsall (1980).
584. Sept. 1939 proposal: ibid., p. 2.
585. one Delivery group physicist: David L. Anderson interview, Oberlin, Ohio, 1981.

PAGE

585–586. "safe . . . of two": Groves (1962), p. 286.
586. Tibbets: besides previous references cf. also Thomas and Witts (1977).
586. "It didn't . . . homeland": LeMay (1965), p. 322.
586. "I'll tell . . . fighting": quoted in Powers (1984), p. 60.
587. "I wanted . . . creature": LeMay (1965), p. 14.
587. "truancy . . . mania": ibid., p. 16.
587. "I had . . . activities": ibid., p. 17.
587. "When the . . . penetrate": ibid., p. 30.
587. "General Arnold . . . system": ibid., p. 338.
588. "The city . . . of it": Guillain (1981), p. 174.
588. Hansell target directive: cf. Birdsall (1980), p. 107.
589. "I did . . . surprise": quoted in ibid., p. 144.
589. blockbluster meeting: on Dec. 19, 1944. Cf. Capt. Derry to LRG, Jan. 9, 1945. MED 4, Trinity test.
590. Parsons memorandum: WSP to LRG, Dec. 26, 1944. MED 51, Memos from Parsons (misc).
590–591. "Suddenly there . . . blossom": Guillain (1981), p. 176.
591. "urgent . . . future planning": quoted in Birdsall (1980), p. 131, whose argument I follow here.
591. "LeMay is . . . it": quoted in ibid., p. 143.
591–592. "General Arnold . . . lives": LeMay (1965), p. 347.
592. "another month . . . this": ibid., p. 345.
592. Churchill instigated: cf. Irving (1963), p. 90ff.
592. "I did . . . done": quoted in ibid., p. 92.
593. "The first . . . underground": I was the interviewer. Rhodes et al. (1977), p. 213ff.
593. Iwo Jima: cf. esp. Wheeler (1980).
594. "We would . . . can": quoted in ibid., p. 28.
594. "I am . . . can": quoted in ibid., p. 29.
594. "They meant . . . homeland": Manchester (1980), p. 339.
594. poison gas: cf. Wheeler (1980), p. 13.
595. "The invaders . . . flesh": Manchester (1980), p. 340.
595. "We shall . . . dying!": quoted in Costello (1981), p. 546.
596. "The Japanese . . . fast": LeMay (1965), p. 346; his italics.

PAGE

596. Tokyo raid: cf. *United States Strategic Bombing Survey* (1976) (hereafter *USSBS*); Birdsall (1980); Guillain (1981); Kennett (1982); Overy (1980).
596. "All the . . . kids": LeMay (1965), p. 349.
596. 87.4 percent: *USSBS* #96, p. 105.
596. "No matter . . . killed": LeMay (1965), p. 352; his ellipses.
596. "the entire . . . target": quoted in Kennett (1982), p. 176.
597. "outstanding strike": quoted in Birdsall (1980), p. 180.
597. Arnold informed: LeMay remembers otherwise, but cf. ibid.
597. "You're going . . . seen": quoted in Costello (1981), p. 548.
597. "grim . . . grubby": Brines (1944), p. 292.
597. "We will . . . possible": ibid., p. 9.
597. "American fighting . . . way": ibid., p. 11.
598. "The inhabitants . . . everywhere": Guillain (1981), p. 184.
598–599. "The fire . . . spectacle": ibid., p. 182.
599. "The chief . . . wind": *USSBS* #96, p. 96ff.
599. "the most . . . known": quoted in Birdsall (1980), p. 195.
599. "probably more . . . man": *USSBS* #96, p. 95.
599. CONGRATULATIONS . . . ANYTHING: quoted in Birdsall (1980), p. 196.
600. "Then . . . Literally": LeMay (1965), p. 354.
600. 32 sq. mi.: *USSBS* #96, p. 39.
600. "I consider . . . command": quoted in Overy (1980), p. 100.
600. "In order . . . mud": quoted in Johnson and Jackson (1981), p. 19.
600. 100 gms., etc.: these numbers and dates from M. L. Oliphant to J. Chadwick, Nov. 2, 1944. MED 201, Chadwick, J.
600–601. "This loss . . . management": ibid.
601. "the output . . . expected": MLO to LRG, Nov. 13, 1944. MED 201, Oliphant, M. L.
601. Jan. 1945, data: Brobeck and Reynolds (1945).
601. Conant notes on Jan. 6: "Notes on history of S-1." Bush-Conant File, f. 19.
601. U235 critical mass: Conant cites 13 ± 2 kg in JBC to VB, Oct. 18, 1944; King (1979) cites 15 kg for U235 surrounded by a thick U tamper.

PAGE

602. "on the . . . received": quoted in Hewlett and Anderson (1962), p. 301.
602. Groves' U235 farm: toured on a visit to Oak Ridge in 1975, when the bluffside bunker had been converted to an air-pollution sampling station.
603. 250 ppm: Seaborg (1958), p. 16.
603. "Originally eight . . . concrete": Groves (1962), p. 85.
604. "When the . . . aloft": Libby (1979), p. 174.
604. "The yields . . . 1945": Seaborg (1958), p. 50ff.
605. "the astonishing . . . date": Goldschmidt (1964), p. 35.
605. "the unfortunate . . . [them]": Groves (1962), p. 186.
605. "but I . . . it": ibid., p. 191.
606. "his thorough . . . me": ibid., p. 193.
606. "The ALSOS . . . Paris": Lt. Col. G. R. Eckman to Chief, Military Intelligence Service, Sept. 1, 1944. MED 371.2, Goudsmit mission.
607. "It is . . . form": Goudsmit (1947), p. 70ff.
607. "Washington wanted . . . Union": Pash (1969), p. 191.
608. "We outlined . . . oxide": JL, "Capture of material," draft report, July 10, 1946. MED 7, War Dept. special operations (tab E-F).
608–609. "Many of . . . started": ibid. Note that Groves (1962), p. 237, remembers these paper bags as fruit barrels and invents a two-week plant run in the midst of contending armies to manufacture them. Such is memory; JL's is the eyewitness account, confirmed by his contemporary report JL to LRG, May 5, 1945. MED 7 (tab A-C).
609. "Haigerloch is . . . pile": Pash (1969), p. 206ff.
609–610. Haigerloch pile: cf. Irving (1967), p. 244ff.
610. "The fact . . . Alsos": Pash (1969), p. 157ff.
610. "By successively . . . reactions": Hawkins (1947), p. 229.
610–611. "At that . . . flicker": Frisch (1979), p. 161.
611. "The idea . . . so": ibid., p. 159.
611. Feynman named it: cf. ibid.
611–612. "It was . . . mid-morning": ibid., p. 159ff.
612. "These experiments . . . alone": Hawkins (1947), p. 230.
613. "In 1940 . . . war": LRG to GCM, April 23, 1945. MED 7 (tab E-F).

PAGE

613. "Sunday morning . . . ours": quoted in Smith and Weiner (1980), p. 287.
613-614. "When, three . . . death": ibid., p. 288.
614. "I kept . . . struck!": quoted in Bishop (1974), p. 598.

Chapter 18: Trinity

617. "Stimson told . . . details": Truman (1955), p. 10.
618. "The chief . . . distrust": Stimson and Bundy (1948), p. 544.
618. "assistant President": quoted in Byrnes (1958), p. 155.
618. "Jimmy Byrnes . . . world": Truman (1955), p. 11.
618. "that in . . . war": ibid., p. 87.
618-619. "A small . . . geniality": Joseph Alsop and Robert Kitner, quoted in Mee (1975), p. 2.
619. "a vigorous . . . politics": quoted in ibid.
619. "Had a . . . *sometimes*": Ferrell (1980), p. 39.
619-620. "I freely . . . action": Byrnes (1958), p. 230.
620. "We proposed . . . in": "Memorandum of conference," Dec. 8, 1944. Bush-Conant File, f. 20a.
621. "I told . . . me": "Extract from notes made after a conference with the President, December 31, 1944." MED 24, Memos to file by LRG covering two meetings with the President.
621. "it would . . . S-1": quoted in Sherwin (1975), p. 136.
621. "the fear . . . messages": Truman (1955), p. 72.
621-622. "barbarian invasion . . . affairs": ibid., p. 71.
622. "I ended . . . government' ": ibid., p. 72.
622. "go ahead . . . organization": ibid.
622. "He felt . . . Hell": Charles Bohlen, quoted in Giovannitti and Freed (1965), p. 46. Note Truman's nearly identical language, sans cuss word and imperative, at Truman (1955), p. 77.
622. "In the . . . promised": Truman (1955), p. 77.
622. "He said . . . serious": ibid., p. 79.
623. "I replied . . . like that": ibid., p. 82.
623. "one of . . . House": ibid., p. 85.
623. April 24 message from Stalin: quoted in full in ibid., p. 85ff.
624. Stimson memorandum: "Memo discussed with the President," April 25, 1945. MED 60, S-1 White House.

PAGE

625. "Mr. Truman . . . all": quoted in Giovannitti and Freed (1965), p. 80.
625. "a great . . . project": "Report of meeting with the President," April 25, 1945. MED 24.
625-626. "I listened . . . service": Truman (1955), p. 87.
626. first Target Committee meeting: Groves (1962), p. 268, dates this occasion May 2, 1945, but cf. "Notes on initial meeting of target committee" dated April 27, 1945, from which all indicated quotations following are extracted. MED 5D, Selection of targets.
627. "I had . . . bomb": Groves (1962), p. 267.
628. May 1 Harrison memorandum: Bush-Conant File, f. 20A.
628. "The President . . . shut": quoted in Giovannitti and Freed (1965), p. 54.
629. "and late . . . accepted": ibid.
629. "were . . . death": quoted in Sherwin (1975), p. 170.
629. "when secrecy . . . Commission": HLS to VB, April 4, 1945. Bush-Conant File, f. 20b.
629. "I have . . . scene": Eisenhower (1970), IV, p. 2673ff.
629. "I tried . . . accomplished": quoted in ibid., p. 2696.
630. "The mission . . . 1945": ibid.
630. deaths: from Elliot (1972) except for Holocaust victims; that number from Dawidowicz (1975), p. 544.
630. "We all . . . committee": quoted in Giovannitti and Freed (1965), p. 56.
630. Stimson introducing Byrnes: cf. R. Gordon Arneson, "Memorandum for the files," May 24, 1946. Bush-Conant File, f. 6.
631. "A. Height . . . Program": J. A. Derry and N. F. Ramsey, "Summary of Target Committee meetings on 10 and 11 May 1945." MED 5D.
633. "very frank . . . one": VB to JBC, May 14, 1945. Bush-Conant File, f. 20B.
633. Stimson's agenda: copy (misdated May 12, 1945) with notes in HLS's hand in Bush-Conant File, f. 100.
633. "I . . . said . . . September": VB to JBC, May 14, 1945.
633-634. "Mr. Byrnes . . . test": JBC to VB, May 18, 1945. Bush-Conant File, f. 12.
634. "Mr. Byrnes . . . one": quoted in Giovannitti and Freed (1965), p. 62.

PAGE

634. "This question . . . argument": JBC to VB, May 18, 1945.
634. "Some of . . . matter": ibid.
634. Conant told Byrnes: cf. ibid.
635. "the feeling . . . back": quoted in Giovannitti and Freed (1965), p. 116ff.
635. "the wisdom . . . bombs": Weart and Szilard (1978), p. 182.
635. "many hours . . . nights": quoted in Giovannitti and Freed (1965), p. 115.
635. "The only . . . President": Weart and Szilard (1978), p. 181.
635. "I am . . . history": quoted in Clark (1970), p. 685.
636. "Elated by . . . House": Weart and Szilard (1978), p. 182.
636. "I see . . . City": ibid., p. 183.
636. "We did . . . know": ibid.
636. "I have . . . judgment": ibid., p. 205.
636. Szilard memorandum: although Document 101 in ibid., p. 196ff, is usually cited as the memorandum Byrnes read, his memory of the contents—discussed below—makes it clear that he read the enclosure given as part of Document 102, p. 205ff, which Weart and Szilard describe as an "enclosure to Einstein's letter."
637. "essentially due . . . armaments": ibid., p. 198.
637. "Szilard complained . . . me": Byrnes (1958), p. 284.
638. "When I . . . Russia": Weart and Szilard (1978), p. 183.
638. "He said . . . already?": ibid., p. 184.
638. "Byrnes thought . . . manageable": ibid.
638. May 28 Target Committee meeting: minutes at MED 5D.
639. "for thirty . . . hated": Stimson and Bundy (1948), p. 632.
640. "I am . . . weapons": diary, quoted in Steiner (1974), p. 473.
640. May 30: on the evidence of LRG to Lauris Norstad, May 30, 1945, reporting Stimson's decision "this AM." MED 5B.
640–641. "I was . . . Kyoto": quoted in Giovannitti and Freed (1965), p. 40ff.
641. "The Joint . . . Japan": quoted in Feis (1966), p. 7.
641. "with the . . . lives": quoted in ibid., p. 8.
641. 31,000 casualties: cf. ibid., p. 8ff.
642. Groves had doubted: cf. VB to JBC, May 14, 1945.
642. "I told . . . well": Weart and Szilard (1978), p. 185.

PAGE

642. May 31 Interim Committee meeting: cf. notes at Bush-Conant File, f. 100.
642. "S.1 . . . World Peace": handwritten notes "To the Four," May 31, 1945. Bush-Conant File, f. 100.
643. "a terrible . . . breached": Oppenheimer (1961), p. 11.
643. "As I . . . weapon": the deleted phrase is "and industrialists." Byrnes would not meet with the industrialists until the next day and presumably merges the two meetings in memory. The context is the May 31 meeting. Byrnes (1958), p. 283.
644–645. "Bush and . . . panel": Oppenheimer (1963), III (Los Alamos version), p. 15.
647. question mentioned during morning: according to E. O. Lawrence; cf. Sherwin (1975), p. 207.
647. "[Stimson emphasized] . . . harm": Oppenheimer (1961), p. 12.
647. "You ask . . . know": quoted in Giovannitti and Freed (1965), p. 104.
647–648. "We feared . . . surrender": Byrnes (1958), p. 261.
648. "The President . . . knows": quoted in Feis (1966), p. 47.
648. "number of . . . raids": quoted in Sherwin (1975), p. 207ff.
648. 20,000 deaths: Compton (1956), p. 237.
648. "a city . . . lives": ibid. AHC locates this discussion A.M. P.M. is likelier; much else in his memory of this meeting is misplaced.
649. "We were . . . done": LeMay (1965), p. 384.
649. "secret intelligence . . . intentionally": unsigned memorandum dated June 1, 1945, on War Dept. stationery; Top Secret classification authorized by LRG. MED 12.
649. June 1 Interim Committee meeting: minutes at MED 100.
650. "I concluded . . . bomb"; quoted in Feis (1966), p. 44.
650. "sternly questioned": Stimson and Bundy (1948), p. 632.
650. "I told . . . understood": Stimson's diary, quoted in Giovannitti and Freed (1965), p. 36.
650. "Mr. Byrnes . . . weapon": quoted in ibid., p. 107.
650–651. Byrnes to White House: ibid., p. 109.
651. "I told . . . recommend": quoted in ibid., p. 110.
651. "said that . . . done": quoted in ibid.

PAGE

651. "I was . . . happen": Oppenheimer (1963) III (Los Alamos version), p. 15.
651. "Do you . . . No": MED 19, Bohr, Dr. Niels.
651-652. Trinity: cf. esp. Badash (1980), Bainbridge (1945), Else (1980), Lamont (1965), Szasz (1984), and Wilson (1975).
652. "was one . . . desert": Hawkins (1947), p. 271.
652. "followed unmapped . . . winds": Wilson (1975), p. 210.
652. "almost to . . . 1945": Bainbridge (1945), p. 5.
654. "people were . . . explosion": Else (1980), p. 16.
655. Pu critical mass: cf. King (1979), p. 7.
655. "Most troublesome . . . lot": Kistiakowsky (1980), p. 20.
655. June 27 LRG/JRO/WSP meeting: JRO/WSP to LRG, June 29, 1945. MED 50. Preparations and movement of personnel to Tinian.
656. "on purpose . . . time": quoted in Sherwin (1975), p. 193.
656. "What are . . . propagandist?": quoted in Szasz (1984), p. 65.
656. ANY . . . DAYS: quoted in Groueff (1967), p. 340.
656. July 9: Bainbridge (1945), p. 39.
656. "In some . . . spheres": Kistiakowsky (1980), p. 20.
657. "You don't . . . it": interview with G. B. Kistiakowsky, Cambridge, Mass., Jan. 15, 1982.
657. "The castings . . . charge": Bainbridge (1945), p. 39.
657. "That last . . . life": Badash (1980), p. 46.
657. "Very shortly . . . hysteria": JRO to ER, May 19, 1950. JRO Papers, Box 62.
657. nickel: Bill Jack Rodgers, LANL, personal communication.
657. "beautiful to . . . threatened": Smith (1954), p. 88.
658. "Right in . . . this?": quoted in Szasz (1984), p. 72.
658. "on the . . . insignificant": Bainbridge (1945), p. 39.
659-660. "when you . . . rabbit": Libby (1979), p. 171.
660. "We were . . . way": quoted in Johnson (1970), p. 11.
660. "The [high- . . . mistake": Wilson (1975), p. 185ff.
661. "So of . . . me": Badash (1980), p. 59.

PAGE

661-662. "Everybody at . . . work": Kistiakowsky (1980), p. 21.
662. "a. 1 box . . . bomb": J. A. Derry to Adm. W. S. DeLany, July 17, 1945. MED 50.3, Shipment of special materials (bomb).
662. "We drove . . . dud": Badash (1980), p. 75ff.
663. "His was . . . him": Bush (1970), p. 148.
663. "Sunday morning . . . society": Badash (1980), p. 59.
663. "What about . . . whimsical": quoted in Lamont (1965), p. 184.
663-664. "*Gadget complete* . . . there?": Bainbridge (1945), p. 43.
664. "in less . . . sacrifices": quoted in Szasz (1984), p. 75.
664. JRO climbed tower: Lamont puts this visit at 1600, when JRO was in conference with Hubbard. Lamont (1965), p. 190.
664. "Funny how . . . work": quoted in ibid., p. 193.
664. "I had . . . possible": Groves (1962), p. 296ff.
664. "thoughtless bravado": Wilson (1975), p. 225.
665. "Trying to . . . whiskey": Teller (1979), p. 147.
665. "On the . . . Zero": Wilson (1975), p. 227.
665. "Soon after . . . tower": ibid.
666. "the night . . . seen": Lawrence (1946), p. 5.
666. "It was . . . Oppenheimer": Else (1980).
666. 0200 weather conference: details from Szasz (1984), p. 76ff., who finds them in Hubbard's contemporary journal.
666. "What the . . . weather": quoted in ibid., p. 76.
666. "or . . . you": quoted in ibid., p. 77.
666-667. "Sporadic rain . . . tower": Wilson (1975), p. 228.
667. "But my . . . water": Segrè (1970), p. 146.
667. "Hubbard gave . . . = 0": Wilson (1975), p. 228.
667. "I drove . . . S 10,000": ibid., p. 228ff.
667. "I unlocked . . . 5:09:45 a.m.": ibid., p. 229.
668. "With the . . . unendurable": Lawrence (1946), p. 6.
668. "We were . . . eye": Teller (1962), p. 17.
668. "I wouldn't . . . lotion": Teller (1979), p. 148.

PAGE

668. "It was . . . flash": Lawrence (1946), p. 7.
668. "personally nervous . . . fault": MED 319.1, Trinity test reports (misc.).
668. "only of . . . happened": Groves (1962), p. 296.
668. "groups of . . . point": MED 319.1.
669. "Lord, these . . . heart": quoted in Lamont (1965), p. 226.
669. "The control . . . safe)": GBK to Richard Hewlett (n.d.), JRO Papers, Box 43.
669. "I put . . . point": Teller (1979), p. 148.
669. "Dr. Oppenheimer . . . ahead": quoted in Groves (1962), p. 436.
669. "I watched . . . rise": MED 319.1.
669. "but at . . . excited!)": ibid.
670. "Now the . . . zero": quoted in Giovannitti and Freed (1965), p. 196.
670. "the very . . . distance": Bethe (1964), p. 13.
671. "The shock . . . center": ibid.
671. "because higher . . . maximum": ibid., p. 14ff.
671. "any further . . . seconds": ibid., p. 92ff.
671–672. "The expansion . . . screw": Bainbridge (1945), p. 60.
672. "We were . . . nature": Rabi (1970), p. 138.
672. "was like . . . sunlight": Teller (1962), p. 17.
672. "We had . . . back": quoted in *Los Alamos: beginning of an era 1943–1945* (n.d.) (hereafter *LABE*), p. 52.
672. "Just as . . . surprise": MED 319.1.
673. "it looked . . . seconds": quoted in *LABE,* p. 53.
673. "At the . . . breath-taking": MED 319.1.
673. "The most . . . possible": Segrè (1970), p. 147.
673. "From ten . . . sunrise": quoted in Terkel (1984), p. 512ff.
673–674. "The flash . . . yards": D. R. Inglis, MED 319.1.
674. "Most experiences . . . anybody": quoted in *LABE,* p. 53.
674. "the overcast . . . sunrise": MED 319.1.
674. "the path . . . clouds": ibid.
674. "When the . . . ball": ibid.
674. "About 40 . . . T.N.T.": ibid.
674. "From the . . . measurement": Segrè (1970), p. 147ff.
674. "He was . . . noise": L. Fermi (1954), p. 239.
675. "And so . . . worked": Else (1980).
675. "No one . . . display": Wilson (1975), p. 230.
675. "personally thought . . . it": Groves (1962), p. 439.
675. "I slapped . . . dollars": Badash (1980), p. 60.
675. "It's empty . . . wait": quoted in Lamont (1965), p. 237.
675. "I finished . . . test": Wilson (1975), p. 230.
675. "Our first . . . worried": quoted in Szasz (1984), p. 91.
675. "Naturally, we . . . was": Rabi (1970), p. 138.
676. "We waited . . . another": quoted in Giovannitti and Freed (1965), p. 197.
676. "When it . . . it": Oppenheimer (1946), p. 265.
676. "He was . . . it": Else (1980).
676. "When Farrell . . . you": Groves (1962), p. 298.
676. "the best . . . philosophy": quoted in Davis (1968), p. 184.
676. "My faith . . . restored": quoted in Szasz (1984), p. 89.
677. 21 KT, 18 KT: cf. telephone notes of 7:55 A.M. LRG to Jean O'Leary, July 16, 1945. MED 319.1.
677. 18.6 KT: Bainbridge (1945), p. 67.
677. "For the . . . driven": L. Fermi (1954), p. 238.
677. "You could . . . future": quoted in Szasz (1984), p. 91.
677. "Partially eviscerated . . . permanently": SW to LRG, July 21, 1945. MED 4, Trinity test.
677. Frank Oppenheimer experiment: Bainbridge (1945), p. 48.
678. "He applied . . . reality": quoted in Terkel (1984), p. 513.
678. 0836 PWT: Ethridge (1982), p. 81.

Chapter 19: Tongues of Fire

679. Kirkpatrick reported to Groves: cf. handwritten reports dated March 31, April 11, and May 10, 1945, at MED 5C, Preparation and movement of personnel and equipment to Tinian.
680. "Tests showed . . . carburetors": Tibbets (1946), p. 133.
681. "The performance . . . theater": Ramsey (1946), p. 146.
681. eleven B-29's: Peer DeSilva to John Lansdale, Jr., June 28, 1945. MED 371.2.

PAGE

681. "looked . . . Paradise": quoted in Craven and Cate (1958), V, p. 707.
681. "Tinian is . . . landed": Morrison (1946), p. 177.
682. "The first . . . Tinian": Ramsey (1946), p. 147.
682. "Jimmy . . . Bible": quoted in Messer (1982), p. 6.
682. "a warning . . . capitulate": Stimson and Bundy (1948), p. 621.
682. "from the . . . Department": quoted in Giovannitti and Freed (1965), p. 180.
682–683. "Secretary Byrnes . . . there?": quoted in ibid.
683. "We reviewed . . . knows?": Ferrell (1980), p. 41.
683. "Proposed Program for Japan": cf. Stimson and Bundy (1948), p. 620ff.
684. "the statement . . . Japan": quoted in Giovannitti and Freed (1965), p. 185.
684–685. "The foreign . . . concerned": quoted in Feis (1966), p. 67.
685. "It is . . . homeland": quoted in ibid., p. 68.
685. "terrible political . . . war?": quoted in Giovannitti and Freed (1965), p. 203.
685–686. "Operated on . . . posted": MED 5E, Terminal cables.
686. "Well . . . Leavenworth": quoted in Bundy (1957), p. 57.
686. "The following . . . Emperor": quoted in Giovannitti and Freed (1965), p. 203.
686. "Neither the . . . test": quoted in ibid.
686. a year's supply of ammunition: production, that is, "which is estimated to equal 350 division months of defensive fighting from fixed positions." *Effects of Strategic Bombing* (n.d.), cover memorandum dated July 25, 1945, p. 5. MED 319.2, Misc.
686. "Subject to . . . government": quoted in Feis (1966), p. 81.
686. "all your . . . city": MED 5E.
686. "aware of . . . it": ibid.
686. "If any . . . rapidly": ibid.
687. "the imminence . . . August": ibid.
687. Groves' narrative: cf. Groves (1962), p. 433ff.
687. "tremendously pepped . . . confidence": quoted in Feis (1966), p. 85.
687. October 1: Arnold (1949), p. 564.
687. "In order . . . cities": ibid.
687. fifty-eight cities: Overy (1980), p. 100.
687. "practically identical . . . out": quoted in Wolk (1975), p. 60.

PAGE

687–688. "We regarded . . . lives": quoted in Mosley (1982), p. 337ff.
688. "We'd had . . . dropped": quoted in "Ike on Ike," *Newsweek*, Nov. 11, 1963, p. 108.
688. "Doctor has . . . farm": MED 5E.
688. "The cable . . . problem": "Ike on Ike."
688. "Believe Japs . . . homeland": Ferrell (1980), p. 42.
689. "Operation may . . . 10": MED 5E.
689. "always . . . authority": ibid.
689. "Hiroshima . . . here": ibid.
689. Official Air Force historians: i.e., Craven and Cate (1958), V; cf. p. 710.
689. "First one . . . sound": MED 5E.
690. "As a . . . once": Feis (1966), p. 101.
690. Stalin knew of Trinity: according to a secret U.S. intelligence agency history of the Soviet atomic bomb program reported in Szulc (1984), p. 3.
690. "I casually . . . Japanese": Truman (1955), p. 416.
690. "That . . . far": Oppenheimer (1963), III (Los Alamos version), p. 16.
690–691. "We have . . . useful": Ferrell (1980), p. 42.
691 the historic directive: WAR 37683, MED 5E.
691. "in order . . . possible": ibid.
692. C-54's: cf. J. A. Derry to Adm. W.S. DeLany, Aug. 17, 1945. MED 5C.
692. Potsdam Declaration: cf. Truman (1955), p. 390ff.
692. "We faced . . . Declaration": Byrnes (1947), p. 262.
693. Japanese response: this discussion follows Feis (1966), p. 107ff.
693. "I believe . . . war": quoted in ibid., p. 109ff.
693. "In the . . . weapon": Stimson and Bundy (1948), p. 625.
693. three B-29's: J. A. Derry to Adm. W. S. DeLany, Aug. 17, 1945.
693. *Indianapolis:* cf. esp. Ethridge (1982).
694. "I took . . . more": Hashimoto (1954), p. 224.
695. "Those who . . . drowned": quoted in Ethridge (1982), p. 89.
695. "We . . . men": quoted in ibid.
696. "so sweet . . . life": quoted in ibid., p. 92.
696. "at length . . . tinned)": Hashimoto (1954), p. 226.
696. HIROSHIMA . . . THEM: MED 5B.
696. "My chief . . . face": Stimson and Bundy (1948), p. 632.

PAGE

697. "our obligation . . . use": cf. report at MED 76.
697. "badly . . . equalizer": quoted in Giovannitti and Freed (1965), p. 237.
697. "First of . . . thing": ET to LS, July 2, 1945. MED 201, Leo Szilard.
697. "To avert . . . deliverance": Churchill (1953), p. 639.
698. "It was . . . end": Anscombe (1981), p. 64.
698. "It was . . . people": Moyers (1984).
698. "impatience to . . . ordeal": Feis (1966), p. 120.
698. A JAP BURNS: *Life*, Aug. 13, 1945, p. 34. This issue appeared on Aug. 6, postdated as is customary to extend newsstand life. Luis Alvarez suggested to me this exercise in examining the popular mood.
699. cordite charge: not, as some have written mistakenly, its bullet. Cf. "Check list for loading charge in plane. . . ." MED 5B.
699. precaution prepared at Los Alamos: cf. Hawkins (1947), p. 225.
699. orders to bring bomb back: Craven and Cate (1958), V, p. 716.
699. "With the . . . completed": Ramsey (1946), p. 149.
699. Farrell telexed Groves: Feis (1966), p. 114.
699. August 2: J. A. Derry to Adm. W. S. DeLany, Aug. 7, 1945.
700. one Fat Man for drop test: cf. Ramsey (1946), p. 150.
700. "By August . . . busy": Tibbets (1973), p. 55.
700. Spitzer diary: quoted in Thomas and Witts (1977).
701. "At 1400 . . . 6": Ramsey (1946), p. 151.
701. bomb-loading procedure: cf. Harold S. Gladwin, Jr., to Boeing Service Dept., Eng. Div., Aug. 20, 1945. MED 5B.
701. "an elongated . . . fins": Jacob Beser, quoted in Thomas and Witts (1977), p. 216.
702–703. "This radar . . . altitude": Hawkins (1947), p. 225ff.
703. "The operation . . . it": H. S. Gladwin, Jr., to Boeing Service Dept., Aug. 20, 1945.
703. "Through the . . . paper": quoted in Marx (1967), p. 98ff.
703. "paint that . . . big": quoted in Thomas and Witts (1977), p. 232.

PAGE

704. "What . . . plane?"; quoted in ibid., p. 233.
704. "By dinnertime . . . poker": Tibbets (1946), p. 135.
704. "Final . . . 6": Ramsey (1946), p. 151.
704. "to be . . . enemies": quoted in Thomas and Witts (1977), p. 237.
704. "amid . . . premiere)": Ramsey (1946), p. 151.
705. "It was . . . ready": Tibbets (1946), p. 135.
705. "The B-29 . . . airborne": ibid.
706. course, altitude, etc.: cf. navigator's charts printed as end papers to Marx (1967).
706. cordite loading: cf. "Check list for loading charge in plane. . . ." MED 5B. For times cf. Parson's log at Cave Brown and MacDonald (1977), p. 522ff.
706. "At forty- . . . runs": quoted in Lawrence (1946), p. 220.
707. "The colonel . . . 'George' ": quoted in variant forms in Marx (1967), p. 78, and Lawrence (1946), p. 220.
707. "A chemist's . . . guess": quoted in Marx (1967), p. 106, and Lawrence (1946), p. 220ff.
707. "Attention! . . . puzzle": quoted in Talk of the Town (1946), p. 16.
707. "At 4:30 . . . spell": quoted in Lawrence (1946), p. 220.
708. "After leaving . . . Away": quoted in ibid. and in Marx (1967), p. 135ff.
708. "The bomb . . . Tinian": quoted in Marx (1967), p. 136.
708. "Well . . . now": quoted in Lawrence (1946), p. 221.
708. "Our primary . . . Hiroshima": quoted in ibid.
708. "It's Hiroshima": quoted in Marx (1967), p. 143.
708. "As we . . . target": quoted ibid., p. 157.
709. "Twelve miles . . . plane": Tibbets (1946), p. 136.
709. perfect aiming point: Thomas and Witts (1977), p. 220.
709. "Ferebee had . . . goes": Tibbets (1946), p. 136.
709–710. "The radio . . . lead": ibid.
710. "Fellows . . . history": according to Jacob Beser, quoted in Marx (1967), p. 173.
710. "[It was] . . . plane": quoted in Giovannitti and Freed (1965), p. 250.
710. "I don't . . . mountains": quoted in ibid.

PAGE

711. "If you . . . home": quoted in ibid.
711. "I kept . . . smoke": quoted in Marx (1967), p. 171ff.
711. "That city . . . me": quoted in ibid., p. 174.
711. 8:16:02: cf. The Committee for the Compilation of Materials on Damage Caused by the Atomic Bombs in Hiroshima and Nagasaki (1981)—hereafter cited as Committee—p. 21. All statistics from this source unless otherwise indicated. The official time according to Hiroshima City is 8:15.
711. "It . . . impersonal": Tibbets (1973), p. 55.
711. "If I . . . mind": quoted in Marx (1967), p. 221.
711. Hiroshima: cf. in particular Cave Brown and MacDonald (1977); Committee (1981); Hachiya (1955); Liebow et al. (1949); Liebow (1965); Lifton (1967); NHK (1977); Osada (1982); *USSBS* (1976), X.
712. Hiroshima history: cf. Kosaki (1980).
712–713. "Hiroshima was . . . harbor": Cave Brown and MacDonald (1977), p. 554.
713. "The hour . . . garden": Hachiya (1955), p. 1.
713. "Just as . . . leaves": Osada (1982), p. 8.
713–714. "Shortly after . . . delirium": ibid., p. 305.
714. "Accompanying the . . . explosion": Liebow (1965), p. 68.
714. "Because the . . . miles]": Cave Brown and MacDonald (1977), p. 570.
714. "The temperature . . . life": Committee (1977), p. 119.
714. "severe thermal . . . viscerae": ibid.
715. "Doctor . . . he?": Hachiya (1955), p. 92.
715. "The inundation . . . fatalities": Lifton (1967), p. 21.
715. "There was . . . dead": quoted in ibid., p. 27.
716. "I asked . . . impossible": Hachiya (1955), p. 114.
716. "Father Kopp . . . hand": Cave Brown and MacDonald (1977), p. 542.
716. "Ah, that . . . around": Osada (1982), p. 352.
716. "The vicinity . . . arms": ibid., p. 305.
716. "That boy . . . that": ibid., p. 194.
717. "My body . . . ending' ": quoted in Lifton (1967), p. 22.

PAGE

717. "I just . . . world": quoted in ibid., p. 23.
717. "Within the . . . sound": Hachiya (1955), p. 164.
717. "When I . . . ruins": Osada (1982), p. 224.
717. "The shortest . . . hysterically": Hachiya (1955), p. 2.
717–718. "The appearance . . . them": quoted in Lifton (1967), p. 27.
718. "I heard . . . burned": NHK (1977), p. 12ff.
718. "On both . . . sleepwalkers": Osada (1982), p. 313.
718. "Everything I . . . about": quoted in Lifton (1967), p. 29.
718. "That day . . . rags": Osada (1982), p. 10.
719. "The people . . . them": ibid., p. 258.
719. "People came . . . sight": ibid., p. 97.
719. "The flames . . . looks": ibid., p. 234.
719. "Screaming children . . . blood": ibid., p. 305.
719–720. "It was . . . flames": Liebow et al. (1949), p. 856ff.
720. "The whole . . . alive": Osada (1982), p. 8ff.
720. "I really . . . walking": ibid., p. 65ff.
720. "I was . . . her": ibid., p. 122ff.
720–721. "I left . . . her": quoted in Lifton (1967), p. 40.
721. "Beneath the . . . flames": Cave Brown and MacDonald (1977), p. 544.
721. "I was . . . thing": Osada (1982), p. 137ff.
721. "A woman . . . help": NHK (1977), p. 49.
721. "There were . . . up": Osada (1982), p. 43.
721. "Nearby . . . trousers": ibid., p. 364.
721. "I walked . . . felt": quoted in Lifton (1967), p. 50.
722. "I was . . . striking": NHK (1977), p. 39.
722. "a man . . . ankles": quoted in Mary McGrory, "Hiroshima Horrors Relived," *Kansas City Times,* March 24, 1982. p. A13.
722. "A man . . . up": quoted in Lifton (1967), p. 42.
722. "In front . . . blackness": quoted in ibid., p. 49ff.
722. "The corpse . . . hand": NHK (1977), p. 96.
722. "There was . . . blindly": Osada (1982), p. 154.
722. "I saw . . . be?": NHK (1977), p. 52.

PAGE

722. "A streetcar . . . tremble": Osada (1982), p. 55.
723. "The more . . . get": ibid., p. 77.
723. "Since just . . . having": ibid., p. 83.
723. "I went . . . eyes": quoted in Lifton (1967), p. 36.
723. "I and . . . agonies": Osada (1982), p. 230.
723. "At the . . . help": ibid., p. 352ff.
723–724. "Near the . . . Hell": ibid., p. 79ff.
724. "We came . . . flame": ibid., p. 62.
724. "The fire . . . heads": ibid., p. 72.
724. "I had . . . faces": ibid., p. 237.
724. "Between the . . . water": Hachiya (1955), p. 19.
725. "While taking . . . him": NHK (1977), p. 48.
725. "There were . . . me": Hachiya (1955), p. 101.
725. "Men whose . . . sea": Osada (1982), p. 178.
725. "We . . . around": ibid., p. 94.
725. "Bloated corpses . . . earth": ibid., p. 334.
725–726. "I had . . . shore": quoted in Trumbull (1957), p. 76.
726. "I got . . . place": Osada (1982), p. 173.
726. "The river . . . terrible": ibid., p. 219.
726. "There was . . . back": Hachiya (1955), p. 15.
726–727. "Hundreds of . . . drowned": ibid., p. 77ff.
727. "Along the . . . walk": ibid., p. 184.
727. "Night came . . . heaven": NHK (1977), p. 44.
727. "Everybody in . . . legs": Osada (1982), p. 280.
727. "If you . . . burns": ibid., p. 99ff.
727. "Hiroshima . . . land": ibid., p. 54.
727–728. "The bright . . . collapse": Cave Brown and MacDonald (1977), p. 546.
728. "The streets . . . height": Hachiya (1955), p. 8.
728. "Nothing . . . view": ibid., p. 31.
728. "I climbed . . . exist": quoted in Lifton (1967), p. 29.
728. "I reached . . . heart": quoted in ibid., p. 86.
728. "It is . . . instantaneously": Committee (1977), p. 61.
728–730. "In Hiroshima . . . destroyed": ibid., p. 379.
730. "[She was] . . . child": NHK (1977), p. 70.
730. "We gathered . . . out": interview with Sakae Itoh, Hiroshima, Aug. 5, 1982.

PAGE

730. "After a . . . mouths": Hachiya (1955), p. 164.
730. "On the . . . mountain": Osada (1982), p. 72ff.
731. "Towards evening . . . Hiroshima": Hachiya (1955), p. 32.
731. "Survivors began . . . death": Lifton (1967), p. 57.
731. "atomic bomb . . . irradiation": Committee (1977), p. 115.
731. "Following the . . . recover": Hachiya (1955), p. 97.
731. gamma radiation: cf. Hempelmann et al. (1952), p. 286ff.
731–732. anti-clotting factor: cf. Liebow et al. (1949), p. 927.
732. "Hemorrhage was . . . cases": Hachiya (1955), p. 147ff.
732. "found . . . autopsied": ibid., p. 145.
732. "evidence of . . . eye": Liebow et al. (1949), p. 923.
732. "the bodies . . . living": quoted in Lifton (1967), p. 66.
732. "We were . . . cancer": quoted in ibid., p. 61.
732. "Mother was . . . cry": Osada (1982), p. 227.
732–733. "in the . . . instant": Committee (1977), p. 6.
733. "The whole . . . foundations": ibid., p. 336.
733. "Such a . . . nothing": quoted in Lifton (1967), p. 79.
733. "the total . . . dead": quoted in Liebow (1965), p. 82.
733–734. "How many . . . explosion": Cave Brown and MacDonald (1977), p. 549.
734. Standardized Casualty Rate: cf. Liebow (1965), p. 235.
734. "Those scientists . . . it?": Osada (1982), p. 264.
734. "This is . . . home": Truman (1955), p. 421.
734–735. "Gen G . . . time": Aug. 6, 1945, transcript, MED 201, Groves, L. R., telephone conversations.
735. "The greatest . . . earth": quoted in Truman (1955), p. 422.
735. "I suppose . . . on": LS to GW, Aug. 6, 1945. Egon Weiss, personal communication.
735. "At first . . . asleep": Hahn (1970), p. 170.
735–736. "Then one . . . enemies": Frisch (1979), p. 176.
736. "the importance . . . all": "From the Rubble of Okinawa: A Different

PAGE

View of Hiroshima." *Kansas City Star,* Aug. 30, 1981, p. I1.

736. propaganda effort: cf. J. F. Moynahan to L. R. Groves, May 23, 1946. MED 314.7, History.

736-737. "What we . . . longer": quoted in Mosley (1982), p. 340.

737. "a certain . . . airplane": J. F. Moynahan to L. R. Groves, May 23, 1946.

737. "the equivalent . . . weapons": ibid.

737. Nagasaki leaflets: ibid.

737-738. "was originally . . . schedule": Ramsey (1946), p. 153.

738. "With the . . . orders": O'Keefe (1983), p. 97.

738-739. "When I . . . backward": ibid., p. 98.

739. "nothing that . . . resolder them": ibid., p. 99.

739. "My mind . . . finished": ibid., p. 100ff.

739. 0347: Ramsey (1946), p. 154.

740. "The night . . . us": Cave Brown and MacDonald (1977), p. 557.

740. Ashworth changed plugs: cf. his log at Ramsey (1946), p. 154.

740. "Two . . . seen": quoted in ibid., p. 155.

740. "the Japs . . . ocean": quoted in Marx (1967), p. 202.

742. "A smell . . . gates": William C. Bryson, Capt., USN, Sept. 14, 1945. *Bul. Atom. Sci.* Dec. 82, p. 35.

742. surrender offer: this discussion relies in part on Bernstein (1977).

742. "does not . . . Ruler": quoted in Butow (1954), p. 244.

742. "taking a . . . hands": quoted in Bernstein (1977), p. 5.

742. "I cannot . . . war": quoted in ibid., p. 6.

742. "crucifixion . . . President": quoted in ibid., p. 5.

742. "willingness to . . . accomplished": quoted in ibid., p. 6ff.

742-743. "From the . . . people": quoted in Feis (1966), p. 134.

743. "We would . . . bomb": quoted in Bernstein (1977), p. 9.

743. "Truman said . . . kids": quoted in Herken (1980), p. 11.

743. "Provided there . . . August": LRG to Chief of Staff, Aug. 10, 1945. MED 5B.

744. "It was . . . now?": quoted in Scott-Stokes (1974), p. 109.

744. "placing . . . officials": quoted in Bernstein (1977), p. 13.

PAGE

744. "I have . . . unusual": quoted in ibid., p. 15ff.

744. "a plan . . . attack": quoted in Feis (1966), p. 205.

744. "evidence of . . . ancestors?": quoted in ibid., p. 208.

745. "the . . . placed": quoted in Bernstein (1977), p. 13.

745. "Flash! . . . soon": quoted in Feis (1966), p. 209n.

745-746. "Despite the . . . generation": quoted in ibid., p. 248.

746. "If it . . . mad": quoted in Scott-Stokes (1974), p. 109.

746. "An atomic . . . slaughter": Committee (1977), p. 335.

746. "By the . . . identity": Elliot (1972), p. 138ff.

746. "The experience . . . mankind": Committee (1977), p. 340.

747. "The night . . . pounding": Hachiya (1955), p. 114ff.

Epilogue

749. "that I . . . escape": ET to LS, July 2, 1945. Weart and Szilard (1978), p. 209.

749. "The development . . . scale": ibid., p. 211.

750. special prayer: cf. ibid., p. 230.

750. second petition: cf. ibid., p. 231.

750. "and told . . . subject": ibid., p. 223

750. Oppenheimer wondered: according to EOL's memory of the weekend as reported in Childs (1968), p. 366.

750-751. "You will . . . despair": Smith and Weiner (1980), p. 297.

751. "the circumstances . . . it": quoted in Else (1980).

751. "terrible swift . . . wars": quoted in Childs (1968), p. 365.

751. "In one . . . bomb": Ulam (1976), p. 170.

751. "With the . . . upon": Snow (1981), bound galleys, p. 89. I do not find this comment in the published book.

751. letter on postwar planning: Scientific Panel to the Secretary of War, Aug. 17, 1945. Smith and Weiner (1980), p. 293ff.

752. "There is . . . working": quoted in Childs (1968), p. 366.

752. "I emphasized . . . following": Smith and Weiner (1980), p. 301.

752-753. Lilienthal notes: cf. Lilienthal (1964), II, p. 637ff.

753-754. "The wisest . . . itself": Weart and Szilard (1978), p. 223.

PAGE

754. Szilard to biology: Wigner (1964) dates this conversion from 1949, the year of LS's first biology paper, but LS himself says 1946. Cf. ibid., p. 16.
754. "The end ... TNT": Hawkins (1947), p. 214.
754. "was unclear ... it": quoted in Clark (1980), p. 262.
754. "[lost] ... work": Teller (1962), p. 22.
754. "We all ... peacetime": Bethe (1982), p. 45.
754–755. "He expressed ... again": quoted in Blumberg and Owens (1976), p. 185.
755. "that I ... days": Smith and Weiner (1980), p. 308.
755. "In the ... it": Bradbury (1948), p. 10.
755. "Here was ... use": ibid., p. 7.
755–756. "I said ... interested": Teller (1962), p. 22ff.
756. "To demand ... least": Bethe (1982), p. 45.
756. "seeking his ... thing": Teller (1962), p. 23.
756. "Now that ... way": ibid.
757. Teller letter: ET to EF, Oct. 31, 1945. MED Harrison and Bundy File, f. 76.
758. "Under a ... ceremony": Smith and Weiner (1980), p. 210.
758. "It is ... humanity": quoted in Hawkins (1947), p. 294.
758–759. "The limitations ... lot": *Time,* Oct. 29, 1945, p. 30.
759. "Good Uncle ... hit": Brode (1960), XI, p. 8.
759–760. "I was ... way": NB to JRO, Nov. 9, 1945. JRO Papers.
760. "that the ... produced": Feis (1966), p. 173n.
760. "the U.S.S.R. ... enterprise": quoted in Clark (1980), p. 208.
760. "To most ... suitcase": York (1970), p. 107.
760. Oppenheimer ALAS speech: Smith and Weiner (1980), p. 315ff.
764. "to review ... were planned": Teller et al. (1950), p. 1.
764–765. "It is ... national policy": Teller et al. (1980), pp. 44–46.
765. numbers of weapons stockpiled: cf. Norris et al. (1985), p. 107; Rearden (1984), p. 439; Rosenberg (1982).
765. half the T Division's time: Mark (1974), p. 3.
765. "The years ... scientifically": Wigner in Mark and Fernbach (1969), p. 4.
765. "purely scientific data": Teller (1946a), p. 10.
765. "ingenious ... sound": Teller (1946b), p. 13.
765. "Nothing ... world-union": Teller (1946c), p. 13.
765. He saw no defense: Teller (1947a), p. 85.
765–766. "One is ... of man": ibid.
766. "agreement with ... and peace": Teller (1947b), p. 356.
766. "a good ... idealist": Dyson (1979), p. 89.
766. "World government ... Government": Teller (1948b), p. 204.
766. "Due to ... incomplete": Teller (1948a), p. 5.
767. Oppenheimer sought him out: USAEC (1954), p. 714.
767. Bradbury sent delegation: Alfred P. Sloan Foundation H-bomb history symposium, videotape transcript (n.d.), Part I, p. 15.
767. "the best ... results": Coughlan (1963), p. 91.
767. Teller had not expected: according to S. Ulam, Sloan transcript I, p. 40.
767. "Keep your shirt on": quoted by Teller, USAEC (1954), p. 714.
767. "did not ... bomb": ibid.
767–768. "Especially ... all time": York (1976), p. 45ff.
768. "intolerable": quoted in Hewlett and Duncan (1969), p. 395.
768. So did Acheson: cf. Acheson (1969), p. 349: "The American people simply would not tolerate a policy of delaying nuclear research in so vital a matter."
768. "Edward ... about it": Sloan transcript II, pp. 30–31.
768. The GAC recommended: the report and annexes are reproduced in York (1976), p. 150ff.
769. two arguments: for an informed and thorough discussion cf. York (1976).
769. 200 bombs: estimated from Rosenberg (1982), p. 26.
769. "The [H-bomb] ... it": USAEC (1954), p. 251.
770. "Edward was ... United States": Sloan transcript II, p. 26.
770. "The reasons ... should be": Coughlan (1954), p. 65ff.
770. "Truman's words ... bomb": interview with J. Carson Mark et al., *Los Alamos Science* 4: 7 (Winter/Spring 1983), p. 36.
771. "no conclusive ... unsuitable": Mark (1974), p. 10.

PAGE

771. "vastly larger . . . computers": Ulam (1976), p. 213.
771. "I think . . . out": quoted in Mark (1974), p. 9.
771. "We started . . . work": Ulam (1976), p. 214.
771. By February 1950: Mark (1974), p. 8.
771. "the result . . . fizzle": quoted in Hewlett and Duncan (1969), p. 440.
771. more tritium: Mark (1974), p. 8.
771. "seemed rather . . . difficulties": Ulam (1976), p. 217.
772. "He was . . . today": quoted in Hewlett and Duncan (1969), p. 440.
772. Teller refused to believe: cf. Teller (1955), p. 272: "I felt at the time that these calculations, which seemed to be in conflict with earlier results obtained on machines, were hard to believe."
772. "reassurances . . . constructive": Hewlett and Duncan (1969), p. 440.
772. "In the . . . dejectedly": Ulam (1976), p. 212. Ulam places this work at Princeton, confusing it in memory with the later calculation carried out on the MANIAC there. The MANIAC was not built in the summer of 1950; when it was, it calculated the hydrodynamics of the successful Teller-Ulam design: icicles did not form.
772. "turned out . . . explosions": ibid., p. 219.
772. "That Ulam's . . . 1949": Bethe (1982), p. 47.
772. "was desperate . . . solution": ibid., p. 48.
772. the laboratory's time: ibid.
772. Teller told the GAC: cf. Hewlett and Duncan (1969), p. 530.
773. few friends: Bethe (1982), p. 48.
773. "The new . . . 1939": ibid., p. 49.
773. Lothar Nordheim: undated draft letter in Teller-Strauss correspondence file, Strauss Papers.
773. Herbert York: interview with Herbert York, La Jolla, Calif., June 27, 1983.
773. "Two signs . . . de Hoffmann": Teller (1955), p. 273.
773. "It is . . . behavior": Coughlan (1954).
773. "I used . . . while": Bernstein (1980), p. 95.
774. "The first . . . transport": *Los Alamos Science* 4:7 (Winter/Spring 1983), p. 112.
774. Ulam studying fission fireball: cf. Bethe (1982), p. 48: "Ulam . . . made his discovery while studying some aspects of fission weapons."

PAGE

774. Those designs blow apart: cf. Ulam (1966), p. 597: "For the wartime schemata for the 'Super,' the hydrodynamical disassembly proceeded faster than the buildup and maintenance of the reaction."
774. Teller-Ulam invention: Howard Morland, who pieced together odds and ends of information that had escaped the security system and reached the right conclusion, deserves a vote of thanks in any discussion of the mechanism of the H-bomb. His book—Morland (1981)—is invaluable. Further confirming the role of plastic foam in the thermonuclear is a description in Cahn (1984) of an inertially confined fusion reactor fuel pellet, which is effectively a miniature spherical H-bomb brought to fusion temperature (its designers hope) by laser pulse: "The essential parts consist of a glass microballoon coated internally with a layer of deuterium-tritium fuel and externally with a dense 'pusher' layer, the whole separated by plastic foam from an external dense metal pusher layer which itself is coated by an ablative plastic layer which creates the 'rocket' effect." In a conference room at the Lawrence Livermore Laboratory in 1984 I saw relegated to a dusty window ledge three display spheres labeled "plutonium," "lithium" and "foam." Cf. also Allred and Rosen (1976); Bell (1965); Bethe (1982); DeVolpi et al. (1981); Mark (1974); Teller (1980); Teller et al. (1950); Ulam (1966).
776. "On March . . . program": Bethe (1982), p. 48.
776. "technically . . . sweet": USAEC (1954), p. 251.
776. "Dr. Oppenheimer . . . it": ibid., p. 720.
776. "Since I . . . who?": JRO Papers, Box 205.
777. "I watched . . . hand?": Teller (1955), p. 274ff.
777. "This thing . . . Thebes": Davis (1968) reports the remark but oddly attributes it to Lewis Strauss' birthday party.
777. "It's a boy": York interview. This response is frequently misplaced to the

PAGE

earlier George shot on May 8, 1951, which tested thermonuclear feasibility, one of the Greenhouse tests. Teller's response on that earlier occasion—which he attended—was silently to hand Ernest Lawrence a five-dollar bill to pay off a bet that the shot would not succeed.

777-778. "The fireball . . . reef": Libby (1979), p. 303.

778. August 1953 Soviet shot yield: York (1976), p. 85.

778. "This was . . . [fallout]": Bethe (1982), p. 53.

778. "the very . . . bomb": York (1976), p. 85.

778. Nov. 23, 1955: ibid., p. 93.

778. "his first . . . be": Oppenheimer (1963), III (Los Alamos version), p. 8.

778. "It did . . . country": Oppenheimer (1946), p. 265.

PAGE

779. "We know . . . expansion": Elliot (1972), p. 187.

779. "The lesson . . . are": Rabi (1970), p. 70.

780. "The hundred . . . race": Elliot (1972), p. 5ff.

780. "The moral . . . time": ibid.

780. "The one . . . death": ibid., p. 24.

781. "a doctrine . . . contradictory": Kedourie (1960), p. 9.

781. "its essential . . . river": Ward (1966), p. 14.

782. "Once men . . . checks": ibid., p. 56.

785. "Nationalism . . . history": ibid., p. 99.

786-787. "Our societies . . . himself": Elliot (1972), p. 8.

787. "The eradication . . . purpose": Henderson (1976), p. 33.

788. "Of course . . . states": JRO to GD, May 16, 1957. JRO Papers, Box 43.

Bibliography

Abelson, Phillip. 1939. Cleavage of the uranium nucleus. *Phys. Rev.* 56:418.
———, et al. 1943. *Progress Report on Liquid Thermal Diffusion.* Naval Research Laboratory report No. 0-1977.
Acheson, Dean. 1969. *Present at the Creation.* W. W. Norton.
Alexandrov, A. P. 1967. The heroic deed. *Bul. Atom. Sci.* Dec.
Allardice, Corbin, and Edward R. Trapnell. 1955. *The First Pile.* U.S. Atomic Energy Commission.
Allison, Samuel K. 1965. Arthur Holly Compton. *Biog. Mem. Nat. Ac. Sci.* 38:81.
Allred, John, and Louis Rosen. 1976. First fusion neutrons from a thermonuclear weapon device. In Bogdan Maglich, ed., *Adventures in Experimental Physics.* World Science Education.
Alperovitz, Gar. 1985. *Atomic Diplomacy.* Penguin.
Alvarez, Luis W. 1970. Ernest Orlando Lawrence. *Biog. Mem. Nat. Ac. Sci.* 41:251.
Amaldi, E. 1977. Personal notes on neutron work in Rome in the 30s and post-war European collaboration in high-energy physics. In Charles Weiner, ed., *History of Twentieth Century Physics. Academic Press.*
Anderson, Herbert L., et al. 1939a. The fission of uranium. Phys. Rev. 55:511.
———. 1939b. Production of neutrons in uranium bombarded by neutrons. *Phys. Rev.* 55:797.
———. 1939c. Neutron production and absorption in uranium. *Phys. Rev.* 56:284.
Andrade, E. N. da C. 1956. The birth of the nuclear atom. *Scientific American.* Nov.
———. 1957. The birth of the nuclear atom. *Proc. Roy. Soc. A.* 244:437.
Anscombe, G.E. M. 1981. *The Collected Philosophical Papers.* v. III. University of Minnesota Press.
Arendt, Hannah. 1973. *The Origins of Totalitarianism.* Harcourt Brace Jovanovich.
Arms, Nancy. 1966. *A Prophet in Two Countries.* Pergamon Press.

Arnold, H. H. 1949. *Global Mission.* Harper & Bros.
Aston, Francis. 1920. Isotopes and atomic weights. *Nature* 105:617.
———. 1927. A new mass-spectrograph and the whole number rule. *Proc. Roy. Soc. A.* 115:487.
———. 1938. Forty years of atomic theory. In Joseph Needham and Walter Pagel, eds., *Background to Modern Science.* Macmillan.
Axelsson, George. 1946. Is the atom terror exaggerated? *Sat. Even. Post.* Jan. 5.
Bacher, R. F., and V. F. Weisskopf. 1966. The career of Hans Bethe. In R. E. Marshak, ed., *Perspectives in Modern Physics.* Interscience.
Bacon, Francis. 1627. *The New Atlantis.* Oxford University Press, 1969.
Badash, Lawrence, et al. 1980. *Reminiscences of Los Alamos.* D. Reidel.
Bainbridge, Kenneth T. 1945. *Trinity.* Los Alamos Scientific Laboratory, 1976.
Barber, Frederick A. 1932. *The Horror of It.* Brewer, Warren & Putnam.
Batchelor, John, and Ian Hogg. 1972. *Artillery.* Ballantine.
Bauer, Yehuda. 1982. *A History of the Holocaust.* Franklin Watts.
Bell, George I. 1965. Production of heavy nuclei in the Par and Barbel devices. *Phys. Rev.* 139: B1207.
Belote, James and William. 1970. *Typhoon of Steel.* Harper & Row.
Benedict, Ruth. 1946. *The Chrysanthemum and the Sword.* New American Library, 1974.
Bentwich, Norman. 1953. *The Rescue and Achievement of Refugee Scholars.* Martinus Nijhoff.
Bernstein, Barton J. 1977. The perils and politics of surrender: ending the war with Japan and avoiding the third atomic bomb. *Pacific Historical Review.* Feb.
Bernstein, Jeremy. 1975. Physicist. *New Yorker.* I: Oct. 13. II: Oct. 20.
———. 1980. *Hans Bethe: Prophet of Energy.* Basic Books.
Bethe, Hans. 1935. Masses of light atoms from transmutation data. *Phys. Rev.* 47:633.
———. 1953. What holds the nucleus together? *Scientific American.* Sept.
———. 1964. *Theory of the Fireball.* Los Alamos Scientific Laboratory.
———. 1965. The fireball in air. *J. Quant. Spectrosc. Radiative Transfer* (GB) 5:9.
———. 1967. Energy production in stars. Nobel Lecture.
———. 1968. J. Robert Oppenheimer. *Biog. Mem. F. R. S.* 14:391.
———. 1982. Comments on the history of the H-bomb. *Los Alamos Science.* Fall.
Beyerchen, Alan D. 1977. *Scientists Under Hitler.* Yale University Press.
Bickel, Lennard. 1980. *The Deadly Element.* Stein and Day.
Biquard, Pierre. 1962. *Frédéric Joliot-Curie.* Paul S. Eriksson.
Birdsall, Steve. 1980. *Saga of the Superfortress.* Doubleday.
Bishop, Jim. 1974. *FDR's Last Year.* William Morrow.
Blackett, P. M. S. 1933. The craft of experimental physics. In Harold Wright, ed., *University Studies.* Ivor Nelson & Watson.
Blumberg, Stanley A., and Gwinn Owens. 1976. *Energy and Conflict.* G. P. Putnam's Sons.
Bohr, Niels. 1909. Determination of the surface-tension of water by the method of jet vibration. *Phil. Trans. Roy. Soc.* 209:281.
———. 1936. Neutron capture and nuclear constitution. *Nature* 137:344.
———. 1939a. Disintegration of heavy nuclei. *Nature* 143:330.
———. 1939b. Resonance in uranium and thorium disintegrations and the phenomenon of nuclear fission. *Phys. Rev.* 56:418.
———. 1958. *Atomic Physics and Human Knowledge.* John Wiley.
———. 1963. *Essays 1958–1963 on Atomic Physics and Human Knowledge.* Interscience.
———. 1972. *Collected Works,* v. I. North-Holland.
———. 1981. *Collected Works,* v. II. North-Holland.
———, and J. A. Wheeler. 1939. The mechanism of nuclear fission. *Phys. Rev.* 56:426.

Bolle, Kees. 1979. *The Bhagavadgītā.* University of California Press.
Bolton, Ellis. n.d. A few days in January 1939. Unpublished MS.
Booth, Eugene, et al. 1969. *The Beginnings of the Nuclear Age.* Newcomen Society.
Born, Max. 1971. *The Born-Einstein Letters.* Macmillan.
Bothe, W. 1944. Die Absorption thermischer Neutronen in Kohlenstoff. *Zeitschrift für Physik* 122:749.
———. 1951. Lebensbeschreibung. In Ruth Drossel, *Walther Bothe, Bemerkungen zu seinen kernphysikalischen Arbeiten auf Grund der Durchsicht seiner Laborbucher.* Max-Planck-Institut für Kernphysik, Heidelberg. Unpublished. 1975.
Bradbury, Norris E. 1949. Peace and the atomic bomb. *Pomona College Bulletin.* Feb.
Bretall, Robert, ed. 1946. *A Kierkegaard Anthology.* Modern Library.
Brines, Russell. 1944. *Until They Eat Stones.* J. B. Lippincott.
British Information Services. 1945. Statements relating to the atomic bomb. *Rev. Mod. Phys.* 17:472.
Brobeck, W. M., and W. B. Reynolds. 1945. *On the Future Development of the Electromagnetic System of Tubealloy Isotope Separation.* MED G-14-74.
Brode, Bernice. 1960. Tales of Los Alamos. *LASL Community News.* June 2 and Sept. 22.
Brode, Harold L. 1968. Review of nuclear weapons effects. *Ann Rev. Nucl. Sci.* 18:153.
Brown, Louis. n.d. *Beryllium-8.* Unpublished MS
Bundy, Harvey H. 1957. Remembered words. *Atlantic.* Mar.
Bundy, McGeorge. 1969. To cap the volcano. *Foreign Affairs.* Oct.
Burckhardt, Jacob. 1943. *Force and Freedom.* Pantheon.
Burns, E. L. M. 1966. *Megamurder.* Pantheon.
Bush, Vannevar. 1954. Lyman J. Briggs and atomic energy. *Scientific Monthly.* 78:275.
———. 1970. *Pieces of the Action.* William Morrow.
Butow, Robert J. C. 1954. *Japan's Decision to Surrender.* Stanford University Press.
Byrnes, James F. 1947. *Speaking Frankly.* Harper & Bros.
———. 1958. *All in One Lifetime.* Harper & Bros.
Cahn, Robert W. 1984. Making fuel for inertially confined fusion reactors. *Nature* 311:408.
Canetti, Elias. 1973. *Crowds and Power.* Continuum.
Carnegie Endowment for International Peace. 1915. *The Hague Declaration* (IV, 2) *of 1899 Concerning Asphyxiating Gases.*
Cary, Otis. 1979. Atomic bomb targeting—myths and realities. *Japan Quarterly* 26/4.
Casimir, Hendrick. 1983. *Haphazard Reality.* Harper & Row.
Cave Brown, Anthony, and Charles B. MacDonald. 1977. *The Secret History of the Atomic Bomb.* Delta.
Chadwick, James. 1932a. Possible existence of a neutron. *Nature* 129:312.
———. 1932b. The existence of a neutron. *Proc. Roy. Soc.* 136A:692.
———. 1935. The neutron and its properties. Nobel Lecture.
———. 1954. The Rutherford Memorial Lecture. *Proc. Roy. Soc.* 224:435.
———. 1964. Some personal notes on the search for the neutron. *Proceedings of the Tenth Annual Congress of the History of Science.* Hermann.
Chandler, Alfred D., Jr., ed. 1970. *The Papers of Dwight David Eisenhower.* Johns Hopkins Press.
Chevalier, Haakon. 1965. *The Story of a Friendship.* Braziller.
Childs, Herbert. 1968. *An American Genius.* E. P. Dutton.
Chivian, Eric, et al., ed. 1982. *Last Aid.* W. H. Freeman.
Church, Peggy Pond. 1960. *The House at Otowi Bridge.* University of New Mexico Press.
Churchill, Winston. 1948. *The Gathering Storm.* Houghton Mifflin.
———. 1949. *Their Finest Hour.* Houghton Mifflin.

———. 1950. *The Grand Alliance.* Houghton Mifflin.
———. 1950. *The Hinge of Fate.* Houghton Mifflin.
———. 1951. *Closing the Ring.* Houghton Mifflin.
———. 1953. *Triumph and Tragedy.* Houghton Mifflin.
Claesson, Claes. 1959. *Kungälvsbygden.* Bohusläns Grafiska Aktiebolag.
Clark, Ronald W. 1971. *Einstein.* Avon.
———. 1980. *The Greatest Power on Earth.* Harper & Row.
Cline, Barbara Levett. 1965. *The Questioners.* Crowell.
Cockburn, Stewart, and David Ellyard. 1981. *Oliphant.* Axiom Books.
Coffey, Thomas M. 1970. *Imperial Tragedy.* World.
Cohen, K. P., et al. 1983. Harold Clayton Urey. *Biog. Mem. F. R. S.* 29:623.
Cohn, Norman. 1967. *Warrant for Genocide.* Harper & Row.
Colinvaux, Paul. 1980. *The Fate of Nations.* Simon and Schuster.
Collier, Richard. 1979. *1940.* Hamish Hamilton.
The Committee for the Compilation of Materials on Damage Caused by the Atomic Bombs in Hiroshima and Nagasaki. 1977, 1981. *Hiroshima and Nagasaki.* Basic Books.
Compton, Arthur Holly.1935. *The Freedom of Man.* Greenwood Press, 1969.
———. 1956. *Atomic Quest.* Oxford University Press.
———. 1967. *The Cosmos of Arthur Holly Compton.* Knopf.
Conant, James Bryant. 1943. *A History of the Development of an Atomic Bomb.* Unpublished MS. OSRD S-1, Bush-Conant File, folder 5. National Archives.
———. 1970. *My Several Lives.* Harper & Row.
Condon, Edward U. 1943. *The Los Alamos Primer.* Los Alamos Scientific Laboratory.
———. 1973. Reminiscences of a life in and out of quantum mechanics. *Proceedings of the 7th International Symposium on Atomic, Molecular, Solid State Theory and Quantum Biology.* John Wiley & Sons.
Conn, G. K. T., and H. D. Turner. 1965. *The Evolution of the Nuclear Atom.* American Elsevier.
Costello, John. 1981. *The Pacific War.* Rawson, Wade.
Coughlan, Robert. 1954. Dr. Edward Teller's magnificent obsession. *Life.* Sept. 6.
———. 1963. The tangled drama and private hells of two famous scientists. *Life.* Dec. 13.
Craig, William. 1967. *The Fall of Japan.* Penguin.
Craven, Wesley Frank, and James Lea Cate, eds. 1948–58. *The Army Air Forces in World War II.* University of Chicago Press.
Crowther, J. G. 1974. *The Cavendish Laboratory 1874–1974.* Science History Publications.
Curie, Eve. 1937. *Madam Curie.* Doubleday, Doran.
Dainton, F. S. 1966. *Chain Reactions.* John Wiley & Sons.
Darrow, Karl K. 1952. The quantum theory. *Scientific American.* Mar.
Davis, Nuel Pharr. 1968. *Lawrence and Oppenheimer.* Simon and Schuster.
Dawidowicz, Lucy S. 1967. *The Golden Tradition.* Holt, Rinehart and Winston.
———. 1975. *The War Against the Jews 1933–1945.* Bantam.
de Hevesy, George. 1947. Francis William Aston. *Obituary Notices of F. R. S.* 16:635.
———. 1962. *Adventures in Radioisotope Research.* Pergamon.
de Jonge, Alex. 1978. *The Weimar Chronicle.* Paddington Press.
Demster, Arthur Jeffrey. 1935. New methods in mass spectroscopy. *Proc. Am. Phil. Soc.* 75:755.
DeVolpi, A., et al. 1981. *Born Secret.* Pergamon.
Dickson, Lovat. 1969. *H. G. Wells.* Atheneum.
Draper, Theodore. 1985. Pie in the sky. *NYRB.* Feb. 4.
Duggan, Stephen, and Betty Drury. 1948. *The Rescue of Science and Learning.* Macmillan.

Dupre, A. Hunter. 1972. The *great instauration* of 1940: the organization of scientific research for war. In Gerald Holton, ed., *The Twentieth Century Sciences,* W. W. Norton.
Dyson, Freeman. 1979. *Disturbing the Universe.* Harper & Row.
Eiduson, Bernice T. 1962. *Scientists: Their Psychological World.* Basic Books.
Einstein, Albert, and Leopold Infeld. 1966. *The Evolution of Physics.* Simon and Schuster.
Elliot, Gil. 1972. *Twentieth Century Book of the Dead.* Charles Scribner's Sons.
———. 1978. *Lucifer.* Wildwood House.
Ellis, John. 1975. *The Social History of the Machine Gun.* Pantheon.
———. 1976. *Eye-Deep in Hell.* Pantheon.
Elsasser, Walter M. 1978. *Memoirs of a Physicist in the Atomic Age.* Science History Publications.
Else, Jon. 1980. *The Day After Trinity.* KTEH-TV, San Jose CA.
Embry, Lee Anna. 1970. George Braxton Pegram. *Biog. Mem. Nat. Ac. Sci.* 41:357.
Ethridge, Kenneth E. 1982. The agony of the *Indianapolis. American Heritage.* Aug.–Sept.
Eve, A. S. 1939. *Rutherford.* Macmillan.
Everett, Susanne. 1980. *World War I.* Rand McNally.
Feather, Norman. 1940 *Lord Rutherford.* Priory Press.
———. 1964. The experimental discovery of the neutron. In *Proceedings of the Tenth Annual Congress of the History of Science.* Hermann.
———. 1974. Chadwick's neutron. *Contemp. Phys.* 6:565.
Feis, Herbert. 1966. *The Atomic Bomb and the End of World War II.* Princeton University Press.
Feld, Bernard. 1984. Leo Szilard, scientist for all seasons. *Social Research.* Autumn.
Fermi, Enrico. 1949. *Nuclear Physics.* University of Chicago Press.
———. 1962. *Collected Papers.* University of Chicago Press.
Fermi, Laura. 1954. *Atoms in the Family.* University of Chicago Press.
———. 1971. *Illustrious Immigrants.* University of Chicago Press.
Ferrell, Robert H., ed. 1980. Truman at Potsdam. *American Heritage.* June–July.
Feuer, Lewis S. 1963. *The Scientific Intellectual.* Basic Books.
———. 1982. *Einstein and the Generations of Science.* Transaction Books.
Feyerabend, Paul. 1975. *Against Method.* Verso.
Feynman, Richard P. 1985. *Surely You're Joking, Mr. Feynman.* W. W. Norton.
———, et al. 1963. *The Feynman Lectures on Physics,* v. I. Addison-Wesley.
Flender, Harold. 1963. *Rescue in Denmark.* Simon and Schuster.
Fredette, Raymond H. 1976. *The Sky on Fire.* Harcourt Brace Jovanovich.
Friedrich, Otto. 1972. *Before the Deluge.* Harper & Row.
Frisch, Otto. 1939. Physical evidence for the division of heavy nuclei under neutron bombardment. *Nature* 143:276.
———. 1954. Scientists and the hydrogen bomb. *Listener.* Apr. 1.
———. 1967a. The life of Niels Bohr. *Scientific American.* June.
———. 1967b. The discovery of fission. *Physics Today.* Nov.
———. 1968. Lise Meitner. *Biog. Mem. F. R. S.* 16:405.
———. 1971. Early steps toward the chain reaction. In I. J. R. Aitchison and J. E. Paton, eds., *Rudolf Peierls and Theoretical Physics.* Pergamon Press.
———. 1975. A walk in the snow. *New Scientist* 60:833.
———. 1978. Lise Meitner, nuclear pioneer. *New Scientist.* Nov. 9.
———. 1979. *What Little I Remember.* Cambridge University Press.
Gamow, George. 1966. *Thirty Years That Shook Physics.* Doubleday.
———. 1969. Origin of galaxies. In Hans Mark and Sidney Fernbach, eds., *Properties of Matter Under Unusual Conditions.* Interscience.
———. 1970. *My World Line.* Viking.

Giovannitti, Len, and Fred Freed. 1965. *The Decision to Drop the Bomb.* Coward-McCann.
Glasstone, Samuel. 1967. *Sourcebook on Atomic Energy.* D. Van Nostrand.
———, and Philip J. Dolan. 1977. *The Effects of Nuclear Weapons.* U. S. Department of Defense.
Goldschmidt, Bertrand. 1964. *Atomic Adventure.* Pergamon.
———. 1982. *The Atomic Complex.* American Nuclear Society.
Goldstine, Herman H. 1972. *The Computer from Pascal to von Neumann.* Princeton University Press.
Golovin, Igor. 1967. Father of the Soviet bomb. *Bul. Atom. Sci.* Dec.
Goodchild, Peter. 1980. *J. Robert Oppenheimer: Shatterer of Worlds.* Houghton Mifflin.
Goodrich, H. B., et al. 1951. The origins of U. S. scientists. *Scientific American.* July.
Goran, Morris. 1967. *The Story of Fritz Haber.* University of Oklahoma Press.
Goudsmit, Samuel A. 1947. *Alsos.* Henry Schuman.
Gowing, Margaret. 1964. *Britain and Atomic Energy 1939–1945.* Macmillan.
Graetzer, Hans G., and David L. Anderson. 1971. *The Discovery of Nuclear Fission.* Van Nostrand Reinhold.
Grew, Joseph C. 1942. Report from Tokyo. *Life.* Dec. 7.
———. 1952. *Turbulent Era.* Houghton Mifflin.
Grodzins, Morton, and Eugene Rabinowitch. 1963. *The Atomic Age.* Basic Books.
Grosz, George. 1923. *Ecce Homo.* Brussel & Brussel.
Groueff, Stephane. 1967. *Manhattan Project.* Little, Brown.
Groves, Leslie R. 1948. The atom general answers his critics. *Sat. Even. Post.* May 19.
———. 1962. *Now It Can Be Told.* Harper & Row.
———. n.d. *For My Grandchildren.* Unpublished MS, U.S. Military Academy Library.
Guillain, Robert. 1981. *I Saw Tokyo Burning.* Doubleday.
Guillemin, Victor. 1968. *The Story of Quantum Mechanics.* Charles Scribner's Sons.
Hachiya, Michihiko. 1955. *Hiroshima Diary.* University of North Carolina Press.
Hahn, Otto. 1936. *Applied Radiochemistry.* Cornell University Press.
———. 1946. From the natural transmutations of uranium to its artificial fission. Nobel Lecture.
———. 1958. The discovery of fission. *Scientific American.* Feb.
———. 1966. *A Scientific Autobiography.* Charles Scribner's Sons.
———. 1970. *My Life.* Herder and Herder.
———. 1975. *Erlebnisse und Erkenntnisse.* Econ Verlag.
———, and F. Strassmann. 1939. Concerning the existence of alkaline earth metals resulting from the neutron irradiation of uranium. *Naturwiss.* 27:11 (Trans., Hans G. Graetzer, *Am. Jour. Phys.* 32:10. 1964.)
Haldane, J. B. S. 1925. *Callinicus.* Dutton.
Harris, Benedict R., and Marvin A. Stevens. 1945. Experiences at Nagasaki, Japan. *Conn. St. Medical Journal* 12:913.
Harrisson, Tom. 1976. *Living Through the Blitz.* Collins.
Harrod, R. F. 1959. *The Prof.* Macmillan.
Harwell, Mark A. 1984. *Nuclear Winter.* Springer-Verlag.
Hashimoto, Mochitsura. 1954. *Sunk.* Henry Holt.
Haukelid, Knut. 1954. *Skis Against the Atom.* William Kimber.
Hawkins, David. 1947. *Manhattan District History, Project Y, The Los Alamos Project,* v. I. Los Alamos Scientific Laboratory.
Heibut, Anthony. 1983. *Exiled in Paradise.* Viking.
Heilbron, J. L. 1974. *H. G. J. Moseley.* University of California Press.
———, and Thomas S. Kuhn. 1969. The genesis of the Bohr atom. *Historical Studies in the Physical Sciences* 1:211.
Heims, Steve J. 1980. *John von Neumann and Norbert Weiner.* MIT Press.

Heisenberg, Elisabeth. 1984. *Inner Exile.* Birkhäuser.
Heisenberg, Werner. 1947. Research in Germany on the technical application of atomic energy. *Nature* 160:211.
———. 1968. The Third Reich and the atomic bomb. *Bul. Atom. Sci.* June.
———. 1971. *Physics and Beyond.* Harper.
Hellman, Geoffrey T. 1945. The contemporaneous memoranda of Dr. Sachs. *New Yorker.* Dec. 1.
Hempelmann, Louis H., et al. 1952. The acute radiation syndrome: a study of nine cases and a review of the problem. *Annals of Internal Medicine* 36/2:279.
Henderson, Donald A. 1976. The eradication of smallpox. *Scientific American.* Oct.
Herken, Gregg. 1980. *The Winning Weapon.* Knopf.
Hersey, John. 1942. The marines on Guadalcanal. *Life.* Nov. 9.
———. 1946. *Hiroshima.* Modern Library.
Hewlett, Richard G., and Oscar E. Anderson, Jr. 1962. *The New World, 1939/1946.* Pennsylvania State University Press.
———, and Francis Duncan. 1969. *Atomic Shield, 1947/1952.* Pennsylvania State University Press.
Hitler, Adolf. 1927. *Mein Kampf.* Houghton Mifflin, 1971.
Hogg, I. V., and L. F. Thurston. 1972. *British Atillery Weapons and Ammunition.* Ian Allan.
Holton, Gerald. 1973. *Thematic Origins of Scientific Thought.* Harvard University Press.
———. 1974. Striking gold in science: Fermi's group and the recapture of Italy's place in physics. *Minerva* 12:159.
———, and Yehuda Elkana, eds. 1982. *Albert Einstein: Historical and Cultural Perspectives.* Princeton University Press.
Hopkins, George E. 1966. Bombing and the American conscience during World War II. *The Historian* 28:451.
Hough, Frank O. 1947. *The Island War.* Lippincott.
Howorth, Muriel. 1958. *Pioneer Research on the Atom.* New World.
Hughes, H. Stuart. 1975. *The Sea Change.* Harper & Row.
Ibuse, Masuji. 1969. *Black Rain.* Kodansha International.
Infeld, Leopold. 1941. *Quest.* Chelsea, 1980.
Irving, David. 1963. *The Destruction of Dresden.* Holt, Rinehart and Winston.
———. 1967. *The Virus House.* William Kimber. (In U.S.: *The German Atomic Bomb,* Simon and Schuster, 1968.)
Iwamatsu, Shigetoshi. 1982. A perspective on the war crimes. *Bul. Atom. Sci.* Feb.
Jaki, Stanley L. 1966. *The Relevance of Physics.* University of Chicago Press.
Jammer, Max. 1966. *The Conceptual Development of Quantum Mechanics.* McGraw-Hill.
Jászi, Oscar. 1924. *Revolution and Counter-Revolution in Hungary.* P. S. King and Son.
Jette, Eleanor. 1977. *Inside Box 1663.* Los Alamos Historical Society.
Johansson, Sigurd. n.d. *Atomålderns vagga stod i. Kungälv.* Unpublished MS.
Johnson, Charles W., and Charles O. Jackson. 1981. *City Behind a Fence.* University of Tennessee Press.
Johnson, Ken. 1970. A quarter century of fun. *The Atom.* Los Alamos Scientific Laboratory. Sept.
Joliot, Frédéric. 1935. Chemical evidence of the transmutation of elements. Nobel Lecture.
———, H. von Halban, Jr., and L. Kowarski. 1939a. Liberation of neutrons in the nuclear explosion of uranium. *Nature* 143:470.
———. 1939b. Number of neutrons liberated in the nuclear explosion of uranium. *Nature* 143:680.
Joliot-Curie, Irène. 1935. Artificial production of radioactive elements. Nobel Lecture.

Jones, R. V. 1966. Winston Leonard Spencer Churchill. *Biog. Mem. F. R. S.* 12:35.
———. 1967. Thicker than heavy water. *Chemistry and Industry.* Aug. 26.
Jungk, Robert. 1958. *Brighter Than a Thousand Suns.* Harcourt, Brace.
Kapitza, Peter. 1968. *On Life and Science.* Macmillan.
———. 1980. *Experiment, Theory, Practice.* D. Reidel.
Kedourie, Elie. 1960. *Nationalism.* Hutchinson University Library.
Keegan, John. 1976. *The Face of Battle.* Viking.
Kennedy, J. W., et al. 1941. Properties of 94(239). *Phys. Rev.* 70:555 (1946).
Kennett, Lee. 1982. *A History of Strategic Bombing.* Charles Scribner's Sons.
Kevles, Daniel J. 1979. *The Physicists.* Vintage.
Kierkegaard, Søren. 1959. *Either/Or.* Doubleday.
King, John Kerry. 1970. *International Political Effects of the Spread of Nuclear Weapons.* USGPO.
Kistiakowsky, George B. 1949a. *Explosives and Detonation Waves. Part I, Introduction.* (LA-1043).
———. 1949b. *Explosives and Detonation Waves. Part IV, The Making of Explosive Charges.* (LA-1052)
———. 1949c. *Explosives and Detonation Waves. Part IV, The Making of Explosive Charges, cont.* (LA-1053)
———. 1980. Trinity—a reminiscence. *Bul. Atom. Sci.* June.
———, and F. H. Westheimer. 1979. James Bryant Conant. *Biog. Mem. F. R. S.* 25:209.
Koestler, Arthur. 1952. *Arrow in the Blue.* Macmillan.
Korda, Michael. 1979. *Charmed Lives.* Random House.
Kosakai, Yoshiteru. 1980. *Hiroshima Peace Reader.* Hiroshima Peace Culture Foundation.
Kruuk, Hans. 1972. The urge to kill. *New Scientist.* June 28.
Kuhn, Thomas S., et. al. 1962. Interview with Niels Bohr.
Kunetka, James W. 1979. *City of Fire.* University of New Mexico Press.
———. 1982. *Oppenheimer.* Prentice-Hall.
Lamont, Lansing. 1965. *Day of Trinity.* Atheneum.
Lang, Daniel. 1959. *From Hiroshima to the Moon.* Simon and Schuster.
Langer, Walter C. 1972. *The Mind of Adolf Hitler.* Basic Books.
Laqueur, Walter. 1965. *Russia and Germany.* Little, Brown.
Lash, Joseph P., ed. 1975. *From the Diaries of Felix Frankfurter.* W. W. Norton.
Lawrence, Ernest O. 1951. The evolution of the cyclotron. Nobel Lecture.
———, and M. Stanley Livingston. 1932. The production of high speed ions without the use of high voltages. *Phys. Rev.* 40:19.
Lawrence, William L. 1946. *Dawn Over Zero.* Knopf.
Lawson, Ted W. 1943. *Thirty Seconds Over Tokyo.* Random House.
Leachman, R. B. 1965. Nuclear fission. *Scientific American.* Aug.
Lefebure, Victor. 1923. *The Riddle of the Rhine.* Dutton.
LeMay, Curtis E., with McKinlay Kantor. 1965. *Mission with LeMay.* Doubleday.
Levin, Nora. 1977. *While Messiah Tarried.* Schocken.
Libby, Leona Marshall. 1979. *The Uranium People.* Crane Russak.
Liebow, Averill A. 1965. Encounter with disaster—a medical diary of Hiroshima, 1945. *Yale Journal of Biology and Medicine* 37:60.
———, et al. 1949. Pathology of atomic bomb casualties. *American Journal of Pathology* 5:853.
Lifton, Robert Jay. 1967. *Death in Life.* Random House.
Lilienthal, David E. 1964. *The Journals of David E. Lilienthal.* Harper & Row.
Litvinoff, Barnet. 1976. *Weizmann.* Hodder and Stoughton.
Lloyd George, David. 1933. *War Memoirs.* Little, Brown.

Los Alamos: beginning of an era 1943–1945. n.d. Los Alamos Scientific Laboratory.
Lyon, Fern, and Jacob Evans, eds. 1984. *Los Alamos: The First Forty Years.* Los Alamos Historical Society.
McCagg, William O., Jr. 1972. *Jewish Nobles and Geniuses in Modern Hungary.* East European Quarterly.
McMillan, Edwin. 1939. Radioactive recoils from uranium activated by neutrons. *Phys. Rev.* 55:510
———. 1951. The transuranium elements: early history. Nobel Lecture.
———, and Philip H. Abelson. 1940. Radioactive element 93. *Phys. Rev.* 57:1185.
Madach, Imre. 1956. *The Tragedy of Man.* Pannonia.
Manchester, William. 1980. *Goodbye Darkness.* Little, Brown.
Mark, Hans, and Sidney Fernbach, eds. 1969. *Properties of Matter Under Unusual Conditions.* Interscience.
Mark, J. Carson. 1974. *A Short Account of Los Alamos Theoretical Work on Thermonuclear Weapons, 1946–1950.* (LA-5647-MS)
Marsden, Ernest. 1962. Rutherford at Manchester. In J. B. Birks, ed., *Rutherford at Manchester.* Heywood & Co.
Marx, Joseph L. 1967. *Seven Hours to Zero.* G. P. Putnam's Sons.
Masefield, John. 1916. *Gallipoli.* Macmillan.
Massie, Harrie, and N. Feather. 1976. James Chadwick. *Biog. Mem. F. R. S.* 22:11.
Mee, Charles L., Jr., 1975. *Meeting at Potsdam.* M. Evans.
Meitner, Lise. 1959. Otto Hahn zum 80. Geburtstag. *Otto Hahn zum 8. März 1959.* Max-Planck-Gesellschaft.
———. 1962. Right and wrong roads to the discovery of nuclear energy. *IAEA Bulletin.* Dec. 2.
———. 1964. Looking back. *Bul. Atom. Sci.* Nov.
———, and O. R. Frisch. 1939. Disintegration of uranium by neutrons: a new type of nuclear reaction. *Nature* 143:239.
Mendelsohn, Ezra. 1970. *Class Struggle in the Pale.* Cambridge University Press.
Mendelssohn, Kurt. 1973. *The World of Walter Nernst.* University of Pittsburgh Press.
Mendes-Flohr, Paul R., and Jehuda Reinharz, eds. 1980. *The Jew in the Modern World.*
Messer, Robert L. 1982. *The End of the Alliance.* University of North Carolina Press.
Middlebrook, Martin. 1980. *The Battle of Hamburg.* Allen Lane.
Moon, P. B. 1974. *Ernest Rutherford and the Atom.* Priory Press.
———. 1977. George Paget Thompson. *Biog. Mem. F. R. S.* 23:529.
Moore, Ruth. 1966. *Niels Bohr.* Knopf.
Moorehead, Alan. 1956. *Gallipoli.* Harper & Bros.
Morison, Elting E. 1960. *Turmoil and Tradition.* Houghton Mifflin.
Morland, Howard. 1981. *The Secret that Exploded.* Random House.
Morrison, Philip. 1946. Beyond imagination. *New Republic.* Feb. 11.
———, and Emily Morrison. 1951. The neutron. *Scientific American.* Oct.
Morse, Philip M. 1976. Edward Uhler Condon. *Biog. Mem. Nat. Ac. Sci.* 48:125.
Morton, Louis. 1957. The decision to use the atomic bomb. *Foreign Affairs.* Jan.
Mosley, Leonard. 1982. *Marshall.* Hearst.
Moyers, Bill. 1984. Meet I. I. Rabi. *A Walk Through the 20th Century.* NET.
Murakami, Hyōe. 1982. *Japan: The Years of Trial.* Kodansha International.
Murrow, Edward R. 1967. *In Search of Light.* Knopf.
Nagy-Talavera, Nicholas M. 1970. *The Green Shirts and Others.* Hoover Institution.
Nathan, Otto, and Heinz Norden, eds. 1960. *Einstein on Peace.* Simon and Schuster.
NHK (Japanese Broadcasting Corporation), eds. 1977. *Unforgettable Fire.* Pantheon.
Nielson, J. Rud. 1963. Memories of Niels Bohr. *Physics Today.* Oct.
Nier, Alfred O. 1939. The isotopic constitution of uranium and the half-lifes of the uranium isotopes. *Phys. Rev.* 55:150.
———, et al. 1940a. Nuclear fission of separated uranium isotopes. *Phys. Rev.* 57:546.

———. 1940b. Further experiments on fission of separated uranium isotopes. *Phys. Rev.* 57:748.
———. 1940c. Neutron capture by uranium (238). *Phys. Rev.* 58:475.
Nincic, Miroslav. 1982. *The Arms Race.* Praeger.
Norris, Robert S., et al. 1985. History of the nuclear stockpile. *Bul. Atom. Sci.* Sept.
NOVA. 1980. *A is for Atom, B is for Bomb.* WGBH Transcripts.
O'Keefe, Bernard J. 1972. *Nuclear Hostages.* Houghton Mifflin.
Oliphant, Mark. 1972. *Rutherford.* Elsevier.
———. 1982. The beginning: Chadwick and the neutron. *Bul. Atom. Sci.* Dec.
———, and Penny. 1968. John Douglas Cockcroft. *Biog. Mem. F. R. S.* 14:139.
Oppenheimer, J. Robert. 1946. The atom bomb and college education. *The General Magazine and Historical Chronicle.* University of Pennsylvania General Alumni Society.
———. 1957. Talk to undergraduates. *Engineering and Science Monthly.* California Institute of Technology.
———. 1961. Secretary Stimson and the atomic bomb. *Andover Bulletin.* Spring.
———. 1963. Niels Bohr and his times. Three lectures, unpublished MSS. Oppenheimer Papers, Box 247.
———, and H. Snyder. 1939. On continued gravitational contraction. *Phys. Rev.* 56:455.
———, et al. 1946. *A Report on the International Control of Atomic Energy.* Department of State.
Osada, Arata, comp. 1982. *Children of the A-Bomb.* Midwest Publishers International.
Overy, R. J. 1980. *The Air War 1939–1945.* Europe Publications.
Pacific War Research Society. 1972. *The Day Man Lost.* Kodansha International.
Pais, Abraham. 1982. *'Subtle Is the Lord. . .'* Oxford University Press.
Parkes, James. 1964. *A History of the Jewish People.* Penguin.
Pash, Boris T. 1969. *The Alsos Mission.* Award Books.
Patai, Raphael. 1977. *The Jewish Mind.* Charles Scribner's Sons.
Paterson, Thomas G. 1972. Potsdam, the atomic bomb and the Cold War: a discussion with James F. Byrnes. *Pacific Historical Review.* May.
Peattie, Lisa. 1984. Normalizing the unthinkable. *Bul. Atom. Sci.* Mar.
Peierls, Rudolf. 1939. Critical conditions in neutron multiplication. *Proc. Camb. Phil. Soc.* 35:610.
———. 1959. The atomic nucleus. *Scientific American.* Jan.
———. 1981. Otto Robert Frisch. *Biog. Mem. F. R. S.* 27:283.
———. 1985. *Bird of Passage.* Princeton University Press.
———, and Nevill Mott. 1977. Werner Heisenberg. *Biog. Mem. F. R. S.* 23:213.
Perrin, Francis. 1939. Calcul relatif aux conditions éventuelles de transmutation en chaîne de l'uranium. *Comptes Rendus* 208:1394.
Peterson, Aage. 1963. The philosophy of Niels Bohr. *Bul. Atom. Sci.* Sept.
Pfau, Richard. 1984. *No Sacrifice Too Great.* University Press of Virginia.
Planck, Max. 1949. *Scientific Autobiography.* Philosophical Library.
Polanyi, Michael. 1946. *Science, Faith and Society.* University of Chicago Press.
———. 1962. *The Republic of Science.* Roosevelt University.
Pound, Reginald. 1964. *The Lost Generation of 1914.* Coward-McCann.
Powers, Thomas. 1984. Nuclear winter and nuclear strategy. *Atlantic.* Nov.
Prange, Gordon W. 1981. *At Dawn We Slept.* Penguin.
Prentiss, Augustin M. 1937. *Chemicals in War.* McGraw-Hill.
The Protocols of the Meetings of the Learned Elders of Zion. 1934. Trans. Victor E. Marsden, n.p.
Purcell, Edward M. 1964. Nuclear physics without the neutron: clues and contradictions. *Proceedings of the Tenth Annual Congress of the History of Science.* Hermann.
Rabi, I. I. 1945. The physicist returns from the war. *Atlantic.* Oct.

———. 1970. *Science: the Center of Culture.* World.
———, et al. 1969. *Oppenheimer.* Scribner's.
Ramsey, Norman, ed. 1946. *Nuclear Weapons Engineering and Delivery.* Los Alamos Technical Series, v. XXIII. Los Alamos Scientific Laboratory.
Rearden, Steven L. 1984. *History of the Office of the Secretary of Defense,* v. I. Office of the Secretary of Defense.
Rhodes, Richard, et al. 1977. Kurt Vonnegut, Jr. In George Plimpton, ed., *Writers at Work.* Viking, 1984.
Roberts, Richard Brooke. 1979. Autobiography. Unpublished MS.
——, et al. 1939a. Droplet fission of uranium and thorium nuclei. *Phys. Rev.* 55:416.
———. 1939b. Further observations on the splitting of uranium and thorium. *Phys. Rev.* 55:510.
———. 1940. Fission cross-sections for fast neutrons. Unpublished MS. Department of Terrestrial Magnetism Archives, Carnegie Institution of Washington.
———, and J. B. H. Kuper. 1939. Uranium and atomic power. *J. Appl. Phys.* 10:612.
Roberts, Stephen H. 1938. *The House that Hitler Built.* Harper & Bros.
Robison, George O. 1950. *The Oak Ridge Story.* Southern Publishers.
Roe, Anne. 1952. A psychologist examines 64 eminent scientists. *Scientific American.* Nov.
Roosevelt, Franklin D. 1939. *The Public Papers and Addresses, VIII.* Russell & Russell.
———. 1941. *The Public Papers and Addresses, IX.* Russell & Russell.
Rosenberg, Alfred. 1970. *Race and Race History.* Harper & Row.
Rosenberg, David Alan. 1982. U.S. nuclear stockpile, 1945 to 1950. *Bul. Atom. Sci.* May.
Rosenfeld, Léon. 1963. Niels Bohr's contribution to epistemology. *Phys. Today.* Oct.
———. 1979. *Selected Papers.* D. Reidel.
Royal, Denise. 1969. *The Story of J. Robert Oppenheimer.* St. Martin's Press.
Rozental, Stefan, ed. 1967. *Niels Bohr.* North-Holland.
Russell, A. S. 1962. Lord Rutherford: Manchester, 1907–19: a partial portrait. In J. B. Birks, ed., *Rutherford at Manchester.* Heywood & Co.
Rutherford, Ernest. 1962. *The Collected Papers,* v. I. Allen and Unwin.
———. 1963. *The Collected Papers,* v. II. Interscience.
———. 1965. *The Collected Papers,* v. III. Interscience.
Sachs, Alexander. 1945. Early history atomic project in relation to President Roosevelt, 1939–40. Unpublished MS. MED 319.7, National Archives.
Sachs, Robert G., ed. 1984. *The Nuclear Chain Reaction—Forty Years Later.* University of Chicago Press.
Sassoon, Siegfried. 1937. *The Memoirs of George Sherston.* Doubleday, Doran.
———. 1961. *Collected Poems 1908–1956.* Faber and Faber.
Saundby, Robert. 1961. *Air Bombardment.* Harper & Bros.
Schell, Jonathan. 1982. *The Fate of the Earth.* Knopf.
———. 1984. *The Abolition.* Knopf.
Schonland, Basil. 1968. *The Atomists.* Oxford University Press.
Scott-Stokes, Henry. 1974. *The Life and Death of Yukio Mishima.* Farrar, Straus & Giroux.
Seaborg. Glenn T. 1951. The transuranium elements: present status. Nobel Lecture.
———. 1958. *The Transuranium Elements.* Yale University Press.
———. 1976. *Early History of Heavy Isotope Production at Berkeley.* Lawrence Berkeley Laboratory.
———. 1977. *History of Met Lab Section C-I, April 1942 to April 1943.* Lawrence Berkeley Laboratory.

———. 1978. *History of Met Lab Section C-I, May 1943 to April 1944.* Lawrence Berkeley Laboratory.
———, et al. 1946a. Radioactive element 94 from deuterons on uranium. *Phys. Rev.* 69:366.
———. 1946b. Radioactive element 94 from deuterons on uranium. *Phys. Rev.* 69:367.
Segrè, Emilio. 1939. An unsuccessful search for transuranic elements. *Phys. Rev.* 55:1104.
———. 1955. Fermi and neutron physics. *Rev. Mod. Phys.* 28:262.
———. 1964. The consequences of the discovery of the neutron. *Proceedings of the Tenth Annual Congress of the History of Science.* Hermann.
———. 1970. *Enrico Fermi, Physicist.* University of Chicago Press.
———. 1980. *From X-Rays to Quarks.* W. H. Freeman.
———. 1981. Fifty years up and down a strenuous and scenic trail. *Ann. Rev. Nucl. Part. Sci.* 31:1.
Semenoff, N. 1935. *Chemical Kinetics and Chain Reactions.* Clarendon Press.
Shamos, Morris H. 1959. *Great Experiments in Physics.* Holt, Rinehart and Winston.
Shapley, Deborah. 1978. Nuclear weapons history: Japan's wartime bomb projects revealed. *Science* 199:152.
Sherrod, Robert. 1944. Beachhead in the Marianas. *Time.* July 3.
Sherwin, Martin J. 1975. *A World Destroyed.* Knopf.
Shils, Edward. 1964. Leo Szilard: a memoir. *Encounter.* Dec.
Shirer, William L. 1960. *The Rise and Fall of the Third Reich.* Simon and Schuster.
Smith, Alice Kimball. 1960. The elusive Dr. Szilard. *Harper's,* Aug.
———. 1965. *A Peril and a Hope.* MIT Press.
———, and Charles Weiner. 1980. *Robert Oppenheimer: Letters and Recollections.* Harvard University Press.
Smith, Cyril Stanley. 1954. Metallurgy at Los Alamos 1943–1945. *Met. Prog.* 65(5):81.
Smith, Lloyd P., et al. 1947. On the separation of isotopes in quantity by electromagnetic means. *Phys. Rev.* 72:989.
Smyth, Henry DeWolf. 1945. *Atomic Energy for Military Purposes.* USGPO.
Snow, C. P. 1958. *The Search.* Charles Scribner's Sons.
———. 1961. *Science and Government.* Harvard University Press.
———. 1967a. On Albert Einstein. *Commentary.* Mar.
———. 1967b. *Variety of Men.* Scribner's.
———. 1981. *The Physicists.* Little, Brown.
Soddy, Frederick. 1913. Inter-atomic charge. *Nature* 92:400.
———. 1953. *Atomic Transmutation.* New World.
Spector, Ronald H. 1985. *Eagle Against the Sun.* Free Press.
Speer, Albert. 1970. *Inside the Third Reich.* Macmillan.
Spence, R. 1970. Otto Hahn. *Biog. Mem. F. R. S.* 16:279.
Stein, Leonard. 1961. *The Balfour Declaration.* Simon and Schuster.
Stimson, Henry L., and McGeorge Bundy. 1948. *On Active Service in Peace and War.* Harper & Bros.
Strauss, Lewis L. 1962. *Men and Decisions.* Doubleday.
Stuewer, Roger H. 1979. *Nuclear Physics in Retrospect.* University of Minnesota Press.
———. 1985. Bringing the news of fission to America. *Phys. Today.* Oct.
Szasz, Ferenc Morton. 1984. *The Day the Sun Rose Twice.* University of New Mexico Press.
Szilard, Leo. 1945. We turned the switch. *Nation.* Dec. 22.
———. 1961. *The Voice of the Dolphins.* Simon and Schuster.
———. 1972. *The Collected Works: Scientific Papers.* MIT Press.
———, and Walter H. Zinn. 1939. Instantaneous emission of fast neutrons in the interaction of slow neutrons with uranium. *Phys. Rev.* 55:799.

Szulc, Tad. 1984. The untold story of how Russia "got the bomb." *Los Angeles Times,* IV:1. Aug. 26.
Talk of the Town. 1946. Usher. *New Yorker.* Jan. 5.
Teller, Edward. 1946a. Scientists in war and peace. *Bul. Atom. Sci.* Mar.
———. 1946b. The State Dep't report— "a ray of hope." *Bul. Atom. Sci.* Apr.
———. 1946c. Dispersal of cities and industries. *Bul. Atom. Sci.* Apr.
———. 1947a. How dangerous are atomic weapons? *Bul. Atom. Sci.* Feb.
———. 1947b. Atomic scientists have two responsibilities. *Bul. Atom. Sci.* Mar.
———. 1948a. The first year of the Atomic Energy Commission. *Bul. Atom. Sci.* Jan.
———. 1948b. Comments on the "draft of a world constitution." *Bul. Atom. Sci.* July.
———. 1955. The work of many people. *Science* 121:267.
———. 1962. *The Legacy of Hiroshima.* Doubleday.
———. 1977. *In Search of Solutions for Defense and for Energy.* Stanford University Press.
———. 1979. *Energy from Heaven and Earth.* W. H. Freeman.
———. 1980a. Hydrogen bomb. *Encyclopedia Americana,* v. XIV.
———. 1980b. *In Pursuit of Simplicity.* Pepperdine University Press.
———. 1983. Seven hours of reminiscences. *Los Alamos Science.* Winter/Spring.
———, et al. 1950. *Report of Conference on the Super* (LA-575, Deleted). Los Alamos Scientific Laboratory.
Terkel, Studs. 1984. *"The Good War."* Pantheon.
Terman, Lewis M. 1955. Are scientists different? *Scientific American.* Jan.
Thomas, Gordon, and Max Morgan Witts. 1977. *Enola Gay.* Stein and Day.
Thompson, Josiah. 1973. *Kierkegaard.* Knopf.
Tibbets, Paul W. 1946. How to drop an atom bomb. *Sat. Even. Post.* June 8.
———. 1973. Training the 509th for Hiroshima. *Air Force Magazine.* Aug.
Toland, John. 1970. *The Rising Sun.* Random House.
———. 1976. *Adolf Hitler.* Doubleday.
Tregaskis, Richard. 1943. *Guadalcanal Diary.* Random House.
Trenn, Thaddeus J. 1980. The phenomenon of aggregate recoil: the premature acceptance of an essentially incorrect theory. *Ann. Sci.* 37:81.
Truman, Harry S. 1955. *Year of Decision.* Doubleday.
Trumbull, Robert. 1957. *Nine Who Survived Hiroshima and Nagasaki.* E. P. Dutton.
Truslow, Edith C., and Ralph Carlisle Smith. 1946–47. *The Los Alamos Project,* v. II. Los Alamos Scientific Laboratory.
Turner, Louis A. 1940. Nuclear fission. *Rev. Mod. Phys.* 12:1.
———. 1946. Atomic energy from U238. *Phys. Rev.* 69:366.
Ulam. Stanislaw. 1966. Thermonuclear devices. In R. E. Marshak, ed., *Perspectives in Modern Physics.* Interscience.
———. 1976. *Adventures of a Mathematician.* Scribner's.
United States Atomic Energy Commission. 1954. *In the Matter of J. Robert Oppenheimer.* MIT Press, 1971.
United States Special Committee on Atomic Energy. 1945. *Hearings pursuant to S. Res. 179.* USGPO.
United States Strategic Bombing Survey, v. X. 1976. Garland.
Urey, Harold C., et al. 1932. A hydrogen isotope of mass 2 and its concentration. *Phys. Rev.* 40:1.
Veblen, Thorstein. 1919. The intellectual pre-eminence of Jews in modern Europe. *Political Science Quarterly.* Mar.
Völgyes, Ivan, ed. 1971. *Hungary in Revolution.* University of Nebraska Press.
von Kármán, Theodore. 1967. *The Wind and Beyond.* Little, Brown.
von Weizsäcker, Carl Friedrich. 1978. *The Politics of Peril.* Seabury Press.
Waite, Robert G. 1977. *The Psychopathic God: Adolf Hitler.* Basic Books.

Ward, Barbara. 1966. *Nationalism and Ideology.* Norton.

Wattenberg, Albert. 1982. December 2, 1942: the event and the people. *Bul. Atom. Sci.* Dec.

Weart, Spencer R. 1979. *Scientists in Power.* Harvard University Press.

———, and Gertrud Weiss Szilard, eds. 1978. *Leo Szilard: His Version of the Facts.* MIT Press.

Weinberg, Alvin M., and Eugene P. Wigner. 1958. *The Physical Theory of Neutron Chain Reactors.* University of Chicago Press.

Weiner, Charles. 1967. Interview with Otto Frisch, AIP.

———. 1967. Interview with Emilio Segrè, AIP.

———. 1969. Interview with James Chadwick, AIP.

———. 1969. A new site for the seminar: the refugees and American physics in the Thirties. In Donald Fleming and Bernard Bailyn, eds., *The Intellectual Migration.* Harvard University Press.

———, ed. 1972. *Exploring the History of Nuclear Physics.* AIP Conference Proceedings No. 7. American Institute of Physics.

———, and Jagdish Mehra. 1966. Interview with Hans Bethe, AIP.

———. 1966. Interview with Eugene Wigner, AIP.

Weisgal, Meyer W., and Joel Carmichael, eds. 1963. *Chaim Weizmann.* Atheneum.

Weizmann, Chaim. 1949. *Trial and Error.* Harper & Bros.

Wells, H. G. 1914. *The World Set Free.* E. P. Dutton.

———. 1931. *What Are We to Do with Our Lives?* Doubleday, Doran.

Wheeler, John A. 1962. Fission then and now. *IAEA Bulletin.* Dec. 2.

———. 1963a. No fugitive and cloistered virtue. *Phys. Today.* Jan.

———. 1963b. Niels Bohr and nuclear physics. *Phys. Today.* Oct.

Wheeler, Richard. 1980. *Iwo.* Lippincott & Crowell.

Wiesner, Jerome B. 1979. Vannevar Bush. *Biog. Mem. Nat. Ac. Sci.* 50:89.

Wigner, Eugene P. 1945. Are we making the transition wisely? *Sat Rev.* Nov. 17.

———. 1964. Leo Szilard. *Biog. Mem. Nat. Ac. Sci.* 40:337.

———. 1967. *Symmetries and Reflections.* Indiana University Press. Reprint OxBow Press, 1979.

———. 1969. An appreciation on the 60th birthday of Edward Teller. In Hans Mark and Sidney Fernbach, eds., *Properties of Matter Under Unusual Conditions.* Interscience.

Wilson, David. 1983. *Rutherford.* MIT Press.

Wilson, Jane, ed. 1975. *All in Our Time.* Bulletin of the Atomic Scientists.

Wolfe, Henry C. 1943. Japan's nightmare. *Harper's.* Jan.

Wolk, Herman S. 1975. The B-29, the A-Bomb, and the Japanese surrender. *Air Force Magazine.* Feb.

Yahil, Leni. 1969. *The Rescue of Danish Jewry.* Jewish Publication Society of America.

Yergin, Daniel. 1977. *Shattered Peace.* Houghton Mifflin.

York, Herbert. 1970. *Race to Oblivion.* Simon and Schuster.

———. 1976. *The Advisors.* W. H. Freeman.

Young-Bruehl, Elisabeth. 1982. *Hannah Arendt.* Yale University Press.

Zuckerman, Harriet. 1977. *Scientific Elite.* Free Press.

PHOTO CREDITS

1. Culver *2.* Egon Weiss *3.* E. Scott Barr Collection, AIP Niels Bohr Library *4.* Courtesy Otto Hahn and Lawrence Badash, AIP Niels Bohr Library *5.* Mary Evans Rhodes *6.* Bibliothek und Archiv für Geschichte der Max-Planck-Gesellschaft *7.* Niels Bohr Institute *8/9.* Bibliothek und Archiv für Geschichte der Max-Planck-Gesellschaft *10.* AIP Niels Bohr Library *11.* University of California Press *12.* Culver *13/14.* Niels Bohr Institute *15–17.* Emilio Segrè *18.* AIP Niels Bohr Library *19.* Picture People *20.* Niels Bohr Institute *21.* Société Française de Physique, Paris/ AIP Niels Bohr Library *22.* W. F. Meggers Collection, AIP Niels Bohr Library *23/24.* Lawrence Berkeley Laboratory *25.* Photo by D. Schoenberg, Bainbridge Collection, AIP Niels Bohr Library *26.* Françoise Ulam *27.* Egon Weiss *28/29.* Rose Bethe *30.* Wide World *31.* Bibliothek und Archiv für Geschichte der Max-Planck-Gesellschaft *32.* Segrè Collection, AIP Niels Bohr Library *33/34.* Bibliothek und Archiv für Geschichte der Max-Planck-Gesellschaft *35.* Mary Evans Rhodes *36.* Argonne National Laboratory *37.* Picture People *38.* Department of Terrestrial Magnetism, Carnegie Institution of Washington *39.* Wide World

40. Lawrence Berkeley Laboratory *41.* Picture People *42.* Rudolf Peierls *43.* Smithsonian Institution Science Service Collection, AIP Niels Bohr Library *44.* UPI/Bettmann Newsphotos *45.* UPI/Bettmann Newsphotos *46.* Alfred O. C. Nier *47.* Photo by P. Ehrenfest, Weisskopf Collection, AIP Niels Bohr Library *48.* Lawrence Berkeley Laboratory *49.* Picture People *50.* UPI/Bettmann Newsphotos *51.* Picture People *52.* Lawrence Berkeley Laboratory *53.* Argonne National Laboratory *54/55.* Martin Marietta *56.* Philip Abelson *57–61.* National Archives *62/63.* Norsk Hydro *64–67.* Los Alamos National Laboratory *68.* Luis W. Alvarez *69.* Niels Bohr Institute *70.* Françoise Ulam *71/72.* Los Alamos National Laboratory *73.* Oppenheimer Memorial Committee *74.* Emilio Segrè *75.* Picture People *76.* Picture People *77.* Mrs. George Kistiakowsky *78–83.* Los Alamos National Laboratory *84.* AIP Niels Bohr Library *85–96.* Los Alamos National Laboratory

97. USAF *98.* Picture People *99.* Henry L. Stimson Papers, Yale University Library *100.* Robert Muldrow Cooper Library, Clemson University *101.* National Archives *102.* Harold Agnew *103–105.* USAF *106.* Yoshito Matsushige *107.* Hiroshima Peace Memorial Museum *108/109.* Hiroshima Peace Culture Foundation *110/111.* Hiroshima Peace Memorial Museum *112.* Hiroshima Peace Culture Foundation *113.* National Archives *114/115.* Issei Nishimori *116.* Hiroshima Peace Memorial Museum *117.* Yosuke Yamabata *118.* Issei Nishimori *119.* Yosuke Yamabata *120.* Peter Wyden *121.* UPI/Bettmann Newsphotos *122–126.* Los Alamos National Laboratory *127.* Dr. Harold E. Edgerton, MIT, Cambridge, MA *128.* Niels Bohr Institute

Index